ABSORPTION SPECTROSCOPY OF ORGANIC MOLECULES

V. M. PARIKH
Notre Dame University of Nelson
Nelson, British Columbia

ABSORPTION SPECTROSCOPY OF ORGANIC MOLECULES

ADDISON-WESLEY PUBLISHING COMPANY
Reading, Massachusetts
Menlo Park, California · London · Amsterdam · Don Mills, Ontario · Sydney

This book is in the
ADDISON-WESLEY SERIES IN CHEMISTRY

 Printed in the United States of America. Published simultaneously in Canada. Library of Congress Catalog Card No. 72-3460.

ISBN 0-201-05708-5
EFGHIJKLMN-AL-8987654321

We may educate ourselves and may reach many precious and coveted goals. But our childhood memories always remain our most precious treasures. During the preparation of this book I lost two of my brothers in the prime of their youth. This book is dedicated to their memory.

PREFACE

Of all the techniques introduced in the past ten years in the teaching of organic chemistry at the undergraduate level, the most significant is spectroscopy. When organic chemists became aware of the value of an early introduction to spectroscopy, the first thing they tried was an empirical approach to give students a working ability to interpret simple spectra with the aid of some reference tables. But the availability of spectroscopic instruments in educational laboratories has now become so widespread that it is necessary to discard the empirical approach and give students a blend of theory and experimental observations, so that they may use these powerful tools with greater confidence.

This book has grown out of my attempts, over several years, to introduce spectroscopy to students in the basic organic chemistry course. These students have been studying to be chemists, biologists, doctors, engineers, pharmacists, and so on. They have had a variety of goals, many interests, and a wide spectrum of backgrounds. To cater to students with such a variety of backgrounds and goals, and to generate their enthusiasm for a deeper understanding of spectroscopy, I found it essential that they begin with a quick review of basic physics. Chapter 1 is therefore a brief introduction to fundamental physics.

Then, in Chapters 2, 3, 4, and 5, the book covers in considerable depth the fields of ultraviolet, infrared, nuclear magnetic resonance, and mass spectrometry. These chapters explain theoretical principles of each technique, give details of the construction of spectroscopes, and discuss the handling of various instruments, the reading and reporting of data, the interpretation of the spectra obtained, and the possible ways in which the technique may be used in analytical, biochemical, medical, pharmaceutical, and other fields. The goals of these chapters is to help the student use these tools most effectively.

Finally, since the effectiveness of spectroscopic tools depends largely on one's ability to use a combination of methods, Chapter 6 is devoted to the integration of various techniques. Here we discuss the spectral characteristics of a number of compounds, in increasing order of difficulty, and present methods by which the student may correlate this information with that obtained by other techniques.

This text is not, however, limited to being a basic teaching manual. It gives enough material and up-to-date references so that it can be used by undergraduates in the senior year, or it may be used for a special one-semester course in spectroscopy. Researchers and practicing chemists will also find the book useful because along with topics which are fundamental to the understanding of each technique I have discussed (and cited pertinent references in the literature) many specialized topics which are currently being avidly investigated. To increase the book's usefulness on this score, I have included, in the appendix, extensive charts of spectroscopic correlations. Thus, while the novice will be enabled to learn the state of the art, the expert will at the same time find the book a source of quick reference.

In choosing the references, I have tried to direct the reader to the next-most-accessible place he should look for in-depth discussion of the topic under study. Books, review articles, and those original papers which I felt to be most readable are therefore included.

My task in preparing this book has been considerably lightened by helpful comments and constructive criticism from colleagues, friends, and students who read the manuscript at various stages of its preparation. Particular mention must be made of the valuable contributions by Dr. Keith Slessor (Simon Fraser University, Burnaby), Dr. David Chandler (University of Saskatchewan, Regina), Dr. George Neville (Department of National Health and Welfare, Ottawa), Dr. Frederick Greene and Dr. William Moore (both of the Massachusetts Institute of Technology, Cambridge, Massachusetts), Dr. Thomas Dunn (University of Michigan, Ann Arbor), Dr. Shelton Bank (State University of New York, Albany), Dr. Michael Gianni (Saint Michael's College, Winooski), and also Professor Francis T. Bonner (State University of New York,

Stony Brook, Long Island) who acted as a consulting editor for the project. The debt I owe to Dr. Neville is particularly great, for our keen friendship and his deep interest in the topic yielded the most penetrating criticism of the whole manuscript and the most generative suggestions as well.

Appreciation and thanks are also due to Mrs. Shirley Black for the NMR spectra, to Mrs. Lynn Jukes for typing parts of the manuscript, and to my family for their patience during the preparation of this book.

Nelson, British Columbia V.M.P.
September 1973

CONTENTS

ONE

INTRODUCTION

1.1 WHAT IS SPECTROSCOPY?

A narrow beam of sunlight, when it passes through a prism, disperses into a colorful spectrum. This suggests that sunlight is composed of many components which can be separated by passing through some medium. Such a separation of the components constitutes the formation of a spectrum; the study of the interaction of these components of light with matter is known as *spectroscopy*. Light from other sources can be similarly dispersed by various means. The spectra thus produced provide an important means of identifying materials and of studying atomic and molecular structures.

Since a spectrum is produced by an interaction between light and matter, let us first discuss what light is and how it interacts with various substances.

1.2 LIGHT AND RADIATION

Light is usually described as "the visible part of a large range of electromagnetic radiation having corpuscular as well as wavelike properties." It was the early physicists who assigned this dual nature to light. For example, James Clerk Maxwell, a Scottish physicist, said in 1864 that whenever an electric or magnetic disturbance takes place, the effects of the disturbance should travel outward with a definite velocity. He postulated further that if the electrical disturbance is of an alternating nature, its effects should travel in the form of electromagnetic waves capable of transporting energy. He deduced the velocity of these waves mathematically to be 2.9979×10^{10} cm/sec (in vacuum), and since this value was numerically equal to the velocity of light, Maxwell suggested that light too might be a radiation of this kind.

A few years later, the German physicist Heinrich Hertz used a rapidly oscillating electric spark to produce an electromagnetic wave disturbance, of exactly the nature predicted by Maxwell, and showed that these waves have many properties—such as reflection, refraction, polarization, etc.—similar to those ascribed to light. Later on, after the electron had been discovered, it became evident that light is produced by

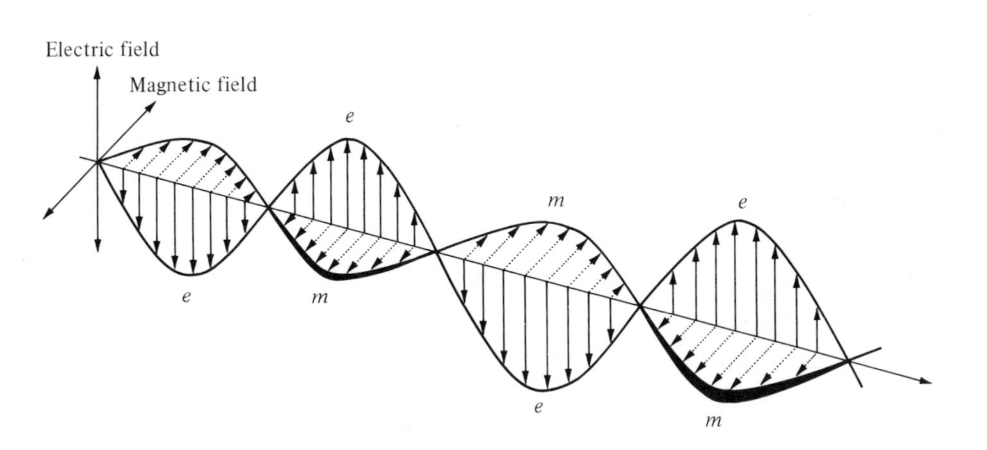

Fig. 1.1 An electromagnetic wave, showing the electric (e) and magnetic (m) components.

Wave picture	Radiation type	Wavelength λ Å	μ	cm	Wavenumber $\bar{\nu} = 1/\lambda$ cm^{-1}	Frequency $\nu = c/\lambda$, Hz	Energy eV	cal/mole	Effects
	Radio	10^{14}	10^{10}	10^{6}	10^{-6}	3×10^{4}	10^{-10}	10^{-6}	Spin orientations
		10^{12}	10^{8}	10^{4}	10^{-4}	3×10^{6}	10^{-8}	10^{-4}	
	Television	10^{10}	10^{6}	10^{2}	10^{-2}	3×10^{8}	10^{-6}	10^{-2}	
	Radar	10^{8}	10^{4}	1	1	3×10^{10}	10^{-4}	2.85	
	Microwave	10^{7}	10^{3}	10^{-1}	10	3×10^{11}	10^{-3}	28.5	Rotational; Molecular transitions
	Far IR	10^{6}	10^{2}	10^{-2}	10^{2}	3×10^{12}	10^{-2}	285	
	Near IR	10^{4}	1	10^{-4}	10^{4}	3×10^{14}	1.24	28.5 k	Vibrational
	Visible red	8×10^{3} Å			1.2×10^{4}	3.7×10^{14}	1.55	35.7 k	Inner shell; Electronic transitions
	violet	4×10^{3} Å			2.4×10^{4}	7.5×10^{14}	3.10	71.4 k	
	Ultraviolet	3×10^{3} Å			3.2×10^{4}	1×10^{15}	4.1	96 k	Valence shell
	X-rays	1	10^{-4}	10^{-8}	10^{8}	3×10^{18}	10^{4}	10^{8}	
	Gamma rays	10^{-2}	10^{-6}	10^{-10}	10^{10}	3×10^{20}	10^{6}	10^{10}	Nuclear transitions
	Cosmic rays	10^{-4}	10^{-8}	10^{-12}	10^{12}	3×10^{22}	10^{8}	10^{14}	

Fig. 1.2 The electromagnetic spectrum.

oscillations of electrons within the atom, and that light is also electromagnetic in nature. It was also realized that the electric and magnetic fields in an electromagnetic wave fluctuate in planes at right angles to each other and to the path of propagation of the wave (Fig. 1.1).

This realization led to the conclusion that light is merely a part of the whole range of electromagnetic radiation, extending from radio waves to cosmic rays (Fig. 1.2). All these apparently different forms of radiant energy can be considered as electromagnetic waves which travel at the same velocity. What makes visible light different from the rest of the radiations is that visible light, with its narrow range of wavelengths and frequencies, is the only part of the spectrum that can be detected by the human eye.

Before we go any further, let us consider the terms *wavelength* and *frequency* that we just used.

1.3 WAVE NATURE OF RADIATION

Radiation is energy in transit displaying the property of continuous waves, somewhat like ripples on a pond. Therefore we must first examine the characteristics of wave motion in general. In wave motion, a disturbance of some kind is propagated through a medium. The particles of the medium oscillate up and down, or back and forth, but they do not progress with the wave. A typical wave motion is one that is set up by regular vibrations in a source (or by some kind of periodic motion in the source). This vibrating center produces a disturbance among the particles of the medium immediately in contact with it; these particles in turn impart *their* disturbance (and thus their energy) to their neighbors. Figure 1.3 illustrates such a wave.

In spectroscopic studies, it is the effects associated with the electrical—rather than the magnetic—component of the electromagnetic wave that are important. Therefore Fig. 1.3 shows the propagation of vibrations in the electric field only. Points x, y, p, and q on this wave

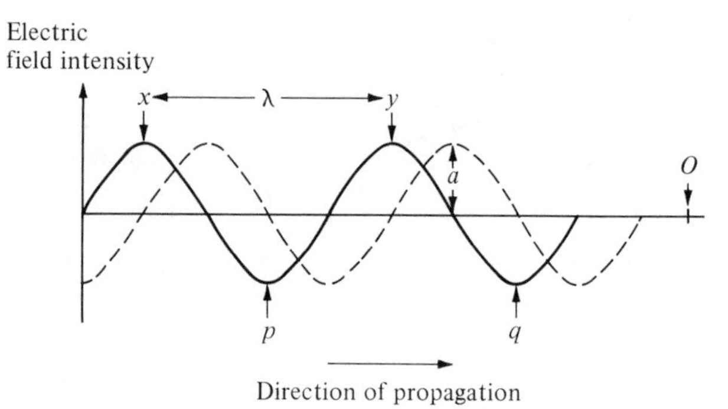

Fig. 1.3 The electric component of a propagating light wave.

represent maximum disturbances in the electric field; distance a is known as the maximum *amplitude* of the wave. The distance from crest x to crest y (or from valley p to valley q) is the wave's *wavelength* λ (Greek lambda). Although the unit commonly used to describe wavelength is centimeters (abbreviated cm), different units are used to express wavelengths in different parts of the electromagnetic spectrum. For example, in the regions of greatest interest to the organic chemist–i.e., in the ultraviolet and visible regions–the units used are *angstrom* (abbreviated Å) and nanometer* (abbreviated nm); 1 Å = 10^{-8} cm and 1 nm = 10^{-7} cm. In the infrared region the unit is *micron* (abbreviated μ). Thus

$$1\ \mu = 1000\ \text{nm} = 10{,}000\ \text{Å} = 10^{-4}\ \text{cm}$$

$$1\ \text{Å} = 10^{-1}\ \text{nm} = 10^{-4}\ \mu = 10^{-8}\ \text{cm}$$

$$1\ \text{cm} = 10^{4}\ \mu = 10^{7}\ \text{nm} = 10^{8}\ \text{Å}$$

Although a wave is frequently characterized in terms of its wavelength λ, people often use terms such as *wavenumber* ($\bar{\nu}$), *frequency* (ν), *cycles per second* (cps) or *hertz* (Hz) for this purpose.

When λ is expressed in centimeters, $1/\lambda$ gives the number of waves per centimeter. This is known as *wavenumber* $\bar{\nu}$ (Greek nu bar) and is expressed in reciprocal centimeters (cm^{-1}). For example, an infrared radiation of

$$50{,}000\ \text{Å} = 5\ \mu = 5 \times 10^{-4}\ \text{cm}$$

is equivalent to

$$1/\lambda = \frac{1}{5 \times 10^{-4}\ \text{cm}} = 2000\ \text{cm}^{-1}.$$

* Nanometer (nm) now replaces the traditionally used term millimicron (mμ).

Note that the longer the wavelength the smaller the wavenumber.

The relation between λ and ν is also very simple. Light travels through space with a velocity c of 2.9979 x 10^{10} cm/sec (approximately 3 x 10^{10} cm/sec). Therefore the number of waves passing by a point O (Fig. 1.3) is c/λ per second. This quantity is known as the *frequency* of the wave motion. In wave motion, when a disturbance has passed from crest x through valley p to crest y, it is said to have completed one cycle. The number of crests to pass a given point in one second therefore represents the number of cycles completed by the wave in one second. The frequency ν is thus expressed in units of cycles per second;* it can also be expressed in units of hertz (Hz). Thus

$$\text{Frequency} = \frac{\text{Velocity of light (cm/sec)}}{\text{Wavelength of radiation (cm)}} = c/\lambda.$$

To understand more fully the relationship between wavelength and frequency, let us consider waves in the infrared region. This region lies between 2 and 16 μ. A wave whose wavelength is 2 microns has

$$\text{Wavelength} = 2\ \mu = 2000\ \text{nm} = 20{,}000\ \text{Å} = 2 \times 10^{-4}\ \text{cm}$$

$$\text{Wavenumber} = 1/\lambda = \frac{1}{2 \times 10^{-4}} = 5000\ \text{cm}^{-1}$$

$$\text{Frequency} = c/\lambda = \frac{3 \times 10^{10}\ \text{cm/sec}}{2 \times 10^{-4}\ \text{cm}} = 1.5 \times 10^{14}\ \text{Hz}$$

* The terms *cycle* and *hertz* are often used with metric prefixes. For example, a kilocycle or kilohertz (kcps or kHz) is 1000 (that is, 10^3) oscillations per second and a megacycle (Mcps) or megahertz (MHz) is 1,000,000 (that is, 10^6) oscillations per second.

Problem 1.1 Use the above relationship between the wavelength and the frequency of a wave to fill in the gaps in the following table.

Radiation	Wavelength λ			Wavenumber $\bar{\nu}$, cm^{-1}	Frequency ν, Hz
	Å	μ	cm		
Visible violet	4500	4.5 x 10^{-1}	-	-	-
Visible red	-	-	7 x 10^{-5}	-	-
X-rays	-	-	-	10^7	-
Television	-	-	-	-	88 x 10^6 (88 MHz)
Ultraviolet	-	3.0 x 10^{-1}	-	-	-
Radio	-	-	-	-	1.6 x 10^7 (16 MHz)

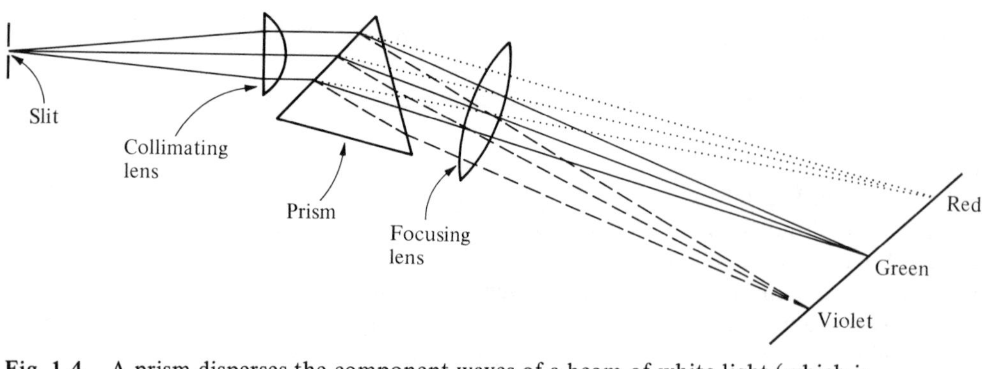

Fig. 1.4 A prism disperses the component waves of a beam of white light (which is polychromatic) into bands of colors known as the visible spectrum.

1.4 ANGULAR SEPARATION OF WAVES TO PRODUCE A SPECTRUM

By now we have become aware of the extensive range of electromagnetic radiation, extending from cosmic rays to radio waves (Fig. 1.2), and of the different classifications arising rather arbitrarily from the wavelengths or frequencies of these radiations. Most of the time the source of radiation—be it the sun or an ordinary light bulb—produces continuous radiation over a wide range of wavelengths. Even a common ray of visible light (such as a sun ray) contains waves with λ ranging from 800 (red) to 400 (violet) nm. These waves travel with a velocity of 3×10^{10} cm in vacuum. They are slowed down when they pass from a vacuum through a transparent medium; the greater the optical density of the medium, the greater the reduction in speed of the waves entering it. As a result of this change in speed, the beam bends or is *refracted* as it passes from one transparent medium to another. *This refraction of a ray depends on the frequency or wavelength of the radiation, and normally increases as the frequency increases or as the wavelength decreases.* Thus, if a narrow beam of white light is passed through a prism (which can be described as a wedge-shaped portion of a refracting medium bounded by two plane surfaces), the violet light, which has the shortest wavelength, is bent most, while the red light, which has the longest wavelength, is bent least (Fig. 1.4). Thus a spectrum of colors is produced.

Basically, production of a spectrum involves dispersion (angular separation) of waves of different wavelengths by some refracting medium. These waves may or may not be from the visible region of the electromagnetic radiation. This process of dispersing radiation into continuous bands of narrow widths of wavelengths is extremely important.

1.5 MONOCHROMATORS

Ordinary light is composed of a large number of wavelengths (therefore it is *polychromatic* light), whereas the light emerging through the prism is in the form of bands of very similar wavelengths. These bands are therefore nearly *monochromatic.* A really monochromatic beam contains radiation of *one* wavelength only. A device which resolves a polychromatic ray into monochromatic (or nearly monochromatic) bands is thus called a *monochromator.**

Although a prism is capable of dispersing light, the spectrum thus obtained is diffuse. In order to get a clearly defined spectrum without too much overlapping of colors, one must use a monochromator system which includes:

a) A fine slit through which the ray may enter
b) A lens or mirror (a collimator) to project the radiation as nearly parallel beams onto the dispersing device
c) A prism or diffraction grating to resolve the light into its component wavelengths
d) An exit slit, which allows a narrow band of the spectrum to pass onto the focusing device
e) A second lens or a mirror beyond the dispersion device to focus the dispersed waves onto a detecting mechanism

---

* Optical glass filters (or interference filters containing colored absorbing materials) allow only a narrow range of wavelengths to pass through; these may also be looked upon as monochromators. So may the cheap colorimeters which contain a series of optical filters for the isolation of spectral regions. However, a monochromator is a device which makes possible a *continuous* selection of wavelengths; a filter does not do this. From this viewpoint, a colorimeter is an instrument with which one cannot continuously vary monochromatic light, whereas a spectrophotometer is an instrument with which one *can* continuously vary it.

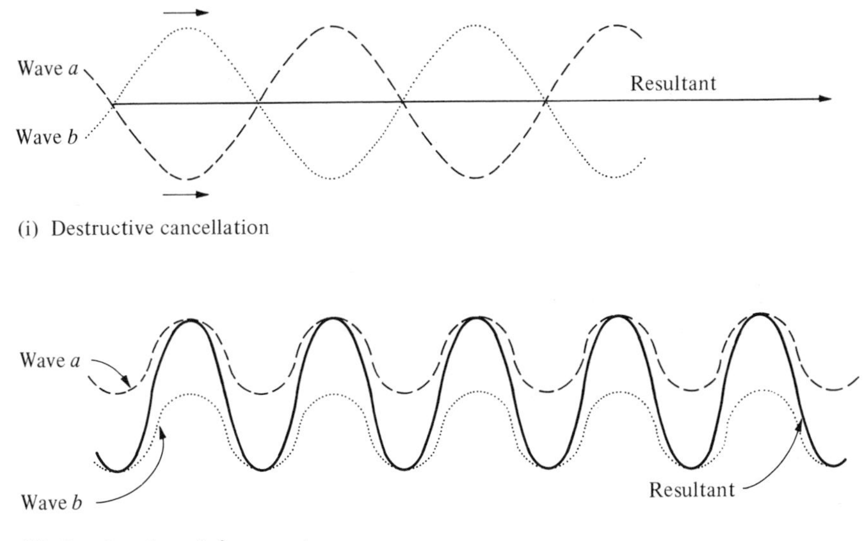

Fig. 1.5 Interference of waves resulting in (i) destructive cancellation and (ii) constructive reinforcement of the effect.

Different monochromators thus differ basically in the types of dispersion devices used, the two main types being prism monochromators and diffraction-grating monochromators. Whereas a prism disperses radiation by means of *refraction,* a grating achieves the same effect through *diffraction by reflection* or *transmission.* To understand how this is achieved, we must first understand what a grating is and how it causes reinforcements and cancellations of waves.

We can prepare a transmission grating simply by drawing a series of very fine equidistant parallel lines, with a fine diamond point, on a polished transparent surface. (Gratings have been made with as many as 40,000 lines to the inch, over a width of 5 to 6 inches, producing 200,000 to 240,000 lines.) Such a plate of glass works like a picket fence, allowing light to pass through the region between the furrows, but blocking its passage through the surface which has been disturbed by diamond etching. Reflection gratings, which are used in ultraviolet, visible and infrared instruments, are prepared in the same manner, but by using a highly reflective aluminized surface instead of a transparent glass.

To understand how such a grating can separate the spectral regions of radiation, we must realize that a combination of two waves (a and b in Fig. 1.5) of exactly the same wavelength and frequency can result either in their complete destruction or in the production of a new wave with double the amplitude. If integration of wave a and wave b is such that troughs of wave a strike the receiver at the same instant as crests of wave b (Fig. 1.5 i), the resultant effect on the receiver is nil, because the resultant wave has zero amplitude. On the other hand, if the crests of both wave a and wave b strike the receiver at the same instant (Fig. 1.5 ii), the effect on the receiver is doubled because the two waves reinforce each other, thus producing a resultant wave with an amplitude equal to the sum of the two component waves. Waves a and b are said to be *exactly out of phase* in case (i) and *exactly in phase* in case (ii); the constructive or destructive combination of waves of the same wavelength is called *interference.*

If ordinary light is collimated by passing it through a lens, the parallel rays thus produced have coherent waves, i.e., all radiations of the same wavelength are in phase and have a common wave front. If such a coherent wave front is reflected by a grating, the reflected waves will no longer all be in phase. This is due to the fact that some waves travel a greater distance from the source to the reflecting surface than others, thus creating a situation in which crests of some waves strike the grating at the same instant as troughs of others (Fig. 1.6).

The reflected waves are not only noncoherent, but are also scattered by the uneven surface of the grating. This causes interference, resulting in both cancellation and reinforcement of reflected waves. *When there is a fine reflection (or diffraction) grating, destructive interference is essentially complete at all angles except those at which constructive reinforcement occurs.* This, in effect, means that when light of a definite wavelength (say red light) is reflected by a fine grating, a pattern con-

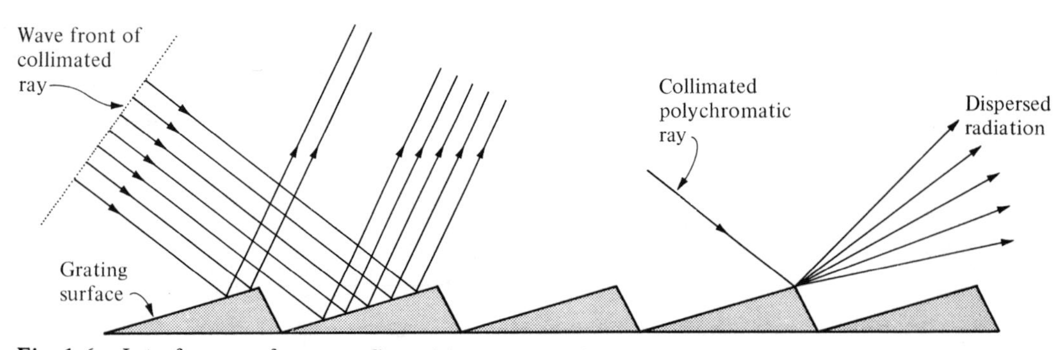

Fig. 1.6 Interference of waves reflected by a grating. (For ease of visualization, the grating is pictured here as a stepped surface.)

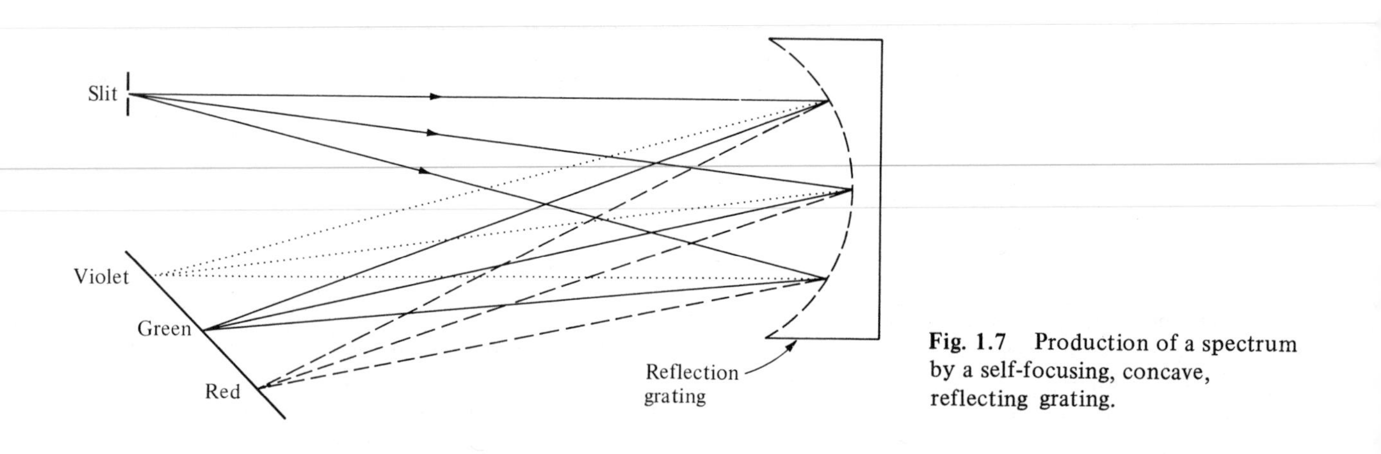

Fig. 1.7 Production of a spectrum by a self-focusing, concave, reflecting grating.

sisting of only a few narrow (sharp) red lines (maxima) is produced on a receiving screen. On the other hand, if white light, with many wavelengths, is incident on the grating, each maximum on the screen becomes a spectrum. Finally, if the grating is in the shape of a concave mirror, maxima of each wavelength can be focused to produce a well-defined spectrum on the screen (Fig. 1.7).

1.6 PHOTONS

In Section 1.2 we said that light has both *corpuscular* (or particle-like) and *wavelike* properties. We have so far considered only the wave nature of radiation. However, the concept that radiation consists of a stream of energy packets, moving in the direction of the beam with the same velocity as that of light, is also important, particularly when we want to understand how molecules of organic substances absorb radiation. In fact, this phenomenon—the absorption or emission of radiation by the molecules of samples through which the radiation passes—is the very basis of organic spectroscopy.

The idea that light beams may be streams of rapidly moving particles is some three centuries old now, and experiments have been performed to support it and also to disprove it. This particle theory had its ups and downs for a couple of centuries, and was almost ready to bow out to the wave theory when, in 1900, it got an opportunity to make a strong comeback. The physicist Max Planck introduced, in that year, the concept of *light quanta* (packets or bundles of light energy) and his colleague Albert Einstein used the concept to explain the so-called *photoelectric effect.*

An account of the controversy between the two theories makes very interesting reading, but because of the limitations of space, we shall simply look at the outcome of the argument, which was that physicists accepted the idea of a dual nature of radiation, wherein the so-called corpuscles are looked upon not as material particles but as "energy bundles" called *photons.* The basic relation between the two approaches allows us to interpret a monochromatic beam of radiation (with a given wavelength or frequency) as being also a train of quanta of a given energy. According to this Einstein-Planck relation, *in any beam of radiation of frequency ν, each photon carries (or is a bundle of) an energy E,* where

$$E = h\nu.$$

Here h is the universal constant, known as *Planck's constant.* It has the value

$h = 6.62 \times 10^{-34}$ joule-sec or $h = 6.62 \times 10^{-27}$ erg-sec.

1.7 THE EINSTEIN-PLANCK RELATION

From our viewpoint, the relation $E = h\nu$ is very important. First, it tells us that *the energy of a monochromatic radiation depends only on the frequency or wavelength of its waves and not on the intensity of its beam.* In other words, a beam of radiation is more or less intense depending on the number of photons per unit time, per unit area, but the energy per photon (i.e., the *quantum energy*) is always the same for a definite frequency of the radiation.

Since we know the value of Planck's constant ($h = 6.62 \times 10^{-27}$ erg-sec), we can readily calculate the energy of individual photons in any kind of radiation. For example, to find out whether ultraviolet radiation is more or less energetic than infrared radiation, consider the frequencies of waves in these regions. Ultraviolet radiation has a wavelength of about 3000 Å or 3×10^{-5} cm. Its frequency is

$$\begin{aligned} \nu &= c/\lambda \\ &= \frac{3 \times 10^{10} \text{ cm/sec}}{3 \times 10^{-5} \text{ cm}} \\ &= 1 \times 10^{15} \text{ cycles/sec (or Hz)}, \end{aligned}$$

and the energy E is:

$$\begin{aligned} E &= h \times \nu \\ &= (6.62 \times 10^{-27} \text{ erg-sec})\,(1 \times 10^{15} \text{ Hz}) \\ &= 6.62 \times 10^{-12} \text{ erg.} \end{aligned}$$

In the same manner one can calculate the energy of the infrared radiation of say 2 microns (20,000 Å) or 2×10^{-4} cm as

$$\begin{aligned} E &= h \times \nu \\ &= h \times c/\lambda \\ &= (6.62 \times 10^{-27} \text{ erg-sec}) \left(\frac{3 \times 10^{10} \text{ cm/sec}}{2 \times 10^{-4} \text{ cm}} \right) \\ &= (6.62 \times 10^{-27} \text{ erg-sec})\,(1.5 \times 10^{14} \text{ Hz}) \\ &= 9.73 \times 10^{-13} \text{ erg.} \end{aligned}$$

Thus we can see that the photons in ultraviolet radiation (which has a shorter wavelength) carry more energy than those in infrared. In general, *the shorter the wavelength (i.e., the greater the frequency) the greater the energy of the photons and the more powerful the radiation.*

Problem 1.2 Calculate the energies of the photons of various radiations described in Problem 1.1 (Section 1.3).

1.8 INTERACTION OF PHOTONS WITH MATTER

We have seen how waves are reflected, refracted, or diffracted when they interact with matter. Our chief interest in the study of these phenomena was to see how light of different wavelengths can be resolved. In effect, we studied the methods for sorting photons of different energies. Organic spectroscopy is essentially the study of how photons are absorbed by organic molecules, and which ones are absorbed. By knowing the kinds of photons that an organic molecule can absorb, we can often find out about the shape, size, atomic arrangement, etc., of a molecule, without actually being able to see it. Of course, we must first be able to answer questions such as: Why do molecules absorb photons? Do they absorb all kinds of photons, or only a certain specific kind? What happens to the organic molecule when it absorbs photons? What happens to the absorbed energy? And so forth.

To answer these questions, let us recall that, according to the kinetic-molecular theory of gases, molecules are in a continuous state of motion (both in the liquid and the solid state they are constantly vibrating). Although the laws of ordinary mechanics (classical physics) are not entirely applicable to the behavior of small bodies such as atoms and molecules (where quantum mechanics should be applied), we can readily appreciate that molecules in motion possess energy. This energy can be resolved into a number of components. For example, the kinetic energy component due to free motion of molecules through space is called *translational energy.* Such a freely moving particle is also capable of rotating or revolving about an axis through its center of gravity. The energy component due to this rotational movement is called *rotational energy.* (Next time you watch a football game, note how the ball, when it is kicked, rotates about its long axis while it is moving freely through the air.) A molecule is made up of two or more atoms bonded together. Imagine that these atoms are little balls bonded together with tiny springs, and you will readily realize that such a flexible system must vibrate and rotate while simultaneously traveling through space. The energy component due to this vibrational movement is called *vibrational energy.*

We shall study more about these three energy components in the chapter on infrared spectroscopy. But for the moment, let us look into yet another energy component which is typical of all molecules, and which is based

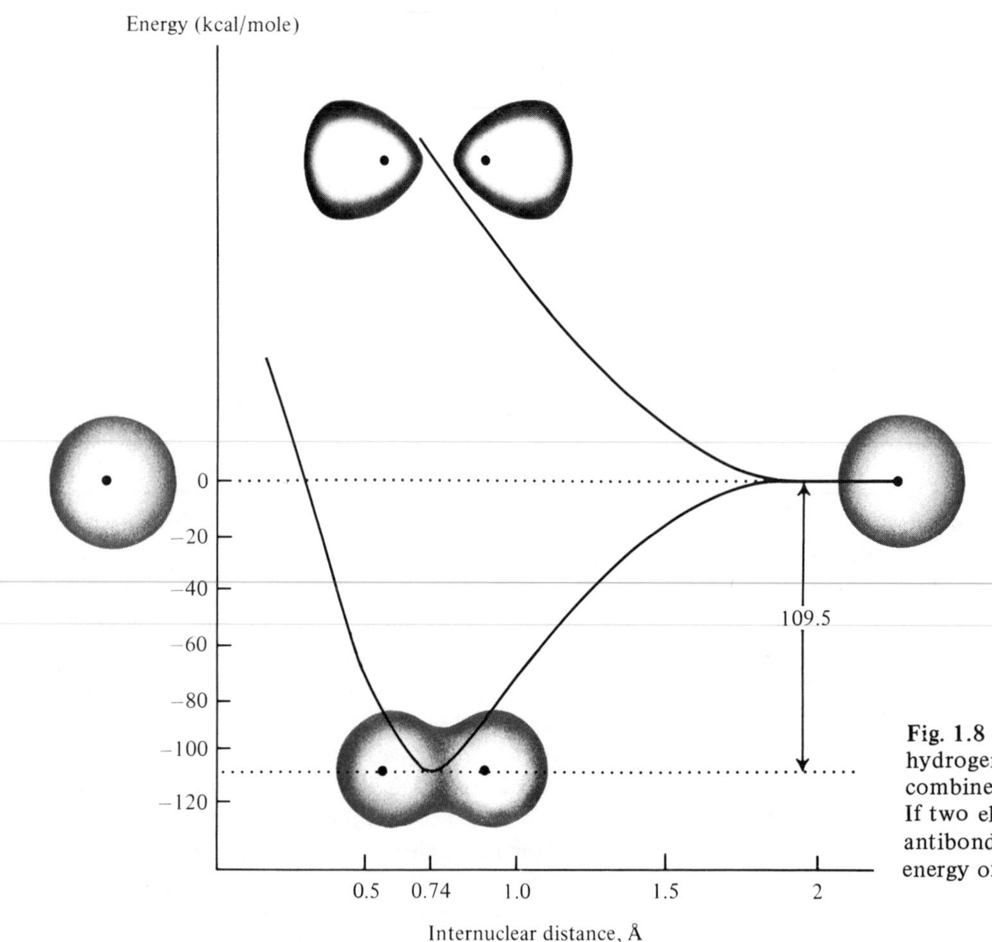

Fig. 1.8 The total energy of hydrogen atoms is lowered when they combine to form hydrogen molecules. If two electrons were to enter the antibonding molecular orbital, the energy of the system would increase.

on their nuclear and electronic structures. Recall that electrons in a given atom occupy discrete energy levels. Recall also that the formation of covalent bonds by the atoms results in an overall lowering of the total energy of the system (Fig. 1.8). This reduction in energy depends on the types of bonds formed (σ, π, etc.), on the distance between the nuclei of the atoms, and, in general, on the degree of delocalization of the electrons. The energy due to the electrons, known as *electronic energy,* also contributes to the total energy content of the molecule. Thus the energy of a molecule can be resolved into at least four major components: (1) *translational energy,* (2) *rotational energy,* (3) *vibrational energy, and* (4) *electronic energy.*

1.9 APPLICATION OF QUANTUM MECHANICS TO SPECTROSCOPY

At this stage we need to understand the contribution of quantum-mechanical principles to spectroscopy. We cannot use the laws of mechanics which apply to the energies of ordinary-sized objects, such as footballs, to understand the behavior of microbodies such as atoms, molecules, electrons, etc. For example, a football can spin or rotate with any number of rotations per minute (depending on how it is kicked). That is, the rotational energy of a football can assume any value on a *continuous* scale, but a rotating molecule cannot rotate freely with any energy it likes. It is subject to what is called *quantum restrictions,* and can have only certain *noncontinuous* values of velocities and energies (Fig. 1.9). The quantization of energy is particularly critical at the atomic and molecular levels.

Thus, for each of the four components of molecular energy, there are certain permissible energy levels. The values of these levels depend on factors such as types of molecular motions and strengths, angles, types and distances of bonds, etc., in the molecule. The quantitative nature of quantum restrictions is beyond the scope of our present study, but we can easily understand a qualitative generalization about them.

Fig. 1.9 Continuous and noncontinuous values for energies of (a) macro- and (b) microbodies.

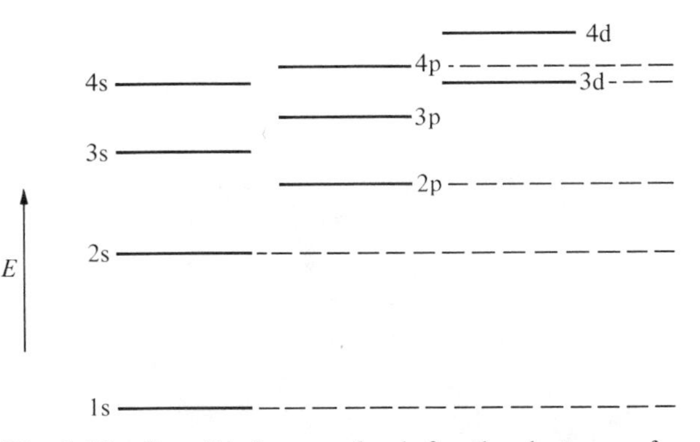

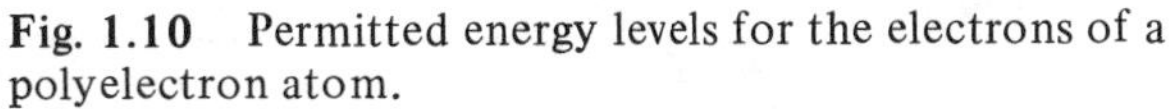

Fig. 1.10 Permitted energy levels for the electrons of a polyelectron atom.

The significance of quantum restrictions on a particular motion of a microbody depends on the space available for such a motion. Naturally, when there is a large space available for a motion, that motion has less significant restrictions on its energy. The energy of such a motion can assume essentially any value on a continuous scale, and a study of energy changes during such motions is therefore of little use to us. Quantum levels of only those movements which are within a very restricted space are important to us here.

Let us once again consider the four components of molecular energy. Now we realize that it is unnecessary to worry about quantum restrictions on the translational energies of molecules because ordinarily the space available for this kind of motion is as big as the volume of the container in which the molecule exists. The three other movements—rotational, vibrational, and electronic—are, however, limited in space to the size of the molecule, and are therefore truly quantized.

To understand how radiation interacts with matter, let us concentrate on, say, the electronic transitions and the permitted energy levels for the electrons of an atom (Fig. 1.10). The diagram shows that the next permitted energy level above the 1s level is 2s, and for a 2s-level electron, the next higher level is 2p. It also shows that the difference in energy between 2p and 2s levels is less than that between 2s and 1s, and that between 4p and 3d is still less. In other words, less energy is required to cause a 3d → 4p transition than is required for 2s → 2p or 1s → 2s transitions.

Let us imagine now that such an atom is exposed to a slowly changing spectrum of electromagnetic radiation. When a long-wavelength radiation, say, from the microwave region, is incident upon this atom, the energy possessed by the photons in this region is so low that the atom that they reach cannot achieve even the weakest electronic transition, 3d → 4p. The atom therefore allows these photons to pass through unabsorbed by the electronic levels. *An atom does not absorb photons unless the photons have enough energy to bring about transition of electrons from one energy level to another, and only those photons which carry the exact amount of energy required for transition of electrons to the next permitted levels can bring about transition.*

Suppose that, instead of microwave radiation, waves from the visible region are allowed to interact with these atoms. Certain waves from this region do have correct frequency (and therefore photons of correct energy) to bring about 3d → 4p transition. The atoms absorb only these photons. Transitions such as 2s → 2p or 1s → 2s do not take place, because these transitions require more energy than any photons in the visible region can provide. Normally, a 2s → 2p transition requires radiation from the far ultraviolet region, and a 1s → 2s transition requires x-ray radiation.

In the case of molecules, electronic transitions occur at the molecular orbital levels. Such transitions require the amount of energy possessed by the photons in the ultraviolet region and become the chief area of interest in ultraviolet spectroscopy. For example, the electrons in

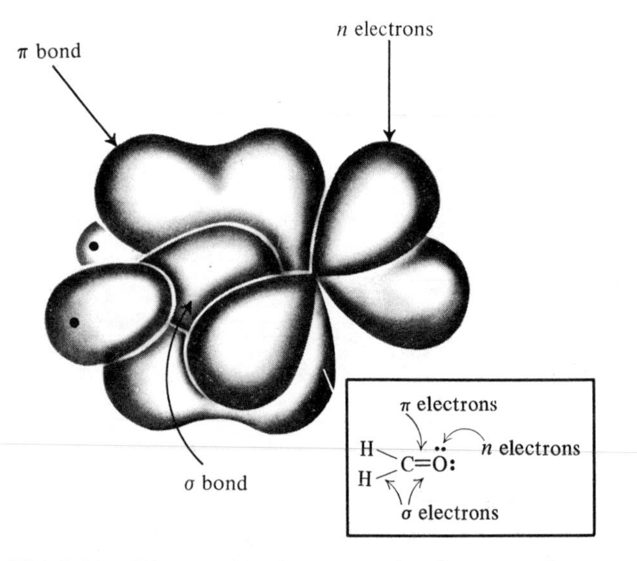

Fig. 1.11 Diagram showing σ, π, and n electrons of formaldehyde.

formaldehyde, HCHO (Fig. 1.11) can be classified intc three main types:

1. Those involved in *head-on* atomic orbital overlap, to form σ bonds (known as σ electrons)
2. Those involved in *parallel* atomic orbital overlap, to form π bonds (known as π electrons)
3. Those which are not participating in bonding (known as nonbonding or n electrons)

The π electrons of formaldehyde have two possible molecular orbital levels: π and π^*. In the *ground state* (that is, the state of lowest energy for the molecule) both π electrons are in the bonding π orbital and have opposite spins, whereas the antibonding π^* orbital is empty. When the molecule absorbs photons of the correct energy, one of these electrons is promoted to the antibonding π^* molecular orbital. This state of a molecule is referred to as an *excited state.* Similar transitions can take place for σ and n electrons also, but we shall postpone further consideration of them until we look at ultraviolet and visible spectroscopy.

Much less energy is required to bring about transitions in molecular *vibrational* and *rotational* states. These energy components are also quantized, but the difference in energy between the permissible levels in the *vibrational* components is much less than that between molecular *electronic* levels. The difference in energy between the quantized *rotational* components is still lower, to a significant degree. Thus, denoting the difference in energy between the quantized levels by ΔE (delta E), we can say, in general, that:

$$\Delta E_{\text{electronic}} \gg \Delta E_{\text{vibrational}} \gg \Delta E_{\text{rotational}}$$

Consequently, rotational transitions are accomplished by considerably weaker photons (longer-wavelength radiations, e.g., microwave) than those required for vibrational transitions (e.g., photons in the infrared region) and electronic transitions need very powerful photons (short-wavelength radiation, e.g., visible and ultraviolet).

We can now see that the passage of electromagnetic radiation through a material can increase the quantized rotational, vibrational, and electronic energy levels of the molecules of the material. This increase takes place only when radiations of the correct frequency are absorbed. Therefore, by exposing a sample of the material to a spectrum of known wavelengths, and investigating *which* frequencies it absorbs, we can learn a great deal about the structure of the molecules of the material.

1.10 SPECTROSCOPIC INSTRUMENTS

Spectroscopic study of organic compounds therefore becomes, in essence, an investigation of the types of waves that can be absorbed by the molecules of the compound. Thus an instrument to carry out such a study (Fig. 1.12) must have the following components:

1. A source of electromagnetic radiation to produce the types of waves required for the study
2. A sample holder placed in such a manner that the radiation from the source can pass through it

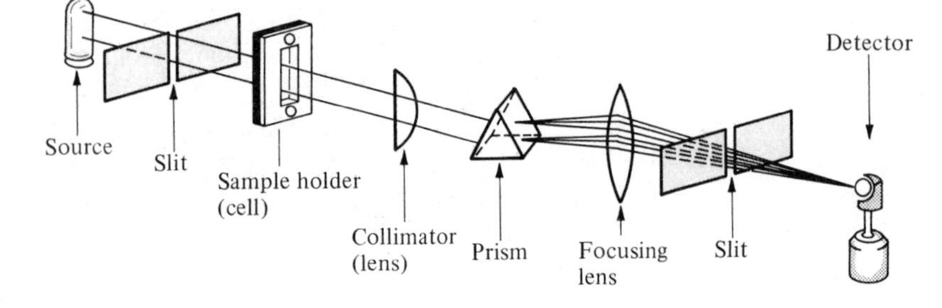

Fig. 1.12 The basic components of a spectrophotometer.

3. A monochromator which receives the waves emerging from the sample and spreads them into a spectrum
4. A device which detects the amount of radiation of various frequencies that reaches it after passing through the sample and the monochromator. This device enables us to detect those waves which are absorbed by the sample and consequently do not reach the detector in full strength.

In practice, various kinds of detectors are used, and instruments are classified accordingly.

1. Instruments in which radiation is detected merely by eye are called *spectroscopes.*
2. Instruments in which photographic plates are used as detectors are called *spectrographs.*
3. In many instruments some electronic devices or phototubes are used to convert electromagnetic radiation into electrical impulses which can be measured by simple galvanometers; such instruments are called *spectrophotometers.*
4. In most cases of group (3), radiation is more accurately detected and measured by electronically amplifying the electrical impulses and recording them on chart paper. Such instruments are referred to as *recording spectrophotometers.*

From the viewpoint of an organic chemist, the four most frequently used types of spectroscopic investigations are as follows.

1. *Ultraviolet and visible spectroscopy,* in which the instruments use a source which can produce radiation of wavelengths between 200 and 800 millimicrons
2. *Infrared spectroscopy,* in which a source capable of producing waves 2 to 16 microns long is used
3. *Nuclear magnetic resonance spectroscopy,* in which the source produces radiation in the radiofrequency region
4. *Mass spectrometry,* in which ions are produced and scanned according to their mass/charge ratio

The principles of nuclear magnetic resonance spectrometry and mass spectrometry are somewhat different from those we have discussed; we shall consider them in Chapters 4 and 5, respectively. In addition to these four types, Raman spectrometry, microwave spectrometry, electron-spin-resonance spectrometry, etc., are being increasingly used in the area of absorption spectrometry. However, these techniques are beyond the scope of our present study.

PROBLEMS

Problem 1.3 Electromagnetic waves 10 Å long are classified as x-rays; classify radiation which has the following properties.

a) 8×10^5 cm
b) 280 MHz
c) 5×10^4 microns
d) 3×10^7 cm^{-1}
e) 6000 Å
f) 8000 nanometers
g) 1.5×10^{15} Hz
h) 1 meter

In each case give at least one source of the radiation and show how it is detected.

Problem 1.4 Calculate the energy per photon for each of the radiations in Problem 1.3. Calculate the energy in terms of ergs for (a), (c), (e), and (g), and in terms of joules for (b), (d), (f), and (h).

Problem 1.5 We have considered electromagnetic radiation in terms of the energy content of a *single* photon in the wave. This energy is expressed in units of ergs or joules, and is very small. The energy of radiation is more conveniently expressed in terms of *calories per mole* and *electron volts per photon*:

1 electron volt (eV) $= 1.602 \times 10^{-19}$ joules,
$= 1.602 \times 12^{-12}$ ergs,

1 calorie = 4.184 joules = 4.184×10^7 ergs.

And the energy per mole of photons* is

$E = Nh\nu,$

where E = energy per mole, in joules,
N = Avogadro's number = 6.02×10^{23},
h = Planck's constant = 6.62×10^{-34} joule-sec,
ν = frequency in hertz.

Thus, on page 7, we calculated the energy of a photon of ultraviolet radiation to be 6.62×10^{-12} erg. The energy per mole of photons of this radiation is therefore

$$E = (6.62 \times 10^{-12})(6.02 \times 10^{23})$$
$$\simeq 4 \times 10^{12} \text{ ergs} \quad \text{or } 4 \times 10^5 \text{ joules}$$

Or

$$E = \frac{4 \times 10^5}{4.184} \simeq 9.5 \times 10^4 \text{ cal/mole}$$

Or

$$E = \frac{6.62 \times 10^{-12}}{1.602 \times 10^{-12}} \simeq 4 \text{ eV/photon}$$

---

* This gives a wrong impression, but chemists use terms such as cal/mole to discuss the energy associated with a process in terms of a mole of a substance undergoing the change.

Calculate similarly the energy of a typical wave in the infrared region, in terms of ergs/mole, joules/mole and electron volts/photon.

Problem 1.6 Fill in the gaps below. (You will need these values for quick reference.)

a) Avogadro's number, $N =$

b) Velocity of light, $c =$ cm/sec

c) Planck's constant, $h =$ erg-sec = joule-sec

d) 1 electron volt, eV = erg

e) 1 joule = erg = cal

f) 1 calorie = joules

g) 1 electron volt = kcal/mole

h) One Å = nm = μ = cm

TWO

ULTRAVIOLET SPECTROSCOPY

2.1 INTRODUCTION

Ultraviolet radiation, as we have said, is that part of the electromagnetic spectrum which bridges the gap between the longest-wavelength x-rays and the shortest-wavelength visible light. This ultraviolet region, which extends from 40 to 4000 Å, is divided into the *near* (2000 to 4000 Å) and *far* (40 to 2000 Å) ultraviolet regions (Fig. 2.1). Far UV radiation is absorbed by air because of moisture, and also because of electronic transitions in atmospheric oxygen, nitrogen, and carbon dioxide. Therefore one must use vacuum apparatus to study far UV radiation. That is why this UV region is also known as the *vacuum region.* Furthermore, glass absorbs radiation of wavelength less than 3000 Å. The instruments used to study far UV radiation must therefore use quartz optics and quartz sample holders. Consequently the region between 2000 and 3000 Å is sometimes also referred to as the *quartz region.*

2.2 THE SHAPE OF UV ABSORPTION PEAKS

We have commented earlier on the fact that it takes radiation of higher energy to effect electronic transitions than to effect rotational or vibrational transitions. Since all these transitions require fixed (quantized) amounts of energy, it is natural to expect that an ultraviolet or visible spectrum of a compound would consist of one or more well-defined, sharp peaks, each corresponding to the transfer of an electron from one electronic level to another. However, a look at a typical ultraviolet spectrum reveals the broad, irregular nature of these peaks (Fig. 2.2).

The peaks are often broad because each electronic state in a molecule is associated with a very large number of vibrational and rotational states. A change in the *rotational* state of a molecule can occur without simultaneous changes in the vibrational and electronic states, because differences between rotational energy levels are small, and weak radiations in the far infrared and microwave regions can readily bring about transitions from one level to another. A far infrared or microwave spectrum of a chemical compound therefore displays several well-defined, sharp peaks.

The differences between the *vibrational* energy levels of a molecule are relatively greater; transitions from one vibrational level to another require more powerful radiations (shorter wavelengths or larger frequencies) such as those in the infrared regions. Transitions in the vibrational states of molecules are thus accompanied by simultaneous changes in the rotational states. The infrared spectra of organic substances therefore do not have peaks as sharp and well defined as those seen in pure rotational spectra of such substances.

The differences between *electronic* energy levels E_1 and E_2 (Fig. 2.3) of two electronic states would be well defined only if the nuclei of the two atoms of a diatomic molecule could be held in fixed positions, i.e., only if no rotational or vibrational motions in the molecule

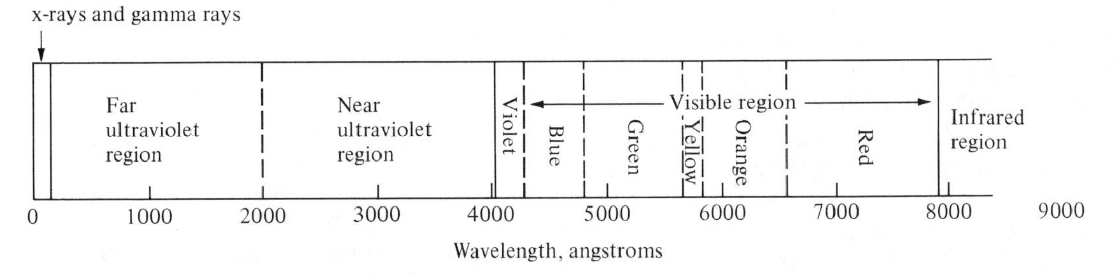

Fig. 2.1 Ultraviolet and visible regions of the electromagnetic spectrum.

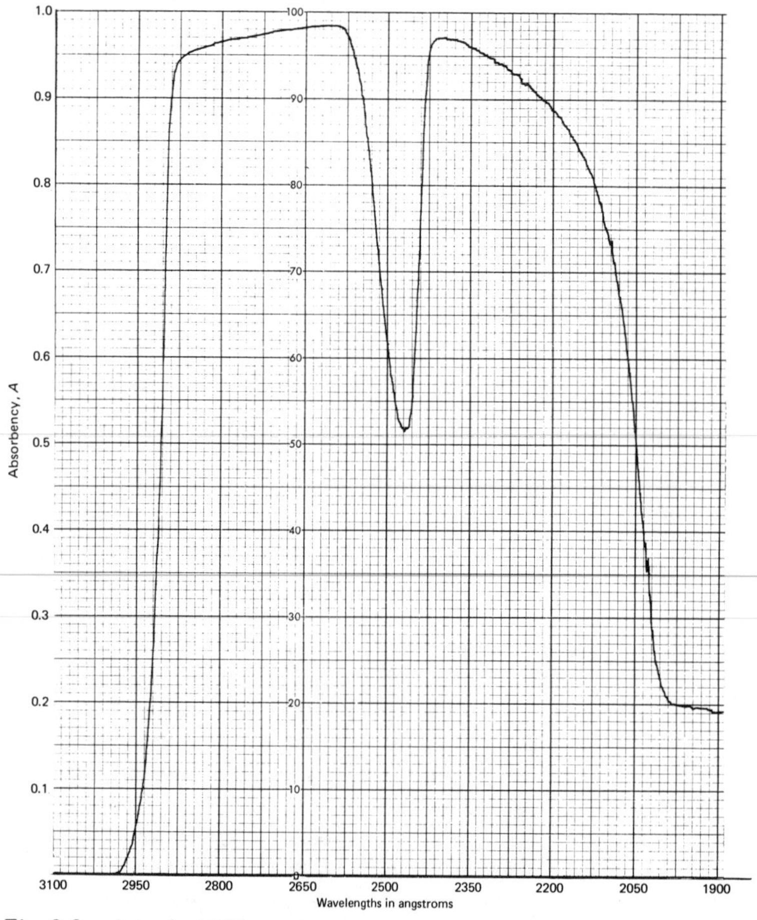

Fig. 2.2 A typical UV spectrum.

occurred simultaneously. However, vibrations and rotations of nuclei do occur constantly. Electronic states in a molecule are therefore always associated with a large number of vibrational and rotational states also (Fig. 2.3). The transition of an electron from one energy level to another is thus accompanied by simultaneous changes in vibrational and rotational states; many waves of closely spaced frequencies are thus absorbed, giving rise to a broad UV peak (Fig. 2.2). This is particularly true in absorption spectra of liquids, of solutions of organic compounds in polar solvents, and of solids. In such spectra, it is almost impossible to distinguish one rotational band from another, and only rarely is it possible to distinguish one vibrational peak from another. Absorption spectra of substances in the gaseous state, or of solutions of substances in nonpolar solvents, often show fine structure.

Figure 2.3 shows a plot of potential energy versus internuclear distance for a light, singly bonded, diatomic molecule. The potential energy of vibration is electronic in nature and is exclusive of the kinetic energy of vibration. The vibrational energy levels, on the other hand,

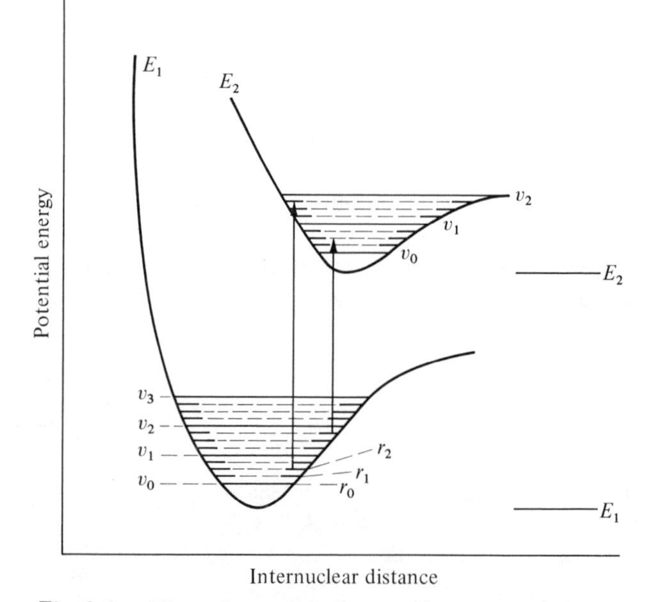

Fig. 2.3 Allowed rotational and vibrational energy levels for two different electronic states, as represented by energy curves for a light, singly bonded, diatomic molecule.

represent the total energy of vibration, both kinetic and potential. In Fig. 2.3, we see the overall energy states; each electronic state is shown to be composed of a series of vibrational states v_0, v_1, etc., and a corresponding series of rotational states, r_0, r_1, etc. The rotational levels are shown in somewhat exaggerated form, since the difference in energy between two adjacent vibrational states is much smaller than the difference in energy between two adjacent electronic levels, and the difference in energy between two rotational states is smaller still. We can obtain an idea of these differences from the fact that powerful UV radiation (energy per photon = 6.62×10^{-12} erg) is required for electronic transitions, but even very weak microwaves (energy per photon = 6.62×10^{-18} erg) can bring about rotational transitions. Electronic transitions require approximately a million times more energy than rotational transitions. This makes it inconvenient to depict rotational energy levels graphically on the same scale as that used for electronic energy levels.

2.3 READING AND REPORTING SPECTRAL DATA

It is obvious that if many molecules jump from one energy level to another, a great amount of energy is absorbed, and this results in a high peak on the absorption spectrum. If there are only a few molecules making the transition, the total absorption of energy is less, and there is consequently a lower peak. This observation is the basis of the *Beer-Lambert law*:

The fraction of incident radiation absorbed is proportional to the number of absorbing molecules in its path.

If the radiation passes through a solution, the amount of light absorbed or transmitted is an exponential function of the molecular concentration of the solute (provided that the solvent is transparent to the radiation in that region) and also a function of the length of the path of the radiation through the sample.

Various forms of the Beer-Lambert law are:

$$I = I_0 \times 10^{-\epsilon cl} \tag{2.1}$$

or

$$-\log \frac{I}{I_0} = \epsilon cl, \quad \text{that is,} \log \frac{I_0}{I} = \epsilon cl \tag{2.2}$$

where I_0 = intensity of *incident* radiant energy,
I = intensity of *transmitted* radiant energy,
c = molar concentration of the solute,
l = internal length of cell (i.e., quartz sample holder) in centimeters.

The ratio I/I_0 is known as the *transmittance T*, and the logarithm of the inverse ratio I_0/I is known as the *absorbance A*. Therefore

$$-\log \frac{I}{I_0} = -\log T = \epsilon cl \quad \text{and} \quad \log \frac{I_0}{I} = A = \epsilon cl.$$

Thus

$$A = -\log T = \epsilon cl \quad \text{or} \quad \epsilon = A/cl \tag{2.3}*$$

where ϵ is the *molar extinction coefficient* of the substance whose light absorption is under investigation; it is a constant,† and is characteristic of a given absorbing molecule or ion in a particular solvent at a particular wavelength. ϵ is numerically equal to the absorbance of a solution of unit molar concentration (c = 1) in a cell of unit length (l = 1). The units of ϵ are thus the same as those of $(c \times l)^{-1}$, namely liters · moles^{-1} · cm^{-1}.

An ultraviolet spectrum is a plot of absorption intensity (ordinate) versus wavelength in Å or nm (abscissa). The absorption intensity is plotted either as ϵ or as $\log \epsilon$ and is reported as ϵ_{max}, meaning that this is the maximum molar extinction coefficient. The wavelength of absorption is usually reported as λ_{max}, meaning that this is the maximum absorption. In case of unknown substances, where molecular weight M and hence molar concentration of the solution cannot be determined, the absorption intensity of a 1% solution of the substance is determined usually in a 1-cm cell. This is then reported as $E_{1\%}^{1\,cm}$, where E is the numerical value of absorbance in a cell of unit length of a solution containing unit mass per given volume of solution. E is thus called the *specific extinction coefficient*. The relation of this expression to the molar extinction coefficient ϵ is given by

$$E_{1\%}^{1\,cm} = \frac{10 \times \epsilon}{\text{molecular weight}}$$

or

$$\epsilon = E_{1\%}^{1\,cm} \times 0.1\ M$$

* When the intensity of the incident radiation I_0 is considered to be 100, then the absorbance A becomes $2 - \log \%T$ in terms of percentage transmittance.

† Since ϵ is a constant and $A = \epsilon \times c \times l$, the absorbance of a solution should remain constant as long as the product of $c \times l$ is constant. That is, if the Beer-Lambert rule is correct, absorbance due to 0.1-molar solution in a cell whose path length is 1 cm should be the same as that due to a 1-molar solution of the same substance in a cell of only 0.1 cm path length. However, this is not always true, and many factors are responsible for this failure of the rule. In practice, therefore, in the quantitative use of UV and visible spectroscopy, one must establish the validity of the Beer-Lambert rule for a given solution over the ranges of concentrations and wavelengths likely to be used.

Terms and symbols used in this text	*Other commonly used equivalent terms and symbols*
Nanometer, nm	Millimicrons, mμ
Hertz, Hz	Cycles per second, cps
Intensity of radiant energy, I_0	Incident radiant power, P_0
Intensity of transmitted radiant energy, I	Transmitted radiant power, P
Internal cell length, l	Internal cell path, b
Transmittance, T	Transmittancy, T
Absorbance, A	Absorbancy, A; Optical density, *od*
Molar extinction coefficient, ϵ	Molar absorptivity, ϵ; Molar absorbancy index, a_M
Specific extinction coefficient, E	Specific absorptivity, a; Absorbancy index, a_s; Extinction coefficient, k

The terms and symbols used in this section are becoming increasingly common, some through recent international agreements. However, several systems of nomenclature and symbols are still in general use, and often create confusion. To avoid such confusion, the list above gives other terms equivalent to those used in this text.

2.4 THE INSTRUMENT

We saw in Chapter 1 that a spectroscopic instrument is made up of three basic components:

1. A source of radiation
2. A monochromator
3. A detector and measuring assembly

1. Radiation source Many chemistry laboratories have visible-ultraviolet recording spectrophotometers. Such an instrument usually contains two different sources of radiation: a tungsten-filament lamp for scanning in the visible region (3500 Å to about 8000 Å), and a hydrogen or deuterium discharge tube for scanning in the UV region (about 1850 Å to about 3900 Å). Some instruments automatically switch from one source to another as the scanning device is changed from one region to another. The discharge tube has a quartz window and contains hydrogen or deuterium gas at low pressure. An electrical impulse of a certain voltage is applied to a pair of electrodes inside the tube, which excites the gas molecules and causes them to emit a continuous radiation in the UV region.

2. Monochromator Either a prism or a grating may be used as a dispersing element in a visible-UV instrument. If a glass prism is used, it is suitable only in the visible region. A low-priced UV instrument may have a prism made of natural quartz; however, the fact that the transmission capabilities of quartz are relatively poor restricts the useful lower limit of such an instrument to wavelengths of about 2050–2100 Å. A prism made of very pure synthetic fused silica extends this range to wavelengths of about 1800 Å. As a matter of fact, in this region the oxygen in the atmosphere absorbs radiation, making it necessary to use special evacuable instruments for scanning radiation with wavelength less than 1800 Å. In such instruments prisms made of calcium or lithium fluoride are often used. An instrument using narrow-band interference filters for the 2000–4000 Å region has recently been described.[1]

3. Detecting and measuring assembly The detectors most commonly employed in visible-UV instruments are photoelectric or photomultiplier tubes. Such a tube has a quartz window to admit UV radiation, an alkaline-earth oxide-coated, semicylindrical cathode, and an anode made of metal wire to collect electrons (Fig. 2.4). When radiation falls on the cathode surface, a very small number of electrons, proportional to the intensity of the incident

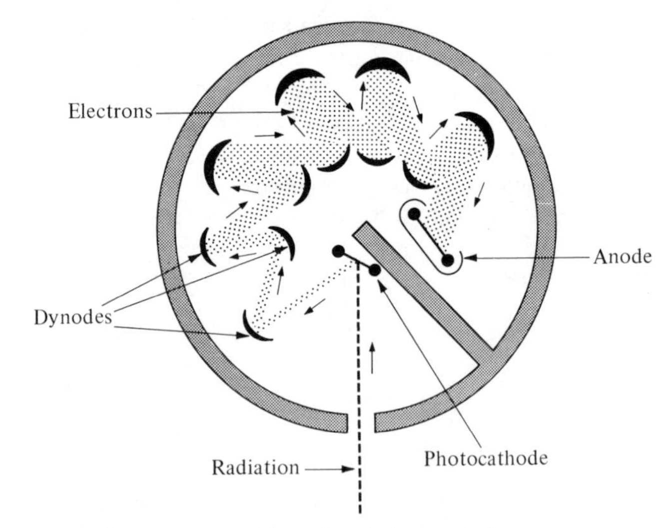

Fig. 2.4 A cross section of a photomultiplier tube, showing how electrons are multiplied by contact with dynodes.

radiation, are emitted from the oxide (which contains electrons that are relatively loosely bound) that is coated on the cathode. If these few electrons were to flow directly to the anode, they would produce only a very small current (about 10^{-10} ampere). But in a photomultiplier tube, the dynodes cause the current to be amplified by a factor of approximately 10^6. The electrons are accelerated as they pass from dynode to dynode. Each time an electron strikes the surface of a dynode it acquires greater energy, strikes the next dynode with a greater power, and causes the dynode to eject several more electrons from its surface.

The three main types of measuring devices used in commercial instruments are: (i) meters, (ii) recorders, and (iii) digital readout units. However, most modern instruments use automatic recorders; this is the type you are most likely to come across in your laboratory. There are a number of different types of recorders available commercially, the most common one being of the type known as a *potentiometric servo-recorder.* With the general trend toward ease of operation, digital readout is becoming available in almost all new manual instruments.

Problem 2.1 On the basis of our discussion in Section 1.10, draw a schematic diagram of the optical path of a double-beam UV spectrometer. Compare your diagram with the one for the instrument available in your laboratory.

Problem 2.2 A solution of a colored substance at a concentration c has a percentage transmittance of 82, which changes to 45.2 when the concentration is increased to $4c$. Establish the validity of the Beer-Lambert law on the basis of this observation. What would the percentage transmittance be at concentration $2c$?

2.5 OPERATING AN UV INSTRUMENT

The actual operation of a spectroscopic instrument to obtain a spectrum for a given compound is surprisingly simple. Often it involves merely turning a few knobs.

Most spectrophotometers today are *double-beam* instruments, in which the primary source of radiation is split into two beams by the use of mirrors. One of them serves as a reference beam, and the other is the one which has the sample interposed in it. Such a procedure has several advantages. For example, since the two beams are mirror images of one another, their relative intensities remain unaltered even through any fluctuations in the light source. Second, this method makes possible the use of an ac amplifier. This is achieved by the use of a mechanical chopper in the form of a rotating disc which alternately chops off the reference beam and the sample beam in such a manner that the frequency at which the two beams alternately reach the amplifier input is the same as that of the power line. This technique also makes it possible to differentiate between the absorption of photons by the solvent in the reference beam and the absorption of photons by the solution in the sample beam. The ac amplifier sends no signals to the recorder so long as the intensities of both the beams are the same. However, should the intensities of the two beams differ, due to greater absorption of photons by the sample, for instance, the imbalance in signal is detected, and this activates the pen on the potentiometric recorder.*

Gases or solutions are usually used as samples in visible–UV study. Sample holders, often known as cells, are made of quartz or fused silica. Those having path lengths of 0.1 mm to 10 cm are often used for gases. For solutions, quartz cells of either fixed or variable path lengths are used (Fig. 2.5).

Fig. 2.5 Variable-path liquid cell. (Courtesy Perkin-Elmer)

Perhaps you recall, from Fig. 1.12, that the sample is placed between the source of radiation and the monochromator. This is the practice in infrared and microwave spectroscopy, in which the radiation is not very powerful. However, the higher energy resulting from polychromatic radiation in the ultraviolet and visible regions may cause photochemical decomposition or fluorescence in the sample. To avoid this, the sample in UV-visible studies is placed on the other side of the monochromator, so that only bands of radiation of narrow wavelengths pass through the sample at any given time. Whenever a solution is used for UV-visible investigation in a double-beam instrument, an identical cell containing pure solvent is placed in the reference beam.

---

* Infrared instruments also operate on the double beam null readout system. To understand this double-beam operation see Fig. 3.1, page 47.

2.6 SELECTION OF A SOLVENT

One has to be careful in choosing a solvent for spectroscopic studies. Polar solvents such as water, alcohol, etc., tend to shift the position and tend to broaden the bands. This is due to the fact that when the electrons of a molecule are in the ground state, the perturbation of the molecule is different from the perturbation which exists when the electrons are in an excited state, due to the nature of the interactions between solvent and solute molecules in the two states. These differences are, however, usually very small and difficult to interpret. Highly pure, nonpolar solvents, such as saturated hydrocarbons, do not affect the absorption bands–i.e., the spectrum of a compound in a pure, gaseous state is usually similar to its spectrum in a nonpolar solvent.

The most important criterion in the choice of a solvent is the ability of the solvent to transmit radiation in the wavelength region under study. The solvent should not absorb radiation in the same region as the solute. In the visible region (wavelengths 4000–8000 Å), any colorless solvent may be employed. In the UV region (3000–4000 Å), solvents with transmission limits below 2500 Å (250 nm) may safely be used, but in the region of wavelengths less than 3000 Å, highly purified saturated alkanes such as hexane and heptane are essential. No suitable solvent is available for the region of wavelengths less than about 1700 Å (170 nm), and spectral study in this far UV region is often carried out on compounds in the gaseous state. Table 2.1 presents the lower cutoff wavelengths in the near ultraviolet region for some of the commonly used solvents. The lower cutoff value for a solvent represents the lower limits of wavelength below which that particular solvent should not be used to determine the UV spectrum of a compound. The wavelengths are given in nanometers; it is presumed that 1-cm cells are used.

After choosing the right solvent, one must decide on the concentration of the solution. The concentration, of course, depends on the quantity of the solute available; anywhere between 0.1 to 100 mg of the substance may be needed. Normally, solutions of 10^{-5} to about 10^{-2} molar concentration may safely be employed. However, for reliable results, the percentage transmittance of the solution should range between about 20 and 65% in most cases. At high concentrations the amount of radiation transmitted is low, increasing the possibility of error.*

2.7 UV ABSORPTION BY ORGANIC MOLECULES

In the early decades of spectroscopy, the spectra of compounds were discussed merely empirically in terms of

* For details on the solvent effects in photometric analysis, see Reference 2.

Table 2.1 Approximate lower cutoff wavelengths for commonly used solvents in UV–visible spectroscopy of organic substances

Solvent	Cutoff wavelength, nm	Solvent	Cutoff wavelength, nm
	200–250		250–300
Acetonitrile	210	Benzene	280
n-Butanol	210	Carbon tetrachloride	265
Chloroform	245	N,N-Dimethylformamide	270
Cyclohexane	210	Methyl formate	260
Decahydronapthalene	200		
1,1-Dichloroethane	235	Tetrachloroethylene	290
Dichloromethane	235	Xylene	295
Dioxane	225		
Dodecane	200		300–350
Ethanol	210	Acetone	330
Ethyl ether	210	Benzonitrile	300
Heptane	210	Bromoform	335
Hexane	210	Pyridine	305
Methanol	215		
Methylcyclohexane	210		350–400
*iso*octane	210		
*iso*propanol	215		
Water	210	Nitromethane	380

chromophores and *auxochromes.* It was recognized that some substances appear colored because they contain functional groups which are capable of absorbing radiation of certain wavelength when an ordinary white light shines on them. For example, a substance in ordinary daylight (λ = 4000–8000 Å) appears blue because it has a functional group (or groups) which absorbs waves of wavelength between 5700 and 5900 Å (yellow light). The transmitted waves, which are now deficient in yellow light, give the effect of blue color. Such functional groups, which confer color on substances, became known as chromophores (*chromo* = color). Most chromophores have unsaturated bonds such as C=C; C=O, N=N, etc. Functional groups such as hydroxyl (—OH), amino ($—NH_2$), halogenic (—Cl, —Br, etc.) cannot confer color on substances, but have the ability to increase the coloring power of a chromophore. Such groups became known as auxochromes.

Today the terms auxochrome and chromophore are no longer limited to absorption of radiation in the visible region only. In the modern sense:

A chromophore is an isolated functional group capable of absorbing visible and/or ultraviolet radiation (800–200 nm).

An auxochrome is a functional group which does not absorb radiation longer than 200 nm, but, when attached to a given chromophore, causes a shift in absorption to longer wavelengths, and also increases the degree of absorption.

For example, pure benzene has an absorption maximum at about 254 nm, but, when an auxochrome (such as a hydroxyl group) is attached to the benzene ring, the degree of absorption increases and λ_{max} shifts to a longer wavelength. Thus phenol,

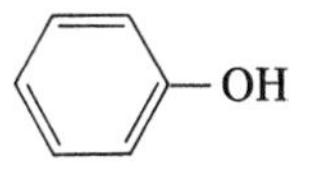

has an absorption maximum at 270 nm with ϵ_{max} (1450) approximately seven times as great as for benzene (ϵ_{max} = 204). A shift in absorption peak of a chromophore from a standard position to:

i) longer wavelength is called the *bathochromic (red) shift*

ii) shorter wavelength is called the *hypsochromic (blue) shift*

iii) greater intensity is called the *hyperchromic shift*

iv) lesser intensity is called the *hypochromic shift*

Knowing that UV and visible radiations cause electronic transitions in the molecules, we are now in a position to correlate the traditional terms, such as chromophores and auxochromes, with the different types of valence electrons present in the organic molecules. Recall that in Chapter 1 we examined three types of valence electrons: sigma (σ), pi (π), and nonbonded (n) electrons. Recall also that without going into any details, we mentioned $\sigma \rightarrow \sigma^*$, etc., transitions. It is now time to look more closely at these transitions and their significance.

Let us start with fundamentals with which we are already familiar. For example, we know that a bond formation between two atoms involves overlap of two orbitals, each containing one electron. Such an overlap gives rise to two new orbitals, known as *molecular orbitals.* One of them, called a *bonding orbital,* has a lower energy than either of the original atomic orbitals, and the other, called an *antibonding orbital,* has higher energy. Figure 2.6 shows how bonds are formed; note the typical potential energy curve.

In the ground state, both electrons of the σ bond are in the bonding molecular orbital and have opposite spins (symbolized by ↑↓). There are no electrons in the antibonding orbital. When the molecule absorbs radiation with energy exactly equal to $\Delta E_{(\sigma \rightarrow \sigma^*)}$, the electronic transition from the bonding to the antibonding orbital occurs with net retention of electronic spins. This excited state is known as a *singlet state.* The excited electron may then undergo spin inversion, which means that the electrons are transposed to the *triplet state,* in which their electronic spins are parallel (↑↑ or ↓↓). Since the amount of energy required for $\sigma \rightarrow \sigma^*$ is very high, this transition is found mainly in the *vacuum* UV region. Saturated hydrocarbons have C—H and C—C single bonds only, and these are sigma bonds. Therefore they do not absorb radiation in the UV-visible regions. (Exception: cyclopropane. Due to ring strain, cyclopropane behaves like an unsaturated compound and absorbs at 190 nm. Generally, cyclic compounds with small ring have strained bond angles. Consequently, they absorb radiation at slightly longer wavelengths than open-chain compounds.)

2.8 EFFECTS OF AUXOCHROMES ON THE POSITIONS OF UV BANDS

Let us now consider the absorption pattern of a molecule which contains one of the functional groups such as $-\ddot{\underset{..}{O}}H$, $-\ddot{N}H_2$, —halogens, etc. Recall that these groups are called *auxochromes.* Note also that a molecule having one of these groups will have a pair of nonbonded elec-

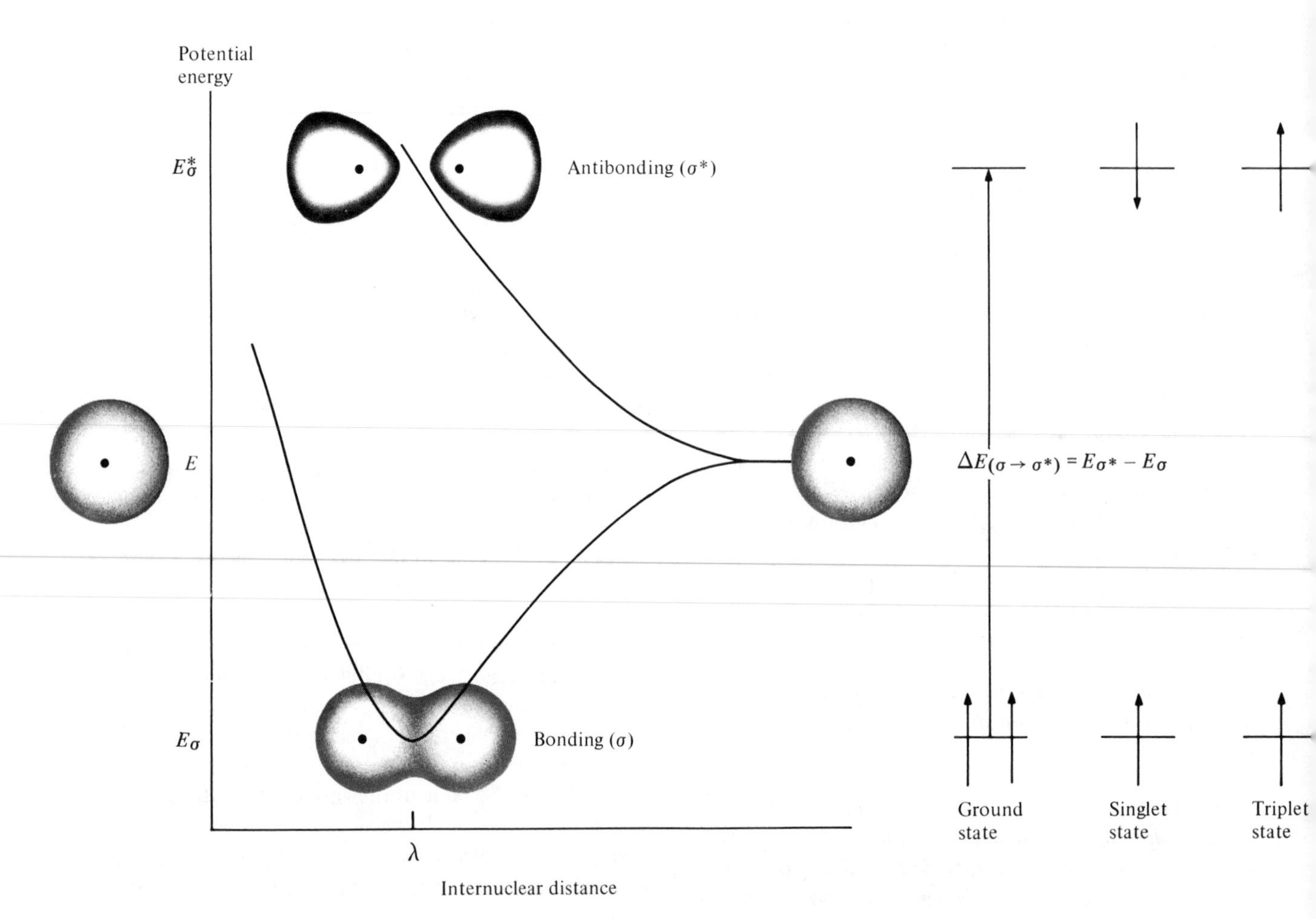

Fig. 2.6 Energy diagram for σ bonds: ground, singlet, and triplet states. Note that the internuclear distance remains unchanged during the transition, so that, even if there is no orbital overlap in the antibonding state, no bond cleavage takes place. This is in accord with *Franck-Condon principle*, which states that electronic excitation occurs so rapidly that the nuclei do not have time to change their positions or their momenta during the transition.

trons in addition to some sigma electrons. The n electrons can be excited to a σ^* energy level, and since they are already at a higher energy level than the σ electrons in the molecule, the energy required for the $n \to \sigma^*$ transition is less than that required for $\sigma \to \sigma^*$. That is,

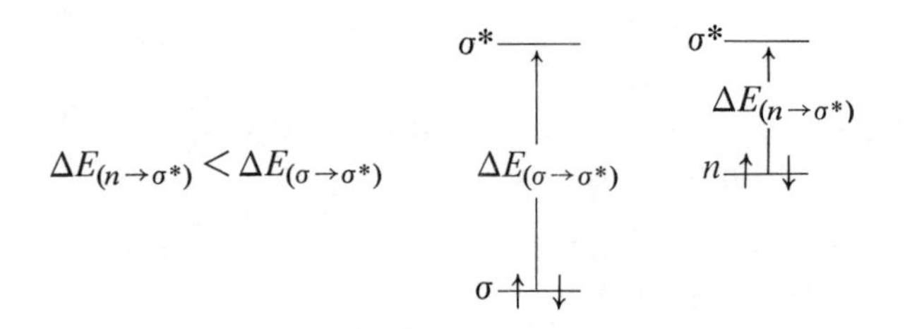

This means that the $n \to \sigma^*$ transition occurs at longer wavelengths than the $\sigma \to \sigma^*$. For example:

Saturated alcohols or epoxides absorb radiation in the region 180–185 nm ($\log \epsilon = 2.5$) due to the $n \to \sigma^*$ transition.

In saturated amines, λ_{max} for $n \to \sigma^*$ is 190–200 nm ($\log \epsilon = 3.4$)

In saturated chlorides, λ_{max} is 170–175 nm ($\log \epsilon = 2.5$)

In saturated bromides, λ_{max} is 200–210 nm ($\log \epsilon = 2.6$)

And so on. We should remember that λ_{max} for the $n \to \sigma^*$ transition depends on the ease with which n electrons can be excited. If in the ground state these n electrons are stabilized by hydrogen bonding, or by some other means, the λ_{max} naturally shifts to a lower wavelength. Such a *hypsochromic shift* (blue shift) occurs when spectra of these substances are analyzed in solvents that lend themselves to hydrogen bonding.

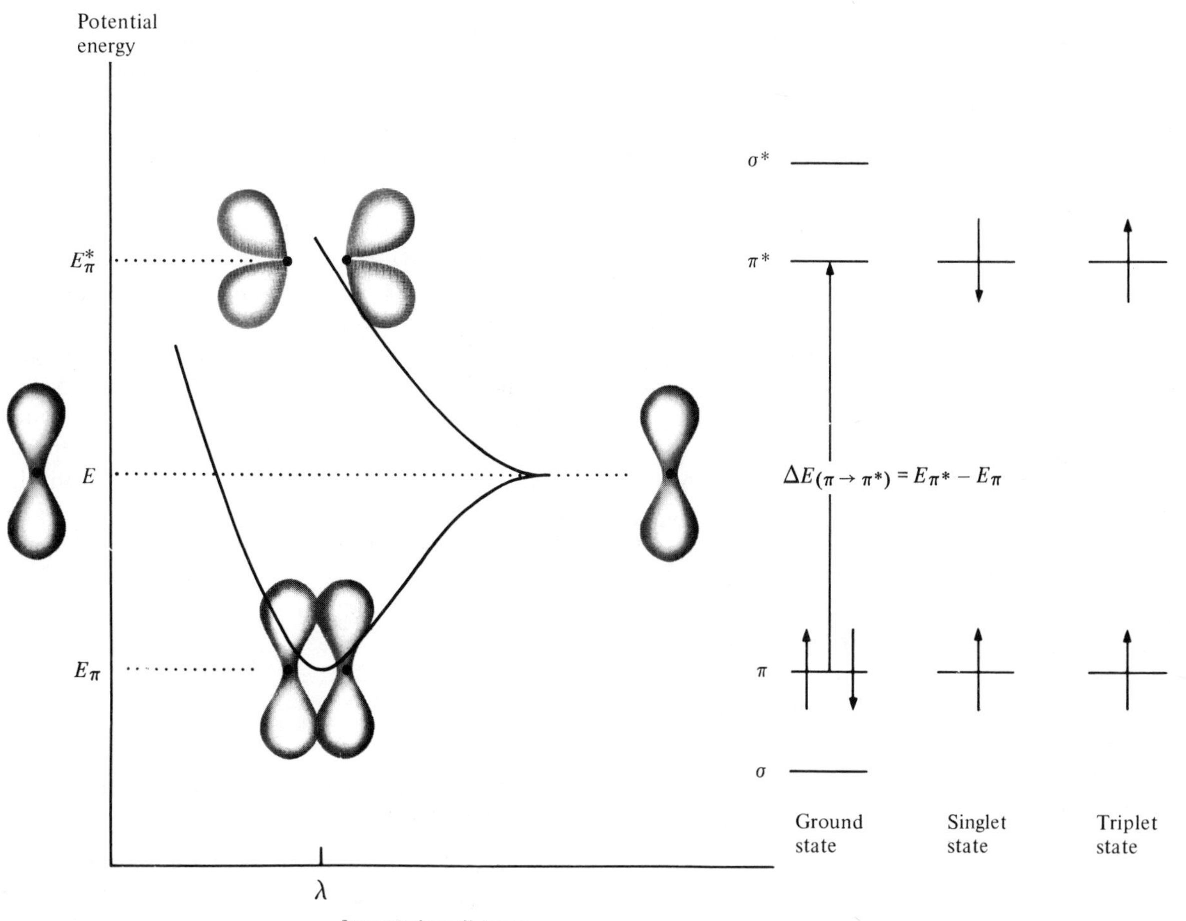

Fig. 2.7 Energy diagram for π-bonding and antibonding molecular orbitals.

Figure 2.7 shows yet another interesting electronic transition in a molecule such as ethylene: $CH_2{=}CH_2$. These molecules contain π electrons in addition to some σ electrons. As we are aware, π bonding involves overlap of parallel atomic orbitals and therefore results in a bond that is considerably weaker than the σ bond. That is, in a molecule, π electrons are at a higher energy than σ electrons. Consequently, $\pi \to \pi^*$ transitions take place at a longer wavelength than $\sigma \to \sigma^*$ and $\Delta E_{(\pi \to \pi^*)} < \Delta E_{(\sigma \to \sigma^*)}$. In simple olefins such as ethylene, propylene, and 1-octene, containing isolated double bonds, $\pi \to \pi^*$ transition occurs in the 170–200 nm region. In such molecules, $\pi \to \sigma^*$ transition can also take place, but for such transitions,

$$\Delta E_{(\pi \to \sigma^*)} \gg \Delta E_{(\pi \to \pi^*)}$$

and therefore they are observed in far UV regions only.

In fact, ethylene displays two peaks of unequal intensity, both due to the $\pi \to \pi^*$ transitions. The strong (ϵ_{max} = 10,000) peak is displayed at λ_{max} = 165 nm, whereas the weak ($\epsilon_{max} \simeq 1000$) peak appears at 210 nm. These peaks undergo *bathochromic shifts* (red shifts) if suitable auxochromes are substituted for the ethylene protons. For example, a basic group, such as $-\ddot{N}R_2$, attached directly to a vinyl carbon, causes the long-wavelength band to shift to 250 nm, whereas a single alkyl substituent shifts this band by about 4–5 nm, to approximately 214 nm (λ_{max} for propylene = 214 nm). The bathochromic shifts due to the substituents are explained on the basis of the resonance interactions. In the case of basic substituents, such as $-\ddot{N}R_2$, the lone-pair *n* electrons of the auxochrome participate in such a resonance, whereas in the case of alkyl groups the electrons of the C—H bond participate in the resonance by means of *hyperconjugation*; an example is:

$$H_3{\equiv}C{-}CH{=}CH_2 \longleftrightarrow \overset{+}{H_3}{=}C{=}CH{-}\overset{-}{C}H_2$$

There are small but significant differences in the absorption spectra of *cis* and *trans* isomers of disubstituted olefins. For a given olefin, both the λ_{max} and ϵ_{max} values of the *trans* isomer are usually greater than those of the *cis* isomer. For example, *trans* diphenylethylene,

$$\begin{matrix} Ph & & H \\ & C{=}C & \\ H & & Ph \end{matrix}$$

displays $\pi \rightarrow \pi^*$ transition at 295 nm (ϵ_{max} = 27,000), which is greater than that shown by its *cis* isomer (λ_{max} = 280 nm, ϵ_{max} = 13,500).

So far we have seen how transitions take place in molecules containing *isolated* auxochromes or π or σ bonds. Almost all these transitions occur below 200 nm, that is, below the region in which most common laboratory instruments can successfully be used. Consequently, had it not been for the interesting effects caused by the associations and conjugations of chromophores and auxochromes, an organic chemist would find very little interest in UV spectroscopy.

Let us now consider a chromophore such as C=O or C=S or N=N, in which both π and n electrons are present. The possible electronic transitions in a simple molecule containing one such group are:

i) $\sigma \rightarrow \sigma^*$
ii) $\sigma \rightarrow \pi^*$
iii) $n \rightarrow \sigma^*$
iv) $\pi \rightarrow \pi^*$
v) $n \rightarrow \pi^*$

Of these five possible transitions, only $n \rightarrow \pi^*$ takes place in the region above 200 nm, and that too with a very low ϵ. For example, the $n \rightarrow \pi^*$ transition in acetaldehyde occurs at 294 nm with log ϵ = 1.08 (cyclohexane solvent); while the $\pi \rightarrow \pi^*$ excitation is at 190 nm with log ϵ = 2.0. As a matter of fact, the presence of a weak absorption band in the 275-295 nm region is considered to be a positive identification of a ketone or an aldehydic carbonyl group.

The $n \rightarrow \pi^*$ transitions are largely affected by the solvent. The bathochromic or hypsochromic shifts in these bands usually depend on the polarity of the chromophore and the electrostatic forces of the solvent. For example, a carbonyl group in the ground state is a resonance hybrid of the two structures;

$$C{=}O \longleftrightarrow \overset{+}{C}{-}\overset{-}{O} = \overset{+\delta}{C}\ \overset{-\delta}{O}$$

The degree of contribution to the hybrid by the polar form depends on the nature of substituents on carbonyl carbon. The electron-releasing substituents make the hybrid more polar and the electron-withdrawing substituents make it less so. The hybrid is therefore stabilized to a greater or lesser extent by hydrogen bonding, or by nonbonding interactions of the carbonyl group with the solvent. The polarity of a structure may also increase or decrease as the molecules go from ground state to excited state. A hypsochromic (blue) shift is therefore observed in a polar solvent if its ground-state structure is relatively more polar (and thus more solvent-stabilized) than its structure in the excited state. For example, acetone shows $n \rightarrow \pi^*$ transition at 279 nm in hexane but at 264.5 nm in water. Conversely, if the structure of a molecule in the excited state is relatively more polar than it is in the ground state, a bathochromic (red) shift occurs. This sort of solvent effect is fairly general, and should be expected in the spectral properties of a variety of compounds.

2.9 EFFECT OF CONJUGATION ON THE POSITIONS OF UV BANDS

It appears that of the five types of electronic excitations that we have considered thus far, only one—the $n \rightarrow \pi^*$ transition—is within the usual range detectable by instruments; therefore this transition might seem to be the only one of praçtical value (Fig. 2.8). Fortunately this is not true, because tremendous shifts in absorption bands are caused by conjugation. We have seen that in a simple olefin a $\pi \rightarrow \pi^*$ transition takes place in the 165-200 nm region, but what happens if the molecule contains two double bonds rather than only one? If there are two or more similar chromophores in a molecule, the transition of electrons depends on how and where these chromophores are located in the molecule. In this case, if the two double bonds are isolated by the presence of at least three single bonds, the $\pi \rightarrow \pi^*$ transition is still in approximately the same region (165-200 nm), although the absorbance of radiation and therefore the ϵ_{max} is a little greater. On the other hand, if the two double bonds are conjugated, absorptions take place at longer wavelengths. This increases the usefulness of UV spectroscopy for the determination of the structure of an organic compound.

The fact that conjugation causes bathochromic (red) shifts suggests that conjugation decreases the energy required for the excitation of electrons. This can be explained on the basis of the *molecular orbital theory.* According to this theory, a linear combination of two

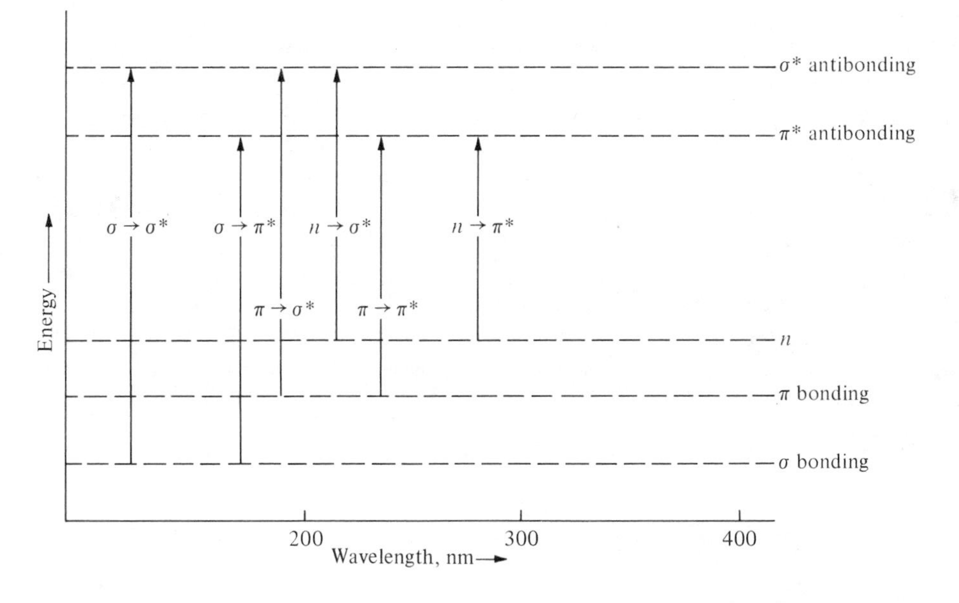

Fig. 2.8 A qualitative representation of different kinds of electronic transitions, the wavelengths at which they take place, and the energy required for them (nonconjugated chromophores only).

atomic orbitals† results in the formulation of two molecular orbitals (MO's) one of which is a low-energy, bonding orbital and the other the high-energy, antibonding orbital. The wave functions of the two MO's are represented as

$$\psi_b = \phi_1 + \phi_2, \qquad \psi_a = \phi_1 - \phi_2,$$

where ψ (Greek psi) denotes a molecular orbital (MO), ϕ (Greek phi) denotes an atomic orbital (AO), the subscripts b and a refer to bonding and antibonding, respectively, and the subscripts 1 and 2 refer to the two atomic orbitals involved in the combination.

In the case of ethylene, in which the two carbon atoms are sp^2 hybridized, there is one partially filled p orbital on each carbon atom. The parallel overlap of these two p orbitals (Fig. 2.9a) produces a bonding orbital (π orbital) whose wave function can be represented as $\psi_b = \phi_1\,(2p) + \phi_2\,(2p)$. Simultaneously an antibonding orbital (π^* orbital) is also produced, whose wave function can be represented as $\psi_a = \phi_1\,(2p) - \phi_2\,(2p)$. The bonding π orbital, which is the low-energy orbital, contains both the electrons of the π bond. One or both of these electrons is promoted to the antibonding π^* orbital upon absorption of ultraviolet radiation by the molecule. The frequency of absorption (i.e., the excitation energy) for this $\pi \rightarrow \pi^*$ transition depends on the difference in energy between the π and π^* levels. Let us denote this difference

$$E_{\pi^*} - E_{\pi} = \Delta E_1.$$

Now consider 1,3-butadiene, $H_2C{=}CH{-}CH{=}CH_2$, which contains two conjugated double bonds. All four carbon atoms in this molecule are sp^2 hybridized, and therefore have one pure p orbital each, so that there is parallel overlap of orbitals. These four p orbitals combine to form four MO's which are delocalized over all four carbon atoms (Fig. 2.9b). If we recognize that in butadiene carbon atom 1 is equivalent to carbon atom 4, and that 2 is equivalent to 3, then we can derive the wave functions of the four MO's by the LCAO method. First we either add or subtract the functions of p_1 and p_4 and then add or subtract the result from the sum (or difference) of the p_2, p_3 combination. Thus we can represent the four MO's by the four wave functions:

$$\psi_1 = (\phi_1 + \phi_4) + (\phi_2 + \phi_3)$$
$$\psi_2 = (\phi_1 - \phi_4) + (\phi_2 - \phi_3)$$
$$\psi_3 = (\phi_1 + \phi_4) - (\phi_2 + \phi_3)$$
$$\psi_4 = (\phi_1 - \phi_4) - (\phi_2 - \phi_3)$$

Figure 2.9(b) shows these MO's and their energy levels. The MO's represented by ψ_1 and ψ_2 have lower energies than the AO's from which they are formed; this lowering is greater for ψ_1 than for ψ_2. Similarly, the MO's ψ_3 and ψ_4 have energies higher than those of the parent AO's, ψ_4 having higher energy than ψ_3. The four

† For a qualitative treatment of molecular orbitals, a non-mathematical, semi-empirical technique known as *linear combination of atomic orbitals* (LCAO)[3] is often used.

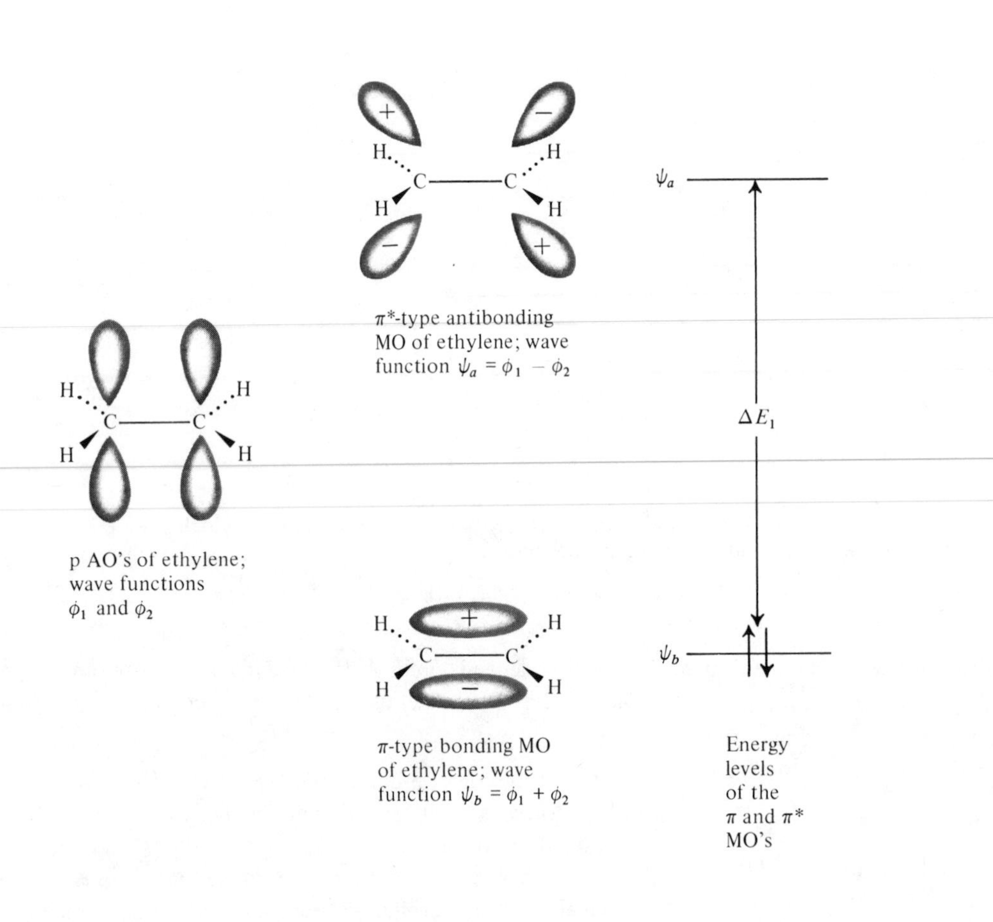

Fig. 2.9 (a) The two molecular orbitals (MO's) formed by the two atomic orbitals (AO's) in ethylene. The difference in energy between the two MO's is given by ΔE_1.

p electrons are therefore in the bonding orbitals ψ_1 and ψ_2, whereas the antibonding orbitals represented by ψ_3 and ψ_4 are vacant.

In 1,3-butadiene, the easiest $\pi \rightarrow \pi^*$ transition involves excitation of an electron from ψ_2 to ψ_3. Figure 2.9(b) shows that the excitation energy required for this transition is:

$$E_{\psi_3} - E_{\psi_2} = \Delta E_2$$

which is less than the ΔE_1 required for the $\pi \rightarrow \pi^*$ transition in ethylene. Consequently, the $\pi \rightarrow \pi^*$ transition in a conjugated system such as butadiene takes place at a longer wavelength than that at which a similar transition occurs in the unconjugated ethylene. In butadiene, this transition occurs at 220 nm in a nonpolar solvent and further it shifts to longer wavelengths (about 40 nm longer for each additional double bond) as the number of coplanar, conjugated double bonds in the molecule increases. The degree of absorption increases simultaneously. For example, 1,3,5-hexatriene, H—(CH=CH)$_3$—H, absorbs radiation at 256 nm (in hexane), whereas α-carotene, which has 10 conjugated double bonds, absorbs radiation in the visible region at 445 nm. A substance which has λ_{max} at 445 nm must be colored. As a matter of fact, substances having λ_{max} even below 400 nm may also be colored because electronic absorption bands are often broad, and may have absorption "tails" that extend into the visible region.

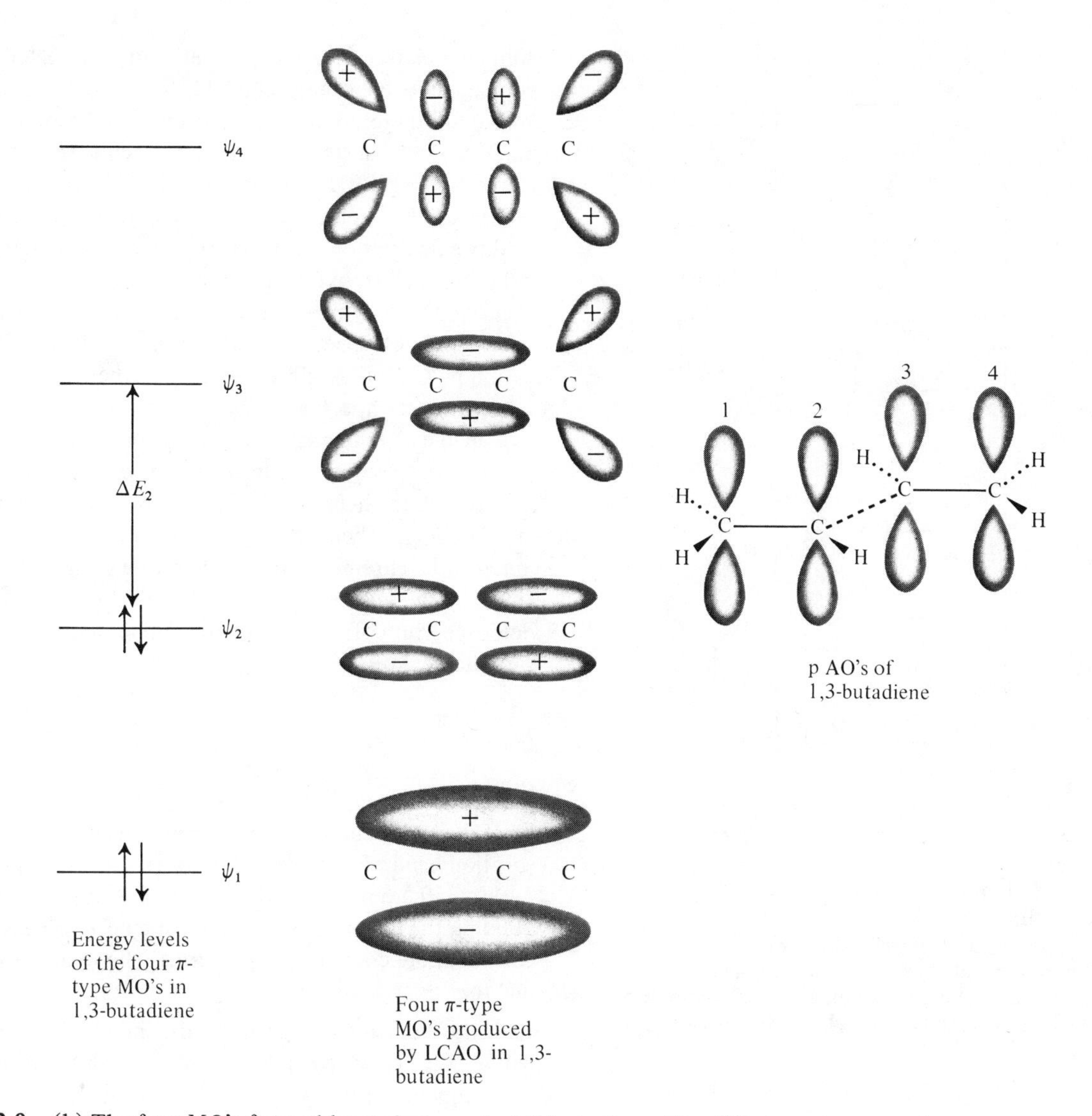

Fig. 2.9 (b) The four MO's formed by π electrons in 1,3-butadiene. The difference in energy between the two MO's involved in the lowest energy transition ($\psi_2 \rightarrow \psi_3$) is given by ΔE_2.

2.10 CONJUGATION OF TWO DIFFERENT CHROMOPHORES

Crotonaldehyde,

```
 H       H
  \     /
   C=C
  /     \
CH3      C=O
        /
       H
```

has an olefinic chromophore conjugated with a carbonyl. An isolated C=C chromophore shows a $\pi \rightarrow \pi^*$ peak at about 165 nm, whereas an isolated carbonyl chromophore shows a strong $\pi \rightarrow \pi^*$ peak at about 170 nm and a weak $n \rightarrow \pi^*$ peak at about 290 nm. When these two chromophores are conjugated, as in crotonaldehyde or in any typical α, β unsaturated carbonyl system (an enone), we have a situation somewhat similar to that in butadiene (see Fig. 2.10).

The conjugated system in a typical enone involves three sp^2 hybridized carbon atoms and an oxygen atom. Each of these four atoms, as in butadiene, contributes one electron to the π-electron system to produce four MO's: ψ_1, ψ_2, ψ_3^*, and ψ_4^*. These four MO's of the enone are similar to those in the butadiene, except that, in the enone (unlike the situation in the diene), nonbonded pairs of electrons (n electrons) exist in the p orbitals of oxygen. In an isolated carbonyl chromophore (Fig. 2.10b), the absorption of UV energy involves transition of one of the lone-pair oxygen electrons from level n to π^*. But, in the conjugated enone, less excitation

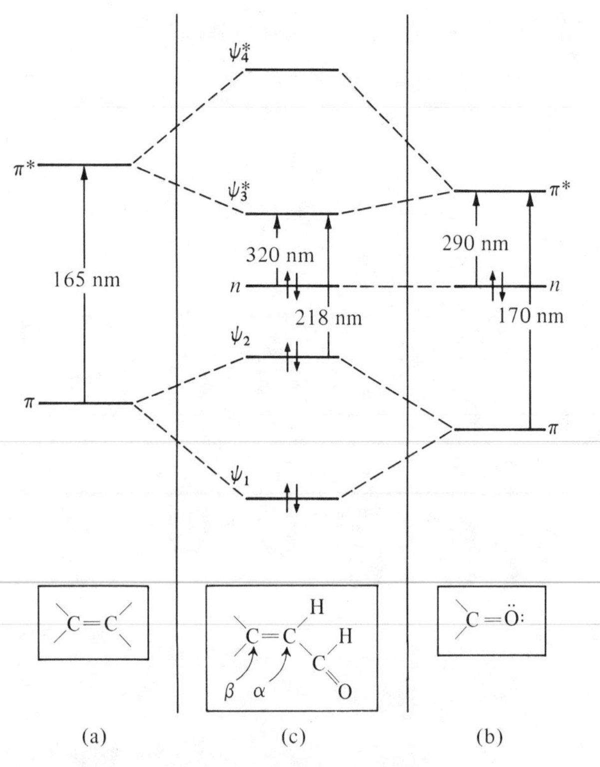

Fig. 2.10 Energy-level diagrams of (a) an isolated C=C chromophore, (b) an isolated C=O chromophore, and (c) an enone (α, β unsaturated carbonyl compound) in which the two chromophores are conjugated. Note that the $n \rightarrow \psi_3^*$ and $\psi_2 \rightarrow \psi_3^*$ transitions require lower energies in enone and thus can occur at longer wavelengths than the corresponding $n \rightarrow \pi^*$ and $\pi \rightarrow \pi^*$ transitions in an isolated carbonyl chromophore.

energy is required because the transition takes place from the n level to the lowest, unoccupied ψ_3^* level. Hence the $n \rightarrow \psi_3^*$ transition in crotonaldehyde occurs at a longer wavelength (λ_{max} = 320 nm, ϵ_{max} = 30 in ethanol) as against the $n \rightarrow \pi^*$ transition in acetaldehyde (λ_{max} = 293.4, ϵ_{max} = 11.8 in hexane). Figure 2.10 shows that the $\psi_2 \rightarrow \psi_3^*$ transition too requires less energy than that needed for the $\pi \rightarrow \pi^*$ transition of an isolated double bond. Hence this band shifts from 170 nm in acetaldehyde to 218 nm (ϵ_{max} = 18,000) in crotonaldehyde.

2.11 DETERMINING THE STRUCTURE OF AN ORGANIC UNKNOWN

Determining the structure of an organic compound is like putting a giant jigsaw puzzle together. Many chemical and instrumental techniques are employed to provide the pieces of this puzzle. No single technique can give us the complete picture of the compound, but each technique complements the others, and adds to the total picture. We should therefore not expect an ultraviolet spectrum to reveal the total structure of the molecule. Instead we should try to get the maximum possible information by relating the electronic spectral characteristics to the structural peculiarities of the molecule. Later, in Chapters 3, 4 and 5, we shall learn how to find out more about a given molecule by studying its infrared, nuclear magnetic resonance, and mass spectra. In Chapter 6, we shall finally learn how to fit these jigsaw pieces together.

The most important role of an ultraviolet spectrum is to enable us to identify a chromophore and to estimate the extent of conjugation in the unknown molecule. We do this with the help of a set of rules by which we can calculate λ_{max}. Then we search for a model system which contains the chromophore and therefore gives a spectrum similar to the one being examined. We examine the ultraviolet spectrum of the unknown compound in terms of the following rules.

λ_{max}

Rule 1

a) If the spectrum of a given compound exhibits an absorption band of very low intensity (ϵ = 10–100) in the 270-to-350 nm region, and no other absorption above 200 nm, the compound contains a simple, *unconjugated* chromophore containing n electrons. The weak band is due to $n \rightarrow \pi^*$ transition.

b) If the spectrum of a given compound exhibits many bands, some of which appear even in the visible region, the compound is likely to contain a long-chain conjugated or polycyclic aromatic chromophore. If the compound is colored, there may be at least 4 to 5 conjugated chromophores and auxochromes. [*Exceptions*: Some nitrogen-containing compounds, such as nitro-, azo-, diazo-, and nitroso-compounds; α-diketones; glyoxal; and iodoform.]

ϵ_{max}

Rule 2

There is a correlation between the intensity of the principal band, the longest-wavelength band, and the length or the area (due to conjugation) of the chromophore.

i) An ϵ value between 10,000 and 20,000 generally represents a simple α,β unsaturated ketone or diene.

ii) Bands with ϵ values between 1000 and 10,000 normally show the presence of an aromatic system. Substitution on the aromatic nucleus by a functional group which extends the length of the chromophore may give bands with $\epsilon > 10{,}000$ along with some which still have $\epsilon < 10{,}000$.

iii) Bands with ϵ less than 100 represent $n \rightarrow \pi^*$ transition.

2.12 CALCULATING ABSORPTION MAXIMA OF UNSATURATED COMPOUNDS

Rule 3

Dienes and trienes: If the compound is suspected to be a conjugated or substituted diene, its wavelength of maximum absorption can be predicted with the help of Table 2.2. To be able to use this table, one must first learn to recognize different types of dienes, conjugations, double bonds, etc. These are as follows.

Table 2.2 Woodward's[4] and Fieser's[5,6,7] rules for calculating ultraviolet absorption maxima of substituted dienes (ethanol solution)

i) Basic λ_{max} for an unsubstituted, conjugated, *acyclic* or *heteroannular* diene	214 nm
ii) Basic λ_{max} for an unsubstituted, conjugated *homoannular* diene	253 nm
iii) Extra double bonds in conjugation (for each C=C)	add 30 nm
iv) Exocyclic double bond (effect is two-fold if bond is exocyclic to two rings)	add 5 nm
v) Substituents on vinyl carbons (for each group given below)	
a) O-acyl (—O—CO—R or —O—CO—Ar)	0 nm
b) Simple alkyl (—R)	add 5 nm
c) Halogen (—Cl, —Br)	add 5 nm
d) O-alkyl (—OR)	add 6 nm
e) S-alkyl (—SR)	add 30 nm
f) N-alkyl$_2$ (—NRR′)	add 60 nm

i) >C=C<C=C>C=C< — A *linear* conjugation; for example, 1,3,5 hexatriene, isoprene, etc.

ii) >C=C<(C=C<)(C=C<) — A *cross* conjugation.

iii) A *cyclic* diene; for example, cyclohexadiene, cyclohepta 1,3-diene, etc.

iv) A *semicyclic* diene; one of the double bonds forms part of a ring and the other is exocyclic, or outside the ring. When only one of the two sp^2 hybridized carbons of a double bond is a part of the ring under consideration, such a double bond is called an *exocyclic double bond.*

v) A *homoannular* diene is one in which the two double bonds are conjugated and are in a single ring; for example, 1,3-cyclohexadiene. Note that both double bonds are exocyclic to ring B.

vi) 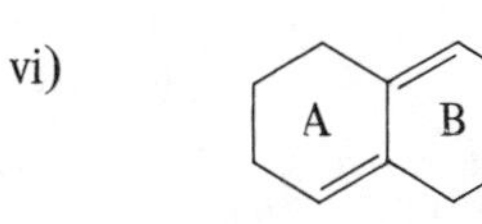

A *heteroannular* diene is a conjugated system in which the two double bonds belong to two different rings. However, these double bonds are also exocyclic, one of them being exo- to ring A and the other exo- to ring B.

Let us now apply the rules in Table 2.2 to a few known compounds, and compare the resulting values of λ_{max} with the values observed experimentally.

1. Abietic acid

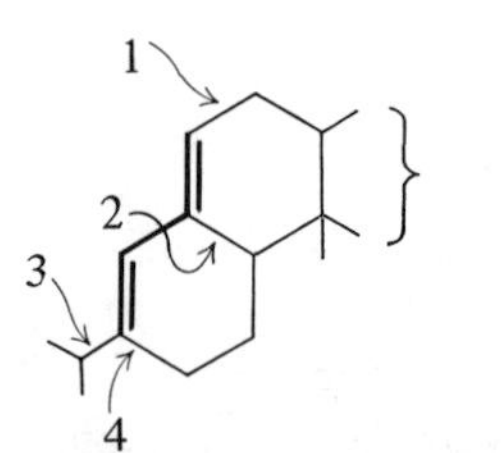

Chromophore is shown by heavy lines; numbers indicate substituents

Basic heteroannular diene		214 nm
Exocyclic double bonds	1 x (5)	5 nm
Substituents R	4 x (5)	20 nm
Calculated λ_{max}		239 nm
Observed[8]		241 nm

2. Ergosterol

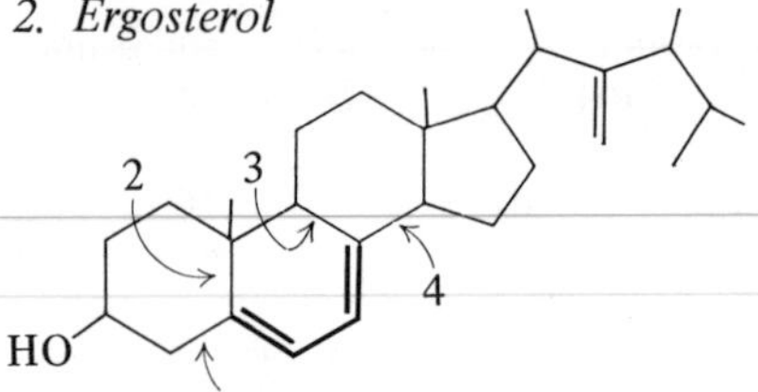

Basic homoannular diene		253 nm
Exocyclic double bonds	2 x (5)	10 nm
Substituents R	4 x (5)	20 nm
Calculated λ_{max}		283 nm
Observed		282 nm

3. 3,β-acetoxyergosta-5,7,14,22-tetraene

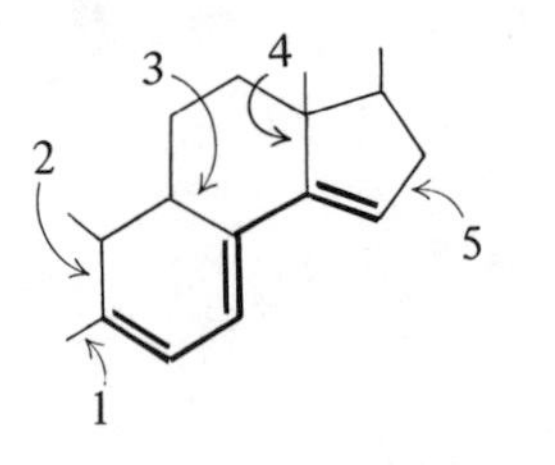

In compounds containing both homoannular and heteroannular double bonds, the diene system which requires *least* energy for excitation (i.e. the one with the longer wavelength of absorption) is used as a base.

Basic homoannular diene		253 nm
Exocyclic double bonds	3 x (5)	15 nm
Substituents R	5 x (5)	25 nm
Extra double bond in conjugation		30 nm
Calculated λ_{max}		323 nm
Observed[9]		319 nm

Steroids are a group of biologically important compounds which have a tetracyclic carbon skeleton similar to that shown above for ergosterol. Woodward's rules are particularly applicable to such dienes, as well as to acyclic dienes. Woodward's rules can also be expected to hold for 5- and 7-membered ring systems, such as cyclopentadienes and cycloheptadienes, provided that the basic λ_{max} values of 228 nm and 241 nm, respectively, are used. In systems containing π bonds only (i.e., no n electrons), UV excitation of the molecule fails to bring about significant change in its polarity, which means that UV excitation also fails to change the molecule's ability to form bonds in the presence of a given solvent. The absorption maxima for such substances, therefore, do not depend on the solvent used (there may be a variation of 1 to 2 nm only). Therefore the spectra of such systems may be measured in any solvent, although ethanol is the one preferred.

One must take care in applying these rules in some cases because steric interactions—as well as certain other special interactions which occur particularly in *cis* conformations of many molecules—cause deviations from Woodward's rules. *In general, positions as well as intensities of long-wavelength bands of* trans *isomers are always greater than those of the corresponding* cis *isomers.* This may be due to the fact that *trans* isomers are normally more elongated than *cis* isomers, and that *cis* isomers are likely to exhibit a greater loss of coplanarity (and therefore greater distortion of the molecules) due to steric interactions of substituents. Forbes and coworkers[10,11] have observed these effects and have suggested modifications to Woodward's rules.

The presence of an absorption peak at a value expected for a particular chromophore may suggest, but does not conclusively prove, the presence of such a chromophore in the molecule. On the other hand, the absence of a peak where it is expected (assuming correct calculations) is fairly strong evidence that the chromophore is absent.

Problem 2.3 Sorbic acid (a) absorbs radiation at 261 nm (ϵ = 25,000) but 2-furoic acid (b) displays a weaker absorption band at 254 nm (ϵ = 11,000). Explain.

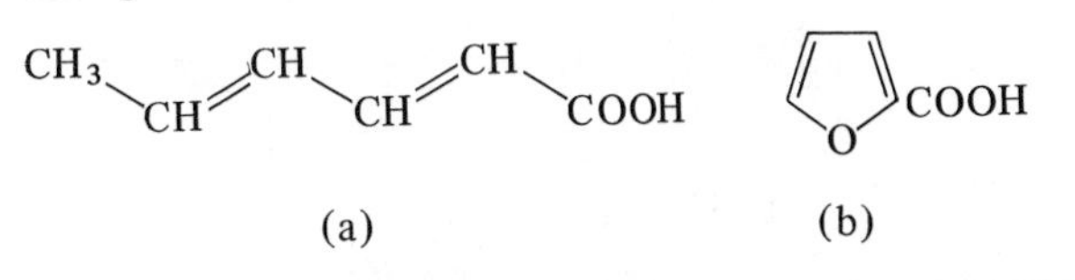

Rule 4

Polyenes: Rule 3 holds fairly well for unsaturated compounds containing up to four conjugated double bonds. However, for systems of extended conjugation, such as those found in carotenoid pigments, Fieser[12] and Kuhn[13] have suggested equations to calculate the basic λ_{max} and ϵ_{max} of UV absorption. (*See right-hand column.*)

$$\lambda_{max} \text{ (in hexane)} = 114 + 5M + n(48.0 - 1.7n) - 16.5R_{endo} - 10R_{exo}$$

$$\epsilon_{max} \text{ (in hexane)} = 1.74 \times 10^4 n$$

where M = number of alkyl substituents
n = number of conjugated double bonds
R_{endo} = number of rings with endocyclic double bonds
R_{exo} = number of rings with exocyclic double bonds

Let us apply these equations to a few compounds.

1. *All trans β-carotene*

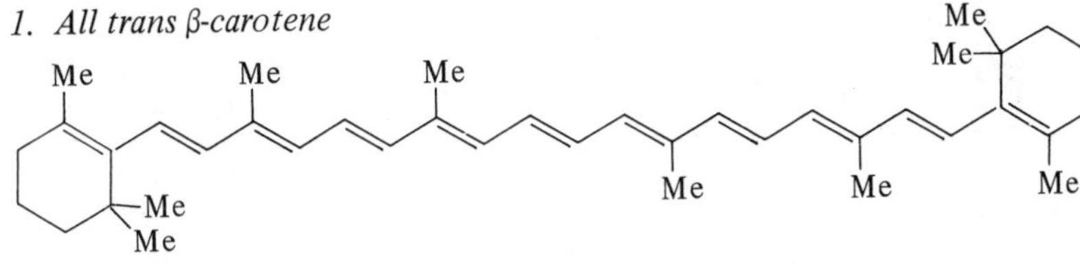

Basic λ_{max} value	114 nm
M = number of alkyl substituents, 5 x 10	add 50 nm
n = number of conjugated double bonds, 11 x [48 – (1.7 x 11)]	add 322.3 nm
R_{endo} = number of rings with endocyclic double bonds, 2 x 16.5	subtract 33 nm
R_{exo} = number of rings with exocyclic double bonds, 0 x 10	subtract 00 nm
Calculated λ_{max}	453.30 nm
Observed[14,15]	452 nm

$\epsilon_{max} = 1.74 \times 11 \times 10^4 = 19.1 \times 10^4$ (calculated)
$= 15.2 \times 10^4$ (observed*)

* The equation for calculating ϵ_{max} is semi-empirical; the value calculated does not always correspond well with the observed value.

2. *All trans lycopene*

Basic λ_{max}	114 nm
M = 5 x 8	add 40 nm
n = 11 x [48 – (1.7 x 11)] (*Note*: Double bonds at ends are not in conjugation with others.)	add 322.3 nm
R_{endo} = 0	subtract 00 nm
R_{exo} = 0	subtract 00 nm
Calculated λ_{max}	476.30 nm
Observed[16]	474 nm

$\epsilon_{max} = 1.74 \times 11 \times 10^4 = 19.1 \times 10^4$ (calculated)
$= 18.6 \times 10^4$ (observed)

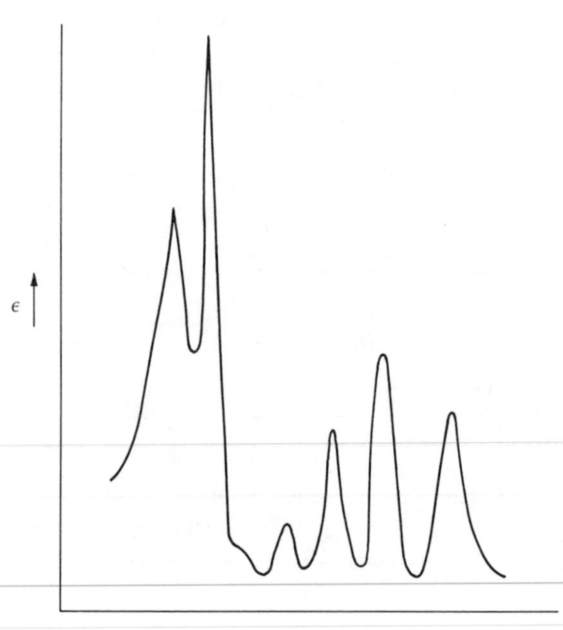

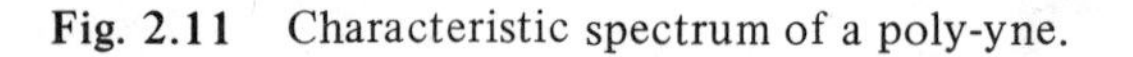

Fig. 2.11 Characteristic spectrum of a poly-yne.

Rule 5

Poly-ynes: Space limitations preclude a detailed consideration of these less-common compounds. However, let us mention one fundamental feature: All compounds containing more than two conjugated triple bonds exhibit spectra similar to that in Fig. 2.11. A poly-yne spectrum always has a series of strong peaks ($\epsilon_{max} = 10^5$) at intervals of 2000 cm^{-1} and a series of weak bands (ϵ_{max} = 200) at intervals of 2300 cm^{-1}. Such a characteristic pattern in the spectrum helps one identify the structure as a poly-yne chromophore.

2.13 CALCULATING ABSORPTION MAXIMA OF CARBONYL COMPOUNDS

Rule 6

Unsaturated carbonyl compounds: We have seen earlier (Section 2.10) that in the case of an α, β unsaturated carbonyl compound such as crotonaldehyde, only $n \rightarrow \pi^*$ (320 nm; weak) and $\pi \rightarrow \pi^*$ (218 nm; strong, ϵ_{max} = 18,000 aproximately) electronic transitions take place in the region above 200 nm. Just as in dienes, the absorption bands in these carbonyl compounds also undergo shifts due to substitution of protons on the carbonyl carbon by functional groups. With the help of Table 2.3, one can predict the value of $\pi \rightarrow \pi^*$ absorption bands in a number of carbonyl compounds. The ϵ values of these transitions are usually above 10,000.

Table 2.3* Woodward's[4] rules modified to facilitate calculation of ultraviolet absorption maxima of enone derivatives (ethanol solution)

Parent enone (acyclic or rings larger than 5 members)		215 nm
Five-membered cyclic enone		subtract 10
Aldehydes		subtract 5
Extended conjugation (for each ene)		add 30
Homoannular component		add 39
Exocyclic double bond		add 5
Substituents		
Alkyl	α	add 10
	β	add 12
	γ and higher	add 18
Hydroxyl	α	add 35
	β	add 30
	γ	add 50
Alkoxyl	α	add 35
	β	add 30
	γ	add 17
	δ	add 31
Acetoxyl	α, β, or δ	add 6
Dialkylamino	β	add 95
Chlorine	α	add 15
	β	add 12
Thioalkyl	β	add 85
Bromine	α	add 25
	β	add 30

* From D. J. Pasto and C. R. Johnson, *Organic Structure Determination*, Prentice-Hall, Englewood Cliffs, N.J., 1969.

The λ_{max} values for $\pi \rightarrow \pi^*$ and $n \rightarrow \pi^*$ transitions in carbonyl compounds depend appreciably on the polarity of the solvent as well as on the nature of the substituents on the chromophoric carbons. In calculating the absorption maxima from Table 2.3, we should therefore remember that a reasonable agreement between calculated and experimental values is to be expected only if ethanol is used as a solvent for the experiment. For any other solvent, we use one of the correction factors given in Table 2.4.

The basic chromophore containing a $>C{=}C<$ (-ene) conjugated with a $>C{=}O$ (-one), as in

$$\overset{\beta}{>C}{=}\overset{\alpha}{C}{-}C{=}O$$

is called an *enone.* If a carbonyl group is conjugated with two double bonds (-diene), such as

$$\overset{\delta}{>C}{=}\overset{\gamma}{C}{-}\overset{\beta}{C}{=}\overset{\alpha}{C}{-}C{=}O,$$

the compound is known as a *dienone.* In the case of cyclic compounds, the ethylenic double bonds conjugated with the carbonyl may be homoannular or heteroannular.

Table 2.4* Solvent correction factor for calculating ultraviolet absorption maxima of enones

For solvents other than ethanol, apply the following correction factors to the value calculated on the basis of Table 2.3.

Solvent		
Water	subtract	8 nm
Methanol		0
Chloroform	add	1
Dioxane	add	5
Ether	add	7
Hexane	add	11

* From D. J. Pasto and C. R. Johnson, *Organic Structure Determination,* Prentice-Hall, Englewood Cliffs, N.J., 1969.

Table 2.5 Nielsen's[19] rules for calculating absorption maxima of α,β-unsaturated carboxylic acids and esters (ethanol solution)

Basic value for acids and esters:	
with α or β alkyl substituents,	208 nm
with α, β or β, β alkyl substituents,	217 nm
with α, β, β alkyl substituents,	225 nm
For exocyclic $=C<$ or any $>C=C<$ endocyclic to a 5- or 7-member ring,	add 5 nm

Let us now apply these rules to a few known compounds and compare the calculated value of λ_{max} with the observed value.

1. *Cholest-4-en-3-one*

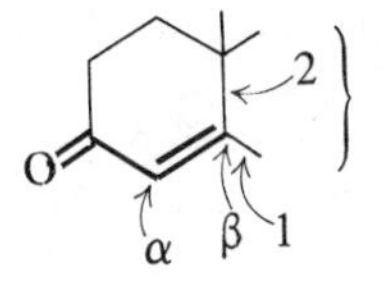

Parent base			215 nm
Substituents β, β	2 × (12)	add	24 nm
Exocyclic $=C<$		add	5 nm
Calculated λ_{max}^{EtOH}			244 nm
Observed[17]			241 nm

2. *Cholesta-2,4-dien-6-one*

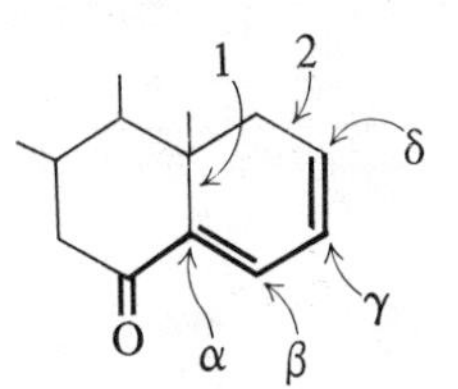

Parent base			215 nm
Extended conjugation		add	30 nm
Homoannular component		add	39 nm
Substituents	α 1 × (10)	add	10 nm
	δ 1 × (18)	add	18 nm
Calculated λ_{max}^{EtOH}			312 nm
Observed[18]			314 nm

3. *3,β-Acetoxy-7-oxolanosta-5,8,11-triene*

Parent base			215 nm
Extended conjugation		add	30 nm
Homoannular component		add	39 nm
Exocyclic double bond		add	5 nm
Substituents	α 1 × (10)	add	10 nm
	β 1 × (12)	add	12 nm
	δ 1 × (18)	add	18 nm
Calculated λ_{max}^{EtOH}			329 nm
Observed[6]			327 nm

Rule 7

Carboxylic acids and esters: When we are dealing with carboxylic acids, we use the rules in Table 2.5, which are in addition to those in Table 2.3, to compute the absorption maxima. The λ_{max} values for α, β unsaturated acids are generally lower than those for corresponding ketones. This is probably because of the n and π electrons resonating as

C=C—C(=Ö:)—ÖH

Such a resonance lowers the electron affinity of the carbonyl group, and hence its capacity to act as acceptor of ethylenic π electrons in excitations involving electron transfer.

A comparison of observed and calculated values illustrates the usefulness of Nielsen's rules.

1. *Cycloheptene-1-carboxylic acid*

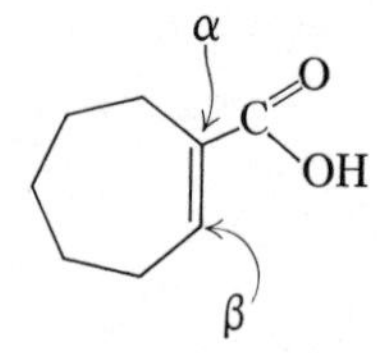

This is an α, β substituted, α, β unsaturated acid. The base value is therefore taken as

therefore taken as		217 nm
C=C endocyclic to 7-member ring	add	5 nm
Calculated λ_{max}^{EtOH}		222 nm
Observed[20]		222 nm

2. *3-Methyl-2-butenoic acid*

$$CH_3-\overset{\overset{CH_3}{|}}{C}=\overset{\overset{H}{|}}{C}-COOH$$

(β at the CH_3-bearing carbon, α at the H-bearing carbon)

β, β substituted unsaturated acid	217 nm
Observed value	216 nm

Problem 2.4 The λ_{max} values of the following enones were found to be: 224 (ϵ_{max} = 9750), 235 (ϵ_{max} = 14,000), 253 (ϵ_{max} = 9550), and 248 (ϵ_{max} = 6890). Assign each spectrum to the appropriate compound.

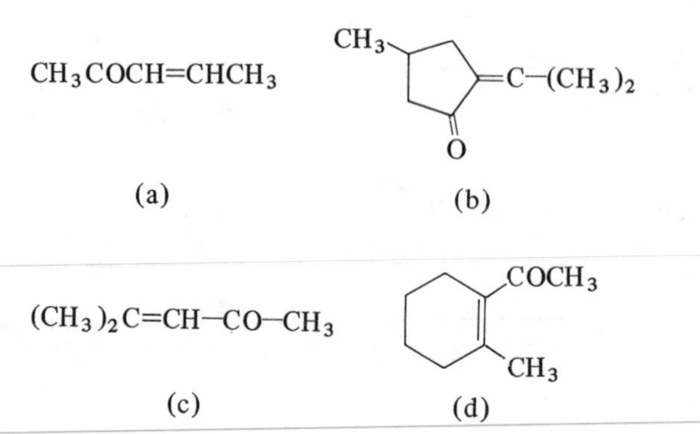

Problem 2.5 The λ_{max} values of the following dienones were found to be: 279 (26,400), 288 (12,600), 348 (11,000). All spectra were determined in ethanol. Assign each spectrum to the appropriate compound.

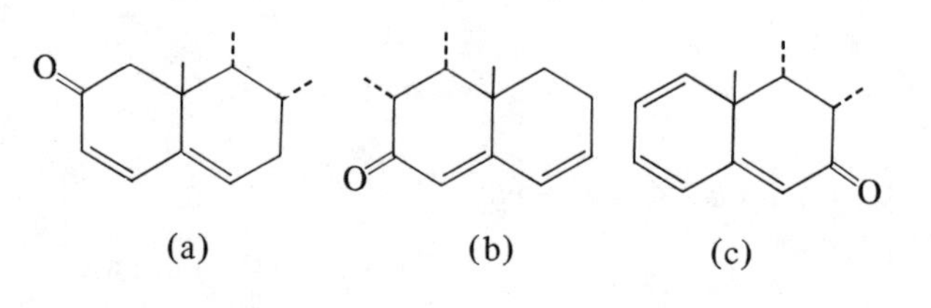

2.14 CALCULATING ABSORPTION MAXIMA OF AROMATIC MOLECULES

There are two types of aromatic molecules: benzenoid and nonbenzenoid. Their spectra show considerable resemblance. In fact, the presence or absence of certain features in UV spectra, such as a low-intensity band (known as a *fine-structure band*) at or about 255 nm, is often used to detect the aromatic character of an unknown substance.

Benzene and monosubstituted derivatives: The simplest aromatic compound is benzene. It has a ring current of π electrons, which shows strong $\pi \rightarrow \pi^*$ absorptions at 184 nm (ϵ_{max} = 47,000) and at 202 nm (ϵ_{max} = 7400). (This is called a *primary band.*) Benzene exhibits a low-intensity band at 255 nm (ϵ_{max} = 230 in cyclohexane) (known as a *secondary* or *fine-structure band*) with a series of fine-structure bands between 230 and 270 nm. Figure 2.12a shows the spectrum of benzene when it is in the

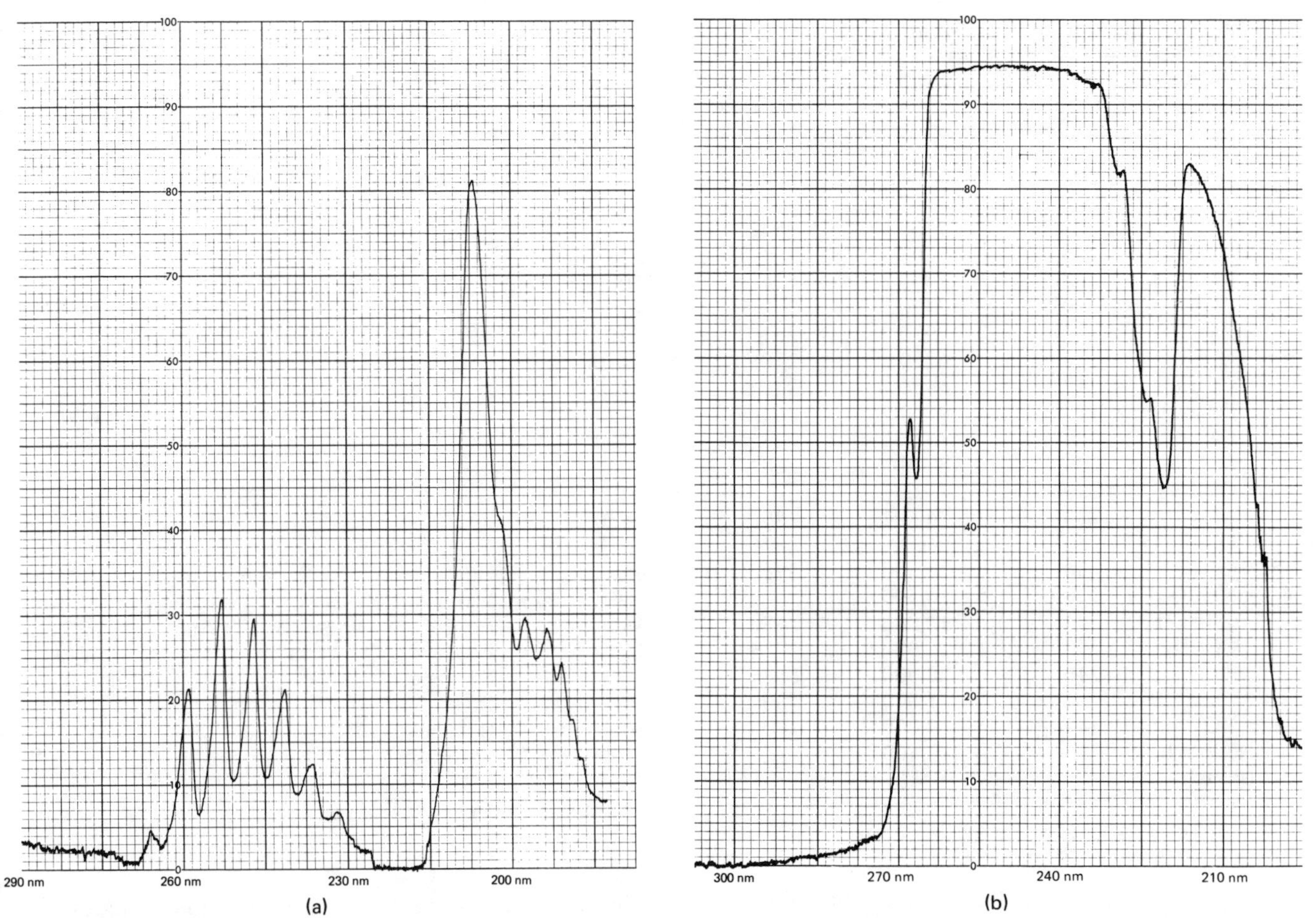

Fig. 2.12 Near UV spectra of benzene in (a) nonpolar solvent, hexane; (b) polar solvent, ethanol.

vapor state or when it is dissolved in a nonpolar solvent. If polar solvents are used, or if a single functional group is introduced on the benzene nucleus, this fine structure is lost, and only broad humps are obtained (Fig. 2.12b). Substituents on the benzene ring also cause bathochromic and hyperchromic shifts of the various peaks. However, unlike the situation that obtains in the case of dienes and unsaturated ketones, one cannot always predict the effects of various substituents on the benzene ring.

Table 2.6[21,22] lists the observed bands in the normally accessible UV region (above 200 nm).

Problem 2.6 Classify the following substituent groups into *ortho-para-directing* and *meta-directing* types and arrange them in the order of increasing $\Delta\lambda$ values of the primary band. Which two groups, when *para* to each other on a benzene ring, would cause maximum bathochromic shift? Why?

$NHCOCH_3$, NH_2, NH_3^+, CH_3, $N(CH_3)_2$, NO_2, OH, $SONH_2$, CN, O^-, CHO, COOH, COO^-, OCH_3, $COCH_3$, Cl, Br

---

To gain a qualitative understanding of the effects of substituents on the UV absorption characteristics of the benzene ring, let us classify the substituent groups into *auxochromes* and *chromophores,* or into *electron-donating* and *electron-withdrawing* groups.

Auxochrome substituents

Auxochromes, such as $-\ddot{\underset{..}{O}}H$ and $-\ddot{N}R_2$, have *n* electrons. When they are excited by UV radiation, one of these electrons is given up to the phenyl nucleus as:

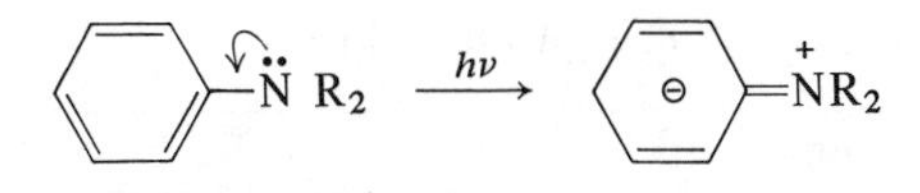

Such an electron transfer introduces the possibility of a new *electron-transfer band* in the near UV region. Apart from this, the interaction between the *n* electrons of the substituent and the π electrons of the ring also causes shifts in the primary and secondary (fine-structure) bands. These shifts, of course, depend on the availability of *n*

Table 2.6 Near-ultraviolet absorptions of some monosubstituted benzenes

Substituent	202-nm or primary band	ϵ_{max}	255-nm or fine-structure band	ϵ_{max}	Solvent
$-NH_3^+$	203	7500	254	169	2% MeOH in H_2O
$-H$	203.5	7400	254	204	,,
$-CH_3$	206.5	7000	261	225	,,
$-I$	207	7000	257	700	,,
$-Cl$	209.5	7400	263.5	190	,,
$-Br$	210	7900	261	192	,,
$-OH$	210.5	6200	270	1450	,,
$-OCH_3$	217	6400	269	1480	,,
$-SO_2NH_2$	217.5	9700	264.5	740	,,
$-CN$	224	13,000	271	1000	,,
$-COO^-$	224	8700	268	560	,,
$-COOH$	230	11,600	273	970	,,
$-NH_2$	230	8600	280	1430	,,
$-O^-$	235	9400	287	2600	,,
$-C{\equiv}CH$	236	15,500	278	650	Heptane
$-NHCOCH_3$	238	10,500	—	—	Water
$-CH{=}CH_2$	244	12,000	282	450	EtOH
$-COCH_3$	240	13,000	278	1100	EtOH
$-Ph$	246	20,000	—	—	Heptane
$-CHO$	244	15,000	280	1500	EtOH
$-NO_2$	252	10,000	280	1000	Hexane
N=N–Ph (*trans*)	319	19,500	—	—	Chloroform

electrons to participate in the p, π interaction. For example, let us consider how the difference in bonding of n electrons with acidic or basic media causes shifts in UV bands (Table 2.7) of compounds like phenol, aniline, etc.

The n electrons of oxygen in phenol interact with the ring electrons. This causes a small shift in the absorption bands (from 203.5 to 210.5 nm in the primary band and from 254 to 270 nm in the fine-structure band). However, the availability of n electrons is greatly enhanced due the negative charge on the oxygen of the phenolate anion; this causes a great shift (from 203.5 to 235 nm and from 250 to 287 nm) in λ_{max} as well as in ϵ_{max}. On the other hand, in the case of the anilinium cation, there are no n electrons available for interaction. The absorption properties of this cation are therefore essentially the same as those of unsubstituted benzene, although a marked auxochromic effect is noticeable in the case of aniline. The bathochromic shifts that take place in phenol-phenolate anion systems in alkaline media are similar to those that take place in enol-enolate anion systems.

Table 2.7 Effects of acidic or basic media on UV absorption characteristics of auxochrome-substituted benzene

Solvent	Compound	Primary band		Fine-structure band	
		λ_{max}	ϵ_{max}	λ_{max}	ϵ_{max}
Water	Benzene	203.5	7400	254	204
Water	Phenol	210.5	6200	270	1450
Alkali (aq.)	Phenolate ion	235	9400	287	2600
Water	Aniline	230	8600	280	1430
Acid (aq.)	Anilinium ion	203	7500	254	160

Table 2.8 Effects of phenyl nucleus and chromophores on ultraviolet absorption properties of each other (all values measured in cyclohexane solution)

Compound	Primary or $\pi \rightarrow \pi^*$ band		Fine-structure band		$n \rightarrow \pi^*$ band	
	λ_{max}	ϵ_{max}	λ_{max}	ϵ_{max}	λ_{max}	ϵ_{max}
Benzene	204	7400	254	204	–	–
Acetaldehyde	180	10,000	–	–	290	17
Benzaldehyde	244	15,000	280	1500	328	20
Nitromethane	201	5000	–	–	271	19
Nitrobenzene	252	10,000	280	1000	330	125
Acetophenone	238	13,000	278	890	320	46

Chromophore substituents

Chromophores contain π electrons. This makes it possible for the electrons to interact through conjugation:

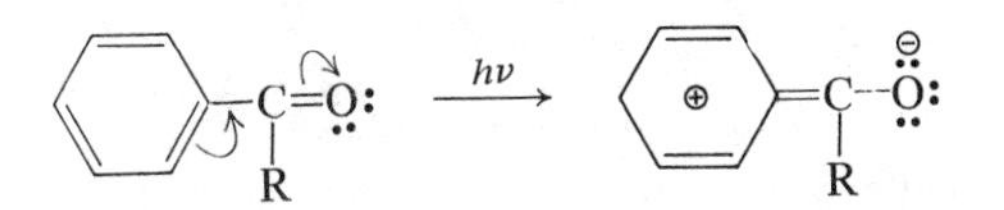

Such an interaction produces a new electron transfer band, as shown in Table 2.8.

Electron-donating and electron-withdrawing substituents

Another way to gain a qualitative understanding of the effects of substituents is to classify the substituent groups according to their electron-donating or electron-withdrawing (inductive) character. For example, note the following from Table 2.6.

i) Any substitution on the benzene ring, irrespective of its electronic character, shifts the primary band (203 nm) to longer wavelengths.

ii) Electron-withdrawing (*meta*-directing) substituents such as $-NH_3^+$, $-SO_2NH_2$, $-CO_2^-$, $-CN$, $-COOH$, $-COCH_3$, $-CHO$, $-NO_2$, etc., have practically no effect on the λ_{max} of the fine-structure band (255 nm) unless the group is also capable of conjugation. Groups that can conjugate with the π electrons of the nucleus are: $-CN$, $-COOH$, $-COCH_3$, $-CHO$, and $-NO_2$.

iii) Electron-donating groups (*ortho-para*-directing) such as $-CH_3$, $-Cl$, $-Br$, $-OH$, $-OCH_3$, $-NH_2$, $-O^-$, etc., or those capable of conjugation, increase the λ_{max} and ϵ_{max} of the fine-structure band. This increase appears to be directly related to the electron-donating effect of the groups; this electron-donating effect becomes more powerful in the groups in the order in which they are mentioned above.

Some controversy[23] surrounds these explanations; therefore one must be cautious in using these correlations.

Disubstituted benzenes and polycyclic aromatic compounds

In the case of disubstituted benzenes, the observations about the effects of electron-donating and electron-withdrawing groups on the spectrum become somewhat useful.[24,25] Examples are given below.

i) Suppose that the two groups are *para* substituted to each other, and that one of the two following conditions hold.

a) The groups are of similar type, i.e., both are electron-withdrawing or both are electron-donating.

The spectral behavior of such disubstituted benzenes is similar to that observed for the most displaced monosubstituted benzene. For example, a $-COOH$ group (an electron-withdrawing group) shifts the primary band (203.5 nm band) of benzene by $\Delta\lambda_1 = 26.5$ nm (to 230 nm in benzoic acid) and a $-NO_2$ group (another electron-withdrawing group) shifts the band by $\Delta\lambda_2 = 65.0$ nm (to 268.5 nm in nitrobenzene). But when these similar groups are *para* to each other in a disubstituted benzene, the shift is only as much as that observed for the most displaced monosubstituted derivative. Thus *p*-nitrobenzoic acid absorbs at 264 nm and *p*-dinitrobenzene absorbs at 266 nm.

b) One of the groups is *ortho-para* and the other is *meta* directing, i.e., the two substituents are of opposing character.

The shift in the primary band of such a disubstituted benzene is usually greater than the sum of the shifts caused individually by the two groups. For example, $\Delta\lambda_1$ for a $-NO_2$ group (electron-withdrawing group) is 65.0 nm and $\Delta\lambda_2$ for a $-NH_2$ group (electron-donating group) is 26.5 nm. Although the calculated sum of the two shifts is

$$\Delta\lambda_1 + \Delta\lambda_2 = 65.0 + 26.5 = 91.5 \text{ nm},$$

the shift observed for the *p*-disubstituted compound is 178.0 nm; that is, the primary band in *p*-nitroaniline appears at 203.5 + 178.0 = 381.5 nm. Several examples of such larger-than-expected bathochromic shifts are given in Table 2.9. Such large shifts in *p*-disubstituted benzenes are attributed to interaction resonance, as illustrated below.

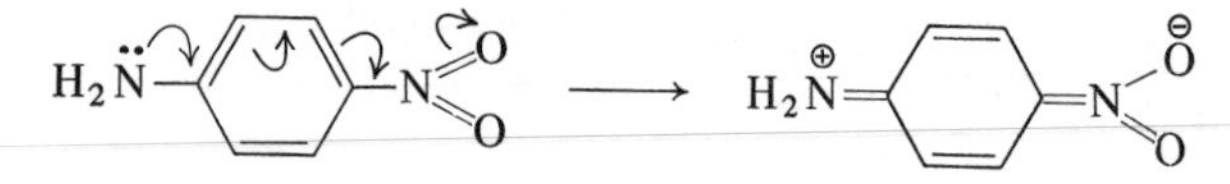

ii) Suppose that the two groups are *ortho* or *meta* to each other. Then the observed shift in Δλ is approximately the same as that calculated by adding the effects of the two individual groups. For example, Δλ due to the —OH group alone is 7 nm and that due to the —COOH group alone is 26.5 nm. The calculated Δλ for *o*-hydroxy benzoic acid is thus 7.0 + 26.5 = 33.5 nm. The observed values for the *ortho* and *meta* compounds are 33.5 and 34.0 nm, respectively. Table 2.9 gives several examples of this type. These small or practically nonexistent deviations from the calculated values in *o* and *m* compounds suggest that there is no interaction resonance of the type generally possible in the *para* disubstituted compounds.

In the case of polycyclic aromatic compounds, not only are the primary and the fine-structural (secondary) bands shifted to the longer wavelengths, but the high-intensity, 184-nm band (second primary band, which is in the vacuum-UV region for benzene) is also shifted to the normally available, near-UV region. These shifts are due to the extended conjugation which becomes possible when there are additional rings. Insignificant shifts occur if conjugation cannot be extended. A list of absorption characteristics of some polycyclic aromatic compounds is given in Table 2.10.

Problem 2.7 Predict the positions of the primary bands in the following disubstituted benzene derivatives in aqueous methanolic solutions.

a) *p*-chlorobenzoic acid b) *o*-nitroaniline
c) *m*-bromotoluene d) 2-hydroxybenzoic acid
e) *p*-hydroxybenzoic acid f) *p*-aminonitrobenzene

Table 2.9 Shifts in the primary bands (203.5 nm bands) of disubstituted ($R_1—C_6H_4—R_2$) benzenes (2% MeOH solution)

R_1	Shift ($\Delta\lambda_1$) for R_1 nm	R_2	Shift ($\Delta\lambda_2$) for R_2 nm	Calculated $\Delta\lambda_1 + \Delta\lambda_2$	Obs. Δλ for *ortho* compound	Obs. Δλ for *meta* compound	Obs. Δλ for *para* compound
CH_3	3.0	CN	20.5	23.5	25.0	26.0	33.5
CH_3	3.0	NO_2	65.0	68.0	62.5	69.5	81.5
Cl	6.0	COOH	26.5	32.5	25.5	28.0	37.5
Cl	6.0	NO_2	65.0	71.0	56.5	60.5	76.5
Br	6.5	COOH	26.5	33.0	—	—	42.0
Br	6.5	$COCH_3$	42.0	48.5	—	—	55.0
OH	7.0	COOH	26.5	33.5	33.5	34.0	51.5
OH	7.0	$COCH_3$	42.0	49.0	49.0	47.0	71.5
OH	7.0	CHO	46.0	53.0	53.0	51.0	80.0
OH	7.0	NO_2	65.0	72.0	75.0	70.0	114.0
OCH_3	13.5	$COCH_3$	42.0	55.5	—	—	73.0
NH_2	26.5	CN	20.5	47.0	—	33.0	66.5
NH_2	26.5	COOH	26.5	53.0	44.5	46.5	80.5
NH_2	26.5	$COCH_3$	42.0	68.5	—	—	108.0
O^-	31.5	$COCH_3$	42.0	73.5	—	—	121.0
O^-	31.5	CHO	46.0	77.5	—	—	127.0
O^-	31.5	NO_2	65.0	96.5	—	—	199.0
$NHCOCH_3$	38.5*	COOH	24.5*	55.0	—	—	84.5*
$NHCOCH_3$	38.5*	NO_2	56.5*	95.0	—	—	112.5*
COOH	26.5	NO_2	65.0	†	—	—	61.0
NO_2	65.0	NO_2	65.0	†	—	38.0	62.5

* In ethanol.
† Not to be added.

Table 2.10 Absorption characteristics of some polycyclic aromatic compounds

Compound	λ_{max}, nm	ϵ_{max}	λ_{max}, nm	ϵ_{max}	λ_{max}, nm	ϵ_{max}
 Benzene	184	47,000	203	7400	255	230
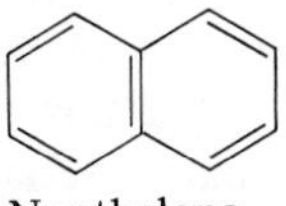 Napthalene	220	110,000	275	5600	314	316
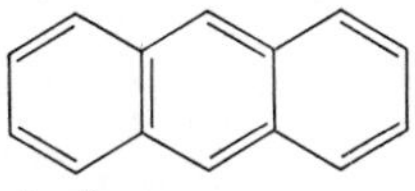 Anthracene	252	200,000	375	7900	*	*
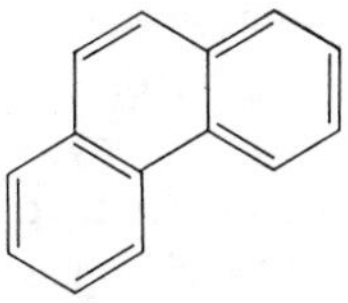 Phenanthrene	252	50,000	295	13,000	330	250
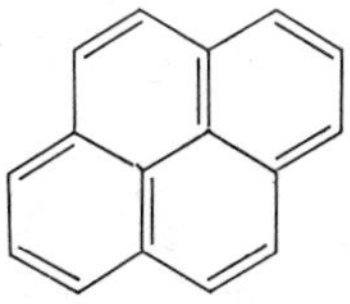 Pyrene	240	89,000	334	50,000	352	630
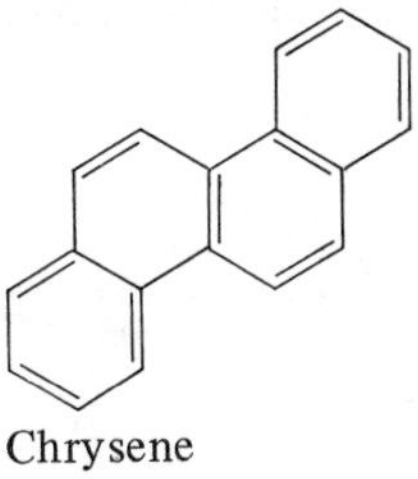 Chrysene	268	141,000	320	13,000	360	630
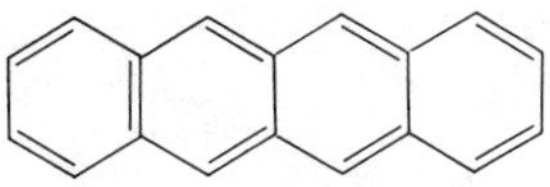 Napthacene	278	130,000	473	11,000	*	*
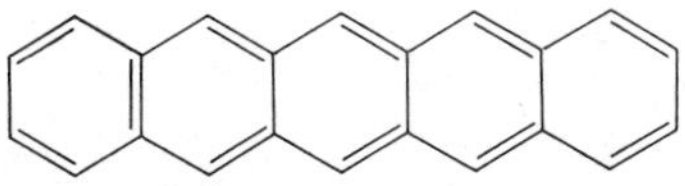 Pentacene			580	12,600		

* These weak bands are usually submerged by strong adjacent bands.

2.15 APPLICATIONS OF UV SPECTROSCOPY

The phenomenon of electronic excitation by UV-visible radiation has been extensively used in chemistry. The entire field of photochemistry, as well as the study of fluorescence and phosphorescence, is based on the principles of electronic excitation. It is these principles which operate when unreactive molecules which are singlets in the ground state are changed into reactive radicals containing unpaired electrons in the excited triplet state. These principles are also usually used to measure the ionization potential of molecules.[26] Spectroscopic techniques, however, are used mainly in determinations of structure and in quantitative estimations.

1. Determination of structure

As mentioned earlier, UV spectroscopy is but one of a number of methods of studying the structure of an organic compound, and may be useful in determining the presence or absence of a chromophore. In fact, UV spectroscopy is often employed to confirm conclusions previously deduced by other methods. The steps are as follows.

i) Collect information on the compound: determine what elements and functional groups are present, perform qualitative tests, etc.

ii) Using the rules given earlier, calculate the possible λ_{max} for the expected chromophore and compare it with the experimental value obtained for the unknown.

iii) Search for a model compound containing a similar chromophore, study its absorption characteristics, and compare the spectrum of the unknown with that of the model compound. For example, model compounds were employed to prove spectroscopically the structures of vitamins such as vitamin A,[27] vitamin B_1,[27] vitamin B_{12},[28] and K_1,[29] and since then UV spectroscopy has replaced the tedious and less rigorous biological assay methods formerly used to detect and estimate these vitamins.

If model compounds are not available, it may be possible to compare the spectrum of the unknown with those for similar compounds listed in the literature.*

2. Determination of configurations

As mentioned earlier (Section 2.12) a *trans* isomer usually has greater λ_{max} and ϵ_{max} values than the corresponding *cis* isomer. For example, in the case of

Ph · CO · CH—CH · Ph (with O bridging the two CH groups)

the *trans* isomer absorbs[30] radiation at 250 nm (ϵ_{max} = 16,800), whereas the *cis* isomer has λ_{max} = 248 nm and ϵ_{max} = 12,900. Such a difference naturally helps us to assign *cis* or *trans* configuration to an isomer. However, take care; various steric interactions between neighboring atoms generally force a *cis* isomer more readily than a *trans* isomer into a nonplanar conformation. When this happens, both the λ_{max} and the ϵ_{max} of the main spectral band are usually affected.

Ultraviolet spectroscopy is of little use in determining the absolute stereochemistry of isomers, particularly when the isomerism is due to only one asymmetric center in the molecule. In such a case all the isomers have the same λ_{max} and ϵ_{max} values. On the other hand, if more than one asymmetric carbon is present in the compound, and if the optical isomers are not mirror images of each other (*diastereomers*), the isomers may not have exactly the same absorption properties. In fact, *optical rotatory dispersion,* one of the recent techniques in studying the structure of materials, is based on this variation of the optical rotation of the plane-polarized radiation with the wavelength of the radiation.

3. Tautomerism

Ultraviolet spectroscopy has been used extensively to detect the tautomers in a mixture. For example, benzoylacetanilide was suspected to exist as an equilibrium mixture of two readily interconvertible keto and enol isomers, I and II.

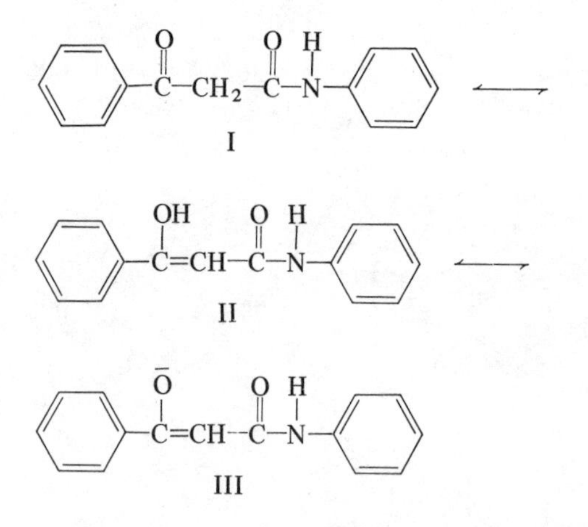

The existence of two forms was confirmed[31] spectroscopically by experiments showing that the compound absorbs maximally at 245 and 308 nm when dissolved in cyclohexane, but the longer-wavelength peak shifts to

* See References 69 through 78.

323 nm when the spectrum is taken at pH 12. The following conclusions can be derived from these observations.

i) The band at 245 nm represents the keto isomer I.

ii) The band at 308 nm represents the enol isomer II, which loses a proton to the strong base and changes to the enolate ion III, which absorbs at 323 nm at pH 12.

The difference between I and II involves the position of one hydrogen atom and one double bond. One of the hydrogen atoms on the central carbon atom in form I migrates to the ketonic oxygen to produce a hydroxyl group in form II. The corresponding electronic redistribution results in the shift of a double bond. The existence of benzoylacetanilide as an equilibrium mixture of a keto (I) and an enol (II) form constitutes an example of *tautomerism,* which is the coexistence of two (or more) compounds that differ from each other only in the position of one (or more) mobile atoms (usually hydrogen) and in the redistribution of the density of electrons.

The phenomenon of tautomerism differs from that of *resonance* in that tautomerism involves change in the σ bond skeleton of the molecule. In resonance, only π or n electrons are moved, and there is no motion of atomic nuclei and no rearrangement of the σ bond skeleton. The characteristic that distinguishes tautomerism from ordinary *molecular rearrangement* is that tautomeric changes are always both rapid and reversible under usual experimental conditions, whereas ordinary rearrangements need be neither rapid nor reversible.

Isomers I and II—which readily interconvert by changing the positions of their atomic nuclei, and which thus exist in a dynamic equilibrium—are known as *tautomers.* The example given here is that of a *keto-enol* tautomerism, although others, such as the *lactam-lactim* tautomerism of amides, the *nitro-aci-nitro* tautomerism of primary or secondary nitro compounds, and the *ring-chain* tautomerism of hydroxyaldehydes are also common.

Another illustration of the use of UV spectroscopy in determining tautomerism is provided by the spectra of β-keto-esters in different solvents. For example, ethyl acetoacetate can undergo tautomeric changes:

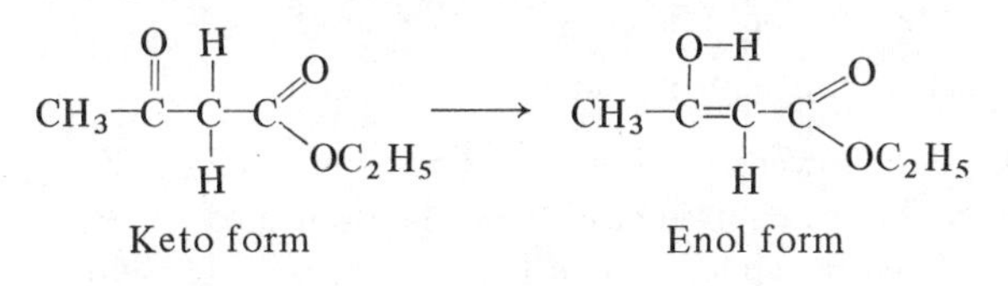

Keto form Enol form

The keto form of this compound is capable of attaining stability by means of the hydrogen bonding with polar solvents such as water. In aqueous solution, therefore, the keto-ester form predominates, as indicated by a weak (ϵ_{max} = 16) band at 272 nm. On the other hand, in nonpolar solvents such as hexane or trimethyl pentane, the hydrogen bonding is not possible. In this situation, however, the enol form can gain stability by means of intramolecular hydrogen bonding. Thus, in nonpolar solvents, a strong absorption at 243 nm (ϵ_{max} = 16,000) indicates the predominance of the enol form. The intramolecular hydrogen-bonded and solvent-bonded forms are as follows:

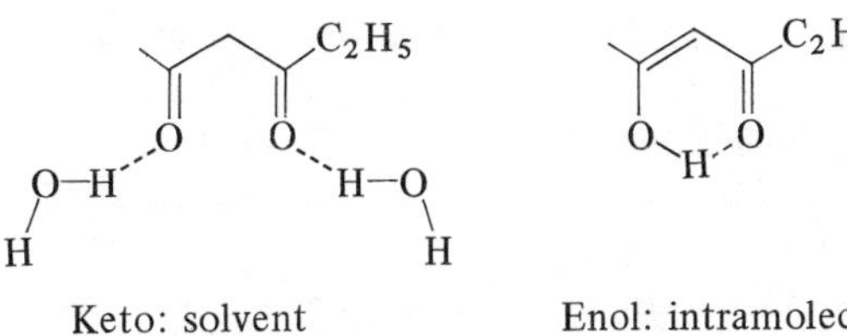

Keto: solvent hydrogen bonded

Enol: intramolecular hydrogen bonded

Since the band intensities in the UV spectra are functions of the concentrations of the absorbing species, one can determine the relative concentrations of the keto and enol forms by UV spectrometry, and also calculate the equilibrium constants.[32,33,34]

Problem 2.8 Predict whether acetylacetone would display a shift in values of λ_{max} and ϵ_{max} on being transformed from polar to nonpolar solvents. Explain.

---

4. Determination of steric interactions

In order for a conjugated system to be an effective chromophore, it must be planar or nearly planar. Only then do the chromophoric groups have maximum interactions or resonance with each other. Any factors that cause steric hindrance or loss of coplanarity in the molecule affect resonance, and thus affect the absorption characteristics of the system. Such effects can be classified into three types.

i) Values of λ_{max} are unchanged, but values of ϵ_{max} are lowered due to *small* steric interactions.

ii) Values of both λ_{max} and ϵ_{max} are lowered, due to *moderate* crowding of atoms.

iii) The molecule absorbs radiation as if it were two distinct entities. This happens when conjugation between the two parts of the molecule is completely inhibited due to twisting of bonds.

An illustration of the moderate steric interactions of type (ii) is provided by biphenyl and its 2,2′-dimethyl derivative.

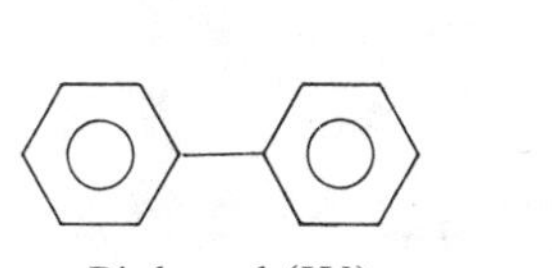

Biphenyl (IV)

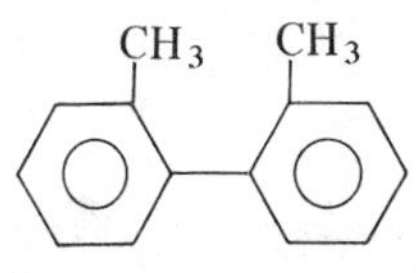

2,2′-Dimethylbiphenyl (V)

Compound (IV) achieves maximum resonance stability by maintaining the two benzene rings in a coplanar relationship. But steric interactions between the methyl groups in (V) twist the central single bond; the two rings thus lose coplanarity, and hence some resonance. This causes a shift to lower wavelength from a broad, intense 245 nm peak in biphenyl to a lower-intensity peak at 222 nm for the dimethyl derivative.

If, in a conjugated system, instead of a single bond, an *essential double bond** is twisted due to steric factors, a bathochromic shift in the spectrum is usually observed.[35]

Problem 2.9 The absorption maxima for the three isomeric methyl derivatives of acetophenone in isopentane are:

ortho isomer 237 (10,700), 245.5 (8360)
meta isomer 241 (12,400), 248.0 (9400)
para isomer 246 (18,000)

Comment on these observations.

---

5. Determination of the hydrogen bond strength

Association of the molecules of the solvent with the molecules of the solute, through hydrogen bonding, can have a profound effect on the spectral characteristics of the solute. For example, in the case of carbonyl compounds, the λ_{max} of $n \rightarrow \pi^*$ transition is highly solvent dependent. When the carbonyl compound is dissolved in polar, protic solvents, such as water, hydrogen bonds are formed between the solvent and the n electrons of the carbonyl oxygen. This lowers the energy of the n orbital in the ground state by an amount equivalent to the strength of the hydrogen bond. However, when the $n \rightarrow \pi^*$ transition takes place, the hydrogen bonds are broken because one of the electrons from the n orbital is removed and the lone, unpromoted electron remaining on the oxygen of the excited molecule is no longer able to sustain the hydrogen bond.[36]

This is probably why the $n \rightarrow \pi^*$ transitions in carbonyl compounds occur at lower wavelengths (blue shift) when the compounds are dissolved in polar, protic solvents than when they are dissolved in nonpolar solvents. For example, the $n \rightarrow \pi^*$ transition in acetone occurs at 279 nm when the spectrum is determined in hexane, but at 264.5 nm when in water. This observation is explained as follows.

In a solution of hexane (a nonpolar solvent), the acetone molecules cannot hydrogen-bond with the solvent molecules either in the ground or in the excited state. Let us assume that the ground-state energy of the acetone molecules is E_{g1} and their energy in the excited states is E_{e1}. The excitation energy, that is, the energy of UV radiation absorbed by the molecules, is therefore

$$E_{e1} - E_{g1} = \Delta E_1.$$

When the measurements are taken in water (a polar solvent) the corresponding energy values for acetone molecules in the two states are E_{g2} and E_{e2}, and the excitation energy is

$$E_{e2} - E_{g2} = \Delta E_2$$

However, E_{g2} is less than E_{g1} by a value approximately equivalent to the degree of hydrogen bonding of acetone molecules with water in the ground state, and $E_{e1} = E_{e2}$ because of the absence of the hydrogen bonds in the excited state (irrespective of the solvent). Therefore

$$\Delta E_2 > \Delta E_1$$

That is, more energy (short-wavelength radiation) is required to bring about $n \rightarrow \pi^*$ transition when acetone is dissolved in water than when it is dissolved in hexane.

On the other hand, $\pi \rightarrow \pi^*$ transitions in carbonyl compounds produce dipolar forms of the type $\overset{+}{C}{-}C{=}C{-}\bar{O}$ in the excited states. These dipolar forms in the excited states are capable of more hydrogen bonding than the nonpolar forms in the ground state. Thus, in the polar, protic solvent, this hydrogen bonding (as well as dipole-dipole interactions) tends to lower the energy of the excited state and thus of the $\pi \rightarrow \pi^*$ transition. Such a lowering is not possible in nonpolar solvents. Consequently, when the polarity of the solvent is increased, red shifts in the $\pi \rightarrow \pi^*$ bands are observed.

People have tried to deduce the strength of hydrogen bonds on the basis of such shifts observed in polar solvents. For example, as mentioned earlier, hydrogen bonding was assumed to be responsible for a shift from 279 nm to 264.5 nm in the $n \rightarrow \pi^*$ transition of acetone when its spectrum was determined in water. The energy associated with these two wavelengths can be calculated as:

for 279 nm, approximately 121 kcal/mole

for 264.5 nm, approximately 126 kcal/mole

Since the lowering in transition energy in this case is considered to be equivalent to the strength of the hydrogen bond, we can conclude that the energy associated with the hydrogen bond is approximately 5 kcal/mole. This value is in good agreement with the known energy value of the hydrogen bond.

We must be careful in evaluating the strength of the hydrogen bond by this method because spectral changes

---

* A double bond is referred to as *essential* if it is double in *all* the principal resonance structures.

are caused not only by hydrogen bonding but also by other factors, such as van der Waals interactions, dipole-dipole interactions, induced dipole orientations, and so forth. All these factors operate in both the ground and the excited states. Also, in keto-enol tautomerism, one tautomer may be more stabilized by a given solvent than another tautomer would be. Such a situation also shows the difference in effects of a nonpolar and a polar solvent.

Intramolecular hydrogen bonding within the absorbing molecule can also be demonstrated by UV spectroscopy. For example, spectroscopic investigation[37] of 5-chloro-7-acetyl-8-hydroxyquinoline (VII) prepared from 5-chloro-8-hydroxyquinoline (VI) revealed that all the bands in the spectrum of (VII) appear at longer wavelengths or are more intense than the corresponding bands in the spectrum of (VI).

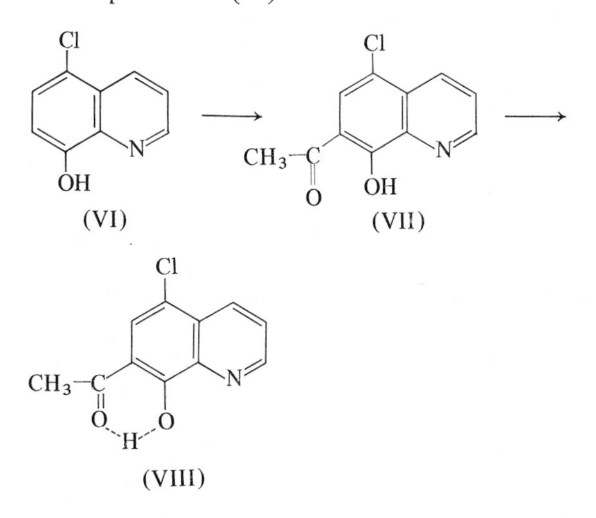

This was presumed to be due to the formation of strong hydrogen bonds between the hydroxy hydrogen and the carbonyl oxygen atoms, as shown in (VIII).

Problem 2.10 The $n \rightarrow \pi^*$ transitions in mesityl oxide $(CH_3)_2C{=}CHCOCH_3$ displayed solvent dependence as follows.

Solvent:	cyclohexane	ethanol	methanol	water
λ_{max}:	335 nm	320 nm	312 nm	300 nm
ϵ_{max}:	25	63	63	112

Assuming these shifts to be due entirely to the hydrogen bonding with the solvent, calculate the strengths of the hydrogen bonds in each solvent.

---

6. Detection of ions and free radicals

Ultraviolet spectroscopy has been used to detect the presence of unstable ionic or free-radical intermediates postulated in organic reaction mechanisms. For example, it was postulated that when triphenylcarbinol, Ph_3COH, is dissolved in concentrated sulphuric acid, the cation $Ph_3\overset{+}{C}$ is produced, as follows:

$$Ph_3COH + 2H_2SO_4 \longrightarrow PH_3\overset{+}{C} + 2H\overset{-}{SO_4} + H_3\overset{+}{O}$$

The formation of the triphenyl carbonium ion was proved by the UV spectrum, which showed a double band at 409 nm and 428 nm with $\epsilon_{max} = 38{,}000$ at each maximum.[38] The formation of the same cation from triphenylmethane was later demonstrated spectroscopically by Leftin.[39]

Similarly, spectroscopic evidence has been obtained[40] to prove that unstable dialkylbenzene radicals are formed by the photolysis of dialkylbenzenes at $-196°K$.

The UV spectra of radical ions are usually similar to the UV spectra of their parent compounds, except that the radical ions show an additional band, often in the visible region. The long-wavelength band is presumed to be due to the odd electron occupying the vacant high-energy orbital. This electron readily undergoes transition to a low-lying excited state, thus producing a band at longer wavelength. For example, 1,1-diphenylethylene reacts with sodium metal in tetrahydrofuran to produce the radical anion $[Ph_2C{-}CH]^{\dot{-}}$, which has a strong band near 600 nm.[41]

7. Quantitative analysis, mixture analysis, and control of purity

UV and visible spectra, though they have only limited scope for qualitative analysis, offer a valuable and simple method of quantitative analysis. The analytical method is based on the fact that substances obey Beer's law (Section 2.3) in the range of concentrations to be studied. The usual practice is to construct a calibration curve, plotting log of transmittance or absorbance versus molar concentration, and to determine the concentration of unknowns by finding the concentration corresponding to the measured transmittance or absorbance of the unknown sample. The use of this technique in vitamin assays was mentioned earlier. Ultraviolet spectrometry is also commonly used in assaying many other pharmaceutical preparations such as alkaloids,[42,43,44,45] antibiotics,[46,47,48,49] and other drugs.

The most common use of UV spectrometry in the field of biological analysis is in the determination of protein at 280 nm. In fact, a simple protein meter utilizing a magnesium lamp which operates at 285 nm has now been developed.[50] Use of ultraviolet spectrometry in the measurement of enzymatic activity is also very common. For example, activity of the enzyme dehydrase can be assayed by measuring the formation of ergosterol at

282 nm.[51] Automated instruments have been developed to determine the activity of protease in laundry washing compounds,[52] and for the enzymatic determination of L(+) lactic acid in biological materials.[53]

If the sample itself does not absorb radiation in the UV region, a suitable absorber is usually obtained by employing a chemical reaction or by the formation of a complex. For example, EDTA–a common chemical used in water analysis–can be determined by measuring the absorbance of its iron complex at 258 and 305 nm.[54]

If the unknown is a mixture of two substances having similar (but not the same) absorption characteristics, the concentration of each can be determined from Beer's law as follows.

Since absorbance $A = \epsilon cl$, at the wavelength λ' the absorbance A' is

$$A' = (\epsilon_1' c_1 + \epsilon_2' c_2)$$

and at a second wavelength λ'' the absorbance A'' is

$$A'' = (\epsilon_1'' c_1 + \epsilon_2'' c_2)$$

where

A' and A'' represent absorbance at λ' and λ'', respectively,
ϵ_1' and ϵ_1'' represent molar extinction coefficients at λ' and λ'', respectively,

the subscripts 1 and 2 refer to the two substances in the mixture,

c_1 and c_2 refer to the molar concentrations of the two substances,

l represents the internal cell length.

The two wavelengths λ' and λ'' are selected such that in one case the ratio ϵ_1'/ϵ_2' is a maximum and at the other $\epsilon_1''/\epsilon_2''$ is a minimum. The concentrations of the two species are then determined as

$$c_2 = \frac{(\epsilon_1'' A' - \epsilon_1' A'')}{l(\epsilon_1'' \epsilon_2' - \epsilon_1' \epsilon_2'')}$$

$$c_1 = \frac{-(\epsilon_2'' A' - \epsilon_2' A'')}{l(\epsilon_1'' \epsilon_2' - \epsilon_1' \epsilon_2'')}$$

From these discussions, you can readily see how UV or visible spectroscopy can be adapted to the control of purity of a substance. If a compound is transparent to UV-visible radiation, traces of impurities can easily be detected, particularly if the impurities themselves have a fairly intense absorption band in the UV-visible region.

8. Study of chemical reactions

Ultraviolet spectroscopy is used extensively in determining rate constants, equilibrium constants, acid-base dissociation constants, etc., of chemical reactions. Information of this type is of considerable importance when one is elucidating structures or determining the reaction mechanisms of organic compounds.

a) Acid-base dissociation constant. The dissociation constant K_a of an acid, HA, in a solvent, SH, according to the equation

$$HA + SH \rightleftharpoons A^- + SH_2^+$$

is given by the expression

$$K_a = \frac{[A^-][SH_2^+]^*}{[HA]}$$

Similarly, K_b, the dissociation constant for a base, B, in a solvent, SH, is given by the expression

$$K_b = \frac{[BH^+][S^-]}{[B]}$$

For substances having pK_a values (that is, $-\log K_a$) within the range of 1 to 12 in aqueous solvents, the term $[SH_2^+]$ represents the pH of the solution, which can readily be determined by a pH meter. Thus, to determine the K_a of the acid, we need only measure the concentrations of HA and A^-. Since spectrophotometry is an accurate method of quantitative analysis, we can use it to determine concentrations of the absorbing species such as HA and A^-.

The same type of data is sufficient to determine the basicity K_b of a base. In practice, K_a or K_b is usually determined by studying the absorption spectra of a series of solutions containing the same total concentrations of acid (or base), but having different pH.[55,56]

In the case of polyfunctional, polyprotic acids, to understand the complete ionization process, you may need a combination of a direct acid–base titration with spectrometric measurements. For example, a biologically important amino acid, cysteine[57] is a trifunctional acid containing three ionizable groups: carboxyl, $-COOH$; amino, $-NH_3^+$; and thiol, $-SH$.

```
          H
          |
HS—CH2—C—COOH
          |
          NH3
          +
```

The carboxyl group–since it is approximately 10^6 times more acidic than the ammonium and thiol groups–is

* Strictly speaking, the quantities in brackets should refer to the activities of the corresponding species. However, we have approximated them to indicate molar concentrations.

readily ionized. Consequently, its ionization constant, K_{COOH}, can be readily determined by a simple base-titration method. However, the ammonium and thiol groups have similar pK_a's, and one cannot identify with certainty which group ionizes first. The routine base-titration technique cannot therefore be employed to determine K_{-SH} and $K_{-NH_3^+}$. Consequently, these ionization constants, called *microscopic ionization constants,* are determined by a series of spectrometric measurements at various pH's.

b) Rate constants. Scientists use spectrophotometry to study reaction kinetics when one of the reactants or the products absorbs radiation in the UV-visible region. The experimenter simply determines the rate at which a particular band appears or disappears as the reaction progresses. As a matter of fact, the latest technique is to have the instrument continuously record the concentrations of a species as a function of time, by fixing the wavelength drive at the desired value and allowing the chart to move at a fixed rate.[58]

c) Equilibrium constant. We have seen how UV spectroscopy is used in determining K_a, the ionization or dissociation or the equilibrium constant of an acid. UV spectroscopy is also used to determine equilibrium constants for a variety of reactions, such as tautomerism, formation of molecular complexes, ionic complexes, etc. The method used is similar to that employed in kinetics to determine orders of chemical reactions. However, the spectrophotometric method is more powerful and fast, and can be used even if the products are not particularly stable. In the case of complex formations, however, the stoichiometry of the reaction must be determined first by other techniques before one can compute an equilibrium constant.[59]

9. Other applications

Techniques based on UV spectroscopic analysis are common in both industrial and government laboratories. For example, in the fish-processing industry, UV devices are used to judge the quality of fish and to estimate the amount of trimethyl amine they contain. This chemical is a product of decomposition in fish and in many meat products. Its concentration increases as the decay of meat takes place. To determine the quality of meat, an extract of the meat is treated with picric acid, which converts the trimethyl amine into its picrate (I). The picrate absorbs radiation at 358 nm.

In the fruit-packing industry, *o*-phenyl phenol (II) is used as a preservative. Citrus fruits are usually either dipped in this chemical or are wrapped in a paper impregnated with it to protect them from mould. This chemical is detected and estimated at its λ_{max} = 245 nm.

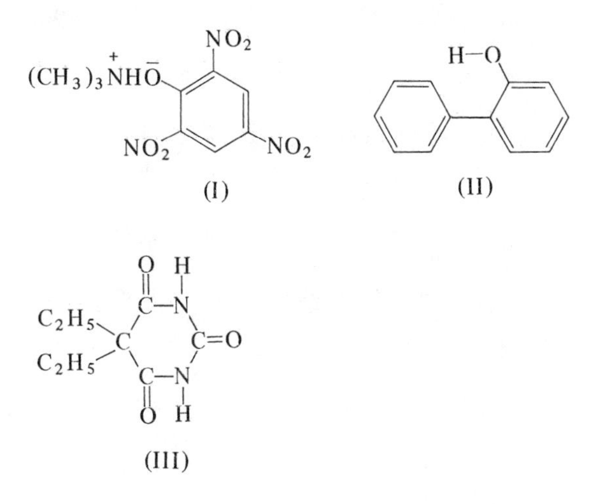

There are innumerable applications of UV spectroscopy in the drugs and pharmaceuticals industry. We have mentioned the use of this technique in vitamin assays. In addition, many drug firms deal in biological materials such as steroids and alkaloids. The steroids usually contain conjugated systems such as dienes, polyenes, conjugated enones, or dienones. All these systems have characteristic absorption properties, and can readily be analyzed spectroscopically. The same is true of the alkaloids. Drugs such as barbiturates can be detected by UV; a number of peaks of barbiturates shift to longer wavelengths with the higher pH. For example, at pH 10, the compound (III) shows two distinct peaks at 209 and 240 nm. These peaks shift to 222 and 255 nm, respectively, at pH 13.4.

2.16 SELECTED REFERENCES

Readers interested in a particular aspect of UV spectroscopy can find more information in the references listed for the corresponding topic. The publications below, which give UV spectral characteristics of several compounds, are useful for comparing the spectrum of the unknown with those listed. In addition, several of them give advanced information on general theory and interpretation.

References cited in the text

1. R. S. Sokolova and E. V. Belkova, *Opt. Spectrosc.* **30**, 303 (1970).
2. E. Sawicki, T. W. Winfield and C. R. Sawicki, *Microchem. J.* **15**, 294 (1970).

3. For details, see: (a) J. D. Roberts, *Notes on Molecular Orbital Theory,* Benjamin, New York, 1961, Chapter 2. (b) H. H. Jaffe and M. Orchin, *Theory and Applications of Ultraviolet Spectroscopy,* Wiley, New York, 1962, Chapter 3.
4. R. B. Woodward, *J. Am. Chem. Soc.* **64**, 72, 76 (1942).
5. L. Fieser, *J. Org. Chem.* **15**, 930 (1950).
6. L. Fieser and M. Fieser, *Steroids,* Reinhold, New York, 1959, pages 15–24.
7. L. Fieser, M. Fieser and S. Rajagopalan, *J. Org. Chem.* **13**, 800 (1948).
8. J. C. Harris and T. F. Sanderson, *J. Am. Chem. Soc.* **70**, 334 (1948).
9. D. H. R. Barton and T. Bruun, *J. Chem. Soc.* 2728 (1951).
10. W. F. Forbes and R. Shilton, *J. Org. Chem.* **24**, 436 (1959).
11. W. F. Forbes, R. Shilton and A. Balasubramanian, *ibid.* **29**, 3527 (1964).
12. L. F. Fieser, *J. Org. Chem.* **15**, 930 (1950).
13. R. Kuhn and C. Grundmann, *Ber. deutschen chem. Ges.* **71**, 442 (1938).
14. Inhoffen *et al., Ann.* **570**, 54 (1950); **571**, 75 (1951).
15. L. Zechmeister, *Fortschr. Chem. Org. Naturstoffe* **18**, 223 (1960).
16. L. Zechmeister, *Chem. Revs.* **34**, 267 (1944).
17. C. W. Bird, R. C. Cookson and S. H. Dandegaonker, *J. Chem. Soc.* 3675 (1956).
18. L. Dorfman, *Chem. Revs.* **53**, 47 (1953).
19. A. T. Nielsen, *J. Org. Chem.* **22**, 1539 (1957).
20. O. H. Wheeler, *J. Am. Chem. Soc.* **78**, 3216 (1956).
21. L. Doub and J. M. Vandenbelt, *J. Am. Chem. Soc.* **69**, 2714 (1947); **71**, 2414 (1949).
22. C. N. R. Rao, *Ultraviolet and Visible Spectroscopy,* Butterworths, London, England (1967), page 60.
23. W. M. Shubert and W. A. Sweeney, *J. Org. Chem.* **21**, 119 (1956).
24. C. N. R. Rao, *J. Sci. Ind. Research (India)* **17B**, 56 (1958).
25. C. N. R. Rao, *Current Sci. (India)* **26**, 276 (1957).
26. K. Watanabe, *J. Chem. Phys.* **22**, 1564 (1954).
27. R. A. Morton, *The Applications of Absorption Spectra to the Study of Vitamins, Hormones and Coenzymes,* Hilger, London, 1942.
28. N. G. Brink and S. K. Folker, *J. Amer. Chem. Soc.* **71**, 2951 (1949).
29. E. A. Doisy, S. B. Binkley and S. A. Thyer, *Chem. Revs.* **28**, 477 (1941).
30. N. H. Cromwell, F. H. Schumacher and J. L. Adelfang, *J. Am. Chem. Soc.* **83**, 974 (1961).
31. C. A. Bishop and L. K. Tong, *J. Phys. Chem.* **66**, 1034 (1962).
32. G. Kjellin and J. Sandstrom, *Acta. Chem. Scand.* **23**, 2888 (1969).
33. J. Pitha, *J. Org. Chem.* **35**, 903 (1970).
34. J. C. Parham, T. G. Winn, and G. B. Brown, *ibid.,* **36**, 2639 (1971).
35. H. H. Jaffe and M. Orchin, *Theory and Applications of Ultraviolet Spectroscopy,* John Wiley, New York, 1966, page 385.
36. V. G. Krishna and L. Goodman, *J. Am. Chem. Soc.* **83**, 2042 (1961).
37. D. R. Patel, C. S. Choxi and S. R. Patel, *Can. J. Chem.* **47**, 105 (1969).
38. G. Branch and H. Walba, *J. Am. Chem. Soc.* **76**, 1564 (1954).
39. H. P. Leftin, *J. Phys. Chem.* **64**, 1714 (1960).
40. T. F. Bindley, A. T. Watts and S. Walker, *J. Chem. Soc.* 4327 (1962).
41. H. P. Leftin and W. K. Hall, *J. Phys. Chem.* **64**, 382 (1960).
42. A. M. Wahbi and A. M. Farghaly, *J. Pharm. Pharmacol.* **22**, 848 (1970).
43. Y. Hammouda and N. Khalafallah, *J. Pharm. Sci.* **60**, 1142 (1971).
44. D. P. Page, *J. Ass. Offic. Anal. Chem.* **53**, 315 (1970).
45. D. J. Smith, M. F. Sharkey and J. Levine, *ibid.* **54**, 609 (1971).
46. D. E. Tutt and M. A. Schwartz, *Anal. Chem.* **43**, 338 (1971).
47. V. Stefanovich and M. Q. Ceprini, *J. Pharm. Sci.* **60**, 781 (1971).
48. R. E. Kling, *J. Ass. Offic. Anal. Chem.* **54**, 21 (1971).
49. L. Angelucci and M. Baldieri, *J. Pharm. Pharmacol.* **23**, 471 (1971).
50. P. A. Bennett, J. V. Sullivan and A. Walsh, *Anal. Biochem.* **36**, 123 (1970).
51. R. W. Topham and J. L. Gaylor, *J. Biol. Chem.* **245**, 2319 (1970).
52. D. A. Detmar and R. J. Vogels, *J. Amer. Oil Chem. Soc.* **48**, 77 (1971).
53. J. D. Cameron and B. J. Francis, *Analyst (London)* **95**, 481 (1970).
54. S. N. Bhattacharyya and K. P. Kundu, *Talanta* **18**, 446 (1971).
55. H. Irving, H. S. Rossotti and G. Harris, *Analyst* **80**, 83 (1955).
56. M. R. Chakrabarty, E. S. Hanrahan, N. D. Heindel and G. F. Watts, *Anal. Chem.* **39**, 238 (1967).
57. G. E. Clement and T. P. Hartz, *J. Chem. Ed.* **48**, 395 (1971).
58. A. D. Banerji and B. Chattopadhyay, *Indian J. Chem.* **8**, 993 (1970).
59. R. L. Moore and R. C. Anderson, *J. Am. Chem. Soc.* **67**, 168 (1945).

Books and general references

60. E. Sawicki, "Photometric Organic Analysis," Part 1, *Chemical Analysis,* Vol. 31, edited by P. J. Elving and I. M. Kolthoff, Interscience, New York, 1971.

61. A. E. Gillam and E. S. Stern, *An Introduction to Electronic Absorption Spectroscopy in Organic Chemistry,* third edition, edited by C. J. Timmons, St. Martin's Press, New York, 1971.
62. L. Lang, *Absorption Spectra in the Ultraviolet and Visible Region,* **XIII**, Academic Press, New York, 1970.
63. R. J. Manning, *An Introduction to Ultraviolet Spectrophotometry,* Beckman Instruments, Fullerton, California (1969).
64. C. N. R. Rao, *Ultraviolet and Visible Spectroscopy,* second edition, Butterworths, London, 1967.
65. W. F. Forbes, "Ultraviolet Absorption Spectroscopy," in *Interpretive Spectroscopy,* edited by S. K. Freeman, Reinhold, New York, 1965.
66. C. Sandorfy, *Electronic Spectra and Quantum Chemistry,* Prentice-Hall, Englewood Cliffs, N.J., 1964.
67. H. H. Jaffe and M. Orchin, *Theory and Applications of Ultraviolet Spectroscopy,* John Wiley, New York, 1962.
68. S. F. Mason, "Molecular Electronic Absorption Spectra," *Quart. Revs.* **15**, 287 (1961).

Compilations of UV spectra

69. J. P. Phillips, L. D. Freedman and J. C. Craig, *Organic Electronic Spectral Data* **VI**, Interscience, New York, 1970.
70. *Sadtler Standard Ultraviolet Spectra,* in 68 volumes, 30,000 spectra, Sadtler Research Labs., Philadelphia, Pa. (1970).
71. L. E. Keuntzel, *Ultraviolet and Visible Spectra Coded on IBM Cards,* American Society for Testing Materials, 1916 Race Street, Philadelphia, Pa.
72. *Selected Ultraviolet Spectral Data* (loose-leaf sheets), 1945 to 1968, Vol. I to III, American Petroleum Institute Research Project 44, Thermodynamics Research Center, Department of Chemistry, Texas A & M University, College Station, Texas.
73. R. G. White, *Handbook of Ultraviolet Methods,* Plenum Press, New York, 1965.
74. A. I. Scott, *Interpretation of the Ultraviolet Spectra of Natural Products,* Pergamon Press, New York, 1964.
75. L. Lang, *Absorption Spectra in the Ultraviolet and Visible Region,* I–XI, Academic Press, New York, compilations since 1961. XII and XIII published by Akademiai Kiado, Budapest, Hungary, 1970.
76. H. M. Hershenson, *Ultraviolet and Visible Absorption Spectra,* Academic Press, New York; index for 1930–1954 published in 1956 and index for 1955–1959 published in 1961.
77. M. J. Kamlet, *Organic Electronic Spectral Data,* Interscience, Vols. I through IV, covering 1946–1959.
78. C. Karr, Jr., *Applied Spectroscopy* **13**, 15–25, 40–45 (1959); *ibid.* **14**, 146–153 (1960); tabulation of the lowest-frequency electronic absorption bands for 408 polycyclic aromatic hydrocarbons and 204 polynuclear heterocyclic aromatics.

Reviews

79. R. C. Gore, R. W. Hannah, S. C. Pattacini and T. J. Porro, *J. Ass. Offic. Anal. Chem.* **54**, 1040 (1971) (UV and IR spectra of pesticides).
80. J. Kracmar and J. Kracmarova, *Pharmazie* **26**, 1 (1971) (Analysis of pharmaceutical and medicinal products).
81. C. N. R. Rao, V. Kalyanaraman, and M. V. George, "Electronic Spectra of Radical Ions," in *Applied Spectroscopic Reviews,* edited by E. G. Brame, Jr. **3**, 153, 1970 (Radical ions).
82. C. N. R. Rao, in *The Chemistry of Nitro and Nitroso Groups,* Part 1, edited by H. Feuer, Interscience, New York 1969, Chapter 2 (Nitro and nitroso compounds).
83. M. Szwarc, *Progr. Phys. Org. Chem.* **6**, 323 (1968) (Radical ions).

PROBLEMS

2.11 Use Nielsen's rules to calculate values of λ_{max} for the following α,β-unsaturated acids in ethanol.

a) *trans*-$CH_3CH{=}C(CH_3)COOH$ b) *cis* isomer of (a)

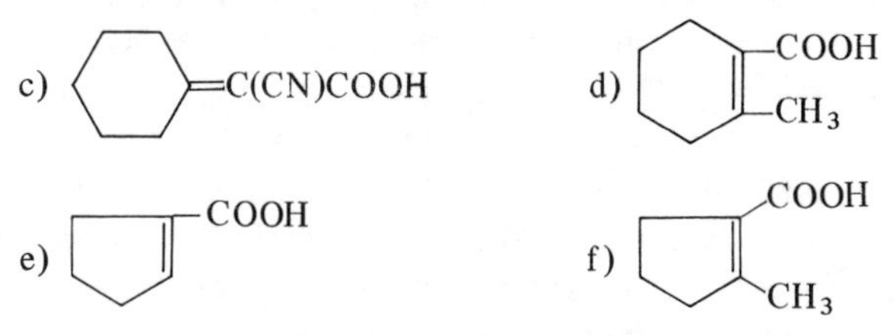

2.12 An unknown compound with the molecular formula $C_{10}H_{14}$ displays a strong absorption band at 268 nm. Draw at least seven different structures corresponding to the above formula. Calculate the expected λ_{max} for each structure and identify the unknown on the basis of the observed value of λ_{max}.

2.13 Three dimethylbiphenyl isomers show the following UV spectral characteristics:

a) λ_{max} = 250.5 (16,000)

b) λ_{max} = 222.0 (5800) and 260.0 (750)

c) λ_{max} = 255.0 (21,000)

Show which of the compounds are the 2,2′, 3,3′, and 4,4′ isomers. Predict the spectral features of 2,2′-di-*t*-butylbiphenyl.

2.14 Calculate the position of the primary band in each of the following aromatic compounds.

a) salicylaldehyde
b) *p*-tolunitrile
c) *p*-tolualdehyde
d) ethylbenzene
e) *p*-hydroxyanisol
f) hydroquinone

THREE

INFRARED SPECTROSCOPY

3.1 INTRODUCTION

We saw in Section 1.8 that the energy of a molecule can be resolved into at least four components: (i) translational, (ii) rotational, (iii) vibrational, and (iv) electronic. In Chapter 2 we considered ultraviolet and visible radiation (200–800 nm) which is energetic enough to affect the *electronic* energy levels in molecules. We shall now consider radiation which has a longer wavelength: *infrared* radiation, which extends beyond the visible into the microwave region, and is capable of affecting both the *vibrational* and the *rotational* energy levels in molecules.

The wavelength of the infrared region extends from approximately 750 nm (0.75 μ) to almost 830 μ, but, in IR spectroscopy, usually only a small range (between 2.5 to about 15.4 μ) is employed. This region is often called the *fundamental* region. The shorter-wavelength region (2.5–0.75 μ) is called the *near infrared* and the longer-wavelength segment (from 15.4 μ up to the microwave region) is called the *far infrared.* The units by which infrared rays are measured are somewhat confusing at first. Since frequency ($\nu = c/\lambda$) is directly related to the energy of radiation through the Planck relation, $E = h\nu$, one would expect that the unit for measuring infrared rays, particularly for theoretical work, would be expressed in terms of frequency. However, physical chemists prefer to give infrared wavelength in angstrom units and analytical chemists in microns and nm units. Organic chemists employ the *wavenumber* ($\bar{\nu} = 1/\lambda = \nu/c$), which is measured in cm^{-1}. Wavenumbers are proportional to frequencies, and therefore to energies, through the Planck relation $E = h\bar{\nu}c$. For this reason, wavenumbers are sometimes referred to as "frequencies in cm^{-1}."

Problem 3.1 The infrared spectrum of an unknown compound displays absorption maxima at 3.27 μ (medium) and 6.90 μ (weak). Calculate the frequency in terms of wavenumbers for these maxima.

3.2 DESIGN FEATURES OF HIGH-RESOLUTION INFRARED SPECTROMETERS

We saw in Chapters 1 and 2 (Sections 1.1 and 2.4) that the basic components of a spectrometric instrument are:

1) A source of radiation
2) A monochromator
3) A detector-recorder assembly

For reasons mentioned in Section 2.5, a modern spectrometric instrument is an *optical-null, double-beam* instrument. The layout of a modern, high-resolution infrared spectrophotometer is shown in Fig. 3.1.

The source of radiation in a typical infrared spectrophotometer is a small ceramic rod, heated electrically to about 1500°C, and made of either silicon carbide (Glowbar) or a mixture of oxides of the rare-earth elements zirconium, yttrium, and erbium. The output from the source is divided and focused by the mirrors (M_1 and M_2) to produce two beams: the reference beam and the sample beam. Reflecting mirrors rather than lenses are used, since it is difficult to obtain achromatic lenses for the infrared region. The sample beam passes through a metal halide cell, for example, a cell made of sodium chloride, containing the specimen (sample films and reflecting surfaces are also used in place of halide cells). The reference beam passes through the reference material, if reference material is present.

The two beams now pass through a chopping mechanism which consists of a rotating segmented mirror (Fig. 3.2). The chopper rotates at a fixed frequency (usually 11 cycles/sec), alternately transmitting the sample beam and the reference beam to the monochromator. This arrangement permits the two beams to follow the same path to the detector, but at different times. Since each beam is chopped twice by every turn of the chopper, the frequency of alternation of the beams is twice the frequency of rotation of the chopper.

The pulsed beam enters the monochromator through an entrance slit and is dispersed either by a grating or by a *Littrow mount prism.* In the Littrow arrangement in an infrared instrument, a plane mirror is placed behind a

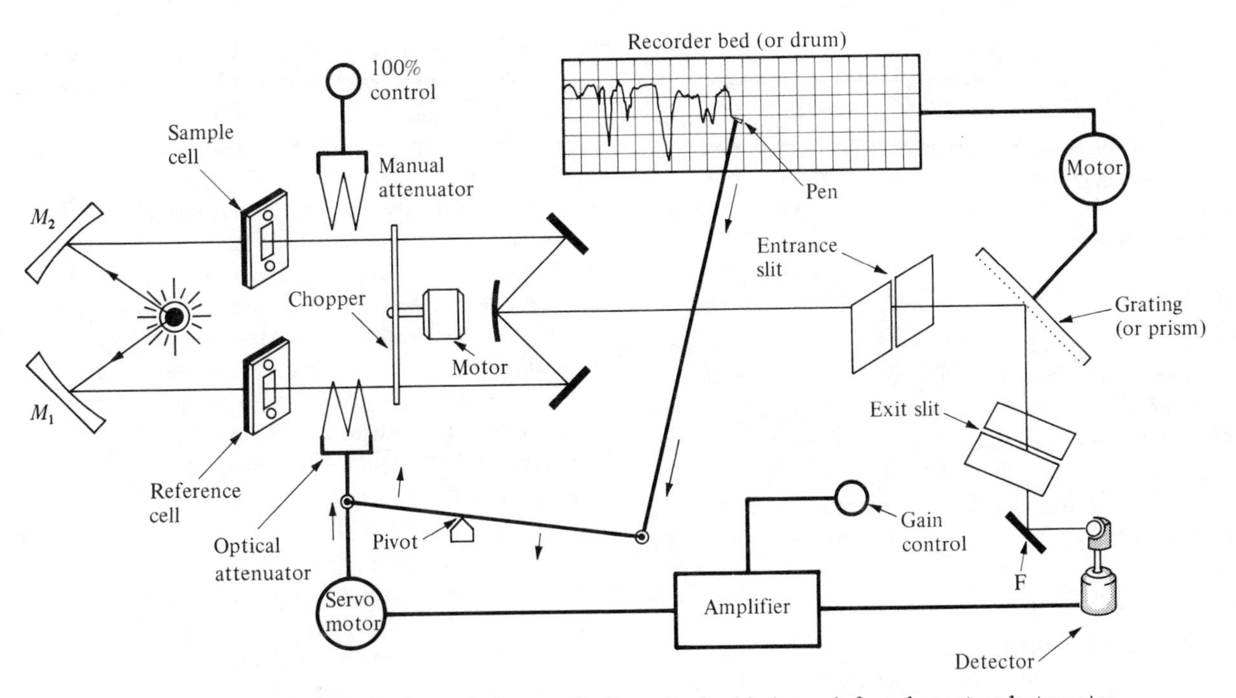

Fig. 3.1 Schematic diagram of a high-resolution, optically null, double-beam infrared spectrophotometer.

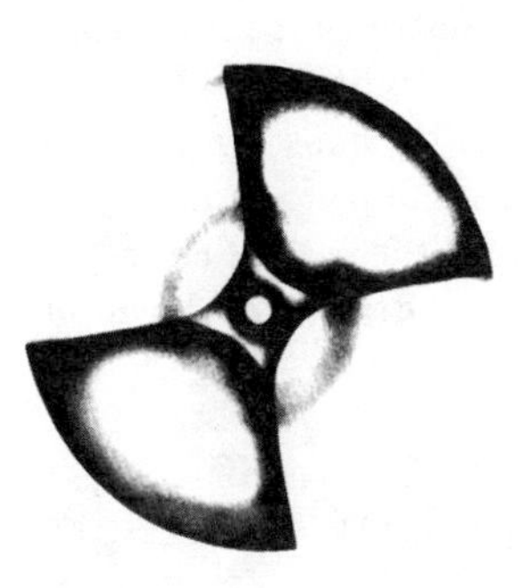

Fig. 3.2 A rotating segmented mirror acts as a beam chopper.

sodium chloride or cesium bromide prism whose apex has an angle of 55–75°. The beam passes through the fixed prism, strikes the slowly rotating mirror, and returns through the prism. The wavelength of the radiation reaching the detector is gradually changed by the slow rotation of the mirror (Fig. 3.3). In high-resolution grating instruments, a filter (*F*) is used to reject radiation of unwanted orders.

The pulsating single beam now emerging through the exit slit is a narrow band consisting of only a very few frequencies. This beam impinges on a detector (a thermocouple) which is connected to an amplifier capable of amplifying only alternating current of the same frequency as that of the two alternating beams. If the sample has preferentially absorbed radiation of this frequency, the difference in beam intensity appears in the detector system as an alternating signal having an amplitude proportional to this difference. If the reference and the sample beams are of equal intensity, i.e., if no radiation has been absorbed, no signal is observed.

The amplified signal drives a servomotor which pushes a wedge (an optical attenuator) into the reference beam to reduce the signal difference to zero, i.e., to restore the instrument to an optically null state. The recorder pen is

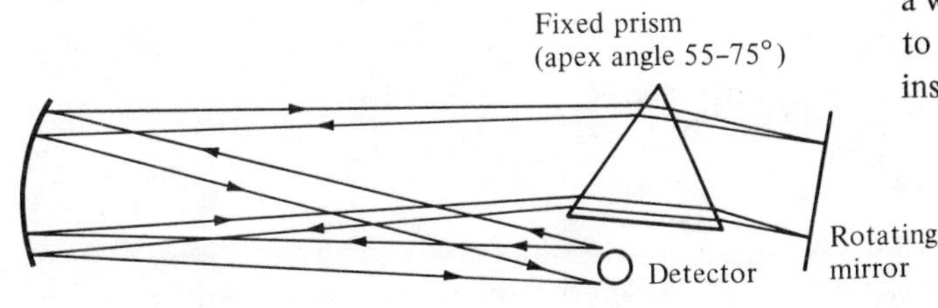

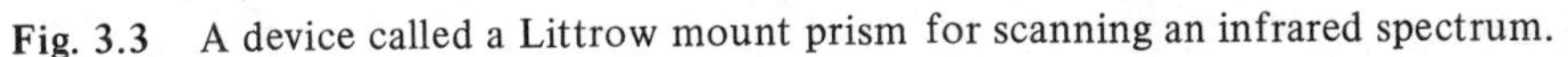

Fig. 3.3 A device called a Littrow mount prism for scanning an infrared spectrum.

coupled mechanically to the attenuator, so that the null position of the wedge is recorded graphically on paper. The resulting chart is a direct indication of the transmittance of the sample. Thus, as the spectrum is scanned by the rotating grating coupled with the recording drum, one obtains a curve showing the percentage of transmission versus wavenumber.

3.3 OBTAINING A GOOD SPECTRUM

To obtain a high-quality spectrum, one must first ensure the effective and accurate performance of the spectrometric instrument. This is achieved by (i) checking the calibration of the scale that measures the frequency, using suitable standard substances such as polystyrene, (ii) frequent checks of gain and noise, and (iii) frequent checks of the beam balance of the instrument.

Frequency calibration

i) The scale that measures frequency (cm^{-1}) or wavelength (μ) is calibrated by placing a polystyrene film in the sample beam, adjusting the instrument for optimum gain at 2500 cm^{-1}, and scanning the entire spectrum. The calibration line [e.g., for polystyrene, absorption maxima at 6.238 μ (1603 cm^{-1})] is recorded and the spectrum of the compound under study is then calibrated against this line. This procedure not only calibrates the wavelength, but also checks the resolving power of the instrument and the response of the pen. For greater accuracy, various calibration frequencies should be used throughout the 4000–600 cm^{-1} range.

Gain and noise control

ii) The amount of power available from electronic amplification is known as *gain.* In optical-null, double-beam spectrophotometers, gain must be precisely controlled if the servomechanism is to operate efficiently. In modern instruments, this control is obtained by means of a gain-control knob and the slit device. Modern instruments also have a mechanism by which one can change the width of the slit automatically as the Littrow prism or the grating rotates. This compensates for the variation of energy from the source as the wavelengths reaching the detector change. Too high a setting of the gain-control knob supplies too much power to the servoloop, causing the pen to overshoot and oscillate, and the optical attenuator (or wedge) to travel considerably farther than is needed to attenuate the energy of the reference beam. This makes accurate determination of λ_{max} for all bands and detection of weak bands very difficult (Fig. 3.4). On the other hand, if the gain-control knob is set too low, the movement of the optical wedge and the recorder pen is too slow, thereby causing loss of spectral details and shifts of absorption bands to lower frequencies.

The optimum setting of the gain-control knob should cause a small amount of overshoot at a frequency at which the sample has no significant absorption, and should allow the pen to respond fast enough to record maximum spectral detail.

Beam balance

iii) One must often check the instrument's *beam balance.* This is done either by completely obstructing both beams simultaneously and scanning, or by adjusting the pen to about 90% transmittance and scanning the entire spectrum without having anything in the path of either beam. In either case, the pen should not drift up or down the percent-transmission scale.

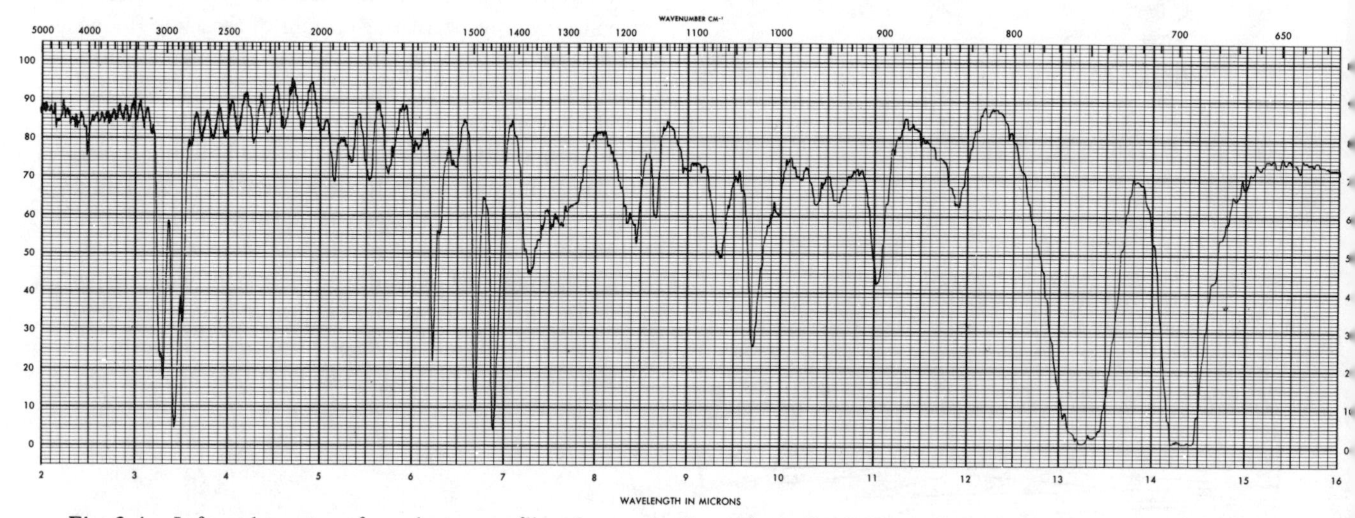

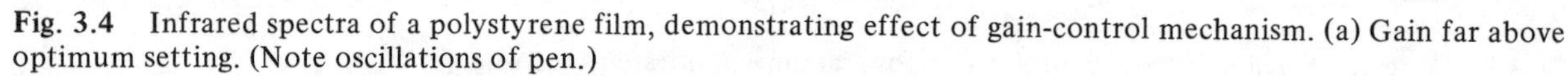

Fig. 3.4 Infrared spectra of a polystyrene film, demonstrating effect of gain-control mechanism. (a) Gain far above optimum setting. (Note oscillations of pen.)

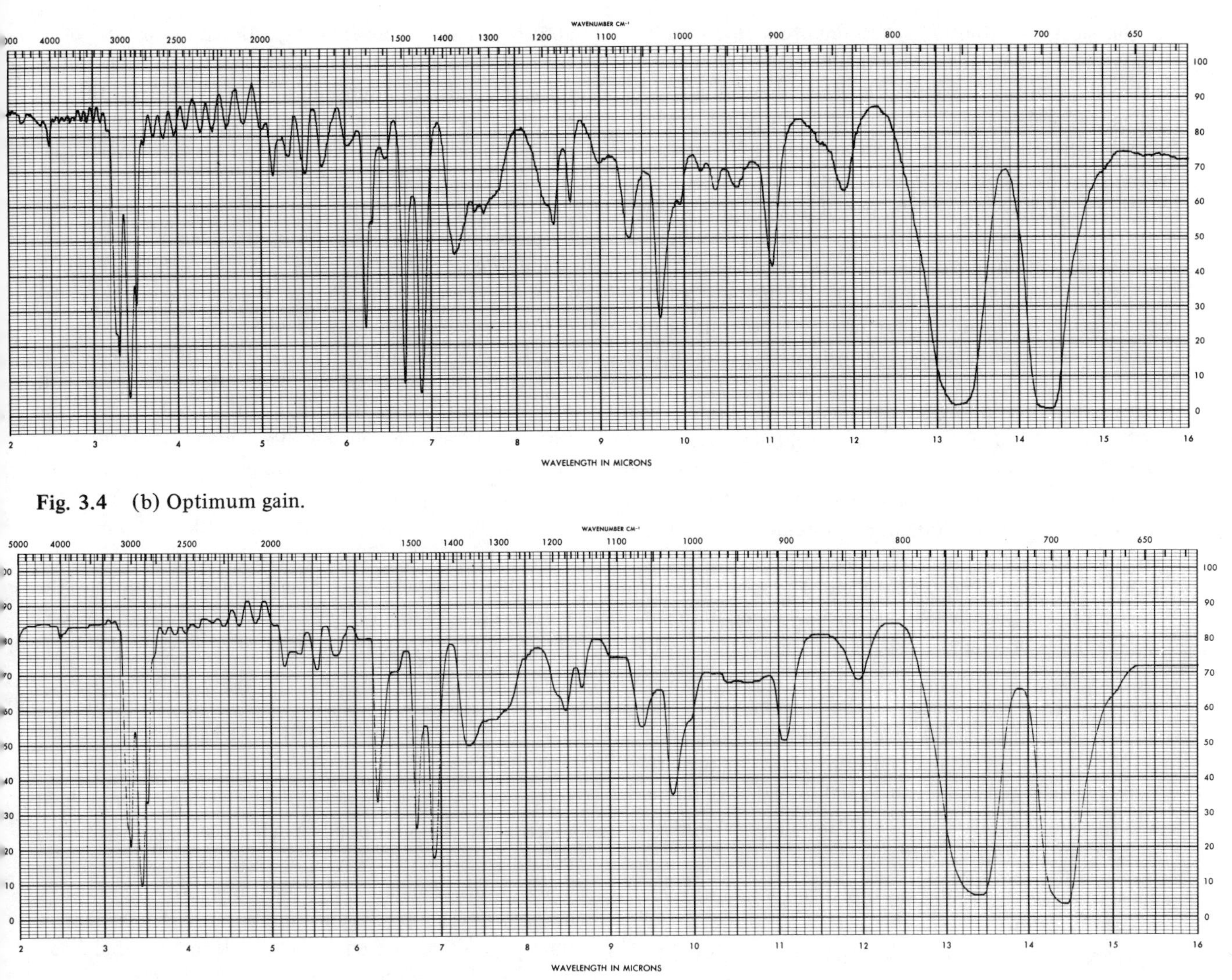

Fig. 3.4 (b) Optimum gain.

Fig. 3.4 (c) Gain too low. (Note shifts in frequency and loss of spectral details.)

Sample preparation

The sample whose spectrum is to be obtained may be a solid, a liquid, a gas, or a solution of any one of these. One determines the spectrum of a solid by making up either (1) a mull or (2) an alkali halide pellet.

To prepare a *mull,* one grinds about 5 mg of the solid to a very fine powder with the smallest possible drop of a suitable mulling agent such as nujol (a mixture of paraffinic hydrocarbons), hexachlorobutadiene, or one of the perfluorocarbons, such as fluorolube. The choice of the mulling agent depends on the spectral region of interest. Nujol (being a hydrocarbon) shows absorption bands only at 3030–2860 (C—H stretching), and 1460, 1374 cm^{-1} (C—H bending) regions. Naturally one cannot obtain information about the absorption by the sample in these regions when one is using nujol. On the other hand, perfluorocarbons are transparent at frequencies higher than 1300 cm^{-1}; thus one can use perfluorocarbons as mulling agents if one wants information in the region between 4000 and 1300 cm^{-1}. The spectrum of the mull is determined by placing it between two sodium chloride plates in the path of the sample beam. Mull spectra are the most reliable device by which to study materials in a solid state, but skill is required in the preparation of mulls in order to obtain high-quality spectra.

To prepare an *alkali halide pellet,* one grinds 1–2 mg of the sample finely, under anhydrous conditions, in an agate mortar. This powder is then thoroughly mixed with 100–200 mg of oven-dried, spectral-grade, 100–200-mesh potassium bromide powder. The mixture is then pressed into a transparent pellet either by pressing under very high pressure (at about 2,500 psi) in an evacuable die, or by using a mini-press. In either case, the KBr pellet is then mounted on a holder and placed in the sample beam of the spectrophotometer. Potassium bromide discs are convenient, and make possible the infrared examination of small quantities of sample. However, they are not as reliable as mull spectra because crystal structures can be altered during the preparation of the pellets. KBr discs should not be prepared from any organic halides other than bromides, because of the risk of halogen interchange.

If the sample is a liquid, its spectrum can be obtained by forming a thin film of the liquid between two sodium chloride or potassium bromide plates. If the sample is gaseous, or if one wants to determine its spectrum in the vapor state, one uses special cells with path lengths up to several centimeters long.

Spectra of solutions are obtained by preparing a 2 to 10% solution of the sample in an appropriate solvent. The cells used in determining the spectra of solutions have path lengths varying from 0.1 mm (macro cells) through 0.002 mm (micro cells) to 0.0002 mm (ultra-micro cells) and are made of a variety of optical materials, such as NaCl, KBr, CsBr, CaF_2, Irtran-2 (ZnS), etc. The solution is introduced into one such cell, and the cell is placed in the sample beam. A reference cell, of the same thickness and material as the sample cell, is filled with pure solvent and placed in the reference beam of the instrument.

The solvent should be able to dissolve the sample readily, should not contain moisture (since cell material is usually attacked by moisture) and should have few absorptions itself. In infrared spectroscopy, the following solvents are commonly used.

i) Carbon tetrachloride (absorbs strongly in 830–670 cm^{-1} region)
ii) Chloroform (strong absorption at 3030, 1220 and 830–670 cm^{-1})
iii) Carbon disulphide (strong absorption at 2350, 2160, and 1650–1350 cm^{-1})

Carbon tetrachloride and carbon disulphide are the preferred solvents, provided enough of the sample can be dissolved to form a solution of reasonable concentration.

Attenuated total reflectance

Recently a technique known as *attenuated total reflectance* (ATR) has been developed for infrared spectroscopy. This technique involves the use of light reflected from the surfaces of materials. When a beam of radiation passes through a prism in such a manner that it is reflected from the back face of the prism, part of the energy of the beam escapes from the face and is returned into the prism. Under proper conditions, this energy can be selectively absorbed by an absorbing substance if the substance is placed on the reflecting surface. The energy absorbed is approximately proportional to the ratio of the refractive indices of the prism and the substance. The refractive index of a substance changes rapidly at the wavelengths at which it absorbs energy. A plot of the energy absorbed by the substance versus the wavelength of radiation would therefore be similar to a conventional infrared spectrum of the substance. If the energy is lost by means other than absorption, the phenomenon is called *frustrated total internal reflection* (FTIR). When several reflectances are utilized, the phenomenon is called *multiple internal reflectance* (MIR).

The ATR technique is particularly useful in analyzing substances, such as polymers, which are difficult to mull, dissolve, melt, etc., so that they can be used in any of the traditional ways described earlier. The ATR technique is also very useful in examining samples whose absorption is too intense for practical transmission work. The spectra obtained by this technique are independent of the thickness of the sample (provided that the sample is thicker than about a wavelength). This enormously simplifies the preparation of samples. Several papers have appeared concerning this technique, and an excellent review has also been published.[1]

3.4 EFFECT OF INFRARED RADIATION ON VIBRATIONAL ENERGY

Molecules are made up of similar or dissimilar atoms held together by chemical bonds. Such a system approximates point masses connected by small springs representing bonds; it has associated with it a number of vibrational motions of the atoms against one another. These vibrations are excited by a change in dipole moment induced by infrared radiation, thus giving rise to the *infrared spectrum.* A similar effect on vibrational energy is achieved when there is a change in polarization caused by visible radiation, in what is known as *Raman spectroscopy.*

Raman spectra occur when molecules of the sample undergo a change in their polarization when they are irradiated with monochromatic* visible radiation. The quanta of incident radiation (with energy $E = h\nu$), when they interact with the molecules, may do one of the following things.

i) They may be reemitted, *without loss of energy*, and scattered in all directions (a process called *Rayleigh scattering*).

* Monochromatic radiation is not essential for Raman activity, but in practice, since very intense radiation is required to induce the excitation, the monochromatic, blue mercury line is used. Almost all the modern instruments now employ the more convenient and energetic sources of laser beams.

ii) They may be absorbed, due to the correspondence of their energy levels with the vibrational and rotational energy levels within the molecule. The absorption of quanta of energy raises the molecules to excited states. However, *nearly all the absorbed quanta* may be released in all directions when the molecules drop back to lower energy states later. This is known as *fluorescence.*

iii) If the energy of the incident photon is much smaller than that required to cause electronic transitions in the molecule, but much larger than that required to cause vibrational transitions in the molecule, a phenomenon known as *Raman scattering* takes place. Such energy is provided by the photons of visible radiation. Visible radiation supplies less energy than ultraviolet radiation (which causes electronic transitions), but more energy than infrared radiation (which causes changes in the vibrational energy levels of the molecule). Thus, when photons given off by visible radiation strike a sample, the molecules of the sample remain in the electronic ground state; transitions occur only on the vibrational and rotational levels associated with the ground state.

Raman effect

To understand the Raman effect, we must first realize that, when a photon, of energy $h\nu$, collides with a molecule initially in a stationary state, of energy E_1, the total energy $h\nu + E_1$ remains constant. After the interaction with the photon, the molecule may achieve a different quantum state with energy E_2 either greater or less than E_1 and release unabsorbed photons with energy $E = h\nu \pm \Delta E$. The term ΔE represents the difference between the two vibrational-rotational energy levels E_1 and E_2. The emerging photons scatter in all directions, in the form of radiation with frequency $\nu \pm \Delta\nu$, and the measurement of $\Delta\nu$ gives the difference between the various vibrational and rotational levels. A Raman spectrum[2] thus consists of lines representing the wavenumber of the incident radiation together with lines, at separation $\Delta\nu$, of both higher and lower wavenumbers, representing the radiation scattered by interaction.

In Raman spectroscopy, the radiation, scattered at right angles to the incident beam, is observed in a spectrograph. The spectrograph registers a spectrum consisting of a strong line (*exciting line*) of the same frequency as the incident radiation together with weaker lines (*Stokes* and *anti-Stokes lines*) on either side of it. The weaker lines are in a symmetrical pattern about the exciting line, and are shifted away from it by frequencies ranging from a few to about 3500 wavenumbers. The shifts in frequencies are due to $\pm\Delta\nu$, and thus, by observing these shifts, we can obtain information about the frequencies of molecular vibrations.

The information obtained from a Raman spectrum of a compound is often complementary to that obtained from its infrared spectrum, and although Raman spectroscopy is not widely used to date by organic chemists, its applications to the problems in organic chemistry are rapidly increasing. Indeed, there are indications that, especially because of lasers, Raman spectroscopy may soon become a routine tool in the hands of organic chemists.

Vibrations of a diatomic molecule

To understand molecular vibrational changes, let us consider a simple diatomic system. Suppose that such a system absorbs infrared radiation and is put in motion by it. The atoms execute a vibrational motion approximated by the *law of simple harmonic motion.* The frequency ν (in Hz) of such a motion is given by

$$\nu = \frac{1}{2\pi}\sqrt{\frac{k}{m_r}} \tag{3.1}$$

where k is the force constant (in dynes/cm) and describes the stiffness of the spring of our model, or the restoring force per unit displacement of point masses, and m_r is the reduced mass, that is, the harmonic mean,

$$\frac{m_1 m_2}{m_1 + m_2}$$

of the masses, m_1 and m_2, of the two atoms (Fig. 3.5).

Equation (3.1) enables us to calculate the frequency of vibration of a diatomic molecule. However, due to quantum restrictions, the vibrational energy levels can assume only special values, given by an expression:

$$E = (v + \tfrac{1}{2})h\nu \tag{3.2}$$

where v is an integer = 0, 1, 2, ... The vibrational energy levels $v_0, v_1, v_2, \ldots$ in Fig. 3.6 are therefore separated from one another by an almost equal amount.

Consequently, given that ν is the frequency of the absorbed radiation which causes a transition from level v_0 to the next-higher level v_1, radiation of frequency 2ν

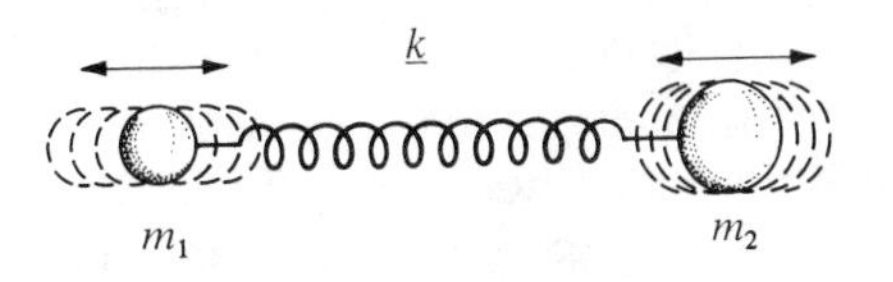

Fig. 3.5 Schematic drawing of a diatomic molecule.

will cause a transition from v_0 to the second level v_2. Frequency ν is called the *fundamental frequency* and 2ν *the first overtone.* The overtones, since they are harmonics of the fundamental frequency, occur at near-integral multiples of the fundamental absorption frequency.

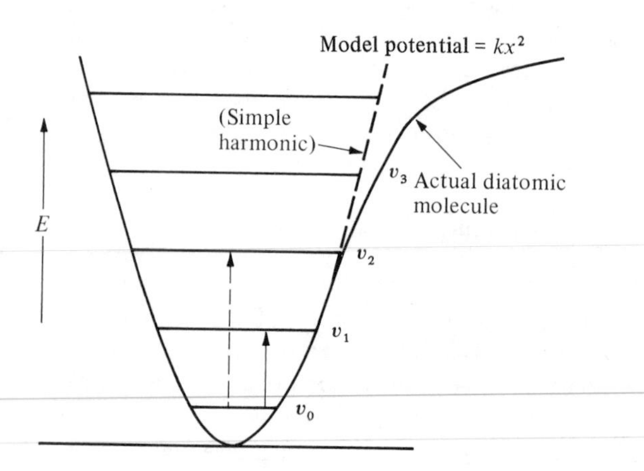

Fig. 3.6 Potential energy curves and vibrational energy levels $v_0, v_1, v_2, \ldots$, for a model and for an actual diatomic molecule. Transition of molecules from one level to the next higher level, for example, $v_0 \rightarrow v_1$ or $v_1 \rightarrow v_2$, is called *fundamental transition,* and that from one level to second- or third-higher levels, for example, $v_0 \rightarrow v_2$ or $v_1 \rightarrow v_3, v_4, \ldots$, is called *overtone transition* (terms derived from musical terminology).

Selection rules

Since the number of energy levels for any molecule is very large, we would expect that an infrared spectrum might be exceedingly complex. However, the observed spectra are relatively simple because of the *selection rules.* These rules simply say that under certain circumstances a band is "allowed" and under certain other conditions a band is "forbidden," and must have zero intensity. For example, according to one selection rule, *only the fundamental transition* ($v_0 \rightarrow v_1$) *is allowed for an idealized system such as a perfectly harmonic vibrator.*

If the molecular vibrations were perfectly harmonic, a vibrational spectrum would have only the fundamental band, with frequency ν, but would contain no overtones of frequency 2ν, 3ν, etc. However, the molecular vibrations are not *strictly* harmonic, and, to the extent that anharmonicity exists, weak overtones of frequency 2ν can often be detected in the spectrum. Strictly speaking, the overtone frequency is usually less than twice the fundamental frequency ν, and the overtone intensity is usually one-tenth to one-hundredth that of the fundamental.*

Types of infrared bands

Infrared spectra of complex molecules, in addition to containing fundamental and overtone absorptions, also contain combination, coupled, and Fermi resonance bands. Relatively weak *combination bands* appear at frequencies equal to the sum of or difference between two or more fundamental frequencies, whereas a *coupling band* results from the interaction of two or more vibration modes of the same portion of the molecule.

When two groups of atoms which are vibrating at similar frequencies are close to each other in the same molecule, a resonance is established between the two vibrating groups and mechanical interaction, or *coupling,* takes place. The coupling between two groups is strongest when their frequencies are nearly identical and when they share a common atom. Strongly coupled groups lose their individual vibrations and vibrate together. In an infrared spectrum, a coupling band appears in a region somewhat removed from the region in which each individual transition would appear.

Sometimes bands which have anomalous intensities and which are displaced by many wavenumbers from their anticipated positions are observed in the spectra of polyatomic molecules. These bands are due to a coupling between a fundamental mode and overtone or combination modes of vibration. Such a perturbation between two modes occurs due to anharmonicity when the modes have almost the same energy and are therefore nearly

---

* Note that the molecular vibrations are not *governed* but *approximated* by the law of simple harmonic motion. Our ball-and-spring model is perfectly harmonic; its potential is given by

$$E = kx^2,$$

where E = potential energy, x = displacement from equilibrium, and k is the force constant. A plot of E versus x gives a parabola such as shown in Fig. 3.6, and the solutions for E and ΔE are as given in Eq. (3.2), where vibrational energy levels are evenly spaced by $h\nu$. In such a case, transitions of more than one quantum are *strictly* forbidden.

The diatomic molecules, however, *do not move* in a perfectly harmonic potential and use the morse curve as shown in Fig. 3.6. At the bottom of the curve (where quantum numbers are low), the curve follows the harmonic potential kx^2, but at higher quantum numbers it does not. This introduces anharmonicity, where $E = k'x^2 + k''x^3 + k'''x^4 + \cdots$, etc. Now the energy levels are not evenly spaced (that is why overtone frequency is not exactly twice the fundamental frequency) and multi-quanta transitions are partially allowed.

degenerate. Interaction of this type between a degenerate overtone and a fundamental mode is called a *Fermi resonance interaction.*

Factors affecting infrared band frequency

Thus, on the basis of the foregoing, we can say that the most important factors determining the frequency of vibration (and thus of absorption) are:

i) Bond elasticity, represented by the force constant k
ii) Relative masses of the bonded atoms, represented by m_r in Eq. (3.1)

However, absorption frequency is influenced by several other effects, both inside and outside the molecule. (1) Electrical effects, (2) the nature, size, and electronegativity of neighboring atoms, (3) hydrogen bonding, (4) phase changes, and (5) steric effects may all cause shifts in frequency. Of these, we shall first consider electrical effects.

In applying the ball-and-spring model to explain molecular vibrations, we have so far ignored the fact that atoms possess electrical properties. These atoms, called *point masses,* may or may not be permanently polarized by their mutual interactions so that the whole molecule has a resultant dipole moment.*

Furthermore, an external field, such as that of a beam of radiation, can electrically polarize these point masses. During molecular vibrations, both the polarizability and the dipole moment of the system may vary as the point masses change their relative positions. If the dipole moment changes as the molecule vibrates, a stationary alternating electric field is produced. The magnitude of this field changes periodically with time at a frequency equal to the vibration frequency. It is this stationary alternating electric field that interacts with the *moving* electric field of electromagnetic radiation. Consequently, *when the frequency of electromagnetic radiation is equal to that of the alternating electric field produced by changes in the dipole moment, vibrational motions are excited in the molecule, and the radiation is absorbed. However, radiation is not absorbed if it fails to produce a change in dipole moment of the molecule. Thus, vibrational modes are infrared-active only if a change in dipole moment occurs during the vibration.*

The operation of this selection rule can be illustrated by considering a symmetrical molecule such as carbon dioxide. For any polyatomic molecule, the number of fundamental absorption bands can be calculated from the number of its atoms and their degrees of freedom. The number of degrees of freedom is equal to the sum of the coordinates necessary to locate all the atoms of the molecule in space. Consider a molecule of N atoms. Such a molecule has $3N$ degrees of freedom because three cartesian coordinates are required to locate each atom in space. However, of these atomic degrees of freedom, three describe the *translation of the molecule as a whole* and three describe the *rotation of the molecule as a whole* if the molecule is nonlinear. These six degrees therefore cannot lead to internal vibrations, and the number of remaining fundamental vibrations is thus $3N - 6$. In the case of a linear molecule, there are only *two* axes (instead of three) about which the molecule as a whole can rotate. The number of molecular vibrations in such a case is therefore $3N - 5$.

The molecule of carbon dioxide, being triatomic and linear, gives rise to $(3 \times 3) - 5 = 4$ fundamental vibrations, two involving changes *along* the bond axis (stretching) and two involving movement of the atoms *out* from the bond axis (bending) as shown in Fig. 3.7. However, in the infrared spectrum of carbon dioxide, one observes an absorption band for the frequency corresponding to asymmetric stretching ν_2 involving change in dipole moment, but *not* for the frequency corresponding to the symmetric stretching ν_1, which involves no change in dipole moment of the molecule. This is in accordance with the selection rule mentioned above.

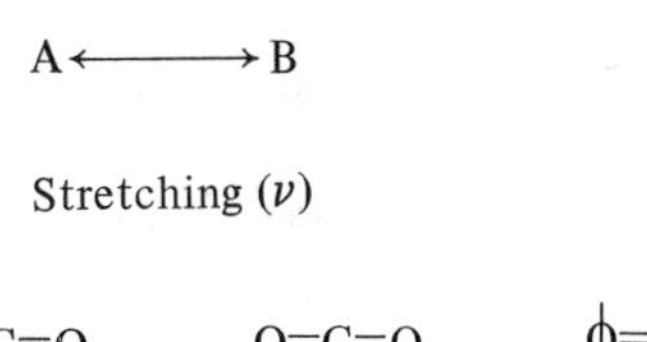

Fig. 3.7 Fundamental vibrations of carbon dioxide. Arrows indicate movement of the atoms along the bond axis, whereas + and − signify vibrations perpendicular to the plane of the paper.

* A molecule which is electrically neutral has a resultant electric *dipole moment* provided that its center of positive charge does not coincide with its center of negative charge. Dipole moment is a vector quantity whose direction is that of a line joining the centers of the two charges, $+q$ and $-q$, and whose magnitude is the length of that line multiplied by the total negative or the total positive charge, these being equal. For example, a proton, $q = \epsilon$, and an electron, $q = -\epsilon$ (where $\epsilon = 4.8 \times 10^{-10}$ esu), which are separated by a distance of one angstrom (10^{-8} cm), have a dipole moment of 4.80×10^{-18} esu cm. The unit 10^{-18} is called the debye (D) and the dipole moment in this example is 4.80 debyes.

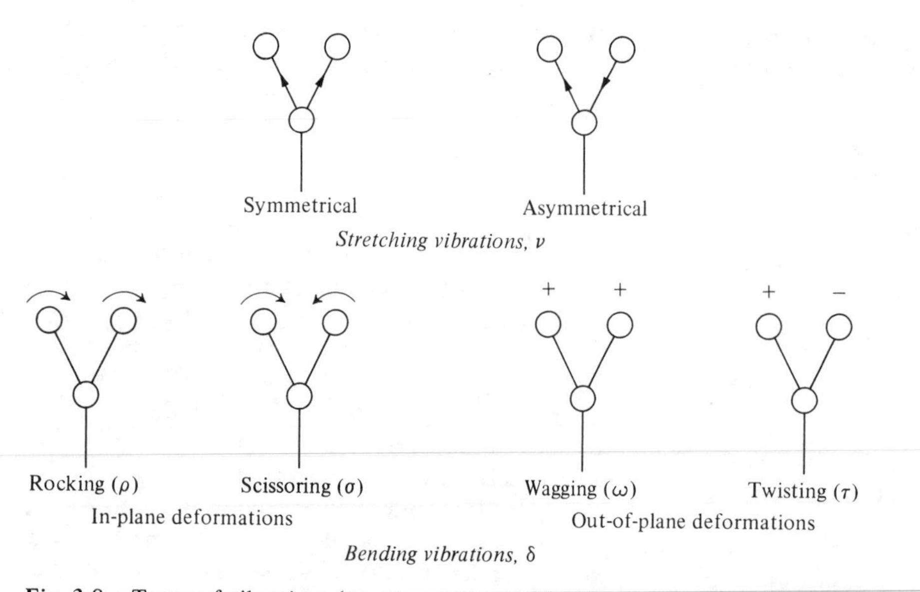

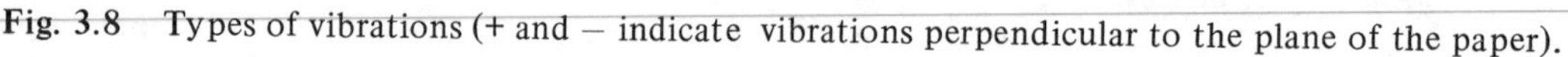

Fig. 3.8 Types of vibrations (+ and − indicate vibrations perpendicular to the plane of the paper).

Always keep in mind this selection rule, which says that there must be a change in dipole moment to accompany a vibrational mode before the vibration becomes infrared-active. It is for this reason that a symmetrical diatomic molecule is completely infrared-inactive, as is the C=C stretching vibration in ethylene or other symmetrical olefins. For Raman activity, however, no change in dipole moment of the bond is necessary, but, in order for vibrational motions to be excited, a change in the polarizability, relative to the ground state, is required. Thus, in molecules which have a center of symmetry, vibrations which are infrared-inactive are generally Raman-active, and vice versa.

Some molecular vibrations (called *fingerprint* vibrations) are characteristic of the entire molecule, whereas others are associated with certain functional groups. The effects of these vibrations on the chemical bonds between the atoms are classified into two major classifications, *stretching* (ν) and *bending* (δ), and are further described by such terms as *scissoring* (σ), *rocking* (ρ), *wagging* (ω), and *twisting* (τ). These are shown schematically in Fig. 3.8.

Frequencies of the stretching and bending vibrations depend largely on the vibrating atomic masses and on the bond orders of the chemical bonds joining them. The lighter the atoms, the higher the frequency of vibration. Similarly, the higher the bond order, the higher the frequency of vibration. Thus a

—C—D (with two vertical bonds on C)

bond vibrates at a lower frequency (lower wavenumber, 2100 cm^{-1}) than a

—C—H (with two vertical bonds on C)

bond (wavenumber 2900 cm^{-1}), and a carbon-carbon triple bond (—C≡C—) has a higher vibrational frequency (2300 to 2100 cm^{-1}, 4.4 to 4.76 μ) than a carbon-carbon double bond (>C=C<), which in turn vibrates at a higher frequency (1700 to 1500 cm^{-1}, 5.88 to 6.7 μ) than a carbon-carbon single bond (1300 to 800 cm^{-1}, 7.7 to 12.5 μ). However, stretching vibrations generally require more energy and thus occur at higher wavenumbers (shorter wavelengths) than the bending vibrations of the same group.

Problem 3.2 What do you understand by the term *reduced mass*? Calculate reduced masses for molecules of HCl, LiH, and N_2, given atomic masses of H = 1, Li = 7, N = 14, Cl = 35.

Problem 3.3 The IR spectrum of an alcohol displays an O—H bond stretching band at 2.77 μ. Calculate the force constant of the O—H bond in this compound.

Problem 3.4 Assume that the force constant for the stretching vibration of a C=O bond in an acyclic ketone is 11.72×10^5 dynes/cm. Calculate the fundamental frequency expected in the IR spectrum for C=O stretching vibration in this compound.

Problem 3.5 What selection rules have you encountered in this section? What is the significance of these rules?

Problem 3.6 Explain the following.

a) Fundamental transition and its dependence on atomic masses and bond strengths

b) Overtone transition and its relation to harmonic motion

c) Combination bands and their origin

d) A coupling band and the frequency of its absorption

e) Fermi resonance interaction

f) The Raman effect and its dependence on polarizability

g) Infrared absorption and its dependence on dipole moment

h) Different types of stretching and bending vibrations

3.5 SPECTRA OF COMPLEX MOLECULES

In the case of diatomic or relatively simple molecules, we can calculate the vibrational frequencies by using Eq. (3.1). For complex polyatomic molecules, however, the mathematical treatment becomes enormously difficult. We must therefore use an empirical approach to study various subparts of the molecule separately.

Certain groups of atoms in a molecule consistently give rise to absorption bands in the same spectral region. For example, all stretching vibrations of groups in which the nucleus of one atom is a proton are observed at wavenumbers above 2250 cm^{-1}. This is consistent with the theory that *vibrational frequency is a direct function of the force constant of the bond and an inverse function of the masses involved.* The hydrogen atom is lighter than any other atom, so hydrogen stretching bands occur at the highest frequencies. Thus C—H stretching is observed in the region 2850 to 3300 cm^{-1} (3.5 to 3.03 μ); O—H stretching at 2500 to 3650 cm^{-1} (4.0 to 2.74 μ); S—H stretching at 2550 to 2600 cm^{-1} (3.92 to 3.85); and N—H stretching at 3030 to 3500 cm^{-1} (3.3 to 2.86 μ). Groups with triple bonds (highest force constant) absorb in the next-highest region of the spectrum. [—C≡C— stretching occurs at 2100 to 2360 cm^{-1} (4.76 to 4.24 μ); —C≡N stretching at 2070 to 2260 cm^{-1} (4.83 to 4.42 μ); and —N≡N stretching at 2080 to 2170 cm^{-1} (4.81 to 4.61 μ).] The only other principal groups to have fundamental absorptions in the high region above 2250 cm^{-1} are those with cumulated double bonds, such as >C=C=O (1930 to 2350 cm^{-1}; 5.18 to 4.24 μ).

Many workers have compiled data concerning characteristic absorption frequencies of functional groups and chemical classes. Correlation charts listing the absorption frequencies (from 4000 to 666 cm^{-1}; 2.25 to 15 μ) of the common classes of organic compounds are given in Appendix 3.1.A, 3.1.B, and 3.1.C.

Problem 3.7 Imagine that an infrared recording chart is arbitrarily divided into four sections according to wavenumbers: (i) 4000 to 2500 cm^{-1}, (ii) 2500 to 1500 cm^{-1}, (iii) 1500 to 1000 cm^{-1}, and (iv) 1000 to 600 cm^{-1}. Predict, without referring to Appendix 3.1, where each of the following modes of vibrations will appear on such a chart:

O—H stretching, C—O stretching, C—H bending, N—H bending, C—C stretching, N—H stretching, C≡C stretching, C—N stretching.

3.6 FACTORS AFFECTING ABSORPTION FREQUENCIES

The correlation charts provide the basic information necessary for a reasonable interpretation of the infrared spectrum of an organic compound. However, assignment of absorption frequencies to specific vibrations should be made only after a careful consideration of the factors causing minor shifts in absorption frequencies. For example, the C=O stretching in a *saturated* acyclic ketone is observed specifically between 1705 and 1725 cm^{-1}; but that for an *unsaturated* acyclic ketone it is normally between 1665 and 1685 cm^{-1}. This effect of the environment of the chromophore on the direction and magnitude of the shift is often of great diagnostic value.

The infrared spectrum of a compound also depends on the physical state of the compound. For example, when the compound is in the *crystalline state,* the absorption frequencies of the groups are modified due to strong interactions, such as hydrogen bonding, between polar groups of adjoining molecules in the orderly arrangement of the molecules in the crystal. When the compound is in the *pure liquid* or *melted solid state,* the same intermolecular forces operate, although the molecules are more randomly oriented. The spectra of compounds in the solid or pure liquid state can therefore sometimes be misleading, and caution should be exercised in using them for precise studies of group frequency correlation. The spectrum of a compound which is in the *gaseous* or *vapor phase* should be ideal for this purpose. However, complexities arise due to (a) the technical difficulties in vaporizing certain organic substances, (b) difficulties in determining vapor-phase spectra, and (c) the fine rotational structure in the vapor phase.

The spectrum of a substance *in solution* ought therefore to be reasonably reliable. However, the solute-solvent interactions in a polar solvent cause considerable shifts, and group frequencies vary appreciably from one solvent to another. Consideration must therefore be given to such possible interactions. Solute-solvent interactions, as well as *intermolecular* interactions, are minimal if the spectrum is determined for an extremely dilute solution in an inert, nonpolar solvent such as *n*-hexane.

A number of *intramolecular* factors, other than those mentioned in Section 3.4 (mass of atoms, force constants, vibrational coupling, Fermi resonance, etc.) are also responsible for shifts in absorption frequencies. Some of these factors are steric interactions, ring strain, resonance, inductive effects, dipolar interactions, intramolecular hydrogen bonding, etc. The effects of these factors are best illustrated by a number of spectra given in the next section.

3.7 REACTIVE GROUPS AND THEIR ABSORPTION FREQUENCIES

An infrared spectrum of an organic compound comprises many bands. This is understandable because organic molecules have many vibration frequencies, according to the $3N - 6$ rule. Consequently, we can find out a lot about the compound from its infrared spectrum, although assigning each band to a particular mode of vibration is rarely possible. As a matter of fact, two nonidentical molecules generally have different infrared spectra, and an IR spectrum therefore is a "fingerprint" of the molecule.

The region most useful for the purpose of "fingerprinting" the compound is that between 650 and 1350 cm^{-1} (15.4 to 7.4 μ). This region comprises a large number of bands due mostly to skeletal vibrations; it also comprises some characteristic frequencies due to bending, and many of unknown origin. We can assign some bands in the 650 to 900 cm^{-1} region to certain particular modes of vibration, but others are too complex, and can be examined only in the light of what we see in the higher- and lower-energy regions of the spectrum. However, in the region below 1350 cm^{-1}, when the spectrum we are studying coincides exactly with the spectrum of a known compound, it can be safely assumed that the two compounds are identical.

The region above 1350 cm^{-1} often provides most of the important information. Characteristic, easily recognizable bands of various functional groups often appear in this region. We can obtain much valuable structural evidence from relatively few of these bands, and a total interpretation of the complete spectrum is seldom required.

Some of the factors that should be taken into consideration when one is trying to determine the structure of a compound on the basis of its infrared spectrum are: shape and position of bands, relative intensities of two bands, weak bands, characteristic patterns, band halfwidths, etc.

The conventional way of observing an infrared spectrum is by correlating band positions and molar extinction coefficients with specific known structural features. The band position is normally cited as the position of maximum absorption ν_{max} in cm^{-1} (Fig. 3.9). In recent years the integrated intensities of the bands are being used increasingly for deriving structural information because of the relationship of this parameter to the polar properties of the molecule.[3,4,5] Correlation has been established between integrated intensities and other important properties, such as hydrogen bonding,[6] presence of functional groups,[7,8,9,10] and chemical reactivities of substituents.[11]

In Fig. 3.9 a typical carbonyl band is replotted in the absorbance mode on the linear wavenumber scale. The integrated intensity of this band is given by the hatched area in the figure.

Although integrated intensities are becoming increasingly significant in determinations of organic structure, the usefulness of these data depends largely on the

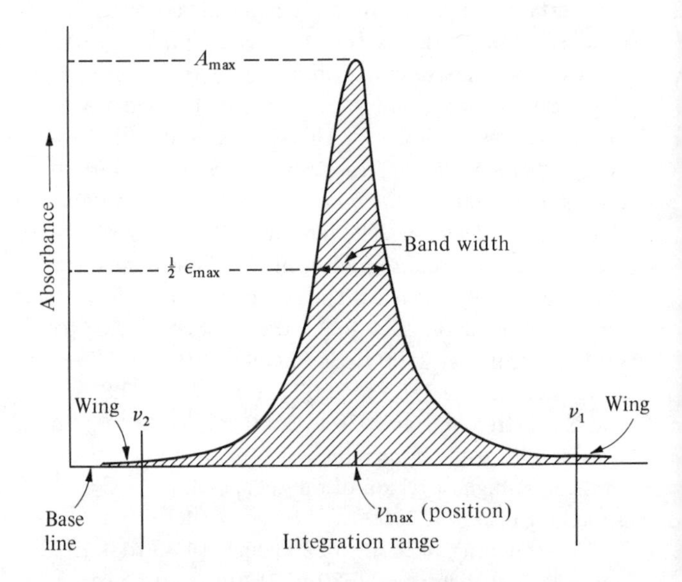

Fig. 3.9 The parameters of a typical infrared band replotted in absorbance mode on a linear wavenumber scale. Note the following parameters. *Position:* given as wavenumber at maximum absorption ν_{max} in cm^{-1}. *Bandwidth:* given in wavenumber as width at half-band height. *Absorbance* = $\log I_0/I$ given at point of maximum absorption A_{max}.
Molar extinction coefficient = $\epsilon_{max} = A_{max}/c \times l$, and

$$\textit{Integrated intensity} = 2.303 \int \epsilon \, d\nu \text{ (in liter mole}^{-1}\text{ cm}^{-2}\text{)} = \frac{\text{hatched area in cm}^{-1}}{\text{moles/liter} \times \text{cm}}$$

accuracy of the instrument, which must have high resolving power and must be so arranged that the width of the slits can be controlled perfectly. We shall therefore follow the conventional way of using the correlation mostly between band position and structure of the compound, and use band intensities only in terms such as strong, weak, medium, variable, sharp, broad, etc., as given in the correlation table. Remember, however, that the mere knowledge of band positions, with some indications of band intensities and shapes, will not suffice for the interpretation of all infrared spectra. To obtain the maximum amount of information from the spectrum, one should combine additional information (chemical and physical properties, empirical formula, etc., of the compound) with the observation of the spectrum. This information, together with that obtained from other spectroscopic techniques, such as UV and NMR spectroscopy, often gives an indication of the structure of the compound. If possible, confirm the structure thus indicated by comparing the spectrum of the unknown with that of an authentic sample.

Hydrocarbons: C—H and C—C stretching and bending vibrations

Let us start with the simple vibrational forms in which equivalent atoms that are close together are treated in equivalent fashion. Consider simple hydrocarbons, a class of compounds in which only two types of atoms (C and H) and only two types of bonds (C—C and C—H) are present. The infrared spectra of three hydrocarbons—*n*-hexane, isooctane, and cyclohexane—are shown in Figs. 3.10, 3.11, and 3.12, respectively.

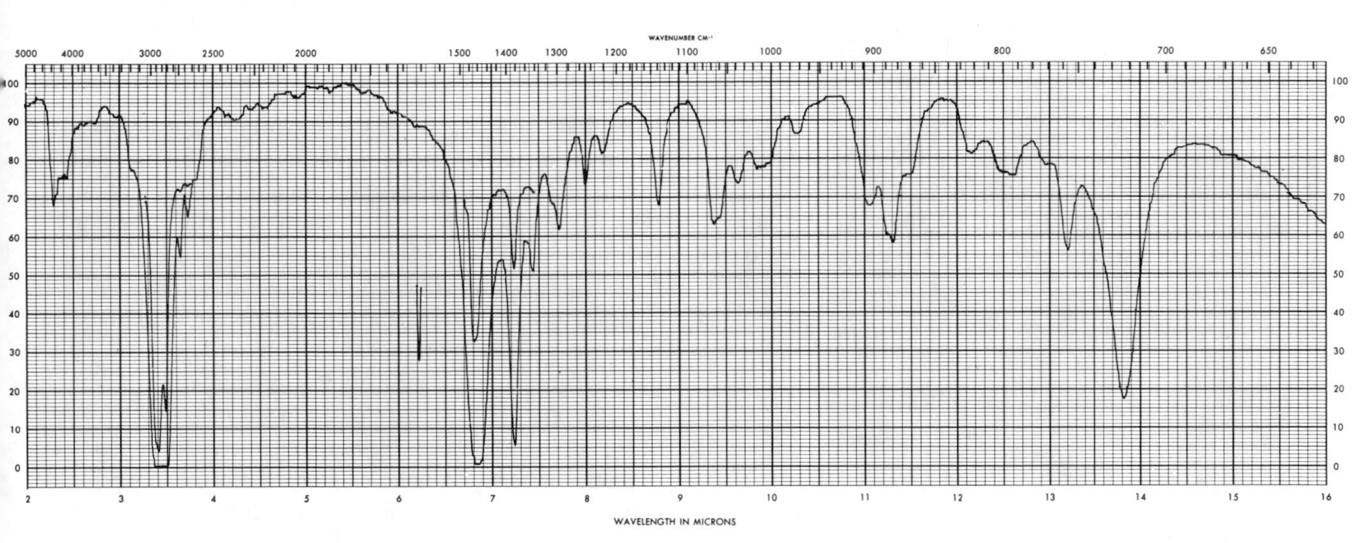

Fig. 3.10 Infrared spectrum of *n*-hexane, liquid film. (Insets at 3.5 and 6.8 μ recorded in 6% solution in CCl_4.)

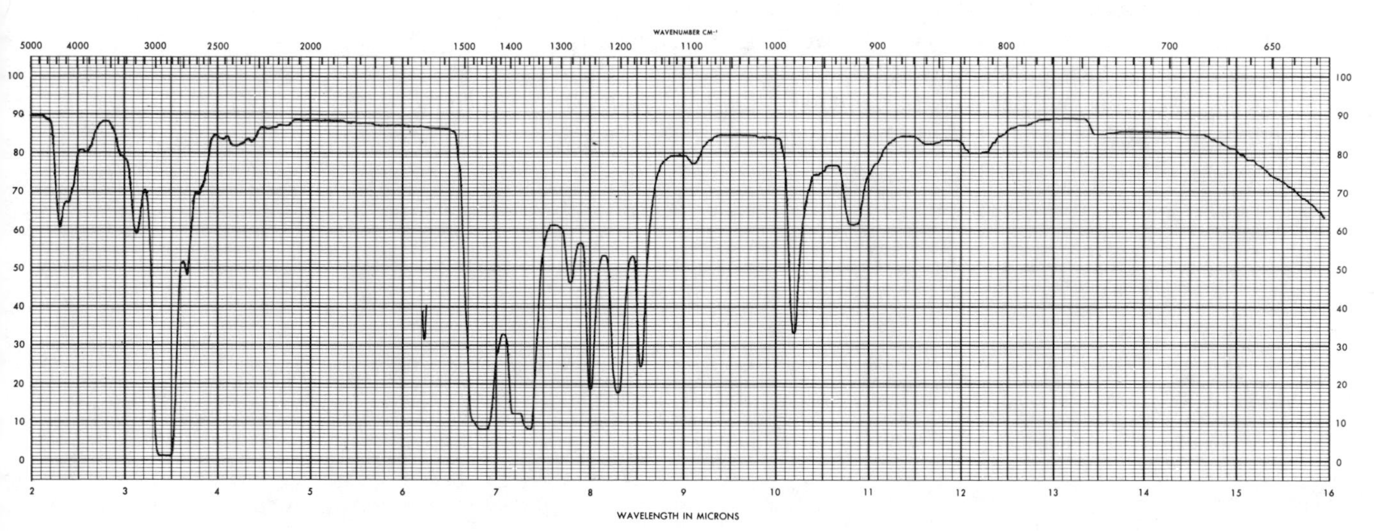

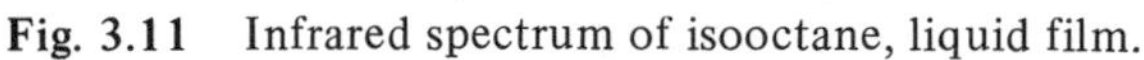

Fig. 3.11 Infrared spectrum of isooctane, liquid film.

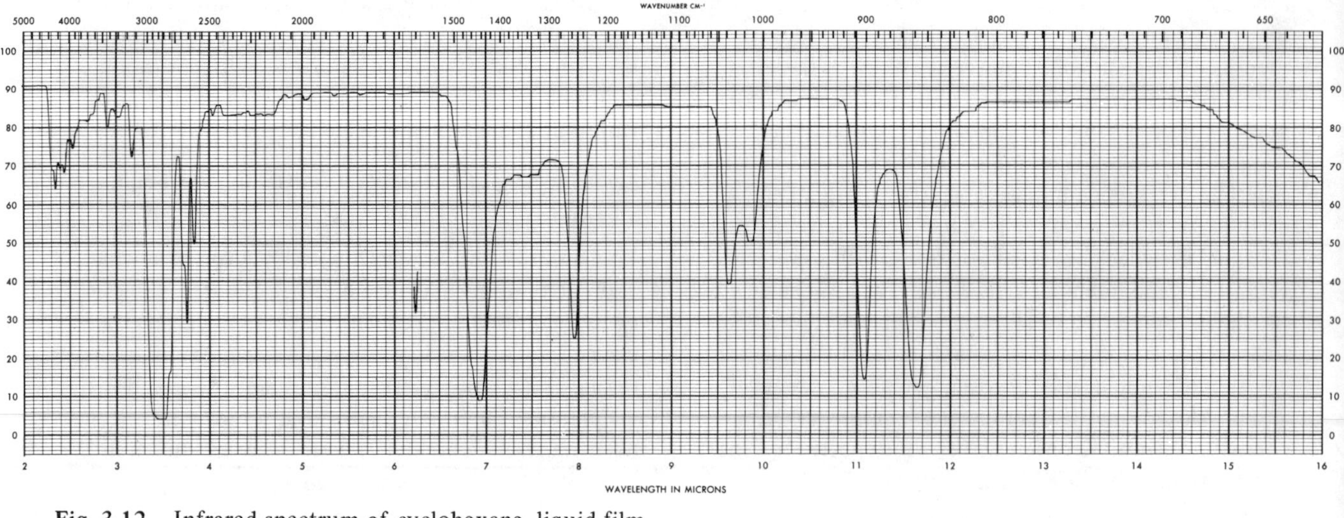

Fig. 3.12 Infrared spectrum of cyclohexane, liquid film.

We saw earlier that all fundamental vibrations with wavenumbers higher than 2500 cm^{-1} may safely be assigned to stretching of bonds between hydrogen and another atom. The C—H stretching absorption usually occurs in the general region between 3300 cm^{-1} (3.0 μ)

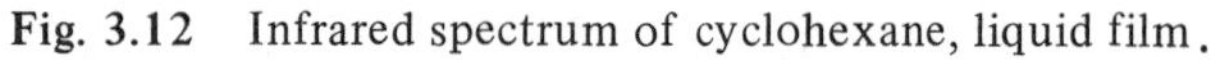

(H—C≡ in alkyne) and 2700 cm^{-1} (3.7 μ) (in H—C(=O)— probably due to an interaction between the C—H stretch and the overtone of the C—H deformation).

Alkanes

In alkanes the C—H stretch region is narrowed to 3000 to 2840 cm^{-1} (3.3 to 3.52 μ). A hydrocarbon containing a methyl group usually shows two distinct bands, one at 2960 cm^{-1} (3.38 μ) [due to an asymmetric stretching of the entire methyl group in which two C—H bonds are extending while the third one is contracting (Fig. 3.13a)] and the other at 2870 cm^{-1} (3.48 μ) [due to symmetric stretching, as in Fig. 3.13b]. The C—H bending deformations of the methyl groups in the hydrocarbons normally occur at 1450 cm^{-1} (6.9 μ) (asymmetrical deformation, Fig. 3.13c) and 1375 cm^{-1} (7.28 μ) (symmetrical deformation, Fig. 3.13d).

The band at 1375 cm^{-1} (7.28 μ) is due only to a methyl on a carbon atom and is quite sensitive to the electronegativity of the substituent attached to the methyl group. It can shift from as high as 1475 cm^{-1} (6.78 μ) (in CH_3—F) to as low as 1150 cm^{-1} (8.7 μ) (in CH_3—Bi). However, this band is extremely useful in detecting the presence of a methyl group in the compound because it is sharp and of medium intensity, and is rarely overlapped by absorptions due to methylene or methine deformations.

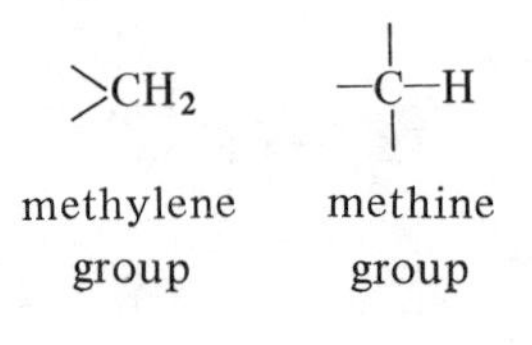

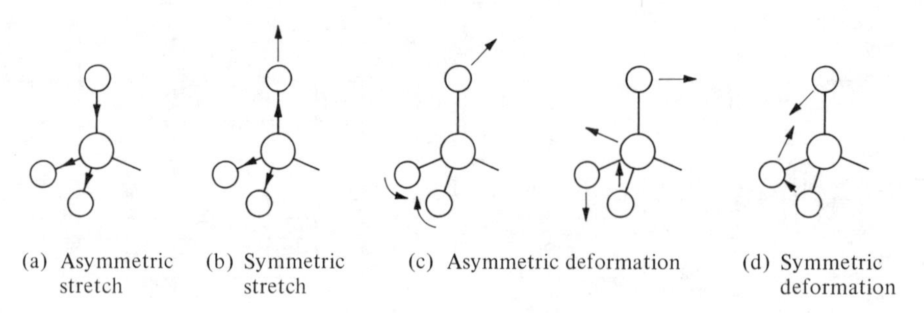

Fig. 3.13 Symmetrical and asymmetrical stretching and bending deformations of a methyl group.

The intensity of this band usually increases with the number of methyl groups in the compound. However, the presence of two or more methyl groups on *one* aliphatic carbon atom (as in isopropyl or *t*-butyl groups) results in splitting of this band due to in-phase and out-of-phase interaction of the two symmetrical methyl deformations. For example, a strong doublet at 1377 cm^{-1} (7.28 μ) and 1360 cm^{-1} (7.26 μ) is noticeable in the spectrum of isooctane (Fig. 3.11). Such a doublet, in which both the bands are of approximately equal intensity, and a band at 1165 cm^{-1} (8.49 μ) are characteristic of all compounds containing an isopropyl group. The band at 1165 cm^{-1} is due to C—C skeletal vibration, about which we shall soon say more.

The C—H bonds in methylene groups undergo a number of stretching and bending vibrations similar to those shown in Fig. 3.8. Consequently, methylene groups give rise to a number of absorption bands. The two stretching vibrations—asymmetrical and symmetrical—occur at 2925 cm^{-1} (3.42 μ) and appear in the spectrum within a range of ±10 cm^{-1}. The bands due to C—H bending deformations, such as scissoring, twisting, wagging, and rocking, normally appear at fairly constant frequencies.

Scissoring and rocking absorption bands are usually strong, and generally appear at wavenumbers of 1465 and 720 cm^{-1} (6.83 and 13.9 μ), respectively (if four or more adjacent $-CH_2-$ are present), whereas the weak bands due to twisting and wagging vibrations appear at wavenumbers of 1250 ± 100 cm^{-1} (7.41 to 8.70 μ). It is easy to see why the scissoring absorption band of methylene, around 1465 cm^{-1}, often overlaps the band due to the asymmetrical bending of methyl at about 1450 cm^{-1}.

In cyclic aliphatic hydrocarbons, the C—H stretching frequencies are the same as in the case of acyclic compounds if the ring is unstrained (for example, cyclohexane, Fig. 3.12); however, the methylene scissoring bands shift slightly to smaller wavenumber. Notice that the 1470 cm^{-1} (6.8 μ) band in *n*-hexane (Fig. 3.10) is shifted to 1450 cm^{-1} (6.89 μ) in cyclohexane. In sterically strained cyclic compounds the C—H stretching normally occurs at slightly higher wavenumbers (for example, 3080 to 3040 cm^{-1} in highly strained cyclopropane).

With the exception of the 1375 cm^{-1} (7.25 μ) band due to C—H symmetrical deformation, all the other C—H absorptions vary only slightly with changes in molecular environments. These slight changes are often beyond the resolving power of ordinary IR instruments, and consequently are of little use for work on structure determinations.

The C—C bond vibrations are also of little value for structural study, since the C—C stretching vibrations appear as weak bands in the 1200 to 800 cm^{-1} region, whereas the C—C bending absorptions occur at wavenumbers less than 500 cm^{-1}, that is, beyond the range of an ordinary infrared instrument.

Alkenes

Like alkanes, simple alkenes contain only carbon and hydrogen atoms. However, the number and types of bonds in an alkene are somewhat different from those in an alkane. An alkene may contain

$$\underset{(sp^3)\,(sp^3)}{\text{C—C}},\quad \underset{(sp^3)}{\text{C—H}},\quad \underset{(sp^2)}{\text{C—H}},\quad \underset{(sp^2)\,(sp^3)}{\text{C—C}}\quad \text{and}\quad \underset{(sp^2)\,(sp^2)}{\text{C=C}}$$

bonds. The vibrational frequencies of some of these bonds are therefore different from those of the bonds in an alkane. For example, a carbon–carbon double bond has a higher force constant than a carbon–carbon single bond, and in a nonconjugated olefin, the frequency of C=C stretching vibration is higher (1680 to 1620 cm^{-1} or 5.95 to 6.17 μ) than that of the C—C stretching vibration (1200 to 800 cm^{-1} or 8.33 to 12.50 μ).

As we saw earlier (Section 3.4), for a vibration to be infrared-active, a change in dipole moment must accompany a vibrational mode. Such a change in dipole moment cannot occur in completely symmetrical molecules. For this reason, the C=C stretching band is absent from the spectra of symmetrical molecules such as ethylene, tetrachloroethylene, etc. On the other hand, the more unsymmetrically substituted double bonds exhibit stronger absorption bands. Mono- and tri-substituted olefins are less symmetrical, and therefore give rise to more intense C=C stretching bands than disubstituted olefins do. Similarly, a *trans* olefin gives only a very weak C=C stretching band, but the intensity of the C=C stretching band is greater in the less symmetrical *cis* isomer. Thus *the absorption bands are more intense for* cis *than for* trans *isomers, for mono- or tri-substituted olefins than for di- or tetra-substituted ones, for terminal than for internal double bonds, and for C=C groups conjugated with certain unsaturated groups (for example, C=C, C=O, etc.) than for nonconjugated ones.*

The absorption frequencies for C=C stretching vibrations as well as for C—H stretching and deformation vibration vary only slightly with the degree of substitutions on the vinyl carbons. These frequencies, listed in Table 3.1, are illustrated by the spectra of three alkenes:

allyl bromide, $RHC{=}CH_2$ (Fig. 3.14),

cyclohexene, $\overset{\text{H}\quad\;\text{H}}{\underset{\text{R}\quad\;\text{R}}{\text{C=C}}}$ (Fig. 3.15), and

2-methyl-2-butene, $R_1R_2C{=}CHR_3$ (Fig. 3.16).

Table 3.1 Infrared absorption frequencies of alkenes

Alkene type	Vibration mode	Frequency, cm^{-1}	Wavelength, μ	Relative intensity
$RHC{=}CH_2$	C—H str (CH_2)	3095–3075	3.23–3.25	m
	C—H str (CHR)	3040–3010	3.29–3.32	m
	Overtone	1850–1800	5.40–5.56	m
	C=C str	1645–1640	6.08–6.10	v
	CH_2 i-p def	1430–1410	7.00–7.10	w
	CH i-p def	1300–1290	7.69–7.75	v
	CH o-o-p def	995–985	10.05–10.15	m
	CH_2 o-o-p def	915–905	10.93–11.05	s
$R_1HC{=}CHR_2$ (*cis*)	C—H str	3040–3010	3.29–3.32	m
	C=C str	1665–1635	6.01–6.12	v
	C—H i-p def	1430–1400	7.00–7.14	w
	C—H o-o-p def	730–665	13.70–15.04	s
$R_1HC{=}CHR_2$ (*trans*)	C—H str	3040–3010	3.29–3.32	m
	C=C str	1675–1665	5.97–6.01	v
	C—H i-p def	1310–1290	7.63–7.75	w
	C—H o-o-p def	980–960	10.20–10.42	s
$R_1R_2C{=}CH_2$	C—H str	3095–3075	3.23–3.25	m
	Overtone	1800–1780	5.56–5.62	m
	C=C str	1660–1640	6.02–6.10	v
	CH_2 i-p def	1420–1410	7.04–7.09	w
	CH_2 o-o-p def	895–885	11.17–11.30	s
$R_1R_2C{=}CHR_3$	C—H str	3040–3010	3.29–3.32	m
	C=C str	1675–1665	5.97–6.01	v
	C—H o-o-p def	850–790	11.76–12.66	m
$R_1R_2C{=}CR_3R_4$	C=C str	1690–1670	5.92–5.99	w
Ar—$HC{=}CH_2$	C=C str	~1625	~6.16	s
C=O or C=C conjugated with C=C	C=C str	1660–1580	6.02–6.33	s

Abbreviations: str = stretching, i-p def = in-plane deformation, o-o-p def = out-of-plane deformation, s = strong, m = medium, w = weak, v = variable.

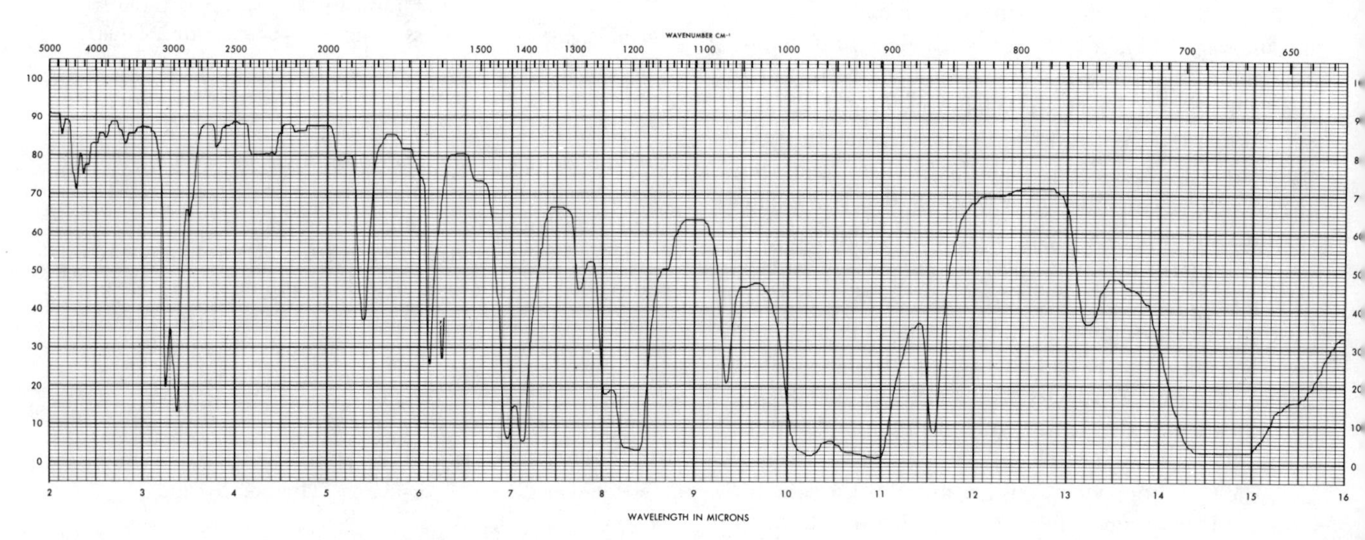

Fig. 3.14 Infrared spectrum of allyl bromide, liquid film.

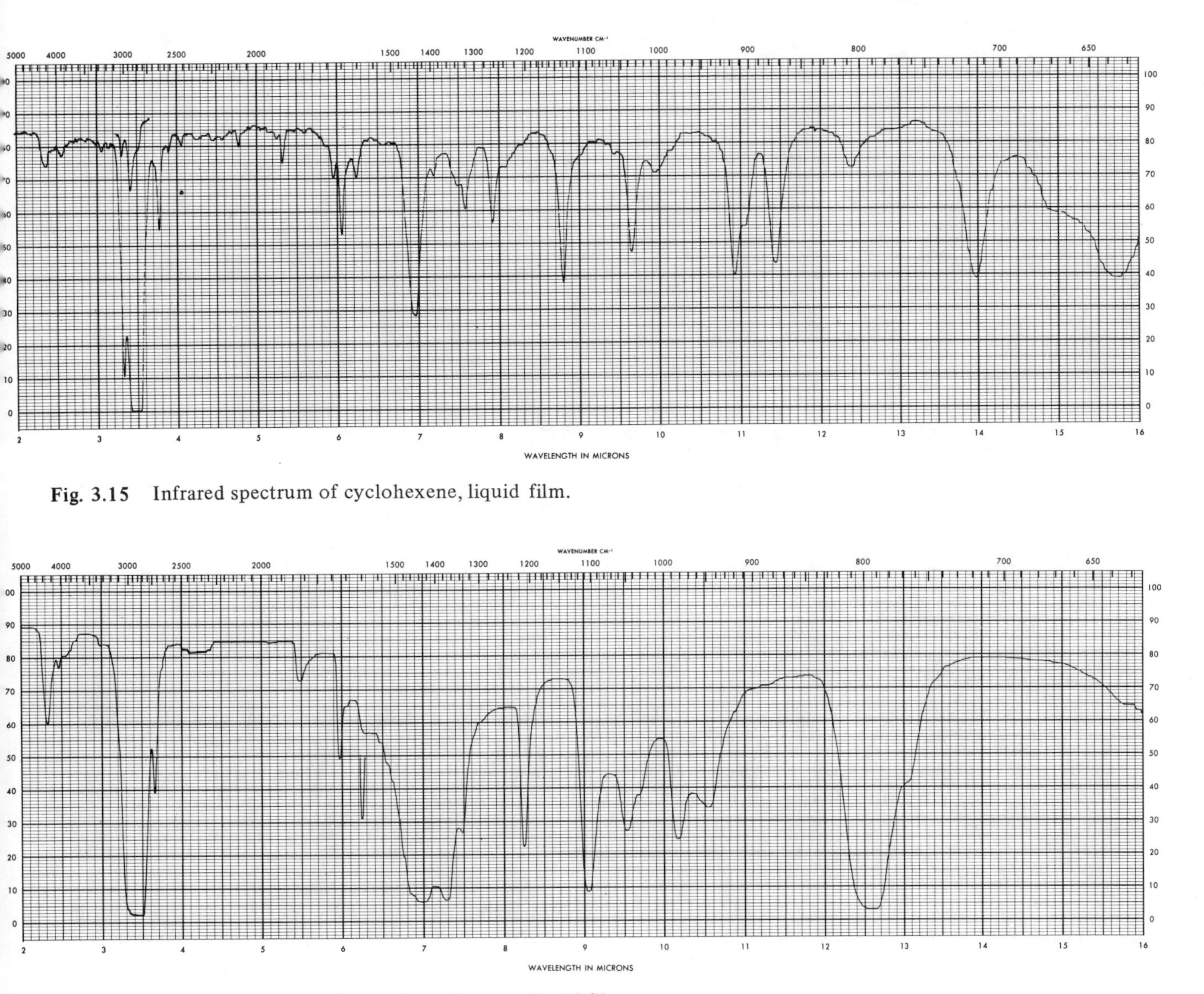

Fig. 3.15 Infrared spectrum of cyclohexene, liquid film.

Fig. 3.16 Infrared spectrum of 2-methyl-2-butene, liquid film.

The position of C=C stretching bands should be the same in cyclic as in acyclic olefins, except for the fact that most rings are strained. In the case of strained rings, the greater the strain, the lower the frequency of C=C stretch. For example, the C=C stretching bands in allyl bromide and 2-methyl-2-butene are observed at wavenumbers of 1640 and 1675 cm^{-1} (6.10 and 5.97 μ), respectively, whereas that in cyclohexene is at 1645 cm^{-1} (6.08 μ). The fact that the C=C stretching occurs at somewhat lower frequency in cyclohexene than in 2-methyl-2-butene may indicate a slight strain in the cyclohexene ring. The C=C stretching in the cyclobutene ring is observed at a still lower frequency (1566 cm^{-1} or 6.39 μ).

The lower frequency of C=C stretching in allyl bromide (1640 cm^{-1}) is caused by the electronegative bromine atom in the molecule. The lowering in frequency is usually greater if the ethylenic carbon is directly attached to chlorine, bromine, iodine, or other electronegative groups.

The weak C=C stretching band is occasionally confused with the carbonyl band if there is a carbonyl group in the compound, especially when there is reason to believe that the frequency of carbonyl absorption may be lowered due to conjugation. There is also the possibility of this band being obscured by other functional groups in the compound. For example, organic nitrites and nitrates absorb radiation in the general region of 1600 to 1650 cm^{-1}. The C=C group absorbs radiation in

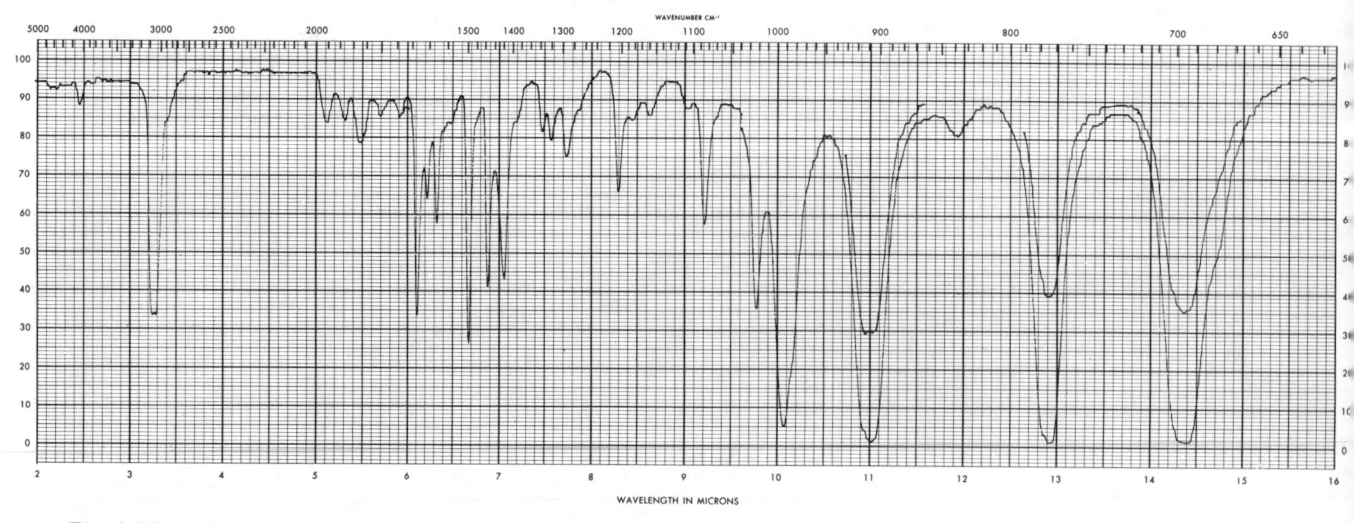

Fig. 3.17 Infrared spectrum of styrene, liquid film. Note an intense C=C stretching absorption band at 1637 cm^{-1} (6.12 μ) and an aromatic C=C stretching absorption band at 1603 cm^{-1} (6.24 μ).

approximately the same region. The bending of the N—H bond gives rise to a broad and strong band in the vicinity of 1500 to 1650 cm^{-1}. And the ionized carboxylic acids and salts absorb around 1600 cm^{-1}, due to the antisymmetric carbonyl stretch.

Conjugation of the double bond with other π electron systems produces interesting results. A conjugated asymmetric diene displays two bands, a weak one near 1600 cm^{-1} and a strong one near 1650 cm^{-1}. If the conjugation is with an aromatic ring, the C=C stretching near 1625 cm^{-1} (6.15 μ) becomes more intense, and an additional band for the aromatic double bonds is observed near 1600 cm^{-1} (6.24 μ). In the case of enones (i.e., compounds containing a double bond conjugated with a carbonyl group), the absorption frequencies of both C=C and C=O stretching are lowered. This lowering is attributed to the resonance:

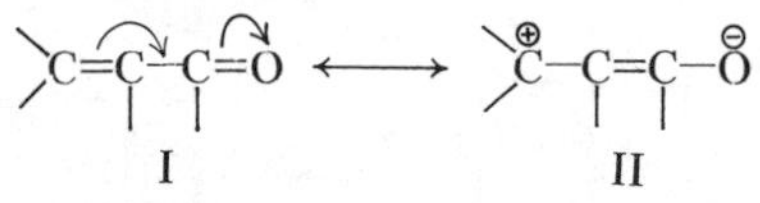

wherein the presence of single bond character in the carbonyl group of structure II can lower carbonyl frequency from about 1720 cm^{-1} to 1680 cm^{-1} and C=C stretching frequency from about 1645 cm^{-1} to as low as 1580 cm^{-1}. The C=C stretching band is usually strong in such compounds, due to the greater polarity of the double bond. These observations are illustrated in the spectra of styrene (Fig. 3.17) (conjugation with an aromatic ring) and of crotonaldehyde (Fig. 3.18) (conjugation with a carbonyl group).

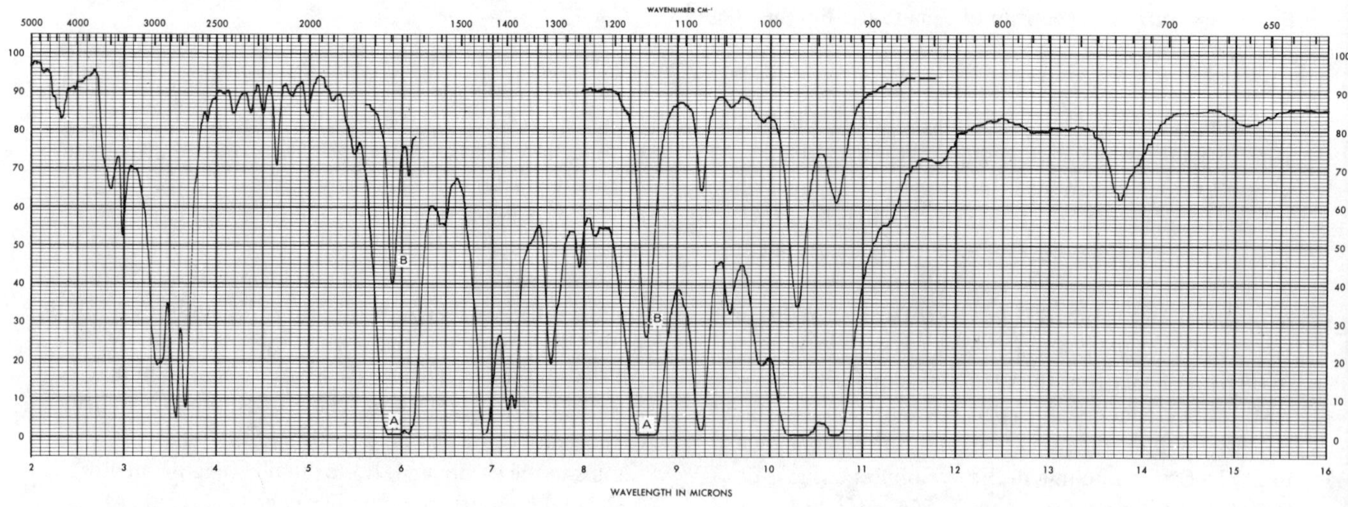

Fig. 3.18 Infrared spectrum of crotonaldehyde, liquid film. Note the carbonyl absorption band at 1695 cm^{-1} (5.90 μ) and the C=C absorption band at 1639 cm^{-1} (6.10 μ).

In compounds containing both olefinic and alkyl C—H bonds, the bands above 3000 cm^{-1} are generally attributed to aromatic or olefinic C—H stretching, whereas those below 3000 cm^{-1} (in the C—H stretching region) are generally assigned to the alkyl C—H stretching. (Exceptions are cyclopropyl compounds). Table 3.1 and Figs. 3.14 through 3.18 clearly illustrate the small effects produced on the C—H stretching absorption by the substituents on the vinyl carbon.

Some of the olefinic and aromatic C—H deformations produce characteristic bands in the region below 1500 cm^{-1} and provide valuable information about the compound. The C—H deformations can be either in the same plane as the C=C bond or perpendicular to it. Table 3.1 shows that most in-plane deformations of the C—H bond take place between 1500 and about 1000 cm^{-1}, and do not show much variation with the substituents. However, the out-of-plane bending vibrations of olefinic and aromatic C—H bonds, occurring between 1000 and 700 cm^{-1}, produce perhaps the most valuable characteristic bands because their pattern depends on the substitution pattern around the ethylenic bond. This dependence makes it possible to distinguish between *cis, trans,* and vinyl structures and provides information about the number of adjacent hydrogens remaining on an aromatic ring. Note in Table 3.1 the values given for various C—H out-of-plane deformations. Of these, those in *cis* disubstituted ethylenes (C—H out-of-plane; 730 to 665) are most affected by the structural changes.

Alkynes and allenes

Alkynes can be terminal or nonterminal, i.e., they can be monosubstituted or disubstituted (R—C≡C—H or R_1C≡C—R_2). All alkynes contain a C≡C triple bond but the terminal alkynes also contain a C—H bond. The force constant for a triple bond is greater than that for a double bond, which in turn is larger than that for a C—C single bond. Consequently, whereas a C—C stretching vibration occurs at 1300 to 800 cm^{-1} and a C=C stretching vibration occurs at 1700 to 1500 cm^{-1}, the C≡C vibrations are observed at the higher frequencies of 2300 to 2050 cm^{-1}. The terminal alkynes show a characteristic but weak triple-bond stretching vibration at 2140 to 2050 cm^{-1} (4.67 to 4.88 μ), whereas the unsymmetrically disubstituted alkynes show a triple-bond absorption at 2260 to 2190 cm^{-1} (4.43 to 4.57 μ) of variable intensity. The C≡C stretching vibration is absent in acetylene or in symmetrically disubstituted alkynes, and the position of this band in terminal alkynes hardly varies at all. The conjugation of a phenyl group with the triple bond hardly lowers triple bond absorption frequency, although the presence of halides somewhat raises the frequency.

The acetylenic C—H stretching vibration is normally observed as a sharp, characteristic band at 3310 to 3200 cm^{-1} (3.02 to 3.12 μ) and the acetylenic C—H bending vibration occurs at 600 to 650 cm^{-1} (16.7 to 15.4 μ). A C—H bending overtone is usually observed between 1200 and 1300 cm^{-1}. If a CH_2 group exists next to the triple bond, as in —CH_2—C≡C—, a weak band at 1336 to 1325 cm^{-1} due to CH_2 deformation indicates the presence of a —CH_2—C≡C— group.[12]

The spectrum of a terminal acetylenic compound is shown in Fig. 3.19, in which C≡C stretching (weak), C—H stretching, and C—H bending vibrations are clearly observed.

Compounds of the allene type (H_2C=C=CH_2) have

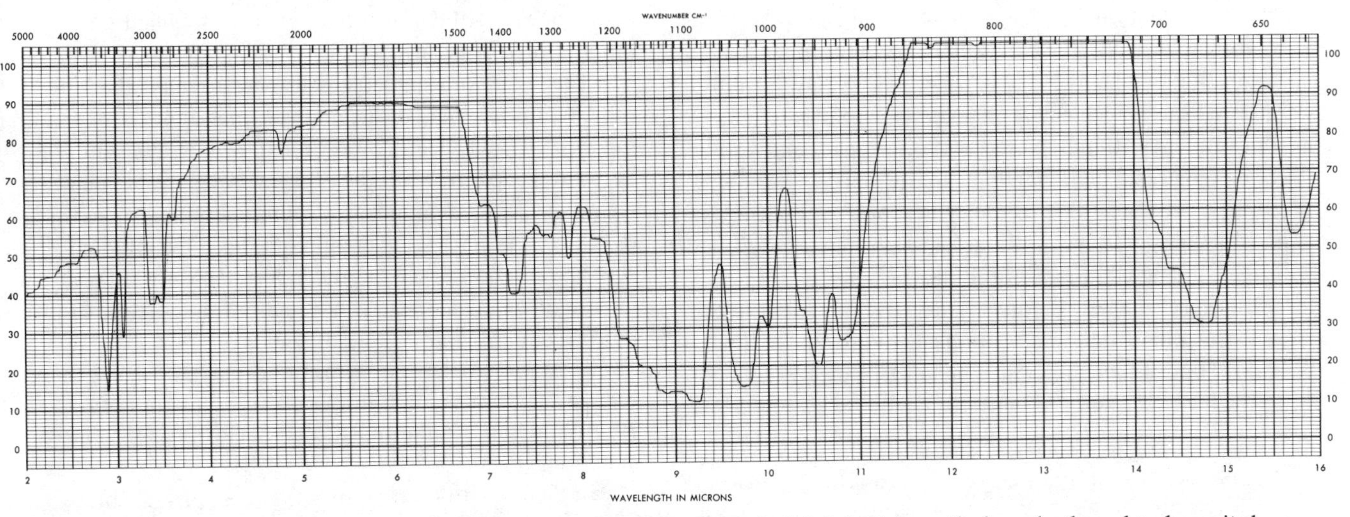

Fig. 3.19 The infrared spectrum of a terminal acetylenic compound: 1,3-2,5-Di-*O*-methylene-4-ethynyl-L-rhamnitol. (Synonym: 1,3-2,5-Di-*O*-methylene-6-deoxy-4-ethynyl-L-lyxo-hexitol, or 2,5-4,6-Di-*O*-methylene-1-deoxy-3-ethynyl-L-arabo-hexitol.) (V. M. Parikh, unpublished synthesis.)

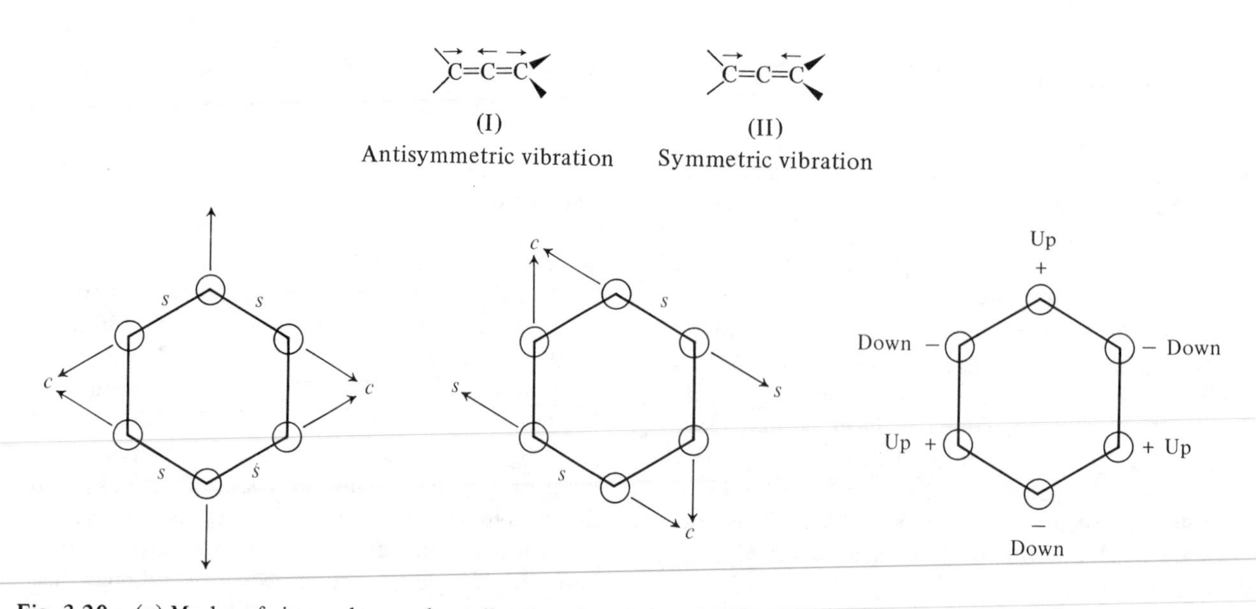

Fig. 3.20 (a) Modes of ring carbon-carbon vibrations in which the center of symmetry is retained. These quadrant stretching and out-of-plane ring bending vibrations become active in some cases, causing absorption of radiation in the region of 1600 cm^{-1} and 700 cm^{-1}. (Legend: *s* = stretching, *c* = compression).

an absorption band in the acetylenic region. This is due to the vibrational coupling of all the three doubly bonded carbons which are involved in an antisymmetric mode (I) absorbing at 1970 to 1950 cm^{-1} (5.08 to 5.13 μ), and in a weakly absorbing symmetric mode (II) which gives a weak band of the C—C type at about 1060 cm^{-1} (9.43 μ).

Aromatic hydrocarbons

Like alkenes and alkynes, aromatic hydrocarbons contain only two types of atoms and mostly single and multiple bonds between carbon-hydrogen and carbon-carbon. However, in aromatic rings, there is usually a fair amount of interaction between C—C and C—H vibrations. This interaction, and a high degree of symmetry, give rise to infrared spectra which are characteristic of aromatic rings and are different from those of ordinary alkenes. For example, in benzene, there are six equal C⋯C bonds, so there are six C⋯C stretching vibrations* of the ring.

In addition, there are several in-plane and out-of-plane bending vibrations of the ring carbons. Benzene also has six equal C—H bonds. These give rise to six hydrogen stretching vibrations, six in-plane hydrogen bendings, and six out-of-plane hydrogen bendings. However, due to the high symmetry of benzene, many modes of vibrations are infrared-inactive. For example, in Fig. 3.20a, carbon atoms which are *para* to each other move in opposite directions in the stretching and in the out-of-plane bending modes shown. These modes retain the center of symmetry, cause no change in dipole, and thus are infrared-inactive. In *para* disubstituted benzenes, these modes are infrared-inactive for the same reason. However, when the groups on the *para* pairs are different, the symmetry is destroyed, and absorption peaks due to these modes are observed in the 1600 to 1585 cm^{-1} and in the 700 cm^{-1} region. The peaks in these regions are particularly noticeable if one substituent is *ortho-para* directing and the other is *meta* directing.

The center of symmetry is also destroyed if the substituents are *meta* or if the ring is mono- or tri-substituted (at 1, 3, and 5 positions). Strong bands in both the 1600 cm^{-1} and the 700 cm^{-1} regions are observed in such compounds. No center of symmetry exists in an *ortho* disubstituted compound. Such a compound absorbs radiation in the 1600 cm^{-1} region, but, if the substituents are identical, it becomes infrared-inactive in the 700 cm^{-1} region.

Thus we see that the forms of normal modes in benzene differ from those in simple alkenes and are not exactly the same in the substituted benzenes as they are in benzene itself.

In the aromatic compounds the most prominent bands are due to out-of-plane bending of the ring C—H bonds in the region of 900 to 650 cm^{-1}. The spectra of aromatic compounds typically exhibit many weak or

* These stretching vibrations in benzene are usually classified as whole-ring stretching, semicircle stretching, quadrant stretching, and sextant stretching. Most in-plane and out-of-plane bendings of the ring carbons are also described as quadrant and sextant bending.

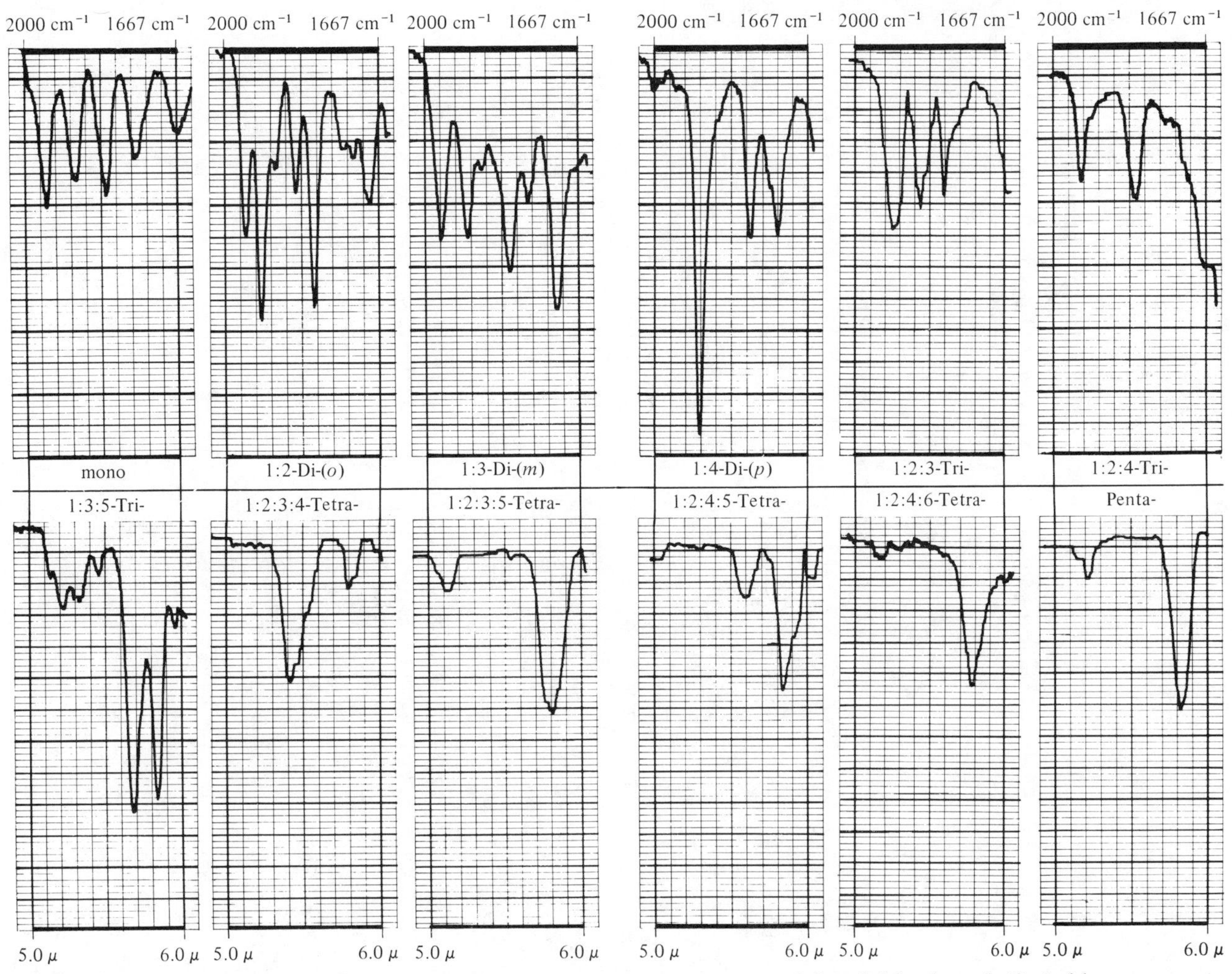

Fig. 3.20 (b) Characteristic patterns of absorption of IR radiation between 5.0 and 6.0 μ by substituted benzenes.

medium peaks in the 3080 to 3030 cm^{-1} (3.25 to 3.30 μ) region. However, the information derived from this region must be used with caution when one is distinguishing between aromatic and olefinic compounds. The bands considered to be of most help in diagnosing the aromatic character of the compound appear in the region between 1650 and 1400 cm^{-1}. There are normally four bands in this region, at about 1600, 1585, 1500, and 1450 cm^{-1} (6.25, 6.31, 6.67, and 6.90 μ), and are due to the C=C in-plane vibrations. Combination and overtone bands in the wavenumber region of 2000 to 1650 cm^{-1} (5.00 to 6.06 μ) are also characteristic of aromatic rings. However, they are very weak, and are obtained only in the case of concentrated solutions of highly symmetric benzenes. Valuable information about the position of substituents on the benzene ring can be obtained by a careful study of the pattern of peaks in this region (Fig. 3.20b). Although these patterns are useful in structural analysis, it is better to prepare a set for each individual instrument.

Table 3.2 gives a list of absorption frequencies typical of aromatic homocyclic compounds. Figures 3.17, 3.21, 3.22, and 3.23 present the infrared spectra of some aromatic homocyclic compounds, showing the effects of substitution.

Alcohols and phenols

When a hydrogen atom from an aliphatic hydrocarbon is replaced by an —OH group, new bands corresponding to the new O—H and C—O bond absorptions appear in the spectrum. A medium-to-strong band, resulting from O—H stretching vibrations in the region from 3700 to 3400 cm^{-1} (2.72 to 2.94 μ) is usually a strong indication that the sample is an alcohol or phenol. It may, however, indicate that the sample is a primary or secondary amine (as N—H also absorbs in this region), or it may indicate that the sample is wet. The exact position and shape of this band depends largely on the degree of hydrogen bonding. A strong, sharp peak may appear in a region as high as

Table 3.2 Absorption frequencies typical of aromatic homocyclic compounds

Aromatic compound	=C—H str (multiple bands)		C=C str (ring stretching)		C—H i-p def (w)	C—H o-o-p def
Monosubstituted	3080–3030 cm^{-1} 3.24–3.33 μ	w–m	1604 ± 3 cm^{-1} 6.25–6.22 μ	v	1177 ± 6 (2,5 *vs* 3,6) 8.45–8.54 μ	751 ± 15 s; 5 adj H wag 13.06–13.58 μ
			1185 ± 3 6.32–6.30 μ	v	1156 ± 5 (3,5 *vs* 4) 8.61–8.69 μ	697 ± 11 s; 5 adj H wag 14.1–14.58 μ
			1510 → 1480* 6.62 → 6.75 μ	v	1073 ± 4 (2,6 *vs* 3,4,5) 9.28–9.36 μ	—
			1452 ± 4 6.87–6.91 μ	v	1027 ± 3 (2,3 *vs* 5,6) 9.71–9.77 μ	—
1 : 2-Disubstituted	3080–3030 cm^{-1} 3.24–3.33 μ	w–m	1607 ± 9 6.26–6.19 μ	v	1269 ± 17 (all clockwise) 7.77–7.99 μ	751 ± 7 s; 4 adj H wag 13.20–13.44 μ
			1577 ± 4 6.33–6.36 μ	v	1160 ± 4 (3,5 *vs* 4,6) 8.59–8.65 μ	—
			1510 → 1460 6.85 → 6.62 μ	v	1125 ± 14 (3,6 *vs* 4,5) 8.78–9.0 μ	—
			1447 ± 10 6.87–6.96 μ	v	1033 ± 11 (3,4 *vs* 5,6) 9.58–9.78 μ	—
1 : 3-Disubstituted	3080–3030 cm^{-1} 3.24–3.33 μ	w–m	1600 → 1620 6.25 → 6.17 μ	v	1278 ± 12 (all clockwise) 7.75–7.90 μ	900 → 860 m; 1 free H 11.12 → 11.63 μ
			1586 ± 5 6.29–6.32	v	1157 ± 5 (2,5 *vs* 4,6) 8.60–8.68 μ	782 ± 10 s; 3 adj H wag 12.62–12.96 μ
			1495 → 1470 6.69 → 6.80 μ	v	1096 ± 7 (4 *vs* 6) 9.07–9.18 μ	725 → 680 m; 3 adj H wag 13.79 → 14.71 μ
			1465 → 1430 6.83 → 6.99 μ	v	1076 ± 7 (2 *vs* 5) 9.23–9.36 μ	—
1 : 4-Disubstituted	3080–3030 cm^{-1} 3.24–3.33 μ	w–m	1606 ± 6 6.20–6.25 μ	v	1258 ± 11 (all clockwise) 7.88–8.02 μ	817 ± 15 s; 2 adj H wag 12.02–12.47 μ
			1579 ± 6 6.31–6.36 μ	v	1175 ± 6 (2,5 *vs* 3,6) 8.47–8.56 μ	—
			1520 → 1480 6.58 → 6.76 μ	v	1117 ± 7 (2,6 *vs* 3,5) 8.90–9.01 μ	—
			1409 ± 8 7.05–7.15 μ	v	1013 ± 5 (2,3 *vs* 5,6) 9.82–9.92 μ	—

1 : 2 : 3-Trisubstituted	3080–3030 cm^{-1} 3.24–3.33 μ	w–m			1175 → 1125 8.51 → 8.90 μ 1110 → 1070 9.01 → 9.35 μ 1070 → 1000 9.35 → 10.00 μ 1000 → 960 10.00 → 10.42 μ	800 → 770 s; 3 adj H wag 12.50 → 12.99 μ 720 → 685 m; 3 adj H wag 13.90 → 14.60 μ — —
1 : 2 : 4-Trisubstituted	3080–3030 cm^{-1} 3.24–3.33 μ	w–m	1616 ± 8 6.15–6.22 μ 1577 ± 8 6.31–6.37 μ 1510 ± 8 6.60–6.66 μ 1456 ± 1 6.86–6.88 μ	v v v v	1225 → 1175 8.16 → 8.51 μ 1175 → 1125 8.51 → 8.89 μ 1125 → 1090 8.89 → 9.18 μ 1070 → 1000 9.35 → 10.00 μ	— 900 → 860 m; 1 adj H wag 11.11 → 11.63 μ 860 → 800 s; 2 adj H wag 11.63 → 12.50 μ —
1 : 3 : 5-Trisubstituted	3080–3030 cm^{-1} 3.24–3.33 μ				1175 → 1125 8.51 → 8.89 μ 1070 → 1000 9.35 → 10.00 μ	900 → 860 m; 1 adj H wag 11.11 → 11.63 μ 865 → 810 s; 1 adj H wag 11.56 → 12.30 μ 730 → 675 s; 1 adj H wag 13.70 → 14.82 μ
1 : 2 : 3 : 4-Tetrasubstituted	3080–3030 cm^{-1} 3.24–3.33 μ					860 → 800 s; 1 adj H wag 11.63 → 12.50 μ
1 : 2 : 3 : 5, 1 : 2 : 4 : 5, and 1 : 2 : 3 : 4 : 5-Substituted	3080–3030 cm^{-1} 3.24–3.33 μ					900 → 860 m 11.11 → 11.63 μ

Abbreviations: str = stretching, i-p def = in-plane deformation, o-o-p def = out-of-plane deformation, w = weak, m = medium, s = strong, v = variable, adj = adjacent, wag = wagging.

* The symbol 1510 → 1480 indicates the range, in cm^{-1}, of absorption of radiation. Usually an *electron-donor* substituent causes absorption near 1510 cm^{-1}, while an *electron-acceptor* substituent causes absorption near 1480 cm^{-1}.

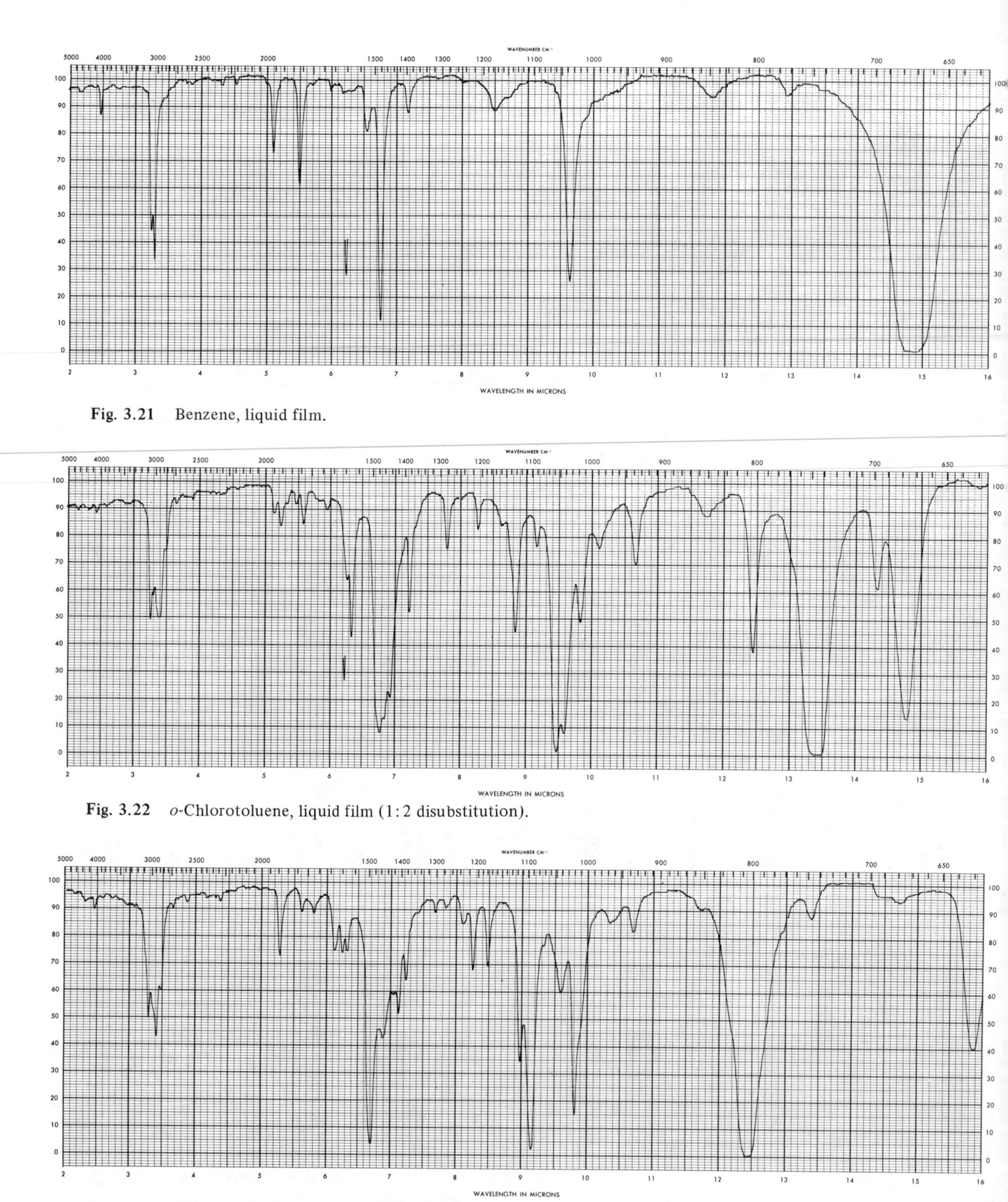

Fig. 3.21 Benzene, liquid film.

Fig. 3.22 *o*-Chlorotoluene, liquid film (1:2 disubstitution).

Fig. 3.23 *p*-Chlorotoluene, liquid film (1:4 disubstitution).

3700 cm^{-1} in gaseous or extremely dilute samples containing unbonded or free hydroxyl groups. But very broad bands may appear in regions as low as 2500 cm^{-1} in chelated compounds. Hydrogen bonding usually seems to broaden the band and shift it to lower frequencies. In fact, as the degree of intermolecular hydrogen bonding in the sample increases, the band corresponding to the free hydroxyl group decreases in intensity and additional bands appear at lower frequencies. The degree of intermolecular hydrogen bonding increases with the concentration of the solution. One can see this effect by recording the absorption bands in the 4000 to 2500 cm^{-1} region for *n*-amyl alcohol solutions of different concentrations (Fig. 3.24).

One cannot observe a similar change in spectral pattern with changes in concentration in solutions of compounds which form only *intramolecular* hydrogen bonds.

The O—H in-plane deformation is normally observed in the region between 1410 and 1260 cm^{-1}. The position of this band depends on the structure of the alcohol. For example, in tertiary alcohols and phenols, this band is observed in the region between 1410 and 1310 cm^{-1}. In secondary and primary alcohols, it is observed in the region between 1350 and 1260 cm^{-1}. The frequency of this band is still lower in the case of α,β-unsaturated alcohols.

The C—O stretching occurs at a lower frequency

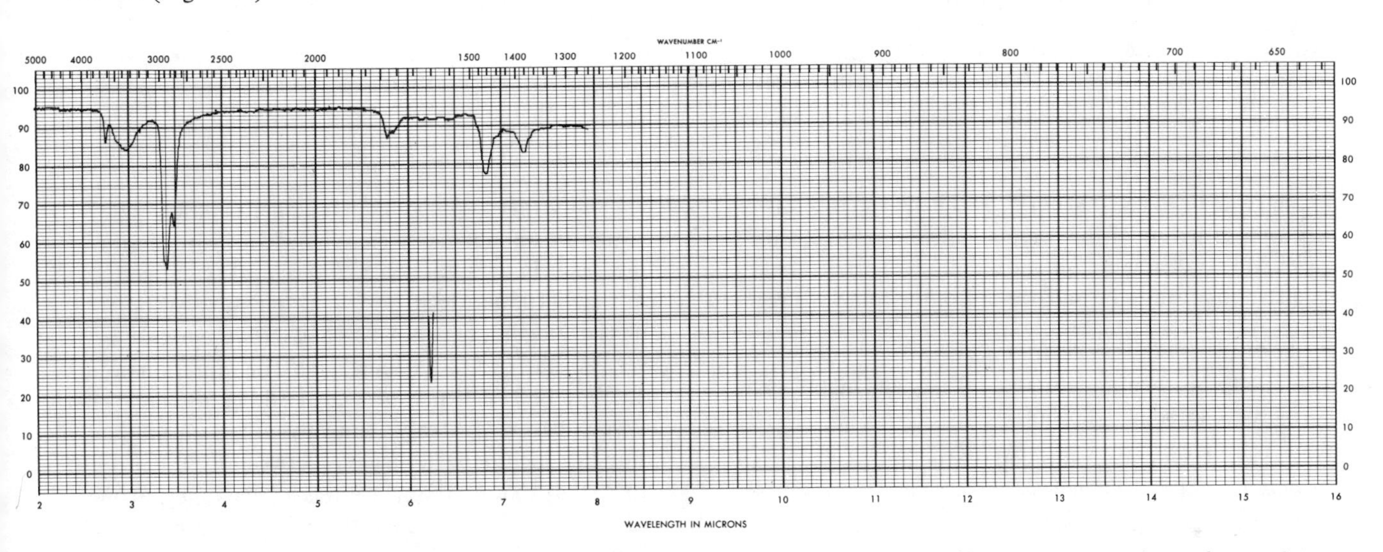

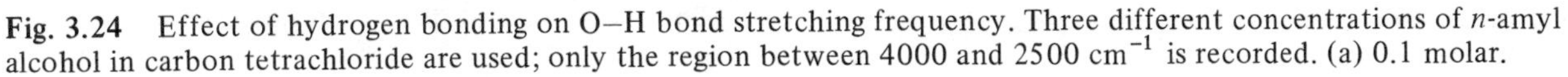

Fig. 3.24 Effect of hydrogen bonding on O–H bond stretching frequency. Three different concentrations of *n*-amyl alcohol in carbon tetrachloride are used; only the region between 4000 and 2500 cm^{-1} is recorded. (a) 0.1 molar.

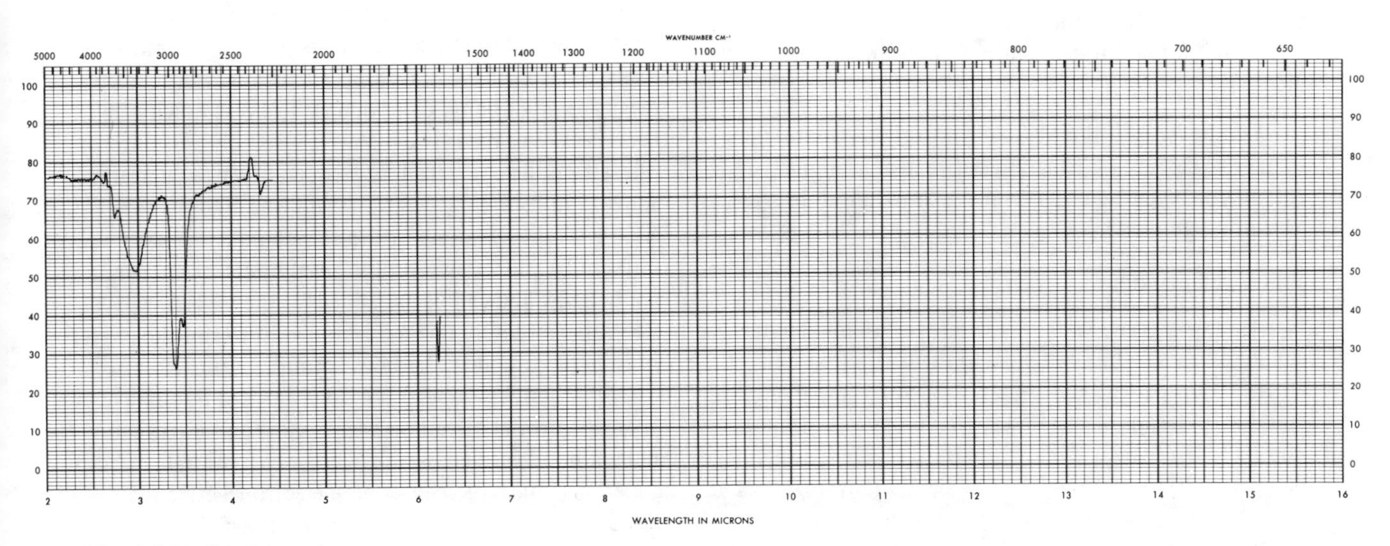

Fig. 3.24 (b) 0.2 molar.

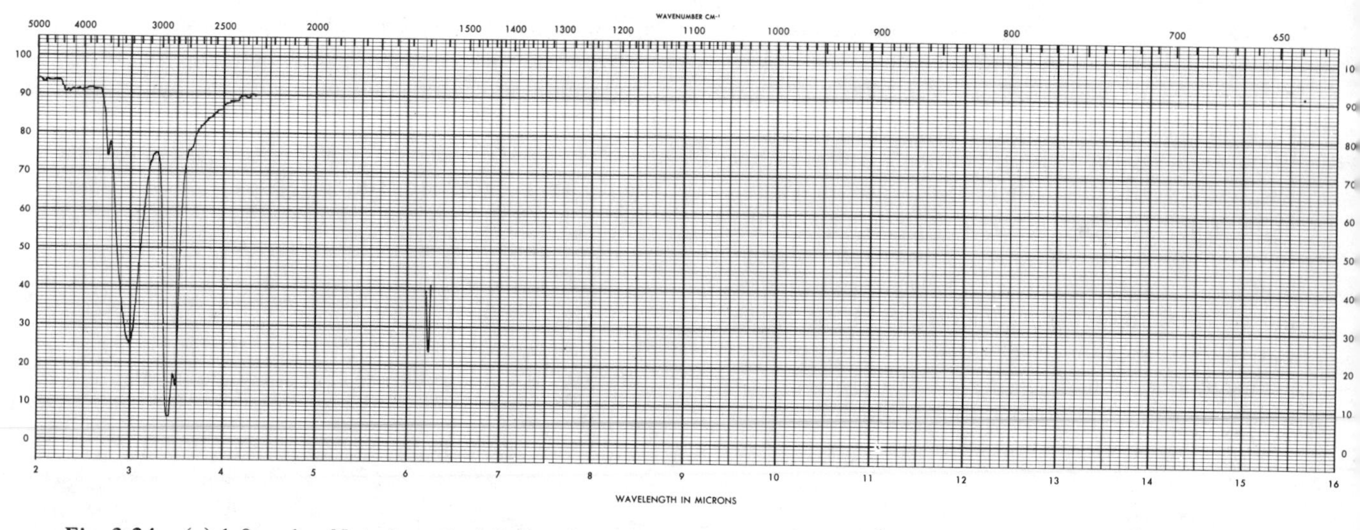

Fig. 3.24 (c) 1.0 molar. Note how the relative sizes of the peaks at 3610 cm^{-1} (2.77 μ) and 3330 cm^{-1} (3.00 μ) change as the concentration changes.

range of 1230 to 1000 cm^{-1}. This band too is sensitive to the structure of the compound being studied, and provides valuable information about it. This band has been observed to be a strong one in the following regions.[13]

1230 to 1140 cm^{-1}, phenols.

1205 to 1124 cm^{-1}, saturated tertiary and highly symmetrical secondary alcohols.

1125 to 1087 cm^{-1}, saturated secondary, cyclic tertiary, or α-unsaturated alcohols.

1085 to 1053 cm^{-1}, saturated primary, 5- or 6-membered ring alicyclic secondary or α-unsaturated secondary alcohols.

Below 1050 cm^{-1}, 7- or 8-membered ring alicyclic secondary, highly α-unsaturated tertiary, di-α-unsaturated secondary, α-unsaturated and/or α-unsaturated primary alcohols.

On the basis of the frequency of this C—O stretching band, one can even distinguish between axial and equatorial hydroxyl groups in complex natural products such as steroids and triterpenes.[14,15]

Ethers and epoxides

Ethers and epoxides contain a C—O—C system which is similar, in its spectral behavior, to the C—C—C system of alkanes. Consequently, it is hard to identify an ether or epoxide by means of infrared spectroscopy. The important band in ethers is the asymmetrical C—O—C stretching band at 1230 to 1000 cm^{-1}, but a band in this region is also observed in other oxy compounds, such as alcohols, aldehydes, ketones, acids, etc. Therefore we consider the possibility that a compound is an ether or an epoxide only if the unknown oxy compound shows no absorption bands in the hydroxyl (3700 to 3300 cm^{-1}) or carbonyl (1850 to 1550 cm^{-1}) regions. We can sometimes distinguish between a *cis* and a *trans* epoxide by means of the C—O stretching band. This vibration occurs at 950 to 860 cm^{-1} in *trans* and at 865 to 785 cm^{-1} in *cis* epoxides. Aryl, alkaryl, and vinyl ethers absorb radiation in a slightly higher region (between 1270 and 1230 cm^{-1}), but so do carboxylic esters and lactones. Figures 3.25, 3.26, and 3.27 present the infrared spectra of an alcohol, a phenol, and an ether, respectively.

Carbonyl compounds

The absorption peak for carbonyl stretching, in the region from 1870 to 1540 cm^{-1}, is perhaps the most easily recognizable band, and is extremely useful in the diagnosis of carbonyl compounds. The reason why this band is easy to recognize is that it is usually very intense; it is within the characteristic frequency range mentioned above, and its precise position can be accurately correlated with the structure of the compound. The band's shifts in frequency, within the given limits, are often predictable, and can be explained on the basis of the electronic structure of the carbonyl group and the environment around the carbonyl group.

A carbonyl group has a σ bond formed by the overlap of an sp^2 orbital of carbon with a p orbital of oxygen and a π bond formed by the overlap of p orbitals of both the atoms. Since oxygen is more electronegative than carbon, the double bond is polar and the carbonyl structure has a fair degree of dipolar character. This dipolar

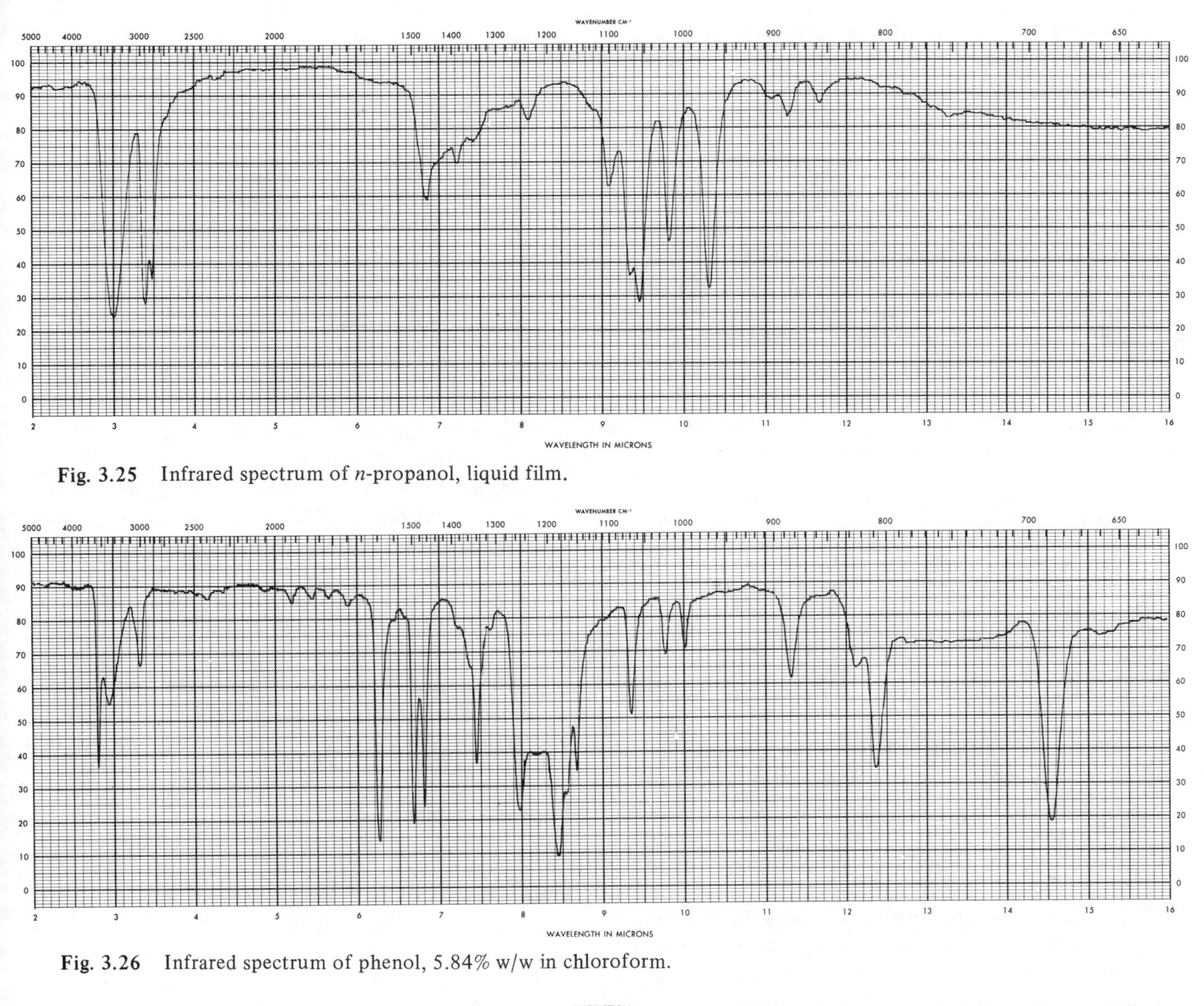

Fig. 3.25 Infrared spectrum of *n*-propanol, liquid film.

Fig. 3.26 Infrared spectrum of phenol, 5.84% w/w in chloroform.

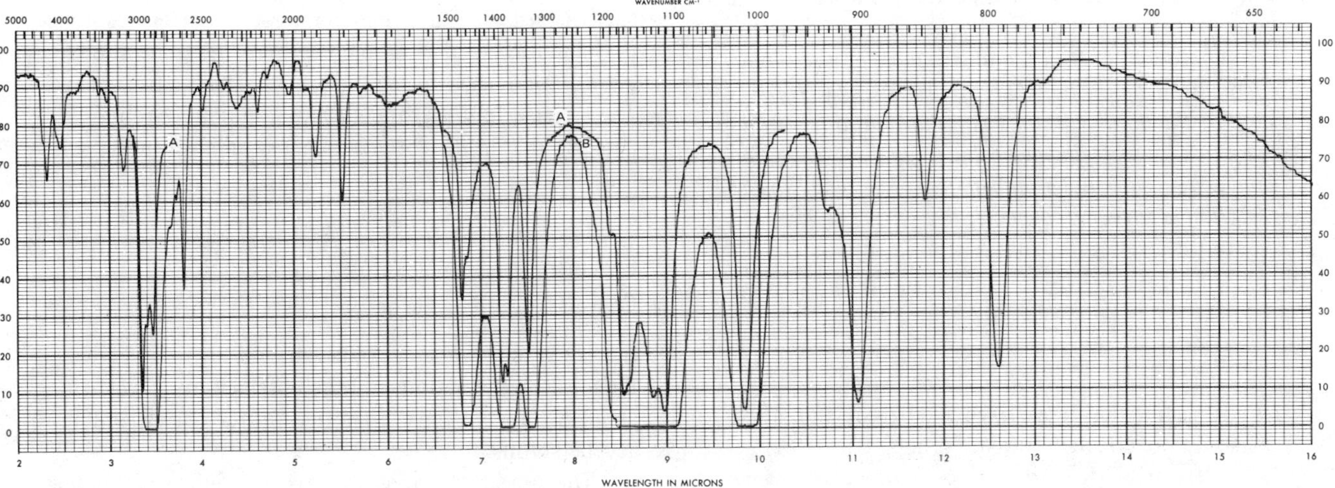

Fig. 3.27 Infrared spectrum of isopropyl ether; liquid film. Inset: 6% solution in CCl_4.

character varies with the changes in the structure of the carbonyl compound. In addition, the carbonyl oxygen has two ($2s^2$ and $2p^2$) nonbonded pairs of electrons.

$$>C{=}\ddot{O}: \longleftrightarrow \; >\overset{+\delta}{C}{-}\overset{-\delta}{\ddot{O}}:$$

These electrons, together with the partial negative charge on oxygen, cause carbonyl groups to form hydrogen bonds easily with solvents capable of acting as hydrogen donors. The carbonyl stretching bands are thus intensely subject to both the *intra*molecular electrical effects (substituents, conjugation, intramolecular hydrogen bonding, nonbonded internal interactions, etc.) and the *inter*molecular factors (physical state, solvent association, intermolecular hydrogen bonding, etc.).

Effects of intramolecular factors on C=O band; variations of group X

To understand the effects of intramolecular factors on carbonyl stretching frequencies, we shall use the simple formula

$$\begin{matrix} X \\ R \end{matrix}\!\!>C{=}O$$

and discuss the effects produced by the changes in the nature of X and R. Such an approach enables us to discuss and compare different types of carbonyl compounds, such as ketones, aldehydes, acids, acid halides, anhydrides, amides, esters, α-halo and conjugated carbonyl compounds, ring compounds, and so forth.

The two primary factors determining the frequency of vibrations are, as we have seen, the *effective masses* of the atoms and the *force constant* of the bonds. The force constant of a double bond is greater than that of a single bond. Thus it is easy to see that *in the case of carbonyl compounds, any factor tending to increase the dipolar character of the carbonyl group decreases the force constant of the bond, and thus shifts the carbonyl stretching band to a lower frequency value (higher wavelength value). Conversely, carbonyl groups that are less polar absorb radiation at higher frequencies.*

In the general formula

$$\begin{matrix} X \\ R \end{matrix}\!\!>C{=}O,$$

let us presume that R is a fixed alkyl group so that we can study the effects of variations in X. If:

X = $-CH_3$ or $-CH_2-$, then the compound is a ketone
X = H, then the compound is an aldehyde
X = OH, then the compound is a carboxylic acid
X = $O^{\ominus}$, then the compound is an anion of carboxylic acid
X = OR, then the compound is an ester
X = $O{-}C({=}O){-}R$, then the compound is an anhydride
X = $-NH_2$, $-NHR$, or $-NR_2$ then the compound is an amide
X = $-Cl$, $-Br$ then the compound is an acid halide

Aldehydes and ketones

The carbonyl stretching in aldehydes generally occurs in the region of 1740 to 1690 cm^{-1} (5.76 to 5.92 μ), whereas in ketones it may be at a slightly lower value of 1730 to 1645 cm^{-1} (5.78 to 6.08 μ)[16]. On the basis of mass effect alone, it has been calculated[17] that the replacement of the carbon atom by a lighter hydrogen atom should cause aldehydes to absorb at about 17 cm^{-1} *lower* frequency than the corresponding ketones. However, the fact that aldehydes absorb at about 15 cm^{-1} *higher* frequency than the corresponding ketones is ascribed to the changes in the force constant.

In the case of ketones, we see a variation in C=O stretching absorption frequency between the cyclic and the acyclic ketones. This variation is due to the difference in interaction between the C=O double bond and the

$$\begin{matrix} C \\ C \end{matrix}\!\!>C{=}O$$

single bonds. This interaction increases as the

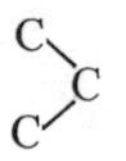

angle decreases. The increased interaction offers resistance to the motion of the carbonyl carbon during the stretching vibration, thus raising the C=O stretching frequency. The increase in this value in cyclic compounds (compared to that in the corresponding acyclic ketone) is as follows:

Six-membered cyclic ketone, 7 ± 14 cm^{-1}
Five-membered cyclic ketone, 37 ± 11 cm^{-1}
Four-membered cyclic ketone, 76 ± 7 cm^{-1}

In a seven-membered cyclic ketone, on the other hand, there is less interaction, and the observed frequency is about 8 ± 3 cm^{-1} lower than that for the corresponding acyclic ketone.

The appearance of weak C—H stretching and bending absorption peaks in the case of aldehydes at 2880 to

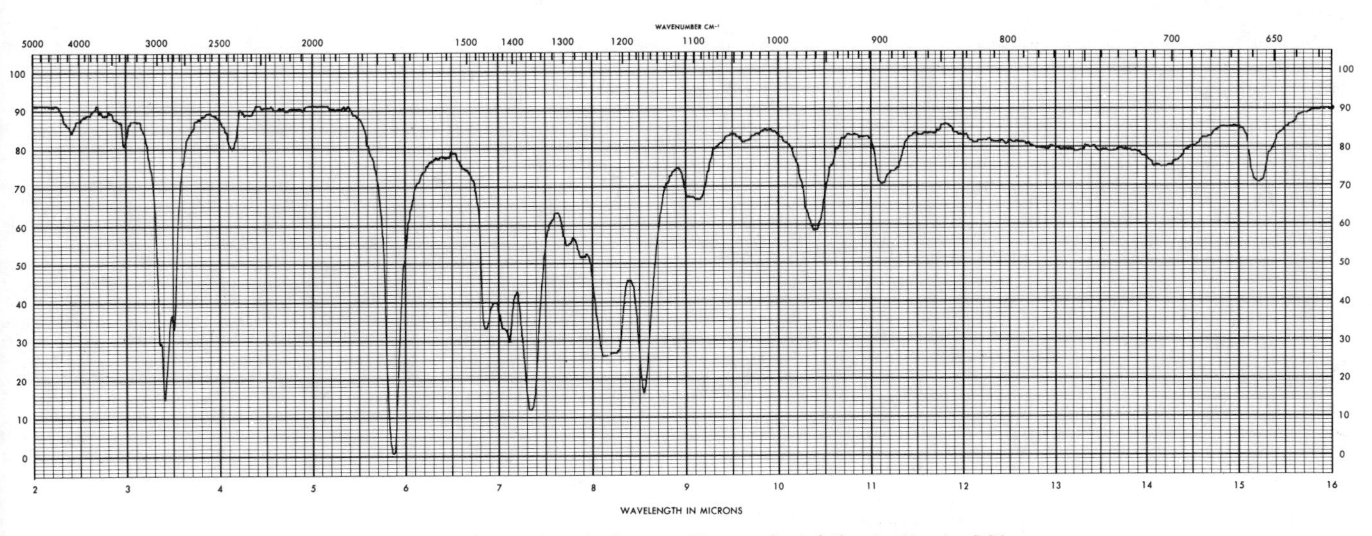

Fig. 3.28 Infrared spectrum of 2-pentanone (methyl propyl ketone); 6.0% solution in CCl_4.

2650 and 975 to 780 cm^{-1}, respectively, sometimes serves to distinguish an aldehyde from a ketone. The spectra in Figs. 3.28, 3.29, and 3.30 for five-membered carbonyl compounds illustrate these points. If a ketone has a methyl group attached directly to carbonyl carbon, as in the case of 2-pentanone, a strong band is normally observed in the region between 1375 and 1350 cm^{-1} (7.26 and 7.41 μ) due to symmetric CH_3 deformation. On the other hand, if one or more methylene $—CH_2—$ groups are directly attached to carbonyl carbon, as in cyclopentanone, the methylene deformation is observed at 1439 to 1400 cm^{-1} (6.95 to 7.15 μ).

Carboxylic acids and their derivatives

Replacement of X in the general formula RCOX by

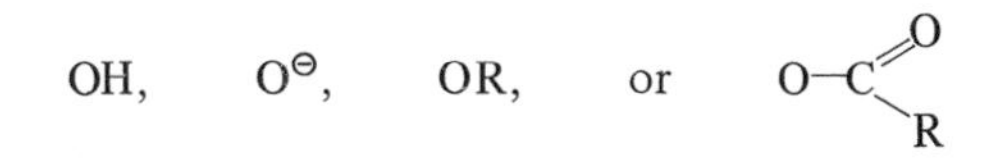

gives us carboxylic acids, acid anions, esters, and anhydrides, respectively. Such a substitution affects the distribution of electrons around the carbonyl group, due either to (a) *resonance* or (b) *inductive effects.* Consequently, the frequency of the carbonyl absorption peaks

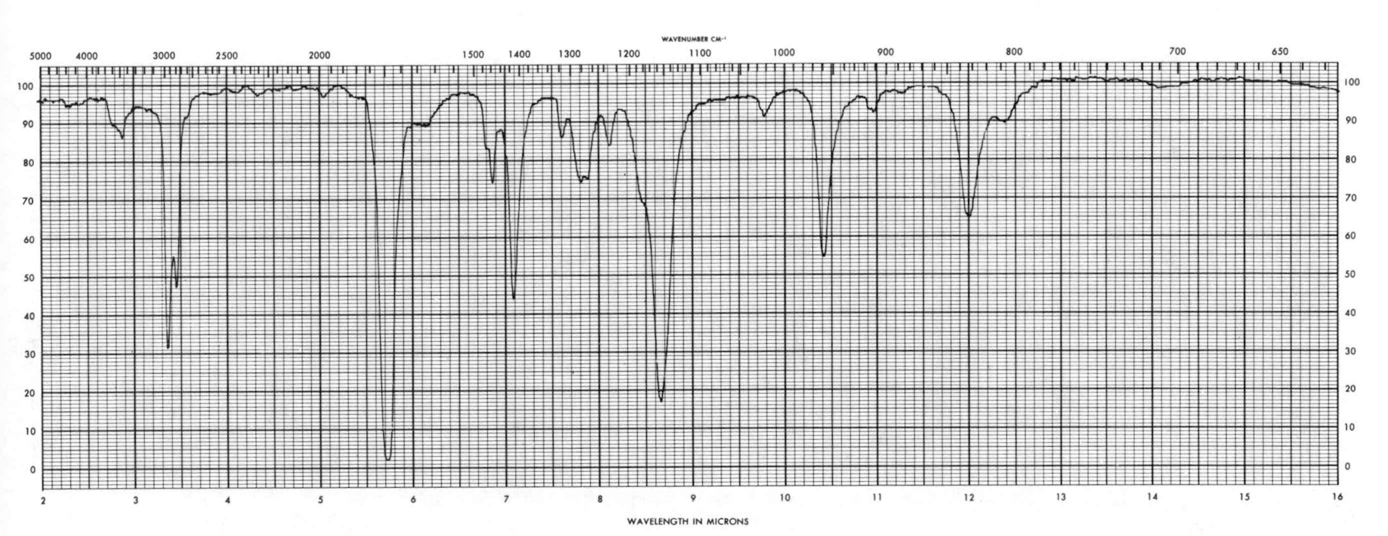

Fig. 3.29 Infrared spectrum of cyclopentanone, liquid film.

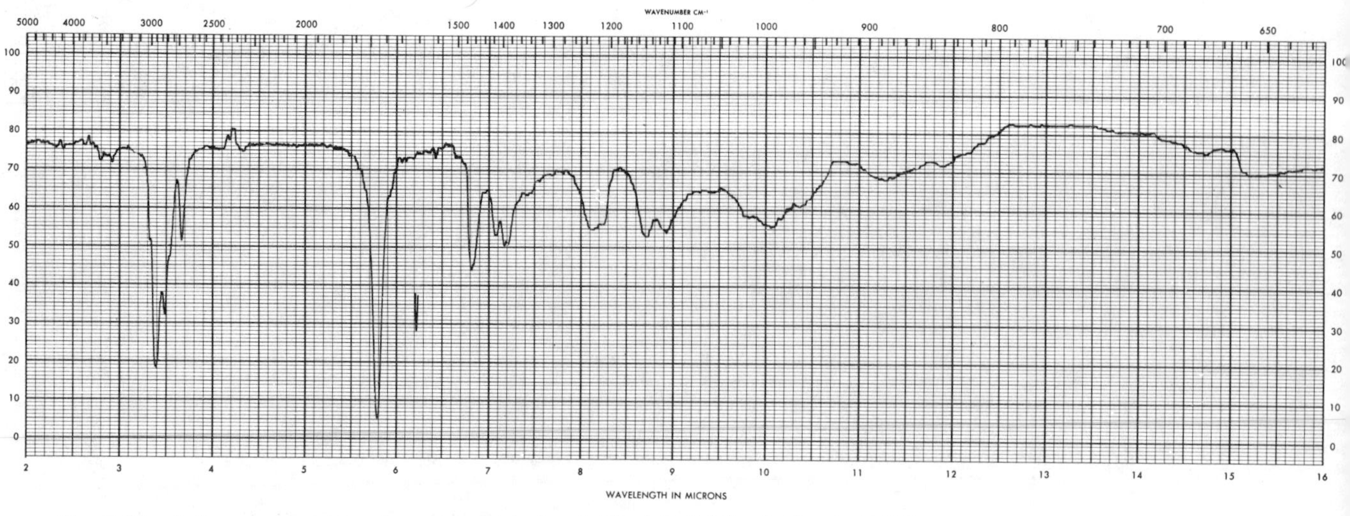

Fig. 3.30 Infrared spectrum of *n*-pentanal (valeraldehyde); 6.0% solution in $CHCl_3$.

is affected. This is also true in acid amides and acid chlorides.

Resonance effect. In the case of amides, esters, acid chlorides, etc., the distribution of electrons is affected by the resonance:

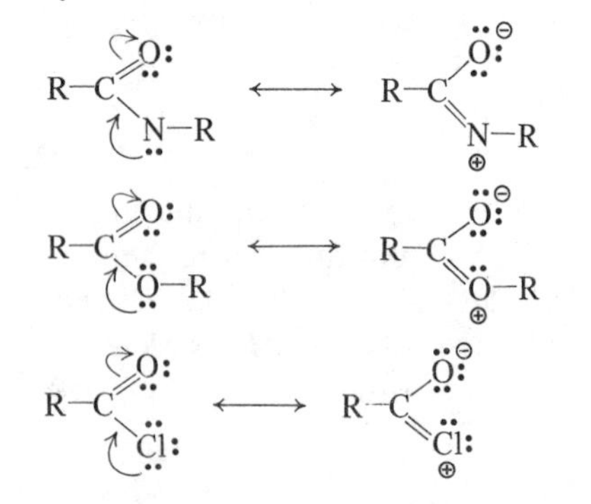

Such a resonance weakens the C=O bond. Resonance of this type may not be important in the case of acids and acid chlorides, but in the case of amides, the dipolar form is dominant. The carbonyl bond in amides is thus weaker than that in a ketone or aldehyde, and consequently the frequency of the absorption peak of amides is between 5 and 55 cm^{-1} lower than that of the corresponding peak of ketones.

A similar effect is observed if X is an aryl; then the carbonyl group is conjugated with the aromatic ring or with the other π electron systems. The frequencies of the carbonyl bands in such cases shift to lower wavenumbers by about 20 to 30 cm^{-1}. Note that the frequency of the carbonyl band in propiophenone (Fig. 3.31) is about 26 cm^{-1} lower than that in 2-pentanone, and that in N,N-dimethyl formamide (Fig. 3.32) is about 34 cm^{-1} lower.

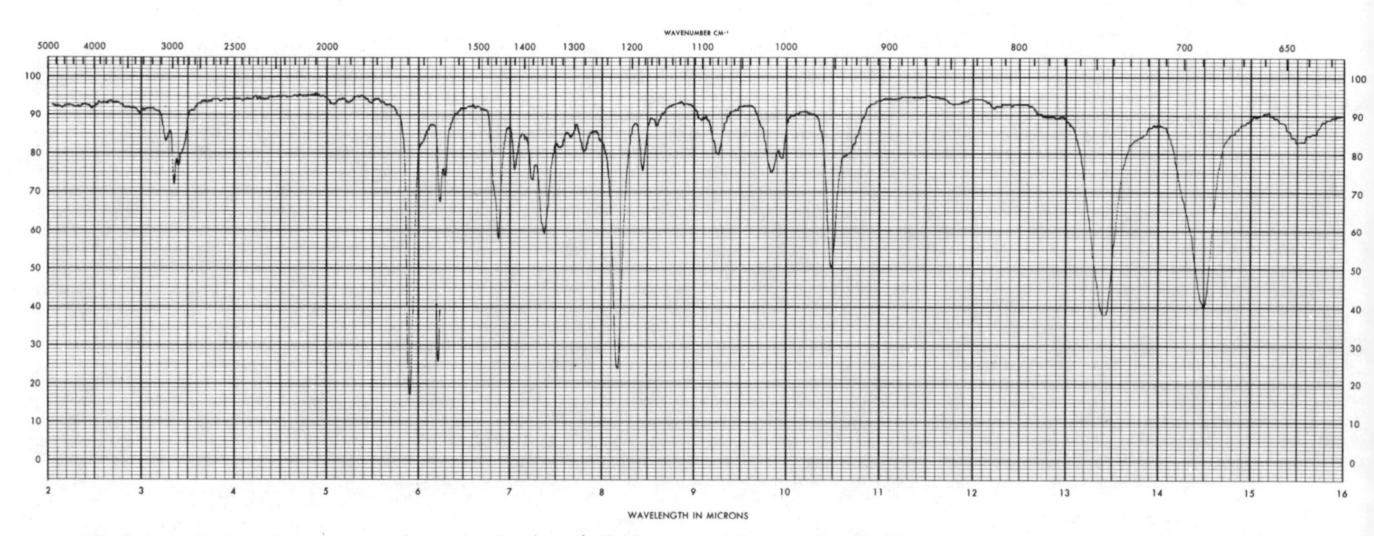

Fig. 3.31 Infrared spectrum of propiophenone (ethyl phenyl ketone), liquid film.

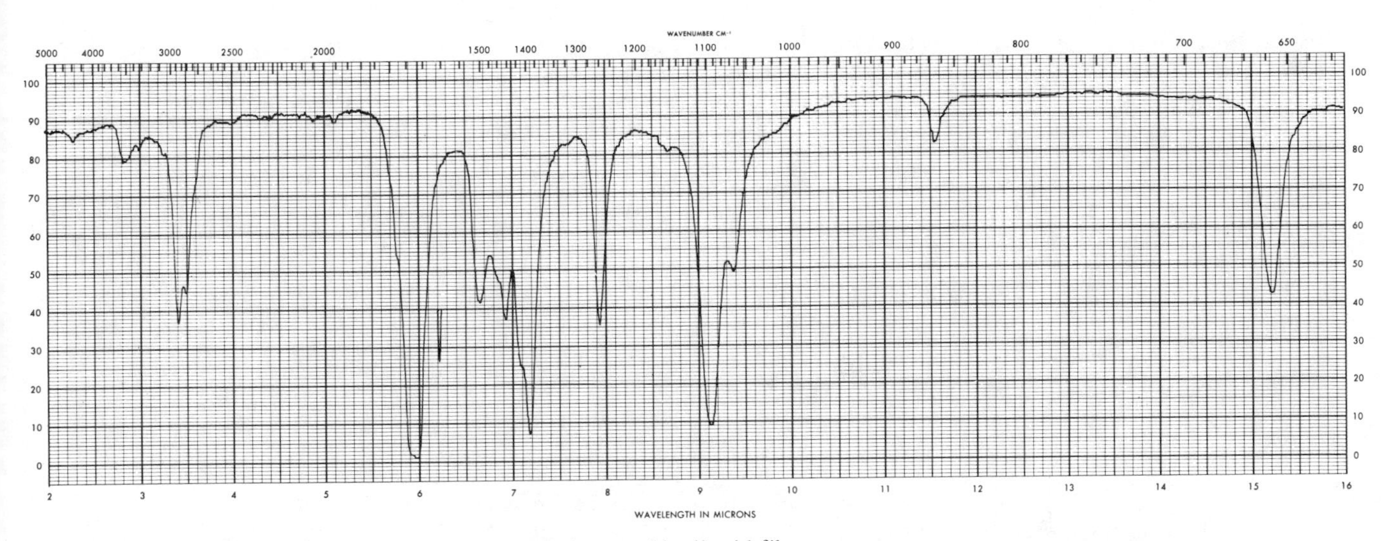

Fig. 3.32 Infrared spectrum of N,N-dimethyl formamide, liquid film.

One must always remember that coplanarity between bonds is essential for maximum resonance, and large band shifts are not observed in cases in which steric effects prevent the coplanarity of the conjugated systems.

Inductive effect. The inductive effect of X is independent of the geometry of the molecule. It is due to the difference in electronegativities of the carbonyl carbon and the X. The inductive effect is usually opposite to the effect of resonance, since the more electronegative X tends to increase the force constant of the C=O bond by resonance:

$$\mathrm{R{-}C(=\ddot{O}{:})X} \longleftrightarrow \mathrm{R{-}C{\equiv}O{:}^{\oplus}}\ \ \mathrm{X^{\ominus}}$$

The increased force constant increases the frequency of the carbonyl absorption peaks by as much as 90 to 100 cm^{-1}. Thus this inductive effect is considered to be responsible for the higher values of carbonyl absorption peaks of aldehydes, esters, acids, acid halides, etc., compared to those of ketones. The actual value for carbonyl absorption peaks, in fact, depends on the net result of the two effects. However, it varies in the order given in Table 3.3.

There are two bands in the carbonyl region of the infrared spectrum that are characteristic of *acid anhydrides.* These bands are at about 1820 cm^{-1} (5.5 μ) due to the *symmetric* and at 1760 cm^{-1} (5.68 μ) due to the *asymmetric* carbonyl stretching vibrations in saturated acyclic anhydrides. The frequencies of these bands are lower (1775 and 1720 cm^{-1}, 5.63 and 5.81 μ, respectively) in conjugated acyclic anhydrides. The relative intensities of these two bands usually furnish important diagnostic clues to enable us to distinguish between acyclic (or *un*strained cyclic) anhydrides and the corresponding strained compounds. Usually, the higher frequency band is more intense in acyclic anhydrides, while the lower frequency band is more intense in strained cyclic (five-membered) compounds.[18,19] The spectra of acetic anhydride (acyclic) and maleic anhydride (cyclic, five-membered) are recorded in Figs. 3.33 and 3.34, respectively.

The high frequency of the absorption peak for carbonyl stretching in *acid chlorides* (1812 to 1790 cm^{-1}, 5.52 to 5.59 μ) may be due to the high electronegativity of chlorine. This high electronegativity makes it hard for the oxygen to draw electrons, thereby weakening the resonance:

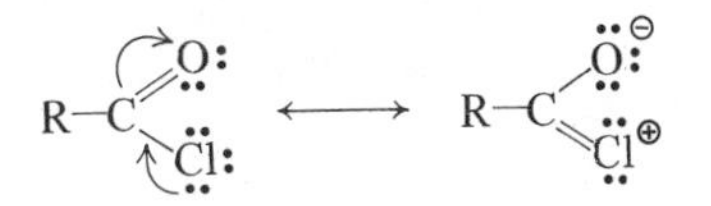

In aromatic acid chlorides, a weak overtone is usually observed as a shoulder on the lower frequency side of the carbonyl band.

It is easy to distinguish *carboxylic acids* from other compounds by infrared spectroscopy, because their C=O bands are generally more intense than those of other carbonyl compounds; carboxylic acids also display very broad O—H stretching bands in the region between 3333 and 2500 cm^{-1} (3.0 to 4.0 μ). This band usually has some fine structure between 2700 and 2500 cm^{-1} (3.71 to 4.00 μ).

Table 3.3 Decreasing order of values of C=O stretching frequency

Type of carbonyl compound	Typical examples	Phase	C=O stretching frequency, cm^{-1}	Value of absorption wavelength, μ
Acid anhydrides (RCOOCOR)			1820 and 1760 (two bands)	5.49 and 5.68
	Acetic anhydride	CCl_4	1825 and 1754	5.48 and 5.70
	Succinic anhydride	$CHCl_3$	1820 and 1776	5.49 and 5.63
	Benzoic anhydride	$CHCl_3$	1818 and 1740	5.50 and 5.75
Acid chlorides (RCOCl)			1812 to 1790	5.52 to 5.59
	Acetyl chloride	CCl_4	1812	5.52
	Isovaleryl chloride	Film	1792	5.58
	Benzoyl chloride	CCl_4	1739	5.75
Carboxylic acids (RCOOH) monomers			1775 to 1750	5.63 to 5.72
	Butyric acid	CCl_4	1775	5.63
	Acetic acid	CCl_4	1760	5.68
Esters (RCOOR)			1780 to 1710	5.62 to 5.85
	Phenyl acetate	Film	1765	5.68
	Vinyl acetate	CCl_4	1765	5.68
	Methyl acetate	CCl_4	1750	5.71
	Methyl propionate	CCl_4	1748	5.72
	Ethyl propionate	CCl_4	1736	5.76
	Propyl formate	CCl_4	1733	5.77
	Ethyl benzoate	CCl_4	1724	5.80
	Benzyl benzoate	Film	1720	5.81
	Ethyl cinnamate	CCl_4	1710	5.85
Aldehydes (RCHO)			1740 to 1690	5.76 to 5.92
	n-butanal	CCl_4	1736	5.76
	Acetaldehyde	CCl_4	1730	5.78
	Valeraldehyde	CCl_4	1730	5.78
	Isovaleraldehyde	Film	1715	5.83

	o-chlorobenzaldehyde	Film	1695	5.90
	p-anisaldehyde	CCl_4	1690	5.92
Ketones (RCOR)			1730 to 1645	5.78 to 6.08
	Norcamphor	Film	1730	5.78
	Butanone	CCl_4	1724	5.80
	Acetone	CCl_4	1720	5.81
	2-Pentanone	CCl_4	1712	5.84
	Methyl isopropyl ketone	Film	1709	5.85
	Methyl phenyl ketone	CCl_4	1695	5.90
	Methyl vinyl ketone	CCl_4	1686	5.93
	Propiophenone	CCl_4	1686	5.93
	Methyl-*p*-tolyl ketone	Film	1675	5.95
	Benzophenone	$CHCl_3$	1669	5.99
	p-Benzoquinone	$CHCl_3$	1645	6.08
Carboxylic acids $(RCOOH)_2$ dimers			1724 to 1665	5.80 to 6.01
	α-chloropropionic acid	Film	1724	5.80
	Butyric acid	Film	1721	5.81
	n-Hexanoic acid	Film	1698	5.89
	Benzoic acid	KBr	1678	5.96
	Salicylic acid	KBr	1665	6.01
Amides $(RCONH_2)$ (RCONHR) $(RCONR_2)$			1678 and shoulder at 1710	5.96 and 5.85
	Acetanilide	$CHCl_3$	1678 and shoulder at 1701	5.96 and 5.88 (sh.)
	N-methylacetamide	CCl_4	1675 and shoulder at 1705	5.97 and 5.86 (sh.)
	N,N-dimethylacetamide	CCl_4	1660 and shoulder at 1710	6.02 and 5.85 (sh.)
	Acetamide	$CHCl_3$	1670 and shoulder at 1716	5.97 and 5.83 (sh.)
	Benzamide	$CHCl_3$	1667 and shoulder at 1709	6.00 and 5.85 (sh.)
	Caprolactam	CCl_4	1667 and shoulder at 1701	6.00 and 5.88 (sh.)

Primary and secondary amides display another band in the carbonyl region between 1650 and 1515 cm^{-1} (6.06 and 6.60 μ). This band is called the amide II band.

Ions and salts of acids $(R{-}COO^{\ominus})$			1570 to 1540	6.37 to 6.49
	Potassium acetate	KBr	1567	6.38
	Potassium stearate	KBr	1543	6.48

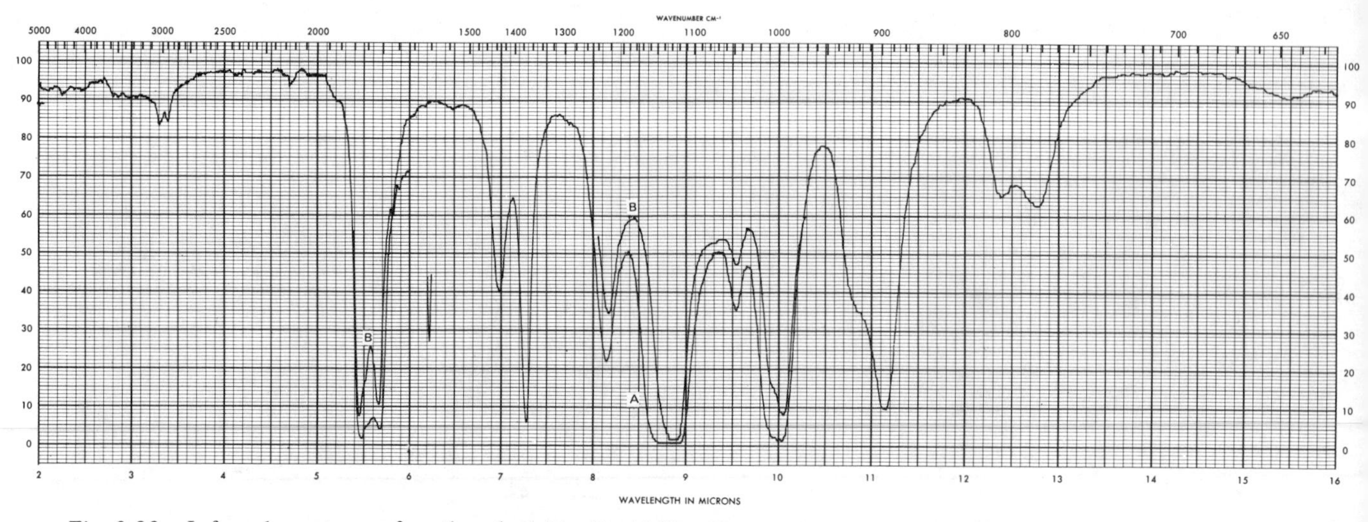

Fig. 3.33 Infrared spectrum of acetic anhydride, liquid film. Inset: 3.5% solution in CCl_4.

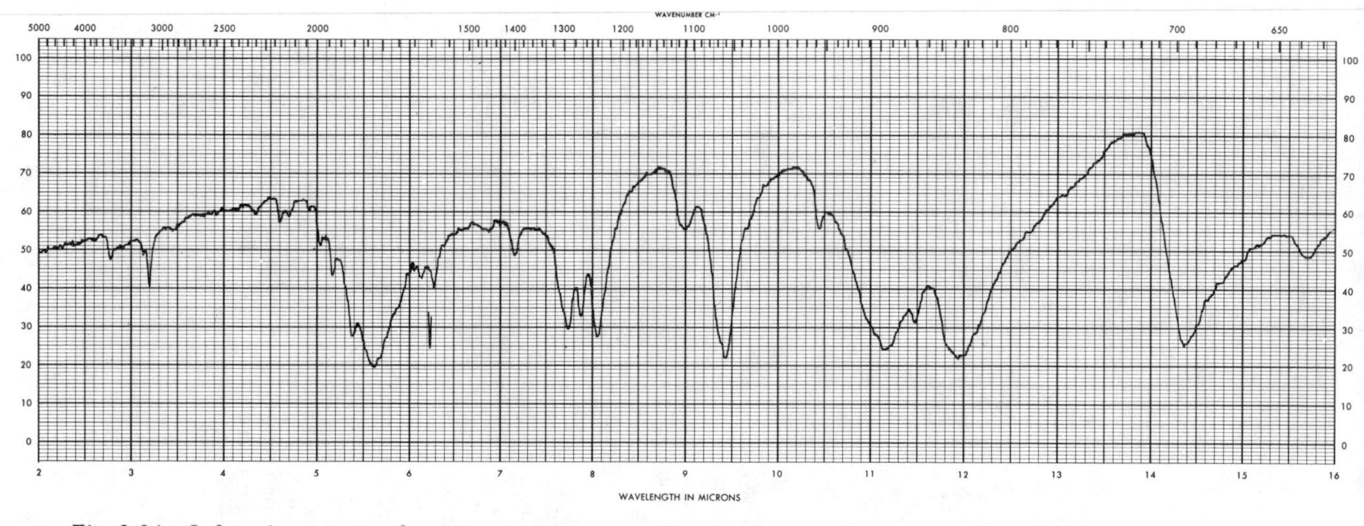

Fig. 3.34 Infrared spectrum of maleic anhydride, crystalline in KBr.

Carboxylic acids in solid or pure liquid form usually exist in the dimeric form due to strong hydrogen bonding:

```
      O···H—O
R—C⟨          ⟩C—R
      O—H···O
```

Only when carboxylic acids are in very dilute solutions, or in the vapor phase, can they exist as acid monomers. The strong hydrogen bonding weakens the C=O bonds, thus lowering the C=O absorption frequency which is a feature of the solid and liquid forms (Table 3.3). Figure 3.35 is a spectrum of a liquid film of *n*-butyric acid.

The appearance of the carbonyl stretching peak at 1721 cm^{-1} (5.81 μ) and a broad hydroxyl peak, with ν_{max} at 2992 cm^{-1} (3.34 μ), indicate that the acid is largely in the hydrogen-bonded, dimeric form.

*Intra*molecular hydrogen bonding causes a greater shift of the carbonyl group to the lower frequency than *inter*molecular bonding does. For example, in *p*-hydroxybenzoic acid, which exists in the dimeric form due to *inter*molecular hydrogen bonding, the carbonyl peak appears at 1678 cm^{-1}. But in salicylic acid (*o*-hydroxyl benzoic acid), in which *intra*molecular bonding occurs, it is at 1665 cm^{-1}. We shall presently discuss the effects produced by substitution of the alkyl group, in R—COOH, by vinyl, aryl, halo-alkyl, and so forth.

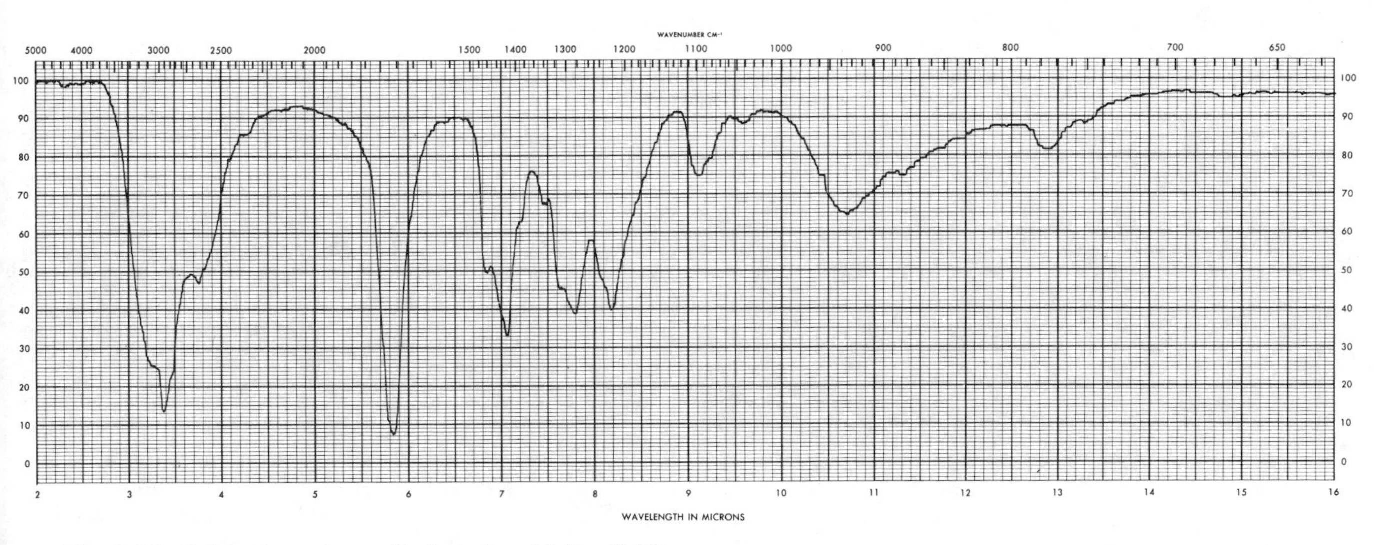

Fig. 3.35 Infrared spectrum of *n*-butyric acid, liquid film.

In the case of carboxylic acids, one usually observes, in addition to the carbonyl and hydroxyl bond stretching absorption peaks, a weak C—O stretching band around 1400 cm^{-1} (7.14 μ) and a strong O—H bending absorption peak around 1250 cm^{-1} (8.00 μ).

Esters also display C—O stretching vibrations in addition to C=O stretching. However, the peculiarity of the esters is that they show two C—O bond absorptions, one at about 1300 to 1160 cm^{-1} (7.69 to 8.62 μ) due to asymmetric

$$\mathrm{C{-}C}\begin{matrix}/\!\!/\\ \diagdown\mathrm{O}\end{matrix}$$

stretching and the other in the region between 1150 and 1035 cm^{-1} (8.69 to 9.65 μ) due to asymmetric O—CH_2—C stretching. These bands are significant in the identification of esters because the high frequency band (near 1250 cm^{-1}) is usually stronger than the C=O stretching band. In aldehydes, ketones, carboxylic acids, amides, etc., on the other hand, the carbonyl stretching band is usually the strongest (Fig. 3.36).

In esters, the carbonyl band appears between 1780 and 1710 cm^{-1} (5.62-5.85 μ). In formates and in alkyl-substituted, saturated, or conjugated esters, the C=O stretching vibration occurs between 1750 and 1710 cm^{-1}. But the frequency of this absorption band is raised to

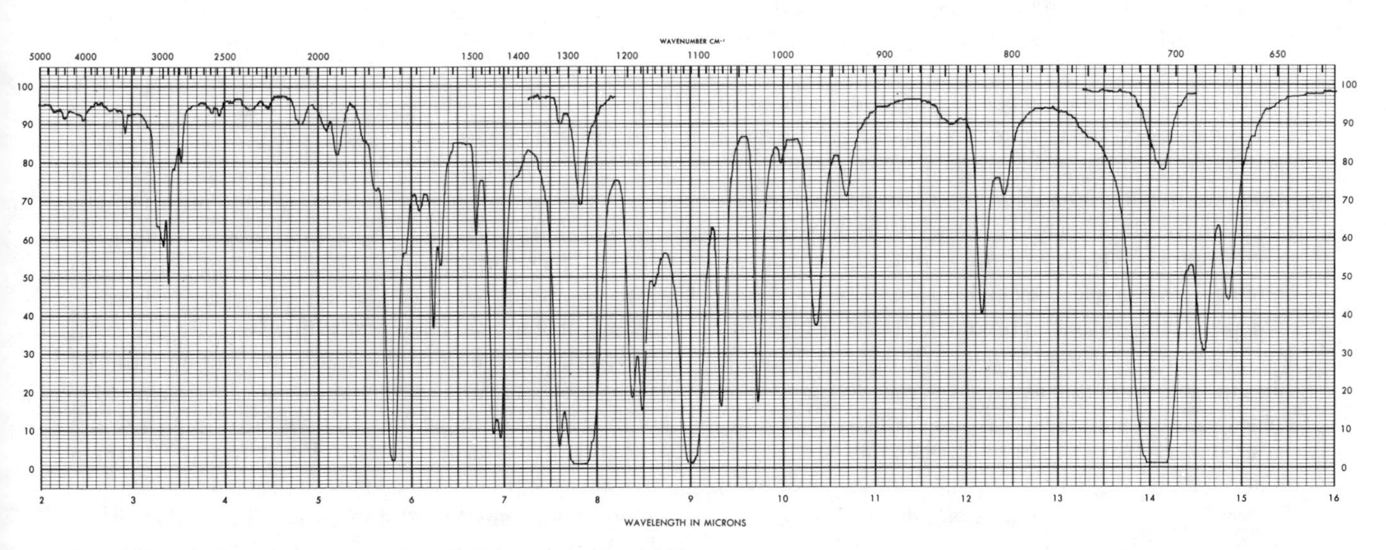

Fig. 3.36 Infrared spectrum of methyl benzoate, liquid film.

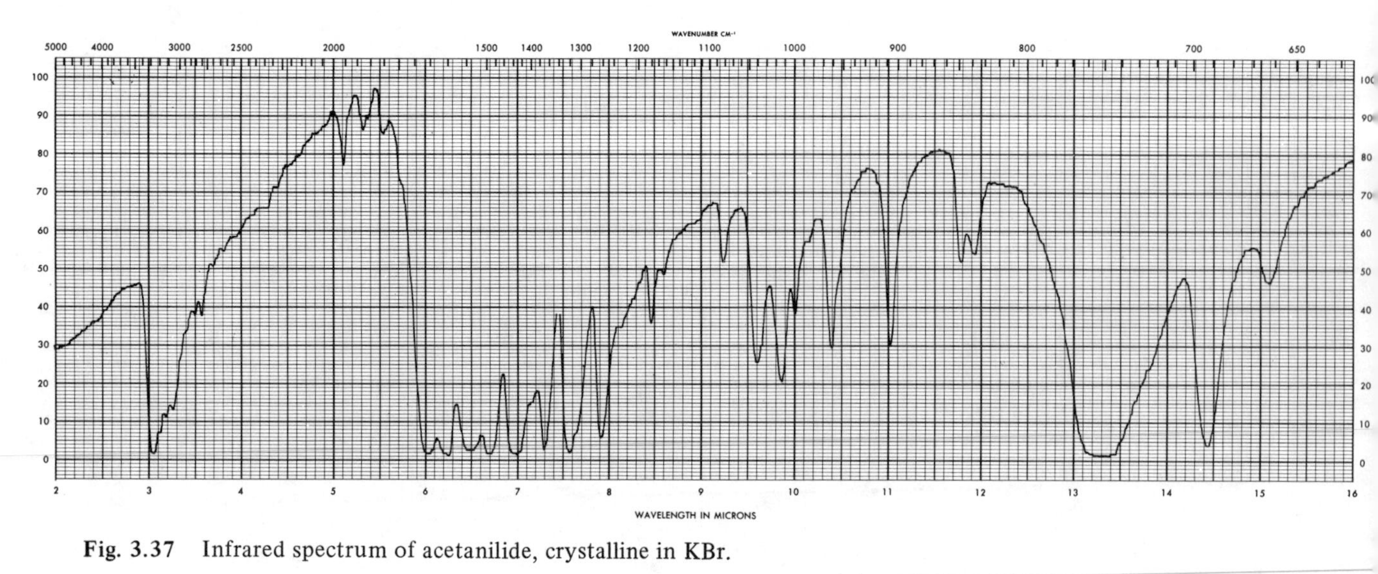

Fig. 3.37 Infrared spectrum of acetanilide, crystalline in KBr.

near 1770 cm^{-1} if the electron-withdrawing groups—such as vinyl, phenyl, etc.—exist on the single-bonded oxygen. This increase is rationalized on the basis of the resonance:

$$\mathrm{R{-}\overset{\displaystyle :\ddot{O}}{\overset{\|}{C}}{-}\ddot{\underset{\cdot\cdot}{O}}{-}C{=}C \longleftrightarrow R{-}\overset{\displaystyle :\ddot{O}}{\overset{\|}{C}}{-}\overset{\oplus}{\underset{\cdot\cdot}{O}}{=}C{-}\overset{\ominus}{C}}$$

This resonance reduces the capability of the ether oxygen to supply electrons to the carbonyl carbon, thus increasing the double-bond character and the force constant of the C=O bond.

The C=O stretching band in the spectra of *amides* appears near 1670 cm^{-1} (5.97 μ), and is known as the *amide* I *band,* so called because all primary and secondary amides display another band in the carbonyl region, due largely to N—H deformation. This second band, between 1650 and 1515 cm^{-1} (6.06 and 6.60 μ), is the *amide* II *band.*

The positions of a number of bands in the spectra of amides depend on the manner in which the spectra are determined, and also on factors such as the size of the rings, substituent groups on nitrogen, the phase, the degree of concentration, etc. These variations in position of absorption bands are partially attributed to hydrogen bonding. For example, the amide I absorption band for solid-phase primary amides (except acetamide) appears near 1650 cm^{-1}, but in very dilute solutions the band shifts to 1690 cm^{-1}. Similarly, the acyclic secondary amides absorb radiation at 1640 cm^{-1} when they are in the solid phase, but at 1680 cm^{-1} when they are in dilute solutions. Tertiary amides are not capable of hydrogen bonding with other amide molecules, and thus exhibit no such variation in position of their absorption bands. Hydrogen bonding with the solvent may, however, affect the amide I band of all amides. The two amide bands are noticeable in Fig. 3.37.

The amide II band shows a similar variation. In primary and secondary amides which are in the solid phase, these bands appear between 1650 and 1620 cm^{-1} and between 1570 and 1515 cm^{-1}, respectively, but shift to about 20 to 30 cm^{-1} lower when these amides are in dilute solutions.

The N—H stretching vibrations of primary and secondary amides also depend on hydrogen bonding. In the solid or pure liquid state, primary amides, which are highly hydrogen bonded, exhibit two N—H stretching bands. One of these occurs at 3350 cm^{-1} (2.99 μ) and is due to asymmetric stretching vibrations. The other occurs at 3180 cm^{-1} and is due to symmetric N—H stretching vibrations. When the primary amides are in dilute solutions, the degree of hydrogen bonding is less, and consequently the absorption bands shift to higher frequencies near 3500 and 3400 cm^{-1}, respectively.

Secondary amides usually show only one band in the N—H stretching region. The exact position of this band depends not only on the degree of hydrogen bonding, but also on whether the N—H group is *cis* or *trans* to the carbonyl group. These two groups are usually *trans* to one another in the N-monosubstituted amides, but can be forced into the *cis* configuration in cyclic structures such as lactams. When the secondary amides are in solid or pure liquid state the bonded *trans* N—H groups absorb radiation in the region from 3333 to 3268 cm^{-1} (3.00 to 3.68 μ), but in dilute solutions this band shifts to 3460 to 3436 cm^{-1} (2.89 to 2.91 μ). The bonded *cis* N—H groups, on the other hand, absorb radiation at a lower frequency of 3185 to 3135 cm^{-1} (3.14 to 3.19 μ) but in dilute solutions the band shifts to 3436 to 3425 cm^{-1} (2.89 to 2.92 μ). Both *cis* and *trans* secondary amides show a weak band near 3100 to 3070 cm^{-1} (3.23 to 3.26 μ) due to an overtone of the 1550 cm^{-1} band.[20]

In the case of *ions and salts of carboxylic acids,* one can explain the considerably lower frequency at which carbonyl absorbs radiation (1570 to 1540 cm^{-1}) on the basis of the near equivalence of the bonds between the two oxygen atoms and the carbon atom. This equivalence is due to resonance:

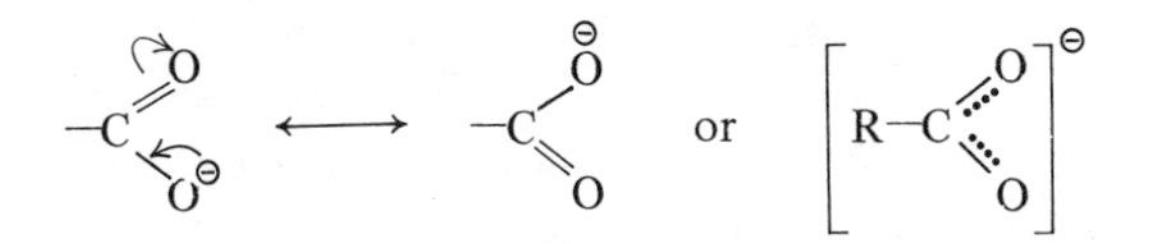

The two oscillators are therefore strongly coupled, displaying a strong asymmetric stretching at 1570 to 1540 cm^{-1} (6.37 to 6.49 μ) and a weaker symmetric CO_2 stretching vibration at about 1440 to 1360 cm^{-1} (6.96 to 7.36 μ).

Intramolecular factors: variations of group R

The frequency at which carbonyl groups absorb radiation in aldehydes, ketones, acids, etc., is affected by variations in group R of the general formula RCOX. Changes in R influence the distribution of electrons in C=O groups. The shifts in frequency thus produced can be explained on the basis of (i) conjugation, (ii) field effects, (iii) strain due to bond angle, and (iv) coupling between carbonyl groups.

i) Conjugation. Table 3.3 presents several values of absorption frequency which show that *the stretching frequencies of carbonyl groups conjugated with other electron systems are generally lower than those of the corresponding nonconjugated compounds.* This lowering is due to the partial loss of the C=O double bond character resulting from the delocalization of π electrons over the entire unsaturated area in the molecule. The greater the degree of delocalization of the electrons, the more the C=O stretching frequency is lowered. For example, vinyl ketones, alkyl phenones, singly conjugated ketones, etc., absorb radiation at frequencies near 1700 to 1670 cm^{-1}, whereas quinones, benzophenones, etc. (doubly conjugated ketones), absorb radiation at lower frequencies of 1680 to 1640 cm^{-1}.

In the case of conjugated aromatic carbonyl compounds, the substituents on the benzene ring often influence the degree of shift of the carbonyl absorption band. *Meta* substituents influence this shift through inductive effect, while *para* substituents do so because of a combination of inductive and resonance effects. It is interesting to note that similar effects also influence the pK values of substituted aromatic acids. In a number of cases, the carbonyl absorption frequencies of aromatic acids have been correlated with Hammett σ constants.[21,22]

The *ortho* substituents exert their influence on carbonyl absorption frequencies of aromatic compounds not only by means of inductive and resonance effects, but also through factors such as chelation, steric effects, field effects, etc. For example, the absorption band for the carbonyl group in benzaldehyde ($CHCl_3$ solution) appears at 1712 cm^{-1}, but, in *o*-hydroxybenzaldehyde, in which intramolecular hydrogen bonding is possible, this band appears at 1698 and 1692 cm^{-1}.

ii) Field effects. A polar group, if oriented properly in relation to the carbonyl group in the molecule, can exert not only field effect (due to the interactions of dipoles through space) but also inductive effect on the frequency of the stretching vibrations of the carbonyl groups. An electron-attracting substituent, such as chlorine, on the carbon which is α to the carbonyl group certainly increases the double-bond character of the carbonyl group through induction. Such an increase in the bond order shifts the C=O stretching frequency to a higher value. Thus the higher positions of the C=O absorption bands in α-chloro ketones, α-chloro acids, α-chloroaldehydes, etc., are easy to explain on the basis of simple inductive effects. However, field effects alone can account for the difference between the rotational isomers of α-chloroacetone which absorb radiation at 1745 and at 1725 cm^{-1}. The fact that an *equatorial* α-halogenated cyclic ketone absorbs radiation at 1745 cm^{-1}, while its *axial* counterpart absorbs radiation at 1725 cm^{-1}, can similarly be explained on the basis of field effects only.

iii) Strain due to bond angle. We have explained (page 72) the difference between the C=O stretching frequencies of acyclic and strained cyclic ketones. The increase in frequency of vibrations of the absorption bands of carbonyls because of strain due to bond angle can be explained on the basis of the increased *s* character of the σ bond of the carbonyl group. The more pronounced *s* character shortens the C=O double bond, and thus increases its force constant. It is possible, by means of a general formula,[23] to predict the difference between the absorption frequencies of cyclic and acyclic carbonyl compounds.

iv) Coupling between carbonyl groups: α and β diketones and other dicarbonyl compounds. α-Dicarbonyl compounds can be either *cis* or *trans*:

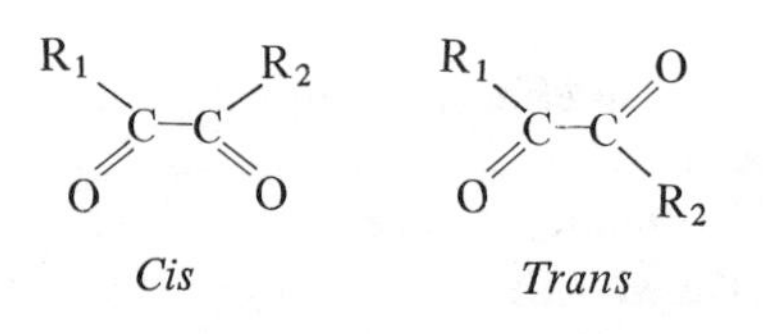

No change in dipole moment is possible in *trans* compounds, but one can observe an interaction between the two carbonyl groups and thus the effect of such interaction on infrared spectra of *cis*-α-dicarbonyl compounds. This effect is particularly noticeable when the high interaction energy of the two polar groups cannot be relieved by enolization. In such cases, two bands at higher frequencies (about 1780 cm^{-1} and 1760 cm^{-1}) are usually obtained.

Enolization is normally possible, and may be dominant in some β-dicarbonyl compounds in which the carbon atom joining the carbonyl groups carries at least one hydrogen atom. For example,

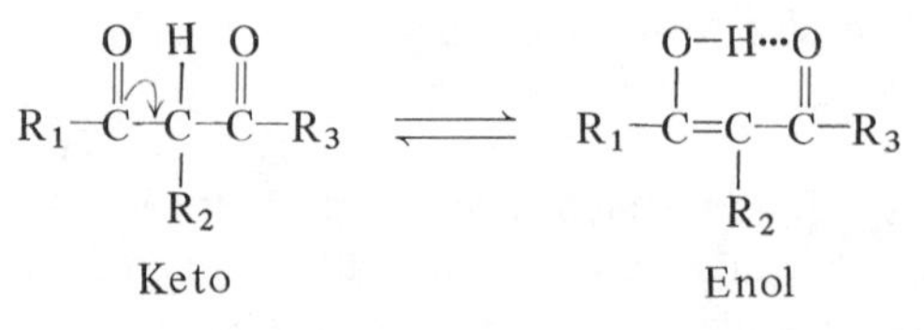

In such cases, the appearance of carbonyl absorption bands depends largely on the extent of enolization. In sterically hindered compounds, and in cases in which enolization is not possible at all, the keto form gives one or two strong absorption bands at about 1700 cm^{-1} (for example, 2-substituted-1,3-cyclohexanediones). On the other hand, in compounds in which keto-enol tautomerism strongly favors enolization (due to hydrogen

Table 3.4 Variation in frequency of vibrations of carbonyl stretching bands due to inter- and intramolecular factors[24] *

Shift to higher frequency	Add cm^{-1}	Shift to lower frequency	Subtract cm^{-1}
Basic value in CCl_4	1720	Neat—solid or liquid state	10
Solvent		*Solvent*	
Hydrocarbon solvents	7	$CHCl_3$, $CHBr_3$, CH_3CN (partially polar)	15
Ring strain		*Ring strain*	
Angle decreases 6 → 5 ring	35	Angle increases 6 → 7 to 10 ring	10
Bridged systems	15		
Substitution on α-carbon		*Substitution on α-carbon*	
(Field and inductive effects)		Each alkyl group	5
Substituent *cis* oriented and coplanar			
—Cl, —Br, —OR, —OH, and —OAc	20	*Alkyl groups substituted by amine*	
		—NH_2 (amide)	5
Substituent *trans* and nonplanar	Nil	—NHMe (monosubstituted amide)	30
		—NMe_2 (disubstituted amide)	55
Alkyl group substituted by electronegative atoms or groups		*Intramolecular hydrogen bonding*	
—H (aldehyde)	10	Weak: α or β-OH ketone	10
—OR (ester)	25	Medium: *o*-OH arylketone	40
—OH (monomeric acid)	40	Strong: β-diketone	100
—O—C=C (vinyl ester)	50	*Intermolecular hydrogen bonding*	
—Cl (acid chloride)	90	Weak: ROH···O=C<	15
—OCOR (anhydrides)	100	Strong: RCOOH dimer	45
		Conjugation (depends on stereochemistry)	
		First C=C	30
		Second C=C	15
		Third C=C	Nil
		Benzene ring	20
		Vinylogous, —CO—C=C—X (X=H or O)	40

* Adapted from J. C. D. Brand and G. Eglinton, *Applications of Spectroscopy,* Oldbourne Press, London, 1965.

bonding, etc.), a peak corresponding to the enol form appears near 1600 cm^{-1} (for example, 2,4-pentanedione). In compounds such as ethyl aceto acetate, in which enolization occurs but the keto-enol tautomeric equilibrium largely favors the keto form, a strong doublet corresponding to the ester and keto carbonyls of the nonenolized form occurs at 1750 and 1730 cm^{-1}, while another doublet representing the enolic form occurs at 1660 and 1640 cm^{-1}. Such a variation in the position of carbonyl absorption bands makes infrared spectroscopy a powerful tool with which to study tautomeric systems.

In the foregoing discussion, we have seen the effects of various intermolecular and intramolecular factors on the carbonyl stretching absorption bands. One or many of these factors may be operative in a given molecule, and we must always remember that the observed position of the carbonyl absorption band in the spectrum is the cumulative effect of all these factors. It is impossible to derive an accurate value of the carbonyl absorption band by taking into account each factor individually. However, one can often estimate the *approximate* position of this band with the help of Table 3.4. In these calculations, a simple dialkyl ketone—acetone—is taken as a reference or parent compound. The carbonyl stretching band occurs at 1720 cm^{-1} (5.81 μ) in acetone. It is also presumed that all the spectra are determined in dilute solutions of carbon tetrachloride. Small variations due to the stereochemistry and the flexibility of the molecule are not taken into account.

From the following examples, we can readily see the usefulness of Table 3.4 in predicting the values of C=O absorption bands. The wavenumbers for C=O stretching vibrations for a number of compounds are calculated as follows.

1. Benzophenone in chloroform

Basic value for C=O absorption	1720 cm^{-1}
Shift due to benzene rings 2 x 20	−40 cm^{-1}
Shift due to solvent $CHCl_3$	−15 cm^{-1}
Calculated	1665 cm^{-1}
Observed	1669 cm^{-1}

2. Vinyl propionate in CCl_4

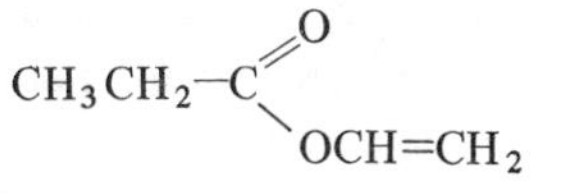

Basic value for C=O absorption	1720 cm^{-1}
Vinyl ester (—O—C=C) shift	+50 cm^{-1}
Alkyl group on a carbon	−5 cm^{-1}
Calculated	1765 cm^{-1}
Observed	1760 cm^{-1}

3. Acetic anhydride in CCl_4

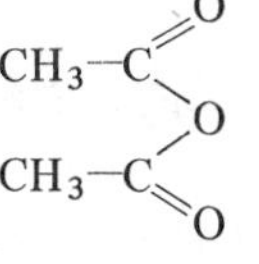

Basic value for C=O absorption	1720 cm^{-1}
Anhydride (—OCOR)	+100 cm^{-1}
Calculated	1820 cm^{-1}
Observed	1825 cm^{-1}

4. Benzamide in $CHCl_3$

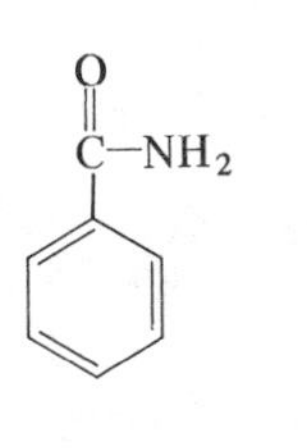

Basic value for C=O absorption	1720 cm^{-1}
Alkyl group substituted by NH_2	−5 cm^{-1}
Alkyl group substituted by benzene	−20 cm^{-1}
Solvent polarity effect	−15 cm^{-1}
Hydrogen bonding	about −15 cm^{-1}
Calculated	1665 cm^{-1}
Observed	1667 cm^{-1}

5. Caprolactam neat liquid

CH_2—CH_2—NH / CH_2 / CH_2—CH_2—C=O

Basic value for C=O absorption	1720 cm^{-1}
Alkyl group substituted by NHMe	−30 cm^{-1}
Alkyl group on α carbon	−5 cm^{-1}
Seven-member ring	−10 cm^{-1}
Neat liquid	−10 cm^{-1}
Calculated	1665 cm^{-1}
Observed	1667 cm^{-1}

Problem 3.8 Suppose that you are investigating a compound with the molecular formula $C_4H_6O_2$. You find that it contains an *ester carbonyl* and an *-ene* group. You determine this on the basis of the following bands in the IR spectrum of a CCl_4 solution of the compound: 3030 cm^{-1} (weak), 1765 cm^{-1} (strong), 1649 cm^{-1} (strong), 1225 cm^{-1} (strong) and 1140 cm^{-1} (strong). Identify these bands, write all the possible structures which include these two groups, and identify the compound.

Amines and amino acids and their salts

Like C—H and O—H bonds, N—H bonds also absorb radiation in the region above 2500 cm^{-1}. Primary amines usually show two bands: one at 3550 to 3350 cm^{-1} (2.82 to 2.99 μ), due to asymmetric stretch, and the other at 3450 to 3250 cm^{-1} (2.90 to 3.08 μ), due to symmetric stretch. Secondary amines exhibit only one band, between 3550 and 3350 cm^{-1} (2.82 to 2.99 μ); tertiary amines do not absorb radiation at all in this region. Aromatic amines absorb radiation in this region at a slightly higher frequency than the aliphatic amines.

Primary and secondary amines are both capable of hydrogen bonding; the bonded N—H groups absorb radiation at lower frequencies than the corresponding unbonded ones. The shift to lower frequency, however, is usually not as great as it is in the case of hydroxyl groups. The NH_3 stretching bands in the spectra of amine hydrochlorides also occur in the region above 2500 cm^{-1}. The spectra of solid salts have absorption peaks at about 3200 cm^{-1}, while the spectra of salts in solution have absorption peaks near 3330 cm^{-1}. Amino acids show a strong band at 3125 to 3030 cm^{-1}, and a weak band at 2760 to 2530 cm^{-1} (primary amino acids only). Table 3.5 presents different modes of vibration of amines and amino acids.

Miscellaneous nitrogen compounds

C≡N groups: nitriles and isonitriles. Due to the triple bond, C≡N groups have a high frequency of stretching vibration. Aliphatic nitriles absorb radiation between 2260 and 2240 cm^{-1}, while aromatic nitriles absorb radiation between 2240 and 2220 cm^{-1}. These bands are usually strong, but variable. The intensity of the vibrations is less if the nitrile group is near an oxygen in the molecule. The isonitriles absorb strongly near 2185 to 2121 cm^{-1}.

C=N groups: oximes, Schiff bases, thiazoles, pyridines, quinolines, pyrimidines, purines, oxazolones, oxazolines, azomethines. The region of C=N stretching vibration is generally 1690 to 1630 cm^{-1}. However, this band can shift as low as 1480 cm^{-1} due to conjugation.

In *oximes* (C=N—OH)—whether aliphatic, aromatic, or amide—the C=N stretching vibration occurs at 1690 to 1620 cm^{-1}. The hydroxyl group in oximes is usually hydrogen bonded; it absorbs radiation at 3300 to 3150 cm^{-1} and the N—O absorbs radiation at 930 cm^{-1}.

The C=N bond in aliphatic *Schiff bases* absorbs radiation at 1670 cm^{-1}, but in aromatic ones it absorbs radiation at 1630 cm^{-1}.

Although compounds containing acyclic, α,β-unsaturated, conjugated C=N bonds absorb radiation at 1665 to 1630 cm^{-1}, corresponding cyclic compounds, such as thiazoles, absorb radiation at frequencies as low as 1660 to 1480 cm^{-1}.

A weak band near 1580 to 1550 cm^{-1} and two medium-intensity bands at 1650 to 1580 and 1510 to 1480 cm^{-1} are attributed to the C=C and C=N stretching vibrations in nitrogenic heterocyclic compounds such as *pyridines* and *quinolines.* The C=N stretching vibration in *pyrimidines and purines* occurs at 1580 to 1520 cm^{-1}.

The C=N bond is weakened by resonance; this indicates that the vibrations are asymmetric stretching vibrations. The intensity of this band thus varies between being strong and being undetectably weak.

Nitro groups: nitro compounds, nitrates, nitramines. A nitro group is a resonance hybrid of the forms:

$$-\overset{\oplus}{N}{\scriptstyle (=O)(-O^{\ominus})} \longleftrightarrow -\overset{\oplus}{N}{\scriptstyle (-O^{\ominus})(=O)} \approx -\overset{\oplus}{N}{\scriptstyle (\cdots O)(\cdots O)^{\ominus}}$$

It shows a strong absorption band due to asymmetric stretching in the region between 1590 and 1500 cm^{-1} and a weak band near 1390 to 1250 cm^{-1}. The large variation in the position of the absorption band is due to the electronegativities of the neighboring groups and also due to the degree of conjugation.

Organic nitrites (R—O—N=O) are capable of rotational isomerism. Consequently, they show two bands: one at 1680 to 1648 cm^{-1}, corresponding to the *trans* isomer, and the other at 1625 to 1600 cm^{-1}, corresponding to the *cis* isomer. Nitrites also show a strong absorption band near 780 cm^{-1} due to N—O absorption.

N=N groups: azo compounds, azoxy compounds, C-nitroso compounds. *Trans,* symmetrical, azo compounds do not show N=N stretching vibration because the center of symmetry prevents changes in dipole moment. However, asymmetric aromatic azo compounds of the type ϕ—N=N—ϕ—X (where X = hydroxyl, amino, etc.) do show N=N stretching vibration near 1410 cm^{-1}. The

Table 3.5 Vibration modes of amines and amino acids (frequencies in cm^{-1})

Compound	N—H stretching		N—H deformation	C—N stretching	C=O stretching	C—O stretching	Other vibrations and remarks
	asym	sym					
Primary amines	3550–3350 (2.82–2.99 μ)	3450–3250 (2.90–3.08 μ)	1650–1580 (6.06–6.33 μ)	1220–1020 ali (8.2–9.8 μ) 1340–1250 aro (7.46–8.0 μ)			Shoulder at 3200 (3.12 μ), overtone of 1610 (6.21 μ) band. N—H wagging strong band at 850–750 (11.76–13.34 μ)
Secondary amines	3550–3350 (2.82–2.99 μ)		1650–1550 (6.06–6.45 μ)	1350–1280 aro (7.41–7.81 μ)			N—H wagging strong band at 750–700 (13.34–14.29 μ). N—H def. band often masked by aromatic band in aromatic compounds.
Tertiary amines				1380–1260 aro (7.24–7.94 μ)			N—Me band in secondary and tertiary amines, observed at 2820–2760 (3.54–3.62 μ)
Amine hydrochloride	About 3380 (2.96 μ) NH_3^+ str		About 1600 (6.25 μ) asym				
Charged amine derivatives	About 3280; (3.05 μ) NH_3^+ str 3350–3150; NH_3^+ str (2.99–3.17 μ) Solid phase		About 1300 (7.7 μ) sym About 800 (12.5 μ) NH_3^+ rock				
Primary amino acids	3125–3030 (3.2–3.3 μ)		1660–1610 band I (6.02–6.21 μ) 1550–1485 band II (6.45–6.74 μ)		1600–1560 ionized (6.25–6.41 μ) carboxyl str.		Weak band at 2760–2530 (3.62–3.96 μ) Medium band at 1300 (7.69 μ)
Dicarboxylic amino acids					1755–1720 (5.7–5.81 μ) unionized carboxyl str		
Amino acid hydrochlorochlorides	3130–3030 (3.2–3.3 μ)		1610–1590 band I (6.21–6.29 μ) 1550–1480 band II (6.45–6.75 μ)		1755–1730 (5.7–5.78 μ) unionized carboxyl	1230–1215 (8.13–8.23 μ)	Series of bands between 3030 and 2500 (3.3 and 4.0 μ)
Amino acid sodium salts	3400–3200 (2.94–3.13 μ) Two bands				1600–1560 (6.25–6.41 μ) ionized carboxyl		

N=N bond in azoxy compounds, ϕ—N=N → O, has no symmetry and absorbs at 1500 cm^{-1} (N=N stretching). N → O stretching vibration is generally observed in azoxy compounds at 1320 cm^{-1}.

3.8 APPLICATIONS

1) Identifying and determining structure. The chief use of infrared spectroscopy is in the determination of structure, the identification, and the qualitative and quantitative determination of organic compounds. No two compounds have exactly the same absorption peaks. Therefore IR spectra can be used as fingerprints to identify unknown compounds. When the spectrum of an unknown compound matches perfectly that of a known compound, one can consider this a positive identification of the unknown. However, you don't always need to match spectra in order to identify an unknown compound. You can determine part or all of the structure of a compound just by careful examination of a spectrum. When you know that certain functional groups or bonds absorb radiation in certain known wavelength regions, you can use correlation charts such as Appendix 3.1, together with your knowledge of factors causing band shifts, to identify most, if not all, infrared bands.

2) Determining purity and quantitative analysis. Using infrared spectroscopy to determine the purity of a sample may, at best, be considered risky because infrared bands are often not very intense, and therefore small amounts of impurities may go undetected. However, if the impurity happens to have some groups which absorb radiation strongly in a region in which the pure sample is not likely to absorb radiation, then IR spectroscopy can be looked on as a feasible method of detecting impurities.

Similarly, in quantitative analysis of the sample, infrared spectroscopy cannot be considered a match for many other analytical techniques, particularly because the base line of an infrared spectrum is usually poorly defined and often does not remain at (or even close to) zero absorbance. Nevertheless, in some areas such as polymer chemistry, the infrared technique is successfully employed in quantitative work; it is used to assay end groups and to determine the degree of polymer branching.

In a recent application of this technique, the acetylation reactions of several primary amines were followed with the aid of the C=O stretching vibration frequencies of their products.[25] The carbonyl groups of the N-monoacetylamines display a strong absorption band at 1640 cm^{-1} and those of the N-diacetylamines show up at 1692 cm^{-1}. This difference in the frequency of absorption of mono and di compounds was used as an analytical probe in the identification of the products. The ratios of the amines were determined by comparing the intensities of the two carbonyl bands.

3) Following chemical reactions. Infrared spectroscopy is often used to follow the progress of chemical reactions, to control chemical reactions, and to study the kinetics of various reactions. These applications are based on the fact that most chemical reactions involve changes in functional groups rather than more fundamental changes in structure. In this technique, one selects a strong band (or bands) in a fairly empty region of the spectrum. Periodically, during the progress of the chemical reaction, one withdraws a small sample of the reaction mixture and scans it over the preselected region to study the appearance or disappearance of this band. A direct plot of the percentage of absorption against time is thus obtained.

For example, the oxidation of an alcohol to an aldehyde or ketone, or the reduction of a ketone to an alcohol can be followed by the appearance or disappearance of the carbonyl band. For accurate work in reaction kinetics, special cells have been designed in which the reaction can be carried out while the cell is placed in the path of the infrared beam of the instrument.

We have already mentioned the application of infrared spectroscopy to the study of the tautomeric equilibria. The tautomeric system may be a keto-enol equilibrium

$$-CO-CH_2 \rightleftharpoons -C-(OH)=CH-$$

or a lactam-lactim equilibrium

$$-CO-NH \rightleftharpoons -C-(OH)=N-,$$

or a mercapto-thioamide equilibrium

$$-N=\overset{|}{C}-SH \rightleftharpoons NH-C=S$$

having a strongly absorbing group. The important bands in the first two equilibria are the carbonyl (at about 1740 cm^{-1}) and the hydroxyl (around 3600 cm^{-1}), whereas in the last equilibrium the C=S band at about 1140 cm^{-1} is often followed.

4) Studying hydrogen bonding. Section 3.7 discussed the effect of hydrogen bonding on a number of band positions. Infrared spectroscopy is a powerful tool in studying such inter- and intramolecular associations. Information about hydrogen bonding is derived not only from the position of the X—H band but also from the bandwidth and from the intensity of the first overtone bands. The bandwidth and the intensity of the fundamental X—H band *increase* in proportion to the strength of the hydrogen bond, whereas the intensity of the first overtone band *decreases* in proportion to the strength of the hydrogen bonding.

5) Studying molecular geometry and conformational analysis. Infrared spectroscopy has been used to study the molecular geometry of very small molecules in the gaseous state. The gaseous molecules undergo vibrational as well as rotational energy changes, thus giving rise to spectra which have a very fine structure. One can derive considerable information about the geometry of these molecules from such a structure, simply because of its fineness.

Since IR spectroscopy provides a powerful tool for the study of the various phenomena of hydrogen bonding, it can also be used to examine conformational preferences in systems in which hydrogen bonding operates.[26] When a series of amino alcohols

Ar—CH—CH—Ar
OH NR′R″

were examined by this technique, the *dl-threo* isomers displayed[27] only a single, intense absorption band in the 3μ region (3350 to 3380 cm^{-1}) assigned to an intramolecular O—H · · · N hydrogen bonding. In contrast, the *dl-erythro* isomers each displayed three bands in the same region (at 3620, 3575 to 3595, and 3485 to 3520 cm^{-1}), attributable, in order of decreasing frequency, to the presence of unassociated O—H, intramolecular O—H · · · π bonding, and intramolecular O—H · · · N bonding. On this basis it was concluded (and verified by NMR studies) that the conformation of the *dl-threo* can be adequately represented by a single rotamer t_A, whereas that of the *erythro* amino alcohols must be represented by an equilibrium mixture of rotamers e_A and e_B.

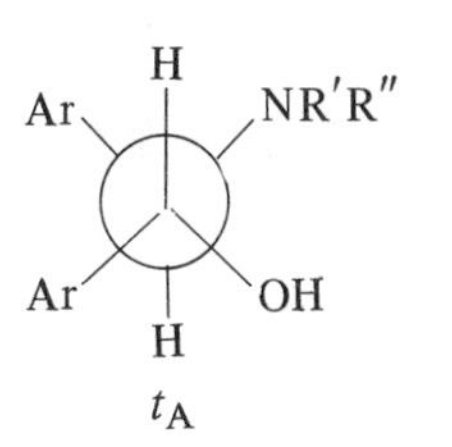

dl-threo amino alcohols

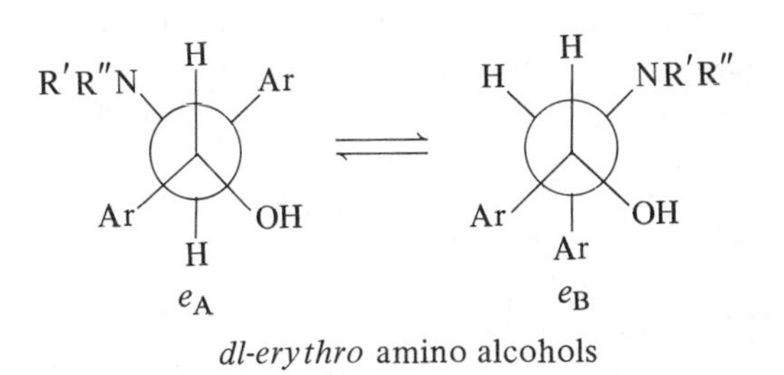

dl-erythro amino alcohols

In another study, a number of cyclohexyl methanesulphonates were examined by IR spectroscopy to determine the equilibrium distribution of conformational isomers.[28] On the basis of the analysis of the bands in the 900 to 1000 cm^{-1} region of *cis* and *trans* isomers of 4-methylcyclohexyl methanesulphonates and 4-*t*-butylcyclohexyl methanesulphonates, it was calculated that the cyclohexyl methanesulphonate is 69% in the equatorial conformation.

Recently the molecular geometry of some hydroxy ketones has been established on the basis of the argument that a strong absorption band around 3400 cm^{-1} is not necessarily due to the overtone of carbonyl vibration, as proposed earlier,[29] but could be due to the strong intermolecular hydrogen bonding in very dilute CCl_4 solutions.[30] Since such intermolecular associations are stereospecific, the appearance of or lack of such bands furnishes clues to the stereochemistry of the hydroxy ketones.

Absorption bands due to C—H stretching vibrations in the 2800 to 2700 cm^{-1} region have also been employed in stereochemical studies. For example, in the case of nitrogen bridgehead compounds, such as yohimbines and related alkaloids, the first correlation between the configuration about the ring fusion and the C—H stretching frequency was proposed in 1956.[31] Since then several papers have appeared describing the infrared correlation of the bands in this region with the stereochemistry of nitrogen bridgehead compounds, and an excellent review of these correlations has recently been published.[32]

Comparisons of Raman and IR spectra are often used to determine the structures and geometries of small molecules. The joint application of these two techniques has also been shown[33] to be fruitful in attacking the complex problems associated with determining the stereochemistry about the cyclobutane ring in the cyclobutane dimers produced by photodimerization of conjugated carbonyl derivatives. Such a technique is based on the selection rule (Section 3.4), according to which, *for centrosymmetric molecules, a vibrationally active Raman transition cannot be active in the IR, and conversely, a vibrationally active IR transition cannot be active in the Raman spectroscopy.* A similar technique is employed in the study of rotational isomerism in phenylethanes, in solid states as well as in solutions.[34]

6) Studying the chemistry of polymers. As mentioned earlier, the field of polymer chemistry has found infrared spectroscopy highly useful. Infrared spectroscopy has provided information which would have been impossible to derive by other techniques. For example, in polymers which have low molecular weight or are highly branched, an infrared spectrum can detect the end groups and give an approximate estimate of the degree of branching in the molecule. Infrared spectra are also used to determine the crystalline–amorphous ratio of the polymer. Such a ratio is helpful in determining the flexibility, plasticity, hardness, softness, etc., of the polymer.[35]

7) Studying reactive species. Infrared spectroscopy is increasingly being used to elucidate the structure of reactive molecules, radicals, or ions, and to study their reactions. The technique used for this purpose involves spectroscopic examination of a reactive species by rapid cooling of a mixture of the absorbing substance (A) and a diluent gas (M) to form a solid matrix. This process avoids the diffusion and the probable reaction of the reactive species (A). The technique is called *matrix isolation.*[36,37,38]

3.9 SELECTED REFERENCES

Infrared spectroscopy has long been established as a technique highly important to the organic chemist. Consequently, a large number of fine books are available to the reader who wants to explore the subject in a greater depth. There are also compilations of several infrared spectra to help the chemist to compare the spectrum of an unknown with those in the compilations. A few recent review articles provide greater information on the subject.

References cited in the text

1. P. A. Wilks, Jr., and T. Hirschfeld, *Applied Spectroscopy Reviews,* edited by E. G. Brame, Jr., Marcel Dekker, Inc., New York, 1968.
2. C. V. Raman, *Indian J. Phys.* **2**, 1 (1928).
3. J. Overend, in *Infrared Spectroscopy and Molecular Structure,* edited by M. Davies, Elsevier, Amsterdam, 1963.
4. D. Steele, *Quart. Rev.* **18** , 21 (1964).
5. A. S. Wexler, in *Applied Spectroscopy Reviews,* edited by E. G. Brame, Jr., Marcel Dekker, Inc. New York, 1968.
6. G. C. Pimentel and L. A. McClellan, *The Hydrogen Bond,* Freeman, San Francisco, 1960, Chapter 3.
7. R. Cetina and J. L., Mateos, *J. Org. Chem.* **25**, 704 (1960).
8. M. St. C. Flett, *Spectrochim. Acta* **18**, 1537 (1962).
9. D. Hummel and E. Lunebach, *Spectrochim. Acta* **18**, 823 (1962).
10. A. S. Wexler, *Spectrochim. Acta* **21**, 1725 (1965).
11. C. N. R. Rao and R. Venkataraghavan, *Can. J. Chem.* **39**, 1757 (1961).
12. J. J. Mannion and T. S. Wang, *Spectrochim. Acta* **17**, 990 (1961).
13. H. H. Zeiss and M. J. Tsutsui, *J. Am. Chem. Soc.* **75**, 897 (1953).
14. T. L. Allsop, A. R. H. Cole, D. R. White, and R. L. S. Willix, *J. Chem. Soc.* 4868 (1956).
15. R. N. Jones and G. Roberts, *J. Am. Chem. Soc.* **80**, 6121 (1958).
16. C. N. R. Rao, G. K. Goldman and C. Lurie, *J. Phys. Chem.* **63**, 1311 (1959).
17. S. Bratoz and S. Besnainou, *Compt. rend* **248**, 546 (1959).
18. W. G. Dauben and W. W. Epstein, *J. Org. Chem.* **24**, 1595 (1959).
19. L. J. Bellamy, B. R. Connelly, A. R. Philpotts and R. L. Williams, *Z. Elektrochem* **64**, 563 (1960).
20. T. Miyazawa, *J. Mol. Spectr.* **4**, 168 (1960).
21. R. N. Jones, W. F. Forbes and W. A. Mueller, *Can. J. Chem.* **35**, 504 (1957).
22. G. Eglinton, *J. Chem. Soc.* 106 (1961).
23. D. Cook, *Can. J. Chem.* **39**, 31 (1961).
24. Adapted by permission from: J. C. D. Brand and G. Eglinton, *Applications of Spectroscopy,* Oldbourne Press, London, 1965.
25. R. P. Mariella and K. H. Brown, *J. Org. Chem.* **36**, 735 (1971).
26. M. Tichy in *Advances in Organic Chemistry, Methods and Results,* edited by R. A. Raphael, E. C. Taylor and H. Wyberg, Volume 5, John Wiley, New York, 1965, page 115.
27. M. K. Meilhan and M. E. Munk, *J. Org. Chem.* **34**, 1440 (1969).
28. D. S. Noyce, B. E. Johnston, and B. Weinstein, *J. Org. Chem.* **34**, 463 (1969).
29. L. Joris and P. von R. Schleyer, *J. Am. Chem. Soc.* **90**, 4599 (1968).
30. J. Pitha, *J. Org. Chem.* **35**, 2411 (1970).
31. E. Wenkert and D. Roychaudhuri, *J. Am. Chem. Soc.* **78**, 6417 (1956).
32. T. A. Crabb, R. I. Newton and D. Jackson, *Chem. Rev.* **71**, 109 (1971).
33. H. Ziffer and I. W. Levin, *J. Org. Chem.* **34**, 4056 (1969).
34. L. H. L. Chia and H. H. Huang, *J. Chem. Soc. (B)*, 1695 (1970).
35. A. Elliott, *Advances in Spectroscopy* **1**, 214 (1959).
36. A. J. Barnes and H. E. Hallam, *Quart. Revs.* **23**, 392, (1969).
37. H. E. Hallam, in *Molecular Spectroscopy, Proceedings of the 4th Institute of Petroleum Hydrocarbon Research Conference,* London, 1968, page 329.
38. M. E. Jacox and D. E. Milligan, *Appl. Optics* **3**, 873 (1964).

Books and general references

39. C. J. Creswell and O. Runquist, *Spectral Analysis of Organic Compounds,* second edition, Burgess Press, Minneapolis, Minn., 1972.

40. G. Herzberg, *The Spectra and Structures of Simple Free Radicals. An Introduction to Molecular Spectroscopy*, Cornell University Press, Ithaca, New York, 1971.
41. T. Shimauchi, *Infrared Absorption Spectrometric Methods*, Nankodo Press, Tokyo, 1971.
42. P. Gans, *Vibrating Molecules: An Introduction to the Interpretation of Infrared and Raman Spectra*, Chapman and Hall, London, 1971.
43. F. S. Parker, *Applications of Infrared Spectroscopy in Biochemistry, Biology and Medicine*, Plenum Press, New York, 1971.
44. M. Avram and G. D. Mateescu, *Infrared Spectroscopy: Applications in Organic Chemistry*, Dunod Press, Paris, 1971.
45. H. Hediger, *Infrared Spectroscopy: Principles, Uses, Interpretation (Methods of Analysis in Chemistry, Volume II)*, Akad. Verlagsges, Frankfurt, 1971.
46. R. T. Conley, *Infrared Spectroscopy*, second edition, Allyn and Bacon, Boston, Mass., 1972.
47. L. M. Sverdlov, M. A. Kovner, and E. P. Krainov, *Vibration Spectra of Polyatomic Molecules*, Nauka, Moscow, 1970.
48. J. E. Stewart, *Infrared Spectroscopy: Experimental Methods and Techniques*, Dekker Publishing Co., New York, 1970.
49. J. H. van der Mass, *Basic Infrared Spectroscopy*, Sadtler Research Labs., Philadelphia, Pa., 1970.
50. N. L. Alpert, W. E. Keiser and H. A. Szymanski, *Theory and Practice of Infrared Spectroscopy*, second edition, Plenum Press, New York, 1970.
51. F. Scheinmann, editor, *An Introduction to Spectroscopic Methods for the Identification of Organic Compounds*, Volume 1, Pergamon Press, Oxford, England, 1970.
52. K. E. Stine, *Modern Practices in Infrared Spectroscopy*, Beckman Instruments, Fullerton, Cal., 1969.
53. L. Little, *Infrared Spectra of Absorbed Species*, Academic Press, New York, 1969.
54. M. Gianturco, "Infrared Absorption Spectroscopy," in *Interpretative Spectroscopy*, S. K. Freeman, editor, Reinhold, New York 1965.
55. R. G. J. Miller, *Laboratory Methods in Infrared Spectroscopy*, Sadtler Research Labs., Philadelphia, Pa., 1965.
56. J. R. Dyer, *Applications of Absorption Spectroscopy of Organic Compounds*, Prentice-Hall, Englewood Cliffs, N.J., 1965.
57. J. C. D. Brand and G. Eglinton, *Applications of Spectroscopy*, Oldbourne Book Co., London, 1965.
58. G. W. King, *Spectroscopy and Molecular Structure*, Holt, Rinehart and Winston, New York, 1964.
59. J. C. P. Schwarz, editor, *Physical Methods in Organic Chemistry*, Oliver & Boyd, Edinburgh, Scotland, 1964.
60. N. B. Colthup, L. H. Daly and S. E. Wiberley, *Introduction to Infrared and Raman Spectroscopy*, Academic Press, New York, 1964.
61. W. D. Phillips and F. C. Nachod, editors, *Determination of Organic Structures by Physical Methods*, Volume 2, Academic Press, New York, 1963.
62. C. N. R. Rao, *Chemical Applications of Infrared Spectroscopy*, Academic Press, New York, 1963.
63. W. J. Potts, Jr., *Chemical Infrared Spectroscopy*, John Wiley, London, 1963.
64. C. E. Meloan, *Elementary Infrared Spectroscopy*, MacMillan, London, 1963.
65. A. R. Katrizky, *Physical Methods in Heterocyclic Chemistry*, Volume 2, Academic Press, New York, 1963.
66. A. Weissberger and K. W. Bentley, editors, *Elucidation of Structures by Physical and Chemical Methods*, Interscience, New York, 1963.
67. M. Davies, editor, *Infrared Spectroscopy and Molecular Structure*, Elsevier, Amsterdam, 1963.
68. A. D. Cross, *An Introduction to Practical Infrared Spectroscopy*, second edition, Butterworths, London, 1963.
69. K. Nakanishi, *Infrared Absorption Spectroscopy*, Holden-Day, San Francisco, Cal., 1962.
70. W. Brugel, *An Introduction to Infrared Spectroscopy*, Methuen, London, 1962.
71. R. E. Dodd, *Chemical Spectroscopy*, Elsevier, Amsterdam, 1962.
72. G. M. Barrow, *Introduction to Molecular Spectroscopy*, McGraw-Hill, New York, 1962.
73. R. P. Baumann, *Absorption Spectroscopy*, Wiley, New York, 1962.
74. G. H. Beavan, E. A. Johnson, H. A. Willis and R. G. J. Miller, *Molecular Spectroscopy, Methods and Applications in Chemistry*, Heywood, London, 1961.
75. K. E. Lawson, *Infrared Absorption of Inorganic Substances*, Reinhold, New York, 1961.
76. L. J. Bellamy, *The Infrared Spectra of Complex Molecules*, second edition, Methuen, London, 1958.
77. R. N. Jones and C. Sandorfy, in *Chemical Applications of Spectroscopy*, edited by A. Weissberger and W. West, Interscience, New York, 1956.
78. R. N. Jones and C. Sandorfy, "The Applications of Infrared and Raman Spectrometry to the Elucidation of Molecular Structure," in *Techniques of Organic Chemistry*, edited by A. Weissberger, Interscience, New York, 1956.
79. G. Herzberg, *Infrared and Raman Spectra of Polyatomic Molecules*, Van Nostrand, New York, 1945.
80. S. Bhagavantam, *Scattering of Light and the Raman Effect*, Chemical Publishing Co., New York, 1942.

Compilations of infrared spectra

81. *Molecular Formula List of Compounds, Names and References to Published Infrared Spectra,* Supplement No. 14, AMD-31-S-14, American Society for Testing and Materials, 1916 Race Street, Philadelphia, Pa., 1971.
82. *Serial Number List of Compound Names and References to Published Infrared Spectra,* Supplement No. 14, AMD-32-S-14, American Society for Testing and Materials, 1916 Race Street, Philadelphia, Pa., 1971.
83. C. J. Pouchert, *The Aldrich Library of Infrared Spectra,* Aldrich Chemical Co., Milwaukee, Wis., 1970.
84. *A.P.I. Research Product 44, Infrared Files,* Petroleum Research Laboratories, Carnegie Institute of Technology, Pittsburgh, Pa.
85. *Manufacturing Chemists Association Research Project, Infrared Files,* Chemical Thermodynamic Properties Center, Department of Chemistry, Texas A. & M. University, College Station, Texas.
86. *Sadtler Standard Spectra,* Sadtler Research Laboratories, Philadelphia, Pa.; 19,000 grating spectra and 39,000 prism spectra (1970).
87. J. Bellanto and A. Hidalgo, *Infrared Analysis of Essential Oils,* Sadtler Research Laboratories, Philadelphia, Pa., 1970.
88. D. Welti, *Infrared Vapor Spectra,* Sadtler Research Laboratories, Philadelphia, Pa., 1970.
89. A.P.I. Research Project 44, *Selected Infrared Spectra Data,* Volumes I to VI (1943–1970) and Volume VII (1970), Thermodynamics Research Center, College Station, Texas.
90. *An Alphabetical List of Compound Names, Formulae and References to Published IR Spectra,* AMD-34, American Society for Testing and Materials, 1916 Race Street, Philadelphia, Pa., 1969 (an index to 92,000 published IR spectra).
91. *An Index of Published Infrared Spectra,* Volumes 1 and 2, Ministry of Aviation, Technical Information and Library Services, H.M. Stationery Office, London, 1960.
92. *Documentation of Molecular Spectra,* the D.M.S. System, Butterworths, London.
93. *Infrared, Raman, Microwave, Current Literature Service,* Butterworths, London, and Verlag-Chemie, Weinheim, Germany.
94. *Tables of Wavenumbers for the Calibration of Infrared Spectrometers,* Butterworths, London, 1961.
95. H. A. Szymanski, and R. E. Erickson, *Infrared Band Hand Book,* second edition, Plenum Publishing Co., New York, 1970.
96. H. M. Hershenson, *Infrared Absorption Spectra Index,* Academic Press, New York, 1959.
97. G. Roberts, B. S. Gallagher and R. N. Jones, *Infrared Absorption Spectra of Steroids, an Atlas,* Volume 2, Interscience, New York, 1958.
98. C. R. Brown, M. W. Ayton, T. C. Goodwin and T. J. Derby, *Infrared–A Bibliography,* Technical Information Division, Library of Congress, Washington, D.C., 1954.
99. K. Dobriner, E. R. Katzenellenbogen and R. N. Jones, *Infrared Absorption Spectra of Steroids, an Atlas,* Volume 1, Interscience, New York, 1953.

Review articles

100. T. Shimanouchi, *Kagaku No Ryoiki* **25**, 89–95 (I), 187–92 (II), 275–80 (III); 377–84 (IV), 506–12 (V), and 611–16 (VI) (1971) (Infrared and Raman spectroscopy).
101. H. J. Sloane, *Appl. Spectrosc.* **25**, 430–41 (1971) (Technique of Raman spectroscopy. State of the art. Comparison to infrared).
102. C. N. R. Rao, S. N. Bhat, and P. C. Dwivedi, *Appl. Spectrosc. Rev.* **5**, 1–170 (1971) (Spectroscopy of electron donor-acceptor system).
103. S. N. Timasheff and L. Stevens, *Trans. Bose Res. Inst., Calcutta* **31**, 75–86 (1968); *Chem. Abstr.* **73**, 135422s (1970) (IR, UV, and circular dichroism studies of proteins and polypeptides).
104. G. Szekely, *J. Chromatogr.* **48**, 313–21 (1970) (Combined application of chromatography and spectroscopy).
105. P. H. Rao, *Indian Chem. Mfr.* **8**, 9 (1970) (Industrial applications).
106. C. Rocchiccioli, *Chim. Anal., Paris* **52**, 273–82 (1970) (Organic analysis).
107. S. F. Orr, *Spectroscopy,* 32–35 (1969) (General).
108. A. Pande, *Chem. Process. Eng., Bombay* **2**, 31–37 (1968); *Chem. Abstr.* **72**, 96204e (1970) (Moisture analysis).
109. H. E. Hallam and C. M. Jones, *J. Mol. Struct.* **5**, 1–19 (1970) (Conformational isomerism of the amide group).
110. J. A. Perry, *Appl. Spec. Revs.* **3**, 229 (1970) (Quantitative analysis).
111. M. Tsuboi, *ibid.,* page 45 (IR of nucleic acids).
112. F. S. Parker and K. R. Bhaskar, *ibid.,* page 91 (Hydrogen–deuterium exchange in biological molecules).
113. P. J. Hendra and P. M. Stratton, *Chem. Revs.* **69**, 235 (1969) (Laser Raman spectroscopy).
114. A. J. Barnes and H. E. Hallam, *Quart. Revs.* **23**, 392 (1969) (Matrix isolated species).
115. H. Levinstein, *Anal. Chem.* **41**, 81 A (1969) (IR detectors).

116. B. R. Kowlski, P. C. Jurs, T. L. Isenhour and C. N. Reilley, *Anal. Chem.* **41**, 1945, 1949 (1969) (Computerized learning machines–interpretation of IR data).
117. C. E. Carraher, Jr., *J. Chem. Ed.* **45**, 462 (1968) (IR spectroscopic practices).
118. R. N. Jones, J. B. Digiorgio, J. J. Elliott and G. A. A. Nonnenmacher, *J. Org. Chem.* **30**, 1822 (1965) (Raman spectroscopy).

PROBLEMS

Problem 3.9. Figure 3.38 shows an infrared spectrum of a liquid film of an unknown compound with the molecular formula $C_6H_{10}O$. The boiling point of the unknown colorless liquid is found to be 155.6°. Identify the compound.

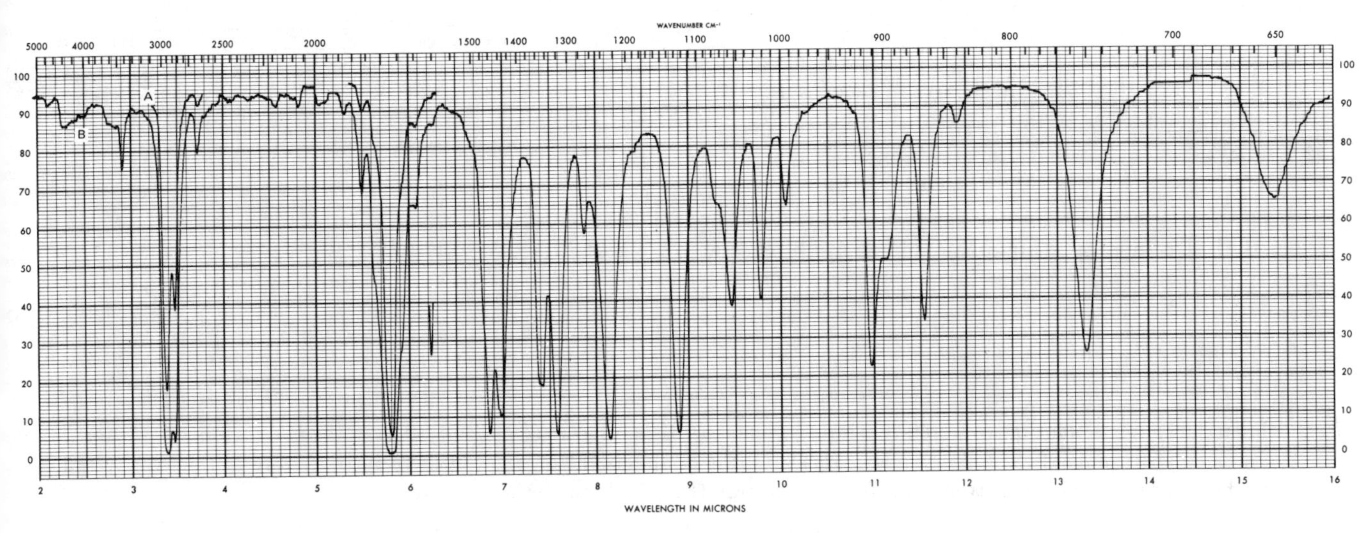

Fig. 3.38 Infrared spectrum of an unknown substance with molecular formula $C_6H_{10}O$.

Problem 3.10. An unknown liquid has a boiling point of 217.7° and a molecular formula of $C_8H_{14}O_4$. An infrared spectrum of a liquid film of the compound is recorded in Fig. 3.39. Identify the compound.

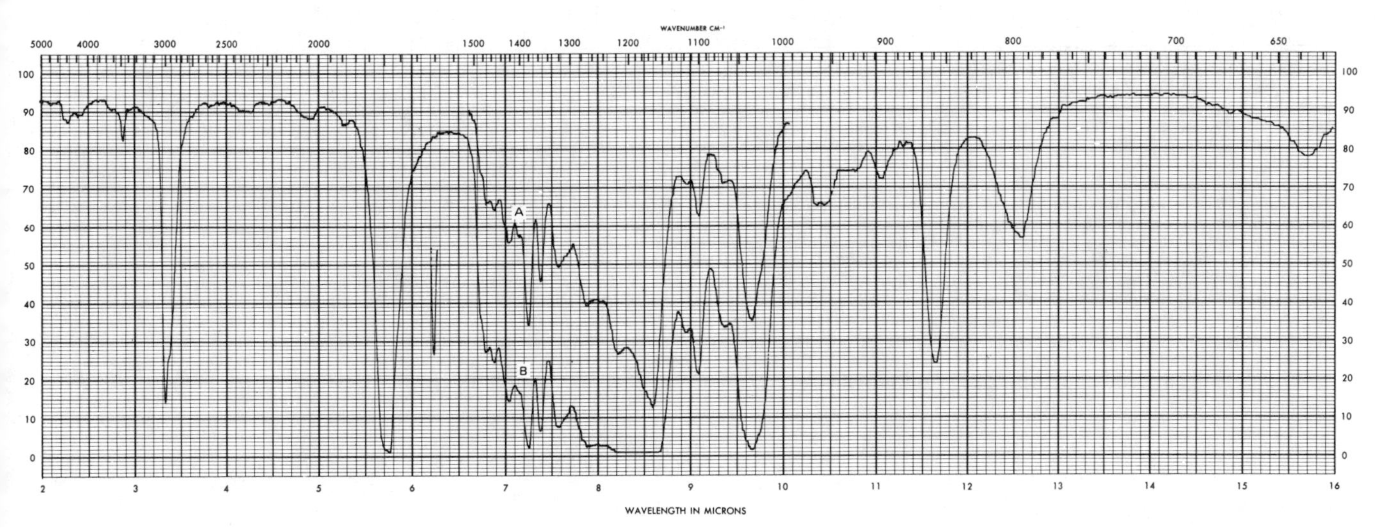

Fig. 3.39 Infrared spectrum of an unknown substance with molecular formula $C_8H_{14}O_4$.

Problem 3.11. Figure 3.40 is the infrared spectrum of a colorless liquid (boiling point 170°). The molecular formula of the compound is $C_7H_{12}O$. Deduce the structure of the compound.

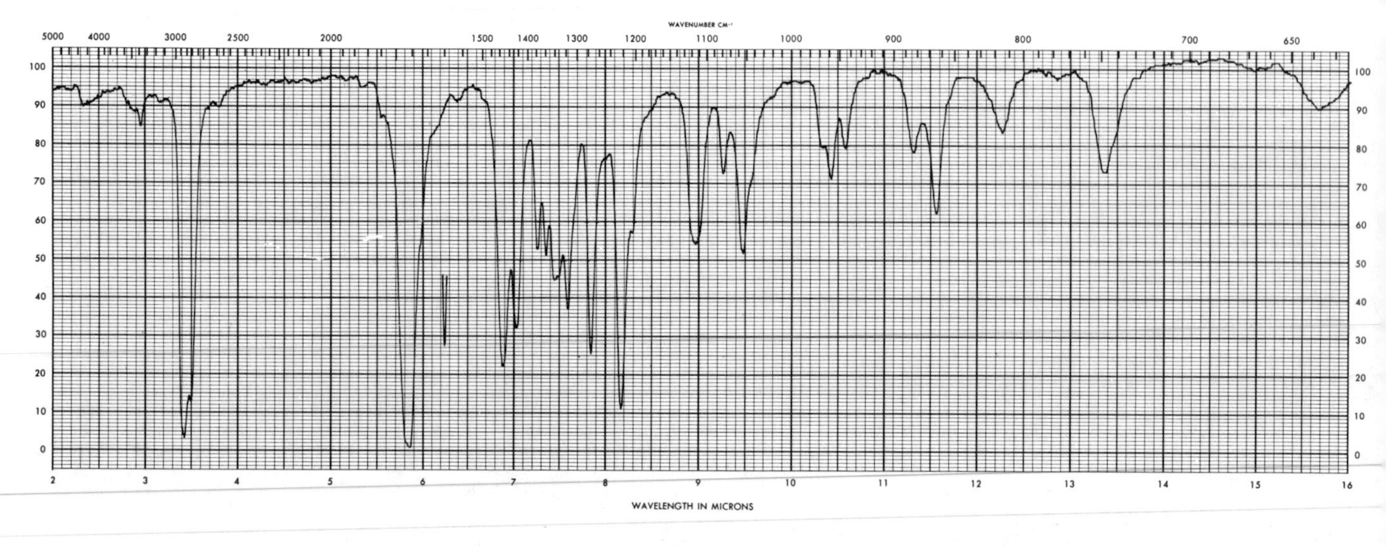

Fig. 3.40 Infrared spectrum of an unknown substance with molecular formula $C_7H_{12}O$.

Problem 3.12. The solution of a colorless liquid in hexane displays a weak peak ($\epsilon = 18$) in the UV region at 290 nm. The molecular formula of the unknown is C_4H_8O and the boiling point = 75.7°. Deduce the structure of this compound, using the information derived from the infrared spectrum (Fig. 3.41) of the compound.

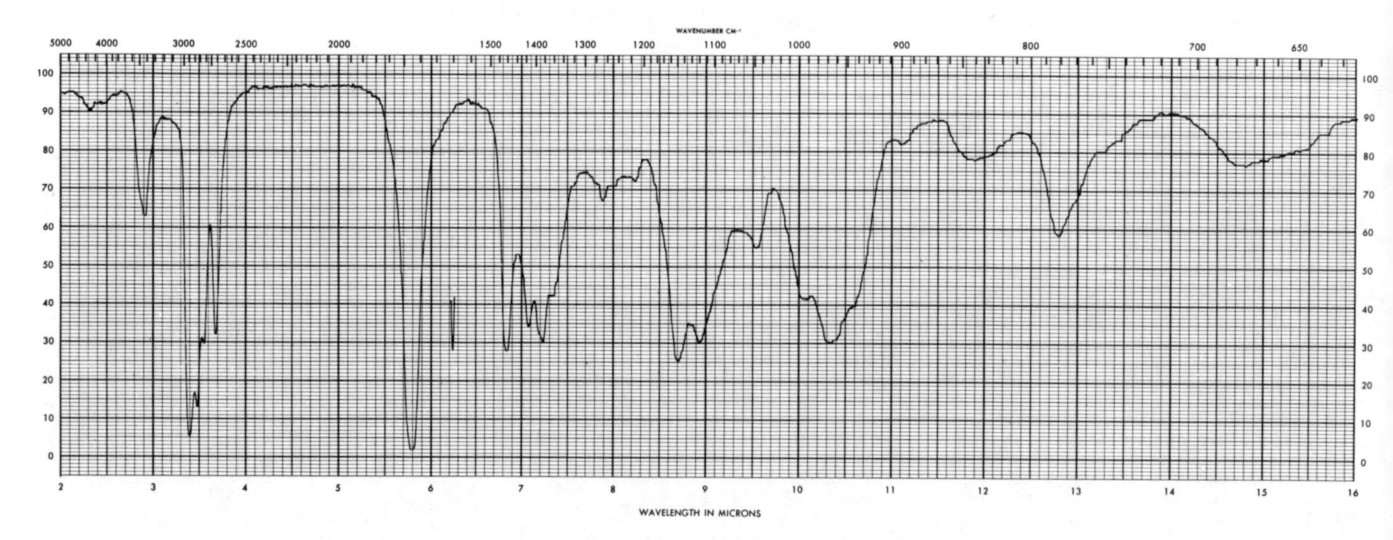

Fig. 3.41 Infrared spectrum of an unknown substance with molecular formula C_4H_8O.

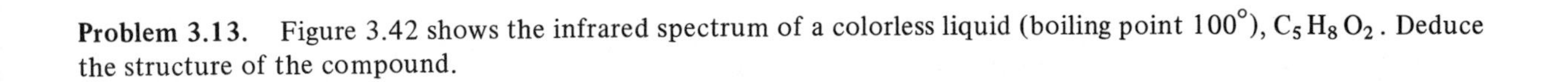

Problem 3.13. Figure 3.42 shows the infrared spectrum of a colorless liquid (boiling point 100°), $C_5H_8O_2$. Deduce the structure of the compound.

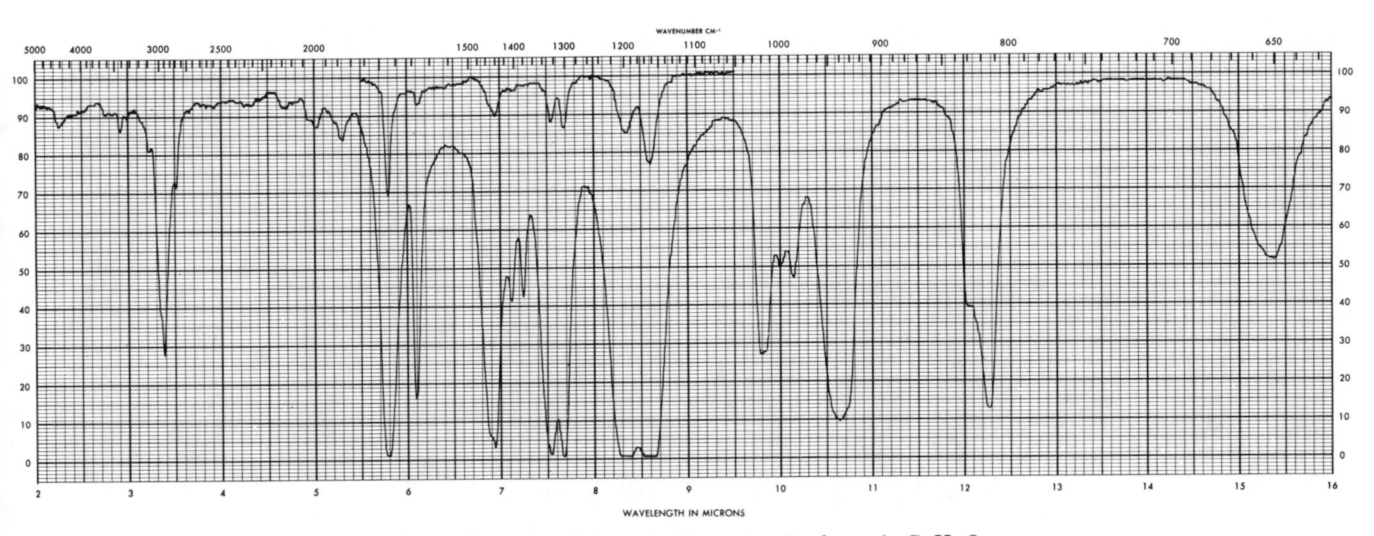

Fig. 3.42 Infrared spectrum of an unknown substance with molecular formula $C_5H_8O_2$.

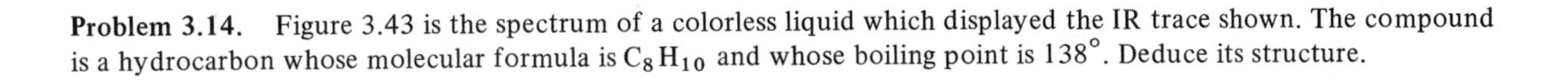

Problem 3.14. Figure 3.43 is the spectrum of a colorless liquid which displayed the IR trace shown. The compound is a hydrocarbon whose molecular formula is C_8H_{10} and whose boiling point is 138°. Deduce its structure.

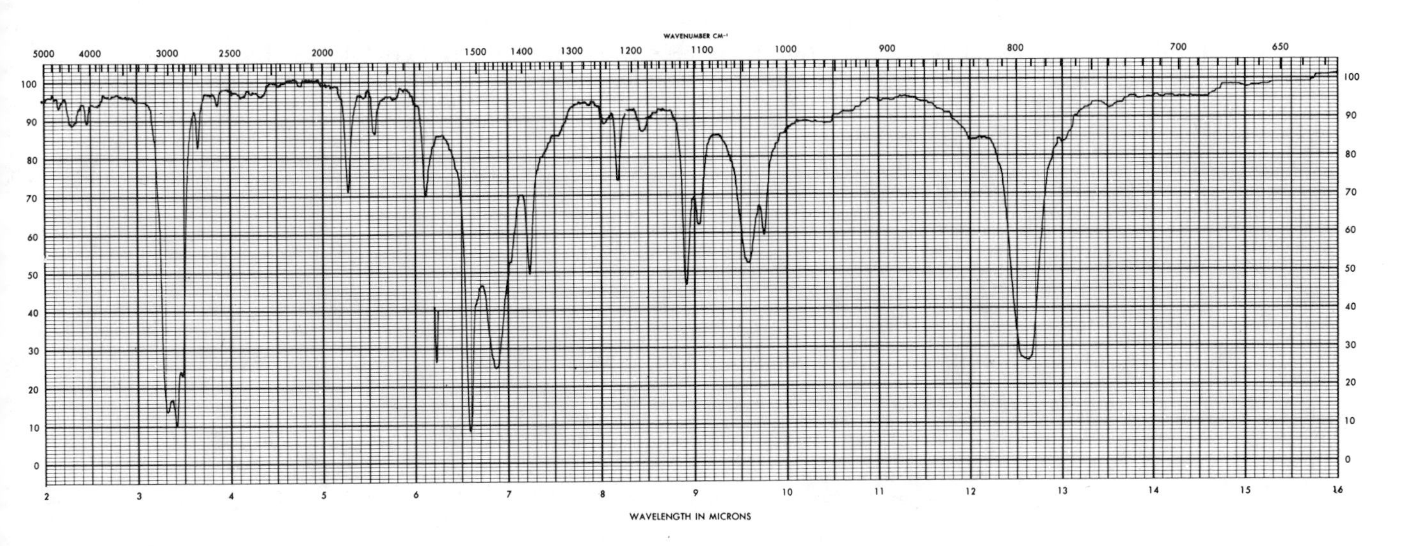

Fig. 3.43 Infrared spectrum of an unknown substance with molecular formula C_8H_{10}.

Problem 3.15. Deduce the structure of the compound whose IR spectrum is recorded in Fig. 3.44, given that the molecular formula of the unknown is $C_8H_8O_2$ and that its boiling point is 249.5°.

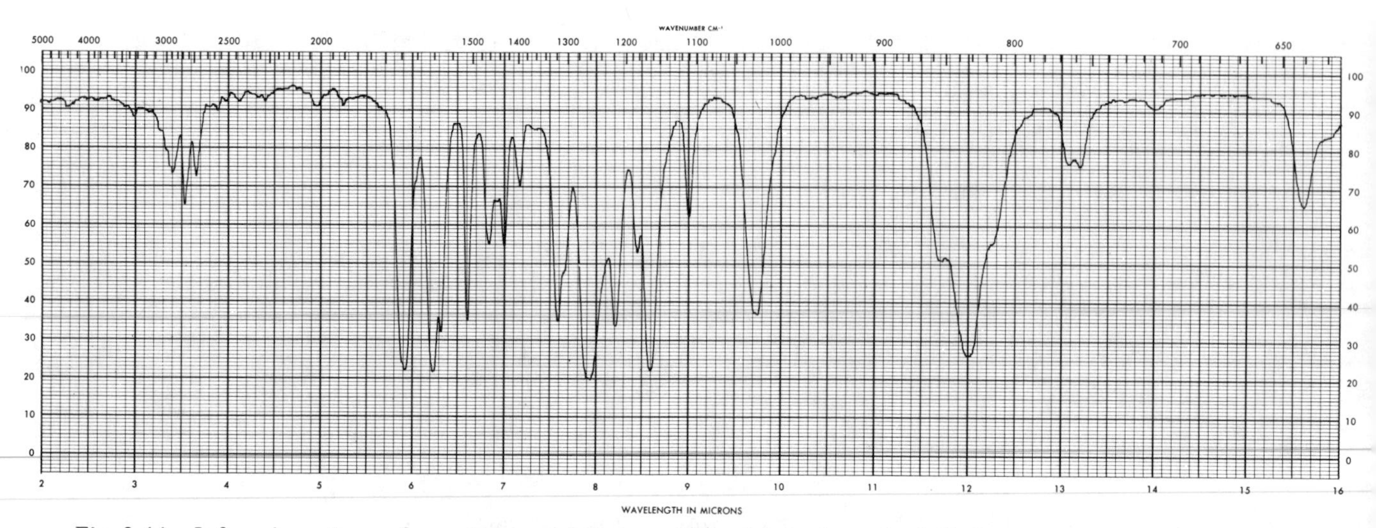

Fig. 3.44 Infrared spectrum of an unknown substance with molecular formula $C_8H_8O_2$.

Problem 3.16. An organic, colorless liquid with a molecular weight of 89.09 and a boiling point of 131° is found to contain C, H, O, and N. The IR spectrum of the pure liquid shows the following characteristics. (All values are in cm^{-1}.)
2950 (m), 1550 (s), 1460 (m), 1438 (m), 1382 (s), 1230 (m), 1130 (w), 896 (w), 872 (w), 800 (s).

Deduce the structure of this compound.

Problem 3.17. The IR spectrum of a yellow crystalline solid (melting point 148°) is determined in the solid state in the form of a KBr pellet. The following important peaks were noted. (All values are in cm^{-1}.)
3500 (m), 3380 (s), 3230 (m), 3100 (w), 1930 (w), 1630 (s, sh), 1600–1590 (s, br), 1480 (s), 1450 (m), 1300 (s, br), 1175 (m), 1008 (s), 840 (s), 750 (s).

Identify the compound.

Problem 3.18. A colorless liquid with the molecular formula $C_6H_{11}BrO_2$ has a boiling point of 177.5°. The IR spectrum of the pure liquid displays the following absorption maxima. (All values are in cm^{-1}.)
2960 (m), 2940 (w), 2880 (w), 1752 (s), 1480 (m), 1380 (m), 1280 (m), 1230 (m), 1170 (s), 1040 (m), 625 (s).

Identify the compound.

FOUR

NUCLEAR MAGNETIC RESONANCE SPECTROSCOPY

4.1 INTRODUCTION

In our study so far we have seen that molecules absorb ultraviolet-visible radiation ($\lambda = 200$ to 800 nm) and infrared radiation ($\lambda = 2000$ to 16,000 nm or 2 to 16 μ). The shorter, more energetic UV waves affect the energy levels of the electrons, whereas the longer, less energetic, IR waves generate mechanical oscillations—e.g., rotation, deformation, and vibration—in the molecule. In NMR spectroscopy we shall be using radiation of very long wavelength ($\lambda = 10^7$ to 10^8 microns, that is, 1000 to 10,000 cm) or extremely low energy: *radiofrequency waves.* These waves naturally are not energetic enough to affect the electronic, vibrational, or rotational energy levels of the molecule. However, they are able to interact with the *nuclei* of certain atoms exposed to a strong magnetic field; hence the name *nuclear magnetic resonance* spectroscopy.

Problem 4.1. Use the λ values given above to calculate the frequency range and energy range (in units of eV and cal/mole) for each kind of radiation: ultraviolet, infrared, radiofrequency, etc.

4.2 MAGNETIC AND NONMAGNETIC NUCLEI

Atomic nuclei contain nucleonic particles such as protons, neutrons, etc. Protons are positively charged, which means that atomic nuclei are positively charged also. Furthermore, like electrons, nucleonic particles also spin on their own axes. Consequently they possess *quantized spin angular momentum* or *spin quantum number.* Individual protons and neutrons have spin quantum numbers of $+\frac{1}{2}$ and $-\frac{1}{2}$ only.

As a result of the angular momentum of the nucleons, most—though not all—nuclei also possess a spin. The total nuclear spin quantum number, I, is a characteristic constant of a nucleus; this constant depends on the number of nucleons and on the symmetry of charge distribution. In general, the following rules apply to nuclear spins.

i) Nuclei which have an odd mass number, and thus an odd number of nucleons, have half-intregal spins such as $\frac{1}{2}, \frac{3}{2}, \frac{5}{2} \ldots$
ii) Nuclei which have odd numbers of protons *and* neutrons (and thus even mass numbers) have integral spins such as 1, 2, 3, . . .
iii) Nuclei which have even numbers of protons and neutrons (and thus even mass numbers) always have zero spin, presumably because there is pairing of oppositely directed spins in the respective nucleons.

Table 4.1 gives values of I for some nuclei of the elements commonly found in organic compounds.

Problem 4.2. Classify the following nuclei according to their spin quantum numbers: spin = half integral, full integral, or zero.

^{23}Na, ^{24}Na, ^{15}N, ^{10}B, ^{17}O, ^{27}Al, ^{24}Mg, ^{35}Cl

---

Let us examine Table 4.1 and note a few facts.

i) Nuclei of phosphorus (P^{31}) as well as protons (H^1) all have the same spin, and thus the same angular momentum, although their masses are different.
ii) Nuclei of sulfur (S^{32}), oxygen (O^{16}), and carbon (C^{12}) all have zero spin; that is, they have no angular momentum.
iii) Spinning of positively charged nuclei causes the distributed positive charge to rotate, giving rise to the equivalent of a current flowing in a circular path. This in turn produces a magnetic field, with a magnetic moment μ directed along the axis of the spin (Fig. 4.1).
iv) Naturally, nuclei such as H^1 and F^{19}, which have spin quantum number $I > 0$, behave like tiny bar

Table 4.1 Some common nuclei and their spins

Nucleus	No. protons, Z	No. neutrons, n	Mass number $Z + n = A$	Spin, I
Hydrogen	1 (odd)	0	1 (odd)	$\frac{1}{2}$
Deuterium	1 (odd)	1 (odd)	2 (even)	1
Tritium	1 (odd)	2 (even)	3 (odd)	$\frac{1}{2}$
Boron	5 (odd)	5 (odd)	10 (even)	3
Boron	5 (odd)	6 (even)	11 (odd)	$\frac{3}{2}$
Carbon	6 (even)	6 (even)	12 (even)	0
Carbon	6 (even)	7 (odd)	13 (odd)	$\frac{1}{2}$
Nitrogen	7 (odd)	7 (odd)	14 (even)	1
Oxygen	8 (even)	8 (even)	16 (even)	0
Oxygen	8 (even)	9 (odd)	17 (odd)	$\frac{5}{2}$
Fluorine	9 (odd)	10 (even)	19 (odd)	$\frac{1}{2}$
Phosphorus	15 (odd)	16 (even)	31 (odd)	$\frac{1}{2}$
Sulfur	16 (even)	16 (even)	32 (even)	0

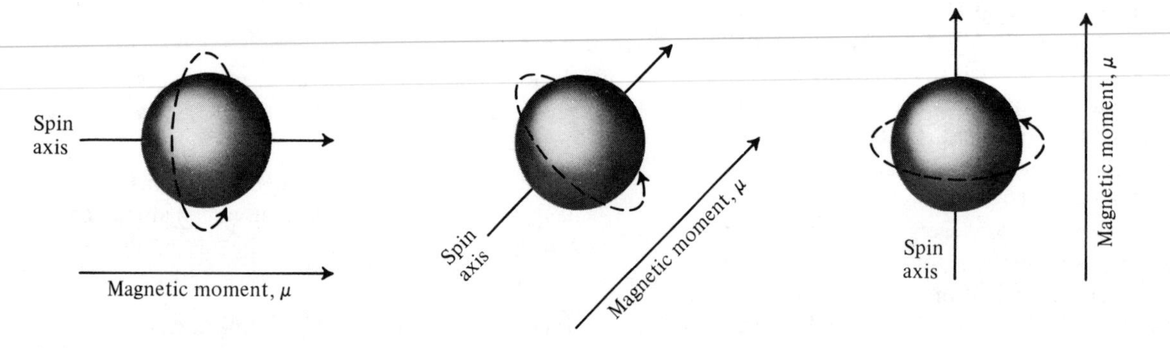

Fig. 4.1 Magnetic moment μ produced by spinning nuclei.

magnets, while those with zero spin ($I = 0$), for example, C^{12}, O^{16}, and S^{32}, are nonmagnetic.

4.3 NUCLEAR PRECESSION

A magnet, when suspended in a magnetic field, comes to rest pointing parallel to the field in which it is situated. Should it be disturbed a little, it oscillates. The frequency of this oscillation depends on the strength of the magnetic field* and on the nature of the magnet itself.

Similarly, a nucleus, which is itself a tiny magnet, should, when placed in an external magnetic field, line itself up with that field. It should also oscillate if disturbed. But here the situation is somewhat complicated. Imagine that the nucleus is spinning in such a manner as to produce a magnetic moment μ (Fig. 4.2). Let us now place our nucleus in a powerful magnetic field, H_0, whose direction is neither exactly parallel or exactly antiparallel to the axis of the nuclear magnet, but makes an angle θ with it. The external field then exerts a torque (i.e., a force at right angles to the magnetic moment) on the nucleus such that the nucleus now precesses around the direction of the applied field, just as a spinning gyroscopic top precesses when tilted with respect to the lines of force of the earth's gravitational field.† The angular frequency (ω_L) of this precession (also known as the *Larmor precession frequency*), or oscillations as in case of a magnetic needle, depends on the strength of the applied magnetic field and on the nature of the nuclear "magnet." Magnetic nuclei of different atoms, therefore, have different *characteristic precession frequencies,* and an instrument which enables us to detect these frequencies thus enables

* A bar magnet (magnetic dipole) will change the orientation of another magnetic dipole placed in its vicinity, and is therefore said to give rise to a *magnetic force.* The region of space in which a magnetic force exists is called a *magnetic field.* A pole which exerts a unit of attractive or repulsive force on another equal pole placed at a unit distance from it in a vacuum is known as a *unit magnetic pole.* The *intensity of the field* at a given point in the magnetic field of a magnet is the force exerted on a unit pole at that point by the magnet, and is expressed in units of *gauss.*

† Actually, the torque should tend to tip μ toward H_0, thus changing θ, but because the nucleus is spinning, μ precesses around H_0 at the same angle θ.

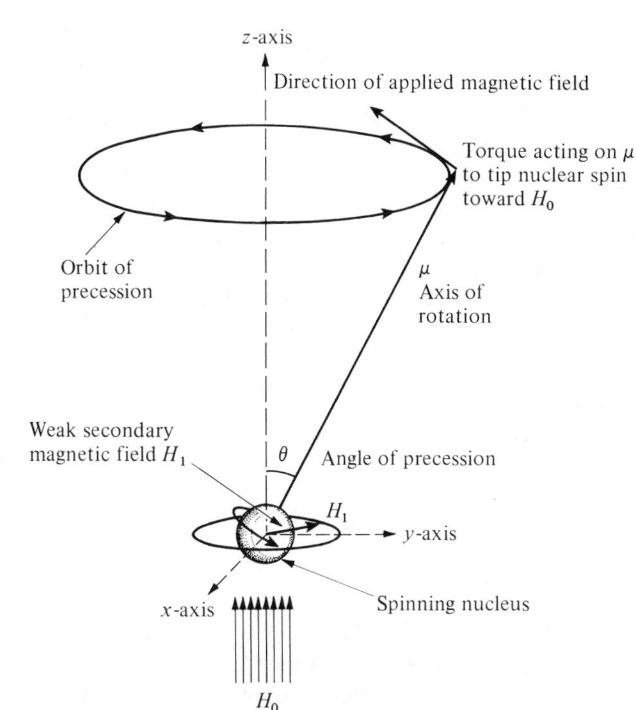

Fig. 4.2 A magnetic nucleus precesses when placed in an external uniform magnetic field (H_0). A set of three coordinates, x, y, z, are shown to originate from the center of the nucleus. The z-axis is parallel to H_0, and makes an angle θ with μ, the magnetic moment or spin axis of the nucleus. Note the weak secondary magnetic field which must rotate in the same plane as the magnetic moment μ, as well as at right angles to H_0, in order to supply necessary resonance energy. This can be achieved by using a plane-polarized radiofrequency signal. When the frequency ν of H_1 equals the linear frequency ν_L of precession, resonance occurs.

us to detect the identities of the corresponding nuclei too! With the help of classical electrodynamics, we can express the above statement in simple mathematical terms as:

$$\omega_L = H_0 \cdot \gamma, \tag{1}$$

where ω_L (Greek omega) is the Larmor angular frequency of precession, H_0 is the applied magnetic field, and γ (Greek gamma) describes the nature of the given nucleus. γ is a ratio of the magnetic moment μ of the nucleus to the angular momentum of the nucleus. It is therefore called the *gyromagnetic ratio.*

The magnitude of the angular frequency ω_L of a motion can be converted to that of its linear frequency ν_L by the following formula:

angular frequency = 2π x linear frequency.

Therefore $\omega_L = 2\pi \times \nu_L$

(Larmor angular frequency) (Larmor linear frequency)

Thus:

$\omega_L = H_0 \cdot \gamma$

or

$2\pi \times \nu_L = H_0 \cdot \gamma$

or

$\nu_L = H_0 \cdot \gamma / 2\pi$

and

$$\gamma = \frac{2\pi \cdot \nu_L}{H_0}. \tag{2}$$

According to Eq. (2), γ is a function of the ratio of the linear frequency of precession to the applied magnetic field.

Just now we compared a spinning nucleus with a spinning gyroscopic top, but the behavior of these two is not quite similar. A gyroscope can spin with any velocity, and its angular velocity can be varied continuously. Also, there are no restrictions on the angle θ that a top may precess at, with respect to the earth's gravitational field. In the case of microbodies like nuclei, however, the restrictions imposed by quantum mechanics must be taken into consideration (Section 1.9). In terms of quantum theory, a precessing nucleus is allowed only certain orientations with respect to H_0. We can calculate the number of these orientations by an easy formula: $2I + 1$, where I is the spin quantum number of the nucleus (Table 4.1). Thus, for a hydrogen nucleus with $I = \frac{1}{2}$, only two precessional orientations are possible. One has a component magnetic moment *parallel* to H_0 and the other a component magnetic moment *antiparallel* to H_0 (Fig. 4.3). The angles θ_1 and θ_2 of these orientations are calculated to be 54°24′ and 125°36′. When the magnetic nuclei assume the allowed orientations relative to the applied field, the energies of the nuclei are either raised or lowered by a value equal to $\mu \times H_0$, depending on their orientations. Those nuclei which are aligned against the

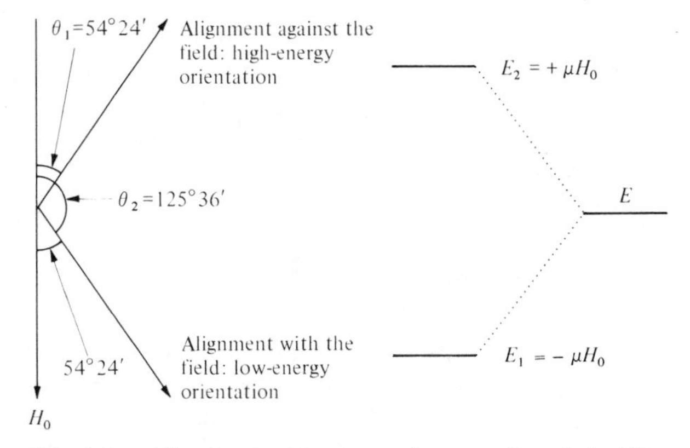

Fig. 4.3 Alignment and energy changes of nuclei with spin $\frac{1}{2}$ in applied magnetic field.

field are in the high-energy state. The energy difference ΔE (that is, the amount of energy required for the transition of protons from spin state E_1 to E_2) between the two states is

$$\Delta E = 2\mu H_0.$$

In general, for a nucleus with spin quantum number I, the difference in energy

$$\Delta E = \mu H_0/I.$$

4.4 NUCLEAR RESONANCE

Earlier we saw that detecting the precession frequency of the nuclear "magnets" serves to identify the nuclei. But how do we go about detecting these frequencies? This is where the principle of nuclear magnetic resonance comes in.

A common example of resonance in everyday life is a radio. Different radio stations send out radio waves of different frequencies, and the air is full of a large number of these signals. In your radio there is a mechanism capable of oscillation. The frequency of this oscillation can be gradually varied by means of the dial. When you are turning the dial (tuning), you are in effect gradually changing the frequency of oscillation of your set. When the value of this oscillation becomes exactly the same as that of the waves transmitted by a nearby radio station, your set responds to the incoming frequency by absorbing energy from the incoming waves. This phenomenon of energy absorption by an oscillating medium, when there is a correspondence between the frequency of the incoming signal and that of the medium, is known as *resonance.* Although the atmosphere is full of numerous frequencies of radio waves, your radio set, when tuned to a particular frequency, is selectively absorbing the waves transmitted by that particular station only.

In the case of our precessing nuclear "magnets," a similar resonance phenomenon can be observed when these "magnets" are exposed to electromagnetic radiation. For example, we have already seen that in the applied magnetic field H_0 the nuclei of hydrogen ($^1H_1$) are precessing in two orientations, which differ in energy by $\Delta E = 2\mu H_0$ (Fig. 4.3). If energy equal to $2\mu H_0$ is supplied to this system, a hydrogen nucleus undergoes transition from one spin state to another. This energy for transition can be supplied by electromagnetic radiation in the radio-frequency range. If the magnetic field H_1 of the radiation is oscillating in a plane at right angles to the applied field H_0 (Fig. 4.2) and if the frequency of radiation ν is the same as the Larmor frequency of precession ν_L of the nuclei, transfer of energy from the radiation to the nuclei can take place. This phenomenon of the transfer of energy is called *resonance.*

The energy of radio waves is given by the Einstein–Planck equation as:

$$E = h \times \nu$$

When resonance takes place, this energy is absorbed to bring about the nuclear transitions. The energy of transition ΔE for a nucleus of spin quantum number I is given by Eq. (3) as:

$$E = \mu H_0/I.$$

Therefore, at resonance,

$$\underset{\text{(energy of radiation)}}{h\nu} = \underset{\text{(transition energy)}}{\mu H_0/I}$$

or

$$\nu = \frac{\mu}{I \cdot h} \cdot H_0.$$

We can therefore place the nuclear "magnets" in a powerful magnetic field H_0 so that they will precess, and then expose the system to radio waves whose frequency we shall deliberately vary gradually (in a manner similar to turning the dial on the radio) through the precession frequency of the nuclei. When the two frequencies (ν_L and ν) match, there will be an exchange of energy between the rotating field H_1 and the precessing nucleus, resulting in the magnetic moment of the nucleus being reoriented opposite to the direction H_0. Knowing the frequency of the radio waves absorbed thus gives us information about the magnetic properties of the nuclei. In practice, however, due to technical problems, the procedure employed is slightly different from the one we have just discussed.

Problem 4.3. The magnetic moment of an electron is expressed in terms of the Bohr magneton, β_0. Similarly, for the nuclei, the magnetic moments, μ_N, are expressed in units of the nuclear magneton, μ_0, which is defined as

$$\mu_0 = \frac{eh}{4\,Mc} = 5.05 \times 10^{-24} \text{ ergs/gauss},$$

where e and M are the charge and the mass, respectively, of the proton. The magnetic moment μ_N of the nucleus of an atom is a multiple of the nuclear magneton. Its value is given in units of μ_0. Magnetic moments of ^1H, ^{13}C, and ^{19}F are 2.793, 0.7022, and 2.627 μ_0 units, respectively. Calculate the frequencies at which these atoms will resonate in a newly developed 70.5 kG NMR instrument operating at 300 MHz.

Problem 4.4. Calculate the gyromagnetic ratio γ for the three atoms given in Problem 4.3.

4.5 ENERGY ABSORPTION AND RELAXATION

The exchange of energy between radio waves and spinning nuclei involves transitions of nuclei from E_1 to E_2 and also from E_2 to E_1 levels (Fig. 4.3). Transitions take place in both directions because, in the radiofrequency field (very-low-energy field), there is a competing process of thermal agitation in the sample. Because of this thermal agitation, there exists a condition known as the *Boltzmann condition,* due to which the probability of nuclear transition upward $E_1 \rightarrow E_2$ (by absorption of energy) is exactly equal to that of nuclear transition downward $E_2 \rightarrow E_1$ (by emission of energy). If this were absolutely true, no NMR phenomenon would be observed because of the cancellation of equal numbers of nuclei of opposite spin states. Normally, however, the population of nuclei in the lower energy state is slightly greater than that in the higher energy state, and this slight excess population, oriented along the direction of the magnetic field H_0, is directly proportional in magnitude to the precessional field strength.

For example, at room temperature and in a very weak magnetic field, such as the earth's magnetic field (about 0.5 gauss), if we could examine the orientations of some 10 billion protons, we would notice that the population distribution in the two orientations was nearly equal, there being an average of only 1 proton more in the lower energy state than in the higher energy state. If the same sample (containing 10 billion protons) were to be placed in a more powerful magnetic field (about 10,000 gauss) the population in the E_1 state would be about 20,000 protons more than that in the E_2 state. It is this excess of nuclei in the lower energy state which is responsible for the net absorption of the radio waves.

It is natural to assume that the absorption of energy from radio waves boosts the nuclei from the E_1 to the E_2 level and thus reduces the equilibrium ratio of low-energy-state to high-energy-state nuclei. If this happened, the high energy state would soon be *saturated,* that is, no more transitions from E_1 to E_2 levels would occur, power absorption would fall, and consequently the intensity of the NMR signal would diminish or disappear. One may also presume that, like electronic transitions, nuclear transitions also involve eventual return of nuclei from the high-energy excited state to the ground state by emitting energy in the form of radiation. However, the radiative transition in the nuclear process is negligible; yet the nuclei do return to the low energy level through a process called *relaxation.*

For nuclei in liquid or gaseous samples, the process of relaxation (i.e., the deactivation of the nuclei, or their return from the perturbed nuclear spin state to the equilibrium or steady state) occurs by two mechanisms:

i) *spin-lattice,* or longitudinal interaction

ii) *spin-spin,* or transverse interaction.

Spin-lattice interaction leads to the nucleus losing its excess energy to the rest of the molecule. To understand this process, we must recall that molecules have translational degrees of freedom which need both kinetic and potential energy. These degrees of freedom together serve as an energy reservoir, usually called a *lattice.* The energy capacity of the lattice is large compared with the energy of interaction of the nuclear magnetic moments with the magnetic field. Thus, when there is an interaction between the spin system and the lattice, the energy from the nuclei at the E_2 level can be transferred to the components of the lattice and the nuclei can undergo transition to the lower E_1 level. Since energy transferred to the lattice is retained by the molecule as additional translational, rotational, or vibrational energy, the total energy of the system is unchanged and no emission of energy takes place.

Spin-spin interaction results when the excited nucleus is deactivated (or dropped to a lower energy state) by the magnetic field of a neighboring spinning nucleus. These so-called relaxation processes strongly affect the widths and shapes of the resonance lines. The width of an absorption band is inversely proportional to the lifetime of the absorbing species in its excited state, i.e., *the more efficient the relaxation process, the broader the resonance band.* For this reason, if maximum resolution is required, paramagnetic impurities such as oxygen, and also magnetic materials, ions, etc., which improve the efficiency of the relaxation mechanism, must be excluded from the sample. The spin-spin relaxation process is more efficient in solids and in very viscous liquids than in gases and dilute solutions. Consequently, when one uses solids or viscous samples for spectral determination, broad signals are obtained.

In samples containing hydrogen atoms attached directly to nuclei, such as ^{14}N, ^{17}O, B, etc., a further relaxation route is possible. These latter nuclei (with spin $>\frac{1}{2}$) have electric quadrupole moments and thus the fields around them are asymmetric. Such nuclei may relax so rapidly, because of interactions of their quadrupole moments with the electrostatic field, that their resonance lines and those of the hydrogen nuclei attached to these nuclei are extremely broad. Such a process of deactivating the nuclei is called *electric quadrupole relaxation.*

4.6 THE NMR SPECTROSCOPE AND ITS USE

After the foregoing discussion, you may already have guessed that for a successful observation of NMR phenomena, we shall need the following basic components: (i) a magnet, (ii) a radiowave generator, and (iii) a device to detect the absorption of radio waves. These requirements are essentially the same as those in a conventional spectrometer (Section 1.10), except that in an NMR instrument a magnet replaces the light source and the sweep generator replaces the monochromator.

We have seen earlier that the magnetic-resonance condition can be achieved by keeping the nuclei in a fixed magnetic field H_0 and varying the radiofrequency. Alternatively, since the resonance phenomenon depends on the characteristic gyromagnetic ratio γ of each nucleus, and since, by Eq. (2),

$$\gamma = \frac{2\pi\nu^*}{H_0},$$

we have defined γ as a function of the ratio of radiofrequency to field, we can achieve the resonance phenomenon even if we maintain a fixed frequency and vary or sweep the magnetic field over a small range. This of course involves automatic conversion, by the recorder, of magnetic field units (gauss) into frequency units (cycles per second, or hertz). The conversion is easily done by the above formula. Figure 4.4 is a schematic diagram of these three fields in an NMR instrument.

* At resonance the Larmor frequency ν_L is equal to the radiofrequency.

In Section 4.7 we shall see that there is usually a very small difference in the resonance frequencies or the strengths of the resonance fields of various nuclei in a molecule. The nuclei are precessing with a frequency of the order of 10^8 to 10^9 Hz. When we sweep the field with a sweep generator, this precessional frequency changes by only *a few* cycles per second. Thus the resolving power of our detecting system—the NMR spectrometer—must be so great that it can detect even so slight a change as 1 or 2 Hz out of 10^8 or 10^9 Hz.

Magnet

Such a requirement makes it essential that the powerful magnet used in the instrument provide a very stable and homogeneous field, which must also remain constant over a long period of time. For this reason the magnets designed for NMR work are large (up to 15 inches in diameter) and thus are capable of producing strong fields (up to 23,500 gauss for 100 MHz work). However, these magnets have a very small air gap between the poles (about 1.5 to 2 inches). Moreover, there is a mechanism which spins the sample tube between the poles, so that inhomogeneity of the field perpendicular to the direction of spinning can be averaged out. If the magnetic field is not homogeneous, the nuclei in different parts of the sample precess with different frequencies, thereby producing broad absorption signals.

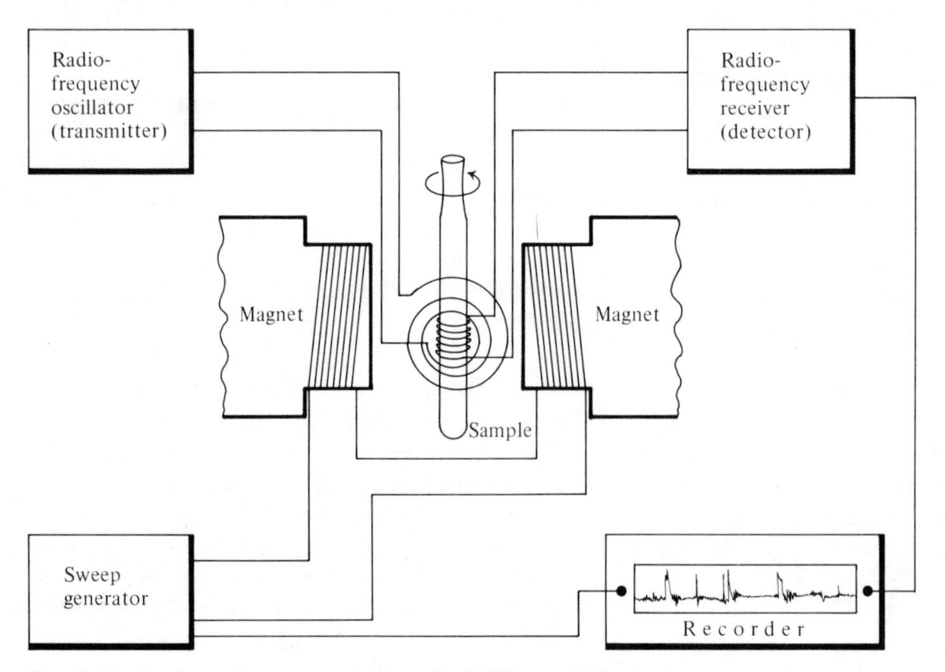

Fig. 4.4 A schematic representation of a NMR spectrometer.

Fundamental limitations in the technology of iron-core electromagnets have in the past limited the extension of the magnetic field to about 23.5 kG. However, superconducting magnet solenoids and improved coils to induce homogeneity of magnetic fields have recently been developed, and commercial instruments operating at fields as high as 70.5 kG (300 MHz) are available.[1] Developments in superconducting materials offer limitless possibilities for extension of magnetic fields, and thus we may expect in the near future to be able to derive considerable information from much smaller samples and more dilute solutions.

Radiofrequency oscillator and sweep generator

The rf oscillator coil is installed perpendicular to the magnetic field, and transmits radio waves of some fixed frequency, such as 60, 100, 220, or 300 MHz (megahertz).

Since the large magnet as well as the rf oscillator both produce fixed fields, a sweep generator is installed to supply a variable dc current to a secondary smaller magnet. This allows us to vary (or sweep) the total applied magnetic field over a small range, and thus to change the precession frequency to that of the rf field. Usually the rate at which the field is swept is of the order of 5 to 10 milligauss per minute. For a typical instrument with a 14,000-gauss magnet, this rate comes out to be about 11 to 12 Hz. If the field is swept too rapidly a characteristic wiggle or ringing pattern is observed on the spectrum.

Detector and Recorder

The coil of the rf receiver or detector is installed perpendicular to both the magnetic field and the oscillator coil, and is tuned to the same frequency as the transmitter. When the precession frequency is matched with the radiofrequency, the nuclei induce an emf (electromagnetic field) in the detector coil by virtue of the change in magnetic flux following nuclear flip-over. This signal is amplified and sent to a recorder.

The recorder gives a spectrum as a plot of the strength of the resonance signal on the y-axis (vertical) versus the strength of the magnetic field on the x-axis (horizontal). The strength of the resonance signal is directly proportional to the number of nuclei resonating at that particular field strength. The area of the peak is therefore a direct measure of the number of resonating nuclei, and for this reason most instruments are equipped with automatic integrators which can record peak areas in the form of a superimposed integration trace on the chart. As far as the x-axis is concerned, accurate measurements of the strengths of continuously sweeping, magnetic fields present a considerable problem. For this reason it is difficult to assign a peak position on an absolute scale, and thus the system employed is to record the position of a peak in relation to the position of an arbitrary standard line. The internal standard used to locate the resonance frequency of most protons is tetramethylsilane (TMS) or $(CH_3)_4Si$, which one adds to the sample before recording the spectrum. We shall later see the nature and purpose of this internal reference.

Sample

To determine the resonance spectrum of the protons of an organic compound, one needs anywhere between 1 and about 30 mg of the sample.* The sample is normally used in the form of its dilute solution (about 2 to 10%) in a solvent which contains no hydrogen atoms of its own. For samples of low polarity, carbon tetrachloride, deuterated chloroform, $CDCl_3$, and deuterated benzene, C_6D_6, are often used. On the other hand, if the sample is soluble only in polar solvents, deuterium oxide, D_2O, acetone-D_6, or dimethyl sulfoxide-D_6 are often employed. TMS cannot be used as an internal reference for substances dissolved in D_2O because TMS is insoluble in D_2O. For aqueous solutions, the standard used is often the sodium salt of 2,2-dimethyl-2-silapentane-5-sulfonic acid, or DSS. The methyl groups in this compound provide a suitable reference peak.

The amount of solution required for NMR work is also very small (0.5 to 1.0 ml). It is placed in a narrow sample tube (OD about 5 mm; ID about 4 mm). A few drops of the material being used for internal reference are then mixed with the solution, and the tube is placed in the

---

* If the quantity of sample available is very small, the electronic noise of the machine may interfere with the weak signal from the nuclei in the sample. This poses a real problem, but it can be overcome by techniques known as *computer averaging of transients* (*CAT*) or Fourier transform. In this technique, the machine scans the resonance region many times and the information from each scan is stored in a computer. The sample signals, appearing at the same frequency each time, add up to a large value, whereas the noise signal, which is random in nature, averages out. The obvious drawback of this procedure is the time required to accumulate several hundred scans. Ernst and Anderson[2a] have developed a technique in which the rf field is applied to the sample as a pulse of 0.1 μs only, and the information obtained from the decay tails of the pulse signals is stored in the computer. This technique permits nearly 500 scans to be made in the time required to run one ordinary spectrum. This method[2b,2c] for improving the signal-to-noise ratio of NMR spectra will probably be much developed in the near future, using Fourier transform treatment for extracting the pulsed data. Such developments will be much appreciated, especially for ^{13}C operations, as the signals generated by these nuclei are very weak.

Good results have also been achieved by using spherical micro cells.[2d] With this technique, only 0.03 to 0.05 ml of solution is sufficient.

narrow gap between the poles of the magnet, where it is spun by an air-driven turbine. In special cases, particularly in biochemical work, large quantities of a sample may be required. Consequently, a low-frequency NMR spectrometer has been developed,[3] capable of holding samples up to 1.25 liters, or the body of a whole animal.

4.7 DISSIMILAR RESONANCES OF SIMILAR ISOTOPES (CHEMICAL SHIFT)

By now you are quite familiar with the expression

$$\nu = \frac{\mu}{Ih} \cdot H_0,$$

and naturally you may also be under the impression that since μ and I are fixed for all the nuclei of a particular isotope, they must all precess at the same frequency ν when a sample containing these nuclei is placed in a uniformly stable magnetic field. For example, a molecule of ethyl alcohol, C_2H_5OH, contains six hydrogen nuclei, all of which have the same nuclear spin,

$$I = \tfrac{1}{2}$$

and the same magnetic moment

$$\mu = 1.41 \times 10^{-23} \text{ erg/gauss.}$$

Therefore you might expect that if a sample of ethyl alcohol were to be placed in a strong, uniform field of, say, 14,000 gauss, all six protons would precess with the same frequency, and resonate at the same frequency:

$$\nu = \frac{\mu}{I \cdot h} \cdot H_0 = \frac{1.41 \times 10^{-23} \times 14{,}000}{\frac{1}{2} \times 6.624 \times 10^{-27}}$$
$$= 60 \times 10^6 \text{ hertz or 60 megahertz (MHz).}$$

* *Note*: Unless stated otherwise, all NMR spectra reproduced here were recorded on a Varian A-60 instrument, using either $CDCl_3$ (internal reference TMS) or D_2O (internal reference DSS) as solvents. The sweep width used is 500 cps.

Problem 4.5. What would be the precession frequency of protons exposed to a more powerful magnet of 23,500 gauss?

Problem 4.6. Fluorine nuclei, ^{19}F, also have spin $I = \frac{1}{2}$ and precess at 56.8 MHz when placed in 14.0 kG magnetic field. Calculate the nuclear magnetic moment, μ_N, for the fluorine nuclei.

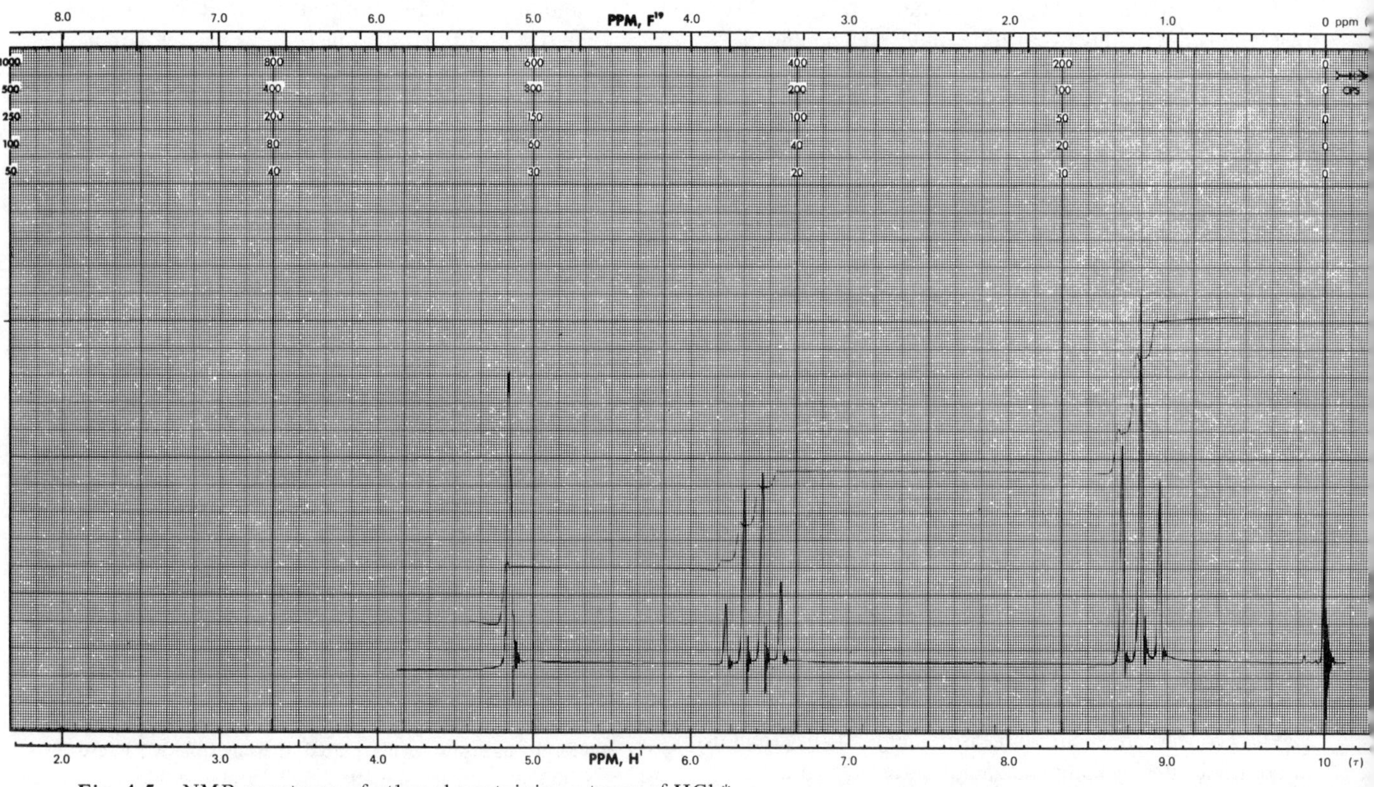

Fig. 4.5 NMR spectrum of ethanol containing a trace of HCl.*

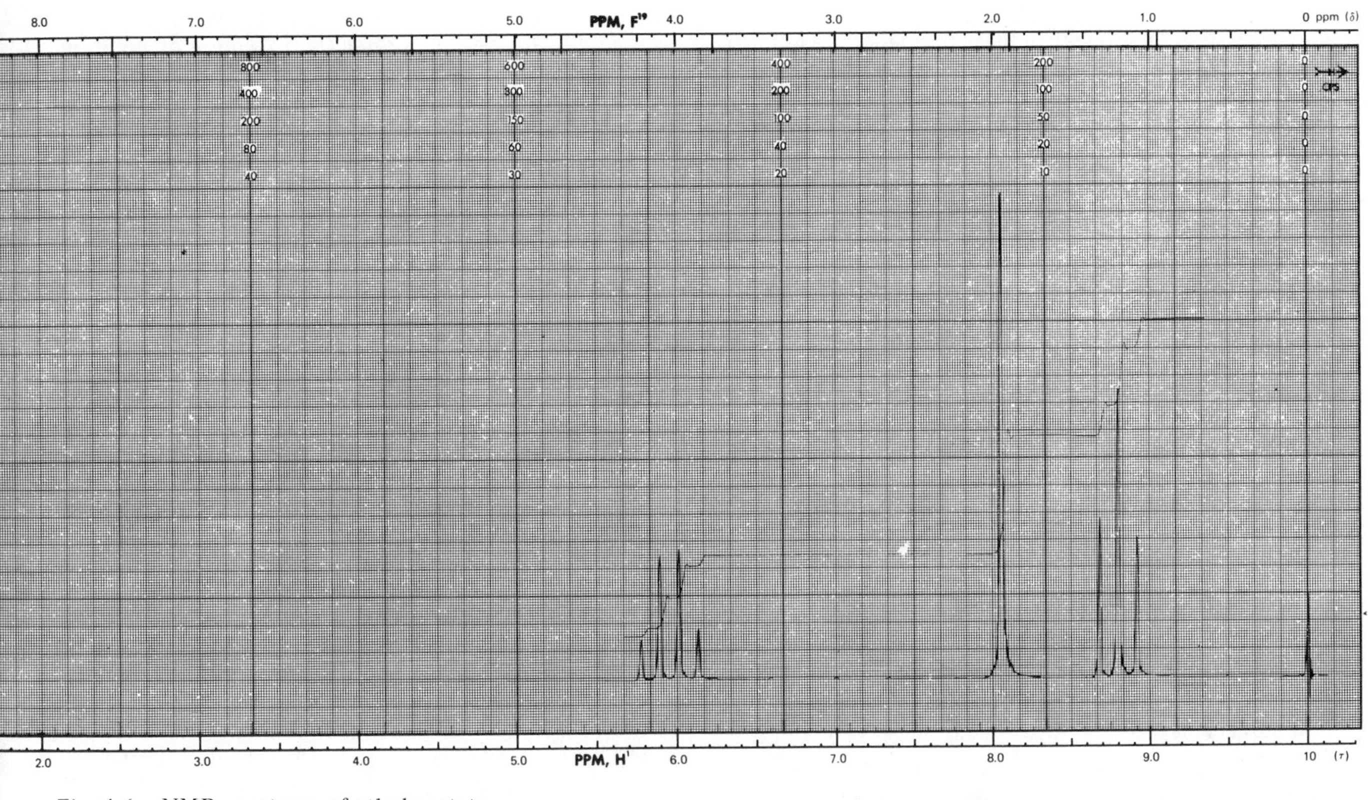

Fig. 4.6 NMR spectrum of ethyl acetate.

But this similarity of precession and of resonance does not occur. Instead of getting one sharp NMR signal representing the six protons, we get a somewhat complex spectrum, shown in Fig. 4.5. Similarly, Fig. 4.6 shows the complex spectrum obtained for the eight protons (hydrogen nuclei) of ethyl acetate:

$$CH_3C\begin{matrix}\nearrow O \\ \searrow OC_2H_5\end{matrix}$$

Here again, some protons resonate at a lower magnetic field strength than others. Note that, in both cases, some three groups of signals appear on the spectrum.

Let us find out why all the protons do not resonate at the same magnetic field strength. As a matter of fact, if all covalently bound protons resonated at identically the same field strength, an organic chemist would have practically no use for NMR spectroscopy because then he could obtain no structural information from it. However, in 1951, when Arnold, Dharmatti and Packard[4] showed that the protons of ethanol resonated with three peaks, it immediately became clear that NMR, discovered only five years earlier, would be very important to organic chemists. Basically, the reason that the hydrogen nuclei in ethanol exhibit three peaks lies in the electronic environment in which these protons exist. *Nuclei in different surroundings differ in their resonance frequencies.* These differences (in peak positions for nuclei which are of the same kind, but located in different electronic environments), from some arbitrarily chosen standard are called *chemical shifts.*

4.8 CHEMICAL SHIFTS AND DIAMAGNETISM

Before we discuss the reasons for chemical shifts, we must understand the terms *diamagnetic* and *paramagnetic.* Molecules of some substances have all their electrons paired, whereas in some other subtances, unpaired electrons may exist. For example, each molecule of NO_2 contains 17 valence electrons, of which 16 are paired and 1, on nitrogen, is unpaired. Electrons in any molecule are spinning at the same time as they are traveling in closed orbits around the nuclei. According to completely classical considerations, both these motions (spinning and orbital) of charged electrons would produce magnetic moments.

In atoms and molecules having paired electrons only, the magnetic moments produced by oppositely spinning electrons balance one another out, so that there is no net magnetic moment; that is, these substances are not *permanently magnetic.* However, when these substances are placed in magnetic fields, the planes of their orbitals are tipped slightly (depending on the strength of the external field) and a small net orbital moment is produced in opposition to the applied magnetic field. As a result of this opposition, these substances are repelled from the applied magnetic field. Such an induced magnetic field, which is in opposition to the applied magnetic field, is called a *diamagnetic field.* Such substances (i.e., those which have paired electrons only) which are repelled by the applied field are called *diamagnetic substances.* The magnitude of the diamagnetic field is directly proportional to that of the applied field.

On the other hand, substances such as NO_2, Mn(II), Cu(II), etc., which contain one or more *unpaired electrons* have a *permanent* magnetic moment because the spin and angular moments of the unpaired electrons are not balanced out. When these substances are placed in an external magnetic field, these permanent atomic and molecular moments align themselves with the field and consequently such substances reinforce the external field. Such a permanent magnetic field, which aligns itself with the applied magnetic field, is called a *paramagnetic field.* Substances which have one or more unpaired electrons, which thus reinforce an applied magnetic field, are called *paramagnetic substances.*

Thus, briefly, a field set up by a given substance *in opposition* to the applied magnetic field is called *diamagnetic,* whereas one which *aligns itself with* the applied field is called *paramagnetic.*

In organic substances, the molecular orbitals usually occupy a large part of the molecule. We may therefore crudely imagine that the component nuclei of a molecule are encased in an electron cloud, containing electrons that are, for the most part, paired. When these molecules are placed in an external magnetic field, the induced electronic circulation produces a diamagnetic field. Since this is opposed to the applied field, you can easily appreciate the fact that, although you are trying to apply an external field of magnitude H_0, the magnitude of the field H *actually experienced* by the nuclei is somewhat less. This protection of nuclei from an external field is called *diamagnetic shielding*; the extent of this shielding depends on the density of the electrons surrounding the nuclei. The shielding parameter σ is defined as follows. The local field H actually experienced by the nuclei is given by:

$$H = H_0 (1 - \sigma).$$

To understand how different protons in a compound are shielded differently by electron clouds, let us look at a simplified molecular orbital (m.o.) picture of a molecule of ethyl alcohol,

```
   H  H
   |  |
H—C—C—O—H
   |② |①
   H  H
```

and try to understand its NMR spectrum (Fig. 4.5). Ethanol has six protons, of which one proton is covalently bonded with oxygen, two protons are covalently bonded with C_1, and three protons are covalently bonded with C_2.

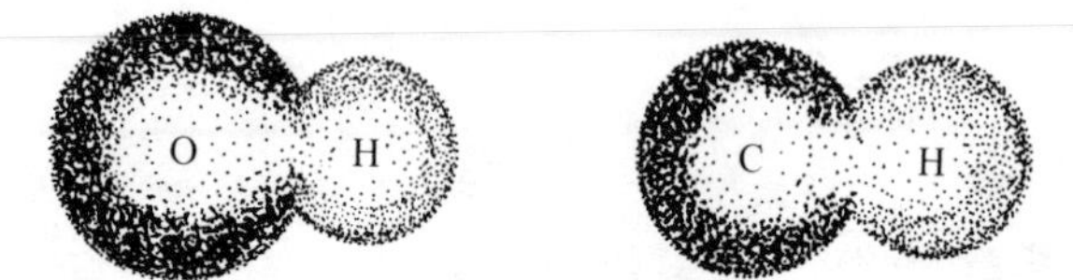

Fig. 4.7 A diagram of a molecular orbital, showing an electron cloud around the hydroxyl proton, which is less dense than that around the methyl or methylene proton.

Oxygen is more electronegative than carbon. This means that if we looked at m.o. pictures of O—H and C—H bonds, they would look roughly like Fig. 4.7, in which the proton attached to the carbon is covered by a denser cloud of electrons than the proton attached to the more electronegative oxygen. Consequently when an alcohol molecule is placed in a uniform magnetic field H_0, the field H experienced by the hydroxyl proton is stronger (due to less shielding) than that experienced by the methyl or methylene protons. It is only natural, then, that as we gradually increase the external field by means of a sweep generator, the less shielded hydroxyl protons resonate first (i.e., at a less-strong field).

Does this mean that the remaining five protons, all of which are attached to carbon, resonate simultaneously?

The fact is that they do not, because, although all of them are bonded to carbon atoms, the two methylene protons bonded to C_1 are slightly *less* shielded than the remaining three methyl protons bonded to C_2. Consider the two carbon atoms: C_2 has its three sp^3 orbitals bonded with hydrogens and the fourth with carbon as

```
   H
   |
H—C—C
   |
   H
```

whereas, in the case of C_1, two of its sp^3 orbitals are bonded with hydrogens, the third with a carbon, and the fourth with the electronegative oxygen. The oxygen naturally pulls the electron cloud of the C—O bond away from C_1, which in *its* turn tries to partially make good the

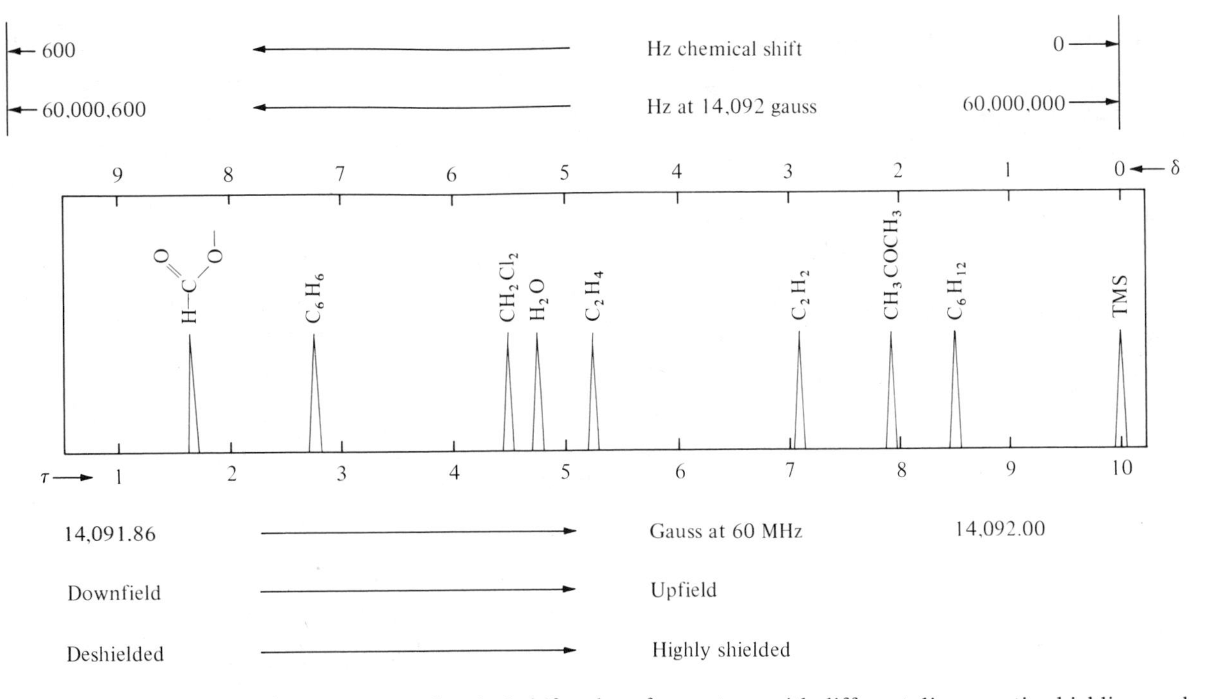

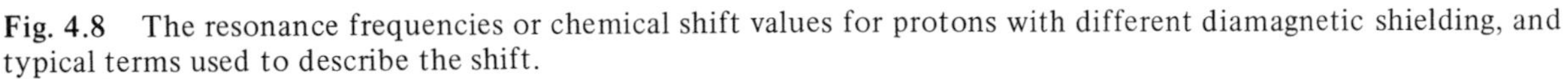

Fig. 4.8 The resonance frequencies or chemical shift values for protons with different diamagnetic shielding, and typical terms used to describe the shift.

loss by pulling the electrons away from the less electronegative hydrogens which are directly attached to it. The decrease in density of the electrons is not experienced as strongly by the protons on C_2. As a result, the two protons at C_1 are somewhat less shielded than the three at C_2, and therefore resonate at a weaker field than the three protons on C_2.

We have thus established, on the basis of inductive effects, a correlation between chemical constitution and resonance of protons. If you now consider the NMR spectrum of ethanol, you will note that three groups of signals appear on the spectrum. On the basis of the integral trace* (the superimposed line indicating the relative numbers of resonating protons) you will notice that the height of the trace jump at the central group of peaks is almost twice that of the smaller peak at the left, and the trace at the extreme right is three times higher. Since the integral traces the peak areas, it is obvious that the three peaks are in the approximate ratio of 1 : 2 : 3, obviously representing 1 proton of —OH, 2 protons of $—CH_2—$, and 3 protons of $—CH_3$, respectively. In a similar fashion we can explain the origin of three sets of peaks in the spectrum of ethyl acetate (Fig. 4.6).

In a conventional NMR spectrometer, when the sweep is started, the recorder pen moves from left to right as the magnetic field gradually increases. The left side on our spectrum (Fig. 4.8) therefore reflects the less-strong field (downfield); this is the side on which peaks from less-shielded or deshielded nuclei appear. The right side is the strong-field (upfield) side, showing the peaks corresponding to the resonance of more shielded nuclei. *Chemical shifts are thus due to diamagnetic shielding effects produced by circulation of both bonding and nonbonding electrons in the neighborhood of the nuclei. The magnitude of the chemical shift in a given compound is directly proportional to the strength of the applied magnetic field H_0.* This is understandable because diamagnetic shielding is the result of the current induced by the electrons in the applied magnetic field. At 60 MHz, the absorption peaks for the majority of hydrogen-containing groupings lie within a range of 600 Hz (Fig. 4.8). It is also worth noting at this stage that *the amplitude of a particular* NMR *signal is proportional not only to the number of equivalent resonating nuclei at that field strength, but is also proportional to the square of the applied magnetic field.* For example, the amplitudes of the signals recorded for a given compound at 60 MHz are all four times as great as those recorded for the same compound at 30 MHz.

* The use and importance of integral trace will be explained in Chapter 6.

4.9 FACTORS AFFECTING CHEMICAL SHIFTS

We have now established a correlation between chemical shifts and chemical constitution, in which the inductive effects due to electronegative atoms (and groups) cause deshielding of nuclei attached to or near them. Inductive effects, and thus also shielding effects, fall off rapidly as the number of intervening bonds increases. It is important, however, to realize that—although in many cases the observed chemical shifts can be explained on the basis of diamagnetic shieldings of protons—such an explanation proves inadequate in many other cases in which intramolecular factors are effective. Such factors are: (i) local deshielding caused by paramagnetic currents about the nuclei, (ii) shielding and deshielding caused by currents in neighboring atoms and groups, (iii) deshielding due to intramolecular van der Waals forces[5] and due to (iv) intramolecular electric fields produced by the dipole moments of functional groups in molecules.[6]

i) *Local paramagnetic currents*

Deshielding due to local paramagnetic currents depends on the existence of low-lying excited states of the magnetic nuclei. In the case of protons, there are no such states, and hence the local paramagnetic effects are negligible in proton resonance spectroscopy. We shall therefore not consider these local currents here. However, bear in mind that these currents are the dominant ones for heavier nuclei.

ii) *Shielding and deshielding caused by currents in neighboring atoms and groups*

Let us consider the resonance of protons in three different compounds: ethane (CH_3—CH_3), ethylene (H_2C=CH_2), and acetylene (H—C≡C—H). On the basis of diamagnetic shielding only, we may expect that the values of the chemical shifts here should follow the order of electronegativity of the carbon to which the hydrogen is attached (electronegativity: $sp > sp^2 > sp^3$). However, the acetylenic hydrogen (attached to the most electronegative sp carbon) has a chemical shift intermediate to that of the methyl protons (attached to the least electronegative sp^3 carbon) on the stronger-field side (extreme right-hand side on an NMR chart) and ethylene protons (attached to the sp^2 carbon) on the weaker-field side (left-hand side on an NMR chart). We may also expect the protons in benzene to resonate at a frequency similar to the frequency of protons in ethylene, since they are all attached to sp^2 carbons only. But, contrary to this expectation, benzene hydrogens appear to be less shielded than ethylenic hydrogens, and consequently resonate at a weaker field strength.

To understand this apparent anomaly, let us carefully reexamine the effects of currents induced by an applied magnetic field. Imagine that we are first considering an isolated hydrogen atom. The proton in this atom, with its associated 1s orbital, has spherical symmetry. The induced shielding *at the nucleus*, therefore, always remains the same no matter how the nucleus orients itself with respect to the applied field; but at a neighboring point *outside the atom* the induced field depends on the orientation of the nucleus. This induced external field averages to zero if there is a large bulk of nuclei capable of having random orientations. Such a case arises, for example, in compounds which are in solution (or in liquid or gaseous states) when the molecules of the solute are free to move and orient themselves in any random pattern. Strictly speaking, this is true only in the case of molecules which are free to move and all of whose electrons have equal freedom to circulate in all directions; that is, in the case of molecules whose electronic environment is *isotropic* (symmetrical). Thus, *in the case of molecules having an isotropic electronic environment, the field arising from local diamagnetic circulations about one nucleus has no effect on the amount of shielding of a neighboring nucleus.*

On the other hand, in those molecules in which the movement of electrons is not completely free, i.e., in which the induced circulation of charge has some preferred orientation with respect to the applied field, the secondary field induced by the circulation of electrons at one point will not average to zero at a neighboring point, but will reinforce or diminish the applied field at that point. Thus, *in the case of molecules whose electronic environment is anisotropic (unsymmetrical), the induced circulation of electrons around one nucleus may cause shielding or deshielding of the neighbouring nucleus.*

We can now readily see why acetylenic protons are more shielded than ethylenic protons, although the former are attached to a more electronegative sp carbon. The acetylene molecule (Fig. 4.9) is linear, and π orbitals of the triple bond are symmetrical about the molecular axis. When acetylenic compounds or acetylene molecules are placed in an external magnetic field, the preferred orientation of the molecule is one in which its axis is parallel to the applied field (because electron currents flow preferentially in a direction perpendicular to the applied field). The π electrons circulate within the axially symmetric orbital, producing a field which is diamagnetic at the place at which the protons are located on the symmetric axis. The acetylenic protons thus experience an anomalously high degree of shielding, although the inductive effect of sp carbon tends to reduce the density of electrons near these protons. A similar explanation also applies to the nitrile grouping C≡N.[7a,7b]

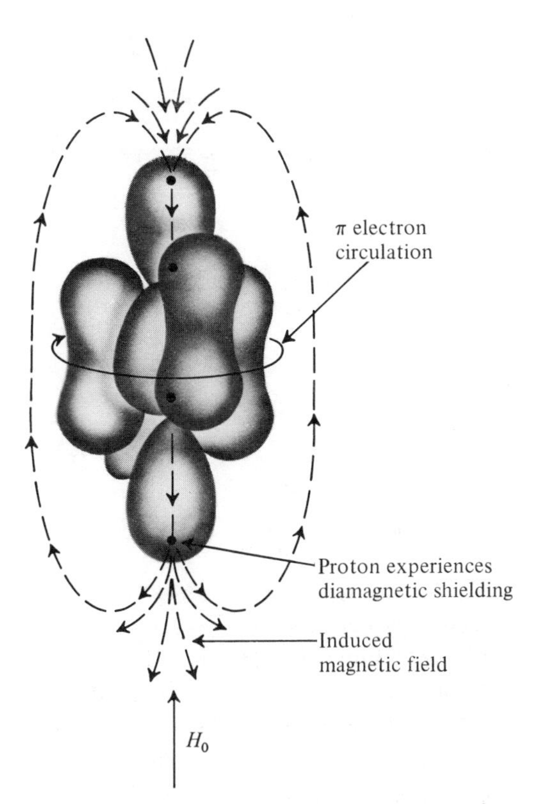

Fig. 4.9 Preferred orientation of the acetylene molecule causes paramagnetic effect at carbons, but diamagnetic shielding of the protons located on the axis. Note that the nuclei in the space surrounding the axis (i.e., in the space on either side of the carbon) are deshielded.

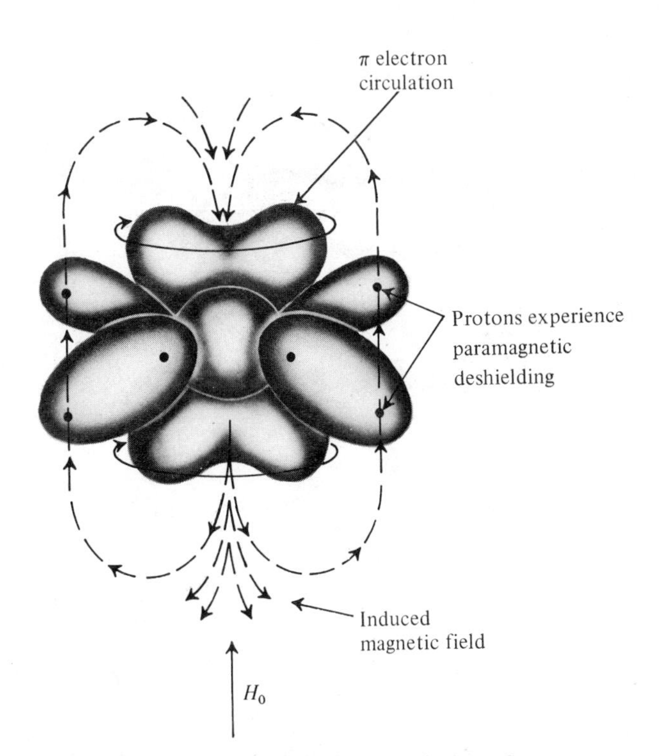

Fig. 4.10 Induced diamagnetic circulation of π electrons in a double bond and the deshielding of ethylenic protons thus caused. Note that the effect is opposite to that experienced by acetylenic protons.

Let us see whether the same argument can be used to explain the fact that aromatic protons of benzene resonate at a less-strong field (and hence are less shielded) than the ethylenic protons. All these protons are bonded to sp^2 carbon, and thus there is no difference in the inductive effects they experience. These molecules contain carbon–carbon π bonds and have an anisotropic electronic environment, but unlike the case for acetylene, the π electron cloud around a double bond is not symmetrical around the molecular axis. In Fig. 4.10 we can see that the π electron cloud of a double bond is concentrated to the left and right of the bond axis in a plane perpendicular to that of the paper (due to the sidewise overlap of the dumbbell-shaped p orbitals). When a molecule containing a double bond, be it a molecule of ethylene or of benzene, is placed in the external magnetic field, it orients itself in such a manner that the electrons circulate perpendicular to the applied field, and thus the π electrons circulate within the π orbital, as shown in the diagram. A secondary induced field, perpendicular rather than parallel to the bond axis, is thus generated, and the ethylenic protons experience a paramagnetic deshielding greater than the deshielding due only to the inductive effect of sp^2 carbon. A similar deshielding would be observed in groups such as

$$>C=O; \qquad >C=N; \qquad >C=S, \qquad \text{etc.}$$

In the case of benzene and other aromatic molecules, a similar paramagnetic deshielding of the protons takes place. However, since π electrons are delocalized, circulation of electrons over a number of atomic centers becomes possible, and so-called *ring currents* are readily developed in the electron cloud loops provided by aromatic nuclei (Fig. 4.11). These currents produce stronger deshielding effects than those produced in simple ethylenic compounds, and hence the aromatic protons resonate at lower (i.e. weaker) fields than simple ethylenic protons.

Those protons which do not lie directly in the field of influence of the ring current are naturally not deshielded. On the contrary, in the space immediately against the face of the benzene ring, a strong induced diamagnetic field exists. Any protons sterically held in this region are therefore abnormally shielded. For

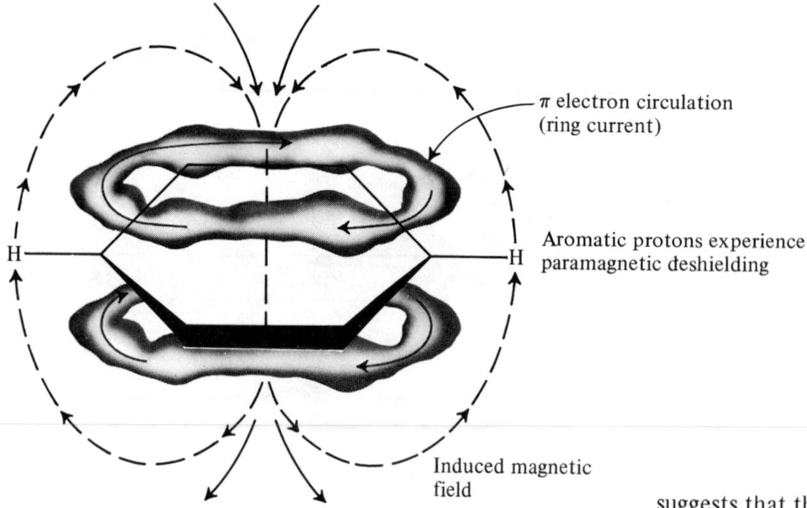

Fig. 4.11 The ring-current effect, causing deshielding of aromatic protons.

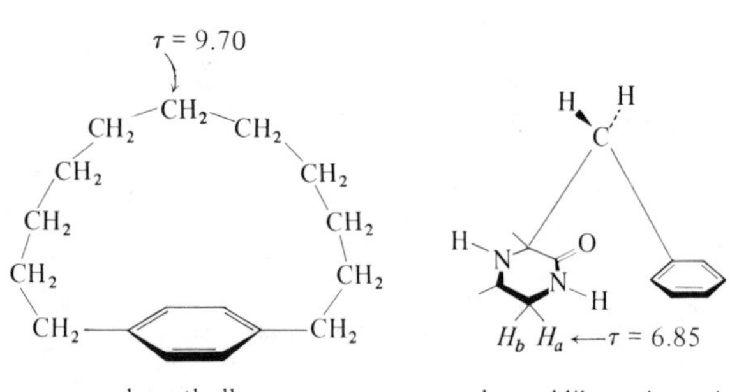

Fig. 4.12 Abnormal shielding of some methylene protons.

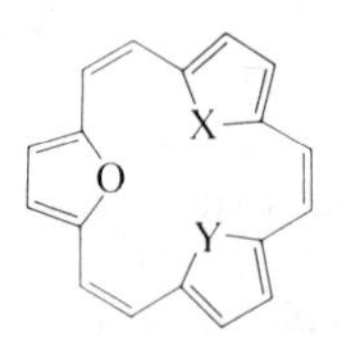

Fig. 4.13

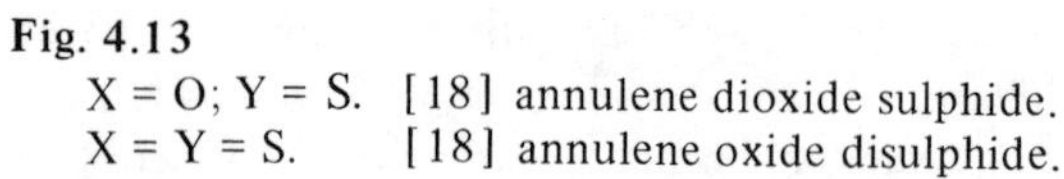
X = O; Y = S. [18] annulene dioxide sulphide.
X = Y = S. [18] annulene oxide disulphide.

example, *p*-polymethylbenzenes[8] (Fig. 4.12) show some methylene signals at abnormally high fields.

An interesting use of this fact is made when one is determining the conformation of monobenzyldiketopiperazine (Fig. 4.12). The observation[9,10] that the methylene proton H_a is more shielded than H_b strongly suggests that the molecule must have a folded conformation.

The concept of magnetically induced ring currents in closed π electron systems is widely used as a criterion of aromaticity. For example, on the basis of low-field chemical shifts, it was shown that [18] annulene dioxide sulphide[11] (Fig. 4.13a) has aromatic character, but [18] annulene oxide disulphide[12] (Fig. 4.13b) is nonaromatic and probably nonpolar.

These *ring-current effects* are used to explain the unusually large chemical shifts even in small-ring, saturated compounds such as cyclopropane.[13]

A carbon–carbon single bond is anisotropic. Therefore the induced secondary field caused by σ electrons at a neighboring point does not average to zero when there are random tumblings of the molecules (Fig. 4.14). This induced secondary field causes shielding or deshielding of the neighboring nuclei. One can observe this in rigid molecules such as cyclohexane, cyclohexanone, etc. In such molecules, axial protons are more shielded than equatorial ones, although the effect is very small. *In all known examples of epimeric cyclic compounds, the axial protons are found to resonate at a slightly higher field than the equatorial protons.* However, some exceptions to this generalization have been reported in the cyclobutane system[14,15] and in some tricyclic sesquiterpenes.[16] Later on we shall see how this small effect can be used to determine the orientations of the ring protons in an unknown structure.

In summary we can say that:

a) Secondary magnetic fields capable of operating through space are induced by circulation of electrons in the molecule.

b) These fields are anisotropic, that is, unsymmetrical about the bond. The phenomenon is therefore called *anisotropic effect.*

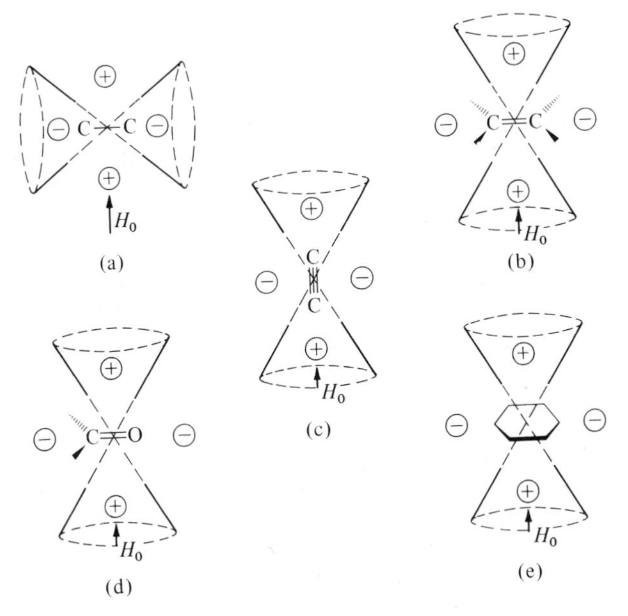

Fig. 4.14 Anisotropic shielding and deshielding zones for: (a) carbon–carbon single bond, (b) carbon–carbon double bond, (c) carbon–carbon triple bond, (d) carbon–oxygen double bond, (e) aromatic molecule. ⊕ Shielding zone; ⊖ deshielding zone.

c) Anisotropic effect is particularly important in molecules containing π bonds; for example, alkenes, alkynes, aromatic systems, and carbonyl groups.

d) Anisotropic effect either reinforces or diminishes the applied magnetic field at any given point in the molecule. Atomic nuclei will therefore experience anisotropic shielding or deshielding depending upon their location within the molecule.

iii) *Deshielding due to intramolecular van der Waals forces*

When atoms which are not bonded to one another, either ionically or covalently, come close to one another, weak forces, called *van der Waals forces,* operate between them, due to attractions between oscillating dipoles in these atoms. The electron clouds in these interacting atoms undergo asymmetrical distortion, causing mutual deshielding of the nuclei. In a sterically congested rigid molecule, therefore, the protons which are subject to van der Waals interactions, due to congestion, resonate at a slightly lower frequency than other similar, but uncongested, protons.

4.10 EFFECT OF HYDROGEN BONDING ON CHEMICAL SHIFTS

In addition to the *intra*molecular shielding factors considered above, some *inter*molecular factors may also cause abnormal shifts. Intermolecular shielding is considered to be the sum of four effects:[17]

$$\sigma^s = \sigma^B + \sigma^W + \sigma^A + \sigma^E$$

where σ^s = total intermolecular shielding,

σ^B = bulk susceptibility contribution (i.e., effect of all the electrons and nuclei of surrounding molecules),

σ^W = effect of the van der Waals forces between solute and solvent,

σ^A = shielding due to the anisotopy of all the neighboring molecules,

σ^E = effects of the distribution of intramolecular charge in the presence of polar molecules.

Among these factors, the most interesting is the effect of the solvent on shielding, particularly with respect to proton exchange and hydrogen bonding.

Hydrogen bonding causes deshielding of protons. The greater the degree of hydrogen bonding of a proton, therefore, the greater the downfield shift (shift to weaker magnetic field) of its resonance. This deshielding is due to the lowering of the effective density of the electrons near the proton by electrostatic attraction. For example, when a X—H molecule hydrogen-bonds with a donor molecule Y to form X—H—Y, the electrostatic field of the hydrogen bond repels the electrons of the X—H bond, as illustrated by the development of partial charges; for example,

$$\overset{\delta-}{X}—H—\overset{\delta+}{O}=C\lt .$$

This reduces the density of the electrons around H and causes deshielding.

4.11 EFFECT OF TEMPERATURE ON CHEMICAL SHIFTS

Usually changes in temperature do not greatly affect proton shifts, but in the case of molecules having intermolecular hydrogen bonds, it is observed that in dilute solutions or at elevated temperatures the proton resonance signal appears at higher fields. This is presumably because inert solvents and high temperatures break up the complexes which are weakly hydrogen bonded. (This is particularly true of the dimers.) Such upfield shifts of hydroxyl protons are observed when one dilutes alcohols, phenols, carboxylic acids, etc., with inert solvents such as carbon tetrachloride. An interesting use of this phenomenon is in studying the disruption of the structure of water by added electrolytes.[18]

Thus we can say in general that *the chemical shifts of protons bonded directly to electronegative atoms (for example, in —OH, —SH, and —NH groups) and thus capable of forming intermolecular hydrogen bonds, show a significant dependence on molecular concentration and on temperature. Protons attached directly to sp^3 carbon show no such dependence, but protons attached to sp^2 and sp carbons show a small upfield shift when solutions containing such protons are diluted.*[19]

Protons attached to electronegative atoms are acidic. The acidic protons sometimes undergo exchange reactions with the solvent as

$$CH_3COOH_{(a)} + HOH_{(b)} \rightarrow CH_3COOH_{(b)} + HOH_{(a)}.$$

NMR spectra of systems in which such exchange reactions occur fairly rapidly display only a *single* peak representing both the protons *a* and *b*. The chemical shift of such a peak is a weighted average of the chemical shifts of the two types of protons, that is,

$$\delta_{obs} = N_a \cdot \delta_a + N_b \cdot \delta_b,$$

where δ_a and δ_b are the chemical shifts of the individual protons (*a*) and (*b*), respectively, N_a and N_b are the mole fractions of (*a*) and (*b*), respectively, and δ_{obs} is the observed chemical shift of the peak representing both the protons.

This phenomenon of proton exchange will be discussed further in Section 4.15 (Fig. 4.25).

Problem 4.7. List briefly the intramolecular and intermolecular factors affecting chemical shifts. On the basis of such a list, can you suggest a relationship between the shift of a proton and its acidity?

4.12 UNITS WITH WHICH TO EXPRESS NMR PEAK POSITIONS

In NMR spectroscopy, we are in reality trying to determine the strength of magnetic field which is required to bring about the resonance of certain nuclei in a given compound. It is natural to expect, therefore, that a NMR spectrum would be a plot of signal strength versus magnetic field, in units of milligauss. Measuring a magnetic field is difficult. However, it is convenient to determine accurately the values of *absorption frequencies of protons relative to those of a suitable standard.* The most logical standard would be a "bare" proton, having no shielding electrons, but experimentally this is not possible, and instead one uses as standards compounds having sharp resonance peaks. The chemical shifts may then be expressed as the difference between the resonance frequency of the protons in the sample (ν_{sample}) and the resonance frequency of the protons in the standard ($\nu_{standard}$).

It has already been mentioned that tetramethylsilane (TMS) is commonly used as an internal reference. Using TMS in the same solution as the compound of interest (i.e., using it as an internal reference) is particularly suitable because:

i) It minimizes the effects of the solvent on the chemical shift, as both sample, reference, and sample tube are subject to the same effects (bulk diamagnetic susceptibilities).

ii) TMS is unreactive (except with concentrated sulfuric acid, with which it should not be used), and it does not associate with the sample.

iii) TMS is symmetrical, and thus gives a sharp peak of 12 equivalent protons.

iv) It is extremely volatile, and thus allows recovery of the pure sample.

v) Methyl protons of TMS are strongly shielded, and therefore resonate at a very strong field. By comparison, proton signals for most of the organic compounds appear at a somewhat weaker field. The TMS peak thus does not obscure any of the other peaks in a spectrum. It also enables us to define a scale for the spectra by arbitrarily assigning a frequency value for TMS protons equal to zero. Thus, in proton magnetic resonance, peaks appearing at a field weaker than the standard are assigned a positive sign. If any peaks appear at all at a field stronger than that of the standard, they are naturally assigned a negative sign.

Thus, if TMS is used as a standard, we can logically express chemical shift as ν_{sample}-ν_{TMS}. However, the magnitude of chemical shift is directly proportional to the applied magnetic field, H_0. So that workers with NMR equipment having different oscillator frequencies and magnetic fields may have a simple basis for comparison of spectra, the chemical shift parameter δ (delta) has been defined as:

$$\delta \text{ (in parts per million)} = \frac{(\nu_{sample} - \nu_{TMS}) \text{ in Hz.}}{\text{spectrometer frequency in MHz}}.$$

Note that δ so defined becomes dimensionless and independent of the value of H_0. For example, if TMS is used as a reference for which $\nu_{TMS} = 0$ and if a 60-MHz spectrometer is used,

$$\delta = \frac{\nu_{sample}}{60} \text{ ppm.}$$

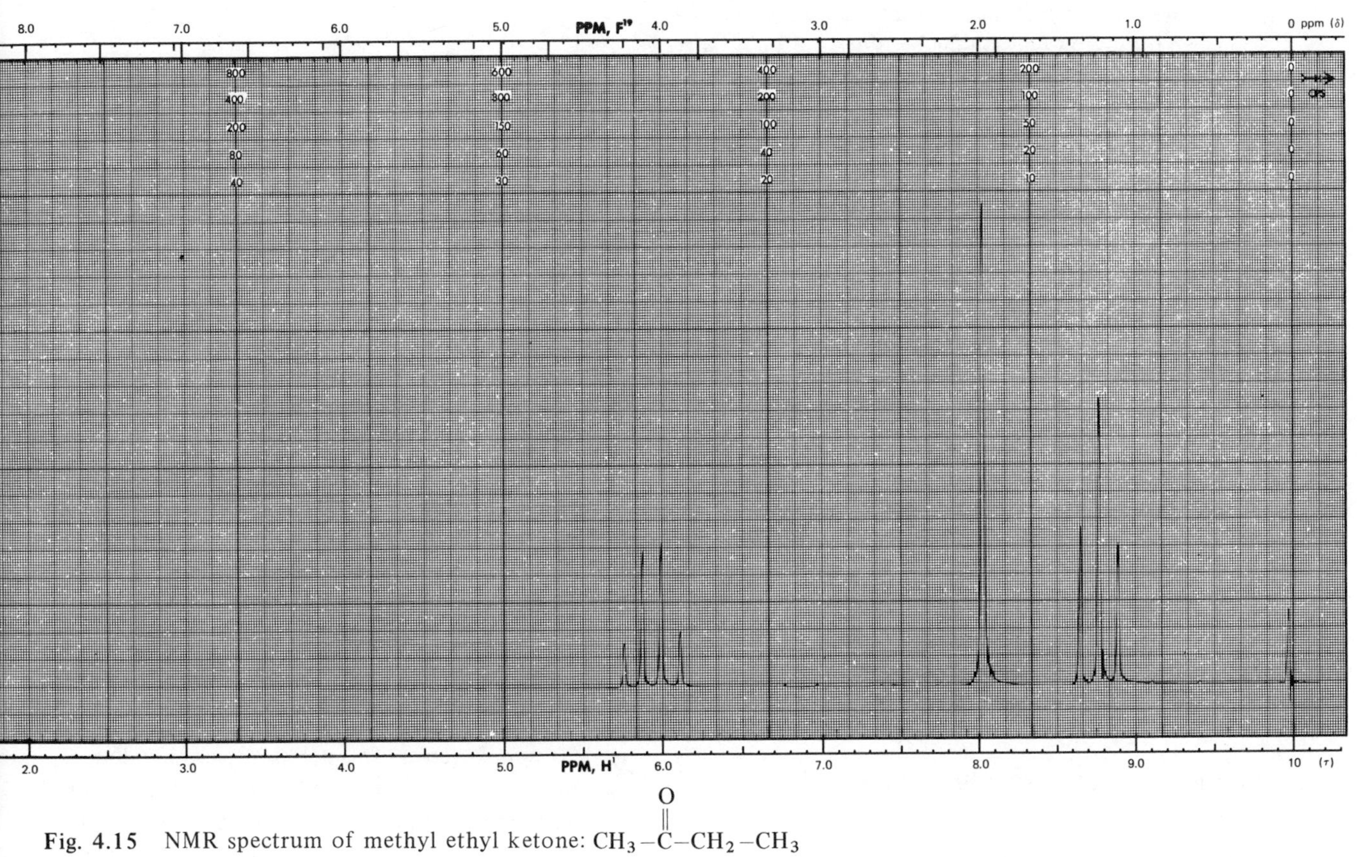

Fig. 4.15 NMR spectrum of methyl ethyl ketone: $CH_3-\overset{\overset{\displaystyle O}{\|}}{C}-CH_2-CH_3$

However, if a 100-MHz instrument is used, the value of ν_{sample} increases correspondingly, and then

$$\delta = \frac{\nu_{\text{sample (increased)}}}{100} \text{ ppm.}$$

Consider, for example, the spectrum of methyl ethyl ketone

$$CH_3-\overset{\overset{\displaystyle O}{\|}}{C}-CH_2CH_3$$

given in Fig. 4.15. This spectrum was made on a 60-MHz machine, with TMS as internal reference. Note that the sharp, tall peak in the spectrum appears at about 130 Hz downfield from the TMS peak. Therefore

$$\delta = \frac{130}{60} = 2.17 \text{ ppm.}$$

The same peak appears at 217 Hz if a 100-MHz machine is used, so that again

$$\delta = \frac{217}{100} = 2.17 \text{ ppm.}$$

Remember that when the δ system is used, TMS is presumed to be the internal reference and need not be specified. However, it is a good idea to get into the habit of always specifying the internal reference anyway! Similarly, since δ is independent of operating frequency, that too need not be specified.

The δ system, although used extensively, is often criticized because values increase in the downfield direction. (That is, on a chart, although the magnetic field increases from left to right, the δ values increase from right to left. This may confuse you in the beginning.) The other system used is the *tau* (τ) scale, in which the values of τ parallel the increasing values of the applied field, and the TMS peak is assigned the value 10.00. Thus

$$\tau = 10.00 - \delta.$$

Most τ values therefore fall in the range 0–10, with a higher τ value indicating greater shielding of the nucleus.

Having studied why chemical shifts take place and how to express them, we are now in a position to find Appendix 4.1 meaningful. Appendix 4.1 lists the chemical shifts of protons in various structural environments.

At this stage of understanding, you should see if you can rationalize the given values on the basis of your knowledge of the factors causing chemical shifts. Also remember that, in addition to the inductive effect due to a functional group in the molecule, factors such as molecular geometry and the degree of contribution by various molecular conformations greatly affect the chemical shift of a given proton.

4.13 WHY SOME NMR PEAKS SPLIT (SPIN–SPIN COUPLING)

Having realized that equivalent protons (i.e., protons in identical electronic and magnetic environments) should show similar chemical shift, you may be expecting, for example, that a spectrum of ethyl acetate

$$CH_3{-}C(=O){-}OCH_2CH_3$$

(Fig. 4.6) should show three sharp peaks, corresponding to

methylene $-O-CH_2-$, acetyl $CH_3-C(=O)-$ and methyl CH_3-C

protons, respectively. However, notice that only the central peak (at 1.96 ppm) due to

$$CH_3{-}C(=O){-}$$

appears as a sharp three-proton peak, the methyl and the methylene group peaks being multiplets. How can we account for this multiplicity?

Coupling

So far we have taken into account only the electronic environment of the proton, and have neglected the small interactions between magnetic nuclei in the same molecule. Recall that magnetic nuclei, such as protons, orient their spins with or against an applied magnetic field. These orientations influence the spin of the nearest bonding electron which, in turn, affects the spin of the other bonding electron in the orbital, and so on through to the next proton. Thus information concerning their spin states is transmitted to one another by neighboring nuclei by means of the intervening bonding electrons. This effect is known as *spin–spin coupling,* and since it causes splitting of peaks in the spectrum, it is sometimes referred to also as *spin–spin splitting.*

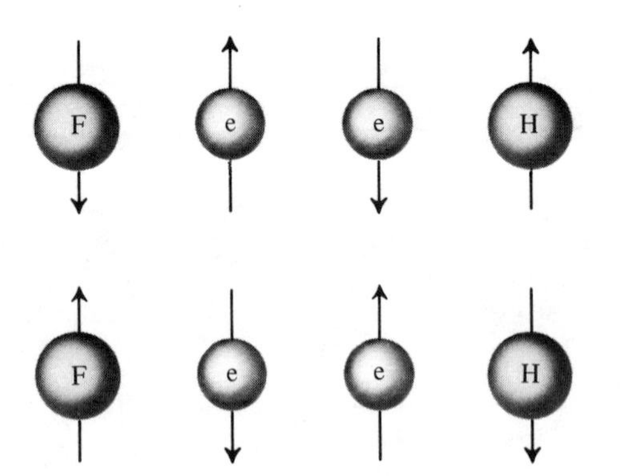

Fig. 4.16 Spin–spin coupling of magnetic nuclei ($I = \frac{1}{2}$) through the bonding electrons.

We can get a better understanding of the phenomenon by considering a simple diatomic molecule such as hydrogen fluoride, HF. When placed in an external magnetic field, each of the two nuclei can assume two orientations: either $+\frac{1}{2}$ (↓ spin) or $-\frac{1}{2}$ (↑ spin). If the fluorine nucleus has ↑ spin, the bonding electrons (of the H—F bond) having favorable ↓ spin tend to remain near the fluorine nucleus (Fig. 4.16). The statistical probability of the other electron (with ↑ spin) being nearer to the hydrogen nucleus is thus greater. This polarization of bonding electrons caused by the spin orientation of the fluorine nucleus thus affects the energy required for a magnetic transition of the proton (because the pairing of neighboring magnets lowers the potential energy of the system*). Since the fluorine nucleus can assume either of two spin orientations, and since there are an enormous number of hydrogen and fluorine nuclei in the sample, the proton peak in the spectrum splits into a doublet.

Another explanation for this phenomenon is that the magnetic moment resulting from $(-\frac{1}{2})$ or $(+\frac{1}{2})$ orientation of the fluorine nucleus either reinforces or diminishes the applied magnetic field. The protons in close proximity to the oriented fluorine nucleus thus experience either increased or diminished magnetic field and the proton peak consequently appears as a doublet in the spectrum.

---

* In general, the potential energy of the system is lowered when two coupled nuclei have paired (antiparallel) spin orientations, and raised when the spins are parallel. However, in some cases, the lower energy state is the one in which the spins of the coupled nuclei are parallel. The magnitude of coupling is given by a coupling constant J, which is considered to be *positive* (that is, $J > 0$) when *antiparallel* coupling produces a low-energy system and *negative* (that is, $J < 0$) when *parallel* coupling produces a low energy system. The relative signs of the various coupling constants within the molecule sometimes influence the appearance of the fine structure of the spectrum. However, detailed discussion of this is beyond our present scope.

Coupling constant

The proton peak in this case splits into two peaks of equal intensity because fluorine nuclei can affect protons through two orientations only, and because there is essentially an equal probability of the two orientations occurring. The separation of two peaks of the doublet depends on the effectiveness of the coupling, and is denoted by a *coupling constant J* (in cps or Hz) which is independent of the applied magnetic field, and is little affected by a change of solvent.

Ordinarily, there is very little effective coupling between nuclei which are separated by more than three covalent bonds, except in cases such as aromatic compounds and conjugated alkenes, in which electrons are delocalized. The magnitude of the value of J is small and can be related to a number of physical parameters, the most important being hybridization, dihedral bond angles, and electronegativity of substituents. However, if one is to observe this *first-order spin–spin splitting*, the separation (in Hz) between the chemical shifts of two coupling nuclei must be at least six times greater than the coupling constant J (that is, $\Delta\nu/J$ must be > 6). As this ratio becomes smaller, i.e., as the value of the chemical shift between the spin-coupled nuclei becomes comparable to or less than J, other magnetic transitions involving both groups of nuclei become more probable, giving rise to a complex splitting pattern (see Section 4.14). However, if the value of $\Delta\nu$ is zero (or nearly so), i.e., if there is no difference in the values of the chemical shifts of the interacting nuclei or if the interacting nuclei are *chemically equivalent,* the multiplets coalesce to give an intense single peak.

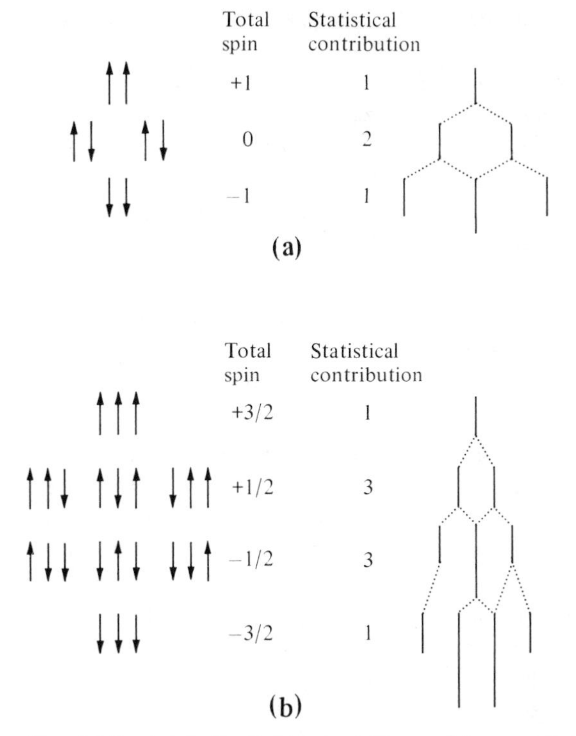

Fig. 4.17 (a) Left: Spin orientations of two methylene protons. Right: Splitting effects of methylene proton orientations on methyl peak. Note the intensity ratio of multiplets (1:2:1). (b) Left: Spin orientations of three methyl protons. The four energy levels correspond to all possible combinations of alignments of the nuclear moments parallel to and antiparallel to H_0. Right: Splitting effects of methyl proton orientations on methylene peak. Note the intensity ratio of the multiplets (1:3:3:1).

Multiplicity

Let us now return to the explanation of the multiplicity of peaks in the ethyl acetate spectrum (Fig. 4.6). This compound, as we have seen, exhibits three groups of peaks in the approximate ratio of 2: 3: 3. Let us number these peaks 1, 2, and 3, from left to right, respectively. Peak number 1, which is a two-proton peak, corresponds to the methylene ($-O-CH_2-$) protons, and is split into a symmetrical quartet. Peak number 2 is unsplit, but peak number 3, corresponding to the $-CH_3$ of the ethyl group of the ester ($O-C-CH_3$), is split into a symmetrical triplet. This splitting pattern is due to the first-order coupling between methylene and methyl protons of the ethyl group. The equivalent methyl protons of the acetyl have no nearby magnetic nuclei to couple with. In the external magnetic field, the two methylene protons orient themselves such that their combinations are grouped into three energy levels, and produce three coupling effects on the neighboring protons (Fig. 4.17a). Similarly, the three methyl protons are oriented in four different energy levels to cause four coupling effects on their neighbors (Fig. 4.17b).

The relative positions of the multiplets (i.e., the values of J) are functions of the separations of the energy levels, and the relative intensities of the multiplets correspond to the orientation probabilities of the neighboring protons. Figure 4.17 shows that in the case of methylene protons the probability of intermediate energy level combination (↑↓ or ↓↑) is twice that of either the low-energy (↓↓) or the high-energy (↑↑) orientations. As a result, the methyl peak is split into a triplet having relative areas 1 : 2 : 1. Similarly, the probability of two methyl protons combining at an intermediate energy level is three times as great as the probability of their combining at the low- or the high-energy levels. The quartet of methyl protons therefore have subareas in the ratio 1 : 3 : 3 : 1. In general, it can be said that the multiplicity of a peak (the number of subpeaks produced by coupling) depends on the number of protons on the adjacent atoms. *Given that n is the number of equivalent protons on adjacent atoms, the multiplicity of a given group is n + 1. The relative areas of a multiplet are symmetric about the midpoint, and are approximately equal to the ratios of the coefficients* (1:1, 1:2:1, 1:3:3:1, 1:4:6:4:1, 1:5:10:10:5:1, etc.) *in the expansion of* $(R + 1)^n$.

For example, in 2-chloro-*n*-butane,

$$\overset{4}{CH_3}-\overset{3}{CH_2}-\overset{2}{CH}(Cl)-\overset{1}{CH_3},$$

the protons on carbon atom 3 have four neighboring protons (one proton on C_2 and three on C_4). Therefore, when we are considering the multiplicity of these methylene protons, we must take $n = 4$. Thus the multiplicity of C_3 protons is equal to $n + 1 = 5$. The intensity ratio of the 5 subpeaks is given by the coefficients preceding each term in the expansion of $(R + 1)^4 = R^4 + 4R^3 + 6R^2 + 4R + 1$. The intensity ratio is then = 1:4:6:4:1.

Similarly, the proton on C_2 has 5 neighboring protons (three on C_1 and two on C_3). Thus now $n = 5$, and the one-proton peak of the C_2 proton will split into a sextet. The relative intensities of the 6 subpeaks can be obtained by the expansion of the expression $(R + 1)^5 = R^5 + 5R^4 + 10R^3 + 10R^2 + 5R + 1$ as 1:5:10:10:5:1.

Although the formula $n + 1$ is used to determine the multiplicity of a given group, one must remember that its applications are restricted to proton–proton interactions only. A more general formula is:

$$\text{Number of subpeaks} = \text{multiplicity} = 2nI + 1,$$

where I is the spin quantum number of the nuclei causing the splitting and n is the number of nuclei causing the splitting. For example, in ammonia, NH_3, the I for nitrogen = 1 and n, the number of nitrogen atoms interacting with protons, is also = 1. The multiplicity of the equivalent protons of NH_3 is thus determined as:

$$\text{Multiplicity} = 2nI + 1 = 2 \times 1 \times 1 + 1 = 3.$$

The three-proton peak in ammonia thus appears as a triplet, with the intensity ratio of 1:2:1.

Of the halogen atoms, chlorine (^{35}Cl), bromine (^{79}Br), and iodine (^{127}I) each has a spin quantum number = $\frac{3}{2}$. Therefore, in compounds containing these halogens, one would expect spin-spin interactions between the halogen atoms and the adjacent protons. However, no proton–halogen or halogen–halogen splitting is normally observed (except in the case of fluorine) because the spin–spin interactions in such cases have such small coupling constants that the interactions may be ignored.*

It must be remembered that, in situations in which adjacent protons are not equivalent, or in which the $\Delta\nu/J$ ratio is small, these simple splitting patterns are not observed. As a matter of fact, a pattern that is neatly split, exactly according to the above rules, is rarely observed experimentally. Note that in case of ethyl alcohol, as well as in ethyl acetate (Figs. 4.5 and 4.6) the triplets and quartets are not quite in the ratio of 1:2:1 and 1:3:3:1, respectively. The small peaks are not even symmetrical around the midpoint, but vary slightly in their intensities, in such a way that those multiplets on the side toward peaks of protons responsible for causing splitting are slightly larger than the corresponding multiplets on the other side. In most cases splitting is even more complex.

In the case of proton–proton interactions also, the multiplicity of a given group cannot always be determined by the application of the simple formula $n + 1$. For example, in a system such as

$$-\underset{H}{\overset{|}{\underset{|①}{C}}}-\underset{H}{\overset{|}{\underset{|②}{C}}}-\underset{H}{\overset{|}{\underset{|③}{C}}}-$$

if the protons on C_1 and C_3 are equivalent, as they are in $Cl_2CH{-}CHCl{-}CHCl_2$, the peak due to the C_2 proton appears as a triplet (according to the $n + 1$ rule). On the other hand, in a compound such as

$$Cl_2CH{-}CHCl{-}C(=O)H$$

in which the three protons are nonequivalent, an unsymmetrical quartet instead of a symmetrical triplet is observed as a C_2 proton peak. Such a pattern, which is due to the so-called *AMX* system, will be discussed in Section 4.14.

Problem 4.8. The NMR spectrum of a deuterochloroform solution of 4-picoline-N-oxide given in Fig. 4.18 was recorded on a 60-MHz instrument using tetramethylsilane (TMS) as an internal reference.

a) How many groups of peaks are present in the spectrum?
b) Record the position of each peak in terms of: (i) Hz value, (ii) δ units, (iii) τ units.
c) What would the positions of these peaks be in terms of Hz, δ, and τ values if a 100-MHz instrument were used?

---

* The very large electric quadrupole moments of these halogen atoms cause spin decoupling of adjacent protons. Consequently, no splitting is observed in HCl, HBr, HI, etc.

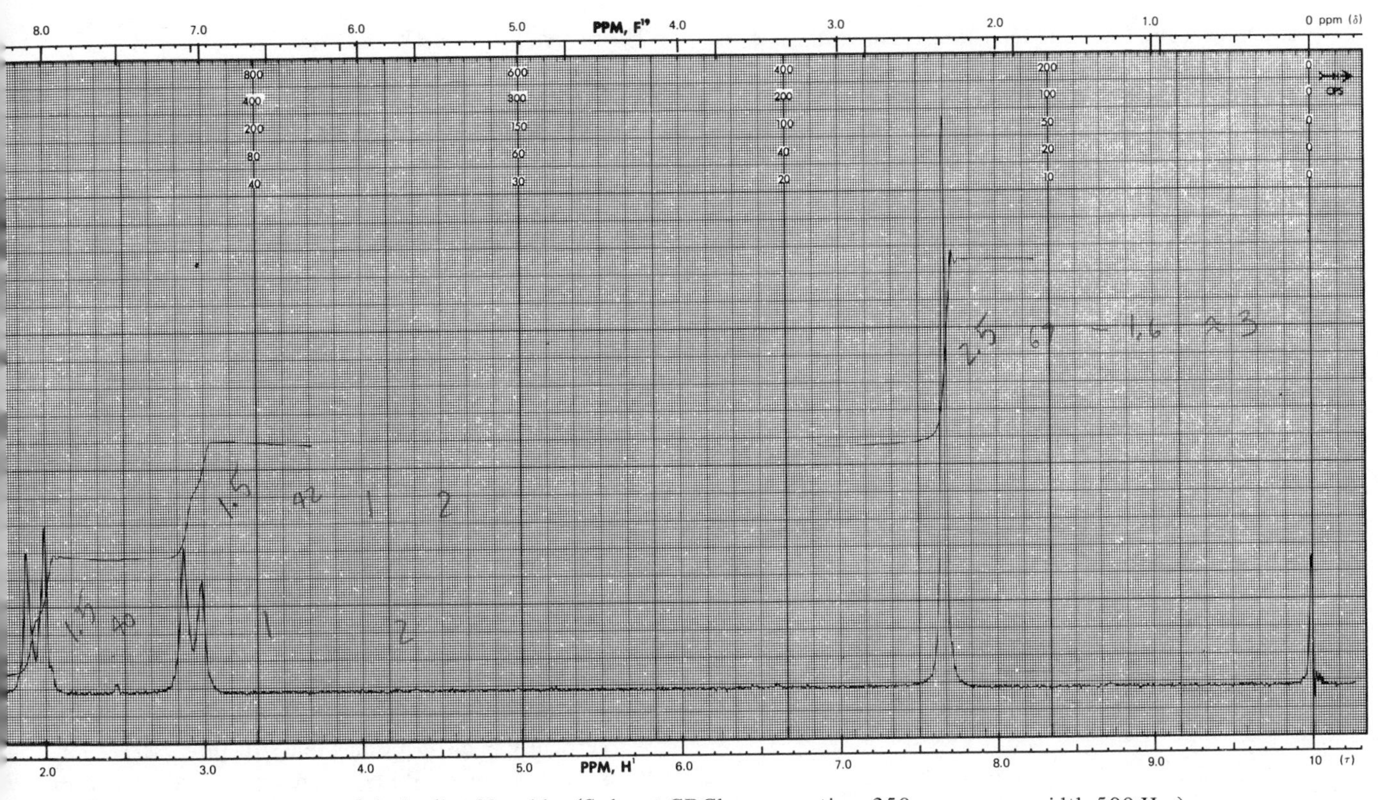

Fig. 4.18 NMR spectrum of 4-picoline-N-oxide. (Solvent $CDCl_3$, sweep time 250 sec, sweep width 500 Hz.)

d) Which is the peak due to the internal reference?

e) How many protons are responsible for the area under each peak? (Disregard the reference peak.)

f) What is the multiplicity of each peak? Consider the structure and try to explain the multiplicity.

g) The subpeaks in each doublet are of unequal intensity. Why?

h) What is the magnitude of the coupling constant J in each multiplet? Is it the same for both the peaks?

i) From the given formula of the compound, decide which protons are most shielded and which are least shielded. Assign the peaks to the proper protons in the structure.

Problem 4.9. Use the NMR correlation chart (Appendix 4.1) to sketch out the spectrum and integral expected at 60 MHz (with TMS as an internal reference), for the following substances.

a) 1-nitropropane, $NO_2—CH_2—CH_2—CH_3$

b) *t*-butyl chloride, $(CH_3)_3—C—Cl$

c) *p*-toluidine, $CH_3—C_6H_4—NH_2$

d) Acetone, $CH_3—CO—CH_3$

e) Methyl-*p*-toluate, $CH_3—C_6H_4—COOCH_3$

f) $(CH_3)_2—C—(OCH_3)_2$

Problem 4.10. The molecular formula of a certain dichloro compound was found to be $C_2H_4Cl_2$. Write two possible isomers for this formula, and show how the two structures could be distinguished by NMR.

4.14 COMPLEX SPLITTING

Before we go on to complex splitting patterns, we should acquaint ourselves with a simple nomenclature often used[20] to discuss proton coupling. According to this system, a group of protons for which the chemical shift difference is comparable to the spin–spin coupling are denoted by the letters *A*, *B*, *C*, etc. (in the order of increasing shielding). Another group of protons which are nearly identical but have resonance positions well separated from *A* or *B*, etc., are denoted by the letters *X*, *Y*, *Z*. The number of protons in each group is denoted by a subscript. The ethyl group in ethyl acetate can thus be described as an A_2X_3 system, but an ethyl group at the end of a long aliphatic chain, for example, in *n*-hexyl bromide, is designated as an A_2B_3 system.

More complex systems such as *ABX, ABC, AMX,* A_2B_2, etc., are also common in organic molecules. For example, in ethyl bromide, CH_3-CH_2-Br, the methyl protons resonate at about 7.97τ and the methylene protons at about 6.75τ. That is, $\Delta\nu$ is not very large. Due to the free rotation of the C—C single bond, the three methyl protons are equivalent, as are the two methylene protons. As a result, the system is A_2B_3. But, in vinyl bromide,

$$(H_A)(H_B)C{=}C(Br)(H_C)$$

the two hydrogens H_A and H_B, though geminal (on the same carbon), are in slightly different electronic environments, due to the restricted rotation of the double bond, and hence are nonequivalent.* The third hydrogen H_C is also nonequivalent with either H_A or H_B. Therefore the system is *ABC*. In an epoxide such as

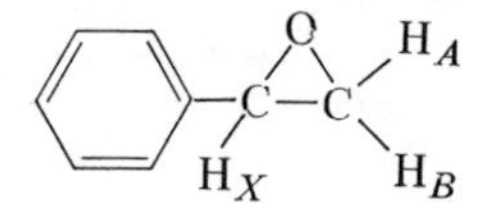

the rotation around the epoxide system is also restricted, making the two geminal hydrogens H_A and H_B nonequivalent. The third hydrogen, H_X, is in a very different electronic environment, and appears at a position considerably downfield from either the H_A or H_B positions. The system is thus *ABX*. In

$$(H_A)(H_M)C{=}C(H_X)-O-\underset{\underset{O}{\|}}{C}-CH_3$$

on the other hand, the three protons differ considerably from each other in their chemical shifts. The system is thus *AMX*.

The fine structure and distortion of the NMR peaks are due mainly to the interactions of protons, of type *AB, ABM,* or *XY,* for which the $\Delta\nu$ is of approximately the same magnitude as the coupling constant *J*. This variation or departure of the peaks from equal intensities is actually of considerable analytic value, and can be used to judge the nature of the system, and thus to determine the structure of the compound. To a certain extent, the values of the coupling constants *J* are also characteristic of the structural and stereochemical relationships of the interacting protons. These values often prove useful in the determination of structures. Some values which are of particular significance are given in Table 4.2.

Difficulty in distinguishing fine structure peaks from those due to chemical shifts is sometimes a problem in the proper use of *J* values. But such a difficulty is often overcome by measuring the spectrum at more than one field strength. Although the separation of peaks due to chemical shifts increases in proportion to the field strength, spin-coupling separations due to fine structure remain constant; therefore a distinction can be made.

A fine illustration of this is provided by the spectra of 1,2,3,4,5-cyclohexane pentol (*proto*-quercitol) recorded[1] at 60, 100, 220, and 300 MHz (Figs. 4.19a and 4.19b). In this compound $\Delta\nu$ for protons on C_1 through C_5 is so small that at 60 and 100 MHz it is impossible to distinguish between the peaks due to fine structure and those due to chemical shifts. At 220 MHz, separate multiplets

* The coupling between two protons H_A and H_B on the same carbon is called *geminal coupling,* whereas that between H_A and H_C or H_B and H_C is called *vicinal coupling.* These will be discussed in Section 4.15(i).

Table 4.2a Geminal proton spin-spin coupling constants *J*, in Hz

Structure	Coupling constant, Hz or cps	Structure	Coupling constant, Hz or cps
Geminal			
$>C(H_A)(H_B)$	12–15	$>C{=}C(H_A)(H_B)$	0.5–3
$(H_A)(H_B)C{=}N-OH$	7.63–9.95 (solvent dependent)	cyclopropane $C(H_A)(H_B)$	3.9–8.8
epoxide $C(H_A)(H_B)$	5.4–6.3	cyclohexane $C(H_A)(H_B)$	12.6

Table 4.2b Vicinal proton spin–spin coupling constants J, in Hz

Structure	Coupling constant, Hz or cps	Structure	Coupling constant, Hz or cps
Vicinal			
CH_3–CH_2–	4.7–9	CH_3CH_2–O–CH_2CH_3	6.97
CH_3–CH_2–X	7.1–7.7	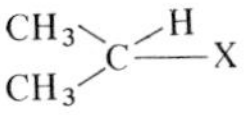 $(CH_3)_2C(H)$–X	6.1–7
where, if X =		where, if X =	
–CH_3	7.26	–OH	6.2
–Cl	7.23	–Cl	6.4
–Br	7.33	–Br	6.5
–I	7.45	–I	6.6
–N–Et_2	7.4	–CH_3	6.8
–CN	7.6	–phenyl	6.9
–phenyl	7.62	–CHO	7.0
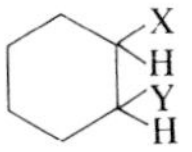 cyclohexane, C(X)(H)–C(Y)(H)	e, e; 2–4 a, e; 2–4 a, a; 5–10	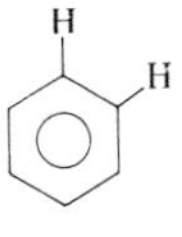 benzene, ortho H	6.5–9.4
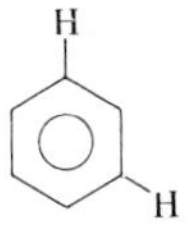 benzene, meta H	0.8–3	 benzene, para H	0.4–1
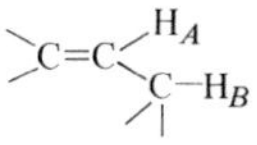 >C=C(H_A)–C(H_B)<	4–10	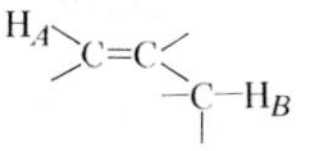 H_A(>C=C<)–C–H_B	0.5–2.5
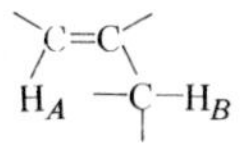 >C=C<, H_A –C–H_B	≈ 0	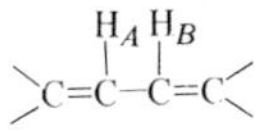 >C=C(H_A)–C(H_B)=C<	9–13
H_AC≡CH_B	9.5–9.8	>CH_A–C≡C–H_B	2–3
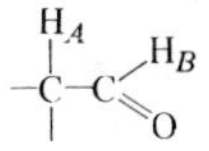 –C(H_A)–C(H_B)=O	1–3	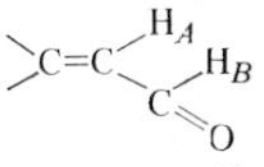 >C=C(H_A)–C(H_B)=O	6–8
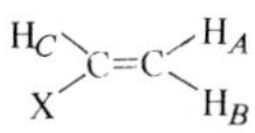 H_C(X)C=C(H_A)(H_B)	*cis* H_A–H_C, 4.6–19.3 *trans* H_B–H_C, 12.7–24.0	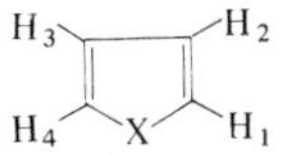 five-membered ring with X, H_1, H_2, H_3, H_4	1–5.5
where, if X =		where, if X =	
–F	H_A–H_C, 4.6 H_B–H_C, 12.7	O	H_1-H_2, 1–2
–Br	H_A–H_C, 7.1 H_B–H_C, 15.2	N	H_1-H_2, 2–3
–Cl	H_A–H_C, 7.4 H_B–H_C, 14.8	S	H_1-H_2, 5.5
–CH_3	H_A–H_C, 9.6–11.1 H_B–H_C, 16.6–17.4	O, N, S	H_1-H_3, 1–2 H_2-H_3, 3–5
–COOH	H_A–H_C, 10.2 H_B–H_C, 17.2		
–phenyl	H_A–H_C, 10.7 H_B–H_C, 17.5		
–H	H_A–H_C, 11.5 H_B–H_C, 19.0		
–CN	H_A–H_C, 11.7 H_B–H_C, 17.9		

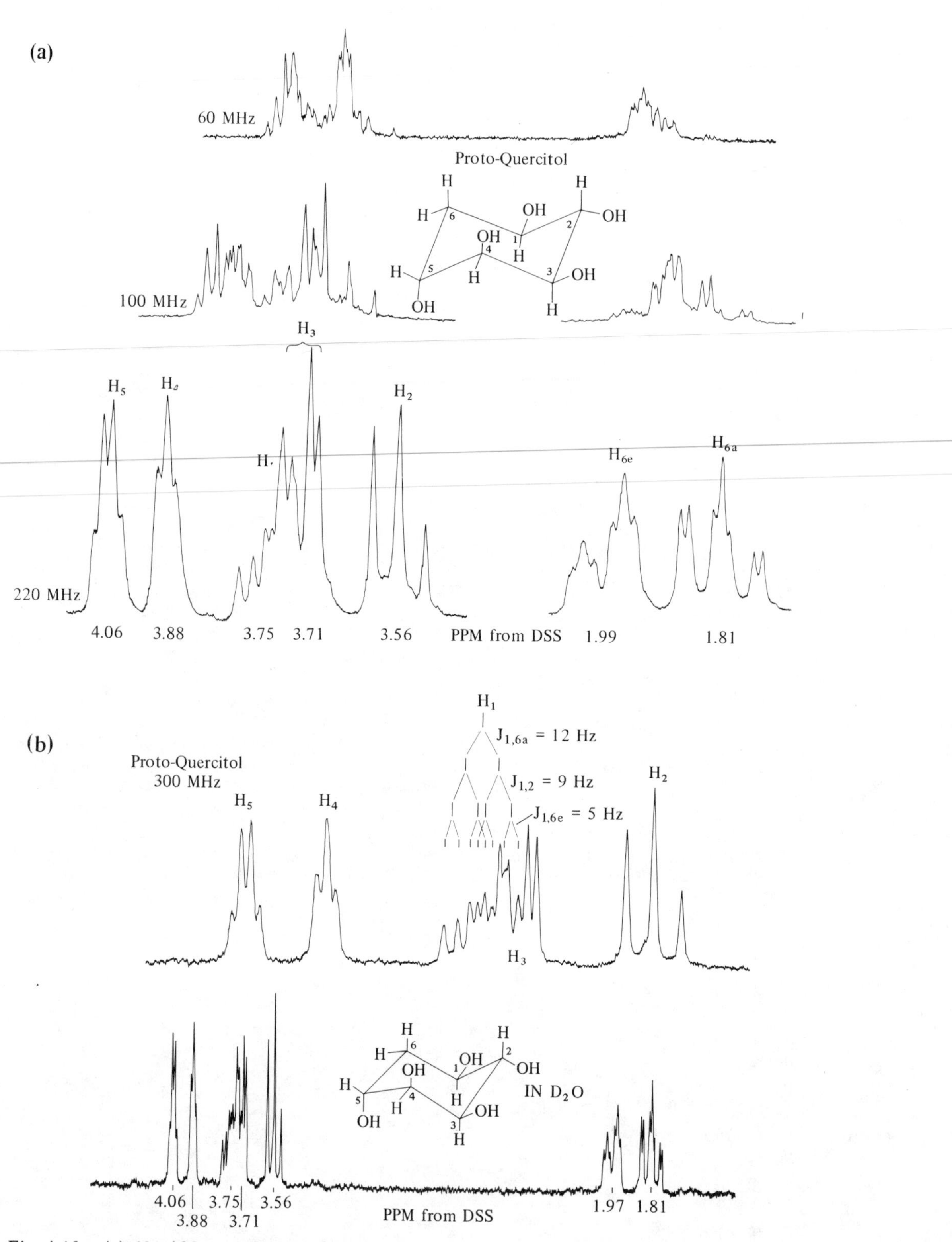

Fig. 4.19 (a) 60-, 100-, and 220-MHz specta of *proto*-quercitol in D_2O. (b) 300-MHz spectrum of *proto*-quercitol in D_2O. [Reproduced from Ref. 1, with permission of the American Chemical Society.]

can be seen for the protons on C_2, C_4, and C_5, as well as for the axial and equatorial protons on C_6. An even better separation of the multiplets from the protons on C_1 and C_3 is achieved at a higher magnetic field of 70.5 kG (that is, 300 MHz). The magnitudes of the observed splittings (that is, the J values), which are characteristic of the structural and stereochemical relationships of the interacting protons, are not changed, although the field strengths *are* changed.

Let us now briefly examine the types of fine spectral patterns to be anticipated for various values of the ratio $\Delta\nu/J$ of the interacting nuclei.

i) Systems with two interacting nuclei: A_2, AB, AX. A system containing two equivalent nuclei* is designated an A_2 system. In such a system, only a single peak is obtained, because the interaction, however strong it is, between such nuclei is not observable.

A system at the other extreme is an AX, which gives rise to a pair of doublets, the subpeaks of which are of equal intensity (Fig. 4.20a), in accordance with the simple splitting rules. In such a system, the chemical shift between proton A and proton X is measured from their centers of doublets.

Let us consider a methylene ($-CH_2-$) group with no protons on adjacent carbon atoms. If such a group is part of an aliphatic chain, the two protons may be chemically equivalent, giving rise to an A_2 system. However, if the methylene group is in a situation in which it is not free to rotate, such as in a ring system, the two methylene protons may be nonequivalent, and thus form an AB system. Under these conditions, the protons exhibit geminal coupling, usually in the range of 12–15 Hz. The spectrum of such an intermediate system AB, in which $\Delta\nu$ (that is, $\delta_B - \delta_A$), between the two nuclei, is comparable to J_{AB}, also consists of two doublets, but the intensities of the four multiplets are no longer equal (Fig. 4.20b), although they are symmetric about the midpoint of the spectrum.

It must be remembered that the general appearance

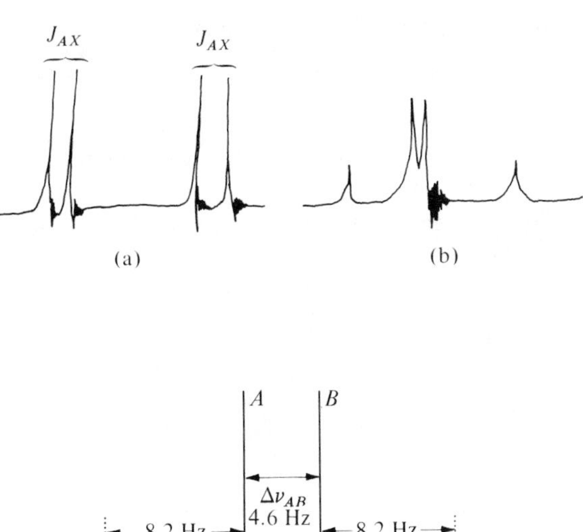

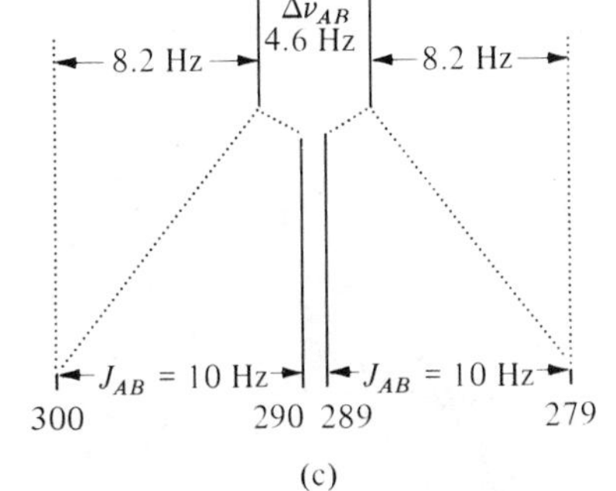

Fig. 4.20 Fine spectral patterns for systems with two interacting nuclei. (a) System AX. (b) System AB. (c) Determining chemical shifts of A and B protons in AB pattern.

of an AB spectrum depends only on the absolute ratio $\Delta\nu/J$. When this ratio is large, the pattern approaches that of an AX system, but as the ratio becomes small, i.e., as the value of $\Delta\nu$ approaches that of J in Hz, or becomes smaller than J, the intensity of the inner pair of lines increases at the expense of the outer pair.

In AB systems, we may still describe the doublet on the left hand qualitatively as having predominantly A lines, although the actual transition is a mixed one in which there is some mixing of spin from the B nucleus. In cases in which $J \geqslant \Delta\nu$, the B signals appear also at lower levels of the magnetic field. The coupling constant J_{AB} and the difference in chemical shifts of the two protons $\Delta\nu_{AB}$ for such spectra can be calculated by using the following relationships, after numbering the multiplets as $\nu_1, \nu_2, \nu_3, \nu_4$, respectively, from left to right (i.e., in order of increasing shielding).

1. The spacing of doublets:

$$J_{AB} = \nu_1 - \nu_2 = \nu_3 - \nu_4$$

2. $\Delta\nu_{AB} = \sqrt{(\nu_1 - \nu_4)(\nu_2 - \nu_3)}$

---

* The nuclei are called equivalent if:

i) They have equal chemical shifts, even though they may occupy quite different environments; e.g., in methylacetylene $CH_3-C\equiv CH$, all four protons appear to be equivalent, since the spectrum shows only a single peak at 8.20τ (using a 60-MHz instrument).

ii) They have equal chemical shifts because they can exchange their positions under the appropriate symmetry operations in a molecule which has elements of symmetry; e.g., protons of benzene.

iii) They have equal chemical shifts because of magnetic equivalence.

3. The relative intensities of multiplets:

$$\frac{i_2}{i_1} = \frac{i_3}{i_4} = \frac{\nu_1 - \nu_4}{\nu_2 - \nu_3}$$

That is, the intensities are inversely proportional to the spacings of the subpeaks.

For the spectral pattern *AB*, once the $\Delta\nu_{AB}$ is determined, the positions of the uncoupled *A* and *B* peaks–i.e., chemical shifts δ_A and δ_B–can be easily ascertained by the formula:

$$\text{Chemical shift of proton } A = \delta_A = \nu_1 - \left(\frac{(\nu_1 - \nu_4) - \Delta\nu_{AB}}{2}\right),$$

$$\text{Chemical shift of proton } B = \delta_B = \delta_A - \Delta\nu_{AB}$$

For example, if an *AB* system gives a four-peak pattern (Fig. 4.20c), such as

$\nu_1 = 300$ Hz, $\nu_2 = 290$ Hz, $\nu_3 = 289$ Hz, and $\nu_4 = 279$ Hz,

the positions of the uncoupled peaks *A* and *B* can be determined as:

$$\Delta\nu_{AB} = \sqrt{(\nu_1 - \nu_4)(\nu_2 - \nu_3)}$$
$$= \sqrt{(300 - 279)(290 - 289)}$$
$$= \sqrt{21} = 4.6 \text{ Hz}.$$

And the position of uncoupled peak *A*, that is, δ_A, is

$$\delta_A = 300 - \left(\frac{(300 - 279) - 4.6}{2}\right)$$
$$= 300 - 8.2 = 291.8 \text{ Hz}.$$

And the position of uncoupled peak *B* is

$$\delta_B = 291.8 - 4.6 = 287.2 \text{ Hz}.$$

Spectral patterns corresponding to the *AB* system are observed as portions of numerous complex spectra, but very few compounds are known to give simple *AB* spectra only. One such compound is 2-bromo-5-chlorothiophene (Fig. 4.21), the spectrum of which was recorded and analyzed by Anderson[21] as:

$\Delta\nu_{AB} = 4.7 \pm 0.2$ Hz, $J_{AB} = 3.9 \pm 0.2$ Hz,

and $\Delta\nu/J = 1.205$.

ii) Systems with three interacting nuclei: A_3, AX_2 (or A_2X), AMX, ABX, ABC, AB_2 (or A_2B).

The six systems with three interacting nuclei are listed above, more or less in the order of increasing complexity of spectra. The systems A_3 and AX_2 are the simplest, because the A_3 system gives rise to only a single peak, whereas the AX_2 system produces two main peaks, in which *A* protons give a 1:2:1 triplet and *X* protons are split into a 1:1 doublet, according to the simple splitting rules.

AMX system. The nuclei of such a system are all nonequivalent, with large $\Delta\nu/J$ ratios. The system therefore shows three chemical shifts and three coupling constants, J_{AM}, J_{MX}, and J_{AX}.

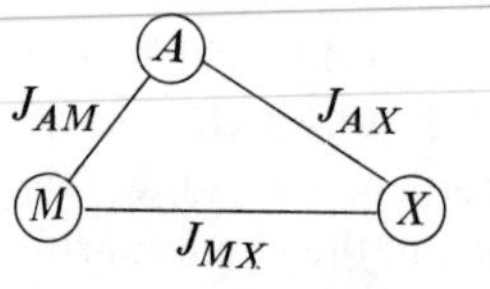

For such a system, an idealized spectrum will show 12 subpeaks of equal intensity arranged in three quartets. The quartets are explained on the basis that the signal due to the *A* proton is first split into a doublet by J_{AM}; and then each of these multiplets is again split into a doublet by J_{AX} (the *M* and *X* signals are similarly split) (Fig. 4.22).

Vinyl protons of

H_A, H_M C=C H_X, O–C(=O)–CH_3

are likely to give fine structure of the *AMX* type, although the spectrum is far from ideal because $\Delta\nu_{AM}$ may not be large enough to give two distinct main peaks.

ABX system. In such a system, the $\Delta\nu$ between two of the three nuclei is comparable to their coupling (*J*) to each other, and both are coupled to a third nucleus well

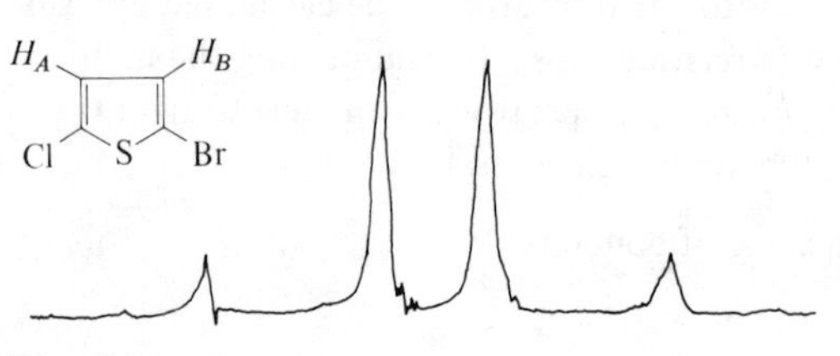

Fig. 4.21 Spectrum of 2-bromo-5-chlorothiophene at 30.5 MHz.

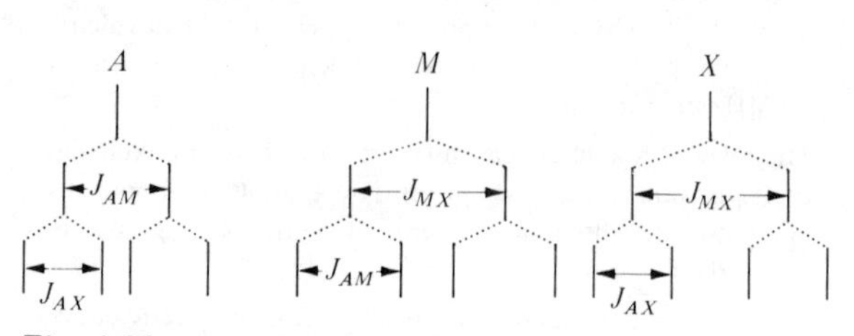

Fig. 4.22 A theoretical *AMX* spectrum.

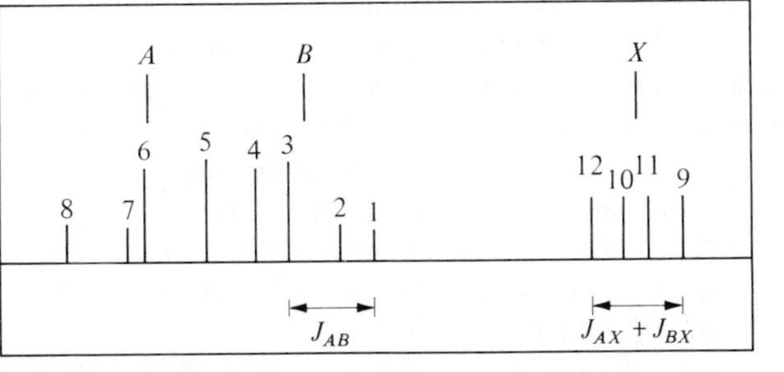

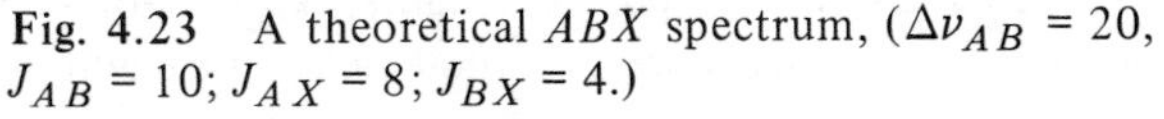

Fig. 4.23 A theoretical ABX spectrum, ($\Delta\nu_{AB} = 20$, $J_{AB} = 10$; $J_{AX} = 8$; $J_{BX} = 4$.)

removed in chemical shift. The spectrum for such a system often exhibits 12 subpeaks. Fifteen are customarily listed, one of these being of zero intensity, and two others usually too weak to be observed, since they represent combination transitions corresponding to the simultaneous excitation of two nuclei. The AB portion of the spectrum consists of two pseudo-AB quartets, whereas the X portion consists of four to six subpeaks (Fig. 4.23).

The total number of subpeaks is sometimes even less than 12, due to partial overlapping or completely degenerate subpeaks. Compounds such as 2-chloro-3-amino pyridine and 5,6-dichloropyridine give spectra of this type.

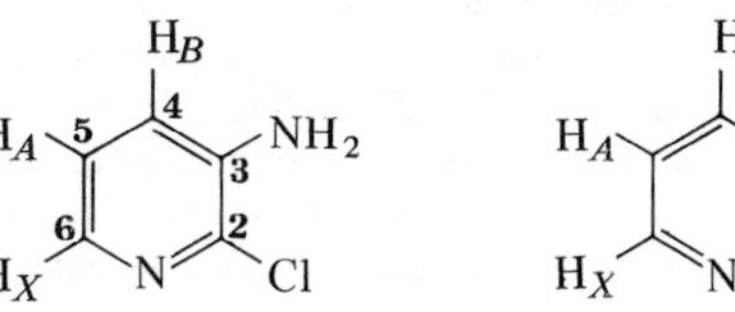

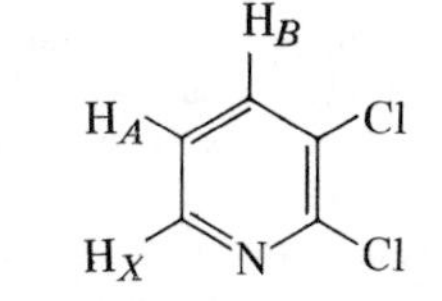

2-chloro-3-amino pyridine 5,6-dichloropyridine

ABC system. In this system, $\Delta\nu_{AB}$ and $\Delta\nu_{BC}$ are of comparable magnitude. That is, $\nu_A - \nu_B \approx \nu_B - \nu_C$. Such systems are quite common in organic molecules, but their spectral patterns are relatively complicated. Some compounds showing such patterns are (i) monosubstituted olefins such as styrene, ethyl acrylate, and vinyl chloride; (ii) trisubstituted benzenes, for example, 3-nitro-*o*-xylene,[22] 2:4-dinitrochlorobenzene, and 3-nitrosalicylic acid; (iii) disubstituted pyridines; and substituted (iv) furans, (v) thiophenes, (vi) epoxides, etc. The ABC spectra also can contain up to 15 subpeaks.

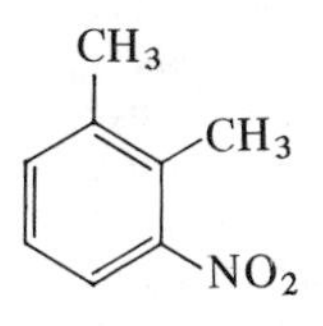

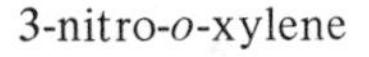
3-nitro-*o*-xylene

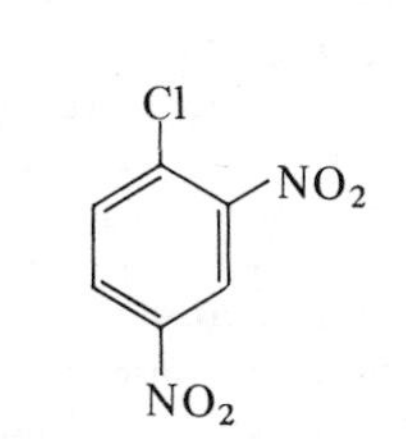

2,4-dinitrochlorobenzene

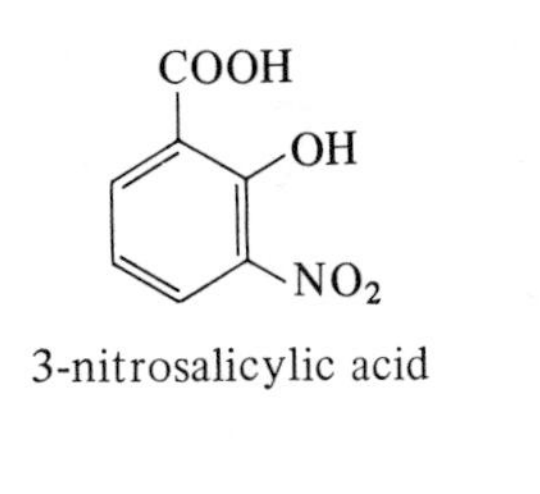

3-nitrosalicylic acid

CH_3

NO_2 CH_3

5-nitro-*m*-xylene

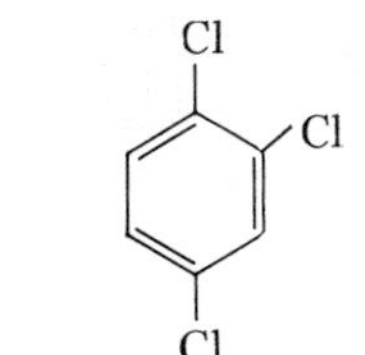

1,2,4-trichlorobenzene

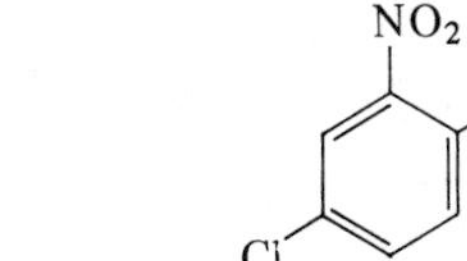

2,5-dichloronitrobenzene

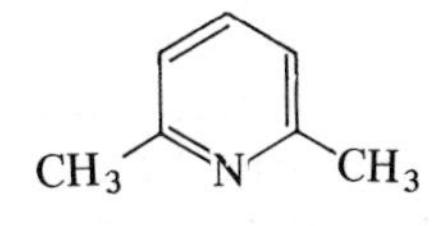

2,6-lutidine

AB_2 system. In a system of type AB_2, two magnetically equivalent nuclei are coupled to a third nucleus, which is not far removed in its chemical shift from the other two.

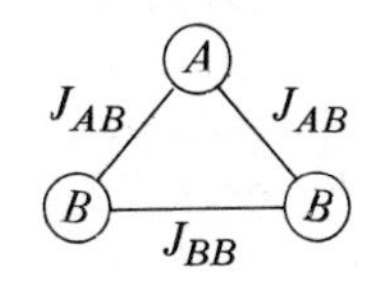

As we have seen earlier, an AX_2 system gives a triplet for A protons and a doublet for X protons, thus producing a total of five subpeaks. But, as $\Delta\nu$ decreases, or as $\Delta\nu/J$ becomes small, each line of the doublet splits into two, and so also the centerline of the triplet splits into two, thus producing a total of eight subpeaks. In addition, the simultaneous change of spin states of all three nuclei produces a weak peak, known as *combination peak.* This peak is usually noticeable only if $\Delta\nu/J$ is extremely small.

Some of the compounds known to give AB_2 or A_2B spectral patterns are trisubstituted benzene derivatives, for example, 5-nitro-*m*-xylene, 1,2,4-trichlorobenzene, 2,5-dichloronitrobenzene, pyrogallol, 2,6-lutidine,[20] 2,6-dichloropyridine, etc. (Fig. 4.24).

In type AB_2 spectra, the τ or δ values for A and B protons can be read directly from the spectrum. If the multiplets are numbered from left to right, the position of peak 3 always gives the chemical shift of proton A and the average of the positions of peaks 5 and 7 gives the chemical shifts of B protons.

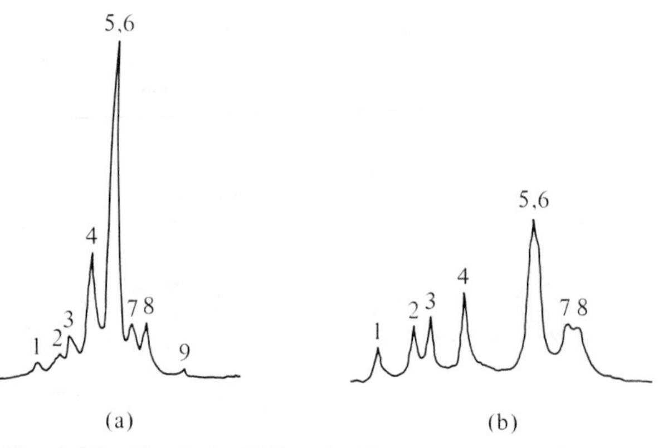

Fig. 4.24 Typical *ABC* and AB_2 type spectra for the aromatic protons of (a) 3-nitro-*o*-xylene (*ABC*), and (b) 2,6-lutidine (AB_2).

iii) Systems with more than three interacting nuclei. There are many possible combinations of four or five or six interacting nuclei. Such combinations may be AX_3, AB_3, A_2X_2, A_2B_2, *ABCD*, A_2B_3, A_2B_2X, $A_2B_2X_2$, . . ., etc. Methods of constructing spectra for such systems from given sets of parameters, or for computing parameters from such spectra, are known, but are beyond the scope of our study. Readers interested in analysis of such systems should consult Reference 100.

Here we merely note that it is not always easy to identify the types of interactions in a complex organic molecule by studying its NMR spectrum. A useful rule to remember is that, among the types of couplings likely to be encountered, *only AB (or AX) and A_2B_2 (or A_2X_2) systems give rise to spectra which are symmetrical about their centers.* Similarly, it should also be remembered that *a gradual transition in splitting pattern occurs from the simple AMX system, through ABX, to the very complicated ABC system.* Analysis of such patterns enables us to assign the absorption positions for the protons observed in the spectrum. This information, together with the knowledge of the chemical reactions of the compound, usually enables us to suggest the possible structure of the unknown.

Problem 4.11. Describe the following spin systems as *AB*, A_2X, etc.

a) CH_3OH b) $CH_2{=}CHCl$
c) $CH_2{=}CHF$ d) $CH_3CH_2NO_2$
e) 2,5-dichloronitrobenzene f) *o*-dichlorobenzene
g) $(CH_3)_2CHOH$ h) $CH_3CH_2CH_3$

4.15 APPLICATIONS OF NMR

Since the first application of NMR to structural problems in organic chemistry in 1953, the growth of this technique has been phenomenal. Within a short period, NMR technique has thrown light on many difficult organic problems, and has continued to solve many intricate problems which remained unresolved when only conventional methods were available to organic chemists.

The importance of this technique lies in the fact that, unlike other physical methods, NMR is mainly concerned with the study of the patterns of protons,* rather than of carbon skeletons.

Chemical shifts, coupling constants, and long-range couplings of protons depend on their electronic environment, which in turn is a function not only of the type of functional groups present, the number and type of intervening bonds, etc., but is also a function of the geometrical disposition of these bonds. As a result, valuable stereochemical information about many types of molecules can be derived from NMR spectra, and the field of conformational analysis has derived special benefits.

i) Geometric isomerism and conformational analysis

In the case of protons on adjacent carbon atoms, the value of the vicinal coupling constant, $^3J_{HH}$ (the superscript 3 indicates the number of intervening bonds be-

---

* Rapid progress is being made in the use of other nuclei such as ^{13}C, fluorine, etc., but the details of this are beyond our present goals. We will, nevertheless, look briefly at a rapidly developing field: the study of the so-called ^{13}*CH satellites.*

The ^{12}C nucleus has spin = 0 and is nonmagnetic, but the heavier isotope of carbon, ^{13}C, has spin = $\frac{1}{2}$, is magnetic, and has a natural abundance of 1.1%. Naturally, in the molecules containing ^{13}C atoms and hydrogens, there is spin–spin coupling between these two magnetic nuclei, the value of $^1J_{^{13}C-H}$ being normally 100 to 200 Hz. This type of coupling can sometimes introduce magnetic nonequivalence between nuclei which otherwise would have been equivalent in molecules containing ^{12}C only. As a result, weak signals given by protons in molecules containing ^{13}C atoms are generally observed in the NMR spectra of a number of compounds. For example, in a molecule such as that of tetrachloroethane $Cl_2HC{-}CHCl_2$, if both carbons are ^{12}C, the two protons are equivalent, and only a single resonance band is observed. But if one of the two carbons is isotopic, the two protons become nonequivalent, the spin coupling system becomes type *ABX*, and bands known as ^{13}*CH satellites* or ^{13}*C sidebands* appear in the spectrum. Since the natural abundance of ^{13}C is only 1.1%, approximately 2.2% molecules of $Cl_2HC{-}CHCl_2$ contain protons that are magnetically nonequivalent. The ^{13}CH satellite bands in the NMR spectrum of the compound are therefore very weak, but they give valuable information about the molecule. Hence scientists seek methods of improving the signal-to-noise ratio in ^{13}C NMR spectra, and use high-magnetic-field spectrometers in order to observe the ^{13}C satellites. Use of 70.5-kG instruments and techniques such as CAT, Fourier transform, etc., have already been mentioned (Section 4.6). For an excellent review of ^{13}C NMR spectroscopy, see Reference 23.

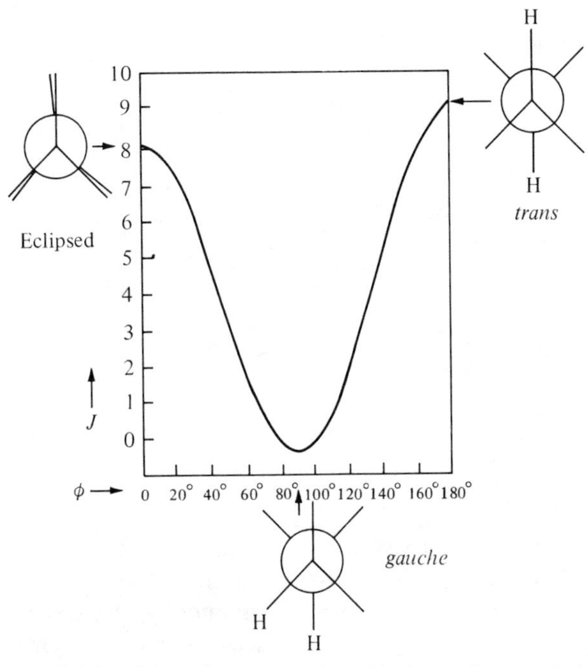

Fig. 4.25 A graph representing the dependence of the vicinal proton-proton coupling constant $^3J_{HH}$ on the dihedral angle ϕ (Karplus' equation).

tween the coupled nuclei) varies with the value of the dihedral angle ϕ (Greek phi) between the two C—H bonds.

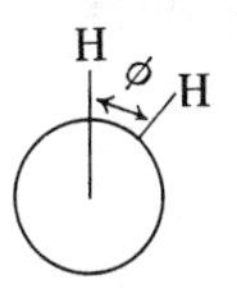

This value is at a maximum when ϕ is 180° (that is, a *trans* conformation) or 0° (eclipsed), and at a minimum when ϕ is about 90° (that is, a *gauche* conformation) (Fig. 4.25). For conformationally rigid systems in which the bond angles are reasonably well known, the value of $^3J_{HH}$ can be calculated by Karplus' relationship,[24] which has been modified by Bothner-By[24a] as

$$^3J_{HH} = 7 - \cos\phi + 5\cos 2\phi.$$

Consequently, the experimentally observed values of $^3J_{HH}$ provide information about ϕ and thus about the conformation of the molecule. However, difficulty arises occasionally in the interpretation of coupling constants, due to the fact that $^3J_{HH}$ values are markedly affected by substituents on the bonded carbons. For example, the $^3J_{HH}$ coupling in cyclopropane[25] is 9.5 Hz for *cis* and 5.5 Hz for *trans* protons, but in cyclopropylamine[26] the *cis* H_A-H_B coupling is reduced to 6.6 Hz and the *trans* H_A-H_C coupling is reduced to 3.6 Hz.

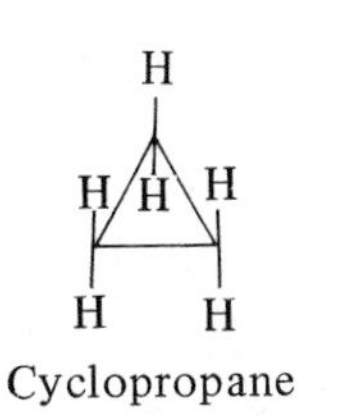

Cyclopropane

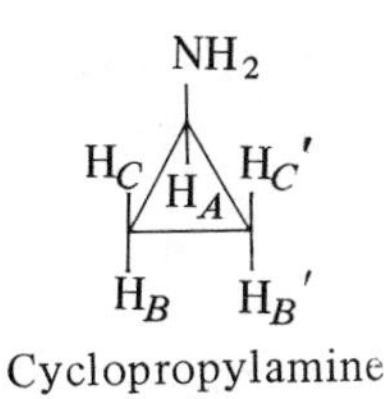

Cyclopropylamine

On the other hand, coupling to protons on other carbons (other than the one carrying the substituent) is found to be increased, since *cis* H_B-H_B' is 12.5 Hz and *trans* H_C-H_B' is 7.5 Hz. Lower $^3J_{HH}$ values are also observed if oxygen, sulfur, or nitrogen is present in the three-membered ring.

Note: *cis* couplings are usually larger than *trans* couplings, the J_{cis}/J_{trans} ratio being about 1.4–1.8. (However, see References 14, 15, 16).

Couplings in six-membered rings have received far more study than those in all other cyclic compounds. The vicinal coupling constants for protons in cyclohexane have been experimentally determined and also have been calculated using Karplus' equation. As a result, three important generalizations are:

i) In a six-membered saturated ring, an axial proton usually resonates at a higher field than the corresponding equatorial proton.

ii) The coupling between two diaxial protons is large: J_{aa} = 8 to 14 Hz (calculations: ϕ_{aa} = 180°, J_{aa} = 11 Hz according to Karplus' equation). Smaller splittings (usually 1 to 5 Hz) are observed with axial–equatorial (J_{ae}) or diequatorial (J_{ee}) interactions (ϕ_{ae} = 60°, J_{ae} = 4 Hz).

iii) In addition to the angular and electronegativity dependence, the gauche vicinal coupling constants in six-membered saturated rings show configurational dependence also,[27,28,29,30]. For example, the gauche coupling constants for a series of steroidal alcohols and their acetates were found to depend on the relative orientation of the C—O bonds, and could be divided into two groups as follows.

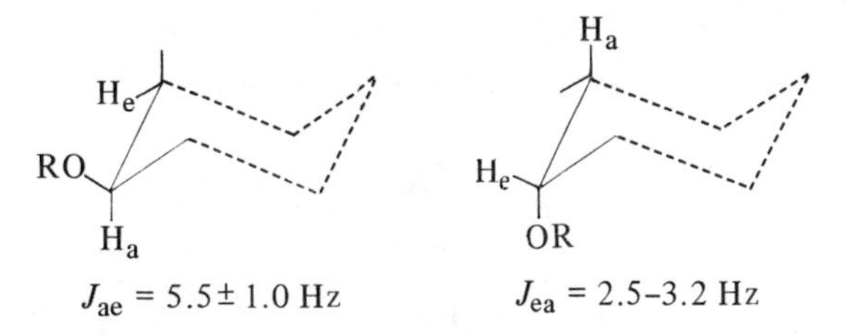

A study by Sundaralingham[31] suggests that this dependence on orientation (with respect to the ring) of the substituent on a ring carbon brings about small changes in the hybridization of that carbon.

Remember that, in cyclohexane itself, the two chair conformations are interchanging so rapidly that the shielding of all protons is averaged to a single value. But in rigid systems, such as *trans* decalin, in which interchange of conformations is impossible, separate signals are observed for axial and for equatorial protons.

Effects of electron-withdrawing substituents, and of the presence of oxygen in the ring, on $^3J_{HH}$ values for six-membered rings are similar to those observed for cyclopropane ring systems. For example, the axial–axial coupling in the following sugar derivative[32] (note O— in the ring) is considerably lower.

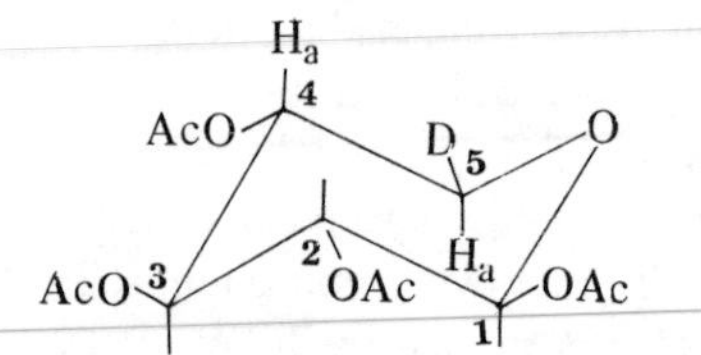

$J_{H_4\text{-}H_5}$, that is, $J_{Haa} = 8.1$ Hz

(It must be remembered, however, that a pyranose ring, such as the one shown here, is flipping, and should not be considered in relation to the fixed or rigid rings encountered in steroids.)

To give us some idea of the power of the NMR technique in solving stereochemical problems, let us use the following reaction. Two products *B* and *C* are theoretically possible if the compound *A* is hydrogenated.

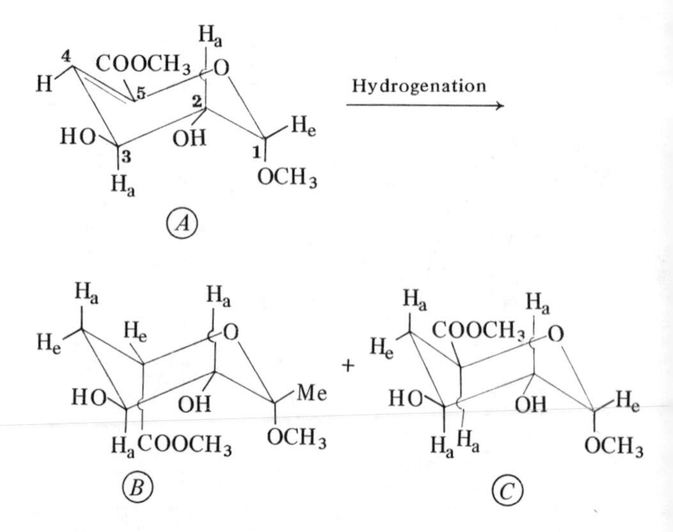

However, only one of the two products was obtained in the crystalline form, on experimentation.[33] The NMR spectrum of the crystalline product in deuterated chloroform was very complicated. A trace of acid was therefore added to the solution to suppress the spin coupling between hydroxyl groups on C_2 and C_3 and neighboring protons. The spectrum thus obtained was better resolved (particularly the multiplets at 5.89τ and 6.50τ) and was analyzed as shown in Table 4.3.

Let us now see how conclusions about the stereochemistry of various protons in the compound can be derived from the given spectral information.

Table 4.3 Analysis of NMR spectrum of a uronic acid ester in acidified deuterated chloroform

Peak no.	Tau (τ) value	No. of protons	Multiplicity	Coupling constants	Inference (see text)
1	5.24	1	Doublet	$J_{1e\text{-}2a} = 3.0$ Hz	H_1 equatorial
2	5.59	1	Symmetrical triplet	$J = 5.0$ Hz	H_5 equatorial
3	5.89	1	Sextet	$J_{3a\text{-}4e} = 4.0$ Hz $J_{3a\text{-}2a}$, $J_{3a\text{-}4a} = 7.2$ Hz	H_3 axial
4	6.23	3	Singlet	—	$-C(=O)OCH_3$
5	6.49	3	Singlet	—	$-C(=O)CH_3$
6	6.50	1	Quartet	$J_{2a\text{-}1e} = 3.0$ Hz $J_{2a\text{-}3a} = 7.2$ Hz	H_2 axial
7	7.58	1	Octet	$J_{gem} = 13.3$ Hz $J_{4e\text{-}5e} = 5.0$ Hz $J_{4e\text{-}3a} = 4.0$ Hz	H_4 equatorial
8	8.21	1	Septet	$J_{gem} = 13.3$ Hz $J_{4a\text{-}3a} = 7.5$ Hz $J_{4a\text{-}5e} = 5.0$ Hz	H_4 axial

Peak 1. This peak at the extreme downfield ($\tau = 5.24$) must be due to the proton that is the most deshielded. H_1 is such a proton because C_1 has two ether linkages. Splitting of the peak into a doublet with $J = 3$ Hz further confirms this conclusion, and indicates that H_1 is coupled with H_2 through J_{ae} interaction. The compound was known to be an α anomer, that is, H_1 was known to be in equatorial position. It was therefore concluded that H_2 is axial.

Peak 2. The proton at C_5 must be the one that is next most deshielded by an ether oxygen and an ester group. The symmetrical triplet ($J = 5.0$ Hz) at 5.59τ must therefore be due to H_5. The fact that a symmetrical triplet is obtained indicates that the two neighboring protons on C_4 must be symmetrically oriented with respect to H_5. This is possible only if H_5 is in equatorial position (hence $J_{4a-5e} = J_{4e-5e} = 5.0$ Hz). If H_5 is axial (as in C), a distorted multiplet would be obtained because $J_{4a-5a} >$ and $\neq$ J_{4a-5e} or J_{4e-5e}.

Peaks 3 and 6. The resolution of these two peaks was particularly improved when the compound was acidified. This indicated that protons at C_3 and C_2 must be responsible for peaks 3 and 6, as they alone would be spin-coupled with the hydroxyl groups in the absence of an acid.

Peak 6 at 6.50τ must be attributed to the H_2 axial proton. The splitting of the peak into a quartet suggests that H_{2a} is coupled with H_{3a} ($J = 7.2$ Hz) on one hand and with H_{1e} ($J = 3.0$ Hz) on the other, thus producing a quartet, according to the theoretical pattern shown here.

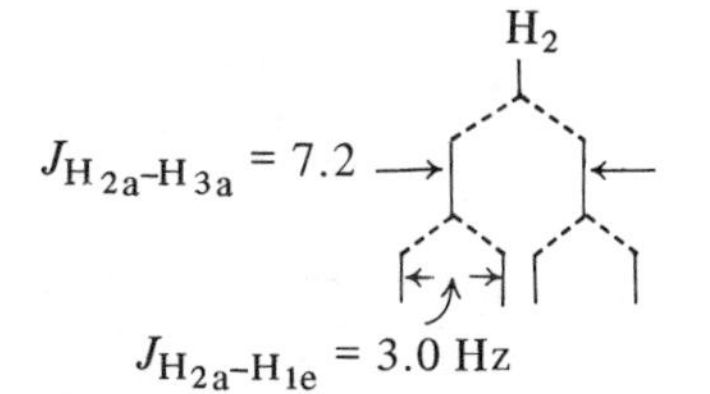

The coupling constant $J = 7.2$ indicates the axial disposition of not only the H_2 proton but also of the H_3 one.

Peak 3, a sextet at $\tau = 5.89$, can similarly be assigned to the H_3 axial proton, which can couple with H_4 axial and H_2 axial, with $J_{aa} = 7.2$ Hz and with H_4 equatorial having $J_{ae} = 4$ Hz, according to the pattern shown for H_3.

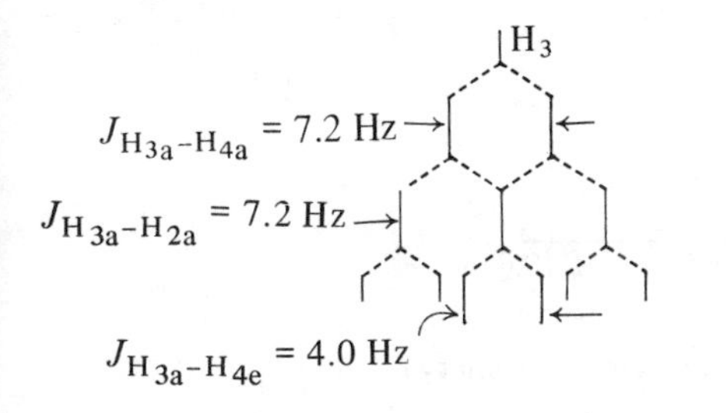

Peaks 4 and 5. These singlets at 6.23τ and 6.49τ, respectively, represent three protons each. Methyl protons of the carbomethoxy

$$-C\overset{\diagup O}{\underset{\diagdown OCH_3}{}}$$

group, however, are less shielded than those of the methoxy

$$C-O\diagdown CH_3$$

group. Hence the lower field signal at 6.23τ is assigned to carbomethoxy protons, while the singlet at 6.49τ must be due to the methoxy group on C_1. These methyl protons are not spin coupled with other protons, and would thus appear as singlets.

Peaks 7 and 8. These two peaks are attributed to the two protons on C_4. Of these, the octet at 7.58τ is assigned to the equatorial proton and the septet at 8.21τ is presumed to be due to the axial proton, because an axial proton usually resonates at a higher field than the corresponding equatorial proton.

The splitting of peak 7 into an octet is due to the three different spin interactions, the strongest of which is at the H_4 axial proton, where $J_{gem} = 13.3$ Hz.* The coupling with the H_5 equatorial proton is 5.0 Hz, while that with the H_3 axial is 4.0 Hz. These interactions with three protons produce an octet, as shown below.

The axial proton at C_4 is also similarly coupled with H_{4e}, H_{3a}, and H_{5e}, with $J = 13.3$, 7.5, and 5 Hz, respectively, and should therefore give an octet. A septet with

---

* The coupling between two nonequivalent protons on the *same* carbon atom is called *geminal coupling*, and the coupling constant, designated J_{gem} or $^2J_{HH}$, may be positive or negative. Its value depends on the state of hybridization of the carbon atom on which the protons are located. For example, for most sp^3-hybridized groups, the $^2J_{HH}$ value is in the range of -10 to -15 Hz; in sp^2-hybridized ethylene, it is +2.3 Hz; whereas in cyclopropanes, in which partial sp^3 as well as sp^2 characters exist, it is -4.8 Hz.

The value of the geminal coupling constant is largely influenced by the nature, position, and orientation of the substituents in the molecule. For example, in formaldehyde,

$$H_2C{=}O,$$

in which the electronegative substituent =O is on the carbon bearing the geminal protons, the coupling is very strong ($^2J_{HH}$ = +42 Hz), but it is weak in

$$H_2C{=}C(H)OCH_3$$

($^2J_{HH} = -2.0$ Hz), in which the electronegative substituent is one atom further removed.

the central subpeak having double the usual intensity is obtained, however, due to the overlap, as shown below.

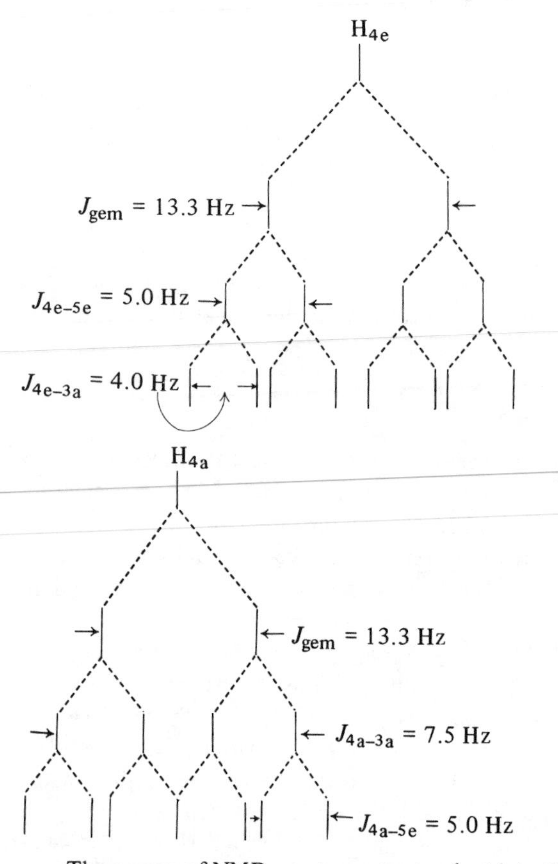

The power of NMR spectroscopy to elucidate the structure and precise geometry of olefinic compounds has been demonstrated by Stehling and Bartz.[19] After having studied some 60 olefins, these workers not only presented a comprehensive correlation of chemical shifts and coupling constants with the structure, but also gave characteristic spectral patterns of olefins.

For example, the structure of a dimer obtained by the oligomerization of 3-methylbutene-1 was determined to be

$$H_a H_b C{=}C(CH_{3d})-C(H_c)(CH_{3f})-C(CH_{3g})_2-CH_{2e}-CH_{3h}$$

on the basis of the spectrum shown in Fig. 4.26.

In this spectrum, signals at 5.26τ and 5.38τ are considered to be characteristic of olefinic protons in the $H_2C{=}CR_2$ system. Of the two protons, the proton *cis* to a

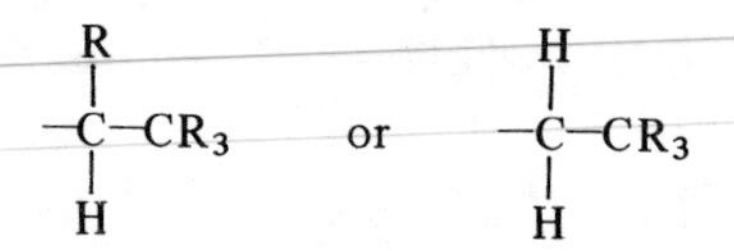

group is considered to give a signal at precisely 5.26τ. The identity of protons *a* and *b* is thus established.

The one-proton quartet centered at 7.95τ is attributed to the α proton *c* on the basis of a precise correlation chart given in Reference 19, and also on the basis of the fact that its coupling with the methyl protons *f* would result in a quartet.

The three-proton multiplet at 8.30τ is claimed to have a pattern characteristic of the CH_3 group in a

$$H_2{-}C{=}C\begin{matrix}R\\CH_3\end{matrix}$$

system.

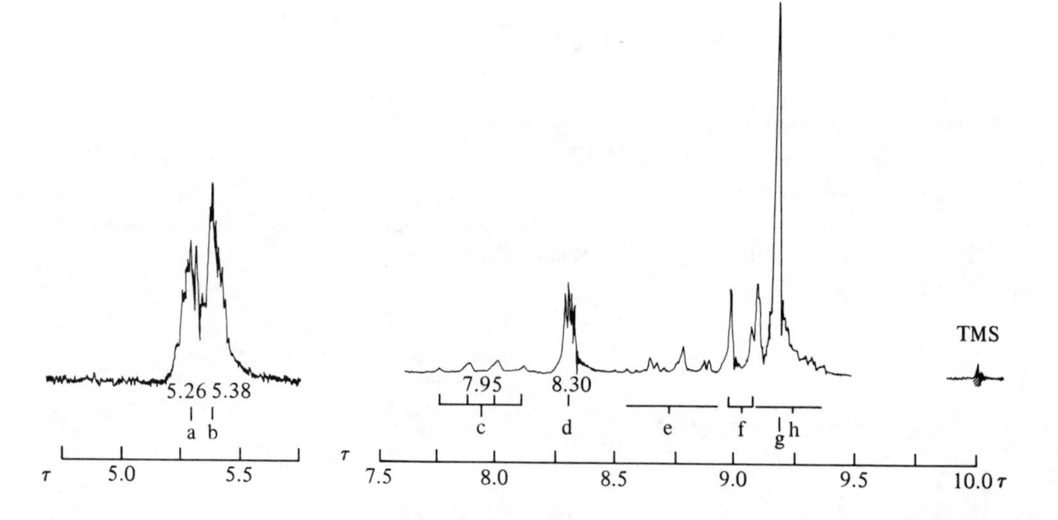

Fig. 4.26 NMR spectrum of a dimer of 3-methylbutene-1 (courtesy American Chemical Society).

Therefore, on the basis of these peaks alone, a structure

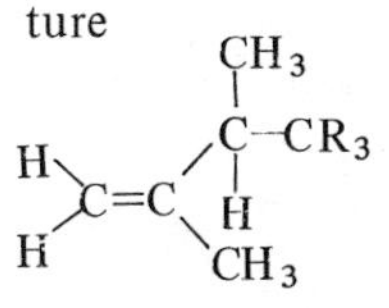

has been established. With regard to the $-CR_3$ part of the molecule, a singlet at 9.18τ is considered to be typical of an "internal" geminal dimethyl group, whereas the multiplets centered at about 8.78τ and 9.18τ have the characteristic "isolated" ethyl group pattern.

ii) Time averaging and its application to rate processes

The electronic environment of nuclei in a substance very often changes because two or more states or subspecies of a molecule undergo rapid interconversion, maintaining the same ratio of one to another, and are thus said to be in a state of rapid equilibrium. Such an equilibrium may be due to chemical exchange between the substance and the solvent, or to tautomerism, or to conformational interchange between axial and equatorial positions, or to formation of different ionic species, or to the formation of complexes, or to different positions of protonation, etc. In short, the equilibrium processes of interest may involve either intramolecular and intermolecular exchange of nuclei or rotations about bond axes. In the ultraviolet or infrared spectrum of such an equilibrium mixture, we observe absorption peaks corresponding to both states, but in NMR spectroscopy, two separate signals are observed if the interchange between two equilibrium positions is sufficiently slow.

Proton exchange. A spectrum of pure alcohol containing only a few drops of water shows peaks corresponding to both alcoholic and aqueous hydroxyl groups. At this concentration of water, the frequency of exchange between the two hydroxyl protons is sufficiently below the frequency of nuclear transitions. However, with the increasing concentration of water, the frequency of proton exchange also increases. As a result, the two peaks first broaden, then coalesce, and finally a single peak is observed at a position between the two separate ones

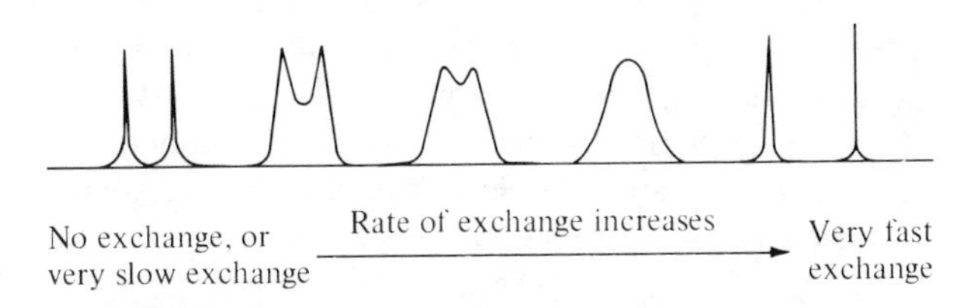

Fig. 4.27 Dependence of shapes of NMR peaks on variations in proton exchange rates in an equilibrium system.

(Fig. 4.27). This phenomenon enables scientists to calculate reaction rates, a value which would be difficult or impossible to measure in any other way[34,35,36].

Rotation about bond axes. Perhaps the best illustration of the application of NMR to the kinetic and thermodynamic study of an equilibrium involving rotation about bond axes is provided by inversion of partially deuterated cyclohexane, C_6HD_{11}.

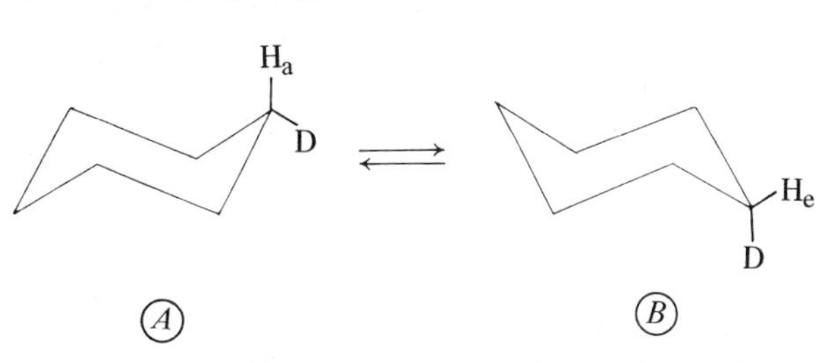

Although signals for equatorial protons normally appear at lower field strengths than the signals for axial protons, at room temperature the exchange between the two forms *A* and *B* is so rapid that only a single sharp peak is obtained in NMR at 8.63τ. On the other hand, at $-89°$, one can observe two sharp peaks, one at 8.88τ corresponding to axial and another at 8.40τ corresponding to equatorial protons. As the temperature is raised, the two peaks become broader, and just merge to a single peak at $-60.3°$.

The spectral behavior in this case is thus similar to that observed in the case of proton exchange. Application of this phenomenon to the study of conformation is easy to understand.

For example, the *cis* and *trans* isomers of decalin can be distinguished because the rigid isomer *trans* decalin contains several differently situated ring hydrogen atoms and a corresponding number of spin-coupled interactions. The resulting spectrum is therefore complex, and a broad region of absorption is observed because most of the peaks fall in the same part of the spectrum. The *cis* isomer, on the other hand, alternates between two conformations, thus rapidly changing the environments of hydrogen atoms from axial to equatorial, and vice versa. The spectral bands for *cis* isomers are thus simpler and narrower because the numbers of nonequivalent ring hydrogens in it are, on a time average, smaller.

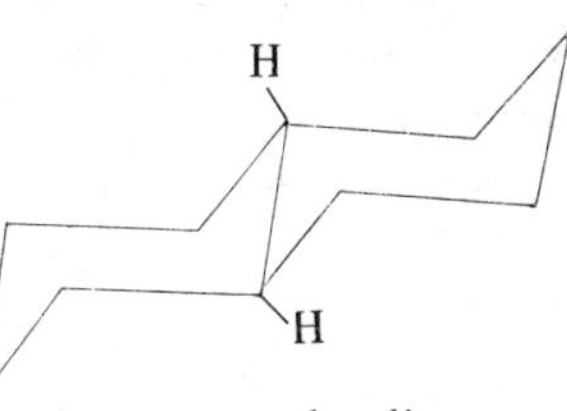

trans decalin

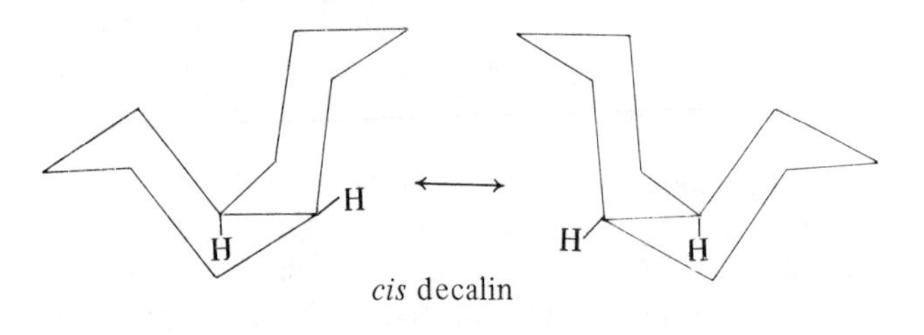

cis decalin

Since the positions and shapes of spectral lines change with changes in reaction rates, it is also possible to use NMR to derive such information as rate constants, free-energy differences, activation enthalpy and entropy, etc.[37] To derive information of this nature, it may be necessary to use computers for the complete analysis of spectra.

Hindered rotation. Isomers which are *cis* and *trans* with respect to a double bond can usually be separated from each other and identified because there is a very high energy barrier between them, which hinders internal rotation. However, in compounds in which isomerism is due to partial double bonds, the energy barrier which hinders rotation may not be high enough to separate *cis* from *trans* isomers. NMR spectroscopy has proved to be an extremely useful tool for the study of hindered internal rotation in such compounds, particularly if the rate of interconversion between the two rotamers is slow enough to allow a difference in chemical shift between signals arising from the two conformers.

Let us consider the case of dimethylacetamide, in which the C—N bond has partial double bond character due to the resonance:

$$O{=}C(CH_3){-}\ddot{N}(CH_3)_2 \longleftrightarrow \bar{O}{-}C(CH_3){=}\overset{+}{N}(CH_3)_2$$

Consequently, rotation around the bond is partially hindered. At low temperatures, the rotation around this bond is so slow that the protons of the N-methyl group closer to the oxygen show a chemical shift different from that shown by the protons of the other N-methyl group. (The environment of the methyl protons depends on whether they are closer to the oxygen or to the third methyl group.) Consequently, two separate three-proton peaks, representing the two N-methyl groups, are observed at low temperature. As the rotation around the C—N bond is only partially hindered, the increase in the temperature of the sample increases the rotation, and when this occurs the two peaks move closer to one another, until at 50° they coalesce, as a result of the frequency of rotation becoming equal to the difference in frequency of the two signals. Since the separation of resonance lines depends on the potential energy barrier to rotation, and since this barrier is overcome when the temperature of the sample is increased, a plot of a function of the peak separation versus the inverse of the absolute temperature enables us to determine the value of the energy barrier to rotation. In the case of dimethylacetamide, this value is about 14 kcal/mole, as determined by this process.

Another example of the application of NMR in such a study can be obtained from papers published on the synthesis of 4-acetamido-4-deoxy-L-erythrofuranoside.[38,39]

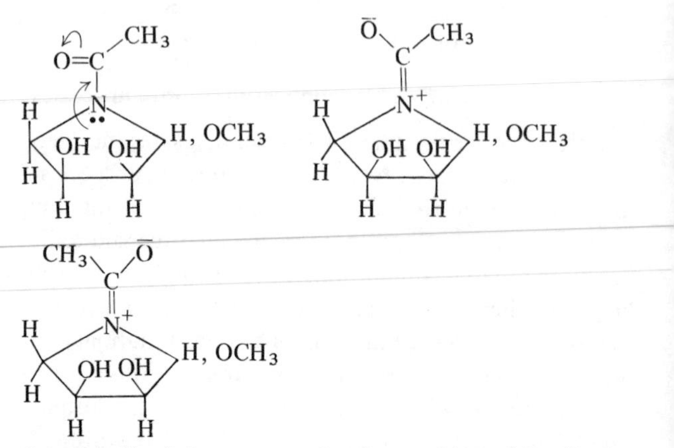

4-acetamido-4-deoxy-L-erythrofuranoside and its dipolar structures

On the basis of the unusually complex NMR spectrum of this furanoside derivative at room temperature, it was initially suggested that either it must have internal hindered rotation or it must be an approximately 1:1 mixture of α and β anomers. Its spectrum showed:

i) Two one-proton doublets at 4.95 and 5.04 ppm in the general region characteristic of anomeric protons
ii) Two three-proton peaks at 3.36 and 3.41 ppm, assigned to the methoxyl protons
iii) Two three-proton peaks at 2.08 and 2.12 ppm, assigned to the protons of the N-acetyl group

However, when the spectrum of the furanoside derivative was subsequently determined at an elevated temperature (80°), it was observed that the two anomeric resonances were broadened and that the *O*-methyl and N-acetyl absorption peaks coalesced to a single peak each. When the furanoside derivative was cooled to room temperature, the doublet pattern was restored. On the basis of this evidence, it was suggested that the compound exists as two rotational isomers and not as a mixture of anomers. The rotational isomerism in this case is due to the partial double-bond character of the C—N bond arising from resonance conjugation between the *p* orbital of nitrogen and the p orbital of the π electron system. This character gives rise to the following two dipolar structures.

Tautomerism. The NMR technique has been found useful in studying keto-enol tautomeric equilibria in β-keto esters, β-diketones and β-ketoaldehydes.

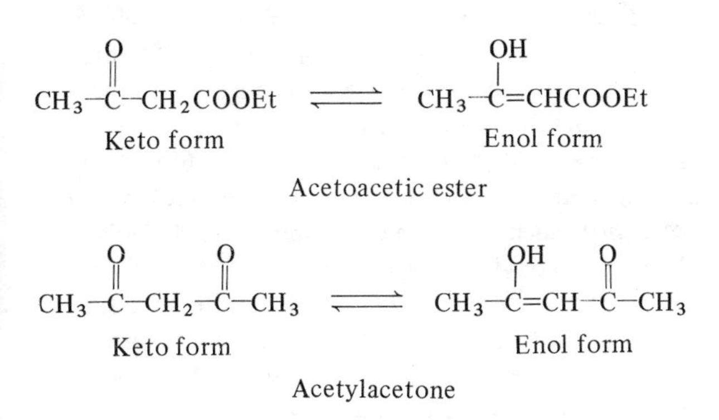

For relatively slow tautomerism, signals from both tautomers occur independently. As is often the case, various signals in such a spectrum are well separated. It is thus possible to determine the keto–enol ratio, and thus the equilibrium constant, by comparing the intensities of the keto ($-CH_2$) and enol ($=CH$) signals. Equilibrium constants for ethylacetoacetate and its mono-, di-, and tri-γ-fluoro derivatives were determined by this technique.[40]

Care should be exercised in detecting tautomerism by NMR techniques, since tautomerism could be masked by the rapid exchange of protons between the tautomers. It is also possible that the concentration of tautomers in equilibrium in the solvents employed could be too small to be detected.[41]

Hydride ion shift. The time-averaging technique has been employed also in the rate and $\Delta H^{\ddagger}$ determinations of 2,3- (and 5,6-) hydride shift.[42] The 2-norbornyl cation in this study was presumed to undergo (i) very fast Wagner-Meerwein rearrangement, (ii) very fast 2,6-hydride shift and (iii) a slow 2,3-hydride shift, at $-120°$, as determined from its NMR spectrum.

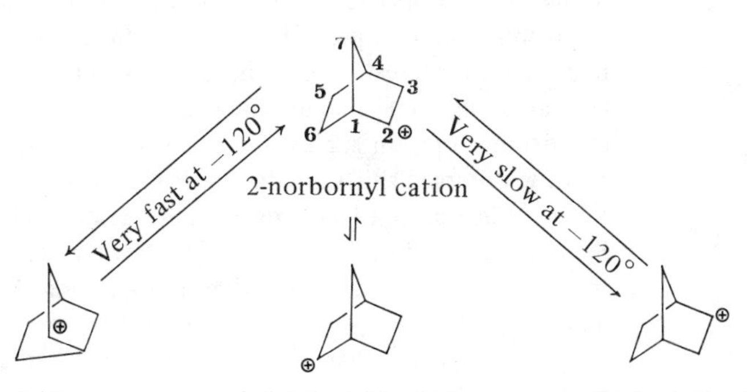

At this temperature three peaks were obtained because fast rearrangement and hydride shift caused three protons at C_1, C_2, and C_6, to average and give a single peak, and also three protons at C_3, C_5, and C_7, to give another averaged single peak, but the proton at the C_4 bridgehead remained unaveraged and gave the third peak. However, as the temperature increased, the 2,3 hydride shift also became fast, and the three peaks began broadening and coalescing until at about 0° only a single peak was obtained, thus showing that the 2,3 shift now had become fast enough to make *all* the hydrogens in the cation equivalent. On the basis of computer programming, the rate constants at various temperatures and the $\Delta H^{\ddagger}$ value for the 2,3-hydride shift were calculated.

Spin decoupling and spin tickling by double resonance. The time-averaging technique has been used in a very interesting manner to study the spectra of relatively complex molecules. In such molecules, if several of the coupling constants have nearly the same value, or if long-range coupling is present, or if a complex absorption multiplet buries one or more nuclei, it is hard to determine which nuclei are spin-coupled. In such a case, more information can be obtained about the spectrum if spin-spin coupling interactions between various nuclei in the compound can be temporarily eliminated.

We have seen earlier that NMR peaks split because the nuclei which cause splitting orient their spins with or against an applied magnetic field and transmit the information concerning their spin states to the neighboring nuclei (Section 4.13). Transmission of this information is possible because the lifetime of nuclei in a given spin state is long in comparison with the reciprocal of the difference ($\Delta\nu$) in chemical shifts, in Hz, of the coupled nuclei.

If, however, the lifetime of a spin state is reduced, i.e., if the exchange between spin states of nuclei is speeded up, the absorbing nuclei can receive only the time-averaged information about the neighboring nuclei. Spin–spin coupling and splitting can thus be eliminated.

In practice this is achieved by techniques such as nuclear magnetic double resonance (NMDR) which involve *spin-decoupling* and *spin-tickling* methods.[43,44,45,46,47]. When these methods are used, the nucleus causing the spin-spin interaction is irradiated at its resonance frequency.

In nuclear magnetic double resonance (NMDR) experiments, formulations such as $^1H - \{^1H\}$ are used to indicate that the proton written outside the braces is being observed, while the other $\{^1H\}$ inside the braces is being irradiated. Such experiments require two radio-frequency fields, whose intensities can be separately controlled. One of them is called the *observing field,* and is used to study the resonance of ^1H, while the other nucleus $\{^1H\}$ is being irradiated by the second field, called the *decoupling field.* The magnitude H_1 of the decoupling field is selected to be at least as great as the total width of the resonance to be irradiated; its frequency is called W_1. The magnitude and the frequency of the observing field are referred to as H_2 and W_2, respectively.

In practice, NMDR experiments can be carried out either by *field-sweep* or by *frequency-sweep* techniques. In the field-sweep arrangement, the spectrum is recorded by scanning the static magnetic field (H_0) while the frequency separation ($W_2 - W_1$) between the two rf fields is set approximately equal to the chemical-shift separation between the nuclei ^1H and $\{^1H\}$. Both W_1 and W_2 are thus scanned during the course of the experiment. In the frequency-sweep arrangement, on the other hand, H_0 is kept constant, W_2 is fixed on the center of the resonance to be irradiated, and the spectrum is observed by slowly sweeping W_1.

In *spin-tickling* experiments, the decoupling field applied to a particular nuclear transition is weak, so that instead of complete decoupling, only the energy associated with the resonance of $\{^1H\}$ is affected. This causes doubleting of the resonance lines of the proton that is being observed. Although spin tickling produces a more complex pattern than that observed for a given proton in simple NMR experiments, the interpretation of the spectrum is usually straightforward, and the position of any particular nuclear transition can be determined very accurately by this technique.

If the two coupled nuclei are not of the same species, the formulation used for such a heteronuclear system is $^1H - \{X\}$, where X represents some nucleus other than a proton, e.g., fluorine. Recently an inexpensive, versatile method has been developed for heteronuclear decoupling, in which an *incoherent radiofrequency field* is used.[48] This is called *noise decoupling.*

Hydrogen bond. As we have said, NMR has become a valuable technique for studying bonding and other molecular interactions. The proton capable of hydrogen bonding experiences different degrees of magnetic shielding in the associated and unassociated states, and its resonance is usually observed at a frequency corresponding to the average shielding for the two states. Since both temperature and degree of dilution affect the degree of hydrogen bonding, this "average" frequency changes when the temperature and the amount of solvent are varied. The difference between the chemical shifts of the proton for the associated and unassociated states is called the *hydrogen bond shift.* Such shifts are very common in alcohols, phenols, carboxylic acids, amines, etc.

NMR has also been used to conduct thermodynamic studies on hydrogen bonding, particularly when the compound under study is in the gas phase. For example, enthalpies of dimerization and tetramerization of methanol have been measured in the vapor phase.[49] These values are especially useful because in the gas phase the hydrogen bond is least modified by the effects of the medium in which the compound is dissolved.

NMR shift reagents. Although it has been known[50] for years that interactions (such as hydrogen bonding) between molecules and solvents cause shifts in the positions of the resonance peaks of protons, the application of this and similar phenomena to improve resolutions of NMR spectra has only just started. Techniques such as NMDR (this page), use of higher frequencies (page 116), deuterium substitution, and formation of derivatives can all be employed with varying success to evaluate chemical shifts of protons whose resonances overlap because of the complexity of molecules under study. However, since 1969—when Hinckley[51] demonstrated that large isotropic shifts are produced for the resonance peaks of the protons of cholesterol when it is placed in solutions of what are now known as *lanthanide-shift reagents*—a great deal of interest has been shown in this *solvent-shift technique* for the simplification of NMR spectra of complex molecules.

Lanthanides are a group of 15 elements (atomic numbers 57 to 71), also known as *rare earths.* The rare-earth complexes are paramagnetic, and have the ability to induce large upfield or downfield shifts in the positions of proton resonance peaks by forming adducts with the molecules of interest. The chemical shifts caused by the lanthanide shift reagents display relatively small line broadening; a recent survey[52] of such reagents suggests that the observed shifts are dipolar in origin. Methods for the preparation of shift reagents are described by Selbin, *et al.*, and by Lyle *et al.*[53,54] A review of these reagents has appeared in the bulletin of a chemical supply house.[55]

Shift reagents other than those of the lanthanide type, such as cobalt(II) complexes and nickel(II) complexes, are also being investigated.[56,57,58,59]

iii) Other miscellaneous applications of NMR

Determination of aromaticity. With the help of NMR, one can now determine experimentally whether or not a compound has a closed ring of π electrons; and aromaticity can now be defined as the *ability of the molecule to sustain an induced ring current.** This ring current greatly deshields the protons on the ring. The aromatic protons therefore appear at low τ. It must, however, be remembered that, in large rings such as [18] annulene, some protons are directed inside the ring and are thus abnormally shielded. For example, in one report, of the 18 protons in [18] annulene, 12 protons appeared at very low $\tau = 1.1$, as expected for aromatic protons, while the remaining 6 gave a signal at abnormally high $\tau = 11.8$, indicating that they were directed inside the ring and were shielded not only more than the normal olefinic protons, but also more than the normal alkyl protons.[61]

Determination of molecular weight. NMR integration provides a simple method for determination of molecular weights.[62,63] This method consists in comparing the integrated intensities of an added standard compound and a recognizable peak (or group of peaks) of the unknown compound in a solution containing known weights of the standard and the unknown compounds.

It is claimed that "this method gives smaller errors than the cryoscopic, etc., methods," since "the values are not affected by dissociation and solvent interaction phenomena or various other effects that cause great errors in methods dependent on ideality in colligative properties of liquids." However, the reliability of this method is questionable, due to inherent differences in relaxation times T_1 and T_2 of both the unknown and the reference material.

Quantitative analysis. NMR technique has been applied in several experiments involving quantitative analysis. The underlying principle that makes such use possible is that the NMR peaks of a compound present in a solvent have areas proportional to the concentration of the solution. Another factor which makes the NMR method attractive is the fact that it offers a rapid, nondestructive analysis.

Moisture analysis by NMR has now become so common that the method has found commercial applications, and automatic control of drying and blending operations of many materials is now being carried out by NMR instruments specially designed for this purpose.[64,65]

NMR technique has also been used in measurements of isotopic abundance ratios[66] such as $^2H/^1H$ and $^6Li/^7Li$, analysis of mixtures,[67] percentages of active hydrogen,[68] proton counting, and so forth.

Pharmaceutical and medicinal chemistry. Because of the advantages of high spectral specificity and rapid analysis with minimal preparation of samples, NMR spectroscopy has found wide acceptance in pharmaceutical and medicinal chemistry. Applications of NMR to this field have been reviewed by a number of authors.[69,70,71,72]

Study of other nuclei. Rapid progress in the field of proton resonance spectroscopy has prompted a flurry of investigations in the use of nuclei other than those of hydrogen. We have already described the extension of resonance spectroscopy to ^{13}C nuclei (page 122). A discussion—even in a perfunctory manner—of NMR spectroscopy of other nuclei is beyond the scope of the present work. However, a list of references to important review articles and research papers is given in Table 4.4.

Table 4.4 Resonance spectroscopy of nuclei other than those of hydrogen

Nuclei	References	Nuclei	References
^7Li	73, 74	^{35}Cl	88
^9Be	75	^{43}Ca	89
^{11}B	76	^{53}Cr	90
^{13}C	77, 78	^{55}Mn	91
^{14}N	79	^{59}Co	92
^{17}O	80	^{77}Se	93, 94
^{19}F	81, 82	^{79}Br	95
^{23}Na	83	^{119}Sn	96
^{27}Al	84, 85	^{129}Xe	97
^{31}P	79, 86	Organometallic comp.	98
^{33}S	87	Halogens	88

4.16 SELECTED REFERENCES

Because of the great importance of NMR technique to organic chemists, a large volume of literature—research papers, books, monographs, compilations of NMR spectra, and review articles—is constantly being published. We have already referred to some of this literature in this chapter and more references are cited here. Butterworths, Inc., of Washington, D.C., even offer a current-literature service by publishing periodically a continuing survey

* This is only one school of thought. According to Musher,[60] "any suggestion that *aromaticity*—whatever the word may mean—be defined in terms of the ability of a molecule to sustain a ring current must be considered unreasonable." On the basis of quantum-mechanical calculations, Musher has shown that "the anisotropic magnetic susceptibility (and chemical shift) of aromatic hydrocarbons, generally attributed to π electron *ring currents*, can be correctly represented as the sum of contributions from localized electrons of both π and σ character. The *delocalization* of the electronic distribution plays no part, whatever is the effect."

and compilation of citations of literature in various areas of nuclear magnetic resonance spectroscopy, electron spin resonance spectroscopy, and nuclear quadrupole resonance spectroscopy.

References cited in the text

1. L. F. Johnson, *Anal. Chem.* **43**, 28A (1971).
2a. R. R. Ernst and W. A. Anderson, *Rev. Sci. Instr.* **37**, 93 (1966).
2b. E. D. Becker, J. A. Ferretti, and T. C. Farrar, *J. Am. Chem. Soc.* **91**, 7784 (1969).
2c. J. S. Waugh, *J. Mol. Spectrosc.* **35**, 398 (1970).
2d. R. A. Flath *et al., Applied Spectrosc.* **21**, 183 (1967).
3. J. A. Jackson and W. H. Wright, *Rev. Sci. Instr.* **39**, 510 (1968).
4. J. T. Arnold, S. S. Dharmatti, and M. E. Packard, *J. Chem. Phys.* **19**, 507 (1951).
5. T. Schaefer, W. F. Reynolds, and T. Yonemoto, *Can. J. Chem.* **41**, 2969 (1963).
6. A. D. Buckingham, *Can. J. Chem.* **38**, 300 (1960).
7a. G. S. Reddy and J. H. Goldstein, *J. Chem. Phys.* **36**, 2644 (1963).
7b. G. S. Reddy, J. H. Goldstein, and L. Mendell, *J. Am. Chem. Soc.* **83**, 1300 (1961).
8. J. S. Waugh and R. W. Fessenden, *J. Am. Chem. Soc.* **79**, 846 (1957); **80**, 6697 (1958).
9. K. D. Kopple and D. H. Marr, *J. Am. Chem. Soc.* **89**, 6193 (1967).
10. F. Bovey, *Nuclear Magnetic Resonance Spectroscopy*, Academic Press, New York, 1969, page 67.
11. G. M. Badger, G. E. Lewis, and U. P. Singh, *Austral. J. Chem.* **19**, 1461 (1966).
12. G. M. Badger, G. E. Lewis, and U. P. Singh, *ibid.* **19**, 257 (1966).
13. D. J. Patel, M. E. H. Howden, and J. D. Roberts, *J. Am. Chem. Soc.* **85**, 3218 (1963).
14. N. Nakagawa, S. Saito, A. Suzuki, and M. Itoh, *Tetrahedron Letters,* 1003 (1967).
15. L. R. Subramanian and G. S. Krishna Rao, *ibid.* 3693 (1967).
16. G. A. Neville and I. C. Nigam, *Can. J. Chem.* **47**, 2901 (1969).
17. A. D. Buckingham, T. Schaefer, and W. A. Schneider, *J. Chem. Phys.* **32**, 1227 (1960).
18. R. E. Glick, W. E. Stewart, and K. C. Tewari, *ibid.* **45**, 4049 (1966).
19. F. C. Stehling and K. W. Bartz, *Anal. Chem.* **38**,1467 (1966).
20. H. J. Bernstein, J. A. Pople, and W. A. Schneider, *Can. J. Chem.* **35**, 65 (1957).
21. W. A. Anderson, *Phys. Rev.* **102**, 151 (1956).
22. R. E. Richards and T. Schaefer, *Mol. Phys.* **1**, 331 (1958).
23. J. B. Stothers, *Quart. Rev. (London)* **19**, 144 (1965).
24. M. Karplus, *J. Am. Chem. Soc.* **85**, 2870 (1963).
24a. A. A. Bothner-By, *Advn. Magn. Resonance* **1**, 195 (1965).
25. S. Meiboom and L. C. Snyder, *J. Am. Chem. Soc.* **89**, 1083 (1967).
26. H. M. Hutton and T. Schaefer, *Can. J. Chem.* **41**, 2774 (1963).
27. D. H. Williams and N. S. Bhacca, *J. Am. Chem. Soc.* **86**, 2742 (1964).
28. H. Booth, *Tetrahedron Letters,* 411 (1965).
29. B. Coxon, *Tetrahedron* **21**, 3481 (1965).
30. J. D. Davis and V. Van Auken, *J. Am. Chem. Soc.* **87**, 3900 (1965).
31. M. Sundaralingham, *J. Am. Chem. Soc.* **87**, 599 (1965).
32. R. U. Lemieux and J. Howard, *Can. J. Chem.* **41**, 308 (1963).
33. H. W. H. Schmidt and H. Neukom, *Tetrahedron Letters,* 2063 (1964).
34. D. E. Leyden and W. R. Morgan, *J. Chem. Educ.* **46**, 169 (1969).
35. S. Brownstein, E. C. Horswill, and K. U. Ingold, *J. Am. Chem. Soc.* **92**, 7217 (1970).
36. N. C. Li, K. C. Tewari, and F. K. Schweighardt, *J. Phys. Chem.* **75**, 688 (1971).
37. F. H. Marquardt, *J. Chem. Soc.* **B**, 366 (1971).
38. W. A. Szarek and J. K. N. Jones, *Can. J. Chem.* **42**, 20 (1964).
39. W. A. Szarek, S. Wolfe, and J. K. N. Jones, *Tetrahedron Letters,* **38**, 2743 (1964).
40. R. Filler and S. M. Naqvi, *J. Org. Chem.* **26**, 2571 (1961).
41. G. A. Neville and D. Cook, *Can. J. Chem.* **47**, 743 (1969).
42. M. Saunders, P. von R. Schleyer and G. Olah, *J. Am. Chem. Soc.* **86**, 5680 (1964).
43. R. Freeman and D. H. Whiffen, *Proc. Phys. Soc. (London)* **79**, 794 (1962).
44. J. D. Baldeshwieler and E. W. Randall, *Chem. Rev.* **63**, 81 (1963).
45. R. A. Hoffman and S. Forsén, *Progress in NMR Spec.*, Vol. 1, edited by J. W. Emsley, J. Feeney, and L. H. Sutcliffe, Pergamon Press, New York, 1966.
46. L. D. Hall and J. F. Manville, *Advances in Chemistry Series,* Number 74, "Deoxy Sugars," Am. Chem. Soc., Washington, D.C., 1968.
47. W. McFarlane, *Chem. Brit.* **5**, 142 (1969).
48. R. Burton and L. D. Hall, *Can. J. Chem.* **48**, 59 (1970).
49. A. D. H. Clague, G. Govil, and H. J. Bernstein, *Can. J. Chem.* **47**, 625 (1969).
50. P. L. Corio, R. L. Rutledge, and J. R. Zimmerman, *J. Mol. Spectrosc.* **3**, 592 (1959).
51. C. C. Hinckley, *J. Am. Chem. Soc.* **91**, 5160 (1969).

52. W. DeW. Horrocks, Jr., and J. P. Sipe III, *ibid.*, **93**, 6800 (1971).
53. J. Selbin, N. Ahmad, and N. Bhacca, *Inorg. Chem.* **10**, 1383 (1971).
54. S. J. Lyle and A. D. Witts, *Inorg. Chim. Acta* **5**, 481 (1971).
55. J. R. Campbell, *Aldrichimica Acta* **4**, 55 (1971); available from The Aldrich Chemical Company, Milwaukee, Wis.
56. M. Ohashi, I. Morishima, and T. Yonezawa, *Bull. Chim. Soc. (Japan)* **44**, 576 (1971).
57. W. A. Szarek, E. Dent, T. B. Grindley, and M. C. Baird, *J. Chem. Soc. D*, 953 (1969).
58. C. A. Cabrera, G. M. Woltermann, and J. R. Wasson, *Tetrahedron Letters*, 4485 (1971).
59. W. A. Szarek and M. C. Baird, *ibid.*, 2097 (1970).
60. J. I. Musher, *J. Chem. Phys.* **43**, 4081 (1965).
61. L. M. Jackman *et al.*, *J. Am. Chem. Soc.* **84**, 4307 (1962).
62. S. Barcza, *J. Org. Chem.* **28**, 1964 (1963).
63. T. F. Page, Jr., and W. E. Bresler, *Anal. Chem.* **36**, 1981 (1964).
64. T. F. Conway and F. R. Earle, *J. Am. Oil. Chem. Soc.* **40**, 265 (1963).
65. H. Reinhardt, *Exp. Tech. Phys.* **17**, 517 (1969).
66. Varian Associates, *NMR at Work* **57**, (1958).
67. A. Mathias and D. Taylor, *Anal. Chim. Acta* **35**, 376 (1966).
68. P. J. Paulsen and W. D. Cooke, *Anal. Chem.* **36**, 1721 (1964).
69. N. W. Rich and L. G. Chatten, *J. Pharm. Sci.* **54**, 995 (1965).
70. D. M. Rackham, *Talanta* **17**, 895 (1970).
71. G. A. Neville, *Can. Spectrosc.* **14**, 44 (1969).
72. H. W. Avdovich and G. A. Neville, *Can. J. Pharm. Sci.* **4**, 51 (1969).
73. L. D. McKeever in *Ions and Ion Pairs in Organic Reactions*, Vol. 1, edited by M. Sware, Wiley-Interscience, New York, 1971.
74. J. Parker and J. A. Ladd, *J. Organometal. Chem.* **19**, 1 (1969).
75. R. A. Kovar and G. L. Morgan, *J. Am. Chem. Soc.* **92**, 5067 (1970).
76. G. R. Eaton and W. N. Lipscomb, *NMR Studies in Boron Hydrides and Related Compounds*, W. A. Benjamin, New York, 1969.
77. A. Allerhand and E. A. Trull, *Ann. Rev. Phys. Chem.* **21**, 317 (1970).
78. J. B. Stothers, *Quart. Rev. (London)* **19**, 144 (1965).
79. *Progress in NMR Spectroscopy* **6**, (1971).
80. D. J. Sardella and J. B. Stothers, *Can. J. Chem.* **47**, 3089 (1969).
81. C. H. Dungan and J. R. Van Wazer, *Compilation of Reported ^{19}F NMR Chemical Shifts: 1951 to Mid-1967*, Wiley-Interscience, New York, 1970.
82. E. F. Mooney, *Introduction to Fluorine-19 NMR Spectroscopy*, Sadtler Research Labs., Philadelphia Pa., 1970.
83. T. Yagi, I. Tatsuzaki, and I. Todo, *J. Phys. Soc. Jap.* **28**, 321 (1970).
84. L. Petrakis and F. E. Dickson, *Appl. Spectrosc. Rev.* **4**, 1 (1970).
85. H. Haraguchi, *Kogaku No Ryoiku* **24**, 802 (1970).
86. M. M. Crutchfield, C. H. Dungan, J. H. Letcher, V. Mark, and J. R. Van Wazer, *^{31}P NMR*, Wiley-Interscience, New York, 1967.
87. N. L. Retcofsky and R. A. Friedel, *Appl. Spectrosc.* **24**, 379 (1970).
88. C. Hall, *Quart. Rev. Chem. Soc. (London)* **25**, 87 (1971).
89. R. G. Bryant, *J. Am. Chem. Soc.* **91**, 1870 (1970).
90. Y. Egozy and A. Loewenstein, *J. Magn. Resonance* **1**, 494 (1969).
91. S. L. Segel, *J. Chem. Phys.* **51**, 848 (1969).
92. R. L. Martin and A. H. White, *Nature* **223**, 394 (1969).
93. B. M. Dahl and P. H. Nielsen, *Acta. Chem. Scand.* **24**, 1468 (1970).
94. M. Lardon, *J. Am. Chem. Soc.* **92**, 5063 (1970).
95. B. Lindman, H. Wennerstrom, and S. Forsen, *J. Phys. Chem.* **74**, 754 (1970).
96. A. G. Davies, P. G. Harrison, J. D. Kennedy, T. N. Mitchell, R. J. Puddephatt, and W. McFarlane, *J. Chem. Soc.***C**, 1136 (1969).
97. A. K. Jameson, C. J. Jameson and H. S. Gutowsky, *J. Chem. Phys.* **53**, 2310 (1970).
98. L. M. Stacey, B. Pass, and H. Y. Karr, *Phys. Rev. L.* **23**, 1424 (1969).

General references

99. A. F. Casy, *PMR Spectroscopy in Medicinal and Biological Chemistry*, Academic Press, New York, 1971.
100. R. J. Abraham, *Analysis of High Resolution NMR Spectra*, Elsevier, New York, 1971.
101. W. W. Pandler, *Nuclear Magnetic Resonance*, Allyn and Bacon, Boston, Mass., 1971, second edition.
102. C. Franconi, *Magnetic Resonances in Biological Research*, Gordon and Breach Science Publishers, New York, 1971.
103. P. Diehl, E. Fluck, and R. Kosfeld, *NMR: Basic Principles and Progress*, Volumes I and II, Springer-Verlag, New York, 1970.
104. B. I. Ionin and B. A. Ershov, *NMR Spectroscopy in Organic Chemistry*, Plenum Press, New York, 1970, second edition.
105. L. M. Jackman and S. Sternhell, *Applications to NMR Spectroscopy in Organic Chemistry*, Pergamon Press, New York, 1969, second edition.
106. R. M. Lynden-Bell and R. K. Harris, *NMR Spectroscopy*, Appleton-Century-Crofts, New York, 1970.

107. *An Introduction to Spectroscopic Methods for the Identification of Organic Compounds,* Vol. 1, edited by F. Scheinmann, Pergamon Press, London, 1970.
108. R. T. Schumacher, *Introduction to Magnetic Resonance,* W. A. Benjamin, New York, 1970.
109. F. A. Bovey, *NMR Spectroscopy,* Academic Press, New York, 1969.
110. E. D. Becker, *High Resolution NMR,* Academic Press, New York, 1969.
111. R. H. Bible, Jr., *Guide to NMR Empirical Method,* Plenum Press, New York, 1967.
112. W. Brugel, *NMR Spectra and Chemical Structure,* Academic Press, New York, 1967.
113. *NMR for Organic Chemists,* edited by D. W. Mathieson, Academic Press, New York, 1967.
114. D. Chapman and P. D. Magnus, *Introduction to Practical High Resolution NMR Spectroscopy,* Academic Press, New York, 1966.
115. P. L. Corio, *Structure of High Resolution Spectra,* Academic Press, New York, 1966.
116. A. Carrington and A. D. McLachlen, *Introduction to Magnetic Resonance,* Harper & Row, New York, 1966.
117. J. W. Emsley, J. Feeney and L. H. Sutcliffe, *Progress in NMR Spectroscopy,* Vols. 1–3, Pergamon Press, New York, 1966–67.
118. J. W. Emsley, J. Feeney and L. H. Sutcliffe, *High Resolution NMR Spectroscopy,* Vols. 1 and 2, Pergamon Press, New York, 1965.
119. A. K. Bose, "Proton NMR Spectroscopy," in *Interpretive Spectroscopy,* edited by S. K. Freeman, Reinhold, New York, 1965.
120. B. Pesce, editor, *NMR in Chemistry,* Academic Press, New York, 1965.
121. R. H. Bible, Jr., *Introduction to NMR Spectroscopy,* Plenum Press, New York, 1965.
122. D. W. Mathieson, editor, *Interpretation of Organic Spectra,* Academic Press, New York, 1965.
123. J. R. Dyer, *Applications of Organic Spectroscopy of Organic Compounds,* Prentice-Hall, Englewood Cliffs, N.J., 1965.
124. N. S. Bhacca and D. H. Williams, *Applications of NMR Spectroscopy in Organic Chemistry,* Holden-Day, San Francisco, Calif., 1964.
125. K. B. Wiberg and B. J. Nist, *The Interpretation of NMR Spectra,* W. A. Benjamin, New York, 1962.
126. J. D. Roberts, *An Introduction to the Analysis of Spin-Spin Splitting in High Resolution NMR Spectra,* W. A. Benjamin, New York, 1962.

Compilations of NMR spectra

127. F. A. Bovey, *NMR Data Tables for Organic Compounds,* Vol. 1, Wiley-Interscience, New York, 1967.
128. *Sadtler Standard NMR Spectra,* Vols. 1–15, 10,000 spectra, Sadtler Research Laboratories, Philadelphia, Pa., 1970.
129. *Selected NMR Spectral Data* (loose-leaf data sheets), Vols. I-III (1947–1969), American Petroleum Institute, Project 44, Vol. II, Chemical Thermodynamic Properties Center, A & M College of Texas, College Station, Texas.
130. *High Resolution NMR Spectra Catalogue,* Vols. 1 and 2, Varian Associates, Palo Alto, Calif.
131. *A Catalogue of the NMR Spectra of Hydrogen in Hydrocarbons and their Derivatives,* Humble Oil and Refining Co., Baytown, Texas.
132. *NMR, EPR and NQR Current Literature Service,* Butterworths, Washington, D.C.
133. H. M. Hershenson, *NMR and ESR Spectra Index,* Acaemic Press, New York, 1965.
134. M. G. Howell, A. S. Kende and J. S. Webb, *Formula Index to NMR Literature Data,* Vols. 1 and 2, Plenum Press, New York, 1966.
135. H. Suhr, *Anwendungen der Kermagnetischen Resonanz im der Organischen Chemie,* Springer-Verlag, New York.

Review articles

136. L. F. Johnson, *Anal. Chem.,* **43**, 28A (Feb.) (1971) [High magnetic field, 300 MHz instrument].
137. T. A. Crabb, R. F. Newton and D. Jackson, *Chem. Revs.* **71**, 109 (1971) [Stereochemical studies of nitrogen bridgehead compounds].
138. H. W. Avdovich and G. A. Neville, *Can. Chem. Ed.* **5** (no. 3), 17 (1970) [General review].
139. J. M. Rowe, J. Hinton and K. L. Rowe, *Chem. Revs.* **70**, 1 (1970) [Application of NMR to biochemistry of biopolymers].
140. H. Kessler, *Angew. Chem., Internat. Edit.* **9**, 219 (1970) [Determining conformational barriers].
141. W. E. Stewart and T. H. Siddall, *Chem. Rev.* **70**, 517 (1970) [Restricted rotation in amides].
142. G. C. K. Roberts and O. Jardetsky, *Adv. Protein Chem.* **24**, 447 (1970) [Protein structure].
143. D. K. Banerjee, *J. Ind. Chem. Soc.* **47**, 199 (1970) [Pharmaceutical applications].
144. H. E. Hallam and C. M. Jones, *J. Mol. Struct.* **5**, 1 (1970) [Restricted rotation in amides].
145. M. Cohn, *Quart. Rev. Biophys.* **3**, 61 (1970) [Biological applications].
146. M. Barfield and B. Chakrabarti, *Chem. Revs.* **69**, 757 (1969) [Spin-spin coupling].
147. W. McFarlane, *Quart. Rev. (London)* **23**, 187 (1969) [Spin-spin coupling].
148. S. Sternhell, *Quart. Rev. (London)* **23**, 236 (1969) [Spin-spin coupling and structure].

149. J. Burgess and M. C. R. Symons, *Quart. Rev. (London)* **22**, 276 (1968) [Ion–solvent and ion–ion interactions].
150. M. Van Gorkom and G. E. Hall, *Quart. Rev. (London)* **22**, 14 (1968) [Equivalence of nuclei].
151. R. E. Lundin, R. H. Elsken, R. A. Flath and R. Teranishi, *Applied Spec. Rev.*, edited by E. G. Brame, Marcel Dekker, Inc., New York, 1968.
152. E. W. Garbisch, Jr., *J. Chem. Ed.* **45**, 311, 402, 480 (1968) [Analysis of complex spectra].
153. J. F. Hinton and E. S. Amis, *Chem. Revs.* **67**, 367 (1967) [NMR of ions in solvents].
154. G. Slomp and G. J. Lindberg, *Anal. Chem.* **39**, 60 (1967) [NMR of nitrogen-containing organic compounds].
155. R. S. Ferguson and W. D. Phillips, *Science* **157**, 257 (1967) [Advances in instrumentation].
156. B. Dischler, *Angew. Chem., Internat. Edit.* **5**, 623 (1966) [Mathematical treatment of NMR spectra].
157. J. D. Swalen, *Progr. in NMR Spectroscopy* **1**, 3 (1966) [Computer technique].
158. R. A. Hoffmann and S. Forsen, *Progr. in NMR Spectroscopy* **1**, 15 (1966) [Double resonance].
159. R. C. Cookson, T. A. Crabb, J. J. Frankel and T. Hudec, *Tetrahedron Suppl.* **7**, 355 (1966) [Geminal coupling].
160. E. Lusting and W. B. Moniz, *Anal. Chem. Annual Rev.* **38**, 331R (1966) [General review].
161. J. E. Anderson, *Quart. Rev. (London)* **19**, 426 (1965) [Rate processes of inversion of cyclic compounds].
162. G. R. Eaton and W. D. Phillips, *Advan. Mag. Resonance* **1**, 103 (1965).
163. A. A. Bothner-By and J. A. Pople, *Ann. Rev. Phys. Chem.* **43**, (1965).
164. L. W. Reeves, *Adv. Phy. Org. Chem.* **3**, 187 (1965) [Reaction velocities and equilibrium constants by NMR].
165. C. S. Johnson, in *Advances in Magnetic Resonance*, edited by J. S. Waugh, Vol. 1, 1965, Academic Press, New York, page 33 [Chemical rate processes].

PROBLEMS

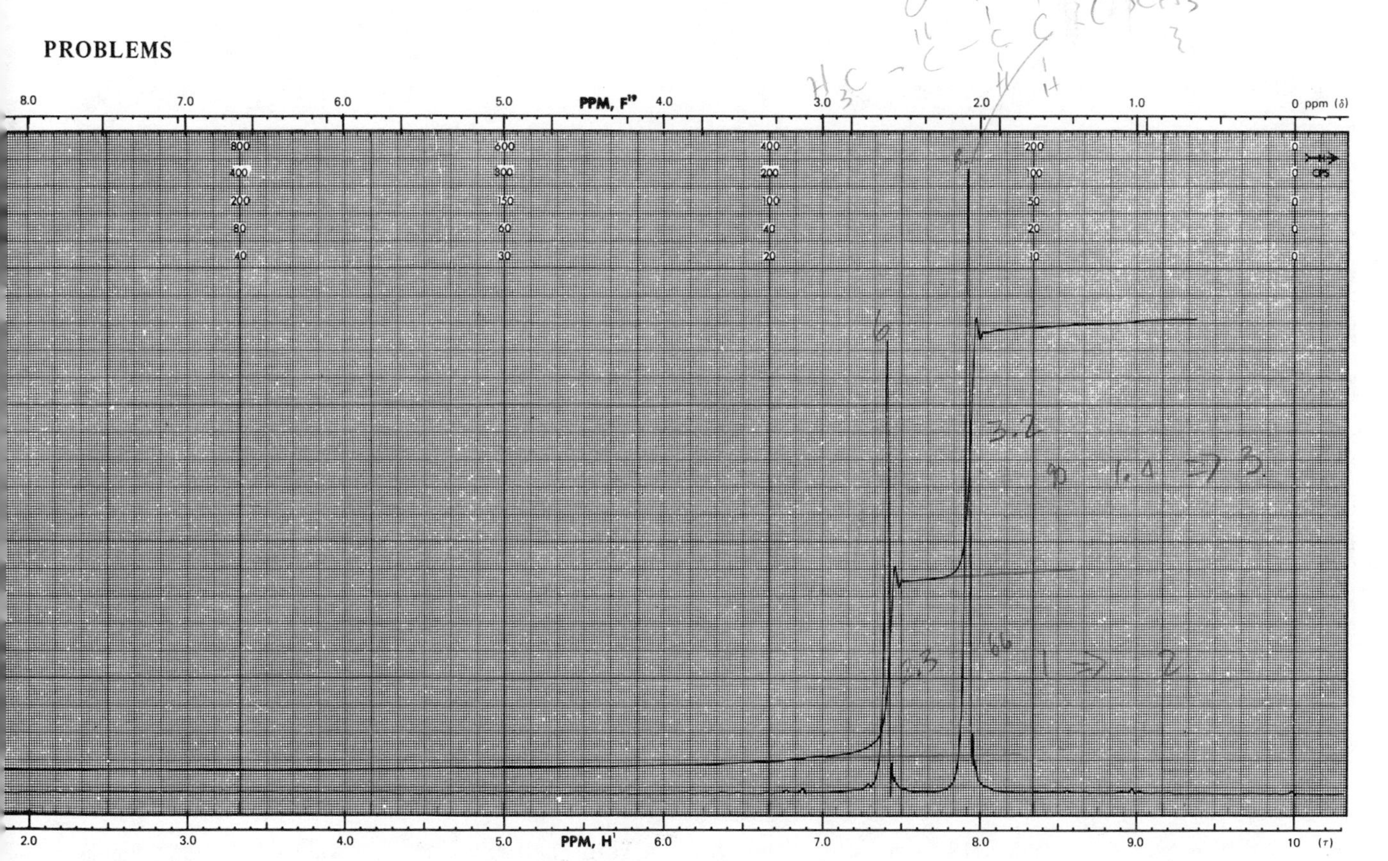

Fig. 4.28 NMR spectrum of the unknown described in Problem 4.12.

Problem 4.12. An unknown liquid has a boiling point of 194° and the molecular formula $C_6H_{10}O_2$. The infrared spectrum of this unknown displays a strong absorption band at 1715 cm^{-1}. The NMR spectrum of the solution of the unknown in CCl_4 is given in Fig. 4.28. Identify the compound.

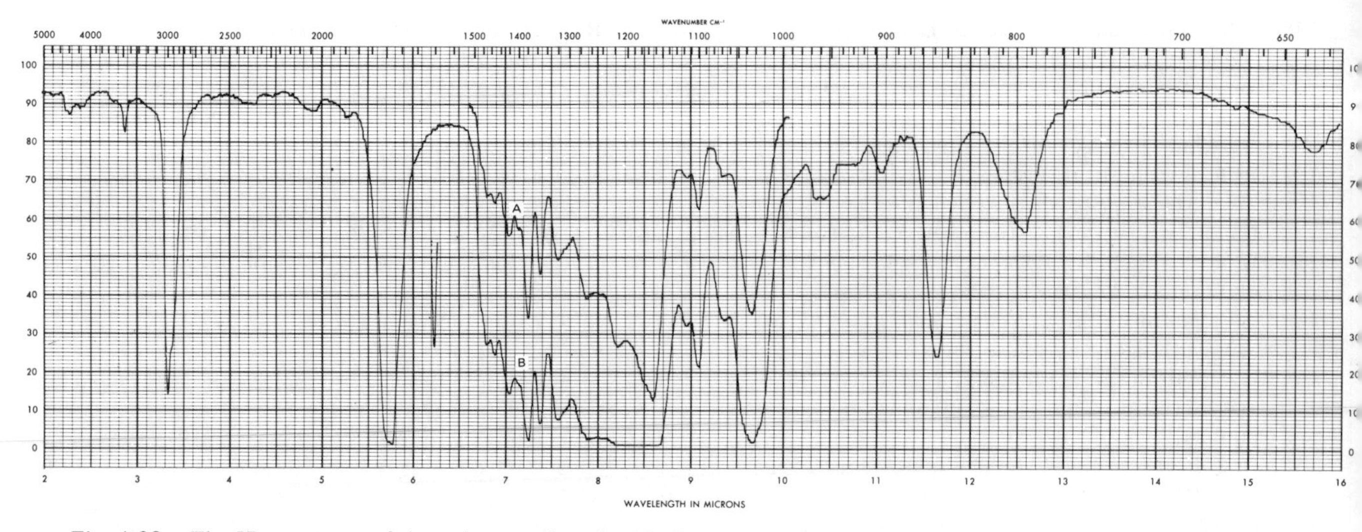

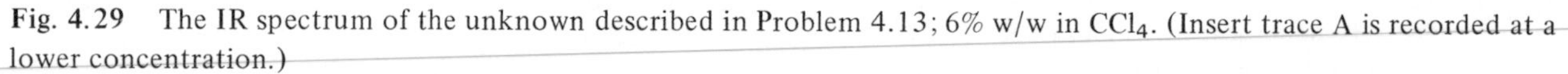

Fig. 4.29 The IR spectrum of the unknown described in Problem 4.13; 6% w/w in CCl_4. (Insert trace A is recorded at a lower concentration.)

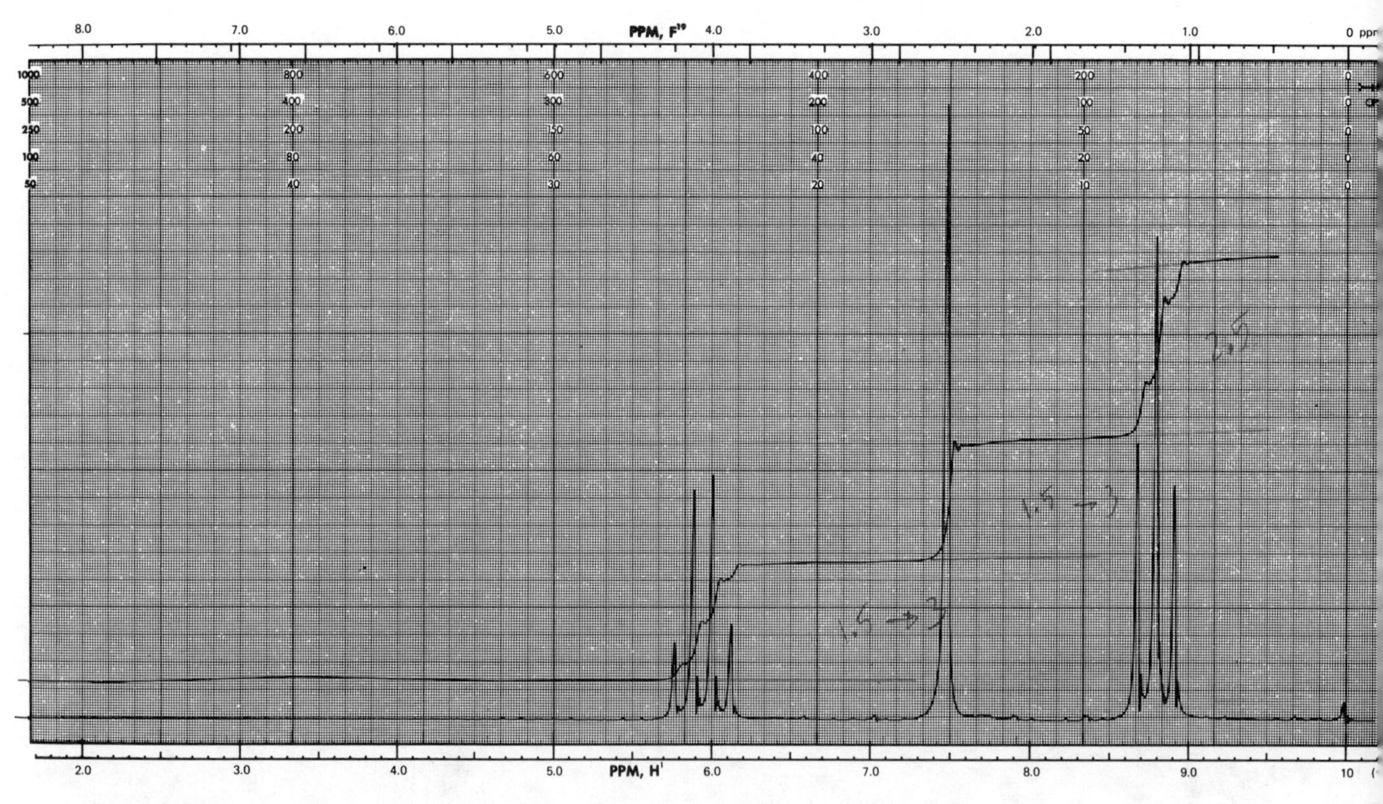

Fig. 4.30 NMR spectrum of the unknown described in Problem 4.13. (Neat liquid.)

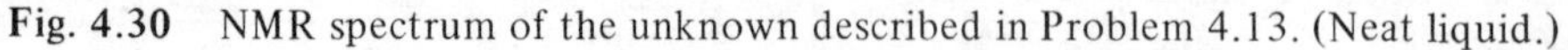

Problem 4.13. The IR and NMR spectra of an unknown liquid with a boiling point of 218° and a molecular formula $C_8H_{14}O_4$ are recorded in Figs. 4.29 and 4.30, respectively. Identify the unknown.

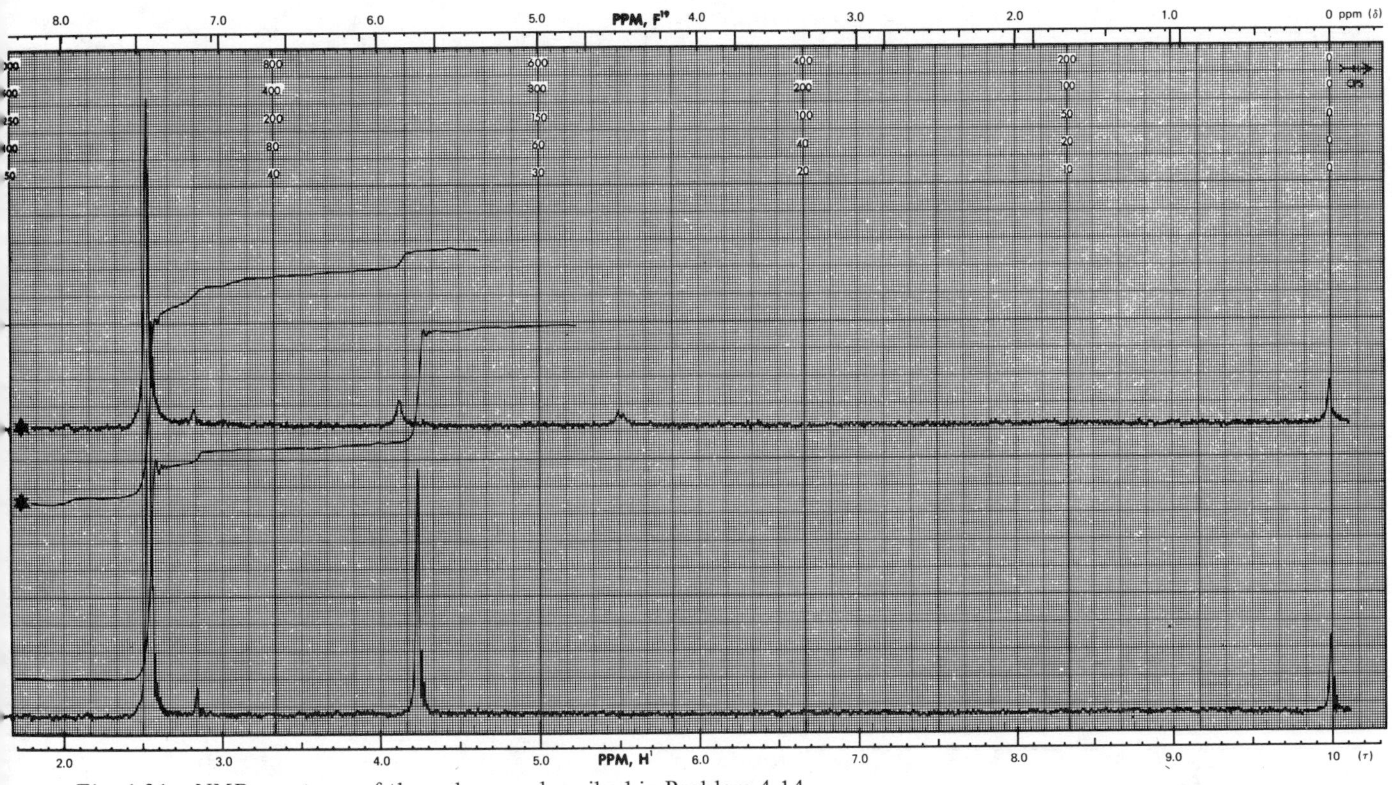

Fig. 4.31 NMR spectrum of the unknown described in Problem 4.14.

Problem 4.14. The NMR spectrum of a crystalline solid which has a melting point of 96° and a molecular formula $C_6H_3OBr_3$ is determined in a solution of $CDCl_3$. A small quantity of D_2O is then mixed with the solution, and the spectrum redetermined. (The upper trace indicated by ★ in Fig. 4.31 is for D_2O solution.) Identify the unknown.

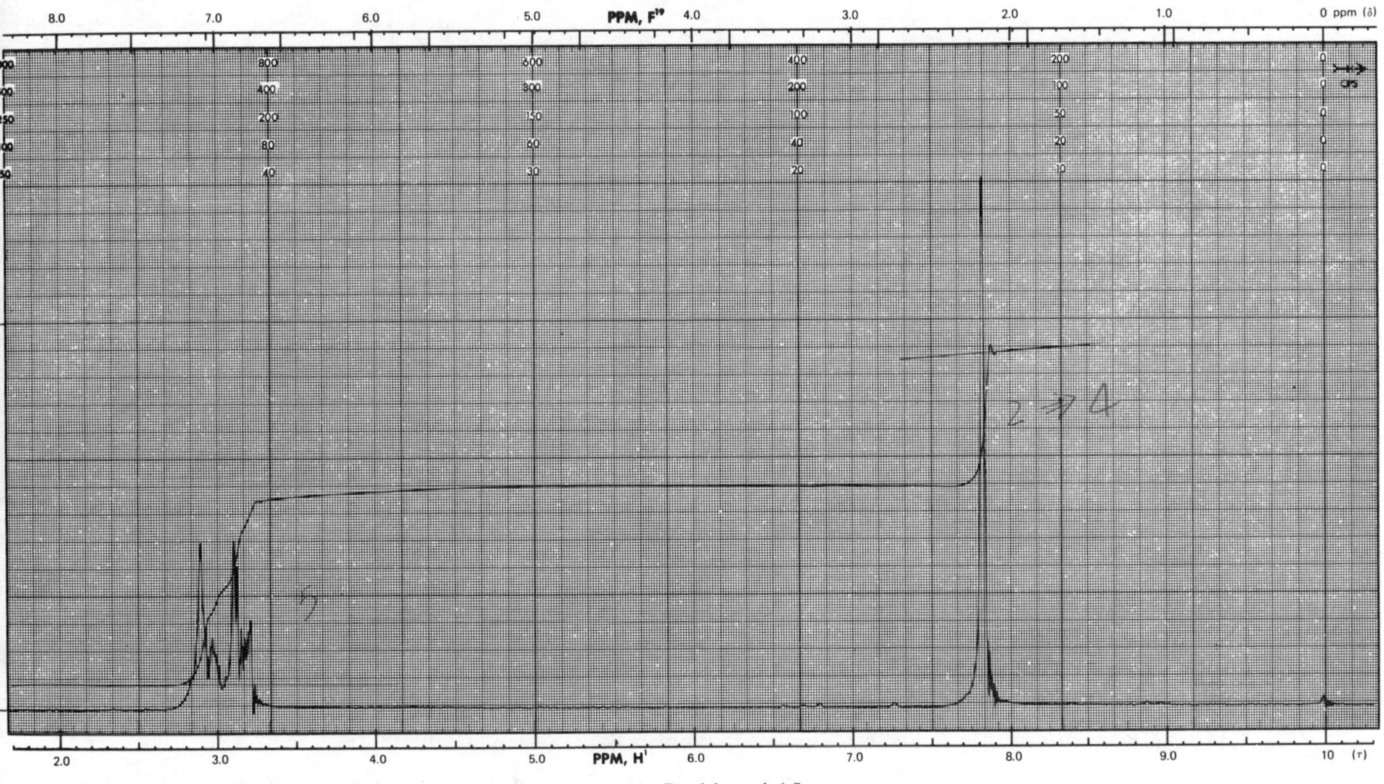

Fig. 4.32 NMR spectrum of the unknown described in Problem 4.15.

Problem 4.15. Identify the unknown which has the molecular formula C_7H_7Br, a boiling point of 184°, and an NMR recorded in Fig. 4.32. The unknown is dissolved in CCl_4 to record the spectrum.

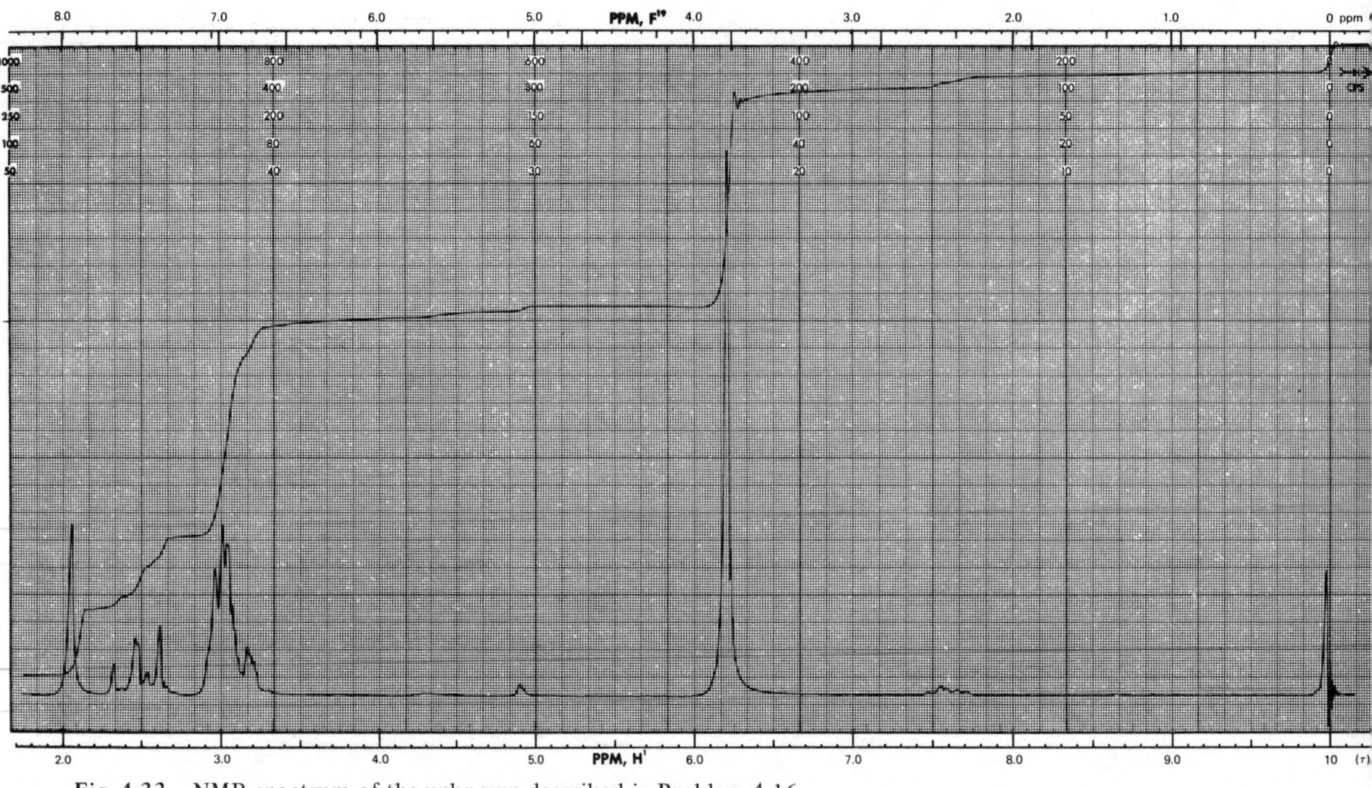

Fig. 4.33 NMR spectrum of the unknown described in Problem 4.16.

Problem 4.16. The molecular formula of an unknown liquid is $C_7H_8O_2$ and its boiling point is 244.3°. Figure 4.33 shows the NMR spectrum of the pure liquid. Determine the structure of the unknown.

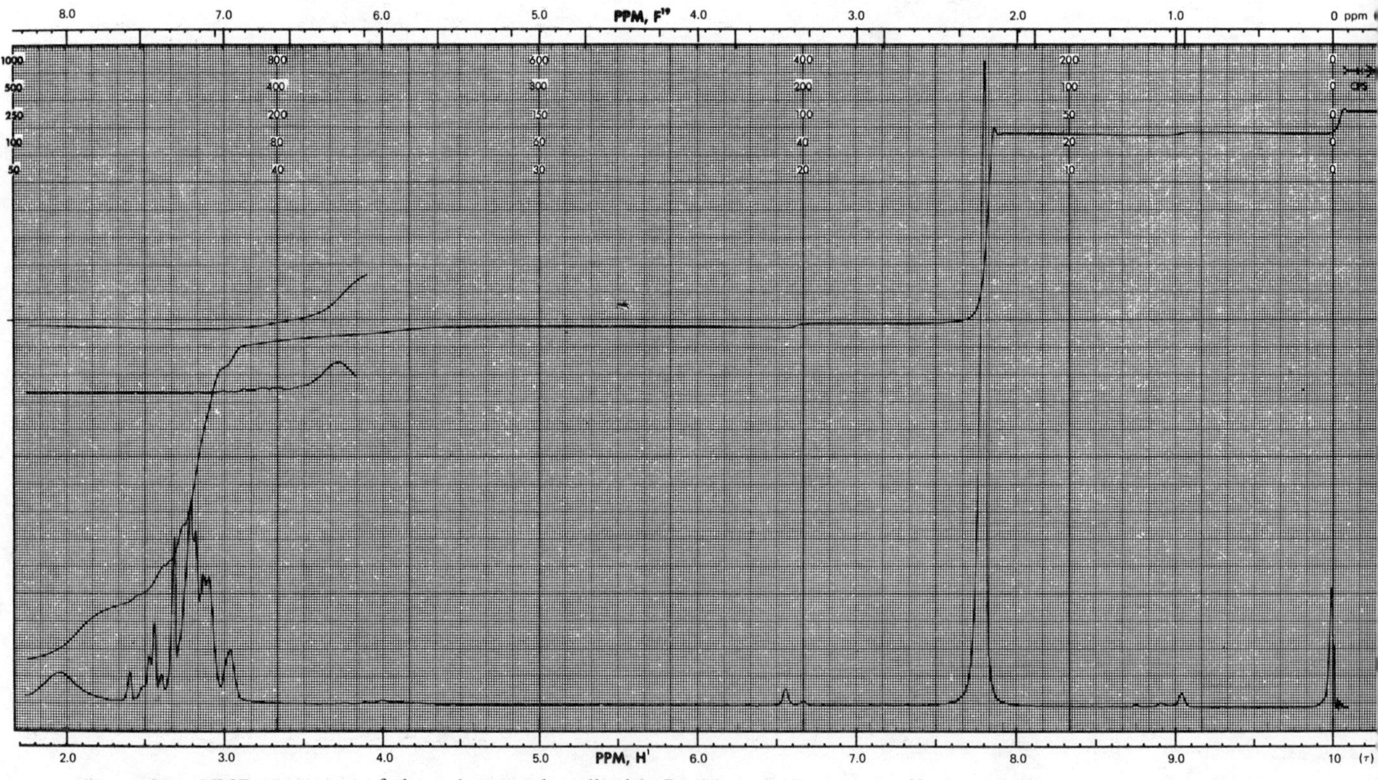

Fig. 4.34 NMR spectrum of the unknown described in Problem 4.17. Sweep offset, 100 Hz.

Problem 4.17. The molecular formula of an unknown liquid is C_7H_8O and its boiling point is 203°. Figure 4.34 shows the NMR spectrum of the pure liquid. Determine the structure of the unknown.

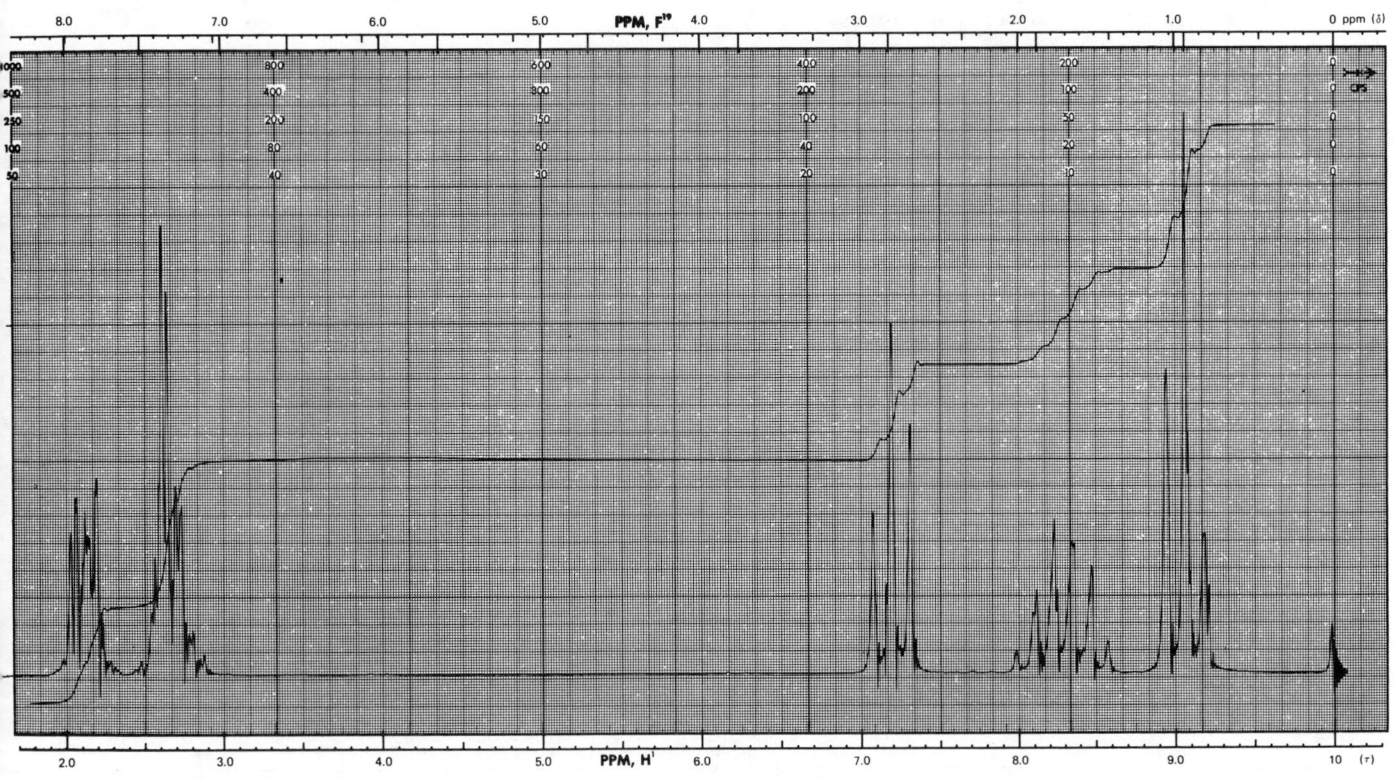

Fig. 4.35 NMR spectrum of the unknown described in Problem 4.18.

Problem 4.18. The molecular formula of an unknown liquid is $C_{10}H_{12}O$ and its boiling point is 229°. Figure 4.35 shows the NMR spectrum of the pure liquid. The unknown also displays a strong band at about 1700 cm^{-1} in the IR spectrum. Determine the structure of this unknown.

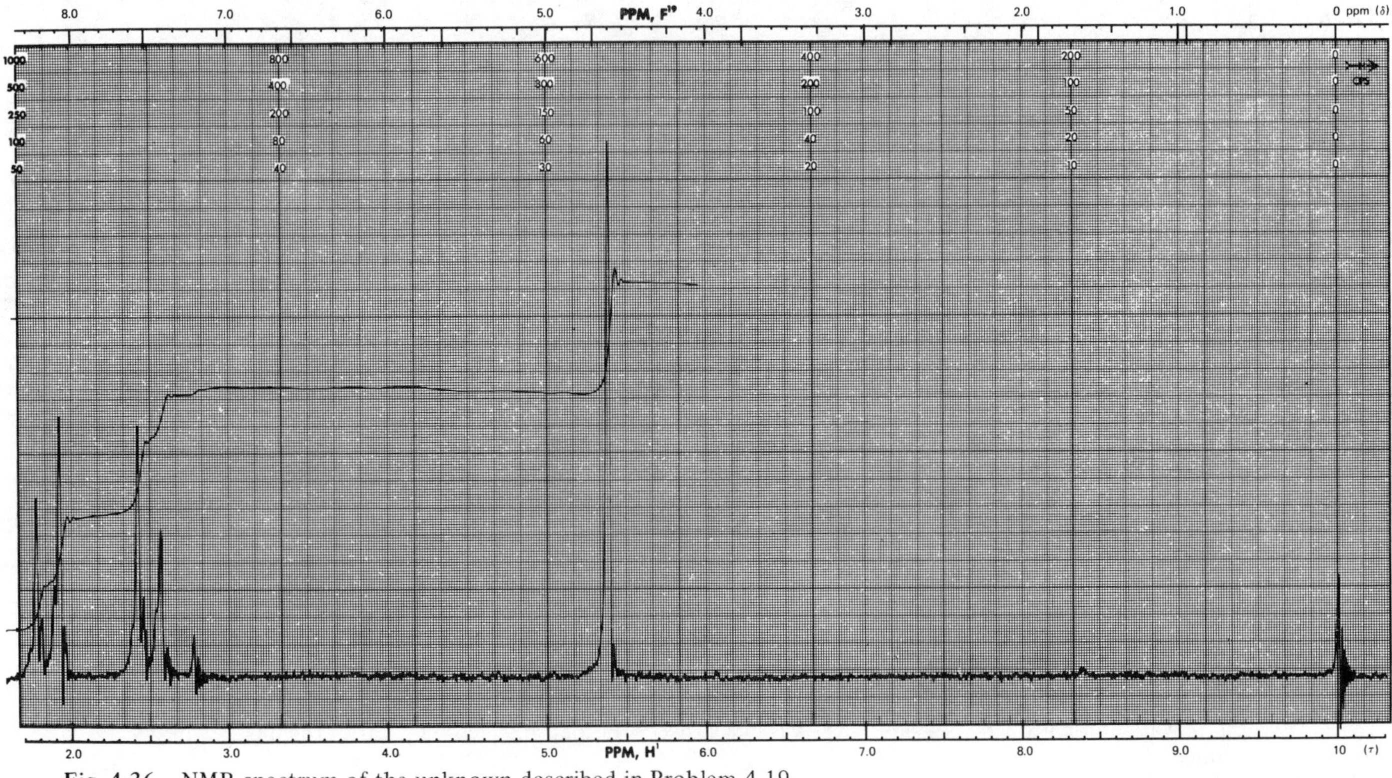

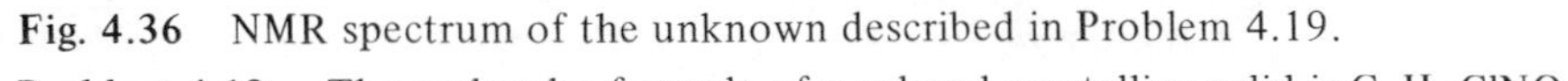

Fig. 4.36 NMR spectrum of the unknown described in Problem 4.19.

Problem 4.19. The molecular formula of a colored crystalline solid is $C_7H_6ClNO_2$. Its melting point is 71°. Figure 4.36 shows the NMR spectrum of a solution of the crystals in $CDCl_3$. Determine the structure of the unknown.

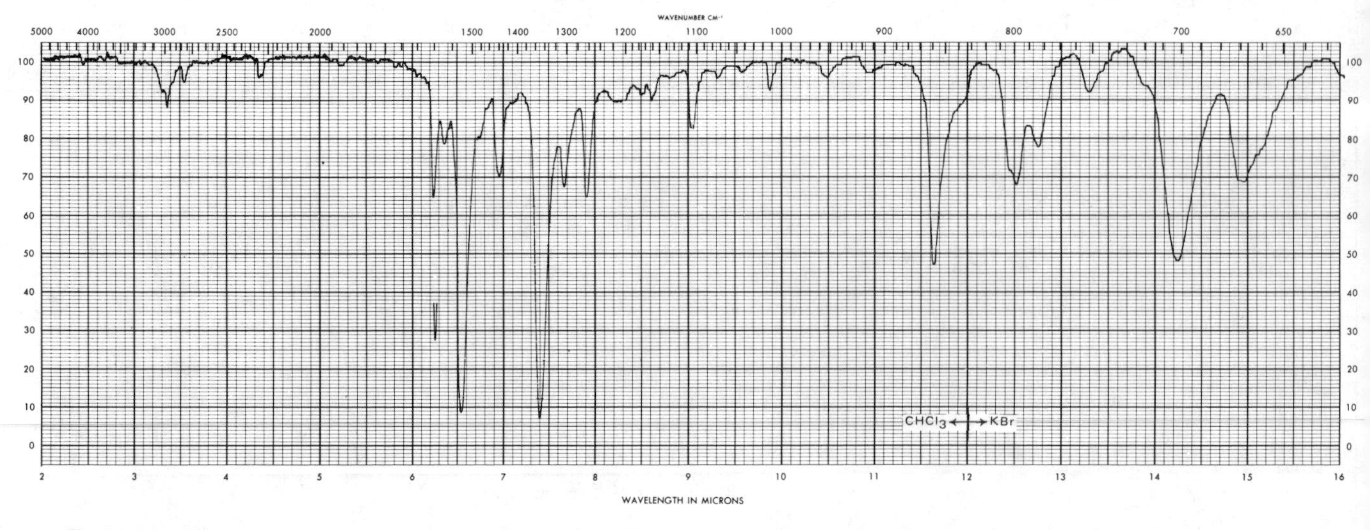

Fig. 4.37 IR spectrum of the unknown described in Problem 4.20. (KBr pellet is used to record the trace between 12μ and 16μ.)

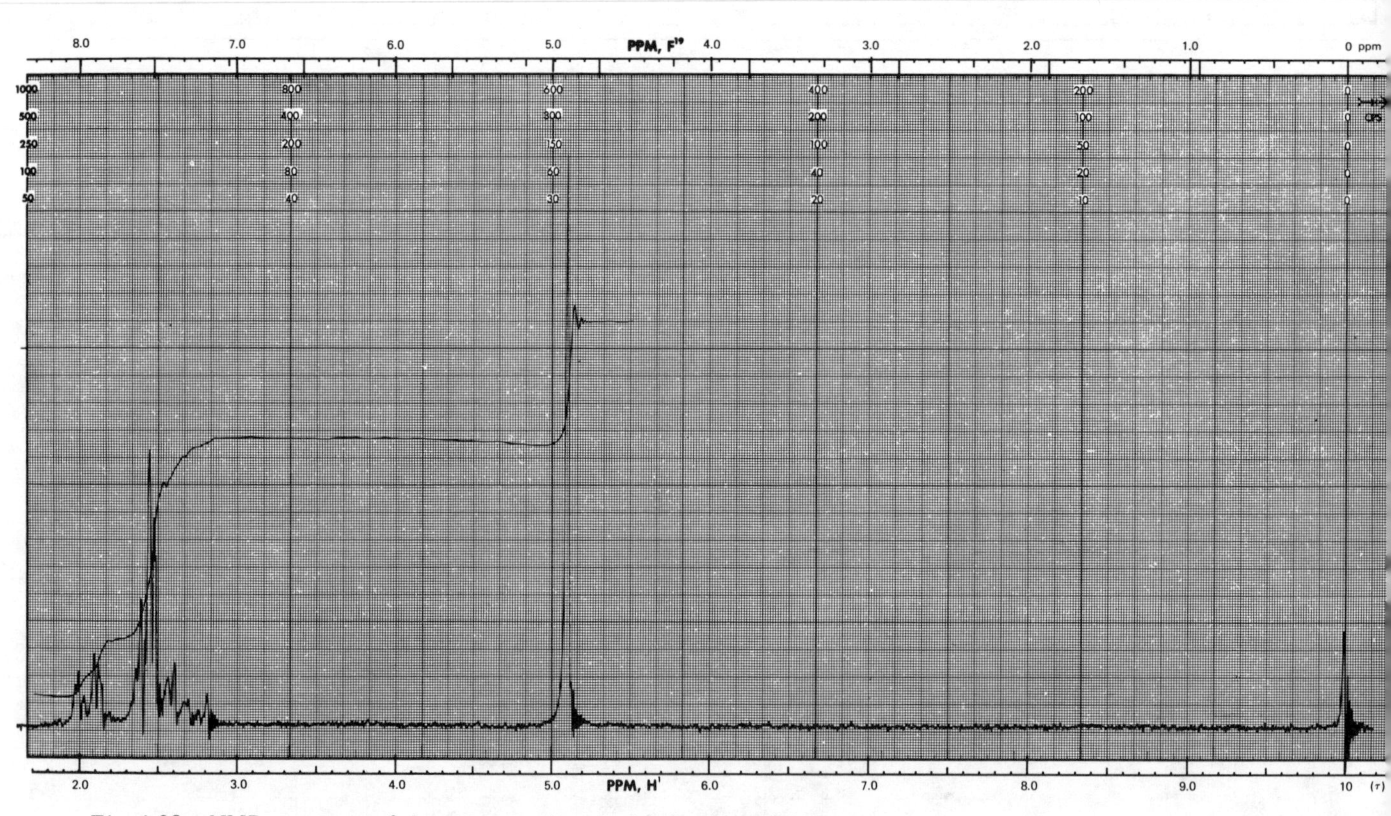

Fig. 4.38 NMR spectrum of the unknown described in Problem 4.20.

Problem 4.20. The IR and NMR spectra of an isomer of the unknown described in Problem 4.19 are recorded in Figs. 4.37 and 4.38, respectively. The IR spectrum was determined using a 6% w/w solution of the crystals, while the NMR spectrum was determined when the unknown was in a $CDCl_3$ solution. The melting point of the unknown is 49°. Determine its structure.

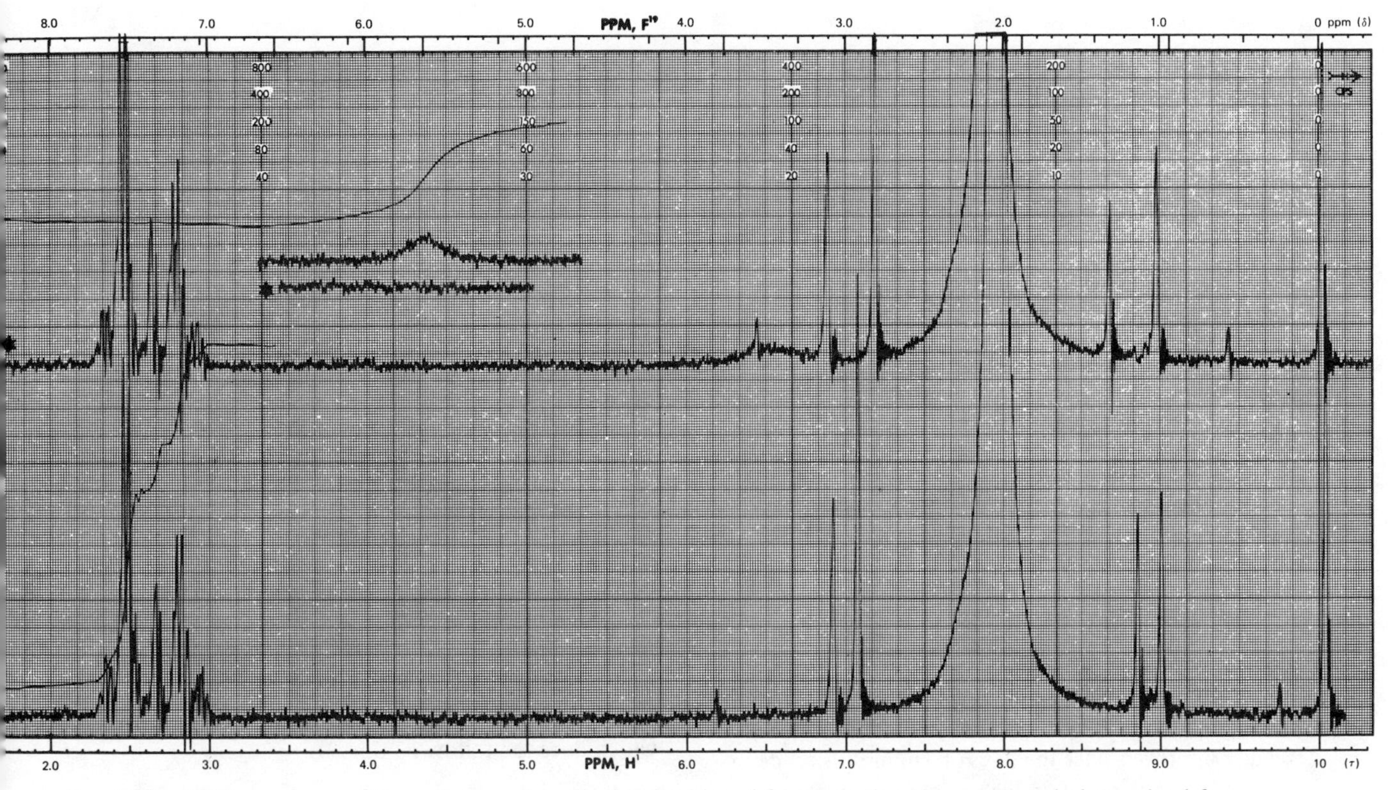

Fig. 4.39 NMR spectrum of the unknown described in Problem 4.21. Note that the spectrum is determined for crystals dissolved in acetone.

Problem 4.21. The yellow monoclinic needles of an unknown melt at 97°. The compound contains C = 51.8%, H = 3.59%, and N = 9.93%. The NMR spectrum of a solution of the crystals in *acetone* is determined before and after the addition of D_2O to the sample. Figure 4.39 shows the traces (indicated by ★) which are determined after the addition of D_2O. The inset is recorded with the sweep offset at 200 Hz. Determine the structure of the unknown.

FIVE

MASS SPECTROMETRY

5.1 INTRODUCTION

Mass spectrometry, unlike NMR spectroscopy, has been used by physicists and chemists for a long time. The development of mass spectrometry began around the turn of the century, when Wien[1] in 1898 first produced crude mass spectra and Thomson[2] in 1905 obtained evidence for the existence of stable isotopes by using apparatus in which positive ions of different charge-to-mass ratios were made to travel in different parabolic traces. The instrumentation and technique of mass spectrometry have steadily improved since the days of Thomson's crude apparatus, and the modern high-resolution spectrometer is a very sophisticated instrument, capable of distinguishing between ions whose masses differ by amounts so minute that differences are measured to the third decimal place or more. There has been an amazingly rapid increase in the number and types of mass spectrometers and mass spectrographs since the second world war, and the range of applications of this technique has broadened so greatly that it has become a routine method for dealing with many analytical problems in organic and inorganic chemistry, determining stable isotopes, studying free radicals in the gas phase, examining ionization phenomena, and it is also used in space research and biomedical research. Organic chemists, however, have been somewhat slow in recognizing the potential of mass spectrometry as a tool with which to elucidate organic structures.

Petroleum chemists have employed mass spectrometry extensively since the late 1940's for quantitative analysis of hydrocarbon mixtures. However it was only in the early 1960's that organic chemists began to recognize the power of this technique for determining organic structure.

5.2 PRINCIPLES OF THE MASS SPECTROMETER

Many different mass spectrometers, which vary considerably in design and operation, are commercially available. However, the basic principle underlying all of them is as follows: *Substances in the gaseous or vapor state, when they are subjected to high-voltage electric current, can be made to lose electrons and form positively charged ions (cations). These cations can be accelerated and deflected by magnetic and/or electrical fields. The deflection of an ion depends on its mass, charge, and velocity. If the charge, velocity, and deflecting force are constant, the deflection is less for a heavy particle and more for a light one.*

A mass spectrometer is therefore an instrument which does the following things.

1) It generates a beam of cations from the specimen under investigation.
2) It disperses this beam into a mass spectrum according to the mass-to-charge ratio of the ions.
3) It detects and records the values of the relative masses and the abundances of these ions.

i. Cation production Most mass spectrometers—particularly those used to measure the abundance of a given isotope or the mass of an atom, to carry out chemical analysis, to elucidate organic structure and to study the ionization and dissociation of molecules—use an *electron-impact (EI) technique* to produce ions. The vaporized sample is introduced under low pressure into a tube called the *ion source*, and the entering molecules of the sample are bombarded with a beam of electrons emitted from a hot tungsten or rhenium wire (Fig. 5.1).

If the energy of the bombarding electrons is less than the *ionization potential** of the molecules (less than about 10 eV in the case of most organic molecules), no apparent change takes place in the molecules. However, if the energy of the electrons is gradually increased just above the ionization potential, the probability also

* The ionization potential of a molecule is the energy required to remove an electron from the highest occupied molecular orbital to form an ion, $M^{+\cdot}$.

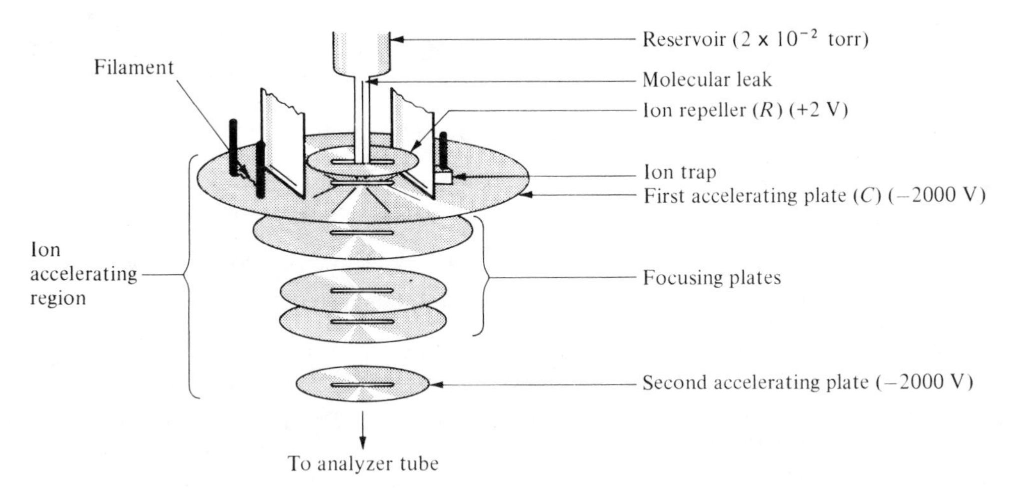

Fig. 5.1 An electron impact ion source in a mass spectrometer.

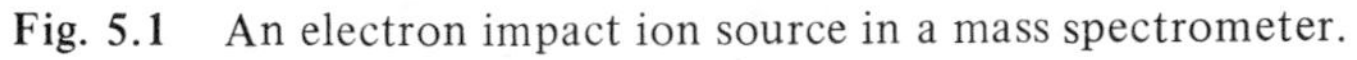

increases that a collision between a molecule and an electron will induce ionization, by the process

$M + e^- \rightarrow M^{\ddagger} + 2e^-$

(for example,

$CH_3OH + e^- \rightarrow CH_3OH^{\ddagger} + 2e^-$).*

If the energy of the electron beam is still further increased, some of this excess energy is transferred to the molecular ions, resulting in the fragmentation of the molecule. The electrical potential of the bombarding electron which is just high enough to initiate fragmentation is called the *appearance potential.* The magnitude of the appearance potential is equal to or greater than the sum of the dissociation energy of the fragmented bond and the ionization potential. When the bombarding electrons have very high energy, more than one bond in the molecule can be broken. In practice, a voltage of about 70 eV is used to accelerate the electrons so that there is enough energy to break any bond in the molecule. Thus, if you introduce vapor into the ion source and bombard it with a 70-eV electron beam, the resulting fragmentation process is:

$$CH_3OH + e^- \rightarrow CH_3OH^{\ddagger} + 2e^-$$
$$CH_3OH^{\ddagger} \rightarrow CH_3^+ + \dot{O}H\dagger$$
$$\rightarrow CH_2OH^+ + \dot{H}$$
$$CH_2OH^+ \rightarrow CHO^+ + H_2$$

The actual fragmentation process occurring in an ion source, however, is rarely so simple. In addition to molecular ions and simple fragmentation ions, the reactions inside the ion source often produce a variety of different ions (Section 5.5).

The electron-impact technique, though used widely, may have only limited usefulness when it comes to determining organic structure, for the following reasons: (1) The necessity for having the sample in the vapor phase increases the possibility of pyrolysis, particularly of less volatile samples. (2) The use of 70-eV acceleration voltage (which is nearly 7 times the ionization potentials of most organic molecules) often destroys the molecular ion. Such destruction due to "hard ionization" poses a serious problem, because a totally absent or a very weak molecular ion peak in the mass spectrum renders the determination of molecular weight and empirical formula very difficult. In addition, the high energy of the "hard-ionization" process increases (i) the number of possible reaction paths for the molecular ion to follow during the decomposition process and (ii) the number of reactions that take place in consecutive steps. Because of these factors, a complex mass spectrum is often produced, and the identification of individual ions often becomes difficult.

Techniques for "soft ionization" of molecules developed in recent years are: field ionization, field desorption, chemical ionization, surface ionization, photoionization, focused radiation or microprobe ionization, etc. Some of these techniques will be discussed in Section 5.6. The "soft" and "hard" ionization techniques, in fact, are complementary to each other, because, although the former produce molecular ions in greater abundance, the latter could provide many specific fragment ions which, if identified, could furnish insight into the overall structure of the molecule. Instruments such as Avco's MS-900, in which the surface ("soft") ionization source can be readily converted to an electron-

* An ion is denoted by the appropriate charge, + or −, whereas a radical is denoted by a dot•. Thus the symbol ‡ signifies a radical cation.

† According to the rule of conservation of charge, only one ion can be formed from the decomposition of a singly charged ion.

impact ("hard") type are therefore now commercially available. A combined electron-impact–field-ionization instrument is also being marketed (manufactured by Varian MAT).

In an electron-impact instrument the ion source also contains ion repellers, ion-acceleration plates and ion-focusing plates (Fig. 5.1). A small electrostatic field applied between the repeller electrode (R) and the first acceleration plate (C) forces the ions to the ion-accelerating region. A potential of about 1000 to 2000 volts is used to accelerate the ions through a slit system, so that a narrow beam of fast-moving ions is formed. This collimated, well-defined beam of positive ions is analogous to the light beam in a spectrophotometer. Given that the mass of the ions is m and their charge e, and that their initial kinetic energy is negligible, then their kinetic energy after they have been accelerated to a velocity v must be equal to the electrostatic energy acquired in passing through the applied voltage V. That is,

$$\tfrac{1}{2}mv^2 = eV \tag{5.1}$$

or

$$v^2 = \frac{2eV}{m}. \tag{5.2}$$

ii. Dispersing the ionic beam into a mass spectrum The fast-moving ionic beam, after leaving the ion source, enters the analyzer tube.

a) Ion-deflection technique. In the analyzer tube the ions are subjected to a uniform magnetic field, H (strength about 500–10,000 gauss), which is generated by an electromagnet and is perpendicular to the direction of the ionic beam. In the magnetic field, the ions are deflected along a circular path of radius r, according to Eq. (5.3):

$$r = \frac{mv}{eH}. \tag{5.3}$$

That is,

$$r^2 = \frac{m^2v^2}{e^2H^2} \tag{5.4}$$

or

$$v^2 = \frac{r^2e^2H^2}{m^2}. \tag{5.5}$$

Thus, from Eqs. (5.2) and (5.5), we have:

$$\frac{2eV}{m} = \frac{r^2e^2H^2}{m^2}$$

or

$$\frac{m}{e} = \frac{H^2r^2}{2V}. \tag{5.6}$$

We can conclude from Eq. (5.6) that:

1) The radius r of the circular path of the ions (i.e., their deflection) is dependent on the following factors.
   a) V, the accelerating voltage
   b) H, the strength of the deflecting magnetic field
   c) m/e, the mass-to-charge ratio of the ions
2) For the ions accelerated by a fixed voltage V, the deflection caused by a fixed magnetic field H is proportional to the mass/charge ratio of the ions.

Since most of the ions produced when one is recording a mass spectrum of an organic compound are singly charged, it follows that at fixed values of V and H all the ions of the same mass (m_1) will follow the same circular path and be recorded at the same mass value (m_1) on the spectrum, whereas another set of ions of identical mass (m_2) will follow a different path and be recorded at a different mass value (m_2). Thus the analyzer tube in a mass spectrometer serves the same function as a monochromator (in an optical absorption spectroscopic instrument) in dispersing a fast-moving beam into a spectrum according to the mass-to-charge ratio of the ions.

In low-resolution, *single-focus* instruments, i.e., in instruments with only one analyzer sector, the dispersed ions arrive at the collector assembly through a fine slit at the end of the analyzer tube and are scanned by a gradual variation either in V or in H. However, a single-focus instrument neither gives a very accurate spectrum nor allows for a good resolution of ions, particularly of those having the same integral mass but differing in elemental composition and therefore in fractional mass. For example, such ions as $C_6H_4N_3O_3$ (accurate mass 166.023264) and $C_9H_2N_4$ (accurate mass 166.026946) cannot be resolved by a single-focus instrument.

Modern high-resolution instruments employ the *double-focus* technique, and are capable of discriminating between the masses of differing particles, even though their differences are so minute. In such instruments, the ions emerging from the source are passed through a radial electrostatic field (electric analyzer) before they enter the magnetic analyzer. The effect of the electrical analyzer is that the ions emerging from it, regardless of their initial kinetic energy spread, follow the same path or are collimated, and the resolution of the spectrum is improved.

Commercial high-resolution spectrometers are basically of two types: those using the Neir-Johnson focusing system and those using the Mattauch-Herzog one.

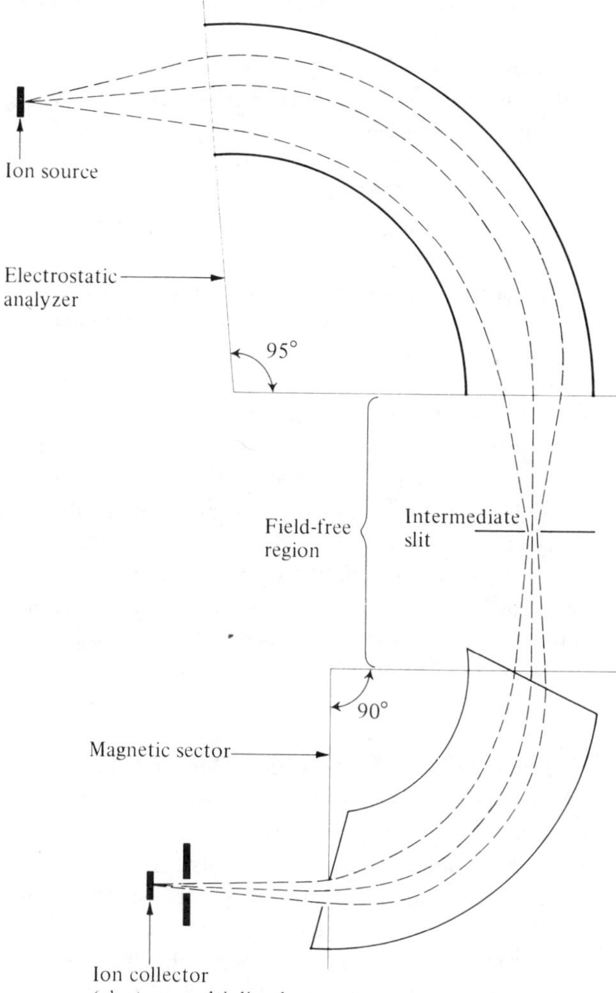

Fig. 5.2(a) A Neir-Johnson double-focusing mass spectrometer.

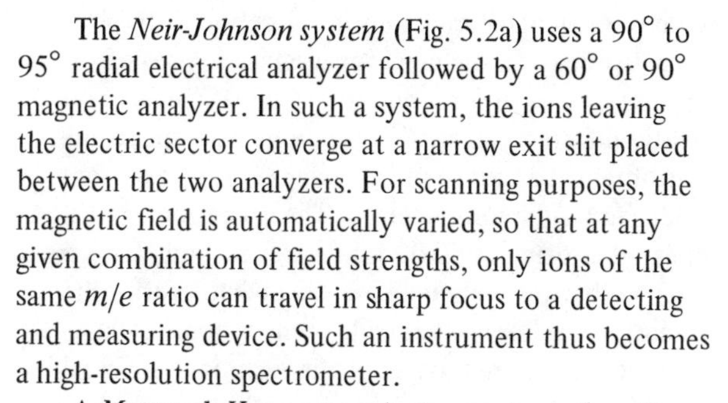
The *Neir-Johnson system* (Fig. 5.2a) uses a 90° to 95° radial electrical analyzer followed by a 60° or 90° magnetic analyzer. In such a system, the ions leaving the electric sector converge at a narrow exit slit placed between the two analyzers. For scanning purposes, the magnetic field is automatically varied, so that at any given combination of field strengths, only ions of the same m/e ratio can travel in sharp focus to a detecting and measuring device. Such an instrument thus becomes a high-resolution spectrometer.

A *Mattauch-Herzog* type instrument, on the other hand, can operate both as a spectrometer and a spectrograph. In such a system (Fig. 5.2b), the electrical analyzer has a 31°50′ geometry which collimates ions of a given m/e ratio. Consequently, all the ions produced in the source can be focused simultaneously by the magnetic analyzer onto a photographic plate, and a spectrograph can be obtained. Alternatively, the spectrum can be scanned across an exit slit, thus making it possible for the instrument to be used as a spectrometer.

In addition to the magnetic-deflection and electrostatic-scanning spectrometers, there are also instruments employing methods such as *time of flight (TOF)*,[3,4,5] *cyclotron or omegatron resonance*,[6] *quadrupole mass filter*,[7,8,9] and *radiofrequency analyzing*,[10,11,12] most of which have been developed in the past few years.

b) Time-of-flight technique. Unlike the techniques of magnetic and electrostatic deflection described so far, the time-of-flight technique does not employ ion-deflection phenomena. Instead, it uses the fact that ions of different masses, when subjected to the same acceleration potential, accelerate to different velocities during the same time interval. In these instruments the ions are produced by a short pulse of electrons in a five-grid electron-beam system, are accelerated by the application of an electric field, collimated by a series of narrow slits and are simply allowed to drift down a 100- to 200-cm, field-free tube to the ion detector. Since the

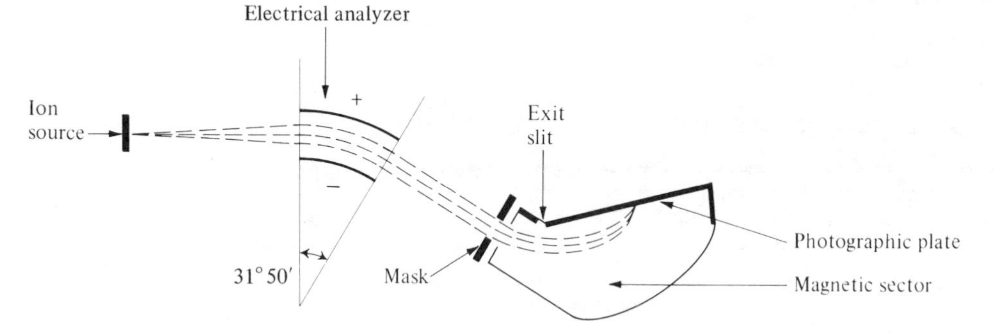

Fig. 5.2(b) A Mattauch-Herzog double-focusing mass spectrometer-spectrograph.

velocity of two differing ions is a function of their respective masses, the detector receives a series of separate groups of ions, one for each mass present, at slightly different times. A continuous mass spectrum is thus obtained as a plot of time elapsed versus current produced by the ions received. The operation can be repeated as often as 50,000 times per second. The spectra obtained by this technique are somewhat inferior in resolution to those obtained by conventional electrostatic-deflection methods, but the technique is valuable for studying fast, gaseous-phase chemical reactions such as explosions, flash photolysis and pyrolysis reactions, etc., and for identifying the components in the effluent stream of a gas chromatograph.

c) Cyclotron-resonance technique. We have seen earlier that moving ions subjected to a deflecting magnetic field H follow a circular path of radius r which, according to Eq. (5.3), is

$$r = \frac{mv}{eH}.$$

The orbital angular velocity ω_c of the ions is thus

$$\omega_c = \frac{v}{r} = \frac{He}{m}. \qquad (5.7)$$

ω_c, which is also known as the *cyclotron velocity*, thus depends on H and on the m/e ratio. In cyclotron-resonance instruments,[6] the ions are exposed to a fixed radiofrequency ω (usually 77 to 770 kHz) and the deflecting magnetic field H is swept so that the orbital frequency ω_c of the ions is gradually brought to the operating radiofrequency ω. When ω_c become equal to ω, resonance occurs, the radiofrequency energy is absorbed, and a signal is generated. Thus, like NMR, the cyclotron mass spectrum is also a plot of magnetic field versus signal intensity. We discussed the mechanism of energy absorption by resonance in the chapter on NMR (Sections 4.3, 4.4, and 4.5).

d) Quadrupole mass filter technique. Quadrupole mass spectrometers are rapidly becoming popular, mainly because of their relatively low price. In the terms of mass range and resolution performance, these instruments are comparable to single-focus spectrometers of very simple design. They are frequently used in analysis of residual gases in evacuated systems, and are therefore also known as *residual gas analyzers.*

Quadrupole-type analyzers behave like mass filters rather than like optical monochromators, since they destroy all ions except those with very precise, predetermined m/e ratios. A quadrupole analyzer consists of a cylindrical tube, 5 to 20 cm long, containing four cylindrical rods placed parallel to one another along the axis of the tube (Fig. 5.3). Rods 1,3 and 2,4 are electrically connected and both dc and rf potential (frequency approximately 500 kHz) are applied between the two pairs. The ions, which enter the tube through a small orifice, react to the applied potential in such a manner that at any given value of dc voltage or radio frequency, only those ions with a certain m/e value are able to travel the linear path along the central axis to the detector, whereas others are diverted and neutralized by contact with one of the four rods. The spectrum is scanned by gradual variation of the dc voltage or the frequency. Instruments operating on a similar principle, but using only a single rod (monopole) which is given both dc and rf potentials are now reported to operate at a satisfactory level of sensitivity.[9]

iii. Detecting and recording mass values and abundances of ions The detection and recording of ions can be done either by photographic plates (spectrographs) or by electron multipliers (spectrometers). A photographic plate detector (used in instruments with the Mattauch-Herzog system) can record all the ions almost simultaneously, and can generally give somewhat better resolution than an electrical detector. However, they are relatively more cumbersome and less accurate in recording relative abundance ratios of ions.

Electrical detectors employed in most modern instruments are either electrometers or electron multipliers. An electron multiplier is similar to the photomultiplier

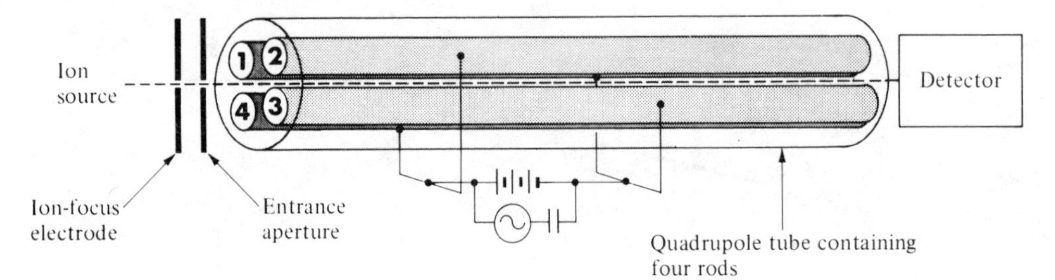

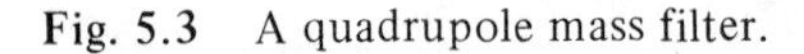
Fig. 5.3 A quadrupole mass filter.

tube (Section 2.4) employed in UV-Vis instruments, except that a beryllium-copper collector plate is used in it as a primary cathode. Such a cathode responds to ions rather than to photons. The very high current gain achieved in an electron multiplier enables the spectroscopist to detect even a single ion arriving at the collector plate.

The pen-and-ink type of recorders employed in UV, IR, or NMR instruments are often too slow for mass-spectrometric work. Most mass spectrometers employ a system of three to five separate galvanometer mirrors which deflect a beam of ultraviolet radiation onto an ultraviolet-sensitive recording paper as ions are collected at the electron multiplier. (This paper does not require wet development.)

5.3 HANDLING OF SAMPLES

As we saw earlier, a mass spectrometer consists, in essence, of an ion source, an analyzer, and a detecting recording mechanism. In addition to these three components, an intricate sample-handling mechanism forms an essential part of a mass spectrometer. The sample may be a gas, a liquid, or a solid, but the fact that it must enter the ionization chamber only at a very low pressure—from 10^{-7} to 10^{-5} torr (1 torr = 1 mm Hg)—causes complications. Naturally, the ions that are produced should not collide with each other or with the molecules, because that would produce inter-ionic, intermolecular or ion-molecule reactions. Also, if the vacuum is not a good one, there is erosion of the electron-producing filament and broadening of the ion beam. Consequently, good instruments need a fast pumping system capable of lowering the pressure rapidly to as low as 10^{-6} or 10^{-7} torr.

There are, basically, three different methods of introducing the sample into the ion source: i) direct insertion, ii) chromatographic eluent insertion, and iii) molecular leak. The choice depends on the volatility of the sample.

i) Direct insertion. Suitable for solids and many liquid organic substances, this method involves placing a sample in a small effusion (Knudsen) cell at the end of a probe and inserting it directly into the evacuated ion source through a vacuum lock system. The sample is then evaporated either by heating the probe or by heating the entire ion source.

ii) Chromatographic eluent insertion. This involves a direct hookup of a gas-liquid chromatography column with the ion source. Submicrogram quantities of the highly purified components eluted from the glc column are passed through specially constructed separators to remove the carrier gas, and the enriched eluent is inserted into the ion chamber through a vacuum lock system. This method of utilizing a glc column is extremely useful and versatile because it can be applied to substances with high molecular weights, but it requires fast scanning and high-speed recording techniques.

iii) Molecular leak. The more commonly used method is the so-called molecular-leak method, which is particularly suitable for highly volatile solids and liquids and for gaseous samples. Insertion of a gaseous sample is simple, and involves a transfer of a gas from a bulb to the inlet sample reservoir via a sensitive metering device. The gas is held in the reservoir (at a pressure of 10^{-1} to 10^{-2} torr) from which it slowly "leaks," through an aperture, into the ion source, where the pressure is always kept below 10^{-4} torr. Solids and liquids must be vaporized completely before the sample reaches the reservoir. To ensure that they are vaporized, the sample-handling systems are heated. Liquids are introduced into such systems by a micro pipette syringe through a silicone rubber septum directly connected to the vacuum system. The size of the sample may vary from about 0.1 mg to several milligrams, but must be such that it can exert a pressure of about 10^{-5} torr in the source region. For compounds unable to meet this requirement or for thermally unstable compounds, it is often necessary to prepare more stable or more volatile derivatives such as acetates, esters, trimethylsilyl ethers, etc.

In some instruments, there is a convenient duplicate inlet system, which makes possible a rapid determination of successive spectra of different compounds, and also enables one to compare the unknown with the standard, because the matched pair of leaks used in this system makes it possible to switch rapidly from one sample to another.

5.4 SPECTRAL PRESENTATION

Under the conditions employed in a mass spectrometer, several molecules of the sample ionize and fragment into different masses. The detector registers these ions and produces signals appropriate to the abundance and the mass/charge ratio of the ions. A complete mass spectrum of a pure compound is thus a graphical record of the fragmentation pattern. It is like a fingerprint of the compound.

As mentioned earlier, a five-element galvanometer recording system can simultaneously record five traces at

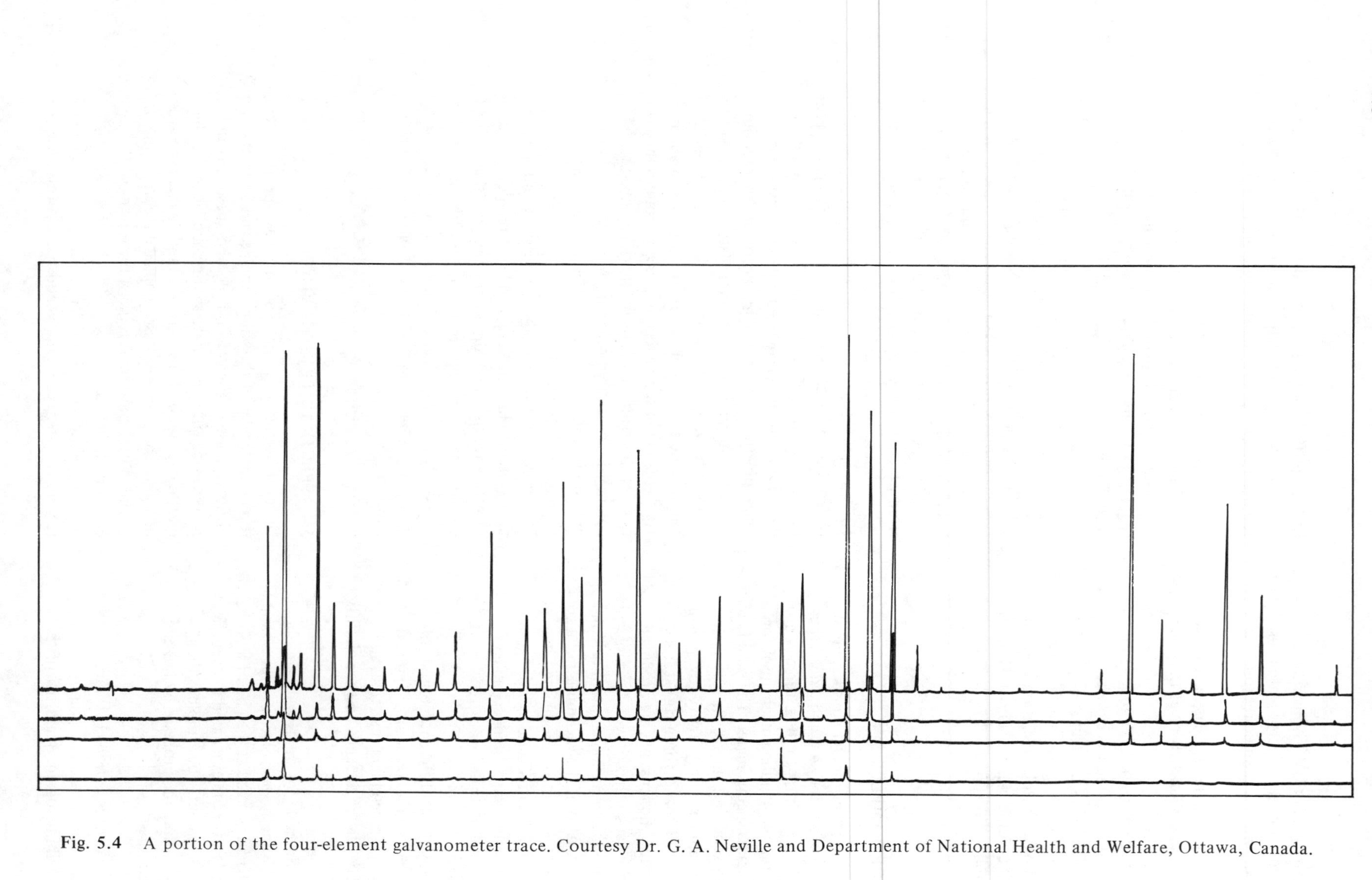

Fig. 5.4 A portion of the four-element galvanometer trace. Courtesy Dr. G. A. Neville and Department of National Health and Welfare, Ottawa, Canada.

five sensitivity levels. However, such a record has limited applications, and is only rarely reported in scientific journals. The commonest method of presentation is either a bar graph or a percentage table.

In the percentage-table system, the most intense peak in the spectrum is given a value of 100% and is called the *base peak.* The intensities of all the other peaks in the spectrum are then reported as percentages of the base peak.

With the bar-graph representation, relative intensity is plotted along the ordinate and mass per unit charge (m/e) along the abscissa. A portion of a mass spectrum traced by a four-element galvanometer recorder is shown in Fig. 5.4. A complete mass spectrum of dimethylketene, $(CH_3)_2{-}C{=}C{=}O$, is reported in tabular form in Fig. 5.5(a) and in bar-graph form in Fig. 5.5(b).

The bar graph, which usually resembles the least sensitive galvanometer trace, is often used because it readily conveys the fragmentation pattern by giving a pictorial impression. The tabular form, on the other hand, provides a quick grasp of mass numbers and percentage abundances, and is therefore used in this book. Remember also that, as long as the ion-source conditions are constant, the fragmentation pattern of a given compound remains unaltered, and any two compounds, even if chemically similar, yield two different fragmentation patterns. However, a variation in temperature or electron energy, or the use of a different type of instrument (e.g. time-of-flight, cyclotron resonance, etc.) produces a very different fragmentation pattern for the same compound.

m/e	% of base peak	m/e	% of base peak	m/e	% of base peak	m/e	% of base peak
25	3	26	10	27	31	29	4
36	2	37	10	38	13	39	58
40	31	41	100	42	35	43	12
53	2	55	3	56	2	68	1
69	0.7	70	46.5M	71	2.4 (M + 1)		

Fig. 5.5(a) A spectrum of dimethylketene in tabular form.

5.5 TYPES OF IONS PRODUCED BY AN ION SOURCE

We mentioned in Section 5.2 that the ion-source reaction often produces a variety of different ions. In addition to molecular ions and simple fragmentation ions, the reaction can produce: i) rearrangement ions, ii) ion-molecule complexes, iii) isotope ions, iv) doubly charged ions, and v) metastable ions.

i) Rearrangement ions. When we are trying to identify an unknown substance, elucidation of its structure by building up a molecule of the unknown from its fragments can get very complicated because of the way the fragments often rearrange themselves. These rearrangements usually involve migration of hydrogen atoms from one part of the ion to another to produce more stable species. McLafferty[13,14,15] has classified such rearrangements as random or specific. Random rearrangements are particularly common among hydrocarbons and perhalocarbons, because the bonds in such compounds require high energy and more time to cleave. The more time

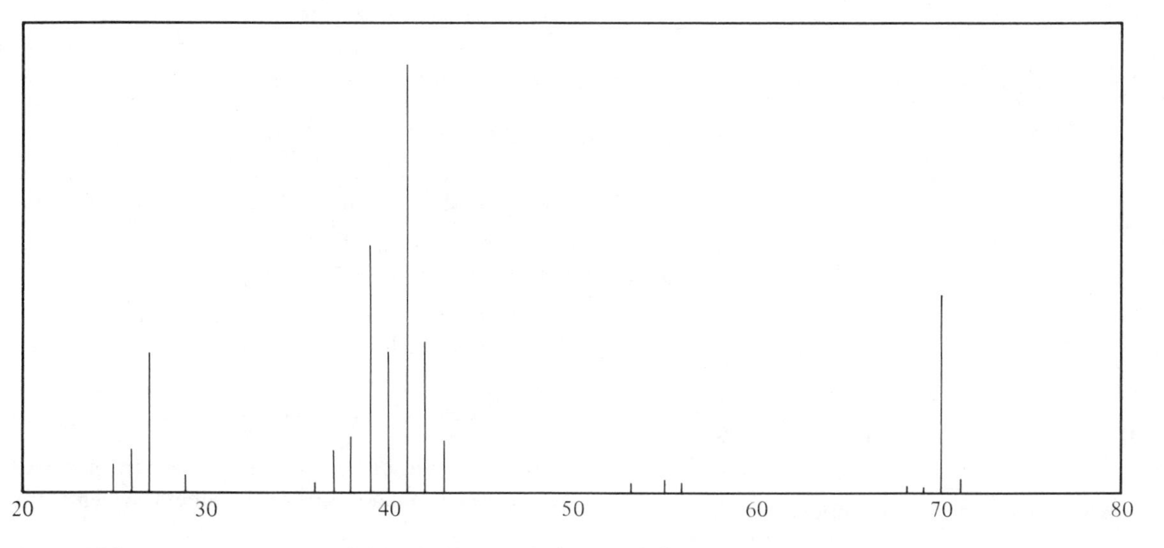

Fig. 5.5(b) A mass spectrum of dimethylketene in bar-graph form.

elapses before cleavage, the more the excited molecular ions isomerize or rearrange their atoms. A typical example of such random rearrangement is afforded by neohexane [$(CH_3)_3CCH_2CH_3$], which produces a fragment ion of mass 43 having the formula $(C_3H_7)^+$. The formation of this ion must involve migration of a hydrogen followed by the rupture of two C–C bonds.

Random rearrangements, which defy simple explanation and pose considerable difficulties when one is trying to elucidate the structure of an unknown are, fortunately, less common than specific ones. Happily enough, the rearrangements that take place most often in molecular ions containing heteroatoms or functional groups usually follow specific mechanisms. Many of these mechanisms are fairly well understood, and the rearranged ions often provide valuable clues to structure. For example, if a ketone molecule (I) containing a gamma hydrogen is bombarded with high-energy electrons, a radical cation (III) and an alkene (IV) are produced. It is believed that the molecular ion initially formed rearranges itself by going through a six-membered cyclic intermediate (II) stage, involving the migration of a hydrogen from a γ carbon to oxygen. The migration of the hydrogen from the γ position is facilitated by simultaneous β cleavage* resulting from two possible modes of electron movements, as follows.

Mode 1: Two-electron movements resulting in heterolysis of the β bond

At the cyclic intermediate stage, the γ hydrogen launches a nucleophilic attack (carrying both the electrons of the C–H bond) on the electron-deficient oxygen. This attack is achieved through "two-electron" transfers, as shown below.

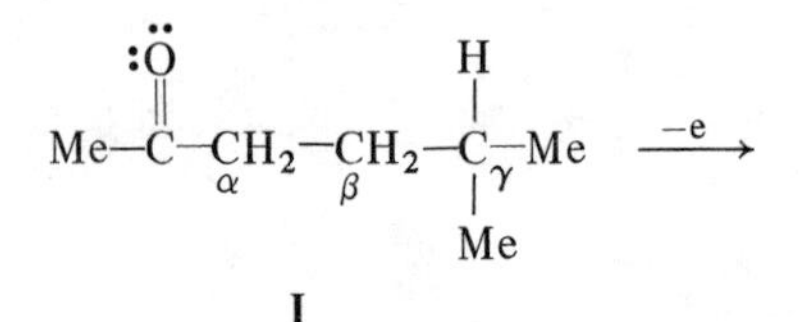

I

5,methyl-2-hexanone

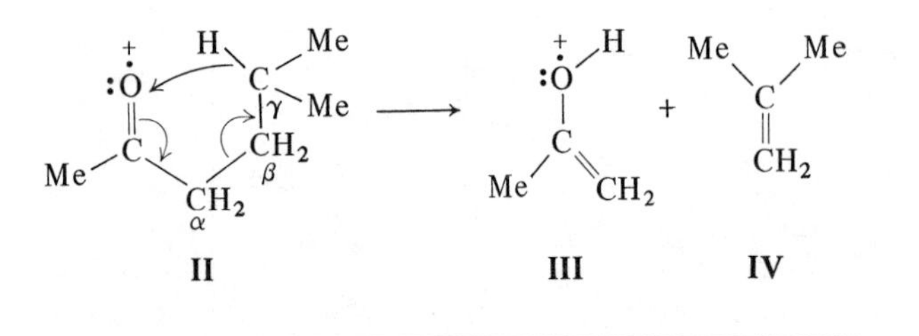

* An α *cleavage* is the rupture of the bond adjacent to a functionalized carbon atom. A β *cleavage* is therefore a fission of the bond between the α and the β carbons.

Mode 2: Single-electron movements resulting in homolysis of the β bond

At the cyclic intermediate stage, the electron-deficient oxygen atom undergoes hybridization to develop some bonding character in the third orbital, so that the "trivalent" O^+ can form an extra bond. The hydrogen atom (not just a proton or a hydride ion) from the γ carbon then migrates to the oxygen through a series of single-electron shifts, as shown below:

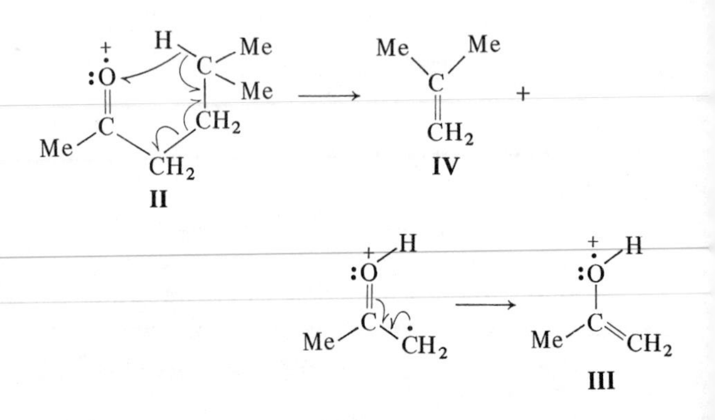

Note the use of a "fishhook" ⇀ to indicate the single-electron shift. We shall follow a recently developed[16] convention to denote the bonding electrons and their shifts during the fragmentation process. According to this convention:

i) A two-electron bond is represented by a line between two atoms, for example, X–Y.

ii) Often a covalent bond loses an electron during ionization by electron impact. For example, the ionization process removes an electron from a C–C bond in alkanes, leaving only *one* electron in the bonding sigma molecular orbital of the ionized species. Such a single-electron, ionized bond is represented by the symbol +• or •+, for example,

$$X{-}Y \xrightarrow{-e} X\cdot{+}\,Y \text{ or } X{+}\cdot\,Y$$

iii) A conventional curved arrow ⟶ denotes a two-electron shift, whereas the shift of a single electron is indicated by a fishhook ⇀.

iv) The cleavage of a covalent bond may occur by:

a) *homolysis*, or transferring one electron of the bond to each fragment:

$$X{-}Y \rightarrow X^{\cdot} + Y^{\cdot}$$

b) *heterolysis*, or transferring both electrons of the covalent bond to one or the other fragment:

$$X{-}Y \rightarrow X^{+} + Y^{-}$$

c) *hemi-heterolysis*, in which the covalent bond is ionized first and the ionized bond is then cleaved:

$$X{-}Y \xrightarrow{-e} X + {}^{\cdot}Y \rightarrow X^{+} + Y^{\cdot}$$

Rearrangements involving migration of γ hydrogens to a polar functional group occur during the mass spectrometry of many olefins, phenyl alkanes, epoxides, aldehydes, ketones, acids, esters, amides, hydrazones, ketimines, nitriles, sulfites, carbonates, and phosphates. This type of hydrogen-transfer reaction as a result of electron impact is now generally referred to as the *McLafferty rearrangement.*[17]

ii) Ion-molecule complexes. Interpretation of mass spectra is further complicated by secondary reactions occurring between the primary ions and the un-ionized gas in the ion source of the spectrometer.[18,19,20] These reactions, although less common under normal operating conditions in modern instruments, take place due to ion-molecule collisions, and are extremely rapid compared to ordinary chemical reactions. In the most extensively studied type of ion-molecule reactions, the molecular ions $M^{\ddagger}$ abstract hydrogen atoms during collisions with the neutral molecules and produce secondary ions having mass-to-charge ratio $(M + 1)^+$ which is one unit above the molecular weight of the sample. Some examples of such ion-molecule reactions involving formation of $(M + 1)^+$ ions are:

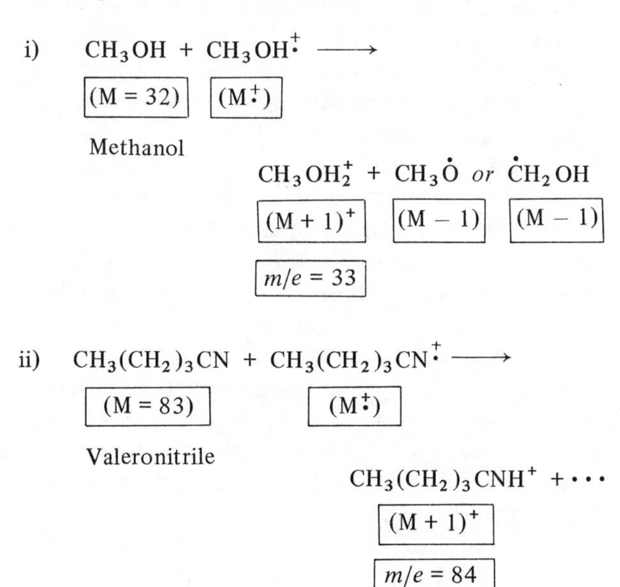

Ion-molecule reactions producing complex ions of mass-to-charge ratio more than $(M + 1)^+$ can also occur. For example, many straight-chain alkylnitriles exhibit a pressure-dependent peak at mass (M + 41). The complex ion responsible for this peak is produced by a combination of the McLafferty rearrangement followed by the ion-molecule reaction. The process can be generally represented as:

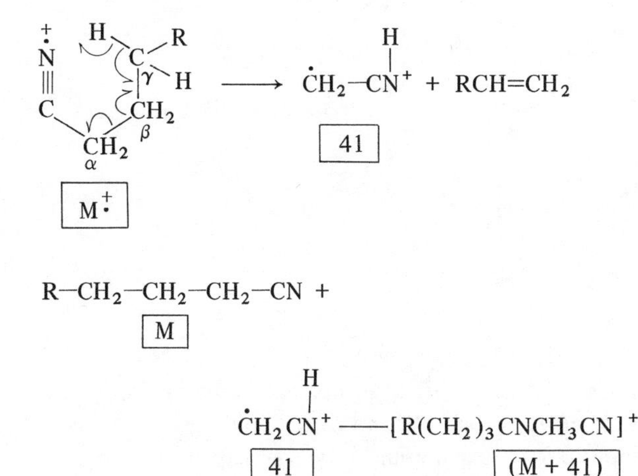

Compounds in which ion-molecule reactions occur are usually those containing heteroatoms such as oxygen and nitrogen, and include such types as alcohols, ethers, esters, aliphatic amines, nitriles, and sulphides. The abundance of ion-molecule complexes in the mass spectra of these compounds depends on the probability of collisions between the ions and the molecules. The higher the number of molecules inside the ion source—i.e., the higher the pressure inside the ion source—the greater the probability of such collisions, and (since many ions and neutral molecules react chemically at every collision) the higher the abundance of $(M + 1)^+$ or $(M + \text{Fragment})^+$ peaks. These peaks are thus pressure-dependent.

The $(M + 1)^+$ and $(M + F)^+$ peaks sometimes cause confusion because, being the peaks corresponding to the highest masses in a spectrum, they may be mistaken for parent ions $(M)^{\ddagger}$. However, although these peaks are sharp, they are weak, and their presence can be detected by varying the spectrometer pressure, or by changing the voltage in the ion repeller. Lowering this voltage causes ions to spend more time in the ion source. The probability of ion-molecule collisions thereby increases, making the corresponding peaks more intense.

Two relatively recent techniques are worthy of some mention here. One, known as *field ionization*,[21] provides spectra which are practically devoid of ion-molecule complexes, whereas the other, called *chemical ionization*, is essentially based on allowing a reaction gas to undergo ion-molecule reactions in the ion source.[22] Both these techniques will be discussed in Section 5.6.

Table 5.1 Some isotopes and their natural abundances

Element	Isotope	Mass	Abundance, %	Isotope	Mass	Abundance, %	Isotope	Mass	Abundance, %
Hydrogen	^{1}H	1.007825	99.9855	^{2}H	2.01410	00.0145	–	–	–
Carbon	^{12}C	12.000000	98.8920	^{13}C	13.00335	1.1080	–	–	–
Nitrogen	^{14}N	14.00307	99.635	^{15}N	15.00011	00.365	–	–	–
Oxygen	^{16}O	15.99491	99.759	^{17}O	16.99914	00.037	^{18}O	17.99916	00.204
Fluorine	^{19}F	18.99840	100	–	– –	–	–	–	–
Silicon	^{28}Si	27.97693	92.20	^{29}Si	28.97649	4.70	^{30}Si	29.97376	3.10
Phosphorus	^{31}P	30.97376	100	–	–	–	–	–	–
Sulfur	^{32}S	31.97207	95.018	^{33}S	32.97146	00.750	^{34}S	33.96786	4.215
Chlorine	^{35}Cl	34.96885	75.537	^{37}Cl	36.96590	24.463	–	–	–
Bromine	^{79}Br	78.9183	50.52	^{81}Br	80.9163	49.48	–	–	–
Iodine	^{127}I	126.9044	100	–	–	–	–	–	–

iii) Isotope ions. Natural carbon ^{12}C contains about 1.1% of its isotope ^{13}C and natural hydrogen contains about 0.0145% deuterium (^{2}H). Consequently, in a fragmentation process, when an ionic species such as methylene ion CH_2^{+} is produced, ions with masses higher than 14, such as $^{13}CH_2^{+}$, $^{12}CHD^{+}$, $^{12}CD_2^{+}$, $^{13}CHD^{+}$, and $^{13}CD_2^{+}$, are also produced in small quantities. A high-resolution spectrum of a compound producing methylene ion does not display a single peak at $m/e = 14$, but gives instead three peaks at masses 14.0156, 15.02, 16.02, and a very weak peak at 17.031.

Like carbon and hydrogen, other elements commonly found in organic compounds are also isotopic mixtures in the natural state (Table 5.1). Due to the scarcity of heavier isotopes of C, H, and O, the isotopic peaks in compounds containing these elements are weak. However, in organic chloro and bromo compounds, the isotopic peaks assume great prominence due to the relatively high percentage of heavy isotopes in chlorine and bromine. For example, a trichloro compound (RCl_3) produces four kinds of molecular ions: $(R{-}^{35}Cl_3)^{+\cdot}$, $(M)^{+\cdot}$; $(R{-}^{35}Cl_2{-}^{37}Cl)^{+\cdot}$, $(M + 2)^{+\cdot}$; $(R{-}^{35}Cl{-}^{37}Cl_2)^{+\cdot}$, $(M + 4)^{+\cdot}$; and $(R{-}^{37}Cl_3)^{+\cdot}$, $(M + 6)^{+\cdot}$. None of these four peaks are really weak. Their relative intensities can be shown statistically to be 29:29:9:1.

Since the isotopic compositions of the elements are reasonably constant, one can use the relative intensity ratios of the molecular ions to determine empirical formulas of organic compounds, providing great care is used in recording such ions. Beynon[23,24] has developed a technique for doing this, and has set forth a very useful table to derive the likely empirical formulas on the basis of measurements of the abundances of particular masses and isotopes. We shall discuss this technique in Section 5.6.

iv) Doubly charged ions. During the ionization process, if some ions are produced which are deficient in two electrons, instead of just one, they appear at half their true masses on the spectrum. Normally the singly charged ions appear at the integral mass values on the scale, but, if a doubly charged ion happens to have an odd mass, it appears at nonintegral mass, according to Eq. (5.6) in Section 5.2. The half-integral mass thus gives away the identity of such ions. Multiple ionization is significant with some aromatic compounds.

v) Metastable ions. Whether they are strong or weak, the peaks produced by the ions we have discussed so far are usually sharp and well defined. However, a mass spectrum occasionally shows some diffused, broad, low-intensity peaks, usually at nonintegral masses. These peaks are due to the so-called *metastable ions.* If an ion m_0, produced in the ion source, is *very stable*, it remains undecomposed during its passage through the accelerator and deflector to the ion collector. On the other hand, if m_0 is *very unstable*, it immediately decomposes (time, 10^{-13} sec) into smaller fragments during the time it spends in the ion source. Each of the fragment ions is deflected and recorded according to its m/e ratio. However, when ions of *intermediate stability* suffer fragmentation (unimolecular decomposition at a rate of roughly 10^{6}/sec) during or immediately after acceleration (but prior to magnetic deflection), these ions display weak and diffuse peaks at nonintegral mass numbers.

The reason for the nonintegral, apparent mass of these metastable ions lies in the energy they possess when they enter the magnetic analyzer. While it is leaving the accelerator, the metastable ion m_0 is accelerated to a velocity determined by its heavier mass. However, the ion deflected by the magnetic deflector is a lighter ion

m^+, which is produced according to the equation

$$m_0 = m^+ + (m_0 - m).$$

Due to this variation between its acceleration mass and its deflection mass, the product ion m^+ is not recorded at its true mass m, but is found at an apparent mass m^*, where

$$m^* = \frac{m^2}{m_0}.$$

The loss of kinetic energy during decomposition and the variation in lifetimes of these ions causes the m^* peak to be diffuse and broad.

Metastable peaks are extremely useful in determining and verifying mechanistic pathways by which ions decompose. Such information can provide valuable insights into the structure of the unknown substance. For example, the spectrum of diisopropylamine[25] shows a metastable peak at $m/e = 22.8$, in addition to the strong peaks at $m/e = 44$ and 86, and a relatively weak peak at 101. The 101 peak at high m/e is due to the ionic diisopropylamine molecule produced by the removal of one of the nonbonded electrons on nitrogen.

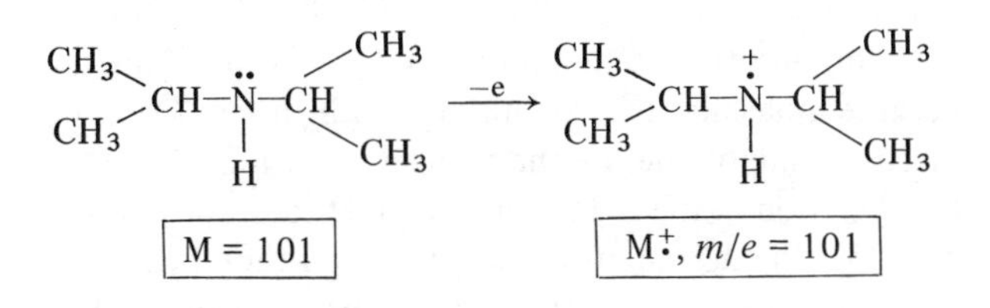

According to the proposed mechanism, the molecular ion M$^{+\cdot}$ immediately undergoes an α cleavage to give the fragment m_0 at $m/e = 86$, which undergoes further fragmentation to an ion m^+ ($m/e = 44$) and neutral propene of mass 42 ($m_0 - m$) by the transfer of a β hydrogen to nitrogen, probably in a homolytic process via a four-membered transition state.

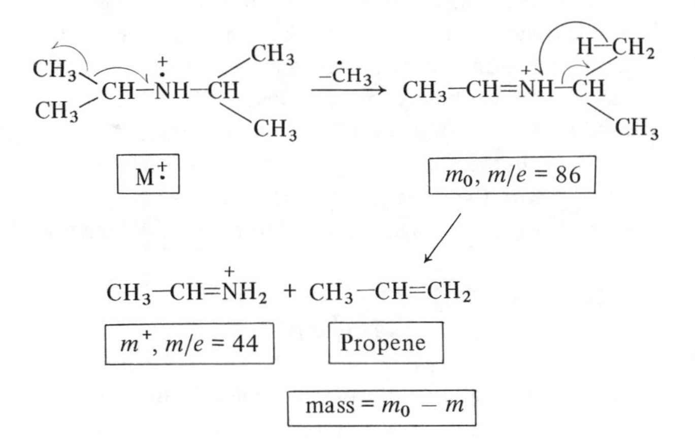

The presence of a metastable peak at apparent mass $m^* = 22.8$ [$m^2/m_0 = (44)^2/86 = 22.5$] is taken as evidence of the proposed fragmentation order.

Metastable transition is also useful when we wish to compare ion structures and determine the amount of kinetic energy of ions in the gas phase. In fact, there has recently been so much interest in the study of metastable transitions that instruments have been produced that are capable of recording both normal and metastable ions or of suppressing the detection of normal ions and amplifying the metastable spectrum.[25a]

Similarly, a very sensitive technique has been developed whereby the metastable ions produced in the flight tube of a time-of-flight instrument can be recorded with highly increased sensitivity by defocusing the interfering normal ions.[25b] This technique, known as *metastable defocusing*, is particularly valuable for the study of collision-induced metastable transitions which produce additional metastable peaks and consequently additional information for the study of reaction mechanisms.[25c,d,e,f,g,h,i]

5.6 DETERMINATION OF MOLECULAR WEIGHT AND MOLECULAR FORMULA

In elucidating the structure of an organic compound, we often want to obtain its molecular weight and molecular formula. In the case of nearly 75% of the organic compounds studied by mass spectrometry, molecular weight can be obtained directly from the spectrum. In these compounds, the molecular ion produced by the initial ionization process is stable enough to reach the detector. Consequently, we can read the molecular weight of such compounds directly by observing the position of the highest-mass peak whose relative intensity is independent of pressure.

The molecular ion peak (parent peak) is intense in the case of those compounds in which factors such as resonance have stabilized the positive charge in the molecular ion. Thus the mass spectra of aromatic and heteroaromatic compounds generally show a strong parent peak. Parent peaks are absent in compounds in which the molecular ion has no stability and is readily susceptible to fragmentation. An approximate order of decreasing stability of molecular ions has been suggested by Biemann[26] as follows: Aromatic and heteroaromatic compounds; cycloalkanes; sulfides and mercaptans; conjugated olefins; straight-chain hydrocarbons; amides; ketones; aldehydes; amines; esters; ethers; carboxylic acids; branched-chain hydrocarbons; nitriles; primary alcohols; tertiary alcohols; acetals.

The logic behind this order is easy to understand. For example, the relative intensities of the parent-ion peaks in the spectra of cyclic compounds are greater than those of their acyclic analogs because the rupture of just one bond in a cyclic compound does not produce smaller fragments, whereas in the case of open-chain compounds, lower mass fractions are produced. Similarly, the higher susceptibility to cleavage of branched-chain compounds causes these compounds to display weaker parent-ion peaks than those exhibited by the corresponding straight-chain compounds.

Thus, in about 75% of the cases, determining molecular weight appears to be simple. For the remaining 25% of compounds, however, the main problem is to ascertain that the peak selected is the parent-ion peak. In these compounds, the parent peaks are very weak or of negligible intensity, and are easily confused with those due to fragment ions and impurities. In such cases it is absolutely essential to ensure the purity of the sample. If the compound can be assumed to be pure, the parent peak can be located by lowering the energy of the bombarding electron beam to as low as the appearance potential. Since the intensity of the peak of a molecular ion depends on its stability under electron bombardment, lowering the energy of the stream of bombarding electrons increases the intensity of the parent peak in relation to the intensities of all other peaks. Yet, if a M^+ is not present at 70 eV, it will still not be present at 10-20 eV either. Therefore, in order to make a positive identification of the parent peak, it is often necessary to utilize other information, such as source and history of the sample, the fragmentation pattern, UV, IR, and NMR characteristics of the compound, and so forth.

Several auxiliary techniques are also useful in the detection of parent peaks and in deducing the molecular formulas of the sample. Some of these techniques are: i) field ionization, ii) chemical ionization, iii) microeffusometry (Graham's law of diffusion), iv) derivatization, v) high resolution, vi) the isotopic-abundance technique, and vii) the valence-requirement rule.

i) Field ionization.[21] Ionization by the electron-impact method requires high energy because the molecule has to *surmount* a high potential-energy barrier, as required in a Franck-Condon transition. In the field-ionization technique, electrons pass from the molecules to the conduction band of a fine metal wire *through* the energy barrier. The passage of an electron "through" a potential-energy barrier which classically it is unable to "surmount" is called the *tunnel effect.*

Field ionization by the tunnel effect is achieved by applying a very high local electric field (of the order of 2×10^8 V/cm or 2 V/Å) on the tip of a very fine metal wire suitably surrounded by earthed surfaces. When the molecule comes near this positively charged conductor, a normal ground-state ($n = 1$) electron of the molecule, if its total energy is the same as the vacant levels in the conduction band of the metal, transfers from the atom to the metal conductor through the energy barrier between the two. The positively charged ion so formed is immediately repelled from the metal surface. Such tunnelling occurs in field ionization because the fields used obviously produce very "thin" energy barriers.

Field ionization is a low-energy process because of tunnelling. Consequently, the fragmentation of the molecular ion is reduced and the intensity of the parent-ion peak is increased during this process. At low energy the possible number of mechanisms for the fragmentation of the molecular ion is also less compared to those available in the electron-bombardment process. Naturally, the number of fragment-ion lines is also considerably less when one is using field ionization. Thus the mass spectra obtained by this technique are often much easier to interpret.

Instruments capable of recording both electron impact and field-ion spectra are now commercially available.

Complications sometimes arise in mass spectrometry because molecules that are unstable at high temperatures often decompose before they ionize at the ion source. A modification in the field-ionization method[27] not only decreases the degree of thermal decomposition of such samples, but further increases the intensity of the peaks of the M^+ ion and the detection sensitivity of the instrument. In this technique, the sample in its saturated solution is applied directly to the field ion emitter, which is activated by benzonitrile. The physically adsorbed molecules undergo evaporation and simultaneous ionization in the ion source at temperatures so low that the molecules do not suffer appreciable decomposition.

ii) Chemical ionization.[22] This technique, although not as effective in detecting the peak of the parent ion as the field-ionization method, is based on ion–molecule reactions. First, highly reactive ions, such as CH_5^+, are produced by introducing a "reaction gas," such as methane, in the ion source. These ions are then allowed to react with the organic molecules under study to produce $(M - 1)^+$ ions, according to the overall equation:

$$CH_4^+ + CH_4 \rightarrow CH_5^+ + CH_3$$
$$CH_5^+ + M \rightarrow (M - 1)^+ + CH_4 + H_2$$

Since the ionization of the organic molecule occurs due to the reaction with CH_5^+ ions, very little energy is transferred to the $(M-1)^+$ ion, and fragmentation of it is reduced. Consequently, this method proves particularly helpful for those compounds whose molecular ions are relatively unstable.

Like the field-ionization technique, the chemical-ionization method also produces spectra which are simpler than those obtained by the electron-bombardment method. The sensitivity of the equipment to important ions in the molecular ion region is also increased, thus making this technique potentially important for the study of complex organic molecules such as carbohydrates, antibiotics,[28] alkaloids, peptides, steroids, and so forth.

iii) Microeffusometry; Graham's law of diffusion. According to Graham's law of diffusion, at constant temperature and pressure, the rates of diffusion of gases, diffusing through the same medium, are inversely proportional to the square roots of their respective densities. In a method based on this law, the streams of the gaseous sample and of a reference compound are allowed to flow from the reservoir to the ionization chamber through a molecular leak. The rates of decay for the unicomponent peaks of the sample and the reference[28a] are then compared so that the approximate molecular weight of the sample can be calculated.

iv) Derivatization. One can often prepare an appropriate derivative of a sample such that the molecular ion of the derivative is stable and shows an appreciable parent-ion peak in the spectrum. For example, in the case of some oligosaccharides, even the field-ionization technique fails to produce significant peaks for the molecular ions. However, the field-ionization spectra of the methylated derivatives of these oligosaccharides often display the molecular ion lines as the most intense lines in the spectra.[29]

v) High resolution. With a high-resolution spectrometer (double-focusing instrument), one can often determine the elemental composition of each peak in the spectrum and thus determine the empirical and molecular formulas of the sample. This determination is based on the fact that the masses of atoms are not, accurately speaking, whole integrals on the basis of the atomic mass standard of carbon: $^{12}C = 12.000000$. For example, according to this standard, the accurate mass of a hydrogen atom ^1H is not exactly 1 atomic mass unit (amu), but is actually 1.0078252. The accurate mass of oxygen ^{16}O is 15.994914 (see Table 5.1). This small fractional difference is due to the nuclear packing fraction of each atom.

Now suppose that we used a low-resolution spectrometer or some other technique, and it indicated that the

Table 5.2 Partial list of elemental compositions and accurate masses of isobaric ions of nominal mass/charge = 166

Accurate mass	Elemental composition
166.004478	$C_{11}H_2O_2$
166.012031	$C_7H_4NO_4$
166.023264	$C_6H_4N_3O_3$
166.026946	$C_9H_2N_4$
166.038864	$C_{12}H_6O$
166.046074	$C_6H_6N_4O_2$
166.057650	$C_7H_8N_3O_2$
166.062994	$C_9H_{10}O_3$
166.074228	$C_8H_{10}N_2O_2$
166.078252	$C_{13}H_{10}$
166.093357	$C_5H_{14}N_2O_4$
166.097038	$C_8H_{12}N_3O$
166.108614	$C_9H_{14}N_2O$
166.120190	$C_{10}H_{16}NO$

nominal molecular weight of an unknown organic sample was 166. This information offers very little clue either to the elemental composition or the molecular formula of the compound because there are some 32 different combinations of, say, C, H, O, and N atoms which could produce molecules of mass 166. Table 5.2 lists 14 of these possible atomic combinations, all of which have the same nominal mass number: 166. The accurate masses of these combinations (as calculated on the basis of atomic masses given in Table 5.1), however, differ slightly from each other. A high-resolution instrument can discriminate between these probable molecular formulas and focus our attention on the exact elemental composition of the sample. For example, although Table 5.2 lists some 14 different formulas ranging from C_5 to C_{13}, if the spectrometer registers the accurate mass of the ion as 166.06299, our attention is immediately drawn to the possible formula $C_9H_{10}O_3$.

We must, however, remember that although it is possible to eliminate most of the 32 probable atomic combinations on the basis of accurate mass, it may not be easy to distinguish between $C_9H_{10}O_3$ (molecular weight 166.062994) and $C_7H_8N_3O_2$ or $C_8H_{10}N_2O_2$. In such cases it is necessary to obtain further clues from the fragmentation pattern of the sample. One technique combines the facilities of high resolution and automatic formula computation to list the ions directly according to their elemental composition.[30a,30b,30c] The details of the technique are discussed by Biemann.[31] Analysis of compounds which are of low molecular weight can be carried out[32] without computer facilities by the use of an extensive table of accurate mass versus elemental composition compiled by McLafferty[33] and Beynon.[23,24] Similar extensive tables are published by Lederberg.[34] The same author has now published[35] a method for rapid calculation of molecular formulas from mass values. This method employs a greatly shortened form of Beynon's tables, but requires the use of a desk calculator.

Calculations may be done by hand, but may prove time-consuming. We shall therefore use Beynon's extensive tables (Appendix 5.1) in our study, although Lederberg's method is explained in Appendix 5.2 for the benefit of those who may have a desk calculator or a small computer readily available.

It must be remembered as well that even if we arrive at the correct molecular formula of the sample, we are still far from actually identifying it. For example, 57 different compounds are presently known to have the same molecular formula $C_9H_{10}O_3$!

Table 5.3 Intensities of M + 1 and M + 2 peaks as percentages of M peak for ions of mass/charge = 166

Elemental composition	M + 1 peak % (mass 167)	M + 2 peak % (mass 168)
$C_5H_{14}N_2O_4$	6.546	0.985
$C_6H_4N_3O_3$	7.809	0.869
$C_6H_6N_4O_2$	8.184	0.698
$C_7H_4NO_4$	8.166	1.094
$C_7H_8N_3O_2$	8.915	0.755
$C_8H_{10}N_2O_2$	9.646	0.818
$C_8H_{12}N_3O$	10.020	0.654
$C_9H_2N_4$	11.283	0.581
$C_9H_{10}O_3$	10.003	1.049
$C_9H_{14}N_2O$	10.751	0.724
$C_{10}H_{16}NO$	11.483	0.800
$C_{11}H_2O_2$	11.997	1.056
$C_{12}H_6O$	13.102	0.989
$C_{13}H_{10}$	14.208	0.933

vi) Isotopic-abundance technique. This method is particularly suitable to identifying compounds which have low to moderate molecular weights and which are capable of producing reasonably stable molecular ions. In Section 5.5 we mentioned that since the isotopic compositions of the elements are reasonably constant, we can utilize the relative intensity of the peaks of the ions to determine empirical formulas of organic compounds. All the elements (except fluorine, phosphorus, and iodine) listed in Table 5.1 contain heavier isotopes, in certain fixed proportions. The mass of a heavier isotope of an element is usually 1 or 2 amu more than that of the lighter isotope. Consequently, molecules containing heavy isotopes appear on the spectrum at *m/e* one or more units higher than the normal. Thus the spectra of most compounds show peaks at M + 1 and M + 2. The intensities of these peaks in relation to the molecular ion peak M is low but fixed (except in those compounds containing chlorine or bromine, in which the percentages of heavy isotopes are quite high).

Since the ratio of the intensities of M, M + 1, and M + 2 peaks is constant for a given molecular composition, the observed values can be compared with those calculated theoretically. A table in which the ratios of the intensities of the M, M + 1, and M + 2 peaks are calculated for all reasonable combinations of C, H, O, and N atoms, up to a molecular weight of 500, has been constructed by Beynon,[23,24] and is extremely useful for this comparison. Appendix 5.1 presents a shortened version of Beynon's table, expressing M + 1 and M + 2 peaks as percentages of the molecular ion peak M.

To illustrate the use of this technique, let us return to Table 5.2, which gives the elemental compositions of 14 ions of nominal mass 166. The ratios of the intensities of M, M + 1, and M + 2 peaks for these 14 possible compositions can be obtained from Beynon's tables (Appendix 5.1). These ratios are reproduced in Table 5.3.

Thus a spectrum showing the intensities of the three peaks at masses 166, 167, and 168 as

Peak	Mass	Intensity
M	166	100
M + 1	167	10.15
M + 2	168	1.1

can be related immediately to the molecular formulas $C_9H_{10}O_3$ or $C_8H_{12}N_3O$. The latter formula can be eliminated on the basis of the nitrogen rule (see below).

Caution is needed in applying this technique because the observed intensities of the M + 1 peaks are usually somewhat higher than the theoretically calculated ones. This is due to the fact that, if the quantity of the sample introduced into the ion source is large, or if the ion repeller potential is low, so that the ions spend more time in the ionization chamber, bimolecular collisions occur. If the sample contains atoms such as O, N, or S, the bimolecular collisions usually produce a species of *m/e* = M + 1 due to the addition of a hydrogen atom to the parent radical ion. Great care must also be taken in the recording and measurement of such ions.

The values given in Appendix 5.1, that is, the percentages of the M + 1 and M + 2 peaks in relation to the peak M, can be calculated as follows:

If the formula is $C_wH_xN_yO_z$, then

The percentage of M + 1 peak =
$(1.08 \times w) + (0.016 \times x) + (0.38 \times y) + (0.039 \times z)$.

Since the multiplication factors for hydrogen and oxygen atoms are small, the approximate percentage of M + 1 peak works out to:

$$\text{Percentage of M + 1 peak} = (1.1 \times w) + (0.38 \times y)$$

and

$$\text{Percentage of M + 2 peak} = \frac{(1.1 \times w)}{200} + (0.20 \times z).$$

For example, for the formula $C_8H_{12}N_3O$:

$$\text{The percentage of M + 1} = (1.08 \times 8) + (0.016 \times 12) + (0.38 \times 3) + (0.039 \times 1) = 10.01\text{* percent.}$$

$$\text{The percentage of M + 2} = \frac{(1.1 \times 8)^2}{200} + (0.20 \times 1) = 0.597\text{* percent}$$

For the formula $C_9H_{10}O_3$:

$$\text{The percentage of M + 1} = (1.08 \times 9) + (0.016 \times 10) + (0.039 \times 3) = 9.997 \text{ percent}$$

and

$$\text{The percentage of M + 2} = \frac{(1.1 \times 9)^2}{200} + (0.20 \times 3) = 1.08\text{* percent.}$$

From the above equations, it follows that in the case of compounds containing C, H, and O only, the number of carbon atoms in the molecule can be calculated as:

$$\text{Number of carbon atoms} = \frac{\text{percentage of M + 1 peak}}{1.1}$$

(to the nearest integer)

vii) Valence requirement rule. When the above techniques suggest several atomic combinations as probabilities for the formula for the molecular ion, one can often eliminate some of these combinations by applying a rule of valence requirement. This rule is the so-called *nitrogen rule*, which applies to all compounds that have only covalent bonds, and that are composed of any possible combination of C, H, O, N, S, P, Si, As, the halogens, and the alkaline-earth metals.

According to this rule, *if the molecular weight of the compound is computed on the basis of the atomic masses of the most abundant isotopes, all organic molecules with even molecular weight must contain an even number (including zero) of nitrogen atoms; those with odd molecular weight must contain an odd number of nitrogen atoms.*

This rule is based on the fact that, for most elements, except nitrogen, the most abundant isotopes of the elements that have even valence have even mass number, while the corresponding isotopes of those with odd valence have odd mass number. Only in the case of nitrogen is the reverse true. That is, although the valence of nitrogen is odd (3), the mass number of its most abundant isotope is even (14).

Thus, on the basis of the nitrogen rule alone, we can eliminate 5 of the 14 possible formulas listed in Table 5.3. Of the two probabilities, $C_9H_{10}O_3$ and $C_8H_{12}N_3O$, on the basis of the isotopic-abundance technique alone, we can now eliminate the latter. And—if we pool the deductions from the nitrogen-rule, isotopic-abundance and high-resolution techniques —we can safely presume $C_9H_{10}O_3$ to be the molecular formula of the sample.

5.7 FACTORS GOVERNING THE FRAGMENTATION PATTERNS

When we are studying a totally unknown sample, its fragmentation pattern often gives us some important clues to its identity. For example, we can sometimes make an educated guess about the type of organic compound on the basis of the intensity of the peak of the molecular ion, M; the intensity of this peak is directly related to the chemical stability of the molecule. An intense M peak may suggest the possibility of a conjugated olefinic, aromatic, or polycyclic compound. On the other hand, a very weak M peak may indicate such compounds as acetals, tertiary alcohols, highly branched hydrocarbons, and so forth. We have already mentioned, earlier in this section, Biemann's listing of the approximate order of stability of molecular ions.

The abundance of a molecular ion naturally depends on the bond strengths of the individual bonds present in the molecule. However, it is hard to draw an exact parallel between bond strengths and ion abundances because the fragmentation process involves molecular ions in the excited state and not molecules in the ground state. Other factors—such as resonance, hyperconjugation, polarizability, inductive and steric factors, energy of transition for fragmentation, and the stability of the fragments produced—also affect the abundances of molecular and other fragment ions.

To enable us to understand the mechanism of ion fragmentation and thus to predict the relative intensities of some of the peaks in the mass spectra, we shall state

---

* In practice, one can never calculate these values with such accuracy.

certain generalizations based on the role of these factors. First, however, it must be emphasized that these statements are only generalizations, and are not to be taken as strict rules. In other words, these statements, though applicable to many compounds, do not hold for all of them.

The prominent peaks in the mass spectrum of a compound are due to the most stable ions produced by the most favorable reaction pathways. The stability of an ion depends largely on its structure. Structural factors can stabilize the ion, and can also lower or raise the free energy of its decomposition products.[36] Such structural factors influencing the intensities of peaks of ions are *electron sharing, resonance,* and *inductive effects.* The following generalizations will help us to understand their role.

1. In the case of alkanes, the relative intensity of the absorption peak of the molecular ion M is greatest for the straight-chain compound, but:

i) The intensity decreases with increased degree of branching.

ii) The intensity decreases with increasing molecular weight.

At higher molecular weights, however, there are many exceptions to this generalization.

It is easy to rationalize these observations on the basis of stabilization of charge of the molecular ions and their fragmentation products. Initial ionization removes a bonding electron from the alkane, producing a net positive charge on the molecule. The stability of the molecular ion $M^{\ddagger}$ thus produced depends on (a) its ability to delocalize this positive charge or (b) the changes in its total free energy during the decomposition reaction. In straight-chain alkanes which have low molecular weight, the stability gained by the fragmentation of the initially formed molecular ion is less than that in the case of branched-chain alkanes. In branched-chain alkanes fragmentation is favored because a cleavage at the branched carbon atom can produce more stable secondary or tertiary carbonium ions. In fact, the initial ionization process itself favors removal of an electron from the most polarizable bond, and the polarizability, in the case of alkanes, increases due to chain branching and to the increased size of the adjacent alkyl groups.

We can illustrate the effect of branching on the fragmentation patterns by looking at some isomers of octane. Since carbon-carbon bonds break more easily than the stronger carbon-hydrogen bonds, and since no particular C–C bond in *n*-octane can be considered to be more polar than any one of the other six, the initial ionization in *n*-octane (I) takes place by the removal of an electron from any one of the seven C–C bonds.

C–C+·C–C–C–C–C–C

C–C–C–C–C–C–C–C (I) → or etc.

C–C–C–C+·C–C–C–C

In 3-methylheptane (II), on the other hand, the bonds adjacent to the branch are more polar than the other C–C bonds, and the polarities of the three C–C bonds of the tertiary carbon also differ because of the sizes of the adjacent alkyl groups. The probabilities of initial ionization of (II) are thus in the following order:

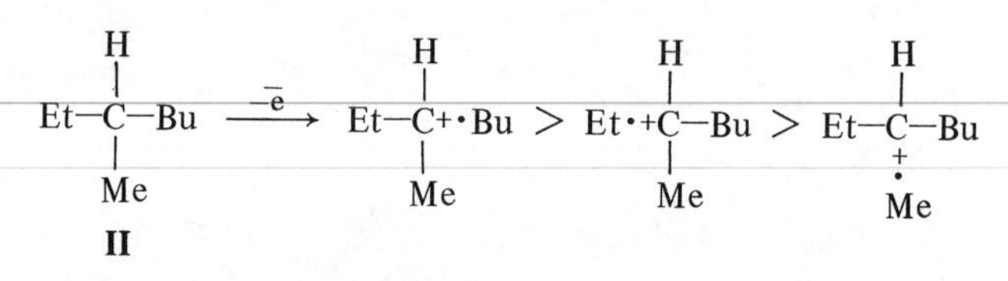

Furthermore, the molecular ion thus produced readily undergoes fragmentation to produce a stable secondary carbonium ion.

Et–CH(Me)+·Bu → Et–CH(Me)$^+$

The same logic applies to 2,2,4,4-tetramethylpentane (III), in which the initial removal of the electron and the subsequent fragmentation produces a large proportion of the stable tertiary carbonium ion, $(C_4H_9)^+$. Consequently, the abundance of molecular ions $M^{\ddagger}$ in the spectrum of (III) is low compared to that of (I) in its spectrum. The intensity of the M peak in (I) is 6.7% of the base peak, which appears at $m/e = 43$ [due to the $(C_3H_7)^+$ ions] in its spectrum, whereas the M peak in (III) is only 0.02% of the base peak, which appears at $m/e = 57$ due to the $(C_4H_9)^+$ ions.

This discussion of the ionization and fragmentation of alkanes illustrates the way such factors as the polarizability of bonds and inductive effects affect the stabilities of alkyl carbonium ions. More generalizations will help us to further understand the role of these factors.

2. Substituted cycloparaffins (saturated cyclic hydrocarbons carrying side chains) tend to lose their side chains through the cleavage of an exocyclic bond. A substituted cycloparaffin is similar to a highly branched hydrocarbon in that the cleavage is favored at the tertiary

carbon in both cases. When the side chain of a substituted cycloparaffin is cleaved, the ring itself behaves as a characteristic entity and retains its identity. Thus a substituted cyclopentyl hydrocarbon often shows a peak at $m/e = 69$ corresponding to the cyclopentyl ion $(C_5H_9)^+$. Generally, the intensity of an M^+ ion peak is greater in the case of an alkane containing a ring than in the case of an n-alkane with the same number of carbon atoms.

3. Certain ions such as $(C_3H_7)^+$ ($m/e = 43$) and $(C_4H_9)^+$ ($m/e = 57$) are usually stable, and contribute strong peaks in the spectra of alkanes. These ions are sometimes formed even if a rearrangement is required for their formation. The reasons for their unusual stability are not yet clearly understood.

4. Formation of a single-carbon fragment is unlikely unless the compound contains a methyl side chain.

5. Except for compounds containing an odd number of nitrogen atoms (recall the nitrogen rule), mass spectra generally have relatively abundant ions with odd mass number. When spectra exhibit prominent peaks at even mass numbers, it generally indicates that the parent compound either (a) contains an odd number of nitrogen atoms or (b) is highly branched, and undergoes fragmentation of two separate side chains, or (c) is cyclic, and suffers ring fragmentation.

6. Molecular ions are usually very abundant in compounds that contain double bonds, or that are cyclic (particularly aromatic or heteroaromatic compounds).

The prominence of the peaks of molecular ions in alicyclic compounds can be explained on the grounds that the removal of carbon atoms from the ring requires cleavage of two carbon-carbon bonds. One reason why aromatic, heteroaromatic, and unsaturated compounds exhibit prominent parent-ion peaks is that the sharing of electrons in a hetero atom, or the π electrons in an unsaturated system, lowers the positive charge at the initial site of ionization and stabilizes the parent carbonium ion.

7. Bonds β to a double bond, β to an aromatic ring, and β to a hetero atom are usually probable sites for ready cleavage because of the resonance stability of the products of such rupture. For example, fission of a bond β to the double bond in an alkene produces a stable allylic carbonium ion.

$$\text{R—}\overset{+\cdot}{\text{CH—}}\text{CH}\underset{\alpha}{\text{—}}\text{CH}_2\underset{\beta}{\text{—}}\text{CH}_2\text{—CH}_2\text{—R} \longrightarrow$$

$$\text{R—}\overset{+}{\text{C}}\text{H—CH=CH}_2 + {}^{\bullet}\text{CH}_2\text{—CH}_2\text{—R}$$

allylic carbonium ion

The product carbonium ion is stabilized by the following resonance:

$$\text{R—CH=CH—}\overset{+}{\text{C}}\text{H}_2 \longleftrightarrow \text{R—}\overset{+}{\text{C}}\text{H—CH=CH}_2$$

$$\approx [\text{R—CH}\text{=-}\text{CH}\text{=-}\text{CH}_2]^+$$

The rupture of a bond β to the aromatic ring gives a resonance-stabilized benzyl ion, $C_6H_5CH_2^+$, of $m/e = 91$:

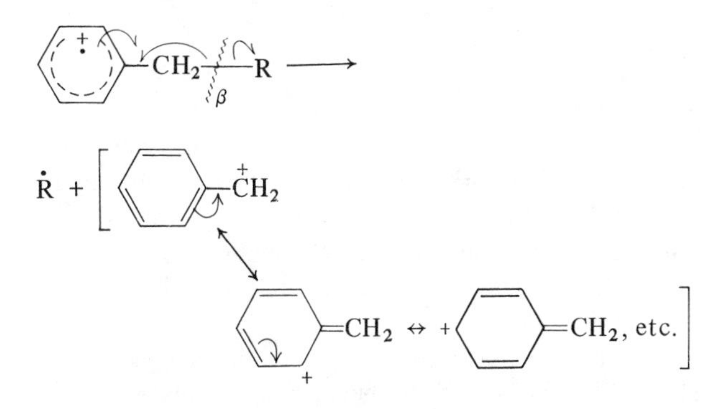

In fact, an intense peak of mass 91 is quite common in the spectra of many alkyl-substituted aromatic compounds. However, Meyerson and co-workers[37] have shown that in most cases this peak is due to the highly stable tropyllium ion $C_7H_9^+$ formed by the rearrangement of the benzyl ion.

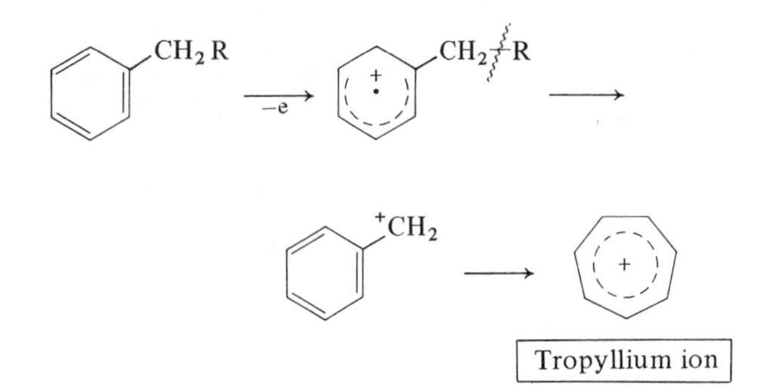

The readiness of formation and marked stability of the tropyllium ion is justified on the basis of Huckel's $4n + 2$ *rule*,[38] according to which a species like the tropyllium ion, containing 6 π electrons in conjugation, can achieve resonance stability and aromaticity in the same way benzene does.

α,β-cleavage is very common in compounds containing a heteroatom because of the resonance stability of the ions produced by the homolytic cleavage of the bonds β to the hetero atom. In some cases the driving force for α,β cleavage is not merely that of resonance stabilization, but also that of subsequent cleavage through the migration of a hydrogen from the β carbon.

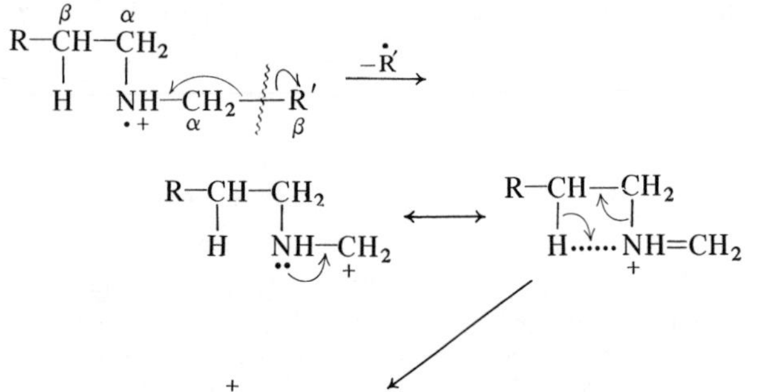

8. Parent ions containing oxygen, sulfur, or nitrogen decompose readily also to eliminate stable molecules such as water, olefins, carbon monoxide, stable alcohols, hydrogen cyanide, nitric oxide, ammonia, hydrogen sulfide, and stable mercaptans. Such an elimination usually entails rearrangement of the metastable ion initially produced.

An example of the elimination of an alkene is given above. Eliminations of this nature also take place in olefins, *n*-alkyl benzenes, vinyl and phenyl ethers, sulfites and carbonyl compounds such as aldehydes, ketones, acids, esters, amides, and carbamates. In these cases fission occurs by concerted heterolysis of bonds in a six-membered transition state (instead of the four-membered one shown above) as:

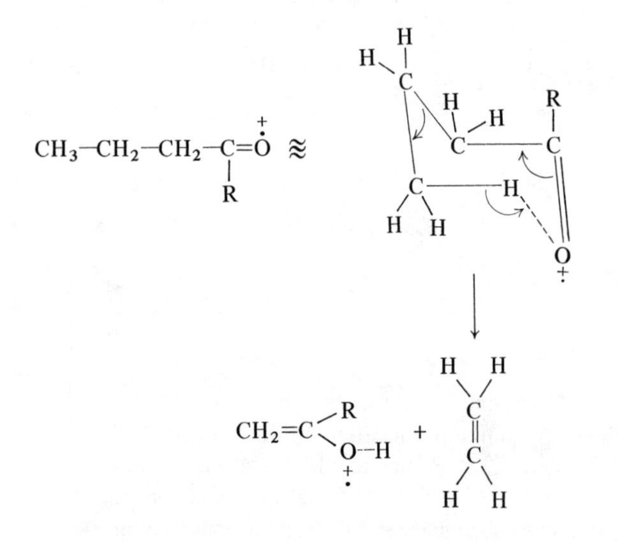

It can thus be said that the factors that play major roles in determining the fragmentation patterns of organic compounds are:

i) Relative bond strengths of various bonds in the parent ion

ii) Relative stabilities of the products which can be formed by the competing degradation paths

iii) Relative probability of certain ion degradation paths

5.8 FRAGMENTATION PATTERNS OF VARIOUS CLASSES OF ORGANIC COMPOUNDS

After our discussion of the factors governing fragmentation patterns, let us now consider the characteristic behavior of various classes of organic compounds in the ionization chamber.

i) Hydrocarbons

a) Alkanes Characteristic features of the alkane spectra are as follows.

1. Straight-chain alkanes show low-intensity molecular-ion peaks. The intensities of these peaks decrease with increasing chain length. At C_{45} the M peak is reduced to an insignificant intensity and is barely detectable. It is very intense in the C_1 to C_5 range (Rule 1).*

2. Since straight-chain alkanes rarely eliminate methyl groups ($-CH_3$), peaks corresponding to (M – 15) are absent, or are of low intensity, in the spectra of long-chain paraffins (Rule 4).

3. Elimination of $-C_2H_5$, $-C_3H_7$, etc., from the molecular ion gives peaks at masses C_nH_{2n+1} ($m/e = 29, 43, 57, 71$, and 85). These peaks are most intense in the C_2 to C_5 range. The peaks at 43 or 57 are usually the base peaks, due to the remarkable stabilities of the propyl ion ($-\overset{+}{C}_3H_7$) and the butyl ion ($-C_4H_9^+$) (Rule 3). The intensities of other peaks 14 mass units ($-CH_2$) apart (71, 85, etc.) decrease with the increasing weight of the fragment. These peaks 14 units apart appear as clusters because each prominent peak is accompanied by a smaller peak one unit higher, due to ^{13}C, and a few small peaks one and two units lower, due to the loss of hydrogen atoms.

4. Low-intensity peaks due to ions produced by random rearrangements are common in the spectra of alkanes. As mentioned earlier, the difficulty in cleavage of C–H and C–C bonds requires that the alkanes spend relatively more time in the ion source. This time factor is responsible for the random rearrangements.

5. In branched alkanes, there is a tendency for the bonds to rupture at the branches, resulting in the formation of relatively stable secondary and tertiary carbonium ions (Rule 1).

---

* These numbers refer to generalizations given in Section 5.7.

6. Molecular ion peaks are more abundant in cycloparaffins than in straight-chain paraffins containing the same number of carbon atoms, although the cyclic compounds tend to lose their side chains (Rules 2 and 6). If the ring undergoes fragmentation, it loses two carbon atoms simultaneously, thus producing an abundance of $\overset{+}{C_2H_4}$ ($m/e = 28$), $\overset{+}{C_2H_5}$ ($m/e = 29$), and M − 28, M − 29 ions in the spectrum. Due to this tendency to lose ethylene C_2H_4 and other even-mass-number fragments, the percentages of fragments having even mass number are usually higher in the spectra of cycloparaffins than in those of acyclic hydrocarbons. For example, the base peak in cyclohexane (Fig. 5.8) is at $m/e = 56$. Compounds containing cyclohexyl rings show prominent peaks at $m/e = 83$, 82, and 81, whereas those containing cyclopentyl rings show an intense peak at $m/e = 69$ (Rule 2).

Mechanisms for fragmentations of different types of alkanes are believed to be as follows.

i) Straight chain alkanes

$$\underset{\boxed{\text{alkane}}}{CH_3-CH_2-CH_2-CH_2-CH_2-CH_3} \xrightarrow[-e]{\text{ionization}} \underset{\boxed{\text{parent ion}}}{CH_3-CH_2\cdot{+}CH_2-CH_2-CH_2-CH_3}$$

$$\underset{\boxed{\text{parent ion}}}{CH_3-CH_2\cdot{+}CH_2-CH_2-CH_2-CH_3} \xrightarrow[\text{followed by heterolysis}]{\text{hemi-heterolysis}} \underset{\boxed{\text{radical}}}{CH_3-\dot{C}H_2} + \underset{\boxed{\text{ion}}}{\overset{+}{C}H_2-CH_2-CH_2-CH_3}$$

$$\xrightarrow{-CH_2} CH_2 + {}^{+}CH_2-CH_2-CH_2$$

ii) Branched chain alkanes

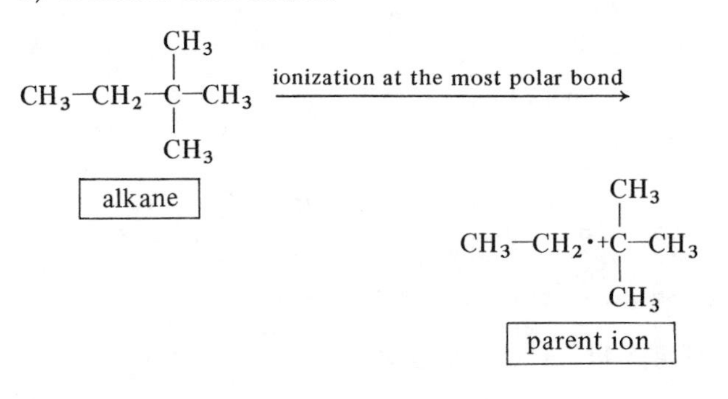

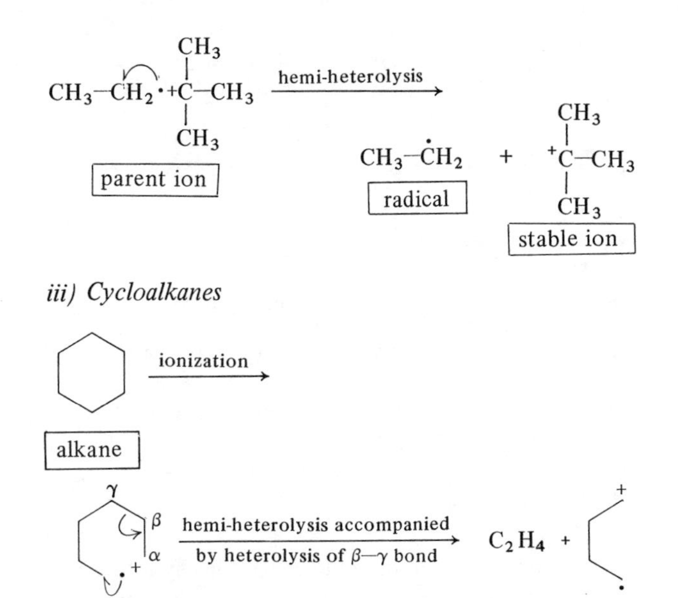

Spectra of n-hexane, 2,2-dimethyl butane, and cyclohexane presented in Figs. 5.6, 5.7, and 5.8 illustrate many of the above characteristics of paraffins.

b) Alkenes Mass spectra of alkenes are usually harder to interpret than those of alkanes or compounds containing other functional groups. Some of the features of the spectra of alkenes are as follows.

m/e	% of base peak	m/e	% of base peak	m/e	% of base peak	m/e	% of base peak
15	6	26	4	27	45	28	11
29	61	30	1	39	20	40	3
41	70	42	41	43	81	44	3
53	2	55	7	56	45	57	100
58	4	71	5	85	0.4	86	15.5 (M)
87	1						

Fig. 5.6 The mass spectrum of n-hexane.

m/e	% of base peak	m/e	% of base peak	m/e	% of base peak	m/e	% of base peak
14	1	15	9	26	3	27	31
28	5	29	48	30	1	39	22
40	3	41	56	42	5	43	100
44	3	53	3	55	12	56	32
57	98	58	4	70	3	71	71
72	5	85	0.2	86	0.1 (M)		

Fig. 5.7 The mass spectrum of 2,2-dimethylbutane.

m/e	% of base peak	m/e	% of base peak	m/e	% of base peak	m/e	% of base peak
15	1.4^R	15.5	0.1*	16	1.0	24.5	0.1**
25.5	0.1*	26	2.3	27	18	28	9.0
29	9	57.5	1*	38	2	39	20.5
40	5	41	57	42	26	43	12.2^R
50	2	51	3	52	1	53	4
54	6	55	34	56	100	67	3
68	2	69	23.4	70	1.4	77	1
78	1	83	4.7	84	73M	85	4.8

^{R} = Rearrangement peak * = Metastable peak
** = Doubly charged peak

Fig. 5.8 The mass spectrum of cyclohexane.

1. Alkenes readily produce molecular ions by losing a π electron. Consequently, spectra of alkenes show low intensity but distinct M peaks. As in alkanes, so in alkenes, the intensity of the M peak decreases with increasing molecular weight.

2. The most intense peak (usually the base peak) in the spectra of olefins is due to the stable, charged species produced by allylic cleavage, that is, by the rupture of the C–C bond β to the double bond (Rule 7). The fragment carrying the double bond is usually the charged species:

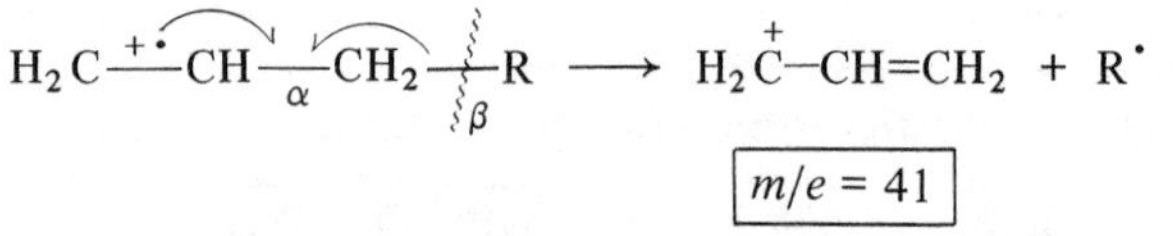

$$H_2C \overset{+\cdot}{—} CH \underset{\alpha}{—} CH_2 \underset{\beta}{—} R \longrightarrow H_2\overset{+}{C}—CH=CH_2 + R^{\cdot}$$

$$\boxed{m/e = 41}$$

3. The natural outcome of allylic cleavage is a series of fragments at masses 41, 55, 69, 83, etc., with the general formula C_nH_{2n-1} (allyl carbonium ions). These peaks are two atomic mass units lower than those in the alkane series. For example, 1-butene displays base peak at m/e = 41 (Fig. 5.9).

4. McLafferty rearrangements are also common in these ions. These rearrangements produce ions of the general formula C_nH_{2n}.

5. Migrations of double bonds through π electron movements make it almost impossible to distinguish between isomers that differ in the position of a double bond in the molecule (*positional isomers*).

c) Aromatic hydrocarbons The mass-spectral data for two aromatic hydrocarbons, benzene and *t*-amylbenzene, are given in Figs. 5.10 and 5.11, respectively. These figures illustrate the following characteristics of the aromatic hydrocarbons.

1. Spectra of aromatic compounds show much stronger molecular ion peaks M than those exhibited by *n*-paraffins of comparable molecular weight (Rule 6). In fact, the M peak of an aromatic compound is usually intense enough to make possible the measurement of M + 1 and M + 2 peaks, and also the determination of the molecular formula of the compound. For example, the molecular-ion peak in benzene is the strongest peak in its spectrum (Fig. 5.10), whereas it is 16.2% of the base peak in the case of *t*-amylbenzene (Fig. 5.11).

m/e	% of base peak	m/e	% of base peak	m/e	% of base peak	m/e	% of base peak
15	2	15.5	0.1*	19	0.1**	19.5	0.1**
20	0.1**	24.5	0.1*	25	1	25.5	1**
26	8	26.5	0.2**	27	25	27.5	0.1**
28	27	29	12.5	35.5	0.1*	37	3
37.5	0.1**	38	4	39	34	40	6
41	100	49	2	50	5	51	4
52	1	53	5	54	2	55	18
56	39 (M)						

* = Metastable peak ** = Doubly charged peak

Fig. 5.9 The mass spectrum of 1-butene.

m/e	% of base peak	m/e	% of base peak	m/e	% of base peak	m/e	% of base peak
26	4	27	3	37	5	37.5	1.2**
38	6	38.5	0.4**	39	14	49	3
50	18	51	20.5	52	20	63	3
73	2	74	5	75	2	76	6
77	14	78	100 (M)	79	6.5		

** = Doubly charged peak

Fig. 5.10 The mass spectrum of benzene.

m/e	% of base peak	m/e	% of base peak	m/e	% of base peak	m/e	% of base peak
27	8	29	7	39	7	41	13
50	5	51	11	55	4	63	4
65	5	77	13	78	6	79	9
91	60	92	6	103	9	104	5
105	26	115	5	117	5	118	4
119	100	120	10	133	4	148	16.2 M
149	2						

Fig. 5.11 The mass spectrum of *t*-amylbenzene.

2. In alkyl benzenes the most probable cleavage is at the bond β to the ring. This gives rise to a base peak at $m/e = 91$, due to the formation of tropyllium ion (Rule 7). However, in compounds containing substitutents on the α carbon atom, the base peaks may have masses higher than 91 by increments of 14, representing substituted tropyllium ions.

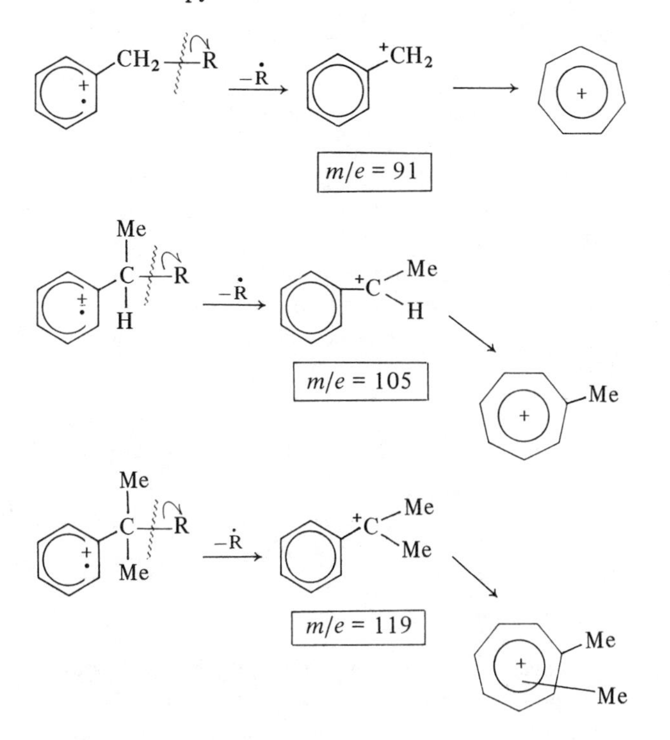

Note the base peak at 119 in *t*-amylbenzene (Fig. 5.11).

3. A strong peak at $m/e = 92$ is observed in the case of many compounds containing a propyl or longer side chain. This peak is due to the $\overset{+}{C}_7H_8$ ion produced by the McLafferty rearrangement.

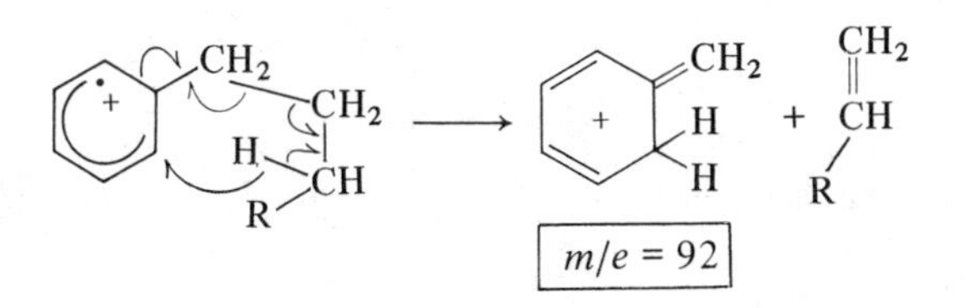

4. If the α-carbon is unsymmetrically substituted, the heaviest substituent on it is preferentially lost. Note the loss of an ethyl group from *t*-amylbenzene.

5. The substituted tropyllium ion can undergo fragmentation so that it exhibits significant peaks at masses 105 and 91 (Fig. 5.11).

6. Fragmentation of a tropyllium ion into a cyclopentyl cation, $\overset{+}{C}_5H_5$, and a cyclopropenium ion, $\overset{+}{C}_3H_3$, results in significant peaks at $m/e = 65$

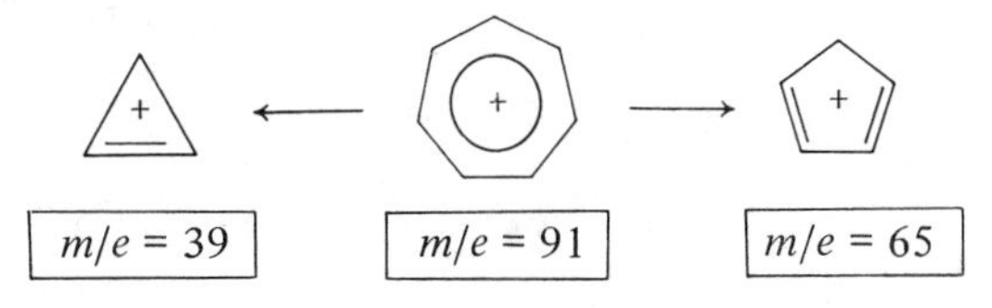

7. α-disubstituted compounds may also produce a tropyllium ion and thus show a base peak at $m/e = 91$, and since many peaks in the spectrum arise from the fragmentation of the tropyllium ion, it is often hard to distinguish between isomers such as *o*-, *m*-, and *p*-xylene.

8. Cleavage of the α bond, although less common, can also take place, producing phenyl ($m/e = 77$), rearranged benzene ($m/e = 78$), and benzene + H ($m/e = 79$) ions.

9. Secondary transitions often give rise to significant peaks at m/e = 107, 106, 105, 93, 63, 52, 51, and 50.

ii) Hydroxy compounds

a) Alcohols The mass spectra of alcohols are considerably harder to interpret for several reasons, such as very low intensity of molecular-ion peaks or total absence of them, metal-catalyzed decomposition of the alcohol in the inlet system, etc.[39] Nevertheless, the following characteristics may be observed in the mass spectra of *n*-octanol (Fig. 5.12), which is a straight-chain alcohol, and ethylene glycol (Fig. 5.13), which is a diol.

1. In the spectra of alcohols, the molecular-ion peak M is usually either very small or totally absent. It is somewhat more abundant in the spectra of secondary than in those of primary and tertiary alcohols. In the primary, straight-chain, or branched alcohols the intensity of the M peak decreases with increasing molecular weight,

m/e	% of base peak	m/e	% of base peak	m/e	% of base peak	m/e	% of base peak
18	5.0	19	2	27	63	28	15.5
29	71	30	3	31	69	32	1
33	0.3	41	100	42	59	43	82
44	7	45	6	55	80.5	56	86
57	40	58	2	59	1	67	4
68	16	69	49	70	55	71	8
73	3	81	1	82	6	83	30
84	44	85	3.5	97	3	112	3
113	0.3	130	0.1 M				

Fig. 5.12 The mass spectrum of 1-octanol.

m/e	% of base peak	m/e	% of base peak	m/e	% of base peak	m/e	% of base peak
14	1	15	8	18	1.5	19	2
26	1.5	27	5	28	7	29	14
30	2	31	100	32	10	33	32
42	3	43	8	44	2.5	45	2.5
60	0.2	61	1.5	62	2.6M	63	0.1
87	0.7	101	1	102	1		

Fig. 5.13 The mass spectrum of ethylene glycol.

disappearing almost totally in alcohols containing more than five carbons. The odd-electron molecular ion is produced initially by the removal of an n-electron from oxygen:

$$R{-}\ddot{\underset{\cdot\cdot}{O}}{-}H \xrightarrow{-\bar{e}} R{-}\overset{+}{\dot{\underset{\cdot\cdot}{O}}}{-}H$$

2. The odd-electron molecular ion readily decomposes into stabler products. One such decomposition involves energetically favored α cleavage. This cleavage leads to more stable, even-electron, oxonium ions. The α cleavage also results in the elimination of α hydrogen and/or a larger substituent.

$$\text{Methanol} \xrightarrow{-\bar{e}} H{-}\underset{H}{\overset{H}{C}}{-}\overset{+}{\dot{O}}{-}H \rightarrow H_2C{=}\overset{+}{O}{-}H + \dot{H}$$

$\boxed{m/e = 31}$

(M – 1)

$$\text{Ethanol} \xrightarrow{-\bar{e}} CH_3{-}\underset{H}{\overset{H}{C}}{-}\overset{+}{\dot{O}}{-}H \rightarrow H_2C{=}\overset{+}{O}{-}H + \dot{C}H_3$$

$\boxed{m/e = 31}$

($M - CH_3$) or (M – 15)

Although ethanol shows an M – 1 peak at $m/e = 45$ by α elimination of a hydrogen radical, the M – 15 peak is more intense because α elimination of a larger substituent is always favored. The larger the neutral radical, the better its chances of being stabilized, either through rearrangement or by further decomposition.[40] The abundance of M – 1 peaks also decreases with increasing molecular weight; but for alcohols, M – 1 peaks are usually more likely to be present than M peaks.

3. In the spectra of primary alcohols, in addition to the M – 1 peak, very low-intensity M – 2 and M – 3 peaks are also observed. These peaks are due to ions such as:

$$R{-}\underset{H}{\overset{H}{C}}{-}\overset{+}{\dot{O}}{-}H \xrightarrow{-H^{\bullet}} R{-}\underset{H}{C}{=}\overset{+}{O}{-}H \xrightarrow{-H^{\bullet}}$$

(M – 1)

$$R{-}\underset{H}{C}{=}\overset{+}{\dot{O}} \xrightarrow{-H^{\bullet}} R{-}C{\equiv}\overset{+}{O}$$

(M – 2) (M – 3)

4. All primary alcohols (except methanol) and high-molecular-weight secondary and tertiary alcohols display peaks at M – 18 due to the loss of H_2O through a cyclic mechanism. In alcohols which have higher molecular weights, dehydration is due to thermal decomposition. Very often, the M – 18 peaks can be mistaken for the M peak.

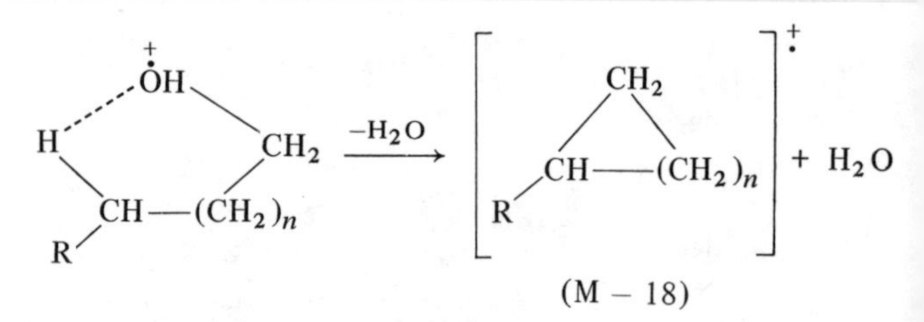

In cyclic alcohols, however, two distinct processes are responsible for the loss of water. The principal dehydration process involves loss of a hydroxyl group, together with a hydrogen atom, from either C_3 or C_4, to produce bicyclo ions.

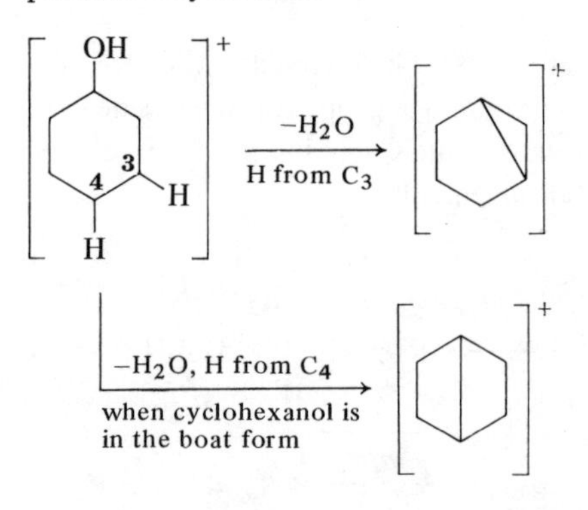

These bicyclo ions may decompose further.[41]

The hydrogen of the hydroxyl is not lost at all in the second, minor process of dehydration. In this mechanism, the oxygen-bonded hydrogen atom migrates during the intermediate free-radical state, producing a hexanal ion. Such a hexanal ion is known[42a] to lose water during the electron impact process, when hydrogens at C_3, C_4, or C_5 are involved. Two recent papers[42b,42c] discuss the mechanism of the elimination of H_2O from *t*-butylcyclohexanols.

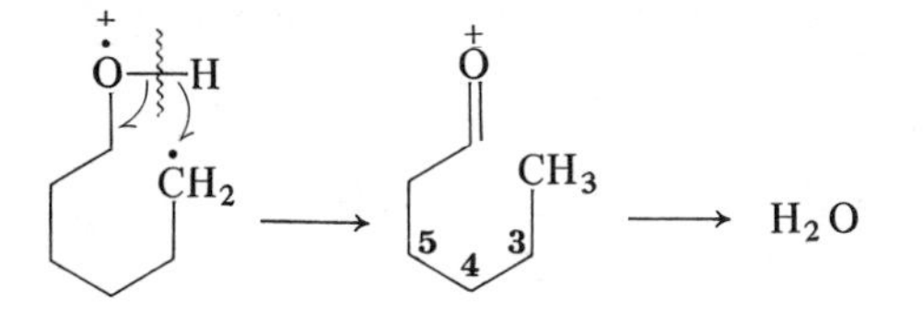

5. Butanol and longer-chain primary alcohols often suffer double elimination. That is, they lose one molecule of water and one of ethylene simultaneously through a six-numbered cyclic mechanism, thus displaying an M – 46 peak.

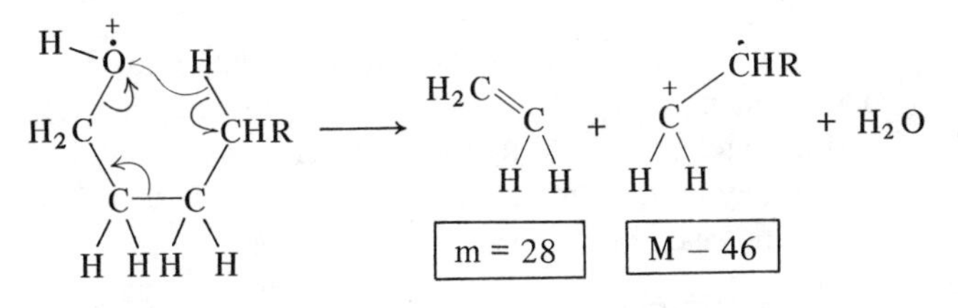

This peak is usually absent in secondary and tertiary alcohols, but could be important in determining the nature of branching on the β carbon. For example, a molecule containing a methyl group on a β carbon loses propylene ($m/e = 42$) instead of ethylene, and displays a peak at M – 60.

6. The most probable site of cleavage, resulting in a very intense peak, is the bond β to the oxygen atom in alcohols (Rule 7). This cleavage gives a strong peak at $m/e = 31$ $(CH_2OH)^+$ in the spectra of primary alcohols, while secondary and tertiary alcohols display analogous peaks at $m/e = 45$ $(MeCHOH)^+$ and $m/e = 59$ $[(CH_3)_2COH]^+$. These peaks are important in the identification of alcohols because very often the fragmentation of the olefinic ion, produced after initial dehydration, causes an alcohol spectrum to look like that of an olefin. The presence of peaks at 31, 45, or 59, however, generally indicates that the compound under investigation is an oxygen-containing compound and not an olefin.

7. Hydrogen abstraction during fragmentation usually produces ions such as H_3O^+ and $CH_3OH_2^+$. Alcohol spectra therefore commonly show peaks at $m/e = 19$ and 33.

8. Unsaturated alcohols generally behave like their saturated analogs. However, the stability of the allyl system decreases the probability of cleavage adjacent to the double bond. The M – 1 peaks are therefore very intense in allyl alcohols, whereas the ions formed by the energetically less-favored cleavage of the bond, β to the oxygen, are less abundant.

$$R{-}CH{=}CH{-}\underset{H}{CH}{-}\overset{+\bullet}{O}H \xrightarrow{-\dot{H}} RCH{=}CH{-}CH{=}\overset{+}{O}H \quad \text{(abundant)}$$

M – 1

$$R{-}CH{=}CH{-}\underset{H}{CH}{-}\overset{+\bullet}{O}H \longrightarrow RCH{=}\dot{C}H + CH_2{=}\overset{+}{O}H \quad \text{(less abundant)}$$

9. Saturated diols also display fragmentation characteristics similar to those of their monofunctional analogs. However, wherever possible, diols fragment so as to produce a more highly substituted oxonium ion. In the case of vicinal diols, the α cleavage of the C—C bond joining the two hydroxyl groups takes place readily. Thus the base peak in propane-1,2-diol[43] is at $m/e = 45$.

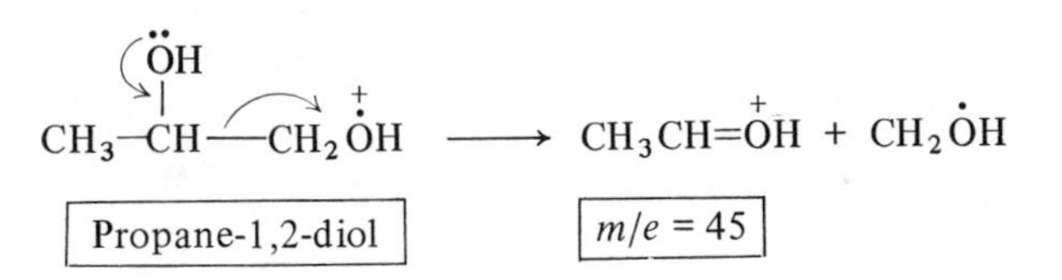

10. The difficulty in identifying alcohols, due to the lack of a significant molecular-ion peak, can be overcome by using derivatives such as trimethylsilyl ethers[44] and esters.[45] The TMS ethers may also not display intense M peaks, but the M – 15 peaks in such derivatives are strong, due to the ready loss of a methyl group.

b) Phenols and other aromatic alcohols

1. As in all aromatic compounds, the molecular-ion peak M in phenols and aromatic alcohols is strong (Rule 6). In phenols it forms the base peak.[46]

2. The M – 1 peak due to the loss of hydrogen is small in phenol, but is very strong in cresols and benzyl alcohols. The production of M – 1 ions is due to the random abstraction of hydrogen bonded to any of the carbon atoms to produce a resonance-stabilized π complex system such as hydroxytropyllium ions.

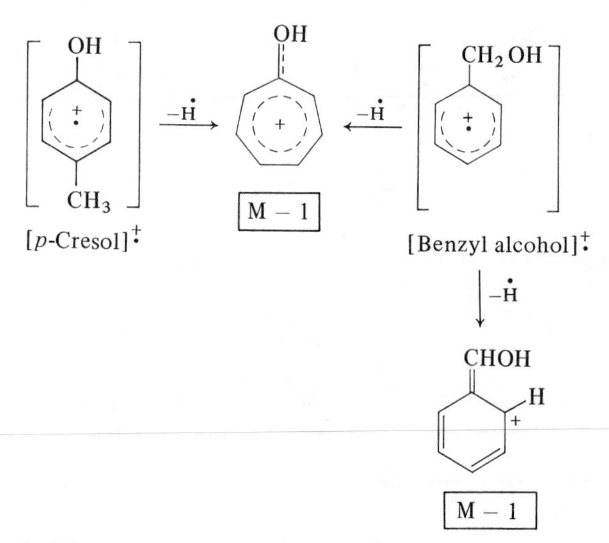

3. The most important fragmentation in phenols and benzyl alcohols is due to the loss of CO and CHO, giving peaks corresponding to M − 28 and M − 29, respectively. This fragmentation too can be attributed to the resonance stability of the π electron system in the resulting ions.

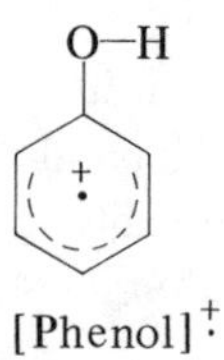

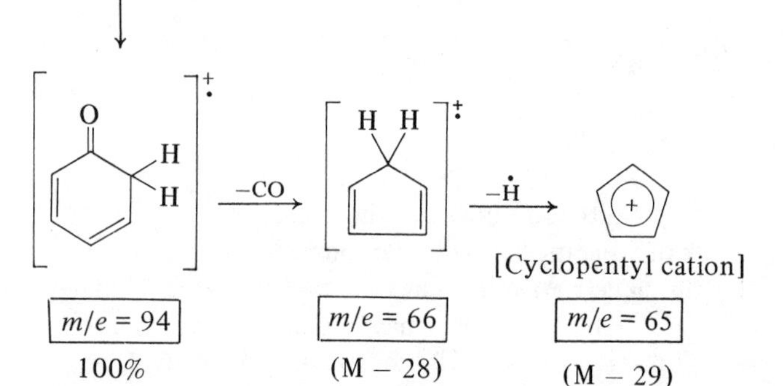

4. Methyl-substituted phenols (cresols), other hydroxybenzenes (catechol, resorcinol, hydroquinone), and methyl-substituted benzyl alcohols display M − 18 peaks due to loss of water. This dehydration is more pronounced if the substituents are *ortho* to each other (the *ortho effect*).

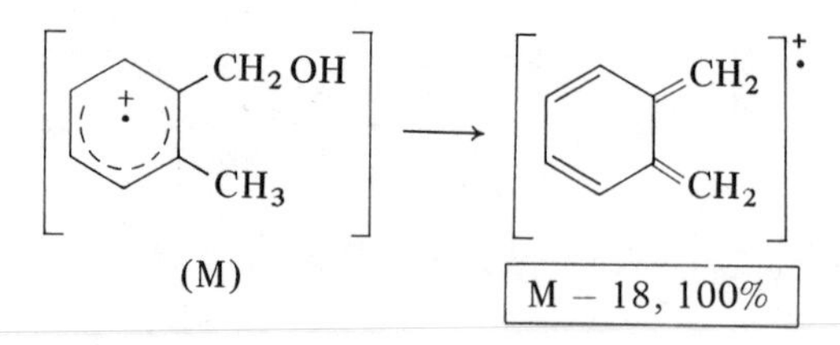

iii) Ethers, acetals, and ketals

Molecular ions of ethers undergo bond fissions similar to those of alcohols.

1. A molecular ion of ether is formed by the removal of one of the nonbonded electrons on oxygen, and, just as with alcohols, the molecular ions of ethers are unstable and readily undergo α cleavage. The M peak of ethers is therefore weak, but can usually be observed.

2. The intensity of the M or the M + 1 peak can be increased by using a larger sample or increasing the operating pressure.

3. In ethers, as in alcohols, α-cleavage (cleavage of a bond β to oxygen) is common. Such a fission favors the loss of a more highly substituted fragment (Rule 7).

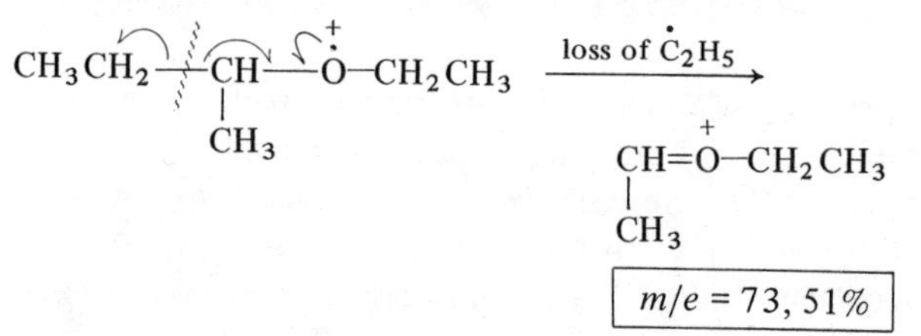

This type of cleavage usually accounts for the base peak and some strong peaks at 45, 59, 73, etc. To a lesser

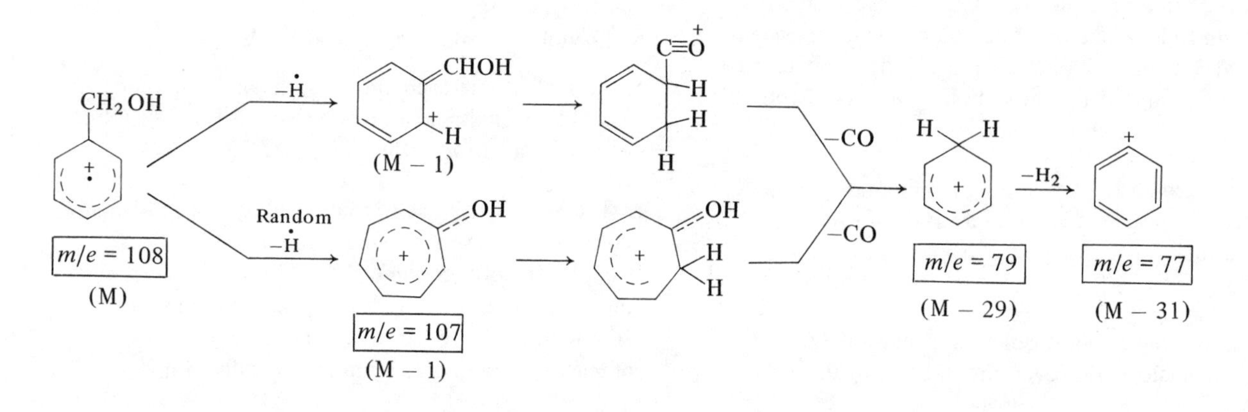

degree, α cleavage—resulting in elimination of the less-substituted fragment—may also take place.

$$CH_3CH_2-CH(CH_3)-\overset{+}{\ddot{O}}-CH_2CH_3 \xrightarrow{\text{loss of } \dot{C}H_3} CH_3CH_2-CH=\overset{+}{O}-CH_2CH_3$$

m/e = 87, 4%

4. In ethers, unlike the case of alcohols, cleavage of bonds α to oxygen can also occur.

$$R-\overset{+}{\dot{O}}-R' \xrightarrow{\text{heterolysis}} R-\dot{O} + \overset{+}{R'}$$

This difference between the behavior of alcohols and ethers may be due to the fact that alkoxyl radicals ($R\dot{O}$) are more stable than hydroxyl radicals ($\dot{O}H$). In such a cleavage, the alkyl portion carries the charge and accounts for hydrocarbon peaks at m/e = 29, 43, 57, 71, etc.

5. Homolytic fission of bonds α to oxygen may often be accompanied by rearrangement of one hydrogen atom to eliminate an olefin.

$$CH_3CH_2-C(H)(CH_3)-\overset{+}{\dot{O}}-CH_2-CH_2-H \longrightarrow CH_3CH_2-C(H)(CH_3)-OH + \dot{C}H_2=\overset{+}{C}H_2$$

In such a case, the charge resides on the olefinic fragment, which displays a peak at one atomic mass unit less than that of the ions produced by the cleavage of a bond α to oxygen without rearrangement, i.e., at 28, 42, 56, 70, etc.

6. Alkyl ethers, like other oxygen-containing compounds and alcohols, exhibit a peak at m/e = 31. Many peaks in the spectrum of diethyl ether (Fig. 5.14) can be identified on the basis of the above generalizations.

Since acetals and ketals are special classes of ethers, their mass-spectral fragmentations fall into the expected pattern of ethers. In these compounds, any bonds of the central carbon atom can be cleaved with almost equal ease to give the corresponding ions:

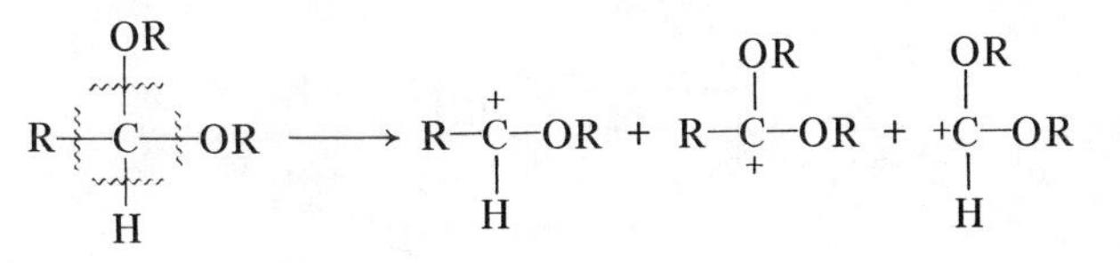

m/e	% of base peak	m/e	% of base peak	m/e	% of base peak	m/e	% of base peak
26	3	27	18	28	6	29	39.5
30	2	31	100	32	1	41	5
42	1	43	7	44	3.5	45	37.5
59	47	60	1	73	3	74	30.5 (M)
75	1.6	76	0.1				

Fig. 5.14 The mass spectrum of diethyl ether.

Aromatic ethers display a somewhat stronger M peak, and show a fragmentation behavior similar to that of aliphatic ethers. Thus:

i) Cleavage of bonds α to oxygen can be homolytic or heterolytic.

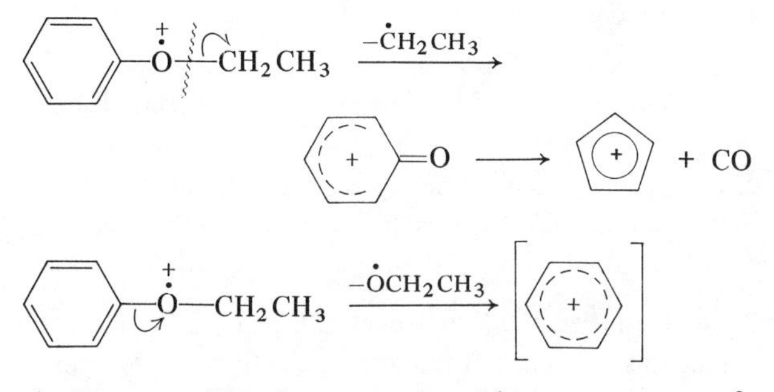

ii) Cleavage of bond α to oxygen with rearrangement of hydrogen to eliminate an alkene

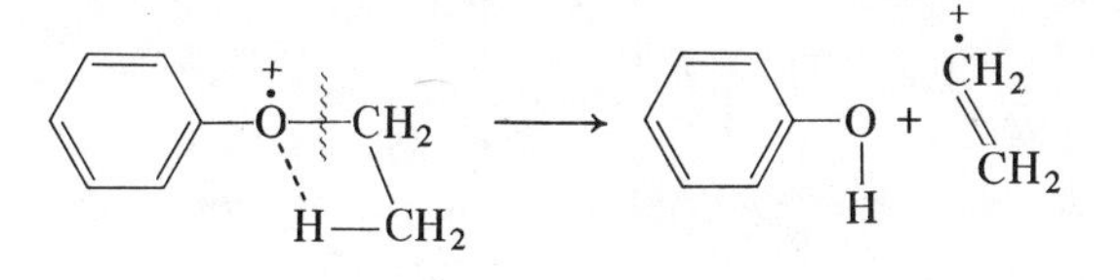

iv) Aldehydes and ketones

Organic carbonyl compounds, like other oxygen-containing compounds, undergo the loss of one of the lone-pair electrons of the oxygen atom. The molecular ion thus produced can undergo fragmentation either through the more favorable α cleavage or, if a concerted migration of γ hydrogen to oxygen is possible, through β cleavage. In α cleavage, the bond between the carbonyl group and an α carbon atom is ruptured by homolytic process, so that the positive charge is retained by the oxygen-containing fragment to produce an oxonium ion, which can stabilize itself by the formation of a triple bond.

$$\begin{matrix} R \\ R_1 \end{matrix}\!\!>C=\overset{+}{\dot{O}} \xrightarrow[\text{homolytic}]{\alpha\text{ cleavage}} R_1-C\equiv\overset{+}{O} + \dot{R}$$

oxonium ion

When a γ hydrogen is available for migration, β cleavage results in the formation of an olefin and a charged enol through the McLafferty rearrangement.

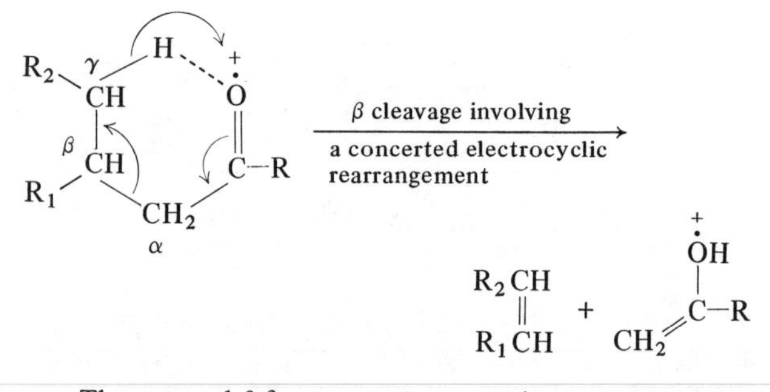

These α and β fragmentations are the most important ones in carbonyl compounds such as aldehydes, ketones, esters, etc.

a) Aldehydes

1. Aliphatic as well as aromatic aldehydes display molecular-ion peaks. The M peak is prominent in aromatic aldehydes due to the resonance stabilization of the molecular ions, whereas the intensity of the M peak in aliphatic aldehydes decreases rapidly in compounds containing more than four carbon atoms.

2. In aldehydes, the M − 1 peak is usually as intense as the M peak. This peak results from the loss of hydrogen through homolytic α cleavage:

$$\mathrm{R{-}\overset{|}{\underset{H}{C}}{=}\overset{+}{\dot{O}}} \longrightarrow \underset{(M-1)}{\mathrm{R{-}C{\equiv}\overset{+}{O}}} + \mathrm{\dot{H}}$$

3. Aldehydes generally display a strong absorption peak at m/e = 29. This peak also results from α cleavage. In lower aldehydes (C_1-C_3), α cleavage results in the formation of the stable formyl ion ($\mathrm{H{-}C{\equiv}\overset{+}{O}}$), which forms the base peak.

$$\mathrm{R{-}\underset{H}{\overset{|}{C}}{=}\overset{+}{\dot{O}}} \longrightarrow \mathrm{H{-}C{\equiv}\overset{+}{O}} + \mathrm{\dot{R}}$$

$\boxed{m/e = 29}$

(base peak)

However, with straight-chain aldehydes of higher molecular weight, this peak is approximately 40% of the base peak which may be displayed at M − 29 due to ions produced through heterolytic cleavage as:

$$\mathrm{R{-}\underset{H}{\overset{|}{C}}{=}\overset{+}{\dot{O}}} \longrightarrow \underset{(M-29)}{\overset{+}{\mathrm{R}}} + \mathrm{\underset{H}{\overset{|}{C}}{\equiv}\dot{O}}$$

This type of heterolytic α fission is so common in some aromatic compounds (in which the resulting ion can be resonance stabilized) that the M − 29 peak sometimes forms the base peak in their spectra.

5. The base peak in butyraldehyde and in many higher aliphatic aldehydes (and also in ketones and esters) results from β cleavage. In butyraldehyde this peak is at m/e = 44, which must be a peak for a rearranged ion because it has an even mass number. In compounds containing C, H, and O, peaks resulting from simple cleavage always appear at odd mass number, but the peaks due to rearranged ions have an even m/e ratio.

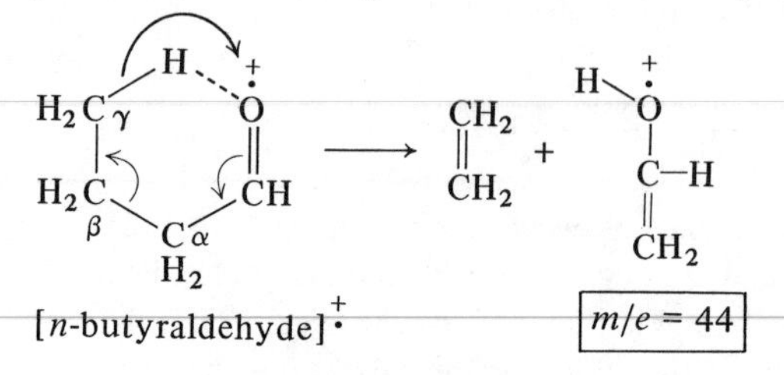

In the case of branched aldehydes, the nature of branching on the α carbon is readily revealed by this β fragmentation. For example, α methyl substitution displays a peak at m/e = 58 (due to $\mathrm{CH_3CH{=}CH\overset{+}{O}H}$), whereas a peak at 72 indicates α-ethyl substitution, etc. Branching at the β carbon and beyond results in the same β-fragmentation product at m/e = 44 as in butyraldehyde.

6. As in α cleavage, so in β cleavage, charged alkyl functions can result in aldehydes containing no α branching. Such fragmentation results in a peak either at M − 44 [$(C_nH_{2n})^{+\cdot}$ if hydrogen migration occurs] or at M − 43 [$(C_nH_{2n+1})^+$ in the absence of such a migration]. Such a fission occurs because the initial ionization process removes a π electron from the carbonyl rather than a n electron from the oxygen.

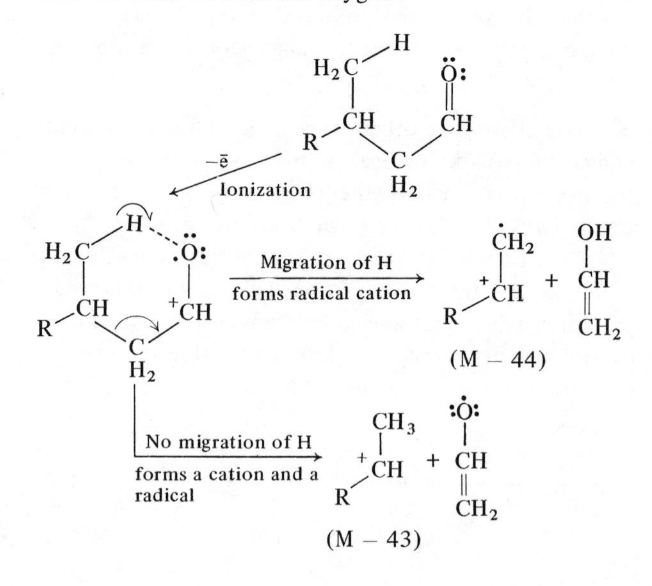

In the spectrum of *n*-hexanal (Fig. 5.15), peaks due to three types of β cleavages are prominent at 44 (enol ion), 56 (M − 44) and 57 (M − 43).

7. Like the parent ions of other oxygen-containing compounds, aldehyde parent ions tend to eliminate stable molecules of water and ethylene (Rule 8), showing peaks at M − 18 and M − 28. In Fig. 5.15 these peaks appear at m/e = 82 and 72, respectively.

m/e	% of base peak	m/e	% of base peak	m/e	% of base peak	m/e	% of base peak
15	2	18	1	26	2	27	27
28	7.5	29	34	30	1.5	31	2
38	1.5	39	17	40	3	41	61.5
42	11	43	62	44	100	45	20
51	1	53	3	54	2.5	55	17
56	87	57	59	58	10.5	60	4
67	9.5	69	2	70	1	71	11.5
72	26	73	3	81	2	82	17.5
83	2.5	99	0.6	100	1.1M	101	0.1

Fig. 5.15 The mass spectrum of *n*-hexanal.

b) Ketones

The fragmentation behavior of ketones is very similar to that of the aldehydes.

1. The parent-ion peak M in ketones is of significant intensity, being greater for the low-molecular-weight (up to eight carbons) than for the high-molecular-weight ketones.

2. Whereas α cleavage in aldehydes produces formyl ions $H{-}C{\equiv}\overset{+}{O}$, similar cleavage in ketones can produce more stable $R{-}C{\equiv}\overset{+}{O}$ ions. The stability of these ions results in the greater importance of α cleavage in ketones than in the corresponding aldehydes.

$$\begin{matrix} R \\ R_1 \end{matrix}\!\!>\!C{=}\overset{+}{\ddot{O}} \xrightarrow{\text{homolysis}} R_1C{\equiv}\overset{+}{O} + \dot{R}\ (\text{if } R > R_1)$$

$$\xrightarrow{\text{homolysis}} RC{\equiv}\overset{+}{O} + \dot{R}_1\ (\text{if } R_1 > R)$$

m/e = 43, 57, 71, etc.

$$\xrightarrow{\text{heterolysis}} R{-}C{\equiv}\dot{O} + \overset{+}{R}_1$$

m/e = 15, 29, 43, 57, etc.

In this respect, ketones show a pattern similar to that of hydrocarbons, in giving a series corresponding to masses C_nH_{2n+1}. This is understandable because the carbonyl group C=O has very nearly the same mass as $-CH_2-CH_2-$. Sharkey *et al.* have proved that α cleavage of larger alkyl groups is favored by demonstrating that in all the aliphatic *methyl* ketones studied,[48] the base peak appeared at m/e = 43 due to the fragment ion $(CH_3CO)^+$. Although α cleavage yielding the $(CH_3CO)^+$ ion can also occur in aromatic ketones, the stability of the phenyl ion is responsible for the fact that the smaller alkyl group is more readily cleaved in alkaryl ketones.

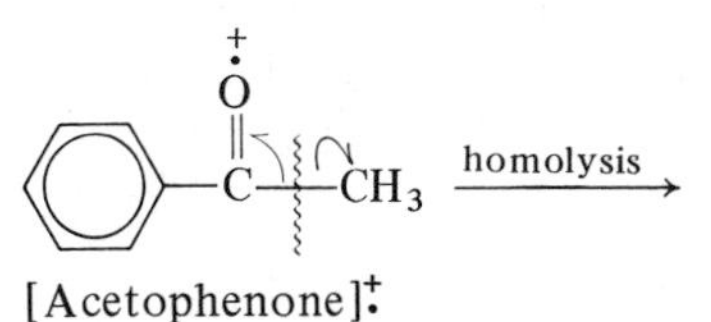

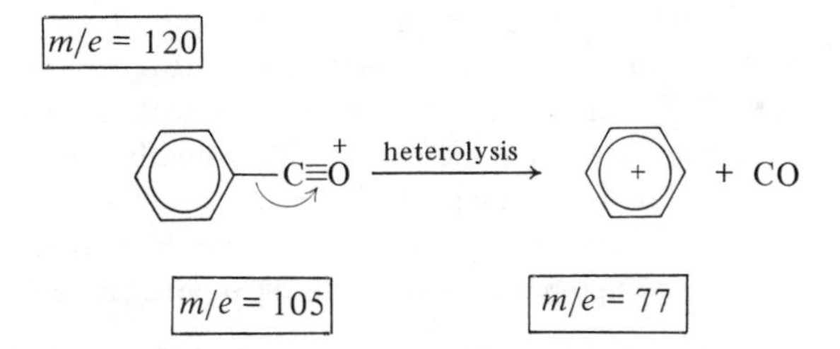

3. McLafferty rearrangements (β cleavage with migration of γ hydrogen) are common in ketones containing a chain of three or more carbon atoms attached to the carbonyl group. However, complications arise due to the following facts.

a) The mass of the ion produced by the migration of a single hydrogen varies with the number of carbon atoms in the alkyl group (R) which is not involved in the rearrangement.

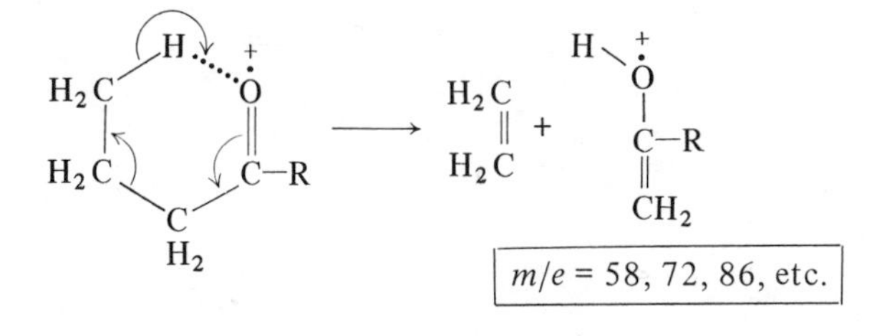

b) If *both* the alkyl groups in the ketone contain three or more carbons, double rearrangement can take place. The first rearrangement produces the enolic cation, which can again undergo β cleavage involving the γ hydrogen on the second alkyl group.

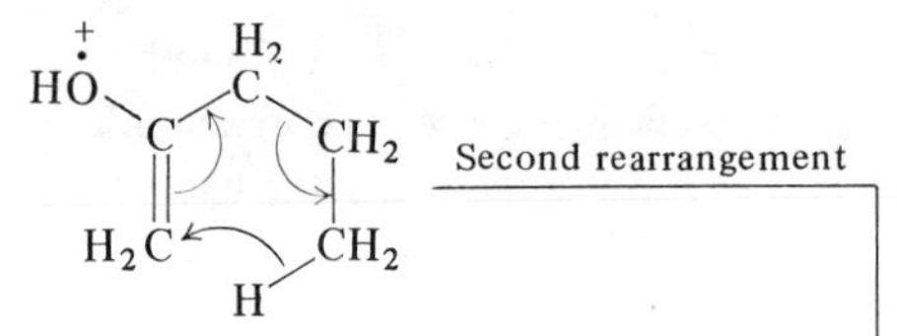

Enol produced after first rearrangement

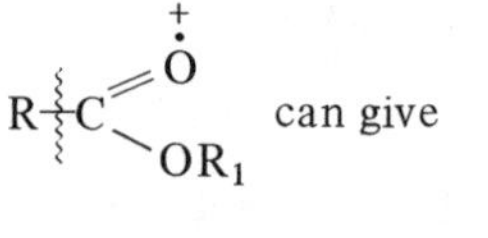

$m/e = 58$

If the γ carbon in either of the alkyl groups is substituted, the order of migration of the γ hydrogen is found to be tertiary > secondary > primary, and the rearrangement preferentially involves the larger alkyl group.

v) Carboxylic acids and esters

1. Aliphatic monocarboxylic acids and their esters generally display a weak but noticeable molecular-ion peak. The intensity of this peak decreases with increasing molecular weight, but increases with the number of multiple bonds in conjugation with the carboxyl group. Aromatic monocarboxylic acids and esters, like other aromatic compounds, display a strong M peak (Rule 6).

2. Like other oxy compounds, cleavage of the bond β to carbonyl (β cleavage), together with the rearrangement of γ hydrogen (McLafferty rearrangement) is the most important mode of fragmentation of carboxylic acids (Rule 7). In monocarboxylic acids, this fragmentation produces ions (I) which exhibit a characteristic peak at $m/e = 60$.

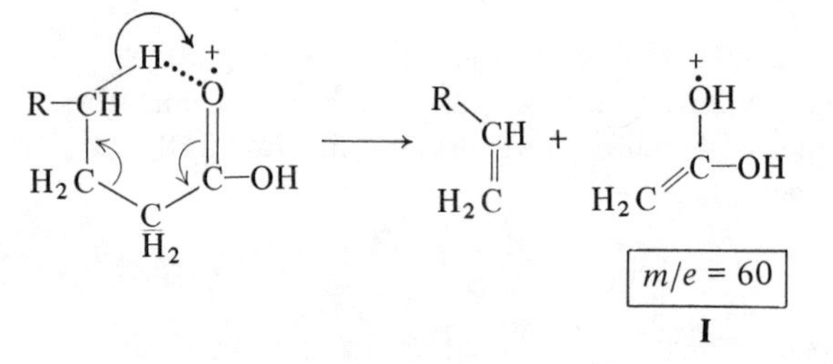

$m/e = 60$

I

A corresponding β cleavage peak in methyl esters appears at $m/e = 74$ due to (II).

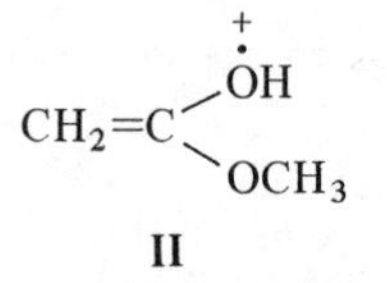

II

Analogous peaks in ethyl and butyl esters appear at 88 and 116, respectively. The importance of this mode of cleavage can be realized from the fact that some aliphatic acids show a base peak at 60, whereas almost all methyl esters of straight-chain carboxylic acids in the C_6-C_{26} range display a base peak at 74.[49]

3. We have seen the importance of α cleavage in aldehydes and ketones. In acids and esters, a similar cleavage of a bond α to carbonyl can produce four kinds of ions.

$R-C(=\overset{+}{O})-OR_1$ can give

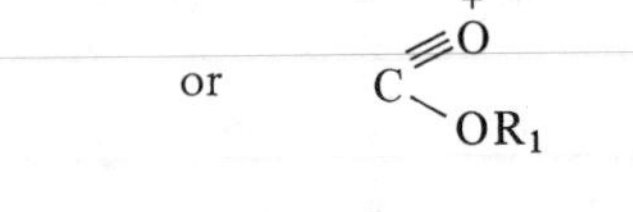

$m/e = 15, 29, 43, 57$	$m/e = 45, 59, 73, 87$, etc.
(M − 45, M − 59, etc.)	(M − 15, M − 29, M − 43, etc.)

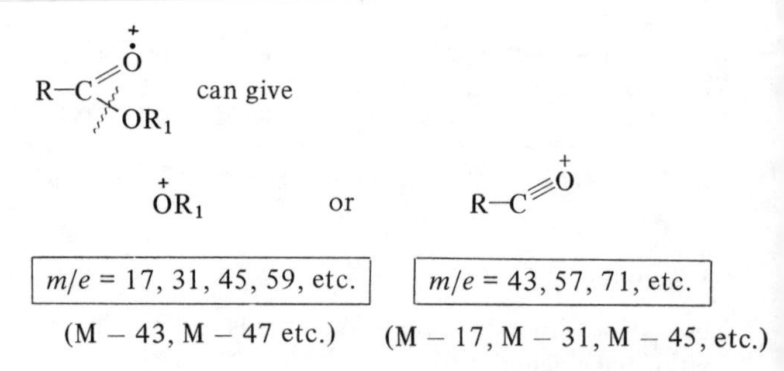

$m/e = 17, 31, 45, 59$, etc.	$m/e = 43, 57, 71$, etc.
(M − 43, M − 47 etc.)	(M − 17, M − 31, M − 45, etc.)

4. Elimination of a stable molecule of water produces a peak at M − 18 (Rule 8). In aromatic acids containing a methyl group *ortho* to the carboxyl, this elimination is so important that in *o*-toluic acid the M − 18 peak is the base peak.

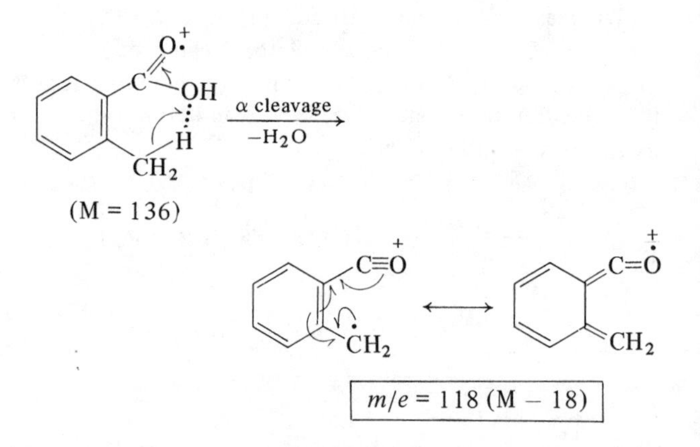

Figure 5.16 shows a mass spectrum of methyl propionate, which illustrates all the typical fragmentation patterns discussed above.

m/e	% of base peak	m/e	% of base peak	m/e	% of base peak	m/e	% of base peak
14	1	15	14	26	6	27	16
28	8	29	59	30	2	31	5
42	2	43	2	45	6.5	55	4
56	3	57	100	58	4	59	31
60	1	87	3	88	34.7 (M)	89	2
90	0.2						

Fig. 5.16 The mass spectrum of methyl propionate.

5. Dibasic acids and their methyl esters display strong M peaks. The intensities of these peaks decrease with the proximity of the carboxylic groups. Dibasic acids show a peak at M − 90 due to the loss of both carboxyls through α cleavage.

6. In the case of hydroxy-acid-methyl esters, and in the case of unsaturated-fatty-acid esters, the fragmentation patterns of their trimethylsilyl derivatives are more useful for elucidation of chemical structures than the esters themselves. Hydroxylation followed by trimethylsilylation is particularly useful in locating double bonds in unsaturated-fatty-acid esters. With TMS derivatives, the technique of combined gas chromatography and mass spectrometry is used.[50,51]

vi) Amines

In amines, as in oxygen-containing compounds such as alcohols, ketones, aldehydes, etc., the initial ionization process occurs through the loss of a nonbonded electron on the hetero atom, nitrogen. However, since the amino group can release electrons more easily than the hydroxyl group can, amines have a lower ionization potential than the corresponding alcohols. In addition, the cleavages characteristic of the hydroxyl group also occur in amines, but to a larger degree. Thus the general fragmentation pattern of amines should be studied in the light of that of alcohols.

1. The molecular-ion peak M is weak to absent in high-molecular-weight aliphatic open-chain amines, but strong in aromatic and cyclic amines (Rule 6).

2. In the case of monoamines, if the molecular-ion peak is present, it appears at an odd mass number in the spectrum. In the case of other nitrogen compounds also, it appears at odd mass number when the compound contains an odd number of nitrogen atoms (the nitrogen rule).

3. A moderate M − 1 peak is observed in many aromatic and low-molecular-weight aliphatic amines due to the loss of a hydrogen radical.

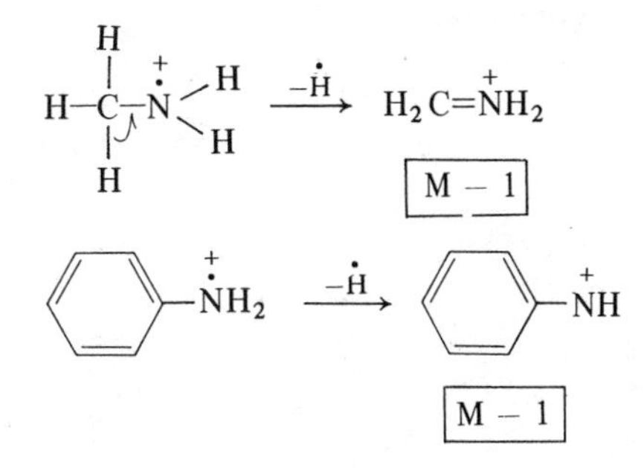

4. In amines, as in alcohols, the most important fragmentation is that of the bond β to nitrogen. In many amines this fragmentation is responsible for the base peak (Rule 7).

$$R{-}C{-}\overset{+}{N}< \longrightarrow \dot{R} + {>}C{=}\overset{+}{N}<$$

m/e = 30, 44, 58, 72, 86, etc.

For primary amines unbranched at the α carbon, this peak is at m/e = 30. A strong peak at 30 may be considered good, although inconclusive, indication that primary amines are present. Secondary and tertiary amines may also display a peak at 30.

5. In cyclic and aromatic amines, both bonds β to nitrogen can rupture.

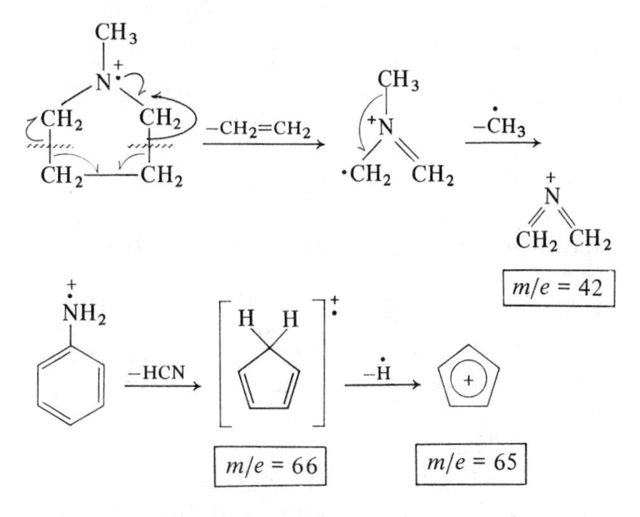

The loss of HCN and H_2CN from aniline is similar to the elimination of CO and CHO from phenol.

6. Like other side-chain aromatic compounds, alkylanilines undergo β elimination of the side chain and show a strong peak at m/e = 106 to the aminotropyllium ion.

7. If the α carbon carries substituents, a complicated fragmentation, involving bonds α and β to nitrogen and a migration of β hydrogen, can take place in secondary and tertiary amines. For example,[52]

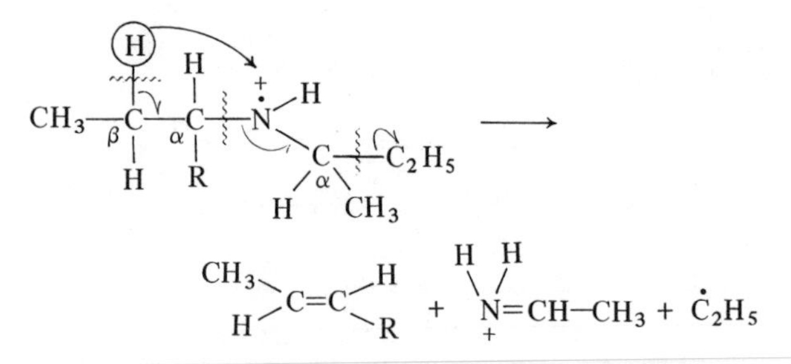

8. Amines display a highly characteristic peak at m/e = 18. Whereas in alcohols this peak is due to H_2O^+, the ions responsible for this peak in amines are ammonium ions, NH_4^+. It is not hard to distinguish an NH_4^+ peak from one due to H_2O^+, since the ratio of mass-18 peak to mass-17 peak is generally much greater for amines than for alcohols.

9. Amines also display rearrangement peaks at masses 31, 45, 59, etc., similar to those in oxygen-containing compounds such as alcohols, aldehydes, etc.

vii) Amides

The mass spectroscopic behavior of amides is generally similar to that of corresponding carboxylic acids.

1. Amides usually display an easily distinguishable molecular-ion peak which, if only one amido group is present, manifests itself at odd mass number in the spectrum.

2. In amides, as in carboxylic acids, cleavage of a carbon-carbon bond β to the carbonyl (β cleavage) is the most important fragmentation process. In amides which have γ-hydrogen available for migration, the McLafferty rearrangement is equally common. In most primary amides, the resulting ion usually forms the base peak.

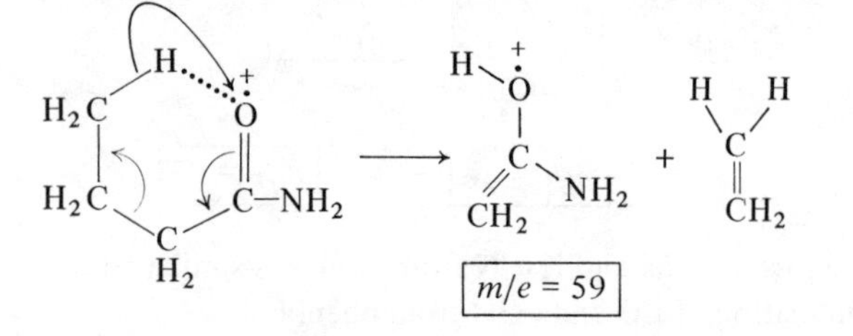

3. In the case of long-chain primary amides, cleavage of the bond γ to carbonyl can also take place, giving minor peaks at mass 72 (without rearrangement) and 73 (with rearrangement).

m/e	% of base peak	m/e	% of base peak	m/e	% of base peak	m/e	% of base peak
15	1	27	8	28	5	29	14.5
30	1	31	1	41	13	42	5
43	11	44	38	45	2	55	6
56	2	57	11	58	1	59	100
60	3	72	19	73	3	85	3
86	3	100	0.6	101	0.8 (M)		

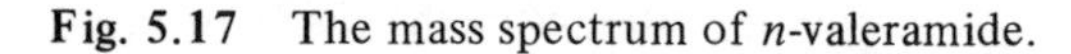

Fig. 5.17 The mass spectrum of n-valeramide.

4. Strong peaks are also displayed at m/e = 44 by primary amides (containing up to four carbon atoms), due to rupture of the bond β to nitrogen (or α to carbonyl). This fragmentation is similar to that observed in amines.

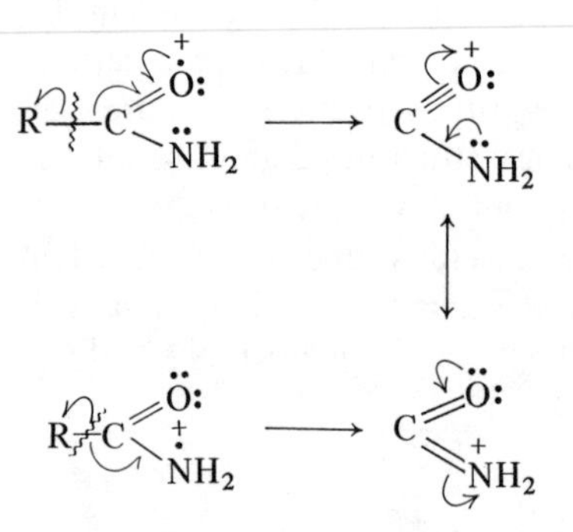

The fragmentation pattern of amides is well illustrated by the spectrum of n-valeramide (Fig. 5.17), in which, due to McLafferty rearrangement, the base peak appears at m/e = 59 and strong peaks are obtained at m/e = 44 and 72. Due to the polarity of the $-CONH_2$ group and its capacity to hydrogen-bond, large molecules containing amido groups are often not very volatile. They also become absorbed by the surface of the inlet system. Difficulties arising from this can be avoided by reducing the amide to an amine with lithium aluminum hydride.

$$R{-}CO{-}NH_2 \xrightarrow{LiAlH_4} R{-}CH_2{-}NH_2$$

viii) Nitriles

1. Methyl cyanide shows a weak molecular-ion peak at odd mass number, but an M peak is usually absent in high-molecular-weight aliphatic nitriles. In such cases, the molecular weight of an unknown can be determined by increasing the quantity of the sample so that the

pressure inside the ion chamber is increased. The ion–molecule reactions are thus encouraged and an M + 1 peak can be observed.

2. Nitriles display a useful M — 1 peak because the product of hydrogen abstraction is stabilized by two resonance forms as:

$$R-\dot{C}H-C\equiv \overset{+}{N} \longleftrightarrow R-CH=C=\overset{+}{\underset{\cdot\cdot}{N}}$$

In aliphatic nitriles containing one nitrile group, this peak is observed at even mass number on the spectrum. Similar even-mass-number peaks are also displayed at 40, 54, 68, 82, etc., in a homologous series, and are produced by a single cleavage at different bonds in the carbon chain.

3. Straight-chain nitriles containing between four and ten carbon atoms display a base peak at $m/e = 41$ due to CH_3CN^+ or $CH_2=C=\overset{+}{N}-H$, which is formed by cleavage, of the bond β to the C≡N group, accompanied by migration of a hydrogen atom (McLafferty rearrangement).[53]

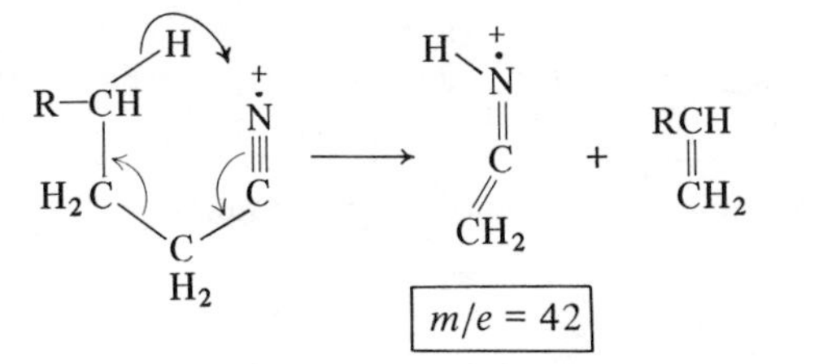

Through a similar mechanism, α-methyl nitriles display a peak at $m/e = 55$.

ix) Nitro compounds

1. Nitro compounds usually do not show an M peak. (A weak M peak is displayed by nitromethane.)

2. Strong peaks are observed at $m/e = 46$ and 30, due to the NO_2^+ and NO^+ ions.

3. The most intense peaks in the spectrum of high-molecular-weight aliphatic nitro compounds are due to hydrocarbon ions produced by C–C cleavage.

4. Aromatic nitro compounds show a strong M peak at odd mass number, a peak at $m/e = 30$ due to NO^+ and M – 30, M – 46, M – 58.

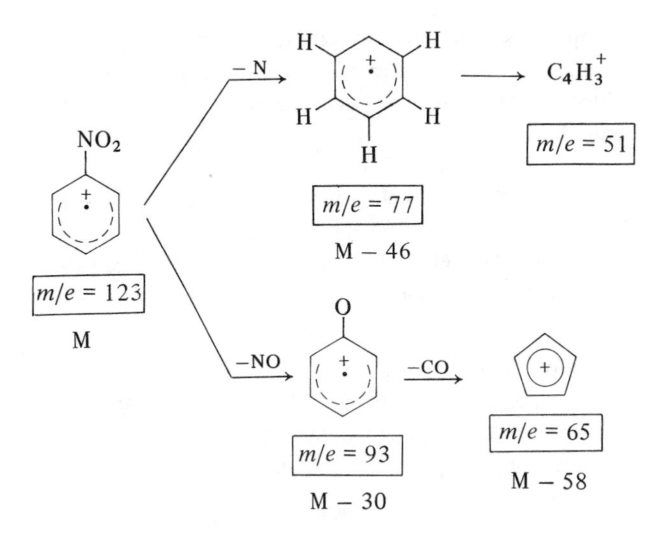

x) Halogen compounds

1. Organic halogen compounds show detectable molecular-ion peaks. The M peaks are strong in aromatic halides. For a given alkyl or aryl group, the intensity of the M peak decreases in the order I > Br > Cl > F, and for a given halogen atom it decreases with increasing size of the alkyl group and extent of α branching.

2. The most noticeable effect in the mass spectra of organic halides is due to the isotopic abundance of halogens. Iodine and fluorine are mono-isotopic, but chlorine and bromide occur as isotopic mixtures of three parts ^{35}Cl to one of ^{37}Cl and equal parts of ^{79}Br and ^{81}Br, respectively. Ions containing one chlorine atom or one bromine atom therefore always show doublets at masses m and $m + 2$. The $m + 2$ peak in chlorides is about a third of the intensity of the peak at mass m, whereas the m and $m + 2$ peaks in bromide ions are of approximately equal intensity. This is also true for molecular-ion peaks M in both compounds. Similarly,

Table 5.4 Isotope ratios (relative to intensity of M peak = 100) for chloro and bromo compounds

Halogen atoms present	Percentage M + 2 peak	Percentage M + 4 peak	Percentage M + 6 peak
Cl	32.6	–	–
Br	97.7	–	–
BrCl	130.0	31.9	–
Cl_2	65.3	10.6	–
Br_2	195.0	95.5	–
Cl_2Br	163.0	74.4	10.4
Br_2Cl	228.0	159.0	31.2
Cl_3	99.8	31.9	3.47
Br_3	293.0	286.0	93.4

compounds containing two chlorine or bromine atoms display peaks at M + 2 and M + 4; those containing three chlorine or bromide atoms show peaks at M + 2, M + 4, and M + 6.

On the basis of the isotopic ratios of chlorine and bromine, one can calculate the relative intensities of M, M + 2, M + 4 and M + 6 peaks for organic compounds containing these elements. Table 5.4 presents these intensities as percentages of M peaks.

3. Chlorides and bromides containing an *n*-alkyl chain of six carbons or more form cyclic chloronium or bromonium ions. Cyclic halonium ions containing four carbon atoms are particularly abundant.

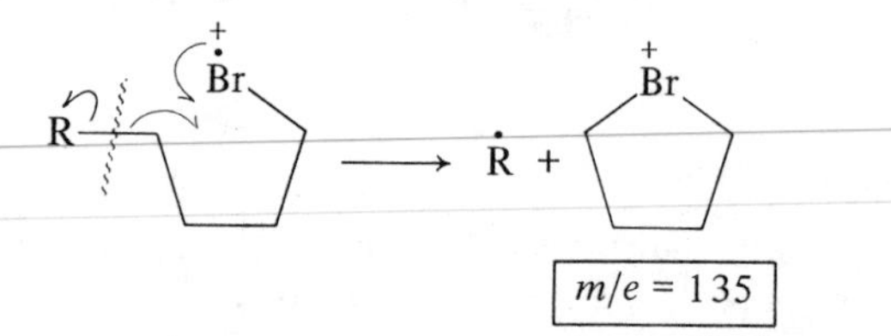

Fluorides, iodides, and branched-chain halides do not form such ions in appreciable abundance.

4. Halides display prominent peaks at M – X, M – HX, M – H_2X and M – R according to the following fragmentation processes.

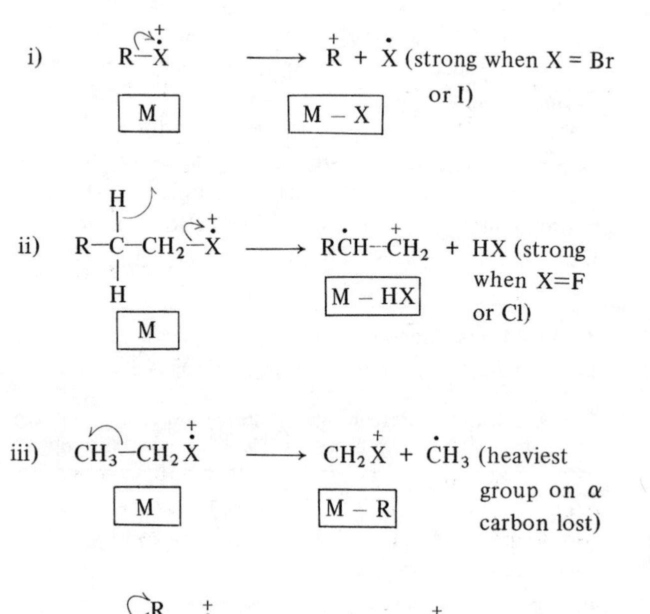

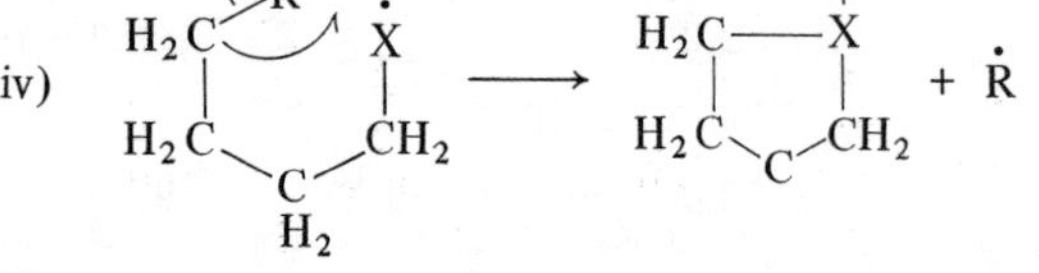

To illustrate the above observations, Figs. 5.18 and 5.19 present spectra of *o*-bromophenol and benzyl bromide.

m/e	% of base peak	*m/e*	% of base peak	*m/e*	% of base peak	*m/e*	% of base peak
18	1	28	4	31	1	32	4
37	2.5	38	5	39	9	46	4
50	2	53	4	61	2	62	3
63	12	64	21	65	28	66	2
74	1	86	5.5	87	5.5	92	5.5
93	10	94	1	117	2	119	2
143	2	144	1	145	2	146	1
171	0.3	172	100 (M)	173	7	174	98 (M + 2)
175	6.5	176	0.4				

Fig. 5.18 The mass spectrum of *o*-bromophenol.

m/e	% of base peak	*m/e*	% of base peak	*m/e*	% of base peak	*m/e*	% of base peak
26	2	27	2	28	2	37	1.5
38	3	39	9.5	41	1.5	43	0.7
44	0.7	45	1.5	49	0.7	50	4
51	5	52	2	61	1.5	62	3
63	7	64	2	65	11	77	3.5
79	3.5	85	1	86	1	89	4.5
90	2.5	91	100	92	8	169	0.3
170	5.3 (M)	171	0.6	172	5.2 (M + 2)	173	0.4

Fig. 5.19 The mass spectrum of benzyl bromide.

5.9 IDENTIFYING AND DETERMINING THE STRUCTURAL FORMULA OF AN UNKNOWN

Knowing the exact molecular formula of an unknown, as well as the fragmentation patterns of various types of compounds, is a great help in elucidating the structure of the unknown. However, the task of determining structure is simplified if one has information and clues from other analytical sources, such as physical properties (e.g., melting or boiling point, etc.), UV, IR, and NMR spectra, behavior of the unknown in a gas-liquid chromatograph, and so forth, before trying to identify the compound on the basis of its mass spectrum. Once you make a list of all probable structures, on the basis of your "other information," with the help of a mass spectrum of the compound, you can quickly narrow down the possibilities.

Once you have determined the exact molecular formula of the compound by the techniques described

Table 5.5 Some ions often produced during mass-spectral fragmentation of organic compounds. (Positive charges on these ions are intentionally omitted.)

m/e	Ion commonly produced	*m/e*	Ion commonly produced
14	CH_2	46	NO_2
15	CH_3	55	C_4H_7
16	O	56	C_4H_8
17	OH	57	C_4H_9, $C_2H_5C{=}O$
18	H_2O, NH_4	58	$C_3H_6NH_2$, $(CH_3COCH_2 + H)$, $(C_2H_5NHCH_2)$
26	CN	59	$(CH_3)_2COH$, $COOCH_3$, $CH_2OC_2H_5$
27	C_2H_3	59	$C_2H_5NO(NH_2COCH_2 + H)$
28	C_2H_4, CO, N_2	60	$C_2H_4O_2(CH_2COO + 2H)$, CH_2NO_2
29	C_2H_5, CHO	61	$C_2H_5O_2(CH_3COO + 2H)$, CH_2CH_2SH
30	CH_2NH_2, NO	69	C_5H_9, C_3H_5CO
31	CH_2OH, OCH_3	70	C_5H_{10}, $(C_3H_5CO + H)$
33	SH	71	C_5H_{11}, C_4H_7O
34	H_2S	73	$OCOC_2H_5$, $C_3H_7OCH_2$
35	Cl	77	C_6H_5
36	HCl	79	Br
41	C_3H_5, $(CH_2CN + H)$	83	C_6H_{11}
42	C_3H_6	84	C_6H_{12}
43	C_3H_7, $CH_3C{=}O$	85	C_6H_{13}, C_4H_9CO, $C_3H_5CO_2$
44	CO_2, $C_2H_4NH_2$, $(CH_2CHO + H)$	91	$C_6H_5CH_2$
45	C_2H_4OH, $(CH_3CHO + H)$, COOH	92	$C_6H_5CH_3$

in Section 5.6, you should determine the number of rings and double bonds in the compound. For a molecule $C_wH_xN_yO_z$, you can obtain the total number of rings and double bonds, that is, the unsaturation sites, R, by the formula

$$R = \tfrac{1}{2}(2w - x + y + 2).$$

For example, the molecule $C_9H_{10}O_3$ (Section 5.6) should contain R, the total number of rings and double bonds:

$$R = \tfrac{1}{2}(2 \times 9 - 10 + 0 + 2) = 5.$$

The presence of one benzene ring in a structure accounts for four unsaturation sites (three double bonds plus one ring). Thus, in the above compound, if one benzene ring is present, the side chain should have only one double bond. This observation is further confirmed from the fact that 55 out of the 57 possible different structures that can be written for $C_9H_{10}O_3$ contain a benzene ring. Thus a quick look at the "other information" immediately enables us to determine the presence or absence of the benzene ring. If the presence of a benzene ring is confirmed, the search is narrowed down to identifying the remaining part of the molecule, consisting of 3 carbon, 3 oxygen, and possibly 5 or 6 hydrogen atoms. Of course, these remaining few atoms can occur in the molecule in 55 different combinations to give rise to aromatic or aliphatic acids, esters, ethers, and hydroxy compounds. We can make further progress by examining the fragmentation pattern of the compound.

Often it suffices to identify a few peaks in the spectrum; indeed, it is frequently impossible to do much more. Thus, in the beginning, list all the *major* peaks together with the molecular-ion peak, and all peaks near the M peak. Also record the mass-to-charge ratio of these ions and calculate the intensities of these peaks in relation to the base peak.

Table 5.5 lists some ions commonly produced in the fragmentation of various compounds. Such a table helps you to identify the important peaks in the spectrum, and should be complemented by additional information from literature.[33]

Important deductions can also be made from knowledge of the *neutral* fragments lost by the molecular or other ions to give certain peaks in the spectrum. Once you identify the molecular-ion peak, it is not too hard to derive the masses of the neutral entities lost. Comparison of these masses with those of the commonly lost fragments (Table 5.6) should reveal the identity of these neutral fragments.

Be careful when you use these tables, because you may reach wrong conclusions if you disregard the fact that various rearrangements may take place during fragmentation. It is useful at this stage to postulate the nature of any functional groups that may be present in the unknown. You can do this by making reasonable guesses, then testing their validity by comparing the fragmentation pattern of the unknown with that of a typical known compound containing such a functional group.

Table 5.6 Some neutral fragments often lost during mass-spectral fragmentation of organic compounds

Fragment number	Mass of lost fragment	Formula of lost fragment	Types of molecular ions fragmented (partial list)
1	1	H	Several types
2	2	H_2	Even-electron ion
3	15	$CH_3^{\cdot}$	A highly branched molecule cleaved at branching site, aldehydes and ketones, esters
4	16	$CH_3^{\cdot} + H$	A highly branched molecule cleaved at branching site
5	16	O	Nitro compounds, sulfoxides, pyridene oxides, epoxides, quinones, etc.
6	17	OH	Alcohols (R⦚OH), acids (RCO⦚OH)
7	18	H_2O, NH_4	Primary alcohols, aldehydes, esters, amines
8	19	F	Fluorides (R⦚F)
9	20	HF	Fluorides (R⦚HF)
10	26	C_2H_2, $\dot{C}{\equiv}N$	Aromatics, nitriles
11	27	$CH_2{=}CH$, HCN	Esters, R_2CHOH, nitrogen-heterocyclics
12	28	CO	Quinones, formates, etc. ($RC⦚\overset{+}{C{\equiv}O}$)
13	28	C_2H_4	$(R\overset{O}{\overset{\Vert}{C}}{-}CH_2C_2H_5)^{+\cdot} \longrightarrow (R\overset{OH}{\overset{\vert}{C}}{-}CH_2)^{+\cdot} + C_2H_4$ (rearrangement), cycloparaffins, alkenes, ethers
14	29	C_nH_{2n+1}, HCO	Same as fragment 3, cycloparaffins
15	30	C_nH_{2n+2}	Same as fragment 4
16	30	NO, NH_2CH_2	Nitroaromatics, amines
17	31	$C_nH_{2n+1}O$	Ethers (R⦚OR′), esters (RCO⦚OR′)
18	31	$\dot{C}H_2OH$, CH_3NH_2	Alcohols, amides
19	32	CH_3OH	Alcohols
20	33	$C_nH_{2n}F$	Fluorides
21	34	H_2S	Thiols
22	35, 36, 37	Cl, HCl, H_2Cl	Chlorides (distinct isotopic peaks for ^{37}Cl are observed)
23	39	C_nH_{2n-3}	Allyl esters
24	40	C_3H_4	Aromatics
25	41	C_nH_{2n-1}, e.g. $CH_2{=}CH{-}CH_2$	Same as 11, alkenes (allylic cleavage)
26	42	C_nH_{2n}	Same as 13
27	42	CH_2CO	Acetamides, unsaturated acetates
28	43	C_nH_{2n+1}, e.g. C_3H_7	Same as 3
29	43	NHCO	Cyclic amides
30	43	$C_nH_{2n+1}CO$	Ketones
31	44	C_nH_{2n+2}, CO_2, $CONH_2$, CH_2CHOH	Same as 4, 16, anhydrides and amides, aldehydes
32	45	$C_nH_{2n+1}O$	Same as 17
33	47	$C_nH_{2n}F$	Same as 20
34	49	$C_nH_{2n}Cl$	Same as 22
35	53	C_nH_{2n-3}	Same as 23
36	55	C_nH_{2n-1}	Same as 25
37	56	C_nH_{2n}	Same as 13
38	57	C_nH_{2n+1}, $C_nH_{2n+1}CO$	Same as 3 and 30
39	58	C_nH_{2n+2}	Same as 4 and 16

Table 5.6 (*continued*)

Fragment number	Mass of lost fragment	Formula of lost fragment	Types of molecular ions fragmented (partial list)
40	59	$C_nH_{2n+1}O$, CO_2R'	Same as 17 and esters R⦚C(=O)—OR′, R′—C(=O)—O⦚R (R′ small and R^+ stable)
50	60	CH_3COOH	Acetates, monocarboxylic acids
51	63	$C_nH_{2n}Cl$	Same as 22
52	67	C_nH_{2n-3}	Same as 23
53	69	C_nH_{2n-1}	Same as 25
54	71	C_nH_{2n+1}	Same as 3
55	72	C_nH_{2n+2}	Same as 4 and 16
56	73	$COOR'$	Same as 40
57	74	$C_3H_6O_2$	Methyl esters of monocarboxylic acids
58	77	C_6H_5	Aromatics
59	79	Br	R⦚Br (equal intensity peak at 81)
60	91	Tropyllium ion	ϕ—CH_2⦚R (aromatic-side-chain compounds)
61	127	I	R⦚I

5.10 REFERENCES

Because of the importance of mass spectrometry in the determination of the structures of organic compounds, a number of excellent books, compilations, and review articles on the subject have been published in recent years. Current interest in the field even sparked the appearance of a new journal, *Organic Mass Spectrometry,* in 1968 (Heydon and Son, Ltd., London). Although this journal deals specifically with the mass spectrometry of organic compounds, three other international journals devoted to mass spectrometry must also be mentioned here. These are:

Mass Spectrometry Bulletin (Mass Spectrometry Data Centre, AWRE, Aldermaston, Berkshire, England)
International Journal of Mass Spectrometry and Ion Physics (Elsevier, Amsterdam)
Mass Spectroscopy (Mass Spectroscopy Society of Japan, Toyonaka, Japan)

You can appreciate the extent of the current interest in this field by noting the fact that *The Mass Spectrometry Bulletin* alone published nearly 20,000 references to the literature on mass spectrometry and related topics in the first four years after its inception, in November 1966.

General References

1. W. Wien, *Ann. Physik.* **65**, 440 (1898).
2. J. J. Thomson, *Phil. Mag.* **10**, 584 (1905).
3. For good reviews, see: (a) D. Price, *Chem. Brit.* **4**, 255 (1968) (b) V. A. Pavlenko, L. N. Ozerov, and A. E. Rafal'son, *Sov. Phys. Tech. Phys.* **13**, 431 (1968).
4. D. B. Wilson, *Vacuum* **19**, 323 (1969).
5. K. A. Lincoln, *Int. J. Mass Spectrom. Ion Phys.* **2**, 75 (1969).
6. (a) C. F. Robinson, *Rev. Sci. Instr.* **27**, 88 (1956). (b) G. Comsa and A. Mircea, *J. Sci. Instr., Ser. 2* **2**, 336 (1969).
7. (a) R. Kohler, W. Paul, K. Schmidt, and U. Von Zahn, in *Proc. Intern. Conf. Nuclidic Masses,* University of Toronto Press, Toronto, Canada (1960), page 507. (b) A. R. Fairbairn, *Rev. Sci. Instr.* **40**, 380 (1969) (c) G. W. Ball, *Vacuum* **19**, 331 (1969).
8. G. I. Slobodenyuk, A. I. Titov, V. S. Voronin, and V. I. Ivashkin, *Instr. Exp. Tech.* 650 (1968).
9. R. F. Herzog, *Rev. Sci. Instr.* **40**, 1104 (1969).
10. G. M. McCracken, *Vacuum* **19**, 311 (1969).
11. H. B. Lall and P. S. Gill, *Nucl. Instr. Methods* **68**, 197 (1969).
12. P. H. Dawson, J. W. Hedman, and N. R. Whetten, *Rev. Sci. Instr.* **40**, 1444 (1969).
13. F. W. McLafferty, *Anal. Chem.* **31**, 82 (1959).

14. F. W. McLafferty, "Decompositions and Rearrangements of Organic Ions," in *Mass Spectrometry of Organic Ions*, edited by F. W. McLafferty, Academic Press, New York (1963), page 331.
15. F. W. McLafferty, *Interpretation of Mass Spectra*, W. A. Benjamin, New York (1967), page 120.
16. (a) H. Budzikiewicz, C. Djerassi, and D. H. Williams, *Interpretation of Mass Spectra of Organic Compounds*, Holden-Day, San Francisco, Calif. (1964), page xii.
(b) J. S. Shannon, *Proc. Royal Austral. Chem. Inst.* 328 (1964).
(c) M. M. Campbell and O. Runquist, *J. Chem. Ed.*, **49**, 105 (1972).
17. For discussion and references on McLafferty rearrangement, see F. W. McLafferty in *Determination of Organic Structures by Physical Methods*, edited by F. C. Nachod and W. D. Phillips, Vol. 2, Academic Press, New York (1962), pages 129–149.
18. D. P. Stevenson, "Ion-Molecule Reactions," in *Mass Spectrometry*, edited by C. A. McDowell, McGraw-Hill, New York (1963).
19. C. E. Melton, "Ion-Molecule Reactions," in *Mass Spectrometry of Organic Ions,* edited by F. W. McLafferty, Vol. 2, Academic Press, New York (1963).
20. J. H. Beynon, "Ions Formed by Intermolecular Processes," in *Mass Spectrometry and Its Applications to Organic Chemistry,* Elsevier, Amsterdam (1960).
21. H. D. Beckey, *Angew. Chem. Internat. Edit.* **8**, 623 (1969).
22. (a) M. S. B. Munson and F. H. Field, *J. Amer. Chem. Soc.* **88**, 2621, 4337 (1966) (b) F. H. Field, *Accounts Chem. Res.* **1**, 42 (1968).
23. J. H. Beynon and A. E. Williams, *Mass and Abundance Tables for Use in Mass Spectrometry*, Elsevier, Amsterdam (1963).
24. J. H. Beynon, "Qualitative Analysis by Mass Spectrometer" and "Masses and Isotopic Abundance Ratios" in *Mass Spectrometry and Its Applications to Organic Chemistry*, Elsevier, Amsterdam (1960), pages 291 and 486.
25. F. W. McLafferty and R. S. Gohlke, *Anal. Chem.* **34**, 1281 (1962).
(a) N. R. Daly, A. McCormick, and R. E. Powell, *Rev. Sci. Instr.* **39**, 1163 (1968)
(b) W. F. Haddon and F. W. McLafferty, *Anal. Chem.* **41**, 31 (1969).
(c) R. G. Cooks, A. N. H. Yeo, and D. H. Williams, *Org. Mass Spectrom.* **2**, 985 (1969).
(d) A. N. H. Yeo and D. H. Williams, *J. Amer. Chem. Soc.* **91**, 3582 (1969).
(e) F. W. McLafferty, D. J. McAdoo, and J. S. Smith, *ibid.,* page 5400.
(f) G. A. Smith and D. H. Williams, *ibid.*, page 5254.
(g) P. F. Donaghue, P. Y. White, J. H. Bowie, B. D. Roney, and H. J. Rodda, *Org. Mass. Spectrom.* **2**, 1061 (1969).
(h) C. C. Fenselau and F. P. Abramson, *ibid.*, page 915.
(i) R. S. Ward and D. H. Williams, *J. Org. Chem.* **34**, 3373 (1969).
26. K. Biemann, *Mass Spectrometry: Applications to Organic Chemistry*, McGraw-Hill, New York (1962).
27. H. D. Beckey, *Int. J. Mass Spectrometry Ion Physics* **2**, 500 (1969).
28. G. P. Arsenault, J. R. Althaus, and P. V. Divekar, *Chem. Commun.,* 1414 (1969).
(a) F. W. McLafferty, "Mass Spectrometry," in *Determination of Organic Structures by Physical Methods*, edited by F. C. Nachod and W. D. Phillips, Academic Press, New York (1962), pages 93–175.
29. H. Krone and H. D. Beckey, *Org. Mass Spectrometry* **2**, 427 (1969).
30. (a) K. Biemann, P. Bommer, and D. N. DeSiderio, *Tetrahedron Letters*, 1725 (1964).
(b) K. Biemann and W. McMurray, *Tetrahedron Letters* 647 (1965).
(c) F. W. McLafferty, *Science* **151**, 641 (1966).
31. K. Biemann, *Pure Appl. Chem.* **91**, 95 (1964).
32. J. H. Beynon, in *Advances in Mass Spectrometry*, Vol. 1, edited by J. W. Waldron, Pergamon Press, London (1959).
33. F. W. McLafferty, *Mass Spectral Correlations,* Advances in Chemistry Series No. 40, American Chemical Society, Washington, D.C. (1963).
34. J. Lederberg, *Computation of Molecular Formulas for Mass Spectrometry*, Holden-Day, San Francisco (1964).
35. J. Lederberg, *J. Chem. Ed.* **49**, 613 (1972).
36. S. Meyerson and A. W. Weitkamp, *Org. Mass Spectrom.* **1**, 659 (1968).
37. P. N. Rylander, S. Meyerson, and H. Grubb, *J. Am. Chem. Soc.* **79**, 842 (1957).
38. Z. Hückel, *Physik* **70**, 204 (1931).
39. U. K. Pandit, W. N. Speckamp, and H. O. Huisman, *Tetrahedron* **21**, 1767 (1965).
40. B. G. Hobrock and R. W. Kiser, *J. Phys. Chem.* **66**, 1648 (1962).
41. R. Srinivasan, *J. Am. Chem. Soc.* **83**, 4923 (1961).
42. (a) J. A. Gilpin and F. W. McLafferty, *Anal. Chem.* **29**, 990 (1957).
(b) M. M. Green, R. J. Cook, W. Rayle, E. Walton, and M. F. Grostic, *Chem. Commun.* 81 (1969).
(c) L. Dolejs and V. Hanus, *Collect. Czech. Chem. Commun.* **33**, 332 (1968).
43. F. A. Long and J. G. Pritchard, *J. Am. Chem. Soc.* **78**, 2663 (1956).
44. A. G. Sharkey, R. A. Friedle, and S. H. Lange, *Anal. Chem.* **29**, 770 (1957).

45. R. Ryhage and E. Stenhagen, *Arkiv Kemi* **14**, 483 (1959).
46. T. Aczel and H. E. Lumpkin, *Anal. Chem.* **32**, 1819 (1960).
47. J. D. McCollum and S. Meyerson, *J. Am. Chem. Soc.* **85**, 1739 (1963).
48. A. G. Sharkey, J. L. Shultz, and R. A. Friedle, *Anal. Chem.* **28**, 934 (1956).
49. R. Ryhage and E. Stenhagen, "Mass Spectrometry of Long-Chain Esters," in *Mass Spectrometry of Organic Ions*, edited by F. W. McLafferty, Academic Press, New York (1963) Chapter 9.
50. P. Capella and C. M. Zorzut, *Anal. Chem.* **40**, 1458 (1968).
51. G. Eglinton, D. H. Hunneman, and A. McCormick, *Org. Mass Spec.* **1**, 593 (1968).
52. R. S. Gohlke and F. W. McLafferty, *Anal. Chem.* **34**, 1281 (1962).
53. R. F. Pottei and F. P. Lossing, *J. Am. Chem. Soc.* **83**, 4737 (1961).

Books

54. M. C. Hamming and N. G. Foster, *Interpretation of Mass Spectra of Organic Compounds*, Academic Press, New York (1972).
55. A. L. Burlingame, *Topics in Organic Mass Spectrometry*, Wiley-Interscience, New York (1970).
56. D. Price and J. E. Williams, *Dynamic Mass Spectrometry*, Sadtler Research Laboratories (1970).
57. J. Roboz, *Introduction to Mass Spectrometry*, Wiley-Interscience, New York (1968).
58. R. Brymner and J. R. Penny, *Mass Spectrometry*, Chemical Publishing Co. (1967).
59. H. Budzikiewicz, C. Djerassi, and D. H. Williams, *Mass Spectrometry of Organic Compounds*, Holden-Day, San Francisco, Calif. (1967).
60. F. W. McLafferty, *Interpretation of Mass Spectra*, W. A. Benjamin, New York (1967).
61. H. C. Hill, *Introduction to Mass Spectrometry*, Sadtler Research Laboratories (1966).
62. R. I. Reed, *Applications of Mass Spectrometry to Organic Chemistry*, Academic Press, New York (1966).
63. R. W. Kiser, *Introduction to Mass Spectroscopy and Its Applications*, Prentice-Hall, Englewood Cliffs, N.J. (1965).
64. H. Budzikiewicz, C. Djerassi, and D. H. Williams, *Interpretation of Mass Spectra of Organic Compounds*, Holden-Day, San Francisco, Calif. (1964).
65. H. Budzikiewicz, C. Djerassi, and D. H. Williams, *Structure Elucidation of Natural Products by Mass Spectrometry*, Vol. I, Alkaloids, Vol. II, Steroids, terpenoids, sugars, and miscellaneous classes; Holden-Day, San Francisco, Calif. (1964).
66. J. Lederberg, *Computation of Molecular Formulas for Mass Spectrometry*, Holden-Day, San Francisco (1964).
67. K. Biemann, "Applications of Mass Spectrometry," in *Techniques of Organic Chemistry*, Vol. 11, edited by A. Weissberger, Wiley, New York (1963).
68. C. A. McDowell, editor, *Mass Spectrometry*, McGraw-Hill, New York (1963).
69. F. W. McLafferty, *Mass Spectral Correlations*, Advances in Chemistry Series, No. 40, American Chemical Society, Washington D.C. (1963).
70. F. W. McLafferty, *Mass Spectrometry of Organic Ions*, Academic Press, New York (1963).
71. J. H. Beynon, *Mass and Abundance Tables for Use in Mass Spectrometry*, Elsevier, Amsterdam (1963).
72. K. Biemann, *Mass Spectroscopy, Organic Chemical Applications*, McGraw-Hill, New York (1962).
73. R. I. Reed, *Ion Production by Electron Impact*, Advances in Chemistry Series, Academic Press, New York (1962).
74. J. H. Beynon, *Mass Spectrometry and Its Applications to Organic Chemistry*, Elsevier, Amsterdam (1960).

Compilations of Mass Spectra

75. American Petroleum Institute Research Project 44, *Selected Mass Spectral Data* (loose-leaf data sheets), Vol. I–VI, Thermodynamics Research Center, College Station, Texas (1947–1970).
76. *Index of Mass Spectral Data*, AMD11, American Society for Testing and Materials, Philadelphia, Pa. (1969).
77. Thermodynamics Research Center Data Project, *Selected Mass Spectral Data* (loose-leaf data sheets) Vols. I and II, Thermodynamics Research Center, College Station, Texas (1959–1969).
78. E. Stenhagen, S. Abrahamsson, and F. W. McLafferty, *Atlas of Mass Spectral Data*, Vols. 1, 2, 3, Wiley-Interscience, New York (1969).
79. F. W. McLafferty and J. Penzelik, *Index and Bibliography of Mass Spectrometry*, 1963–1965, Wiley-Interscience, New York (1967).
80. A. Cornu and R. Massot, *Compilation of Mass Spectral Data*, Heydon and Sons, London (1966).
81. J. H. Beynon, R. A. Saunders, and A. E. Williams, *Tables of Metastable Transitions*, Elsevier, Amsterdam (1966).
82. D. D. Tunnicliff, P. A. Wadsworth, and D. O. Schissler, *Mass and Abundance Tables*, Shell Development Company, Emeryville, Calif. (1965).
83. J. Lederberg, *Tables and Algorithms for Calculating Functional Groups of Organic Molecules in High-Resolution Mass Spectrometry*, NASA, Scientific and Technical Aerospace Report N64-21426 (1964).

84. R. S. Gohlke, *Uncertified Mass Spectral Data*, Dow Chemical Company, Midland, Mich. (1963).
85. American Petroleum Institute and the Manufacturing Chemists Association, *Catalog of Mass Spectral Data*, Chemical Thermodynamics Properties Center, Texas A. and M. University, College Station, Texas.

Review Articles

86. H. D. Beckey, *Angew. Chem. Internat. Edit.* **8**, 623 (1969) [Field Ionization Spectrometers].
87. G. W. Ewing, *J. Chem. Ed.* **46**, A-69, A-149, A-233 (1969) [Instruments].
88. R. G. Cook, I. Howe, and D. H. Williams, *Org. Mass Spectrom.* **2**, 137 (1969). [Ion production].
89. R. G. Cook, *Org. Mass Spectrom.* **2**, 481 (1969) [Rearrangement reactions].
90. J. H. Bowie, B. K. Simons, and S. O. Lowesson, *Rev. Pure Appl. Chem.* **19**, 61 (1969) [Organosulfur compounds].
91. A. B. Foster, *Lab. Practice* **18**, 743 (1969) [Biomedical research].
92. J. A. McCloskey, in *Methods in Enzymology,* Vol. 14, edited by J. W. Lowenstein, Academic Press, New York, 1969, pp. 382–450 [Lipids].
93. J. M. S. Henis, *Anal. Chem.* **41** (10), 22A (1969) [Ion cyclotron resonance].
94. L. I. Yin, I. Alder, and R. Lamothe, *Appl. Spectrosc.* **23**, 41 (1969) [Photoelectron spectroscopy].
95. C. W. Childs, P. S. Hallman, and D. D. Perrin, *Talanta* **16**, 629 (1969) [Computer applications to mass spectrometry].
96. S. Meyerson and A. W. Weitkamp, *Org. Mass Spectrom.* **1**, 659 (1968) [Stereoisomeric effects on mass spectra].
97. M. M. Bursey, *Org. Mass Spectrom.* **1**, 31 (1968) [Substituent effects in mass spectra of aromatic compounds].
98. A. J. D'Eustachio, *Anal. Chem.* **40**, 19R (1968) [Biomedical research].
99. E. Lederer, *Ind. Chim. Belge* **33** (special no.), 60 (1968) [Complex biological compounds].
100. I. A. Pearl and S. F. Darling, *Phytochemistry* **7**, 831 (1968) [Natural glycosides].
101. K. K. Sun and R. T. Holman, *J. Amer. Oil Chem. Soc.* **45**, 810 (1968) [Lipids].
102. J. Lewis and B. F. G. Johnson, *Accounts. Chem. Res.* **1**, 245 (1968) [Organometallic compounds].
103. M. I. Bruce, *Advan. Organometal. Chem.* **6**, 273 (1968) [Organometallic compounds].
104. R. K. Webster, *Proc. Soc. Anal. Chem.* **5**, 117 (1968) [Analytical applications of mass spectroscopy].
105. A. B. Littlewood, *Chromatographia*, 37 (1968) [GLC/MS].
106. H. Ch. Curtius, *Z. Klin. Chem. Klin. Biochem.* **6**, 122 (1968) [GLC/MS].
107. J. Normand and R. Lombre, *Method. Phys. Anal.* **4**, 61 (1968) [GLC/MS].
108. J. D. Baldeschwieler, *Science* **159**, 263 (1968) [Ion cyclotron].
109. J. D. Baldeschwieler, H. Benz, and P. M. Llewellyn, *Advan. Mass Spectrom.* **4**, 113 (1968) [Ion cyclotron resonance].
110. L. Friedman, *Ann. Rev. Phys. Chem.* **19**, 273 (1968) [Ion–molecule reactions].
111. J. H. Futrell and T. O. Tiernan, *Science* **162**, 415 (1968) [Ion–molecule reactions].
112. B. Brocklehurst, *Quart. Rev. (London)* **22**, 147 (1968) [Ion–molecule reactions].
113. F. J. Comes, *Advan. Mass Spectrom.* **4**, 737 (1968) [Photoionization].
114. J. Block, *Advan. Mass Spectrom.* **4**, 791 (1968) [Field ionization].
115. G. G. Wanless, *Advan. Mass Spectrom.* **4**, 833 (1968) [Field ionization].
116. F. H. Field, *Accounts. Chem. Res.* **1**, 42 (1968) [Chemical ionization].
117. D. W. Turner, *Chem. Brit.* **4**, 435 (1968) [Photoelectron spectroscopy].
118. H. Kienitz and R. Kiser, *Z. Anal. Chem.* **237**, 241 (1968) [Use of computers in mass spectrometry].
119. V. A. Pavlenko, L. N. Ozerov, and A. E. Rafal'son, *Sov. Phys. Tech. Phys.* **13**, 431 (1968) [Time-of-flight spectrometers].
120. D. Price, *Chem. Brit.* **4**, 255 (1968) [Time-of-flight spectrometers].
121. K. L. Rinehart and T. H. Kinstle, *Ann. Rev. Phys. Chem.* **19**, 301 (1968) [General review].
122. A. G. Loudon, *Chem. Brit.* **4**, 50 (1968) [Structure determination].
123. J. H. Jones, *Quart. Rev.* **22**, 302 (1968) [Amino acids, peptides].
124. D. B. Chambers, F. Glockling, and J. R. C. Light, *Quart. Rev.* **22**, 317 (1968) [Organometallic compounds].
125. G. W. A. Milne, *Quart. Rev.* **22**, 75 (1968) [Applications to organic chemistry].
126. R. I. Reed, *Quart. Rev.* **20**, 527 (1966) [General mass spectrometry].
127. N. K. Kochetkov and O. S. Chizhov, *Advan. Carbohydrate Chem.* **21**, 39 (1966) [Carbohydrates].
128. K. Biemann, *Fortschr. Chem. Org. Naturst.* **24**, 1 (1966) [Natural products].
129. S. N. Foner, *Advan. At. Mol. Phys.* **2**, 385 (1966) [Free radicals].
130. F. W. McLafferty, *Science* **151**, 641 (1966) [High resolution].
131. G. Spiteller and M. Spiteller-Friedmann, *Angew. Chem. Internat. Edit.* **4**, 383 (1965) [Relation between structure and mass spectra].

132. J. Roboz, *Trace. Anal.*, 453 (1965) [Instrumentation].
133. S. Meyerson, *Record. Chem. Progr.* (Kresge-Hooker Sci. Lib.) **26** (4), 257 (1965) [Origins of mass spectra].
134. M. E. Fitzgerald, *Mass Spectr., New Instr. Tech. Rept.* **13**, 76 (1964) [Instrumentation].
135. F. W. McLafferty and R. S. Gohlke, *Chem. Eng. News* **42**, 96 (May 18, 1964) [Analytical applications].
136. K. Biemann in *Techniques of Organic Chemistry*, Vol. II, edited by A. Weissberger, Wiley, New York (1963) [Applications of mass spectrometry].
137. K. Biemann, *Ann. Rev. Biochem.* **32**, 755 (1963) [General review].
138. E. Kendrick, *Anal. Chem.* **35**, 2146 (1963) [Mass scale based on CH_2 = 14.000000].

SIX

IDENTIFICATION OF ORGANIC COMPOUNDS

In the last four chapters we discussed fundamentals of ultraviolet, infrared, nuclear magnetic resonance, and mass spectrometric methods. Our purpose was to give you enough knowledge of spectroscopy to enable you to understand where and why the various functional groupings and molecular structures show maximum absorption of radiation, or undergo fragmentation. These four techniques, used conjointly, constitute an extremely powerful tool in the hands of the organic chemist. However, no one type of spectroscopy by itself can be used for the precise identification of an organic structure. Often one has to select and correlate relevant data from whatever physical, chemical, and spectroscopic methods there are available.

In this chapter we shall try to show how a composite of all the spectroscopic information about an unknown substance can be used to finally put together the jigsaw puzzle of the molecular structure of the substance. We shall demonstrate this bringing together of information by discussing the spectra of the first five organic compounds presented herein and then, in the rest of the chapter, allowing you, the reader, to practice interpreting the spectra of the next 17 unknowns.

Two features will make spectral interpretation more realistic for you: First, instead of providing predigested abstracts of positions of absorption peaks, we shall present complete infrared and NMR spectra of the organic compounds in question, together with tabular information on mass spectral and UV behavior of each. You are likely to encounter similar situations in research and in routine work on chemical structures; therefore it is important that you understand how to select relevant data. Second, instead of using very sophisticated high-cost instruments and excessively purified chemicals, we have used medium-range instruments, commonly available today in many laboratories, to determine the spectra reported in this chapter. The chemicals used are also ordinary reagent-grade compounds such as are supplied by any manufacturer of chemicals. This is the sort of situation most of you will face during your initial encounters with spectroscopy.

UNKNOWN NUMBER 6.1

Physical data: Colorless liquid, boiling point 144°

Mass spectrum

m/e	% of base peak	m/e	% of base peak	m/e	% of base peak	m/e	% of base peak
27	40	28	7.5	29	8.5	31	1
39	18	41	26	42	10	43	100
44	3.5	55	3	57	2	58	6
70	1	71	76	72	3	86	1
99	2	114	13 (M)	115	1	116	0.06

Isotope ratio

m/e	% of M
114 (M)	100
115 (M + 1)	7.7
116 (M + 2)	0.46

Ultraviolet data

λ_{max}^{EtOH}	ϵ_{max}
275 nm	12

Infrared spectrum: Liquid sample, thin layer

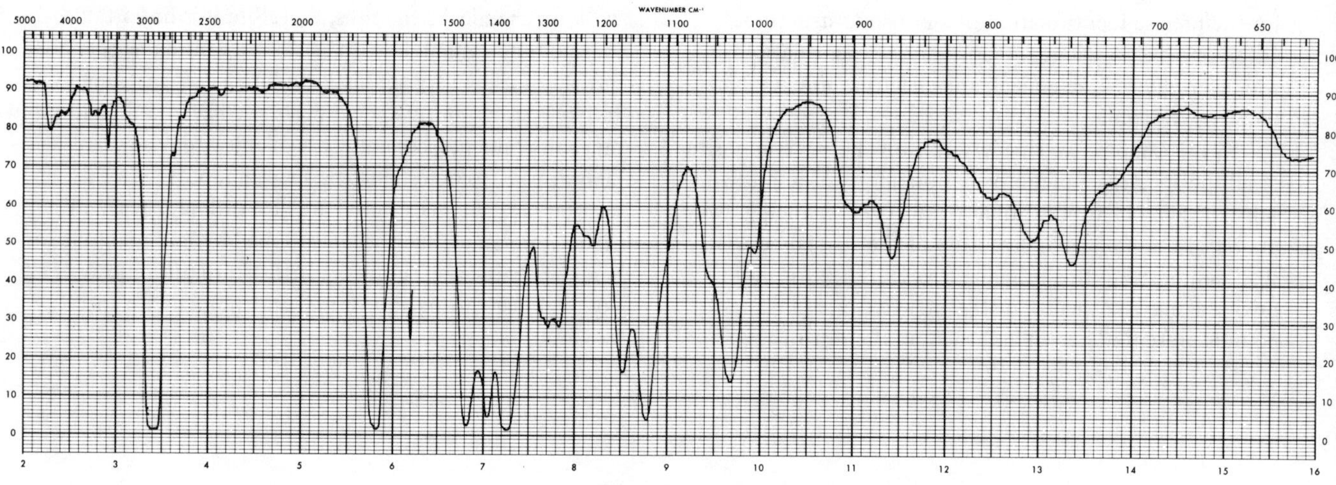

NMR spectrum: Neat, sweep time 250 sec, sweep width 500 Hz

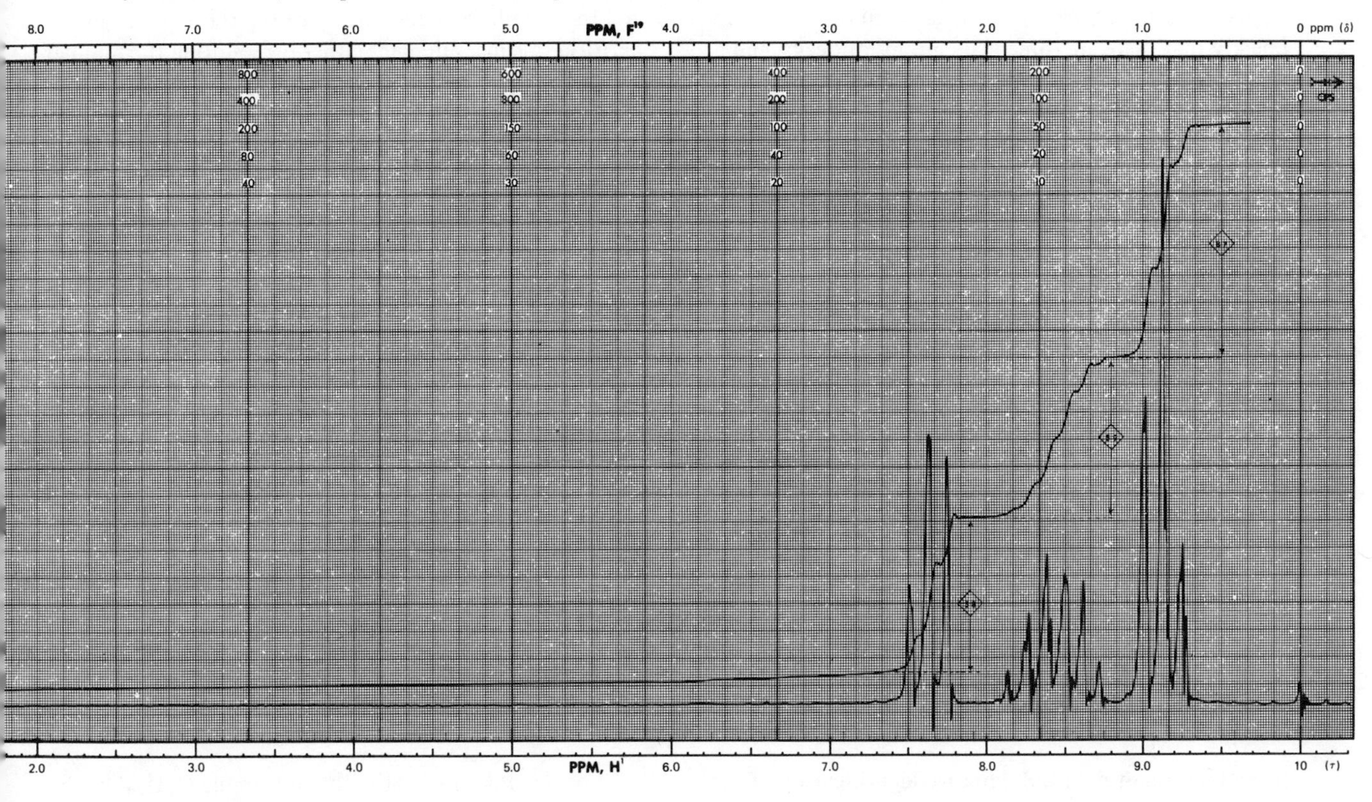

IDENTIFICATION OF UNKNOWN NUMBER 6.1

Summary of relevant data

Before one sets out to identify an unknown, it is a good idea to prepare first a tabular summary of the spectroscopic information, based on the data provided in the problem and in the spectroscopic tables given in the Appendix. Such a summary enables one to quickly correlate the information obtained from different techniques. Let us therefore prepare the summary as follows.

Mass spectrum

i) The five most intense peaks are at masses 43, 71, 27, 41, 39.

ii) The molecular weight of the compound is 114, and the M + 1 peak is 7.7% of the molecular-ion peak. Therefore, from Appendix 5.1, we can select the possible molecular formulas of the unknown within the M + 1 peak intensity range of 6.7 to 8.4%. We can also use the nitrogen rule to eliminate formulas containing an odd number of nitrogen atoms, and thus arrive at the following molecular formulas that might apply in this case.

Formula number	Formula	M + 1	M + 2
M.1	$C_6H_{10}O_2$	6.72	0.59
M.2	$C_6H_{14}N_2$	7.47	0.24
M.3	$C_7H_{14}O$	7.83	0.47
M.4	$C_7H_2N_2$	8.36	0.37

Ultraviolet data. According to ultraviolet rules 1a and 2 (Section 2.11), when a compound exhibits a very low-intensity absorption peak at 275 nm and no other absorption peak above 200 nm, this indicates that it contains a simple, unconjugated chromophore having n electrons ($n \to \pi^*$).

Infrared data. Important infrared peaks are as follows.

Peak number	Wave-length λ, μ	Wave-number $\bar{\nu}$, cm^{-1}	Inference (Appendix 3.1b)
I.1	2.93	3413	C=O str.; overtones
I.2	3.39	2950	C—H str.; alkyl, $-CH_3$ or $-CH_2-$
I.3	5.85	1709	C=O str.; aldehydes, ketones, carboxylic acids
I.4	7.05	1418	$-CH_2$ def. in $-CH_2-CO$
I.5	8.79	1138	C—O str.

NMR data. Several absorption peaks—both large and small—appear on the spectrum. One must learn to recognize quickly that there are three main groups of peaks, and a small peak at 0 ppm (10.00 τ), due to the internal reference TMS. The upper horizontal line is the integral trace. The steps in this line indicate the integrated intensities of the peaks. To measure the size of the step, extend the horizontal portion of the integral trace by drawing a dotted line, and measure the distance between this line and the next horizontal portion of the trace. You can perform this measurement either by counting the number of small squares on the spectrum or by measuring the distance, in millimeters, between the two horizontal lines. The information thus derived is summarized below; possible assignments are suggested on the basis of Appendix tables 4.1 and 4.2.

Signal number	Signal position δ	τ	Integral step size	Rel. no. protons	Multiplicity	Inference (Appendix 4.1)
N.1	2.31	7.63	58	2	Triplet	$-CH_2^*-CO-R$ CH_3CO $CH_3-CO-C=C$
N.2	1.57	8.43	59	2	Sextet	$-CH_2-C-C=C$ $-CH-$, sat. $-CH_2-C-Ar$ $-CH_2-C-O-R$ $-CH_2-C-CO-$ $CH_3-C=C$
N.3	0.86	9.14	87	3	Triplet	CH_3-CN CH_3-C-, sat.

* Protons responsible for the signal are shown by bold type in this chapter.

Interpretation of data and tentative identification of unknown

We can now proceed to correlate the data.

1. The M + 1 peak is 7.7%; the M + 2 peak is 0.46% of the M peak. Of the four possible formulas listed above, the observed values are best suited to the formula M.3; that is, $C_7H_{14}O$. Let us therefore consider it to be the probable molecular formula of the compound. We can support this assumption by calculating the number of carbon atoms as

$$\text{No. of carbons} = \frac{\text{percent M + 1 peak}}{1.1} = \frac{7.7}{1.1} = 7$$

(Section 5.6).

2. The carbon-to-hydrogen ratio of the tentative formula indicates that the compound has an aliphatic structure. This deduction is supported by the following observations.

i) The molecular-ion peak M is small. Aromatic compounds generally exhibit intense M peaks (Section 5.8).

ii) For a molecule $C_wH_xN_yO_z$, the total number of rings and double bonds is obtained by the formula,

$$R = \tfrac{1}{2}(2w - x + y + 2) \qquad \text{(Section 5.9)}.$$

For the unknown 6.1 with the formula $C_7H_{14}O$,

$$R = \tfrac{1}{2}(14 - 14 + 2) = 1.$$

The presence of one benzene ring in the structure requires at least four unsaturation sites (3 double bonds + 1 ring).

iii) The appearance of weak, multiple peaks in the 3080–3030 cm^{-1} (3.25–3.30 μ) region, and three to four bands in the region 1650–1400 cm^{-1} (6.06–7.14 μ) is typical of the infrared spectra of aromatic compounds (Section 3.7). These bands are absent from the infrared spectrum of the unknown.

iv) According to nuclear magnetic resonance theory, aromatic protons resonate downfield at 2 to 3.4 τ (Appendix 4.1). The NMR spectrum of the unknown shows no signals at such a low field.

3. The strong absorption peak for carbonyl at 1709 cm^{-1} (5.85 μ) suggests that the only oxygen in the formula $C_7H_{14}O$ is present in the carbonyl functional group. The compound is thus either an aliphatic aldehyde or a ketone. A very weak C=O overtone absorption peak at 3413 cm^{-1} (2.93 μ) in the infrared spectrum and a weak $n \rightarrow \pi^*$ transition at 275 mμ in the ultraviolet spectrum also support this conclusion.

4. We can rule out the possibility of an aldehyde function,

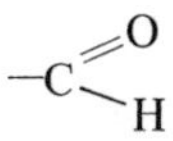

$$-C\begin{smallmatrix}{=}O\\ \diagdown H\end{smallmatrix}$$

because an aldehyde normally shows a weak doublet due to C—H stretching between 3.47 and 3.77 μ and a weak peak at 10.26-12.82 μ due to C—H deformation. Both these peaks are absent from the infrared spectrum of the unknown. More important is the fact that, in NMR, the aldehyde proton being bonded to the carbonyl carbon resonates at a very low τ range of 0.2 to 0.3 (aliphatic aldehydes). No such downfield resonance is observed in the NMR spectrum of the unknown. The formula for the unknown may thus be written as $C_6H_{14}C{=}O$.

5. The NMR spectrum suggests the presence of three types of protons, in the ratio 2:2:3. Assignments such as CH_3CN or $-CH_2-C-Ar$ for the NMR absorption peaks may now be ruled out, for reasons already discussed. The three-proton N.3 peak centered at 9.14 τ is thus due to a methyl group. Since it is split into a triplet with J = 7 Hz, there must be two protons on the adjoining carbon atom. This suggests the structure CH_3-CH_2-.

6. Of the three assignments for the triplet N.1 at 7.63 τ, only $-CH_2-CO-R$ is probable, because the signal is due to two protons only.

7. The two-proton peak at 8.43 τ is a sextet, indicating the presence of five neighboring protons. This enables us to select $-CH_2-C-CO-R$ as the possible assignment, and suggests for the unknown a partial structure

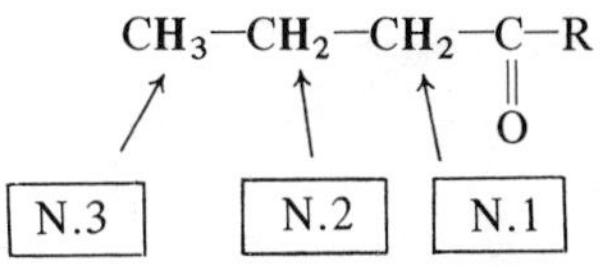

$$\underset{\text{N.3}}{CH_3}-\underset{\text{N.2}}{CH_2}-\underset{\text{N.1}}{CH_2}-\underset{\underset{O}{\|}}{C}-R$$

8. The partial structure now arrived at consists of 7 protons of three different types. Since the compound contains 14 protons, and since there are only three types of protons in the compound, we write the complete structure as:

$$CH_3-CH_2-CH_2-\underset{\underset{O}{\|}}{C}-CH_2-CH_2-CH_3$$

4-Heptanone

Cross checking

1. The suggested structure contains only three types of protons; these are in the ratio 2:2:3.

2. α protons, being closest to the carbonyl group, resonate at the lower field, and exhibit a triplet due to the two neighboring protons on the β carbon.

3. The β protons, with a methyl group on one side and a methylene group on the other, exhibit a sextet.

4. The peak due to γ protons is split into a triplet, due to the coupling of these protons with the two β protons. The triplet is not symmetrical, but the subpeak on the downfield side is more intense than that at the other side of the central subpeak. This indicates that the methyl protons are coupled with the downfield methylene protons. A reverse situation is observed in the case of α protons which are coupled with upfield β protons.

5. The base peak at m/e = 43 in the mass spectrum is due to α cleavage, which is common in ketones, and which produces stable ions $R-C\equiv\overset{+}{O}$. Thus

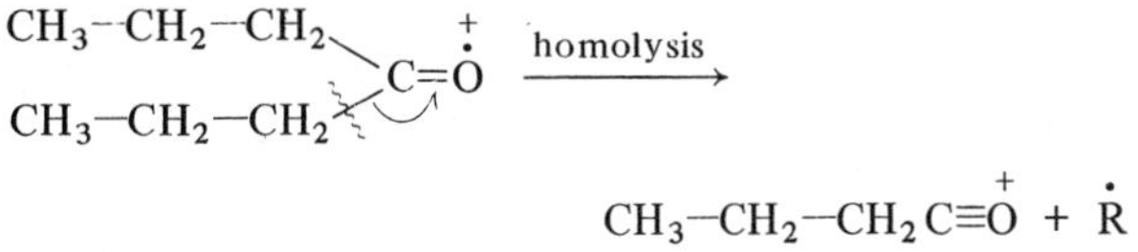

$$\begin{matrix}CH_3-CH_2-CH_2\diagdown\\ CH_3-CH_2-CH_2\diagup\end{matrix}C{=}\overset{+}{\dot{O}} \xrightarrow{\text{homolysis}} CH_3-CH_2-CH_2C\equiv\overset{+}{O} + \dot{R}$$

m/e = 71

$$\begin{matrix}CH_3-CH_2-CH_2\diagdown\\ CH_3-CH_2-CH_2\diagup\end{matrix}C{=}\overset{+}{\dot{O}} \xrightarrow{\text{heterolysis}} CH_3-CH_2-CH_2C\equiv\dot{O} + CH_3-CH_2-\overset{+}{C}H_2$$

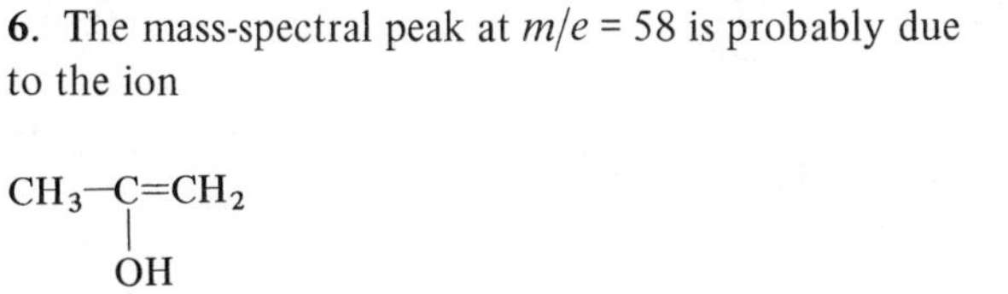

m/e = 43

6. The mass-spectral peak at m/e = 58 is probably due to the ion

$$CH_3-\underset{\underset{\overset{+}{OH}}{|}}{C}=CH_2$$

produced by double McLafferty rearrangement.

UNKNOWN NUMBER 6.2

Physical data: Low-melting needle-shaped crystals from ethanol; melting point 29°; colorless liquid; boiling point 217.6°

Mass spectrum

m/e	% of base peak	m/e	% of base peak	m/e	% of base peak	m/e	% of base peak
15	1.5	26	3	27	6	30.5	0.1**
37	5	38	7	39	16.5	41	1^R
42.5	0.1**	43	1**	43.5	0.3**	44	0.5**
44.5	0.5**	45.5	1**	49	1	50	7
51	8	52	3	55.5	0.2**	56.5	0.3**
57.5	1**	58	0.7**	58.5	3**	61	3
62	6	63	12	64	6.5	65	3
69.5	1*	74	1.5	75	3	76	2.5
77	1	86	1	87	2	88	4
89	17.5	90	26.5	91	4.5	114	1
115	3	116	55	117	100 (M)	118	9

* = Metastable peak ** = Doubly charged ion *R* = Rearrangement peak

Isotope ratio

m/e	% of M
117 (M)	100
118 (M + 1)	9

Ultraviolet data

λ_{max}^{EtOH}	ϵ_{max}
237 nm	17,200
268 nm	750

Infrared spectrum: 9.0% w/w solution in $CHCl_3$ and 5.0% w/w solution in $CHCl_3$, as inset at 12 μ

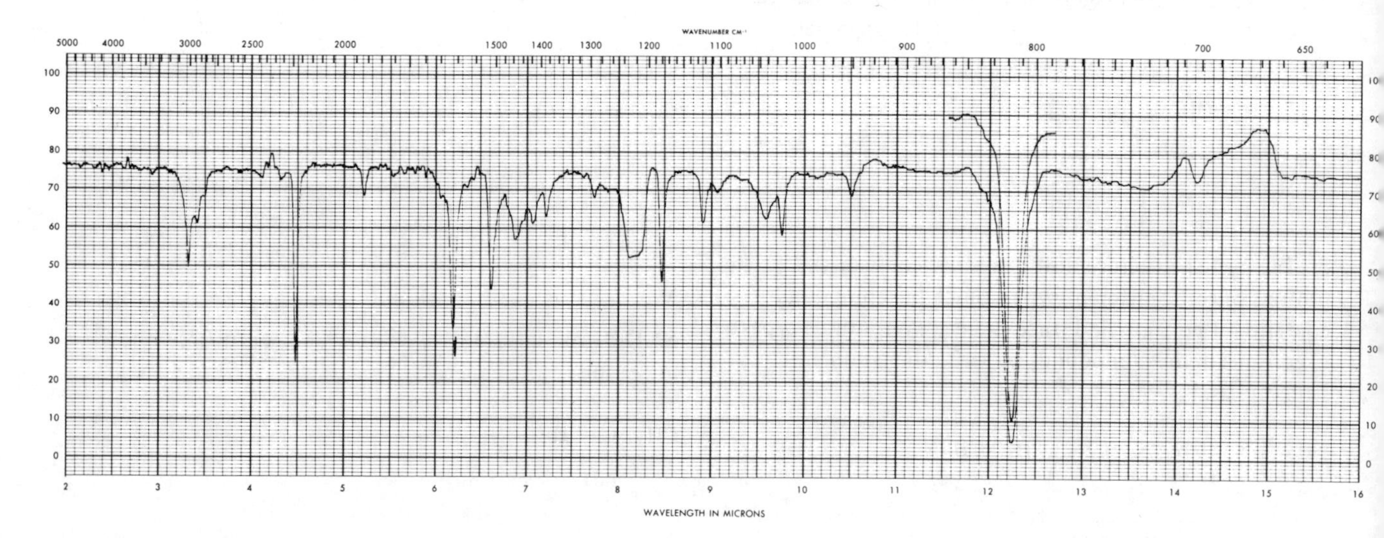

NMR spectrum: Solvent CCl_4, sweep time 250 sec, sweep width 500 Hz

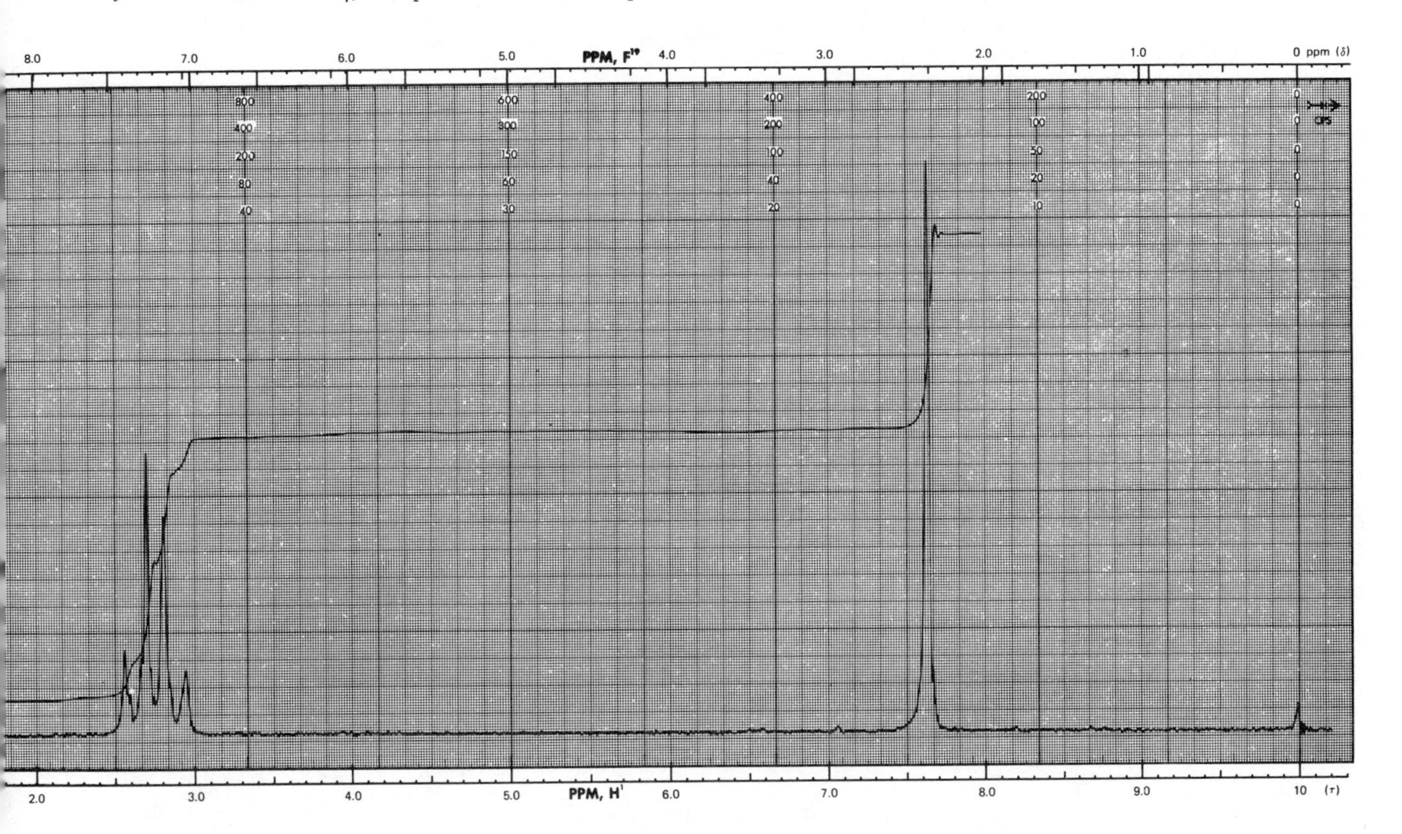

IDENTIFICATION OF UNKNOWN NUMBER 6.2

Summary of relevant data

Mass spectrum

i) The five most intense peaks are at masses 117, 116, 90, 89, 39.

ii) The molecular weight of the compound is 117; the M peak has an odd mass number and is also a base peak. Thus the compound is probably an aromatic structure containing an odd number of nitrogen atoms (recall the nitrogen rule).

iii) The M + 1 peak is 9% of the M peak. The molecular formulas of compounds with mass 117, M + 1 peak intensity between 8.0 and 9.8%, and an odd number of nitrogen atoms are the following. (Appendix 5.1)

Formula number	Formula	M + 1	M + 2
M.1	C_7H_3NO	8.03	0.48
M.2	C_8H_7N	9.14	0.37

Ultraviolet data. A strong absorption peak at 237 nm and a somewhat weaker peak at 268 nm strongly suggests an aromatic structure.

Infrared data. The pattern of absorption in the infrared region between 6 and 7 μ is typical of aromatic compounds. Thus, in preparing our summary, we consult Appendix table 3.3.

Peak number	Wavelength λ, μ	Wavenumber $\bar{\nu}$, cm^{-1}	Inference (Appendix 3.1b and 3.3)
I.1	3.32	3012	=C—H str. (multiple)
I.2	3.45	2899	—C—H str.; $—CH_3$, $—CH_2—$, —CH—
I.3	4.50	2222	—C≡N str.; aryl nitrile or $\alpha:\beta$ unsaturated nitrile
I.4	6.23	1605	—C=C str.; mono or disub. benzenoid compound
I.5	6.64	1506	—C=C str.; mono, 1:2 or 1:4 disub. or 1:2:4 trisub. benzenoid compound
I.6	8.5	1176	—C—H i.p. deformation; mono, 1:4 disub. or 1:2:3, 1:2:4, or 1:3:5 trisub. benzenoid compound
I.7	12.25	816	—C—H o.o.p.; 1:4 disub. or 1:2:3 trisub. benzenoid compound
I.8	14.22	703	

NMR data. There are only two groups of signals in the NMR spectrum.

Signal number	Signal position δ	τ	Integral step size	Rel. no. protons	Multiplicity	Inference (Appendix 4.1)
N.1	7.25	2.75	96	4	Quartet	Ar—H
N.2	2.37	7.63	71	3	Singlet	$Ar—CH_3$

Interpretation of data and tentative identification of unknown

1. Data for the M + 2 peak intensity are not available from the mass spectrum. Therefore it is hard to choose between formulas M.1 and M.2. However, if we try to write reasonable structures for these formulas with the help of the *Handbook of Chemistry and Physics* (edited by R. C. Weast, Chemical Rubber Publishing Co., Cleveland, Ohio, 53rd edition, 1972-73), we arrive at C_8H_7N, because no probable structure has been listed in the handbook under the molecular formula C_7H_3NO. We quickly confirm this conclusion on the basis of the following observations.

i) The compound is an aromatic nitrile. This conclusion is based on the fact that there is a strong molecular-ion peak. Peaks I.3, I.4, I.5, and N.1 support this conclusion.

ii) Nitriles usually show a strong M – 1 peak, due to the loss of hydrogen, and a strong M – 27 peak, due to the loss of HCN. The mass spectrum of the unknown has intense peaks at m/e = 116 (M – 1) and m/e = 90 (M – 27).

iii) The NMR spectrum indicates that the two types of protons are in the ratio 4:3. This suggests that the unknown is likely to contain at least seven protons.

2. Signal N.2 is identified to be due to Ar—CH_3 protons.

3. The presence of four aromatic protons, three methyl protons, and a cyanide group suggests that the compound is a tolunitrile. This enables us to write the possible structure as:

CH_3 —CN

4. We can now readily establish the relative positions of the methyl and the nitrile groups.

i) The pattern formed by peaks I.4 through I.8 is not only characteristic of the benzenoid compounds, but also indicates 1:4 disubstitution. This is further supported by a typical pattern in the 5.0 to 6.0 μ region, which is characteristic of 1:4 disubstituted compounds (see Fig. 3.20b).

ii) The most important indication of 1:4 disubstitution, however, is obtained from the N.1 peak pattern. *Ortho-* or *meta*-disubstituted benzenes may show either a single peak or a complex pattern of peaks, but *para*-disubstituted benzene derivatives nearly always form an A_2B_2 splitting system and display a pattern characterized by four subpeaks symmetrically placed with an intensity relation of weak–strong–strong–weak. The separation between the outer components of such a pattern is usually 8 Hz, and some less intense sub-peaks may also be present in the overall pattern. We note a nearly similar quartet pattern for N.1. Thus we conclude that unknown number 6.2 is *p*-tolunitrile.

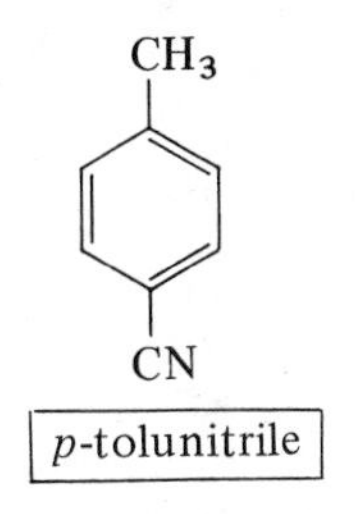

p-tolunitrile

Cross checking. The final confirmation of the proposed structure is obtained as usual, by comparing the physical data of the unknown with the data listed in the handbook for authentic *p*-tolunitrile, and by matching the infrared spectrum of the unknown with the spectrum of an authentic sample of *p*-tolunitrile. The primary band in the UV spectrum of benzene appears at 203.5 nm. According to Table 2.9, *para*-tolunitrile shifts this band to the longer wavelength by approximately 33.5 nm. The observation that unknown 6.2 absorbs UV radiation at 237 nm is a further proof of its identity.

UNKNOWN NUMBER 6.3

Physical data: An oily liquid with odor like cloves; boiling point 254°

Mass spectrum

m/e	% of base peak	m/e	% of base peak	m/e	% of base peak	m/e	% of base peak
18	1	27	3	28	1	29	1
38	1	39	6	40	1	41	3
43	2	45.5	0.2	50	2	51	6
52	2	53	3.5	55	10.5	59.5	0.1
60.5	0.1	62	1	63	2.5	65	5
66	2	77	14.5	78	4	79	4
81	1	89	1	91	10.5	92	2
93	2.5	94	3.5	102	1	103	15
104	10	105	5	107	2	109	1
115	2	117	1	119	1	120	1
121	10	122	6	123	1	131	19
132	7	133	13.5	134	2	135	1.5
137	17.5	138	1.5	147	5	148	1
149	31	150	3	151	2	152	1
163	4	164	100 (M)	165	11.1	166	1

Isotope ratio

m/e	% of M
164 (M)	100
165 (M + 1)	11.1
166 (M + 2)	1

Ultraviolet data

λ_{max}^{EtOH}	ϵ_{max}
282 nm	3000

Infrared spectrum: 6.0% w/w solution in CCl_4

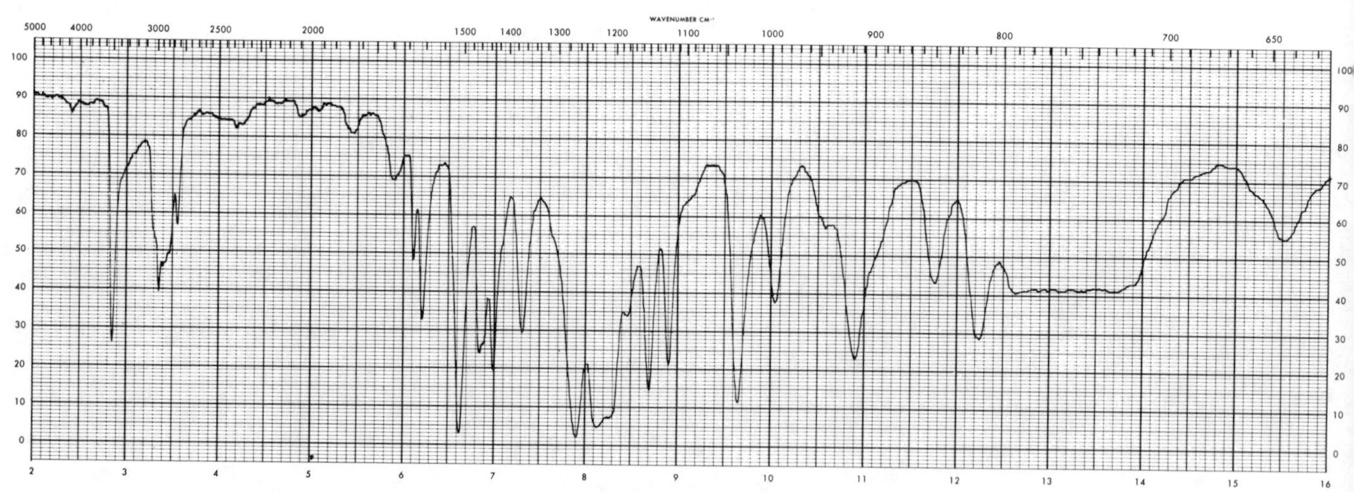

NMR spectrum: Solvent CCl_4, sweep time 250 sec, sweep width 500 Hz; upper spectrum marked with ★ is spectrum after D_2O has been added to the sample.

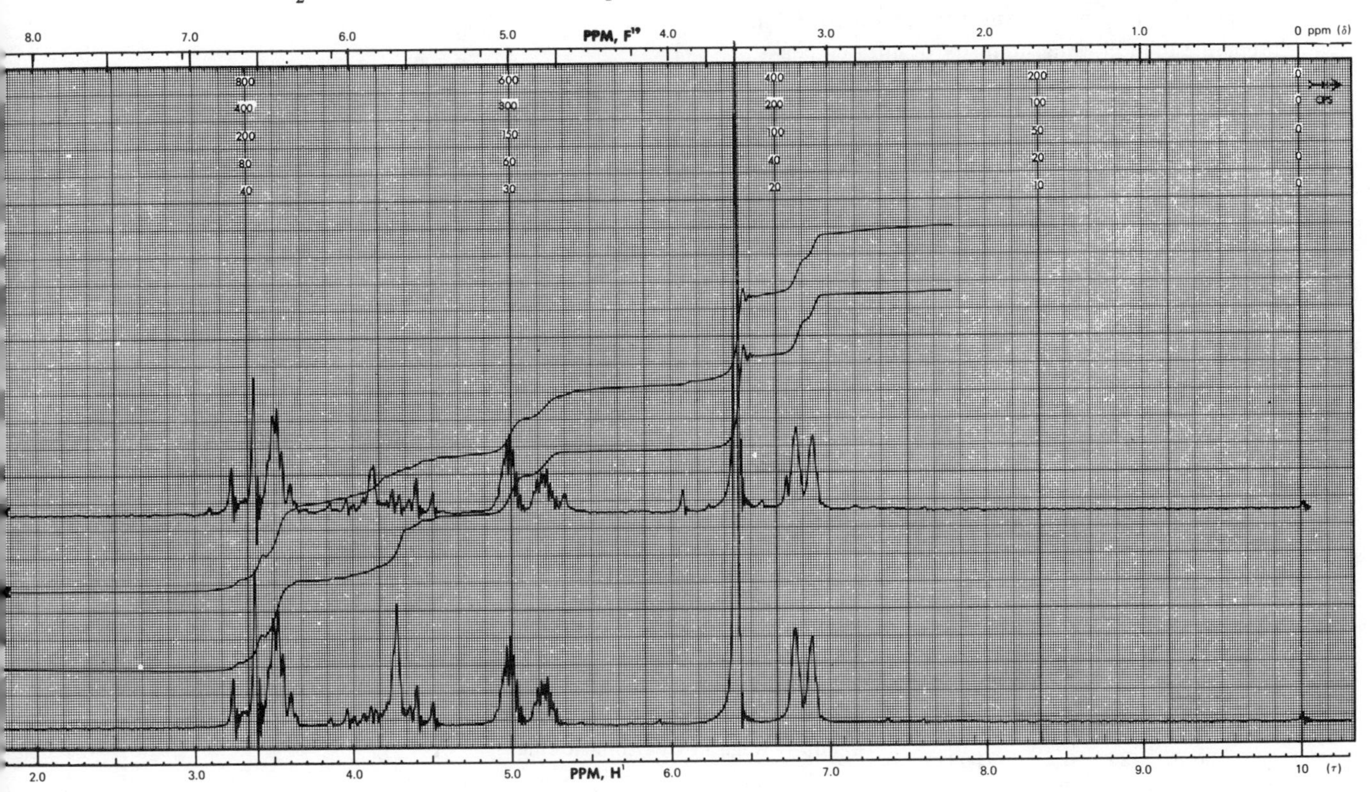

IDENTIFICATION OF UNKNOWN NUMBER 6.3

Summary of relevant data

Mass spectrum

i) The five most intense peaks are at masses 164, 149, 131, 137, 103.

ii) The molecular weight of the compound is 164; the M peak is also the base peak.

iii) The M + 1 peak is 11.1% and the M + 2 peak 1.0% of the M peak. Therefore, from Appendix table 5.1, the possible molecular formulas with M + 1 peak intensities between 10.1 and 11.8 are the following.

Formula number	Formula	M + 1	M + 2
M.1	$C_8H_{12}N_4$	10.36	0.49
M.2	$C_9H_{12}N_2O$	10.72	0.72
M.3	$C_{10}H_{12}O_2$	11.08	0.96
M.4	$C_{10}H_{16}N_2$	11.83	0.64

(Formulas with an odd number of nitrogen atoms are omitted.)

Infrared data. We observe a pattern of three to four peaks in the region between 6 and 7 μ; this is typical of aromatic compounds. Thus we consult Appendix table 3.3 in preparing the following summary of infrared peaks.

Peak number	Wavelength λ, μ	Wavenumber $\bar{\nu}$, cm^{-1}	Inference (Appendix 3.1b and 3.3)
I.1	2.84	3521	O—H str.; alcohols, phenols, carboxylic acids
I.2	3.35	2985	=C—H str.
I.3	3.55	2817	C—H str.
I.4	6.10	1639	C=C str. (too weak to be C=O str.)
I.5	6.21	1610	C=C i.p. vibrations; ring stretch;
I.6	6.62	1511	these values jointly indicate 1:2:4
I.7	6.87	1456	trisub. benzene compound
	and 7.00	1429	
I.8	7.32	1366	O—H def.; phenolic
	and 7.90	1266	
I.9	8.12	1232	C—O str.; phenolic
I.10	8.30	1202	C—H i.p. def. Again these
I.11	8.72	1147	peaks together indicate that
I.12	8.92	1121	the unknown is a 1:2:4 trisub.
I.13	9.65	1036	benzene derivative
I.14	10.09	991	C—H o.o.p. def.; monosub. alkenes
I.15	10.90	917	—CH_2 o.o.p. def.; monosub. alkenes
I.16	11.75	851	2 adj. H wagging in aromatic ring
	and 12.25	816	

NMR data. There appear to be five or possibly six groups of peaks in the NMR spectrum.

Signal number	Signal position δ	τ	Integral step size	Rel. no. protons	Multiplicity	Inference (Appendix 4.1)
N.1	6.6	3.4	33	3	Multiplet	Ar—**H** benzenoid
N.2	5.8	4.2	22	2	Multiplet	—Ar—O**H**, —C**H**=C—CO—R —C=C**H** conj., —C=C**H**$_2$ —C=C**H**—CO—R, —C=C**H** acyclic, nonconj.
N.3	5.04	4.96	11	1	Multiplet	—C=C**H** acyclic nonconj. —C**H**(OR)$_2$
N.4	4.8	5.2	10	1	Multiplet	R—O**H**, —C=C**H**$_2$
N.5	3.6	6.4	33	3	Singlet	—O—C**H**$_3$
N.6	3.2	6.8	22	2	Doublet	ArC**H**$_2$—C=C—

The intensity of the peak numbered N.2 is reduced to nearly half its original intensity when D_2O is added to the sample. Since D_2O brings about an exchange in O—H bond to O—D bond, we conclude that the two-proton peak at 4.2 τ is a combined signal from the two protons, one of which is likely to be from Ar—O**H**.

Interpretation of data and tentative identification of unknown

1. The M + 1 peak is 11.1% and the M + 2 peak is 1.0% of the M peak. Of the four possible formulas listed above, the observed values fit best the formula M.3; that is, $C_{10}H_{12}O_2$. Let us therefore consider it to be the molecular formula of the compound. Furthermore, we can calculate the number of carbon atoms in the compound as

$$\text{No. of carbons} = \frac{\text{percent M + 1 peak}}{1.1} = \frac{11.1}{1.1} = 10.$$

2. The carbon-to-hydrogen ratio of the deduced formula indicates that the compound has an aromatic structure. This deduction is further supported by the following observations.

i) Aromatic compounds generally show large molecular-ion peaks in their mass spectra. In this unknown compound, the molecular-ion peak is the most intense peak (base peak).

ii) For the formula $C_{10}H_{12}O_2$, the total number of rings and double bonds (R) is calculated as

$$R = \tfrac{1}{2}(2 \times 10 - 12 + 2) = 5 \qquad \text{(Section 5.9).}$$

The presence of a benzene ring may account for four of these unsaturation sites. This suggests that the compound contains one double bond (or one saturated ring) in addition to an aromatic benzenoid ring.

iii) The pattern formed by peaks I.5, I.6, and I.7 is not only characteristic of benzenoid compounds, but also indicates a 1:2:4 trisubstituted benzene derivative (also see Fig. 3.20b). This deduction is given additional support by peaks I.10 through I.13 and I.16.

iv) A three-proton signal N.1 at 3.4 τ is a further proof of the presence of a trisubstituted benzene.

3. The presence of a phenolic OH group has already been suggested by the N.2 peak and by the behavior of the compound after the addition of D_2O. This deduction is confirmed by peak I.1. This enables us to assume that peak I.8 is due to phenolic O—H deformation and that peak I.9 is due to C—O stretching.

4. According to Appendix 3.1b, a peak at 1639 cm^{-1} could be due either to the C=O stretching of a carbonyl group or to the C=C stretching of an olefin. The peak is too weak to suggest the presence of a carbonyl group. We therefore turn to Appendix 3.2, and conclude that a monosubstituted alkene, R—CH=CH_2, is possibly responsible for peaks I.4, I.14, and I.15. Olefinic protons resonate in the region 3.6 to 5.4 τ (Appendix 4.1). In this region we see signals from four protons, one of which is already assigned to a phenolic O—H. The remaining three protons could thus be vinylic. This is in agreement with the conclusion from the infrared data. Thus we accept the possibility of a R—CH=CH_2 group in the compound, and assign both peaks N.3 and N.4 to —C=CH_2 protons. We decide that the —R—CH= group is probably responsible for one proton signal at 4.2 τ (peak N.2; part of this signal is already considered to be due to phenolic O—H). Signals N.3 and N.4 can now be considered to be the subpeaks of a two-proton olefinic doublet centered at 5.08 τ. These protons are coupled with the third olefinic proton, which appears at 4.2 τ.

5. Peak N.6 is a two-proton doublet, representing a methylene group placed between an aromatic ring and a monosubstituted alkene group; that is, an Ar—CH_2—CH=CH_2 system.

6. The three-proton singlet at 6.4 τ (signal N.5) is due to the methoxy group, O—CH_3. The position of this peak is best accounted for by putting the methoxy group on the aromatic ring as Ar—O—CH_3.

7. At this point all the available information allows us to write the partial structure of the unknown as

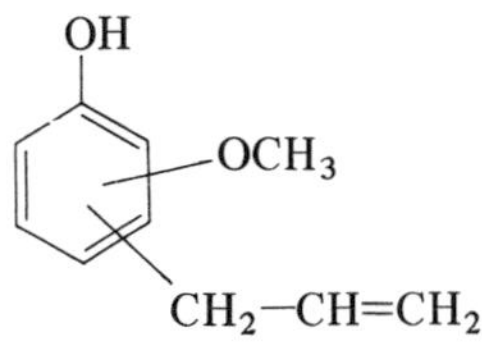

There are six possible arrangements of the functional groups to produce 1:2:4 trisubstituted benzene.

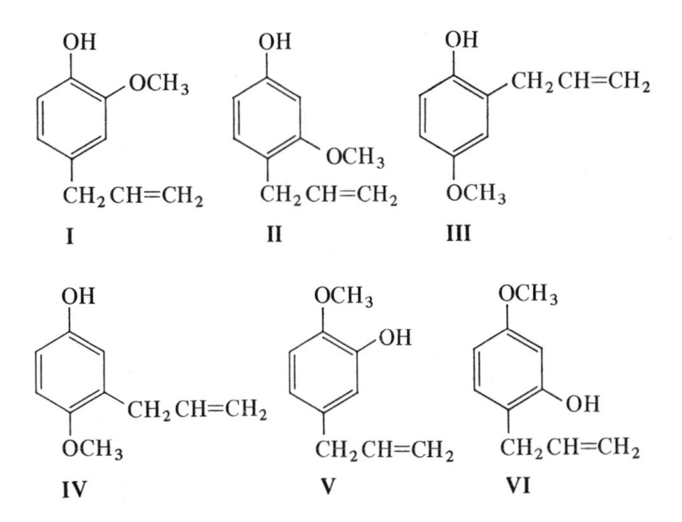

We can now establish the correct structure of the unknown by employing other information such as boiling point, etc., and by surveying the literature. Of the 37 compounds listed under the formula $C_{10}H_{12}O_2$ in the *Handbook of Chemistry and Physics* (edited by R. C. Weast, The Chemical Rubber Co., Cleveland, Ohio, 53rd edition, 1972-73), only eugenol has the right combination of physical properties and reactive groups. We therefore find the infrared spectrum of an authentic sample of eugenol and compare it with that of our unknown compound. A perfect match of the two spectra confirms that unknown number 6.3 is eugenol, with structure I.

OH

OCH_3

$CH_2CH{=}CH_2$

Eugenol

UNKNOWN NUMBER 6.4

Physical data: Colorless liquid, boiling point 197°

Mass spectrum

m/e	% of base peak	m/e	% of base peak	m/e	% of base peak	m/e	% of base peak
12	1	13	1	14	1.5	15	1
25	1.5	26	5	27	7	28	5
29	12	31	1.5	32	1	36	1
37	8	38	13.5	39	40	40	7.5
41	1	42	1	43	1	45	1
46	1.5	47	4	49	2	50	10
51	7	52	1	53	8	55	4
60	1	61	4.5	62	6	63	10
64	5	65	32	66	11.5	67	1.5
72	2	73	1	74	3	75	2
76	20	77	2.5	92	2	93	18.5
94	6.5	104	15.5	105	1.5	120	1
121	96	122	100 (M)	123	7.8	124	0.7

Isotope ratio

m/e	% of M
122 (M)	100
123 (M + 1)	7.8
124 (M + 2)	0.7

Ultraviolet data

λ_{max}^{EtOH}	ϵ_{max}
256 nm	12,600
324 nm	3,400

Infrared spectrum: Liquid sample, 0.01-mm layer; inset, capillary

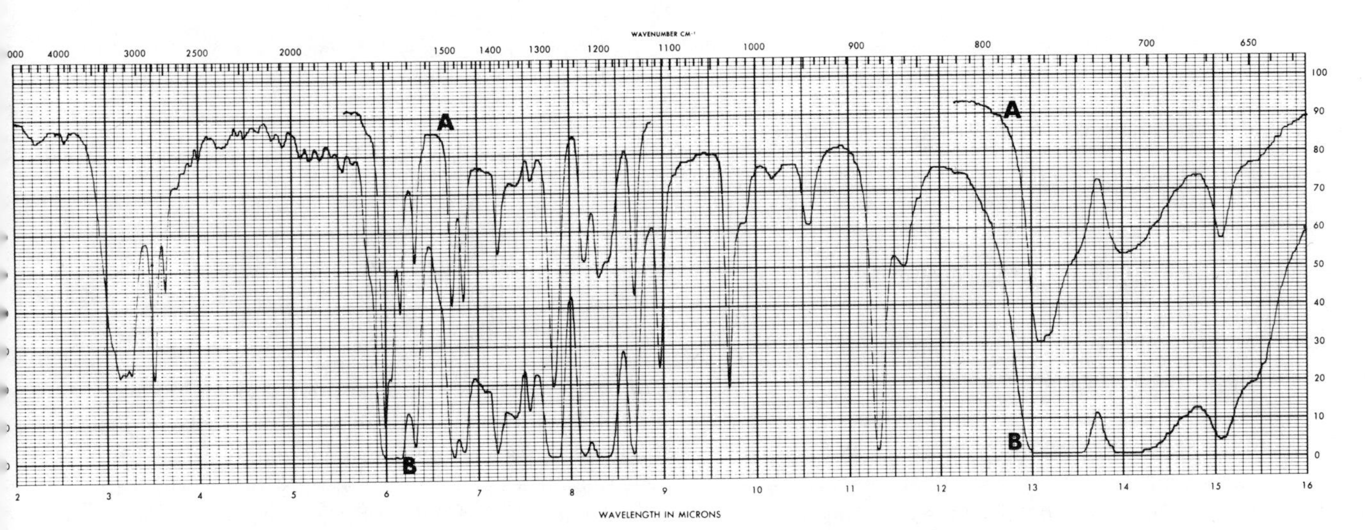

NMR spectrum: Solvent CCl_4, sweep time 250 sec, sweep width 500 Hz, sweep offset 300 Hz; traces marked with ★ are spectrum after D_2O has been added to the sample.

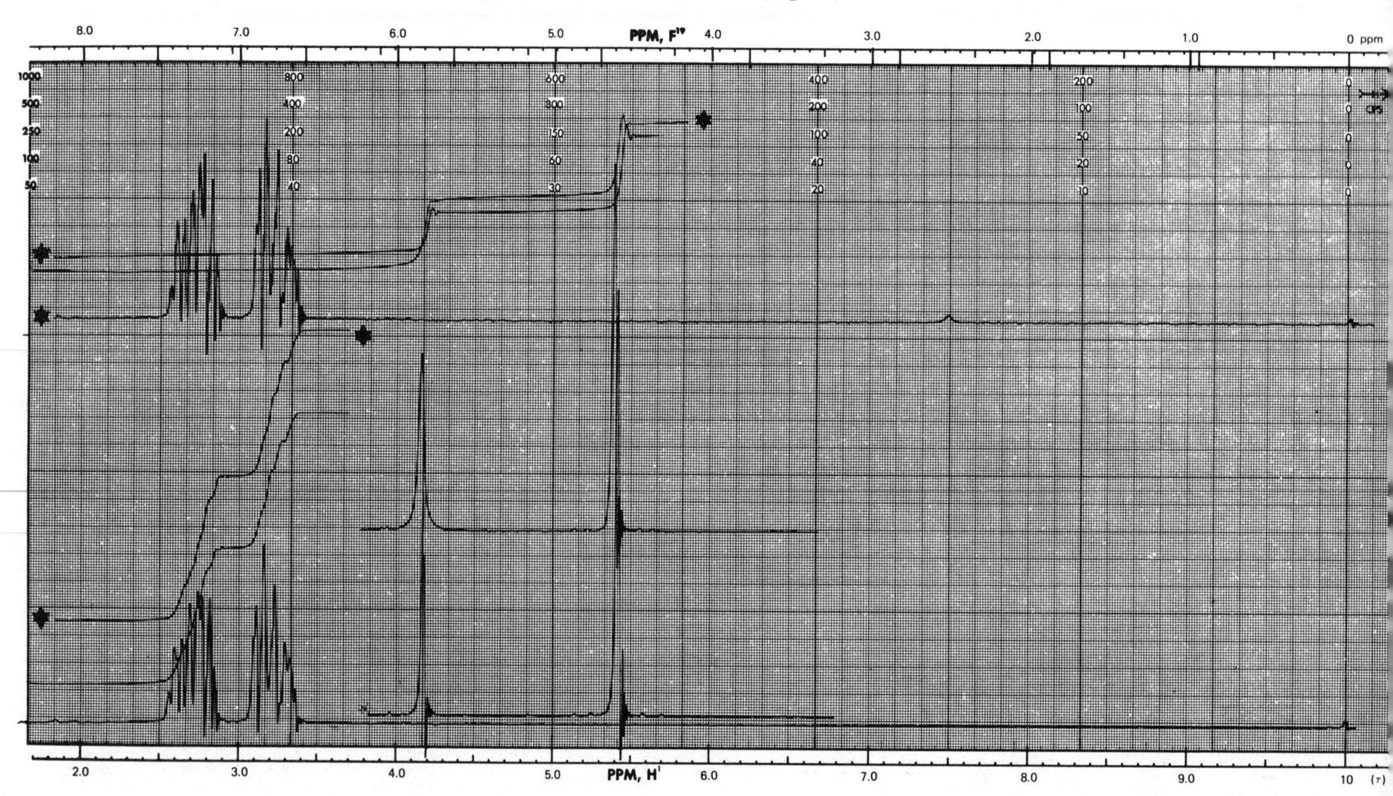

IDENTIFICATION OF UNKNOWN NUMBER 6.4

Summary of relevant data

Mass spectrum

i) The five most intense peaks are at masses 122, 121, 39, 65, 76.

ii) The molecular weight of the compound is 122; the M peak is also the base peak.

iii) The M + 1 peak is 7.8% and the M + 2 peak is 0.7% of the M peak. Therefore, from Appendix 5.1, the possible molecular formulas with M + 1 peak intensities between 6.8 and 8.5 are the following. (Formulas with an odd number of nitrogen atoms are omitted.)

Formula number	Formula	M + 1	M + 2
M.1	$C_5H_6N_4$	7.03	0.21
M.2	$C_6H_6N_2O$	7.38	0.44
M.3	$C_7H_6O_2$	7.74	0.66
M.4	$C_7H_{10}N_2$	8.49	0.32

Ultraviolet data. A strong absorption peak at 256 nm and a weak absorption peak at 324 nm strongly suggest that this unknown is a benzenoid compound. If so, the benzene ring carries a substituent or substituents which are responsible for a red shift of the primary band from 203.5 nm to 256 nm.

Infrared data. The absorption pattern in the region between 6 and 7 μ is typical of aromatic compounds. Therefore we consult Appendix 3.3 in preparing the following summary of infrared peaks.

Peak number	Wavelength λ, μ	Wavenumber $\bar{\nu}$, cm^{-1}	Inference (Appendix 3.1b and 3.3)
I.1	3.15	3175	O—H str.; hydrogen bonded
I.2	3.29	3040	=C—H str. in aromatics
I.3	3.53, 3.65 (Doublet)	2833, 2740	—C—H str.
I.4	6.00	1667	i) —C=O str.; cross conj. dienones or intramolecular hydrogen-bonded aldehydes ii) C=C str.; in alkenes conj. with —C=C— or —C=O
I.5	6.19	1616	—C=C— str.; pattern typical of *ortho*-substituted benzenoid compounds (I.5–I.8)
I.6	6.33	1580	
I.7	6.74	1484	
I.8	6.87	1456	
I.9	7.23	1383	—O—H i.p. def.; phenols
I.10	7.84	1276	—C—H i.p. def.; substituted aromatic
I.11	8.13	1230	—C—O str.
I.12	8.30	1205	—C—H i.p. def.; together these values indicate 1:2 substitution in a benzenoid compound (I.12–I.15)
I.13	8.69	1151	
I.14	8.98	1115	
I.15	9.72	1029	
I.16	11.32	883	—C—H o.o.p. def.; substituted aromatic (883, 760, 702)
	13.10	760	
I.17	14.25	702	

NMR data. There appear to be three or possibly four groups of peaks in the NMR spectrum. We run the spectrum twice, the second time after we have added D_2O. Exchangeable protons such as —**OH**, —**NH**, or activated —**CH** can often be detected by adding a drop of D_2O to the sample, since this causes peaks to either disappear or to be reduced in intensity. Note that the singlet peaks are recorded after a sweep offset of 300 Hz; that is, an offset of 5.00 τ units.

Signal number	Signal position δ	τ	Integral step size	Rel. no. protons	Multiplicity	Inference (Appendix 4.1)
N.1	10.83	−0.83	23(13)*	1	Singlet	Ar—**OH** intramolecularly bonded
N.2	9.59	0.41	23(32)	1	Singlet	—C=C—**C**HO, R—**C**HO, or Ar—**C**HO
N.3	7.31	2.69	50(54)	2	Octet	Ar—**H**
N.4	6.79	3.21	50(54)	2	Sextet	Ar—**H**

* Values in parentheses indicate the integral step size of the signal after D_2O was added.

We have considered peaks N.3 and N.4 to be two separate signals, but, since both represent aromatic protons Ar—**H**, it is possible that there is a single peak centered at 2.95 τ representing four aromatic protons. This peak, which appears as deceptively simple groups of an octet and a sextet, may be a complex 14-subpeak signal, representing a system such as AB_2X, AB_2C, AM_2X, etc.

Interpretation of data and tentative identification of unknown

1. Of the four possible formulas listed for the compound, M.3 (that is, $C_7H_6O_2$) appears to have M + 1 and M + 2 intensity peaks nearest to those observed for the unknown. Let us therefore consider it to be the probable molecular formula of the compound. Furthermore, the number of carbon atoms in the unknown is calculated as

$$\text{No. of carbons} = \frac{\text{percent M + 1 peak}}{1.1} = \frac{7.8}{1.1} = 7.$$

2. The carbon-to-hydrogen ratio of the postulated formula indicates that the compound has an aromatic structure. This deduction is further supported by the following observations.

i) Aromatic compounds usually have very intense molecular-ion peaks. For the unknown we are studying, the molecular-ion peak is also the base peak.

ii) For a formula $C_7H_6O_2$, the total number of rings and double bonds (R) is calculated as

$$R = \tfrac{1}{2}(2 \times 7 - 6 + 2) = 5 \qquad \text{(Section 5.9).}$$

The presence of a benzene ring may account for four of these unsaturation sites. This suggests that the compound contains one double bond (or one saturated ring) in addition to an aromatic benzenoid ring. This additional unsaturation could be of the —C=C— or —C=O type. Strong infrared absorption peak at 1667 cm^{-1} suggests the presence of a carbonyl group.

iii) The ultraviolet absorption spectrum suggests a carbonyl group with an aromatic ring.

iv) The infrared absorption pattern responsible for peaks I.5 through I.8 and I.12 through I.15 is typical of substituted aromatic compounds. Comparison of the peak pattern between 5.0 and 6.0 μ with that in Fig. 3.20b suggests 1:2 disubstitution on the ring.

v) Signals N.3 and N.4 are due to aromatic protons.

3. The presence of a phenolic O—H group has already been suggested because of the N.1 peak. The resonance of the phenolic proton at the extremely low field of −0.83 and the reduction in intensity of this peak when D_2O is added are strong indications of intramolecular bonding. This deduction is confirmed by peak I.1. Thus we can surmise that peak I.9 is due to phenolic O—H deformation and peak I.11 to —C—O stretching.

4. Signal N.2 is due to an aldehydic proton. This enables us to postulate that peak I.4 is due to —C=O stretching of an intramolecularly hydrogen-bonded aldehyde. The low frequency at which this peak appears suggests that it is an aromatic aldehyde. (Confirm by consulting Appendix 3.5.)

5. Let us pull all the above conclusions together. We are now in a position to write the structure for the unknown as

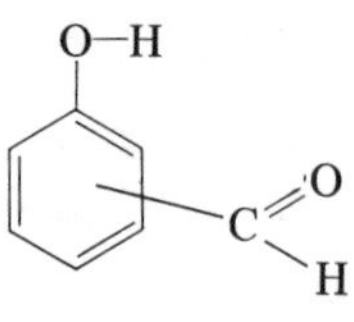

However, the intramolecular bonding of the O—H proton, as suggested by its large chemical shift and I.4 peak, is possible only if the two substituents on the benzene ring are *ortho* to each other. Thus unknown 6.4 is salicylaldehyde, in which the phenolic hydrogen is bonded to carbonyl oxygen.

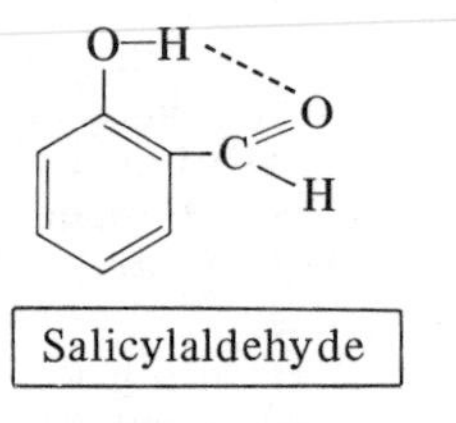

Salicylaldehyde

Cross checking

1. The two substituents are not likely to be *para* substituted. *Para*-disubstituted benzene derivatives usually form A_2B_2 splitting systems, thus displaying a pattern characterized by four subpeaks symmetrically placed with a weak–strong–strong–weak intensity relation (see unknown 6.2). The aromatic protons in the present compound exhibit a 14-subpeak complex pattern.

2. The mass spectrum of this compound displays an intense (96% of the base peak) M − 1 peak. This is common in aldehydes, in which a hydrogen atom is easily lost through an α cleavage, as:

$$\text{Ar—}\underset{\text{H}}{\underset{|}{\text{C}}}\text{=}\overset{+}{\dot{\text{O}}} \longrightarrow \text{Ar—C}\equiv\overset{+}{\text{O}} + \dot{\text{H}}$$

M − 1

3. α cleavage in an aldehyde can also produce peaks at $m/e = 29$ and at M − 29, as:

$$\text{I} \quad \text{Ar—}\underset{\text{H}}{\underset{|}{\text{C}}}\text{=}\overset{+}{\dot{\text{O}}} \xrightarrow{\text{homolysis}} \dot{\text{A}}\text{r} + \text{H—C}\equiv\overset{+}{\text{O}}$$

$m/e = 29$

$$\text{II} \quad \text{Ar—}\underset{\text{H}}{\underset{|}{\text{C}}}\text{=}\overset{+}{\dot{\text{O}}} \xrightarrow{\text{heterolysis}} \overset{+}{\text{A}}\text{r} + \text{H—C}\equiv\dot{\text{O}}$$

M − 29

In the unknown these two peaks are fairly intense.

4. The molecular ions of aldehydes tend to eliminate stable molecules of water and ethylene (Rule 8), and exhibit absorption peaks at M – 18 and M – 28. The peaks at M – 28 and M – 29 can also arise from the loss of CO and —COH. The peaks at m/e = 104 (M – 18), m/e = 94 (M – 28), and m/e = 93 (M – 29) are fairly intense in the mass spectrum of the unknown. In the case of salicylaldehyde, the peaks at M – 28 and M – 29 are due to the loss of CO and CHO, as losses of C_2H_4 and C_2H_5 would yield unusually unsaturated structures.

5. According to Table 2.9, the salicylaldehyde (that is, a disubstituted benzene with —OH and —CHO substituents *ortho* to each other) should display a primary absorption band at 203.5 + 53 = 256.5 nm. Unknown 6.4 displays a strong UV absorption band at 256 nm.

6. Salicylaldehyde is a colorless liquid which boils at 197°.

UNKNOWN NUMBER 6.5

Physical data: Colorless liquid, boiling point 156°

Mass spectrum

m/e	% of base peak	m/e	% of base peak	m/e	% of base peak	m/e	% of base peak
26	1.5	27	18.5	28	3.5	29	18
39	11.5	40	2	41	38	42	14
43	100	44	2.5	51	1	53	2
54	1	55	34.5	56	17	57	22
58	1	69	6	85	49	86	3
93	1	95	1	99	1	107	5
109	5	135	50	136	2	137	49.5
138	2	164	2.3 (M)	166	2.2		

Isotope ratio

m/e	% of M peak
164 (M)	100
166 (M + 2)	95.7

Ultraviolet data

Transparent above 210 nm

Infrared spectrum: Liquid sample, 0.01-mm-thick layer; inset, capillary.

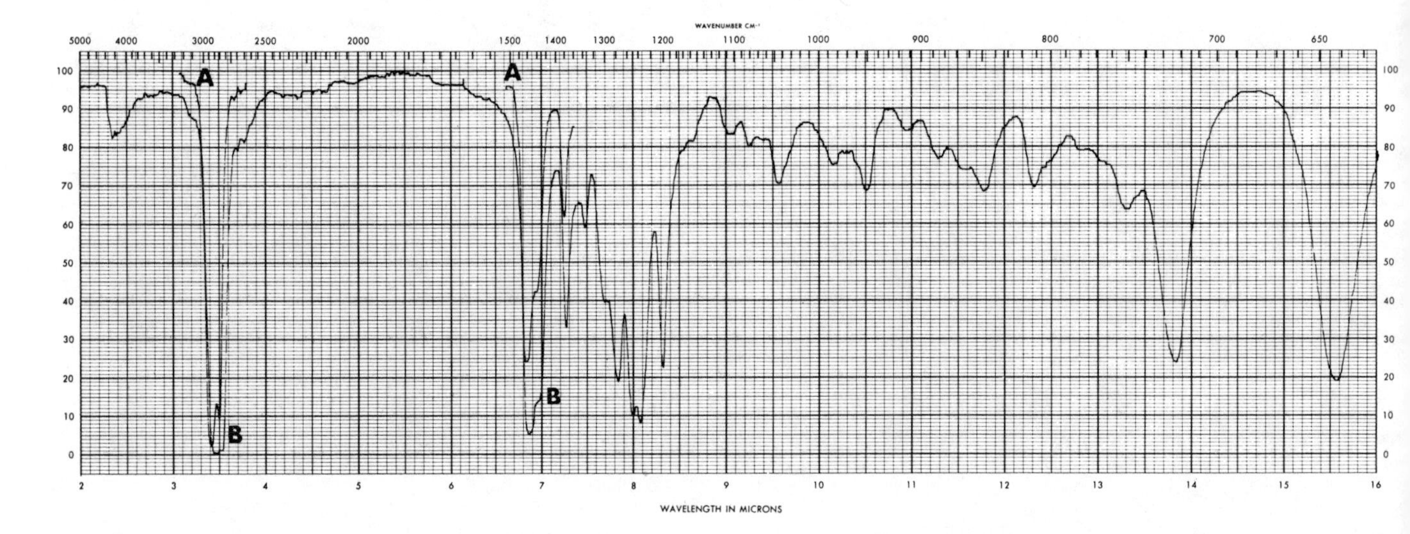

NMR spectrum: Solvent CCl_4, sweep time 250 sec, sweep width 500 Hz

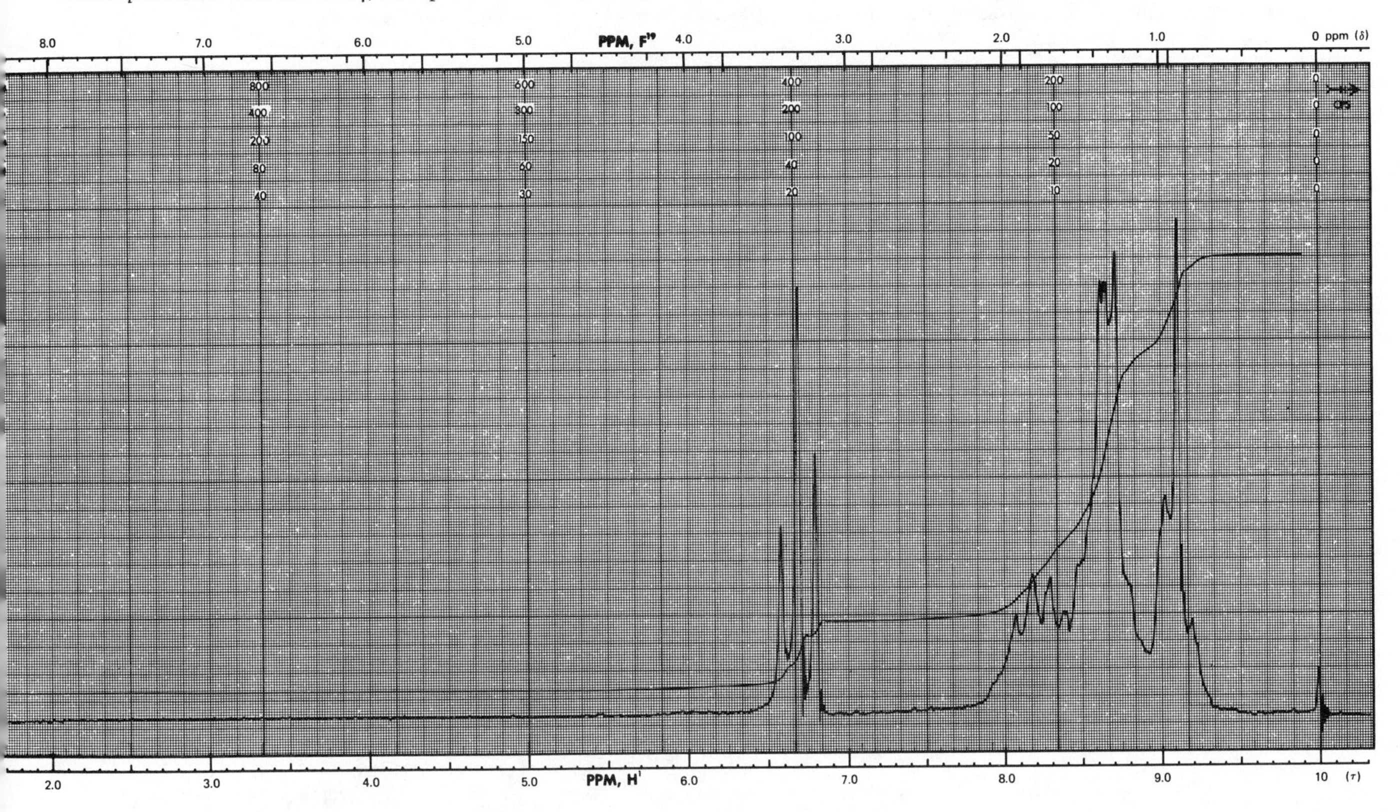

IDENTIFICATION OF UNKNOWN NUMBER 6.5

Summary of relevant data

Mass spectrum

i) The five most intense peaks are at masses 43, 135, 137, 85, 41.

ii) The molecular weight of the compound is 164. The molecular-ion peak is not very intense; therefore the compound is not likely to be an aromatic structure.

iii) The M + 1 peak intensity is not given, but the M + 2 peak is 95.7% of the M peak. The series of peaks with mass numbers 164 and 166 is therefore possibly an isotopic cluster containing halogen atoms. According to Table 5.4, the compound contains one atom of bromine.

Ultraviolet data. The fact that the compound is transparent above 210 nm indicates the absence of carbonyl function, aromatic structure, conjugation of halogen with C=C, etc.

Infrared data. The infrared spectrum of the unknown is a simple one. Only C—H stretching and deformation peaks are present. There are no hydroxyl, carbonyl, aromatic, or alkene absorptions. Some conspicuous infrared peaks are the following.

Peak number	Wavelength λ, μ	Wavenumber $\bar{\nu}$, cm^{-1}	Inference (Appendix 3.1b and Section 3.7)
I.1	3.42	2924	—C—H sym. and asym. str. in a number of methylene groups
I.2	3.50	2857	C—H sym. str. in $-CH_3$
I.3	6.87	1456	Overlap of scissoring absorption
I.4	7.85	1274	of methylene and asym. methyl def.
I.5	8.10	1235	Methylene twisting and wagging
I.6	13.82	724	

NMR data. There are three main groups of absorption peak signals in the spectrum. They are as follows.

Signal number	Signal position δ	τ	Integral step size	Rel. no. protons	Multiplicity	Inference (Appendix 4.1)
N.1	3.32	6.68	24	2	Triplet	$Ar-CH_2C{=}C$, $-CH_2Br$
N.2	1.63	8.37	96	8	Multiplet	—CH (sat.), $-CH_2CBr$, $CH_3C{=}C$, $-CH_2C-Ar$
N.3	0.89	9.11	35	3	Distorted triplet	CH_3CN, CH_3C (sat.)

Interpretation of data and tentative identification of unknown

1. UV, IR, and mass spectral data indicate that:

a) No aromatic structure is present in the compound.

b) Nitrogen is absent, but the compound contains one atom of bromine.

c) C=C is absent.

Therefore we can surmise that signal N.1 is traceable to two protons of $-CH_2Br$, signal N.2 to —CH (saturated) or $-CH_2CBr$, and signal N.3 to a terminal methyl group, CH_3C- (saturated).

2. Signal N.1 is a two-proton peak; therefore the bromine atom must be located on the terminal carbon atom.

3. Signal N.1 is a triplet; therefore there must be two protons on the carbon atom β to bromine. Thus a partial formula for the structure, deduced on the basis of signal N.1, is $-CH_2-CH_2-Br$.

4. Signal N.2 appears to be somewhat complex. However, on careful examination, we can resolve the signal into:

a) a two-proton symmetrical quartet centered at 8.13 τ,

b) a six-proton tall multiplet centered at 8.66 τ.

The quartet at 8.13 τ must result from the two protons β to bromine. This confirms the above partial structure $-CH_2-CH_2-Br$.

5. The remaining four carbons and nine hydrogens could be arranged only in three different formations as:

$CH_3-CH_2-CH_2-CH_2-$ $(CH_3)_2{>}CH-CH_2-$ $CH_3-C(CH_3)_2-$

I *n*-butyl   **II** *iso*-butyl   **III** *tert*-butyl

The *tertiary* arrangement (III) would require a sharp nine-proton peak at high field, and thus can be rejected. The *iso*-butyl arrangement (II), on the other hand, would require the high-field (N.3) signal to be due to six protons and the low-field (N.2b) signal to be due to three protons. The observed peak intensities can be explained on the basis of the *n*-butyl arrangement (I) only.

6. In mass spectrometry, halides usually display a strong peak at M − X according to the following heterocyclic cleavage of the electronegative halogen atom.

$$R\overset{+\bullet}{-X} \longrightarrow \overset{+}{R} + \overset{\bullet}{X}$$

The mass spectrum of the unknown shows a strong peak at $m/e = 85$, which is due to $\overset{+}{R} = M - Br\ (164 - 79 = 85)$. The mass of $\overset{+}{R}$ and the peaks at 29, 43, 57 (C_nH_{2n+1} or $C_nH_{2n+1}CO$) suggest that the R portion of the molecule is alkyl to six carbons or perhaps carbonyl to five carbons (Table 5.6).

7. To find out whether the R is an alkyl or a carbonyl, we look for bromine-containing ions such as $-CH_2\overset{+}{Br}$, $-CH_2CH_2\overset{+}{Br}$, $-CH_2CH_2CH_2\overset{+}{Br}$, etc. Peaks corresponding to these ions indicate that R is an alkyl group. The ion series with mass numbers 93-95 ($-CH_2\overset{+}{Br}$), 107-109 ($-CH_2CH_2\overset{+}{Br}$), and 135-137 $(CH_2)_4\overset{+}{Br}$ suggests the presence of an alkyl chain in the unknown. The molecular formula of the unknown is thus probably $C_6H_{13}Br$. The lack of a carbonyl peak at about 1720 cm^{-1} in the infrared spectrum supports this hypothesis.

8. Bromides containing *n*-alkyl chains of six or more carbon atoms usually produce a large concentration of five-membered cyclic bromonium ions. The spectrum of the unknown shows intense peaks at mass numbers 135 and 137, which can be assigned to the cyclic bromonium ions. This suggests that the unknown is *n-bromohexane,* and the structure of the unknown is as follows.

$CH_3-CH_2-CH_2-CH_2-CH_2-CH_2-Br$

n-hexyl bromide or 1-bromohexane

Cross checking

1. The boiling point of 1-bromohexane is 156°.

2. The fragmentation pattern of 1-bromohexane can be diagrammed as:

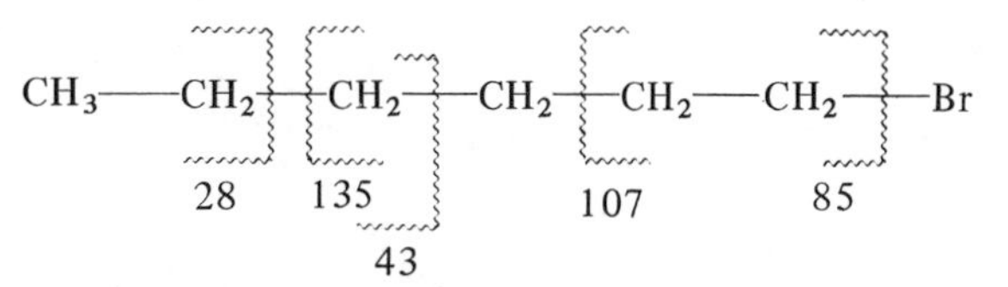

Now we come to the end of our spoon-feeding. From now on, the book will provide only the basic data on the unknown, reproductions of its spectra, and a description of it. You must henceforward tabulate the data, summarize them, and do your own detective work. So from now on, you're on your own. Good luck to you!

UNKNOWN NUMBER 6.6

Physical data: Colorless liquid, boiling point 102°

Mass spectrum

m/e	% of base peak	m/e	% of base peak	m/e	% of base peak	m/e	% of base peak
18	2R	24.5	0.1*	25	1	25.5	0.1**
26	9	27	37	28	11	29	99.5
31	1R	31.5	0.1*	38	1	39	3
41	2.5	42	4	43	4R	44	1R
45	0.5R	53	1	55	2	56	3.5
57	100	58	3.5	74	0.1	86	17.3 (M)
87	1	88	0.1				

* = Metastable peak ** = Doubly charged ion R = Rearrangement peak

Isotope ratio

m/e	% of M
86 (M)	100
87 (M + 1)	5.8
88 (M + 2)	0.5

Ultraviolet data

λ_{max}^{EtOH}	ϵ_{max}
273	20

Infrared spectrum: Liquid sample, thin layer

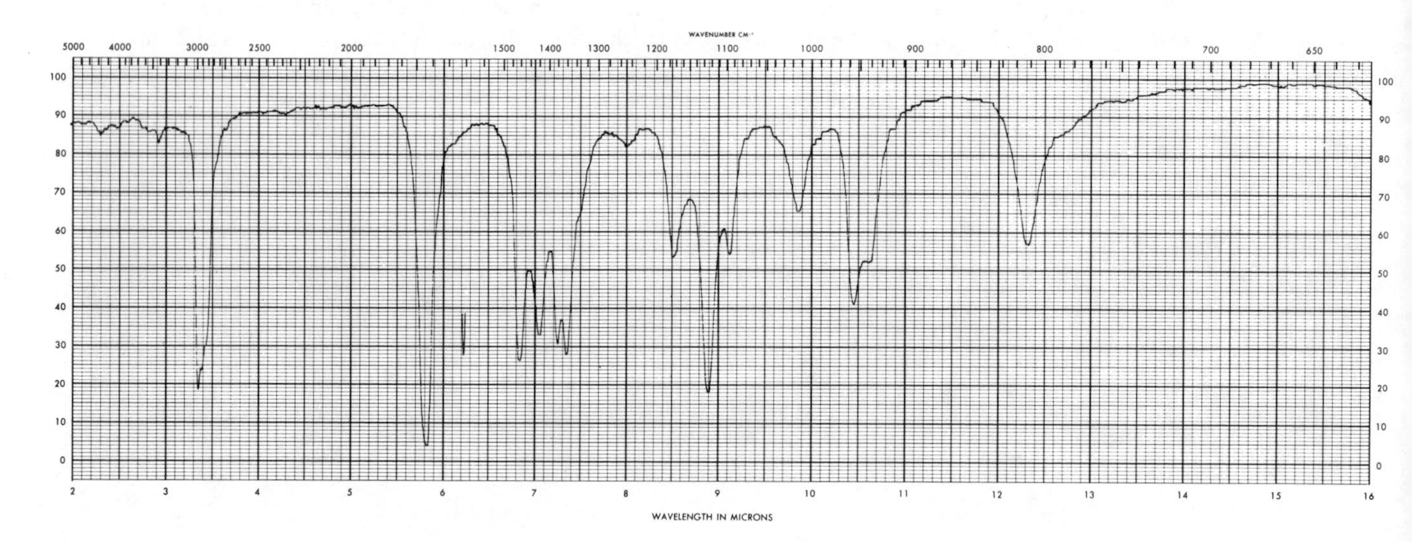

NMR spectrum: Neat, sweep time 250 sec, sweep width 500 Hz

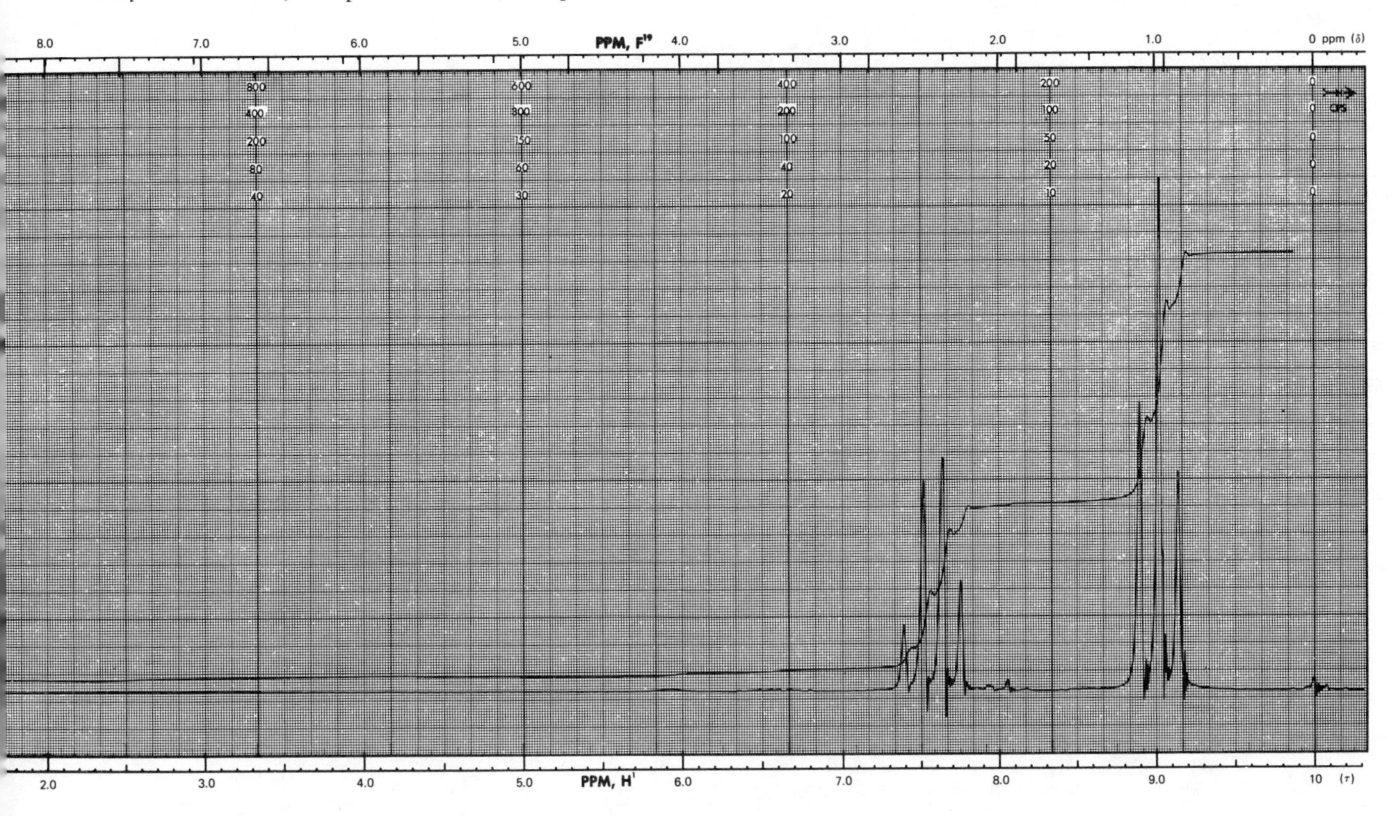

UNKNOWN NUMBER 6.7

Physical data: Low-melting-point crystals, melting point 28.5°, boiling point 184°

Mass spectrum

m/e	% of base peak	m/e	% of base peak	m/e	% of base peak	m/e	% of base peak
27	1	28	2	37	1.5	38	3
39	11	40	1	41	1.5	42.5	0.1
43	2.5	43.5	0.5	44	3	44.5	0.6
45	7	45.5	1	49	1	50	6.5
51	5	52	1	61	2	62	5
63	12	64	3	65	19	66	1
74	2	75	2	76	1	85	4
85.5	0.5	86	4	86.5	0.3	87	1
89	11.9	90	9	91	100	92	8
117	1	119	1	168	0.2	169	11
170	78.6 (M)	171	16.8	172	77.1 (M + 2)	173	6

Isotope ratio

m/e	% of M
170 (M)	100
171 (M + 1)	21.4
172 (M + 2)	98.1

Ultraviolet data

λ_{max}^{hexane}

210 nm

Infrared spectrum: Liquid sample, 0.01-mm layer

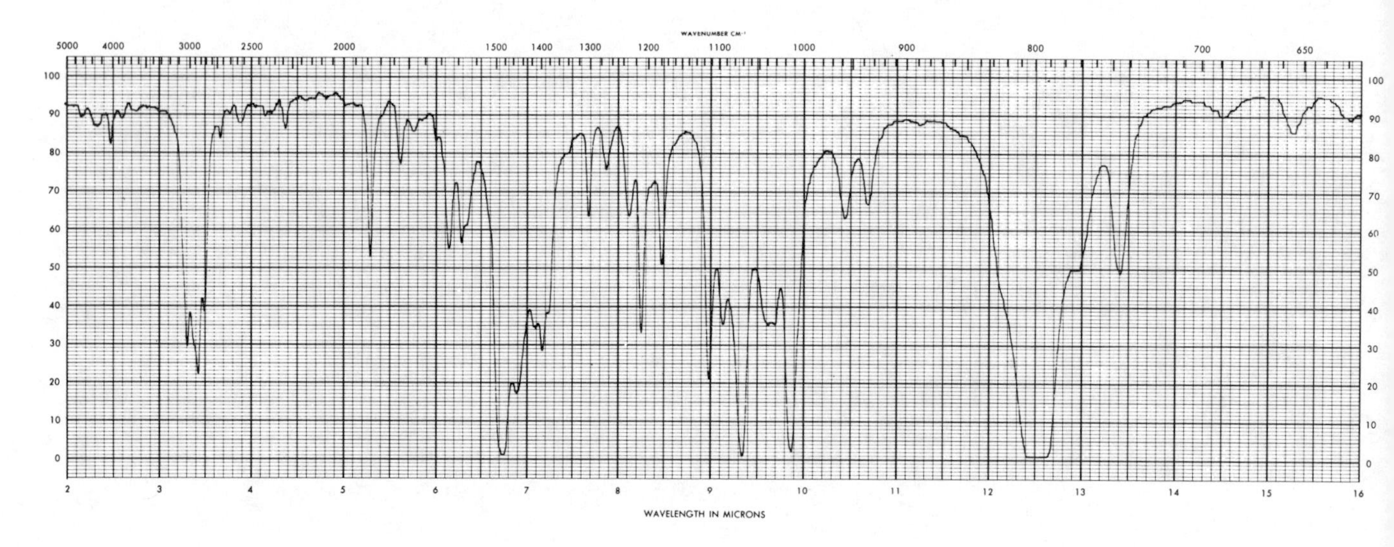

NMR spectrum: Solvent $CDCl_3$, sweep time 250 sec, sweep width 500 Hz

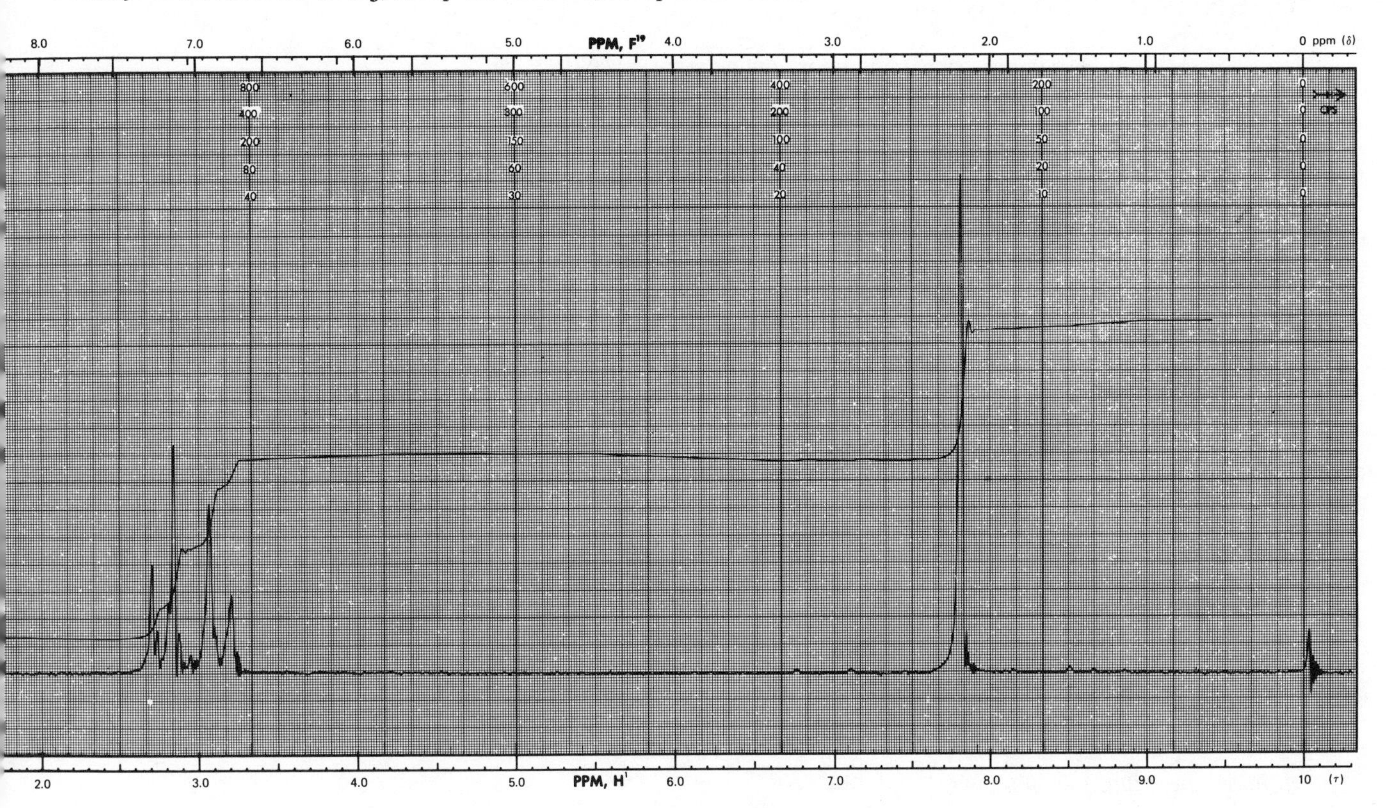

UNKNOWN NUMBER 6.8

Physical data: Colorless liquid, peppermint odor, boiling point 204–205°

Mass spectrum

m/e	% of base peak	m/e	% of base peak	m/e	% of base peak	m/e	% of base peak
38	11	39	35.5	40	3	41	6
43	2	45	1.5	49	3	50	15.5
51	18.5	52	4	53	2.5	59	2
60	2	61	6	62	12	63	24
64	6	65	31	66	2.5	73	1
74	4	75	2.5	76	1.5	77	3
85	1.5	86	2.5	87	2	89	11
90	6	91	100	92	11.5	105	2.5
107	1	118	1	119	99.8	120	83.2 (M)
121	7.4	122	0.7	123	0.1		

Isotope ratio

m/e	% of M
120 (M)	100
121 (M + 1)	8.9
122 (M + 2)	0.8

Ultraviolet data

λ_{max}^{hexane}	ϵ_{max}
251	15,000
258	12,500

Infrared spectrum: Liquid sample, thin film

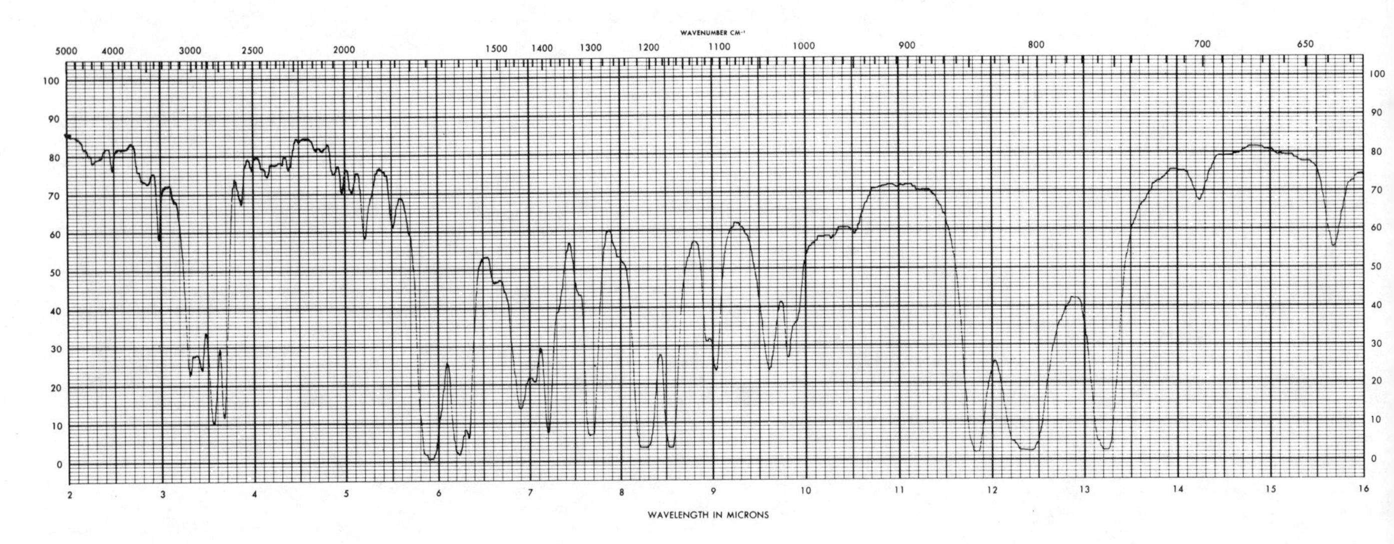

NMR spectrum: Neat, sweep time 250 sec, sweep width 500 Hz, sweep offset 200 Hz

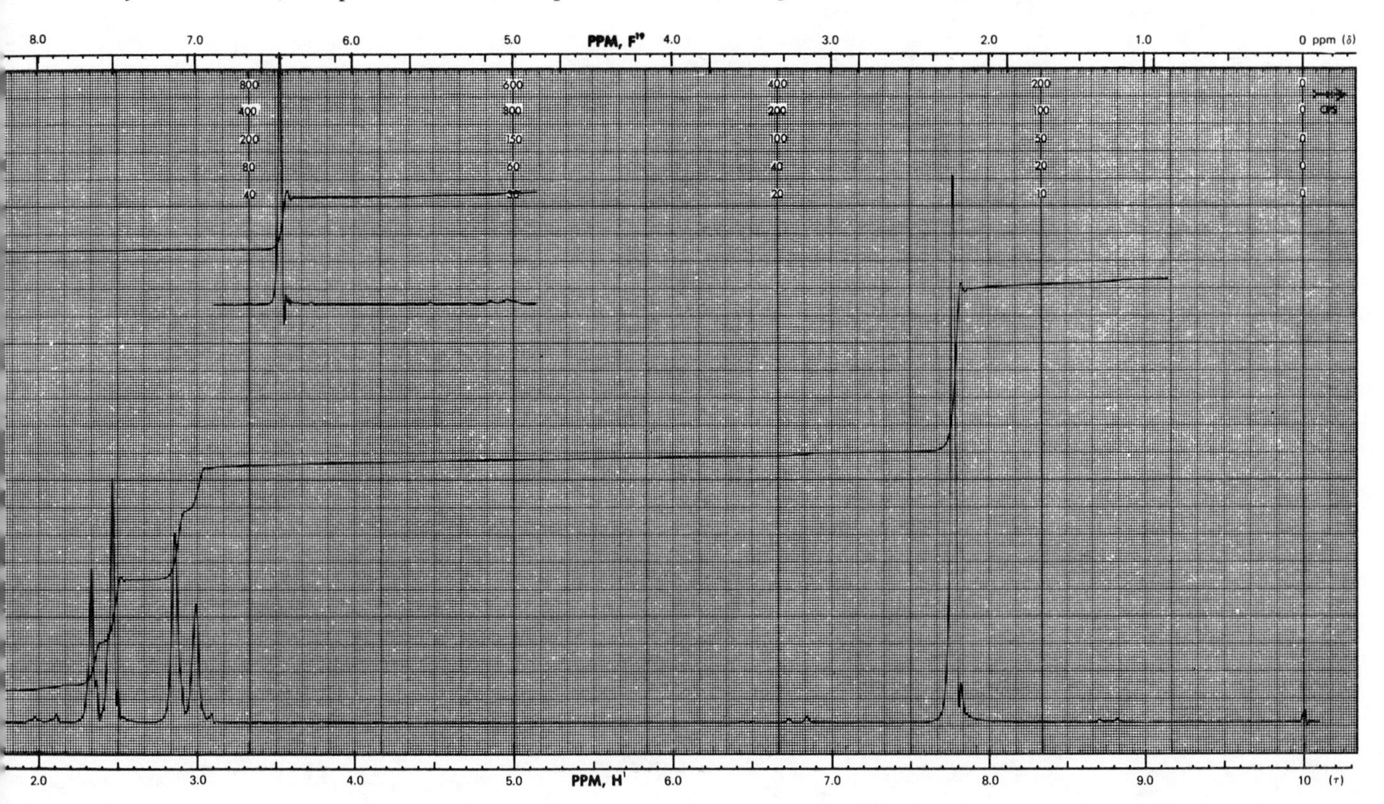

UNKNOWN NUMBER 6.9

Physical data: Colorless liquid, boiling point 210-212°

Mass spectrum

m/e	% of base peak	*m/e*	% of base peak	*m/e*	% of base peak	*m/e*	% of base peak
14	1	15	1	26	3	27	5.5
37	4.5	37.5	0.1**	38	7	39	18
40	2	41	1^{R}	42.5	0.1**	43	1**
43.5	0.4**	44	0.5**	44.5	0.5**	45	0.2**
45.5	1**	46	0.1**	49	1.5	50	7
51	7.5	52	3	56.5	0.3**	57.5	1.5**
58	1**	58.5	3**	59	0.3**	61	3
62	6	63	12	64	6	65	3
69.5	1.5*	74	1.5	75	3.5	76	3
77	1.5	86	1	87	2	88	3.5
89	19	90	29	91	4.5	114	1
115	3	116	54.5	117	100 (M)	118	9
119	0.4						

* = Metastable peak ** = Doubly charged ion *R* = Rearrangement peak

Isotope ratio

m/e	% of M
117 (M)	100
118 (M + 1)	9
119 (M + 2)	0.4

Ultraviolet data

λ_{max}^{EtOH}	ϵ_{max}
229.5	11,000
276	1,280

Infrared spectrum: Liquid sample, thin film

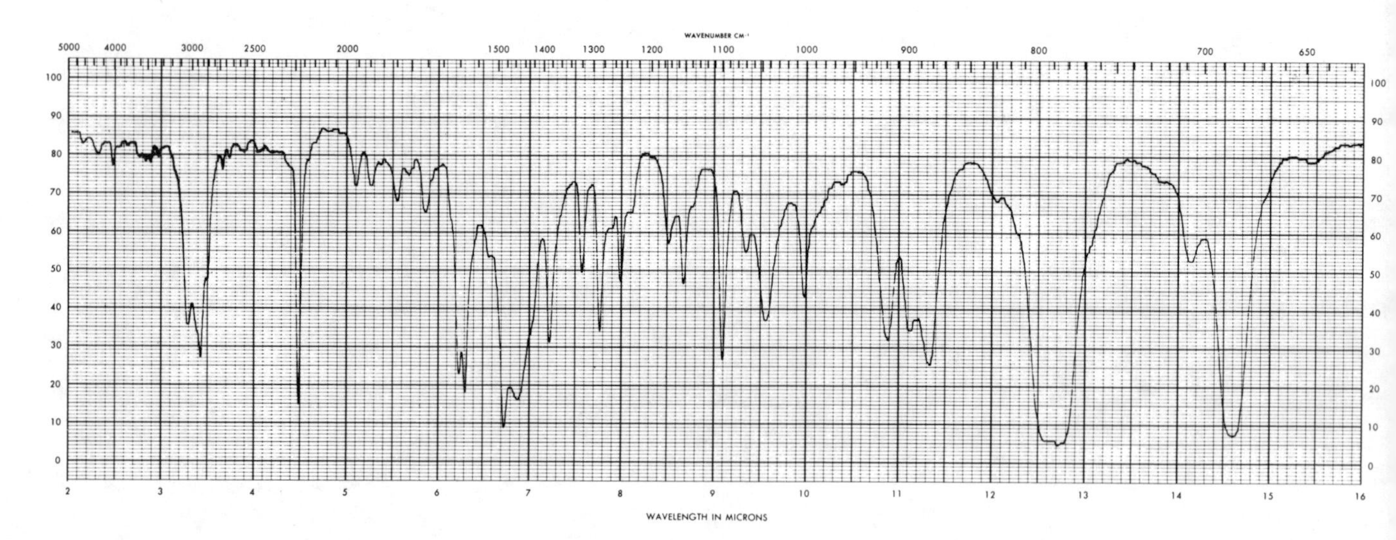

NMR spectrum: Solvent CCl_4, sweep time 250 sec, sweep width 500 Hz

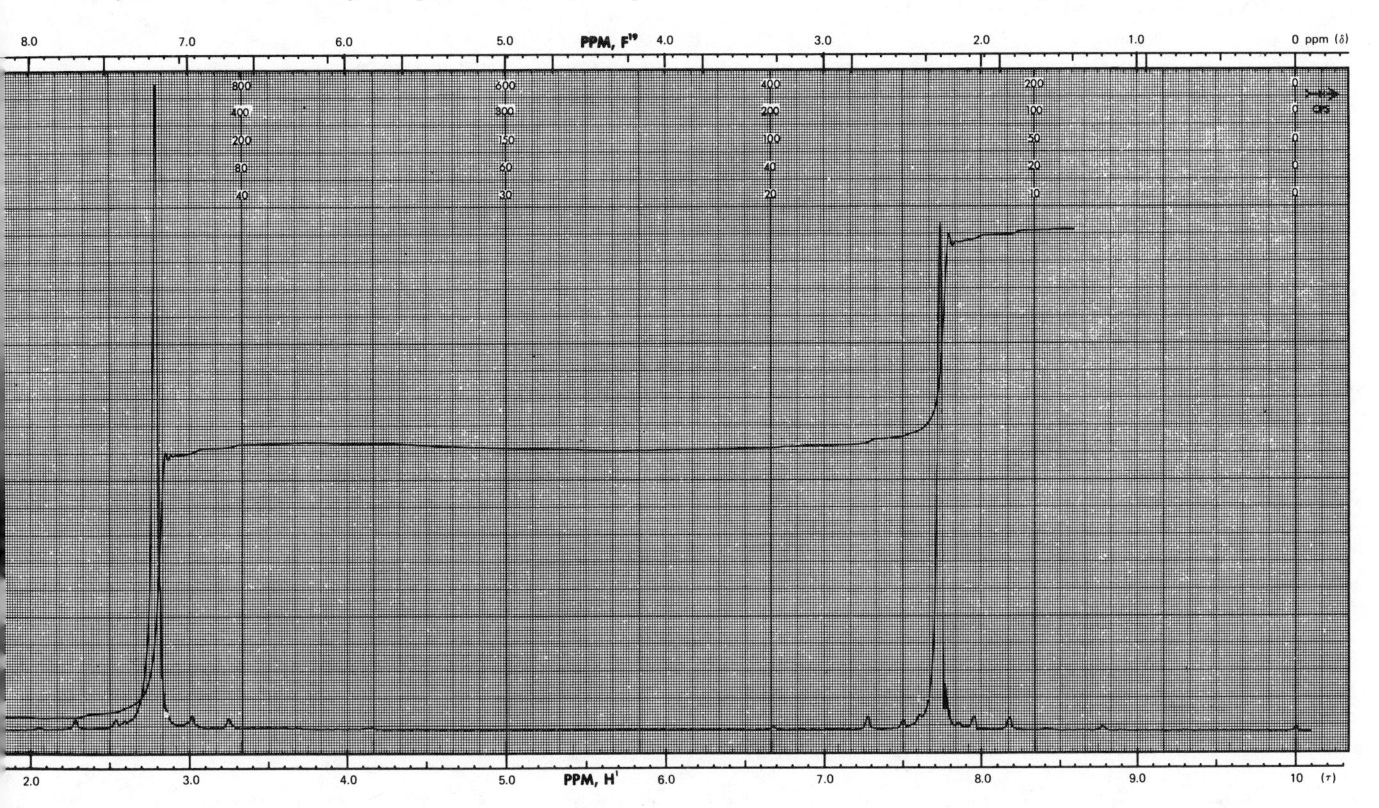

UNKNOWN NUMBER 6.10

Physical data: Colorless water-soluble liquid, smells like ammonia, boiling point 55.5°

Mass spectrum

m/e	% of base peak	m/e	% of base peak	m/e	% of base peak	m/e	% of base peak
13	1	14	4	15	21	17	1
18	13	26	7	27	29	28	37
29	25	30	98	31	1.5	39	1.5
40	3.5	41	6.5	42	14	43	5
44	29	45	2	56	4	57	2
58	100	59	4	72	14	73	21 (M)
74	1						

Isotope ratio

m/e	% of M
73 (M)	100
74 (M + 1)	4.8

Ultraviolet data

Transparent above 210 nm

Infrared spectrum: 6.0% w/w in $CHCl_3$

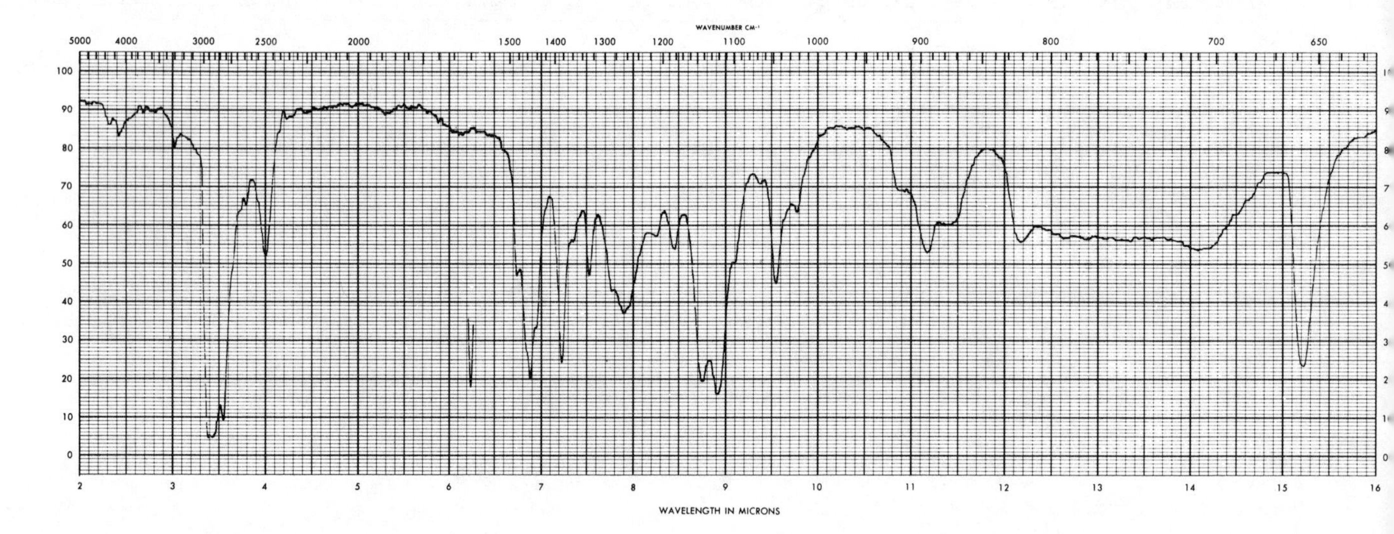

NMR spectrum: Solvent CCl_4, sweep time 250 sec, sweep width 500 Hz

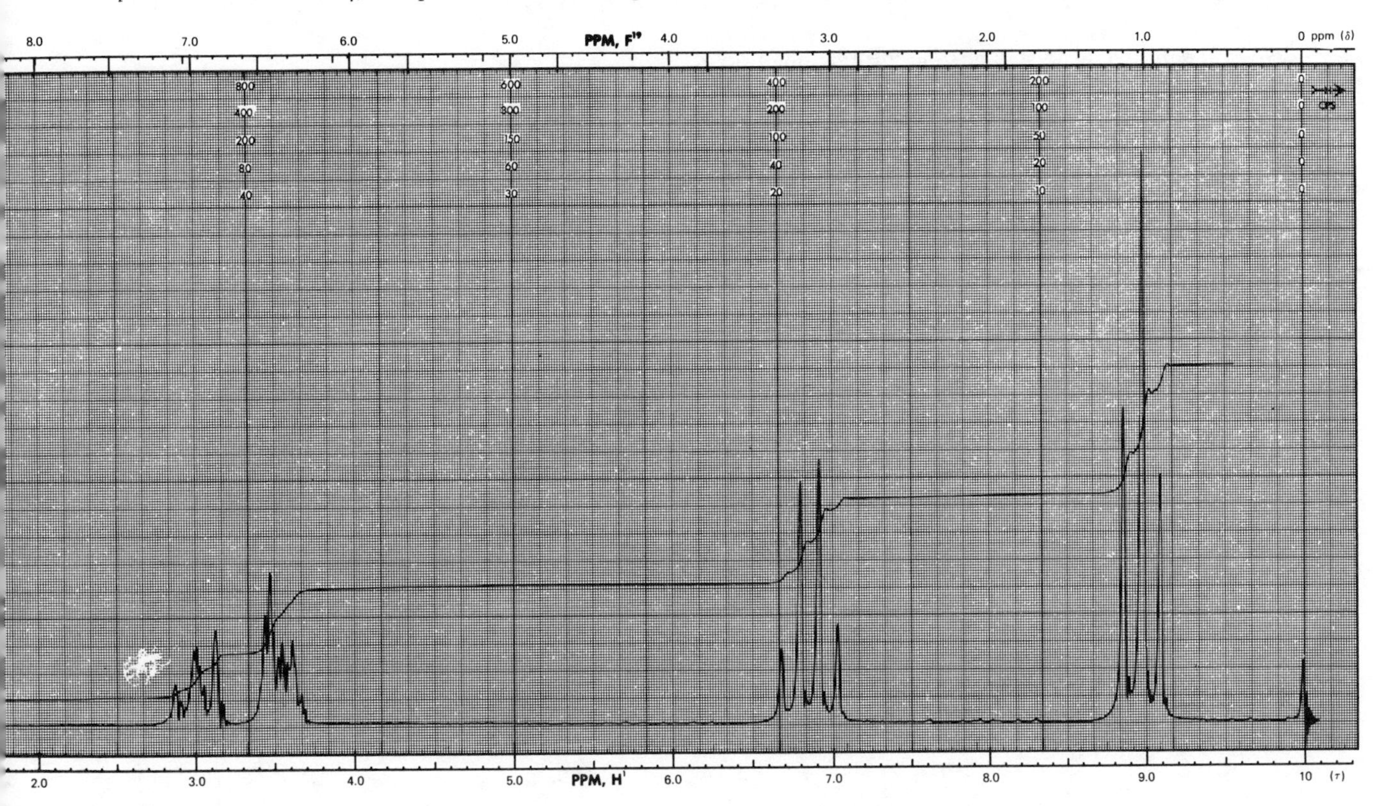

UNKNOWN NUMBER 6.11

Physical data: White crystalline solid, melting point 66.4°

Mass spectrum

m/e	% of base peak	m/e	% of base peak	m/e	% of base peak	m/e	% of base peak
28	2.5	29	1	31	1	31.5	0.3
32	3	37	2.5	38	4.5	39	11
46	3.5	46.5	0.2	50	3	51	1
53	3	61	2.5	62	4	63	7
64	4	65	31	66	2.5	71.5	0.1
73	1	74	2	75	1	86	5
86.5	0.3	87	4.5	87.5	0.3	92	1
93	19.5	94	1.5	117	2	119	2
143	2.5						
145	2.5	171	0.6	172	100 (M)	173	7.2
174	98.5	175	6.5	176	0.4		

Isotope ratio

m/e	% of M
172 (M)	100
173 (M + 1)	7.2
174 (M + 2)	98.5
176 (M + 4)	0.4

Ultraviolet data

$\lambda_{max}^{heptane}$	ϵ_{max}
224.4	9800

Infrared spectrum: KBr disc, 2-mg solid sample in 200 mg KBr

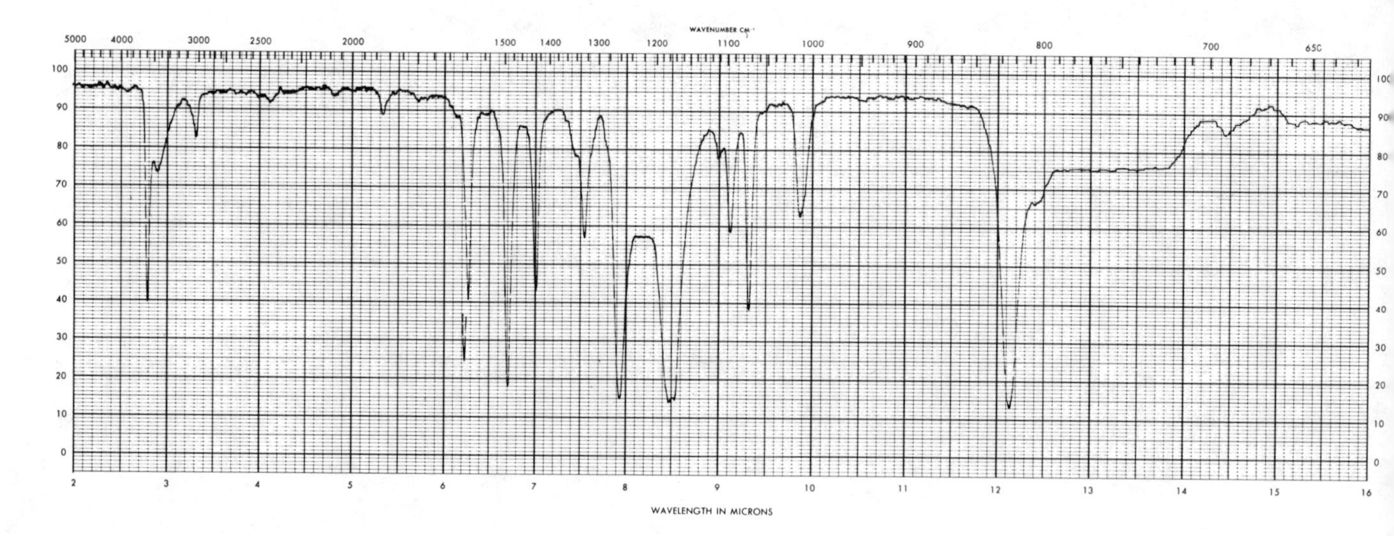

NMR spectrum: Solvent $CDCl_3$, sweep time 250 sec, sweep width 500 Hz; upper trace marked with ★ is spectrum of sample after D_2O has been added

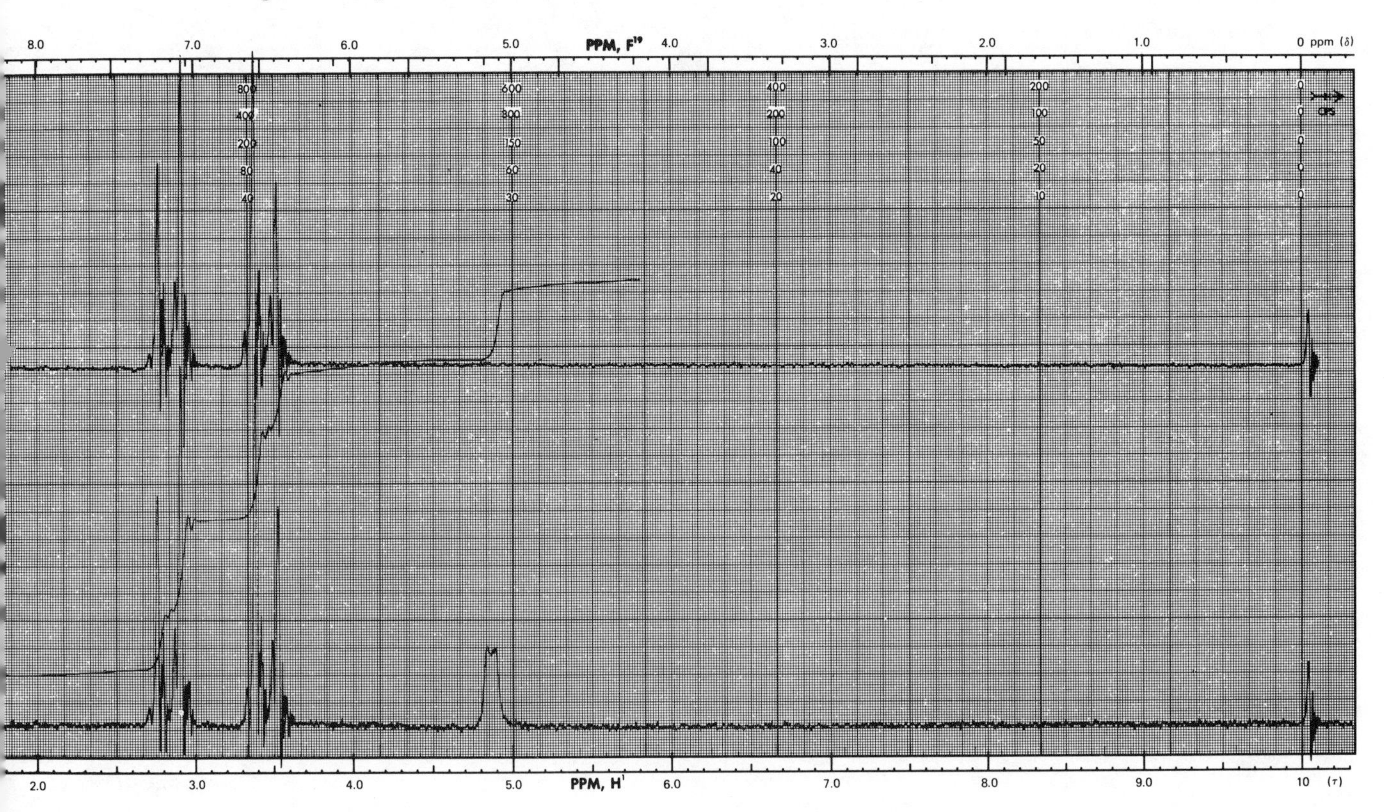

UNKNOWN NUMBER 6.12

Physical data: Colorless plates, melting point 53°, boiling point 243°

Mass spectrum

m/e	% of base peak	m/e	% of base peak	m/e	% of base peak	m/e	% of base peak
25	1	26	7.5	27	23.5	28	4.5
29	13.5	30	2	31	2	35	0.2*
37	6	38	15	39	36.5	40	6.5
41	12	42	4	43	4	45	3
49	2	50	11	51	16	52	13
53	32.5	54	3.5	55	11	61	3
62	8.5	63	15	64	7	65	19
66	22.5	67	6.5	68	4	69	11
71	1*	72	1*	73	1	74	2
75	1.5	76	1	77	10	78	8
79	6	80	5	81	39.5	82	3
90	1	91	1.5	92	2.5	93	7
94	60	95	35.5	96	3	105	2
106	1	107	7	108	5	109	8.5
110	2	121	2	122	1	123	8
124	100 (M)	125	7.8	126	0.7	136	7

Isotope ratio

m/e	% of M
124 (M)	100
125 (M + 1)	7.8
126 (M + 2)	0.7

Ultraviolet data

λ_{max}^{EtOH}	ϵ_{max}
217 nm	Strong

Infrared spectrum: 6.0% w/w in $CHCl_3$

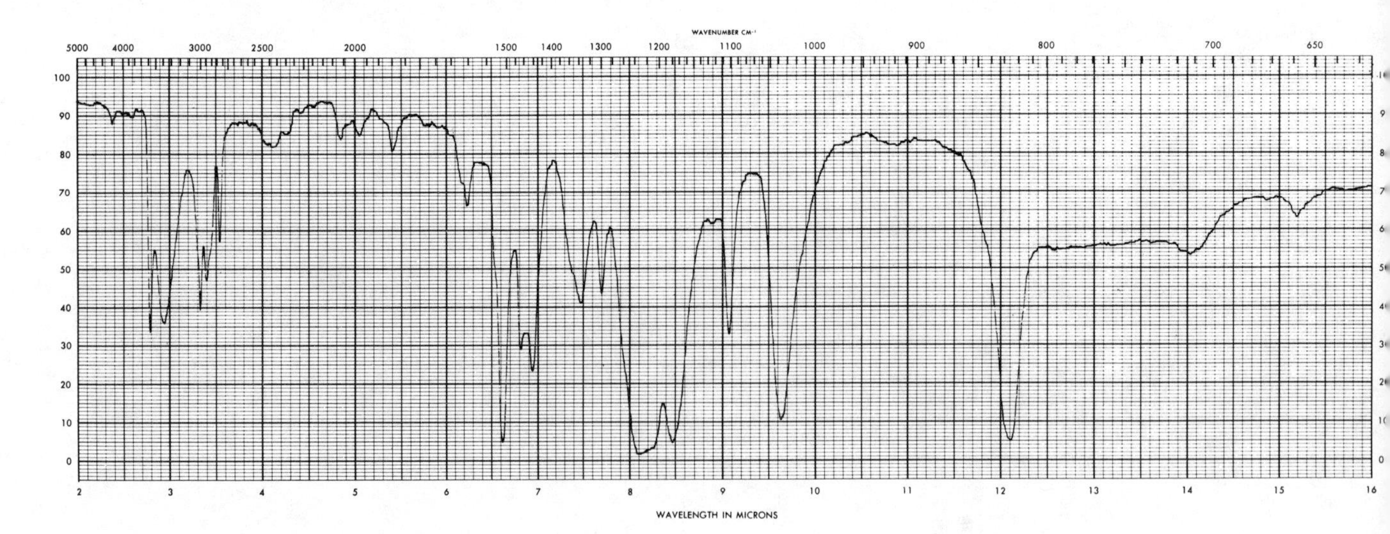

NMR spectrum: Solvent $CDCl_3$, sweep time 250 sec, sweep width 500 Hz; upper trace marked with ★ is spectrum of sample after D_2O has been added

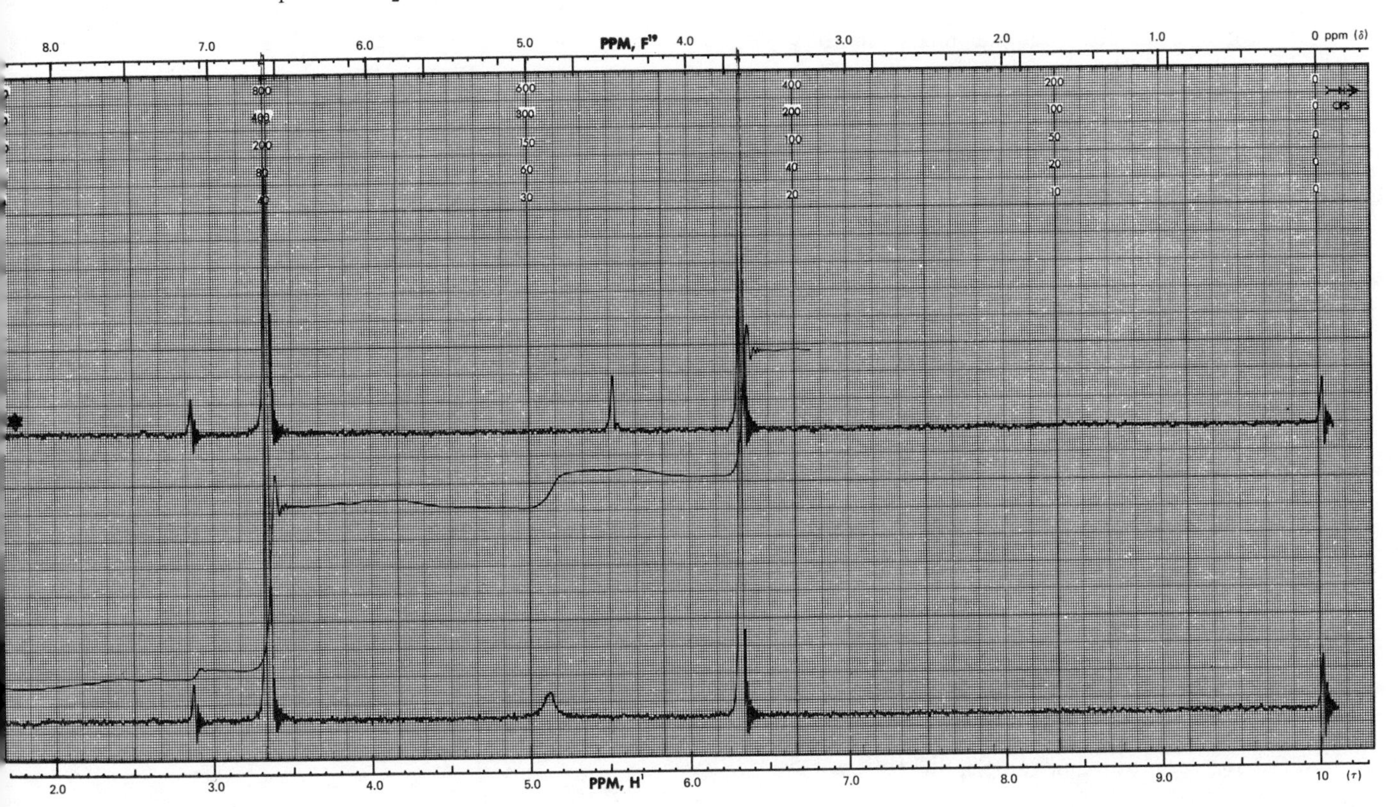

UNKNOWN NUMBER 6.13

Physical data: Colorless liquid, boiling point 205.2°

Mass spectrum

m/e	% of base peak	*m/e*	% of base peak	*m/e*	% of base peak	*m/e*	% of base peak
15	1	26	3	27	5.5	37	5
38	7	39	17	40	2	41	1*R*
43.5	0.3**	45.5	1**	50	7.5	51	8
52	3	55.5	0.2**	56.5	0.3**	57.5	1**
58.5	3**	61	3	62	6	63	13
64	6.5	65	3	69.5	1.5*	74	1.5
75	3	76	2.5	77	1	86	1
87	2	88	4	89	21	90	34
91	4.5	114	1	115	3	116	46.5
117	100 (M)	118	9	119	0.4		

* = Metastable peak ** = Doubly charged ion *R* = Rearrangement peak

Isotope ratio

m/e	% of M
117 (M)	100
118 (M + 1)	9
119 (M + 2)	0.4

Ultraviolet data

λ_{max}^{EtOH}	ϵ_{max}
228.5	11,100
276.5	1,440

Infrared spectrum: Liquid sample, thin layer

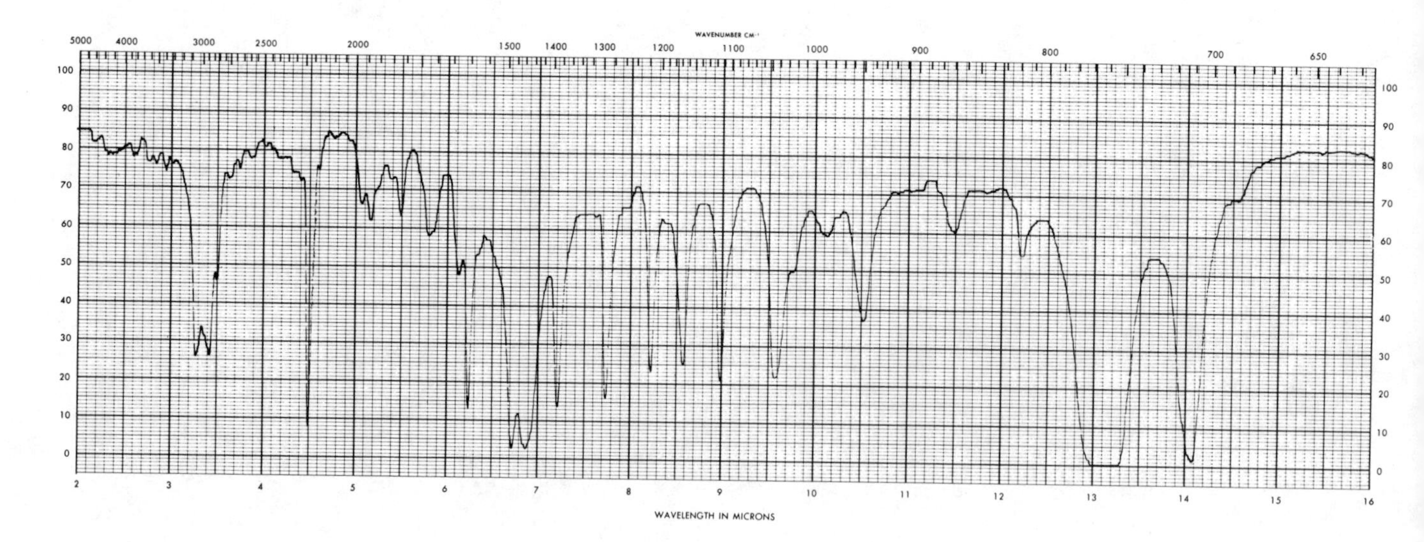

NMR spectrum: Solvent CCl_4, sweep time 250 sec, sweep width 500 Hz

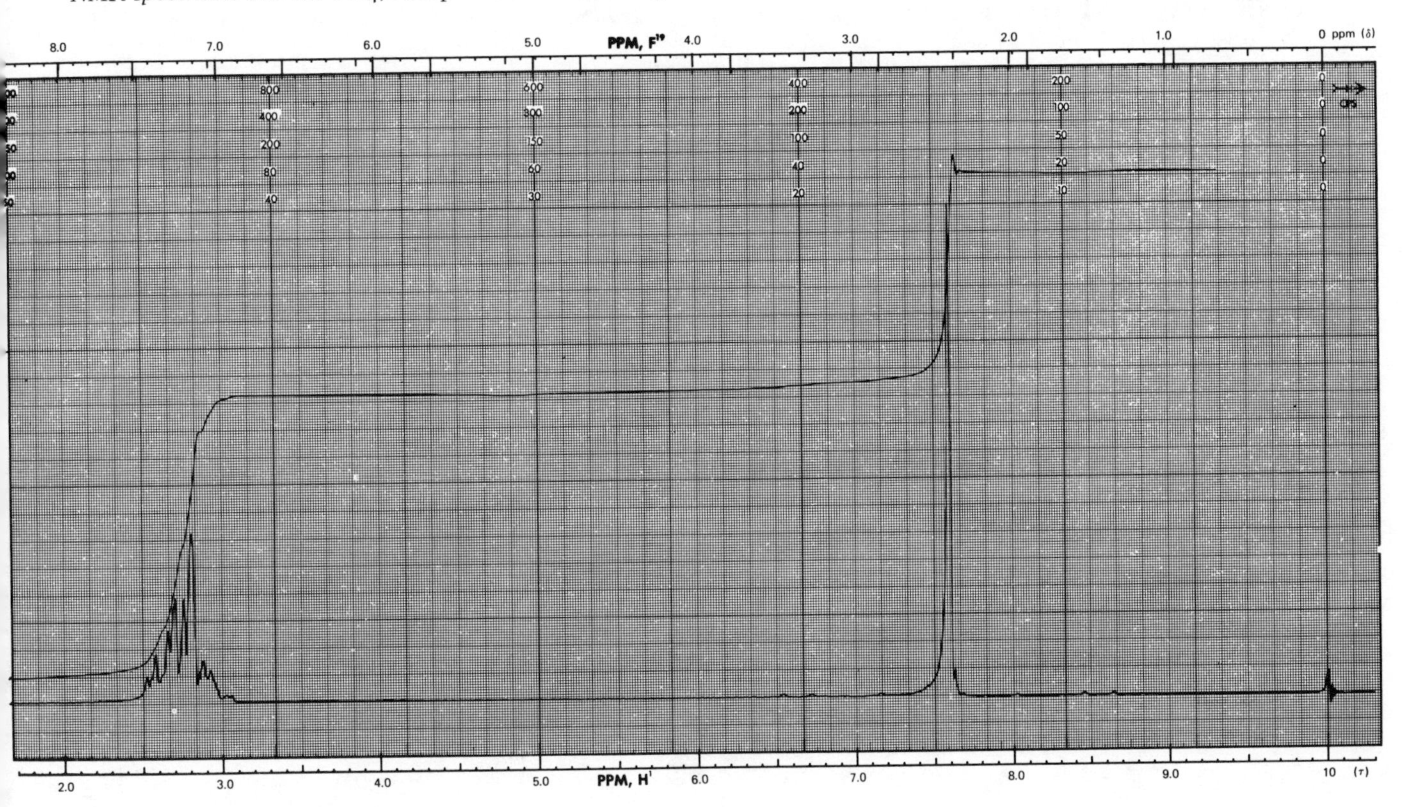

UNKNOWN NUMBER 6.14

Physical data: Colorless liquid, pungent odor, boiling point 105°

Mass spectrum

m/e	% of base peak	m/e	% of base peak	m/e	% of base peak	m/e	% of base peak
14	3	15	9	18	5.5	25	1.5
26	6	27	18	28	4.5	29	30.5
31	1.5	37	7	38	11	39	65
40	11.5	41	100	42	22	43	22
44	19	45	11	50	3	51	3
53	3	55	5.5	57	1.5	67	2
68	3	69	52	70	97.8 (M)	71	4.5

Isotope ratio

m/e	% of M
70 (M)	100
71 (M + 1)	4.4

Ultraviolet data

λ_{max}^{EtOH}	ϵ_{max}
218	18,000
320	30

Infrared spectrum: Liquid sample, 0.007-mm layer; inset, capillary

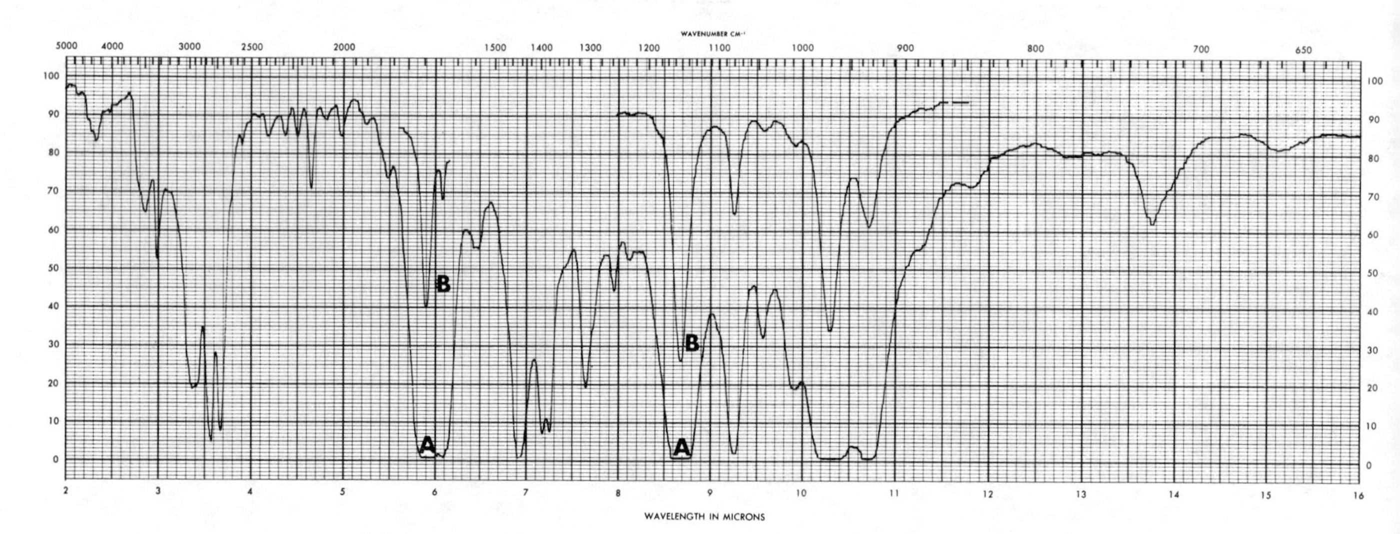

NMR spectrum: Solvent CCl_4, sweep time 250 sec, sweep width 55 Hz, sweep offset 100 Hz

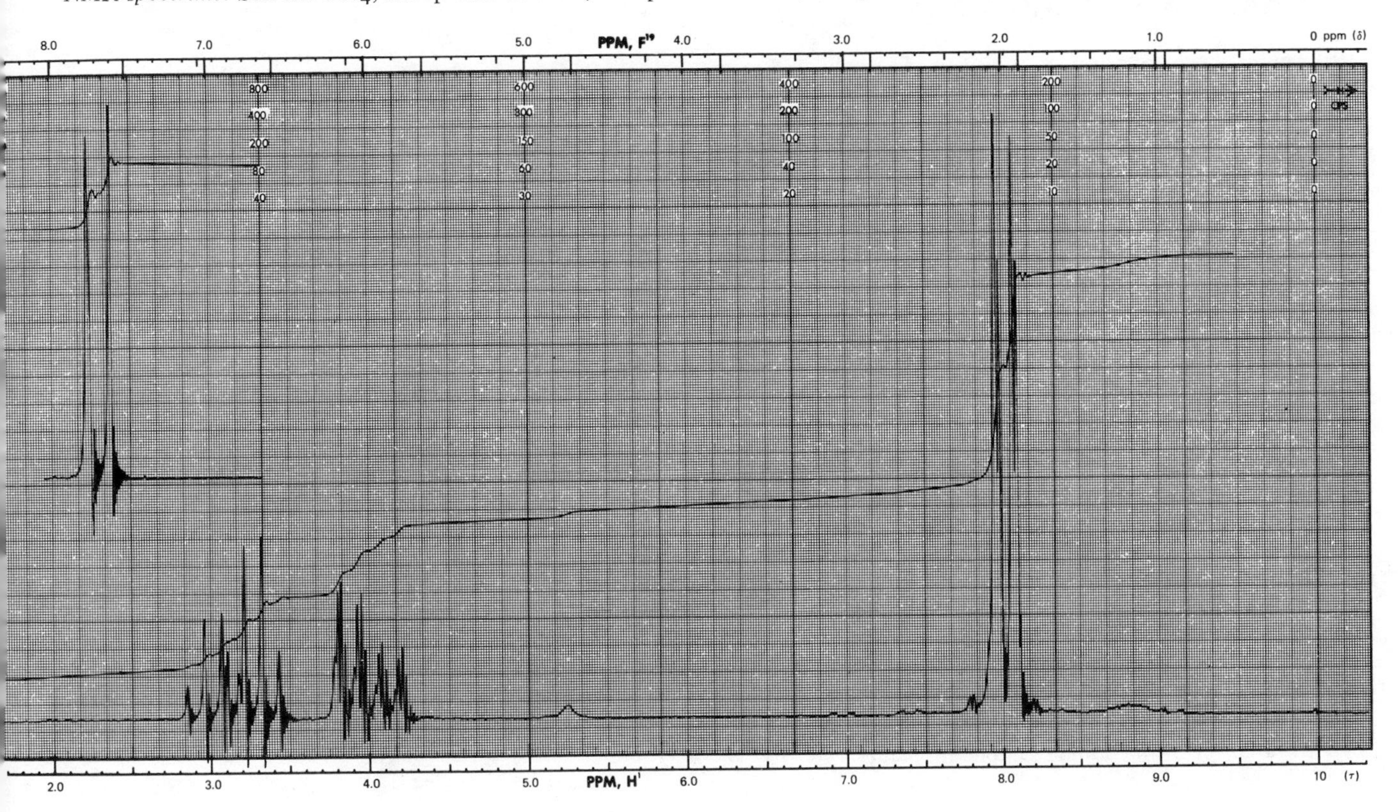

UNKNOWN NUMBER 6.15

Physical data: Colorless liquid, boiling point 181°

Mass spectrum

m/e	% of base peak	m/e	% of base peak	m/e	% of base peak	m/e	% of base peak
28	1	37	1	38	2.5	39	9
40	0.8	41	1	43	2	43.5	0.4
44	2	44.5	0.4	45	5	45.5	0.8
50	4	51	4	61	2	62	4.5
63	10.5	64	3	65	15	66	1
74	1	75	1	85	3	86	3
87	1	89	17.5	90	19	91	100
92	8	117	1	169	7.6	170	89.5 (M)
171	14	172	88.0	173	7	174	0.3

Isotope ratio

m/e	% of M
170 (M)	100
171 (M + 1)	15.8
172 (M + 2)	98.4
174 (M + 4)	0.4

Ultraviolet data

λ_{max}^{hexane}	ϵ_{max}
213 nm	strong

Infrared spectrum: Liquid sample, thin layer

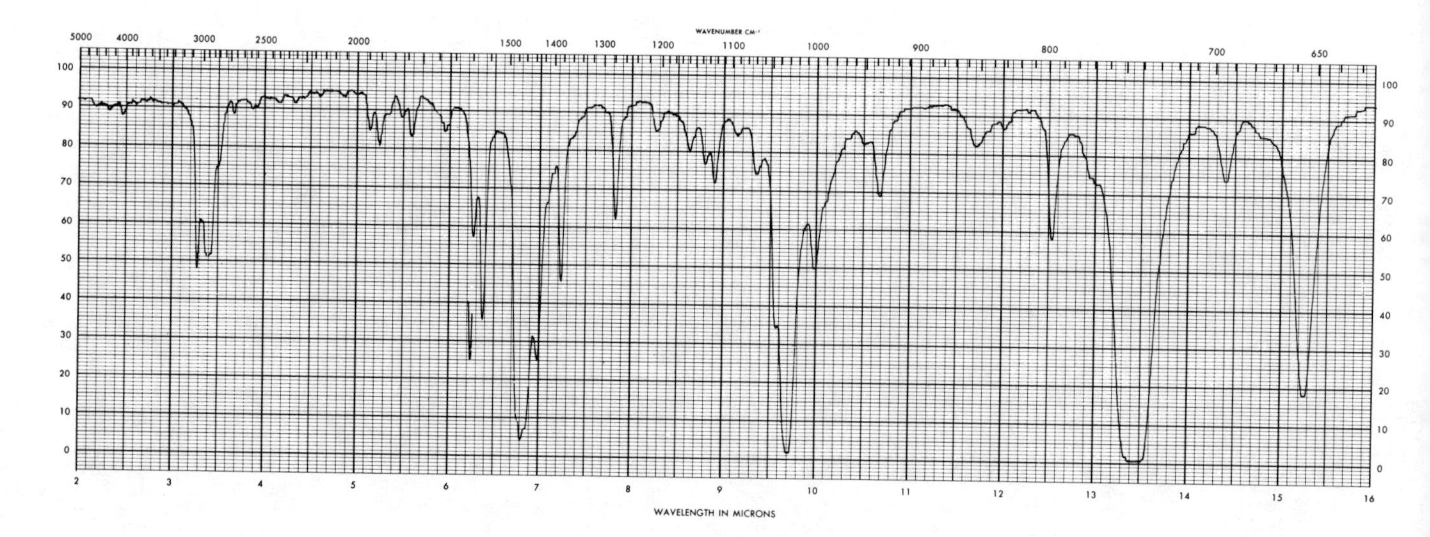

NMR spectrum: Solvent CCl_4, sweep time 250 sec, sweep width 500 Hz

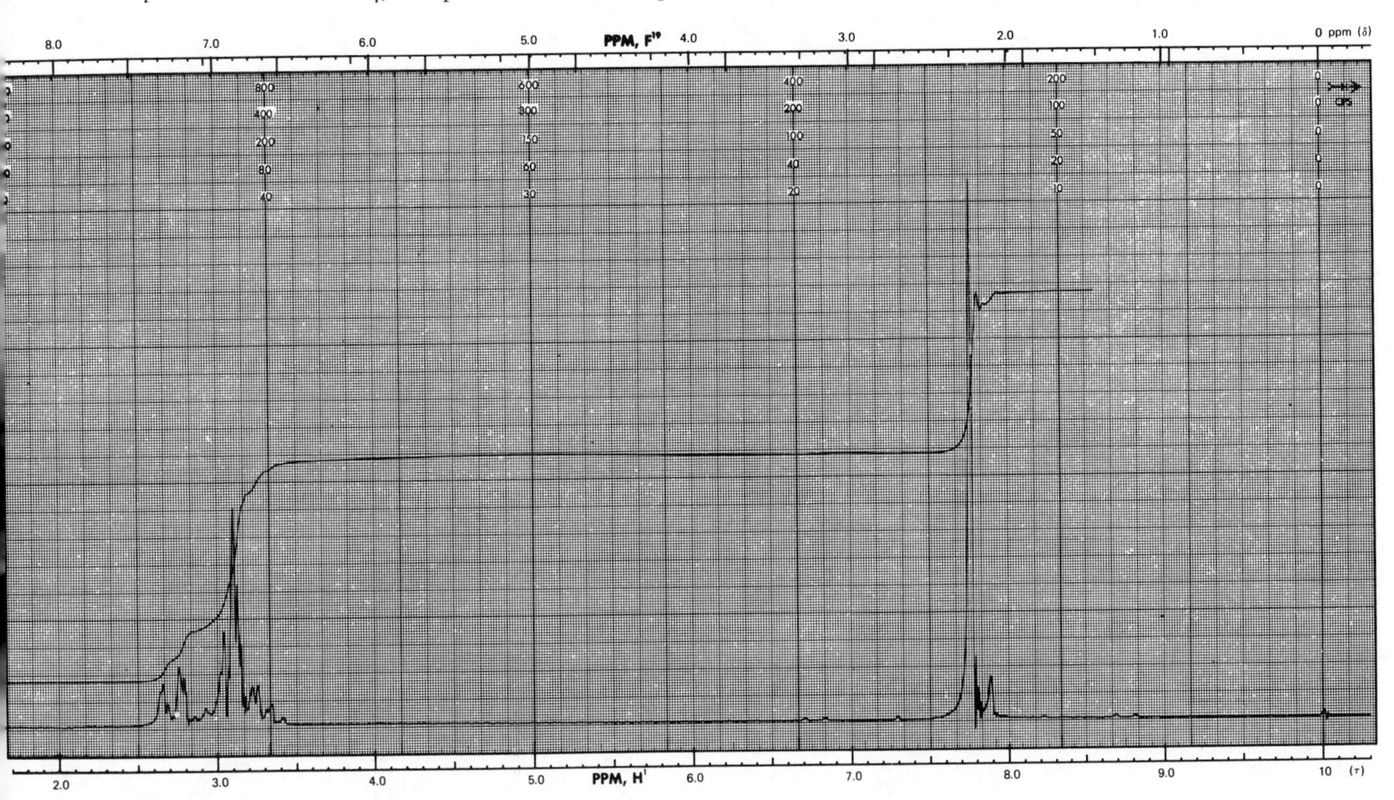

UNKNOWN NUMBER 6.16

Physical data: Colorless liquid, boiling point 151.5°

Mass spectrum

m/e	% of base peak	m/e	% of base peak	m/e	% of base peak	m/e	% of base peak
14	1	15	8	26	2	27	19.5
28	4	29	12.5	38	1	39	10
40	2	41	11.5	42	7	43	100
44	3	53	1	55	4.5	56	1
57	2	58	55	59	8.5	71	13
72	2	85	2	99	2.5	114	5.2 (M)
115	0.4						

Isotope ratio

m/e	% of M
114 (M)	100
115 (M + 1)	7.7

Ultraviolet data

λ_{max}^{EtOH}	ϵ_{max}
275	12

Infrared spectrum: Liquid sample, thin layer

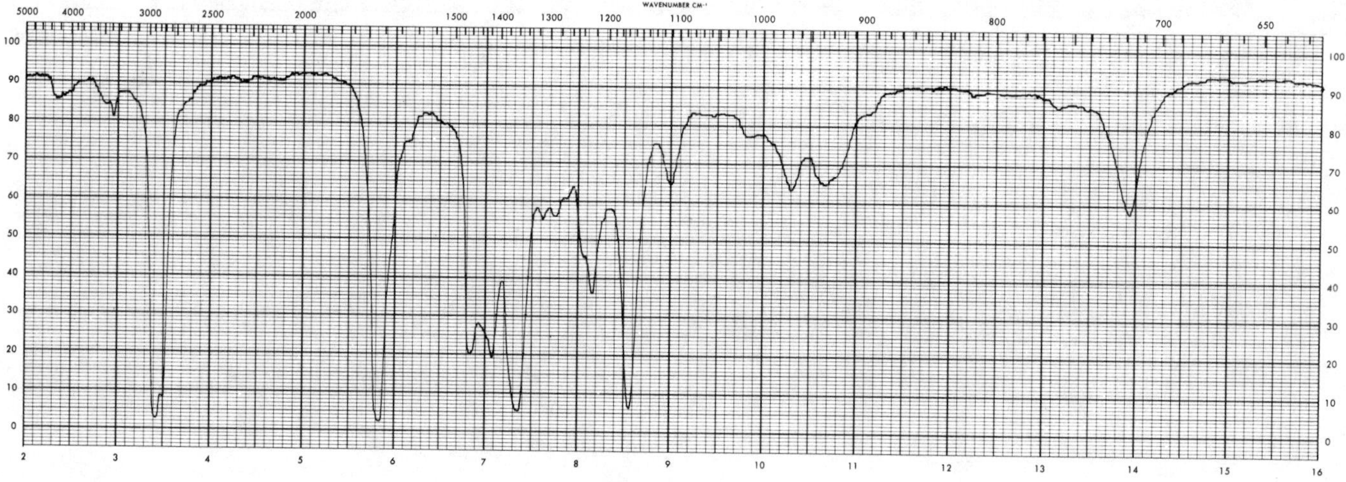

NMR spectrum: Neat, sweep time 250 sec, sweep width 500 Hz

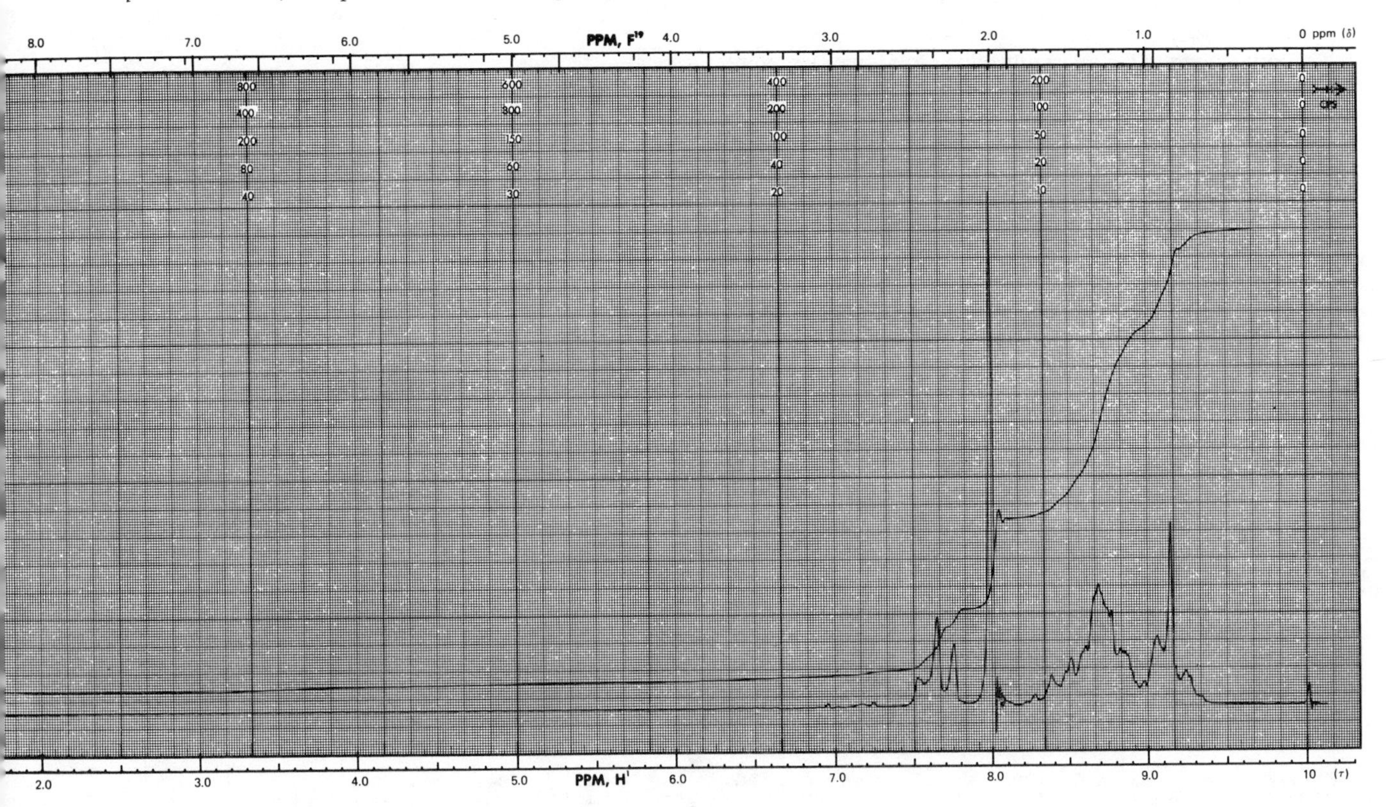

UNKNOWN NUMBER 6.17

Physical data: Colorless crystalline solid, melting point 170.3°

Mass spectrum

m/e	% of base peak	m/e	% of base peak	m/e	% of base peak	m/e	% of base peak
38	6	39	17	41	1.5	42	1
43	2	49	2	50	6	51	7
52	6	53	27	54	13	55	22
56	2	60	1	61	3	62	4
63	7	64	5	65	2	66	1
69	2	71	1	74	1	80	1
81	22.5	82	9.5	91	1	92	1
108	1	109	5.5	110	100 (M)	111	6.7
112	0.6						

Isotope ratio

m/e	% of M
110 (M)	100
111 (M + 1)	6.7
112 (M + 2)	0.6

Ultraviolet data

λ_{max}^{EtOH}	ϵ_{max}
290 nm	2800

Infrared spectrum: KBr disc, 0.5-mg solid sample in 200 mg KBr

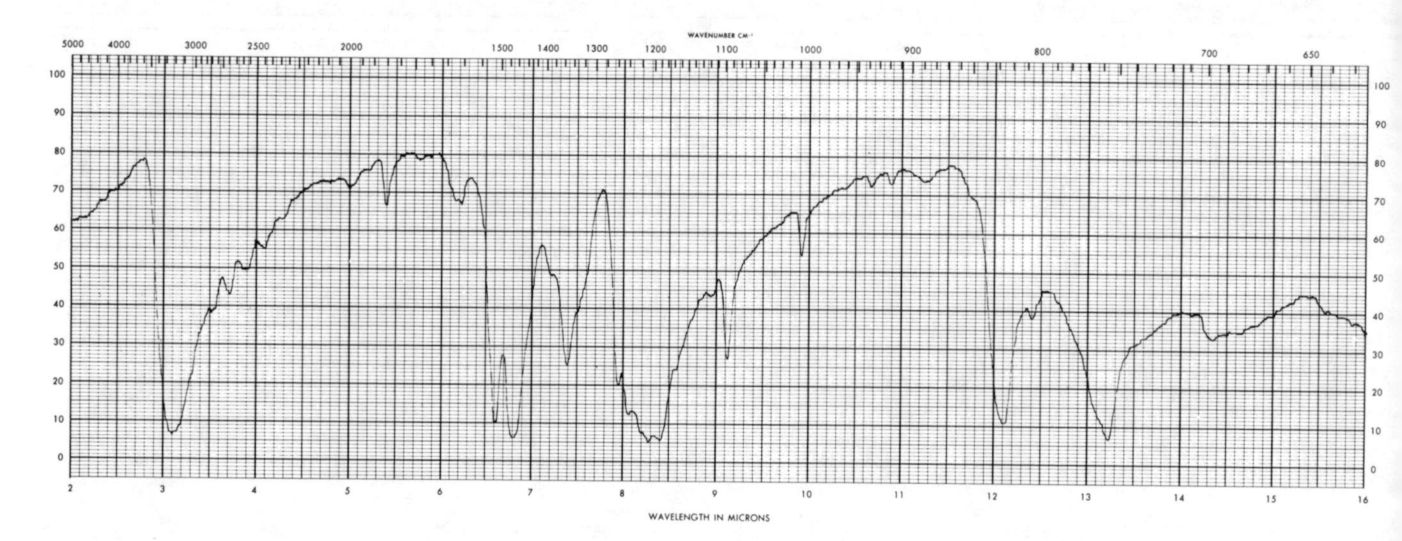

NMR spectrum: Solvent acetone, sweep time 250 sec, sweep width 500 Hz; upper trace marked with ★ is recorded after D_2O has been added to the sample

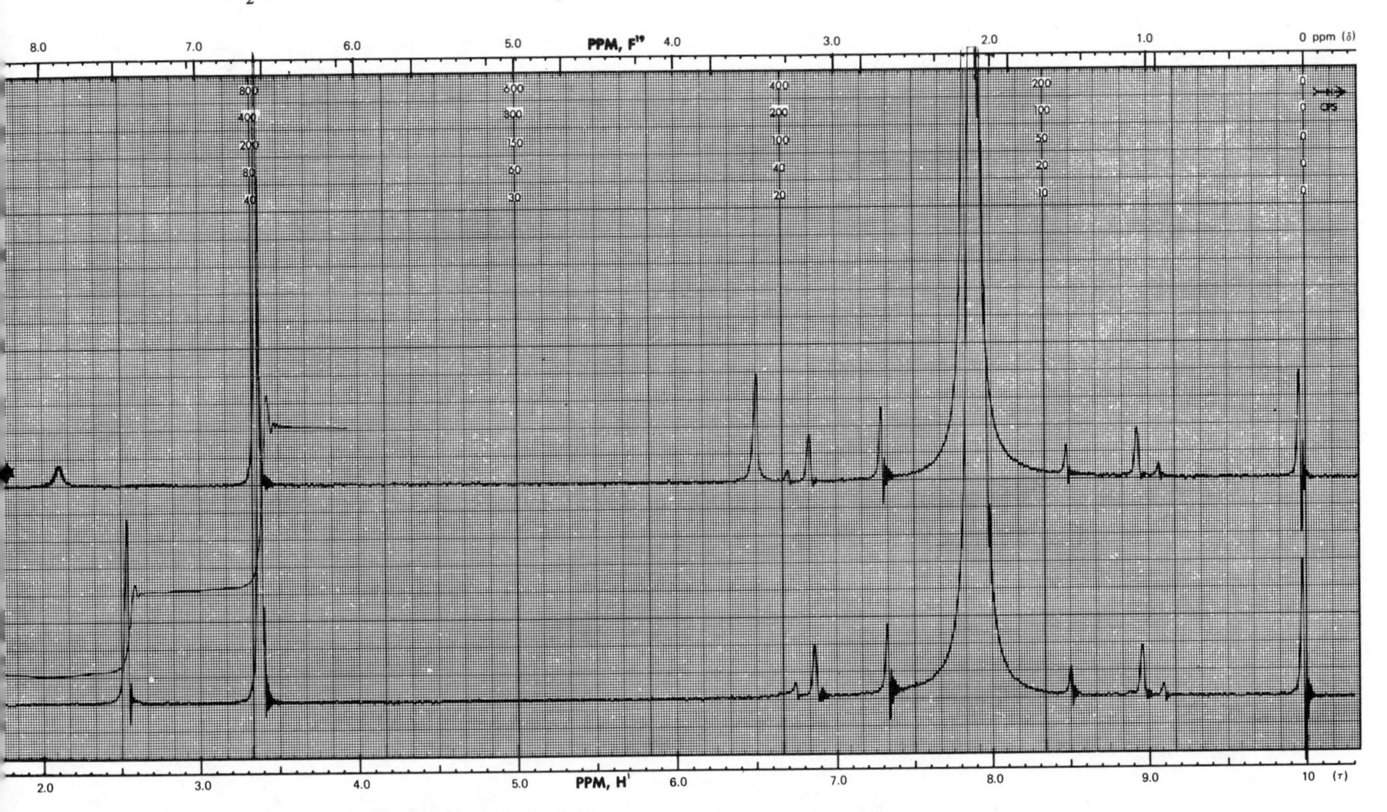

UNKNOWN NUMBER 6.18

Physical data: Colorless liquid, boiling point 120°

Mass spectrum

m/e	% of base peak	m/e	% of base peak	m/e	% of base peak	m/e	% of base peak
18	2	26	7	27	40	28	7
29	31	31.5	0.1	32.5	0.1	37	2
38	4	39	27	40	4.5	41	41.5
42	16	43	100	44	4	48	1.5
50	1.5	51	2	53	3	54	1
55	25.5	56	1.5	57	1.5	69	1.5
70	2.5	71	82.5	72	4.4	79	1
81	1	107	1	109	1	150	0.3 (M)
152	0.3						

Isotope ratio

m/e	% of M
150 (M)	100
152 (M + 2)	100

Ultraviolet data

Transparent above 210 nm

Infrared spectrum: Liquid sample, 0.09-mm layer; inset, capillary

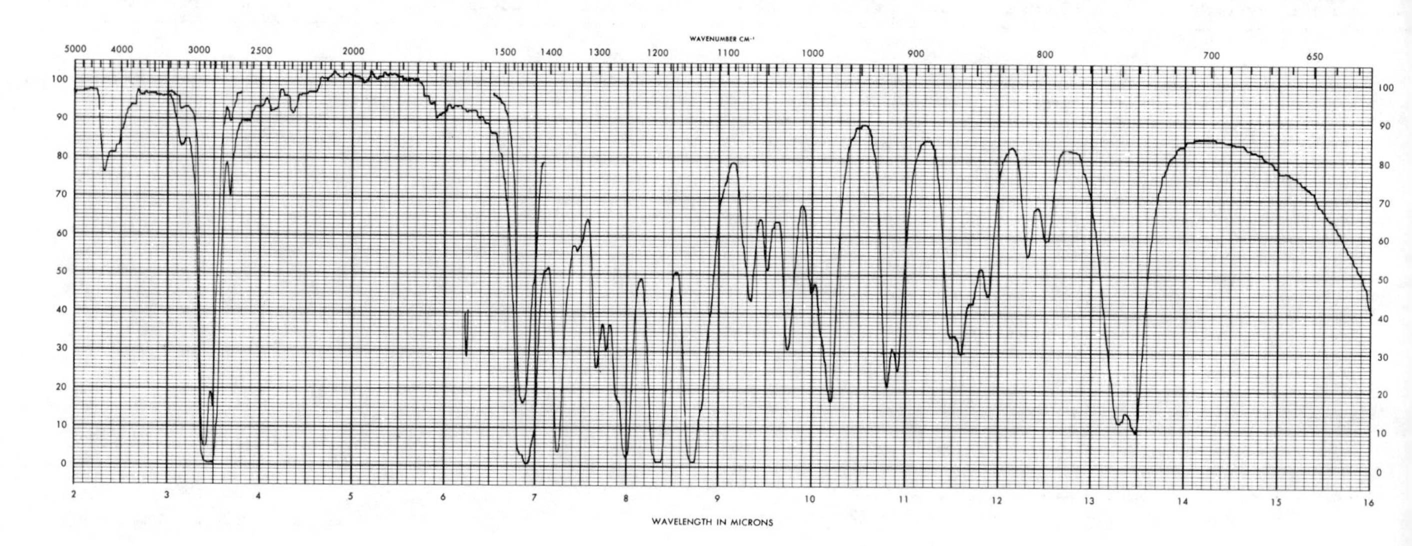

NMR spectrum: Solvent CCl_4, sweep time 250 sec, sweep width 500 Hz

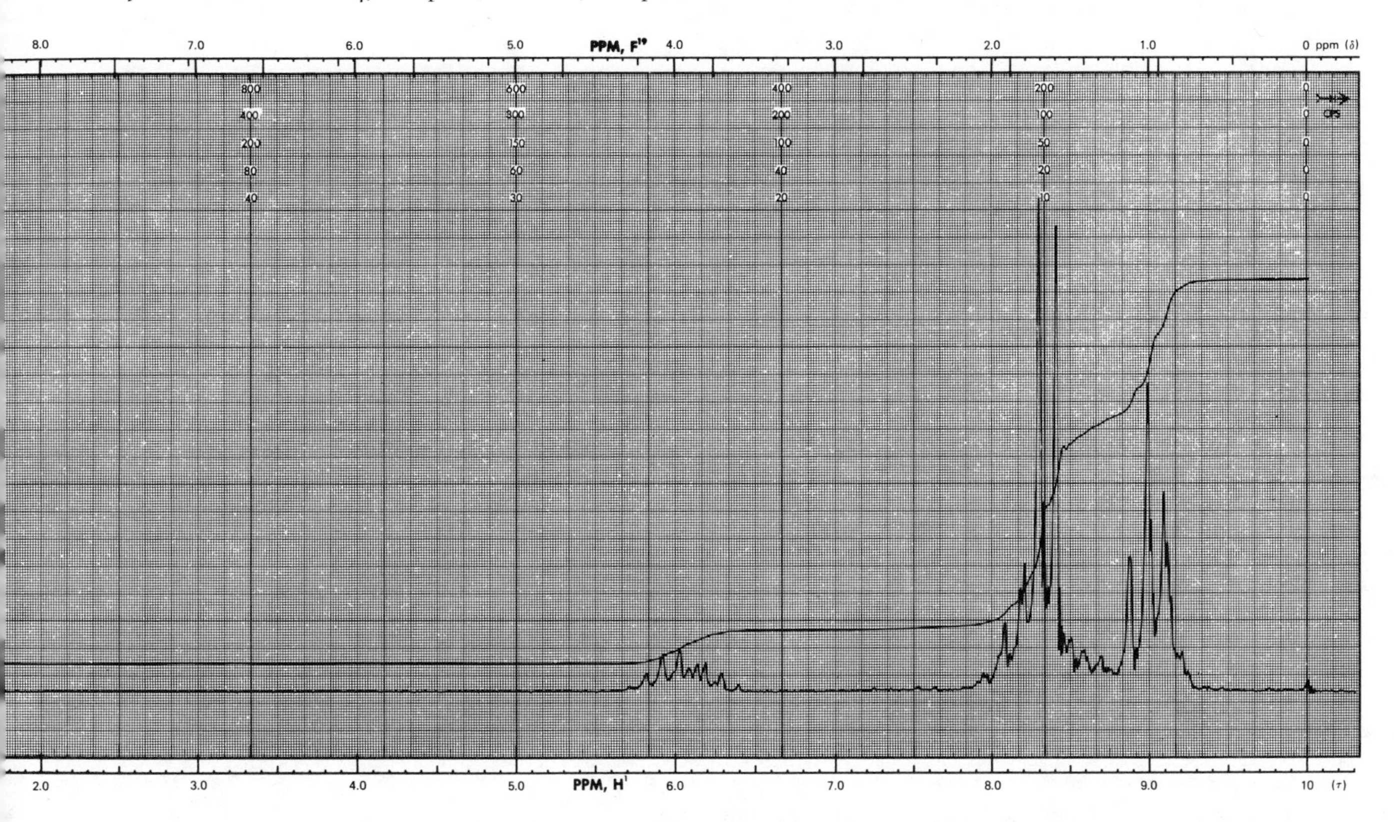

UNKNOWN NUMBER 6.19

Physical data: Colorless liquid, boiling point 146°

Mass spectrum

m/e	% of base peak	m/e	% of base peak	m/e	% of base peak	m/e	% of base peak
26	4	27	9	28	3	37	4
38	9.5	39	37	40	8.5	41	7
42	8.5	49	1	50	4	51	7
52	5	52.5	2.5	53	5.5	53.5	2
54	1	61	1.5	62	3.5	63	8
64	5	65	18	66	22.5	67	5
68	1	75	1	77	5	78	2
79	9	80	3	91	3	92	19
93	2	104	1	105	5	106	27
107	100 (M)	108	8.2	109	0.3	121	1

Isotope ratio

m/e	% of M
107 (M)	100
108 (M + 1)	8.2
109 (M + 2)	0.3

Ultraviolet data

Spectrum determined in alkaline and in acidic solutions

	λ_{max}	ϵ_{max}
0.1N NaOH	266.5	4510
0.1N HCl	269.5	8540

Infrared spectrum: Liquid sample, 0.007-mm layer; inset, capillary

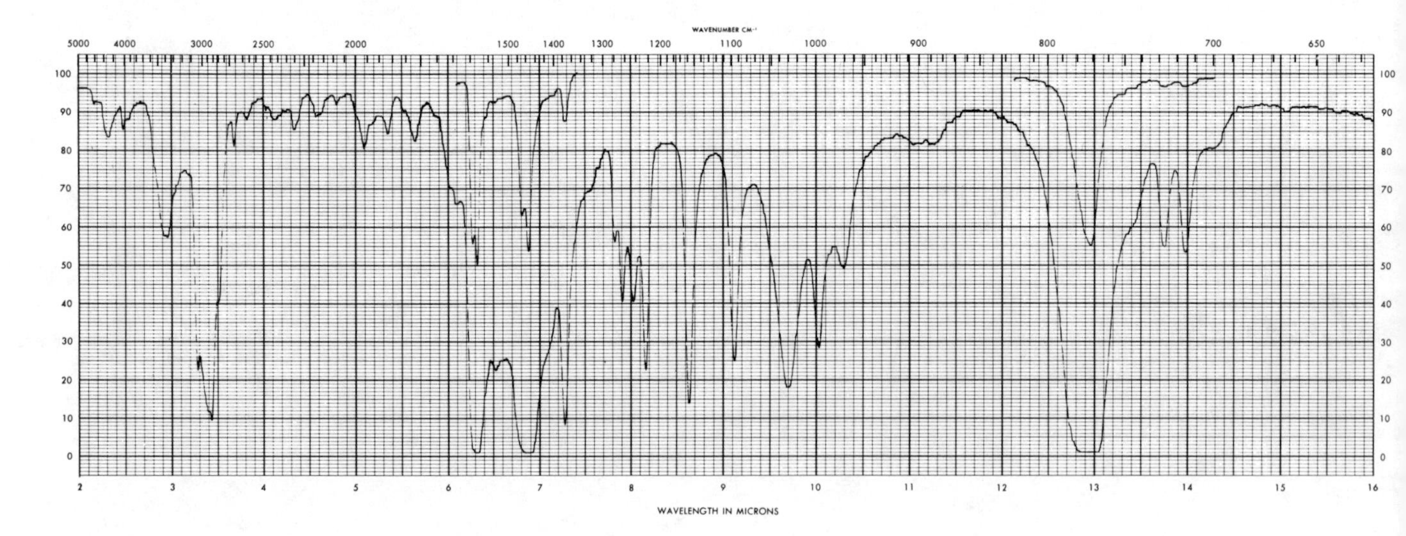

NMR spectrum: Neat, sweep time 250 sec, sweep width 500 Hz

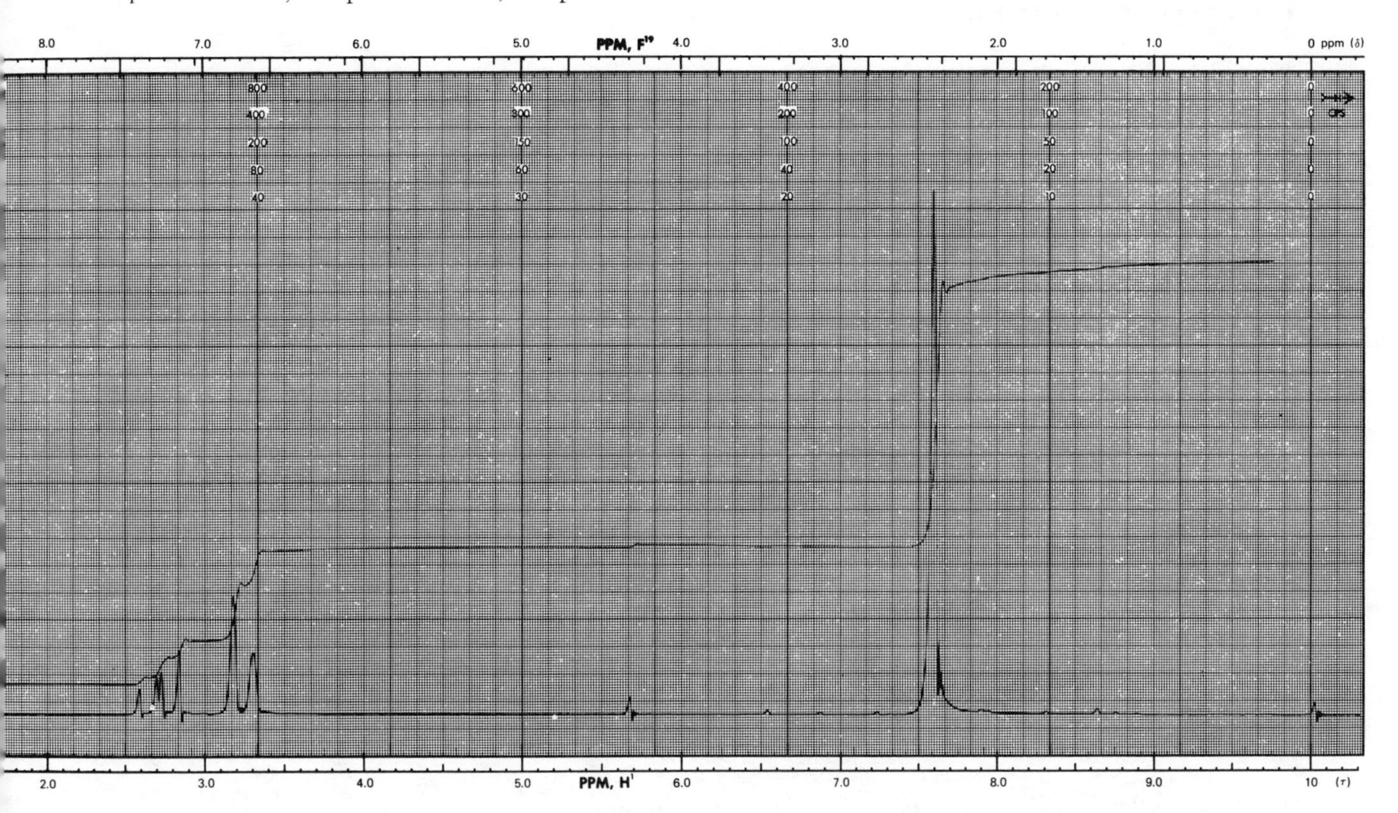

UNKNOWN NUMBER 6.20

Physical data: Colorless liquid, boiling point 252°

Mass spectrum

m/e	% of base peak	m/e	% of base peak	m/e	% of base peak	m/e	% of base peak
39	2	44.5	0.1	45.5	0.1	49.5	0.1
50	1.5	51.5	0.2	52	1	52.5	0.1
53	1	57.5	0.1	58.5	0.1	59.5	0.1
62	1	63	2	64.5	0.1	65	1
65.5	0.1	66.5	0.1	67.5	0.1	70.5	0.1
71.5	0.1	72.5	0.1	73.5	0.1	75	1
76	1	77	5	78	100	79	1.5
80.5	0.5	81	1.5	81.5	0.2	91	1.5
102	1.5	103	7	104	9	105	2
131	10	132	3	133	1	135	4
161	9	162	33.3 (M)	163	3.7	164	0.3

Isotope ratio

m/e	% of M
162 (M)	100
163 (M + 1)	11.1
164 (M + 2)	0.92

Ultraviolet data

λ_{max}^{EtOH}	ϵ_{max}
326	4800

Infrared spectrum: Liquid sample, 0.007-mm layer; inset, capillary

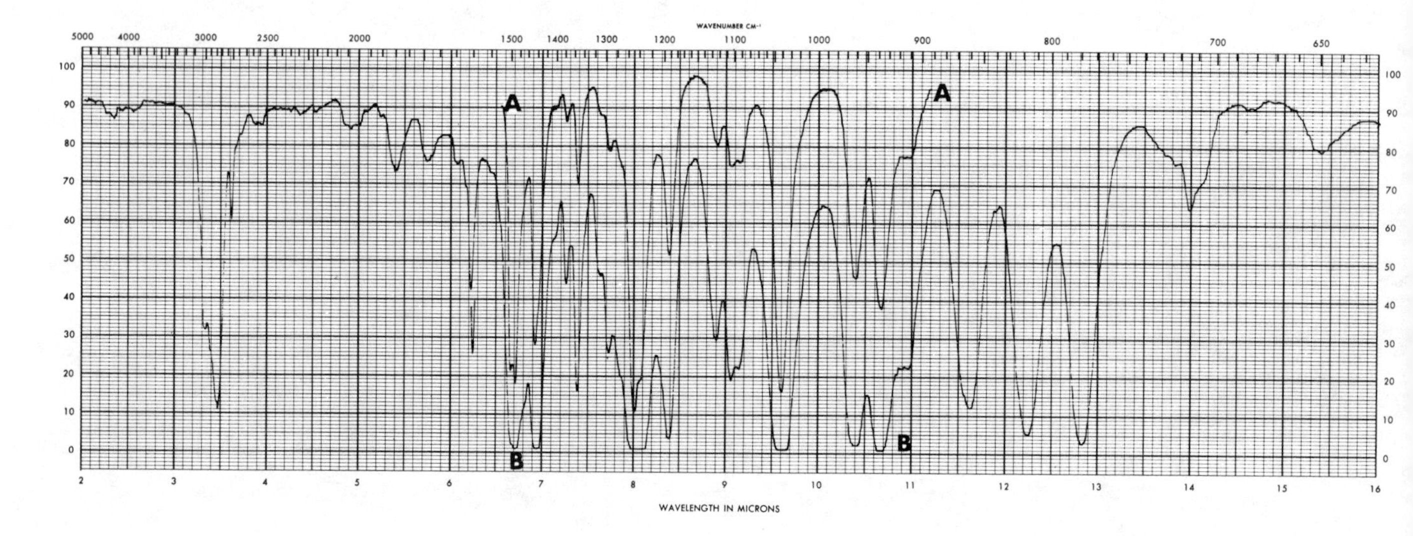

NMR spectrum: Solvent CCl_4, sweep time 250 sec, sweep width 500 Hz

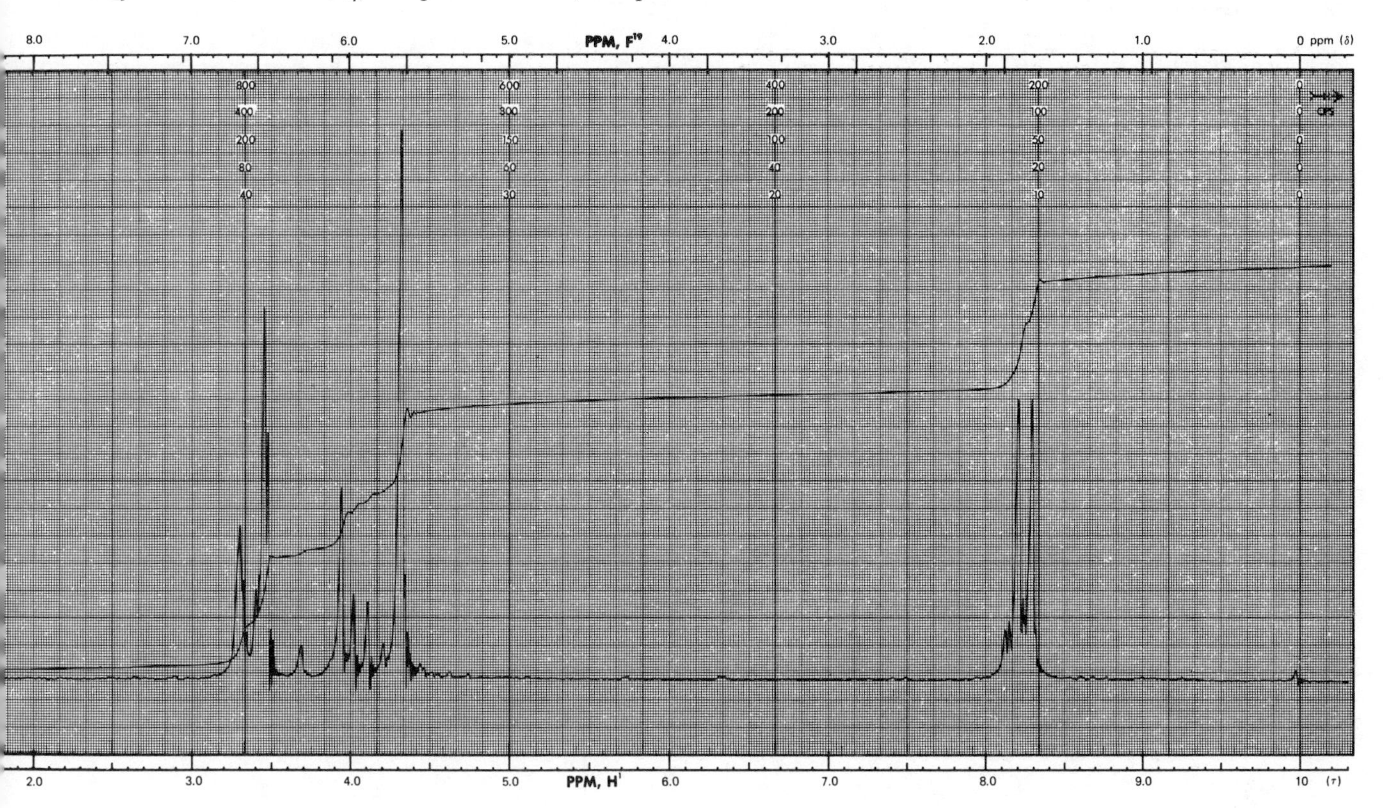

UNKNOWN NUMBER 6.21

Physical data: Colorless liquid, boiling point 165°

Mass spectrum

m/e	% of base peak	*m/e*	% of base peak	*m/e*	% of base peak	*m/e*	% of base peak
15	0.2	27	3	39	6	41	3.5
50	2	51	6	52	2	53	2.5
57	1	57.5	2	58	3	59	3.5
62	1	63	3	65	3.5	77	10.5
78	3.5	79	6	89	1	91	8
92	1	102	1	103	5.5	104	2.5
105	100	106	9	115	3	117	2.5
119	15.5	120	62.5 (M)	121	6.4	122	0.3

Isotope ratio

m/e	% of M
120 (M)	100
121 (M + 1)	10.24
122 (M + 2)	0.48

Ultraviolet data

λ_{max}^{EtOH}	ϵ_{max}
217	7200
267	260

Infrared spectrum: 6.0% w/w in $CHCl_3$

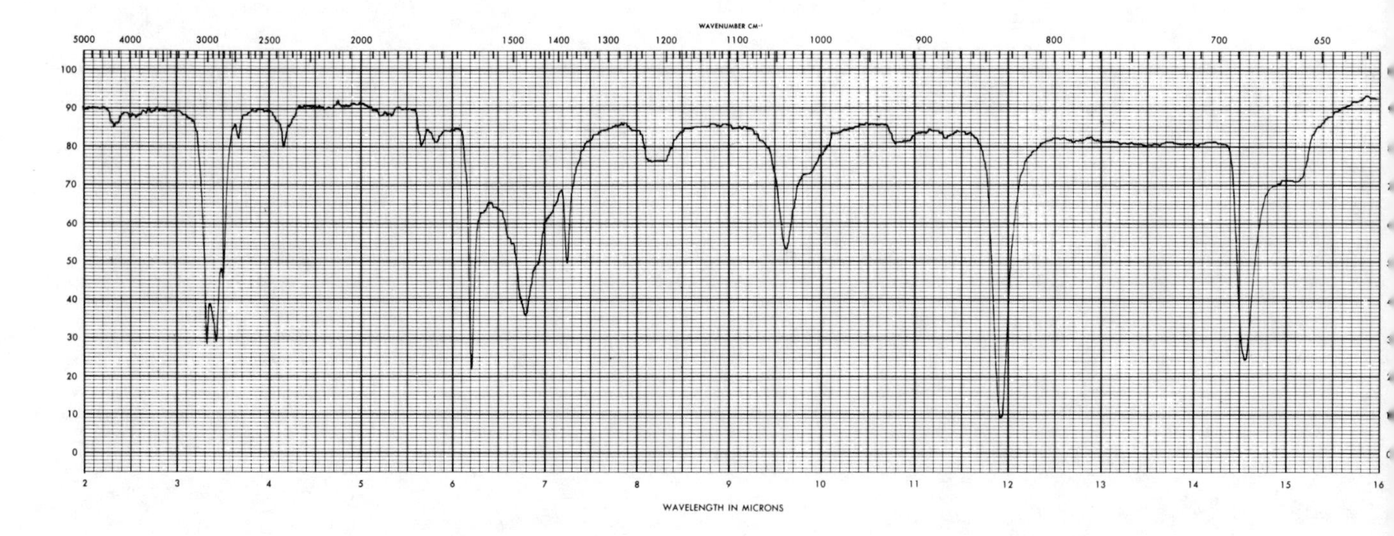

NMR spectrum: Solvent CCl_4, sweep time 250 sec, sweep width 500 Hz

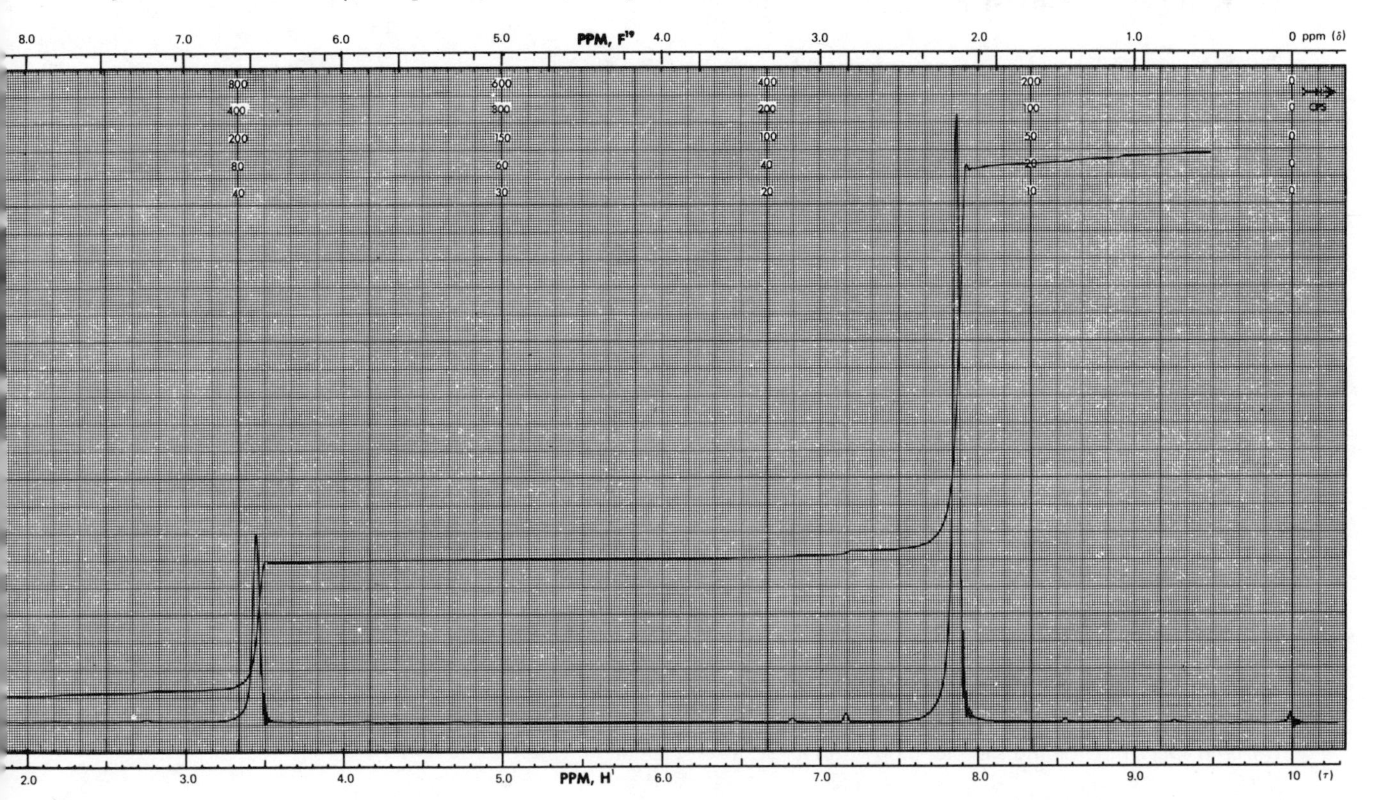

UNKNOWN NUMBER 6.22

Physical data: Colorless liquid, boiling point 83°

Mass spectrum

m/e	% of base peak	m/e	% of base peak	m/e	% of base peak	m/e	% of base peak
15	1R	26	3	27	19.5	28	6
29	2.5R	38	3	39	36.5	40	5
41	36.5	42	2.5	50	5	51	8.5
52	4.5	53	13	54	77	54.5	5.5*
55	5.5	63	1.5	66	2.5	67	100
68	5.5	77	6.0	78	2	79	7.5
80	1.5	81	11.5	82	40.9 (M)	83	2.7
84	0.1						

R = Rearrangement peak * = Metastable peak

Isotope ratio

m/e	% of M
82 (M)	100
83 (M + 1)	6.61
84 (M + 2)	0.2

Ultraviolet data

Transparent above 210 nm

Infrared spectrum: Liquid sample, 0.007-mm layer; inset, capillary

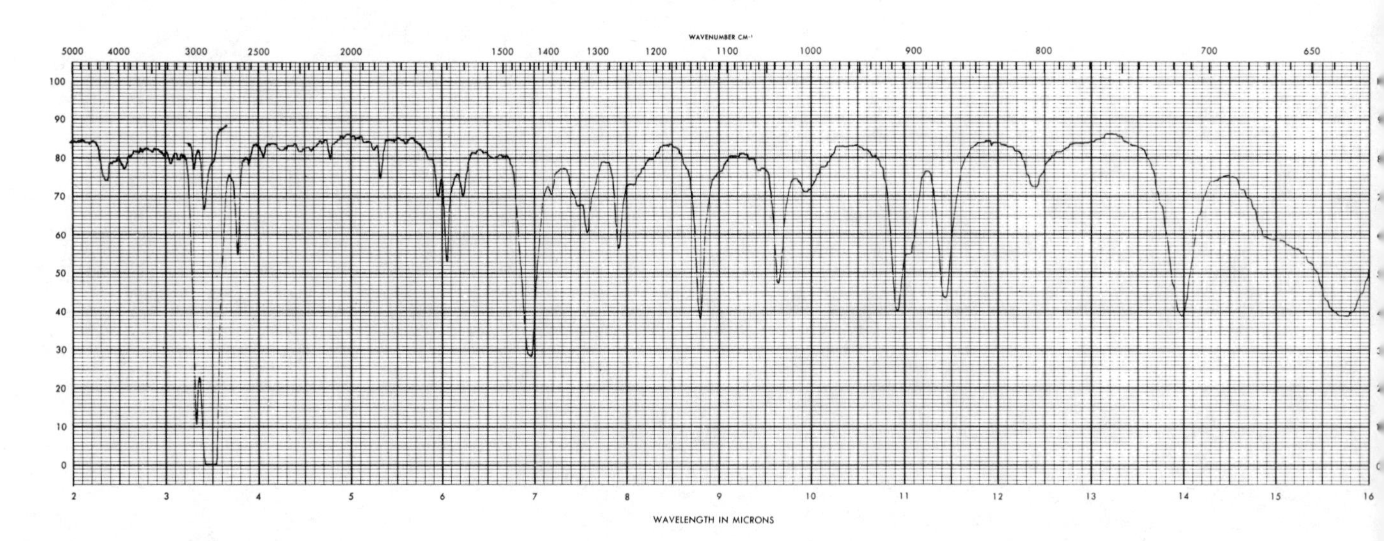

NMR spectrum: Neat, sweep time 250 sec, sweep width 500 Hz

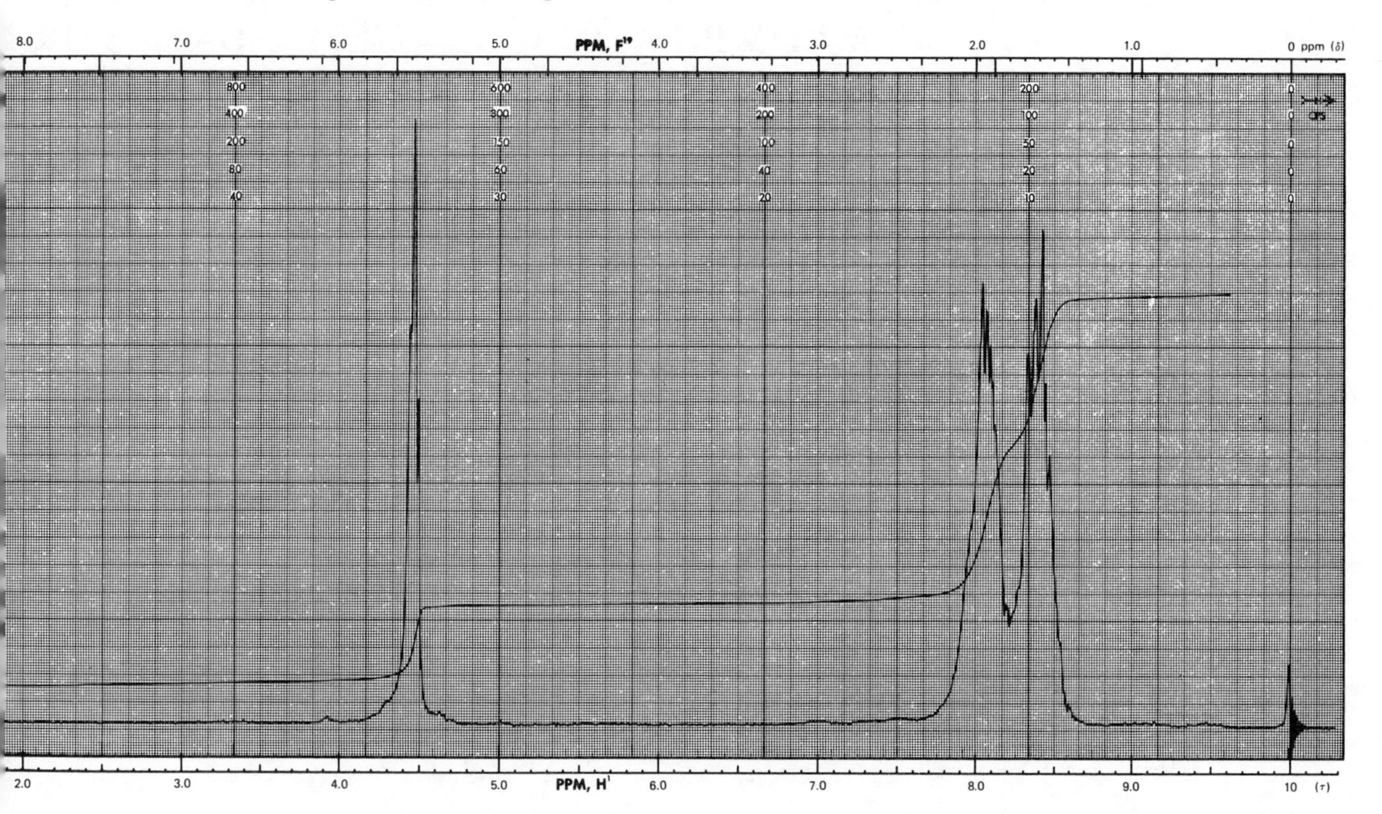

APPENDIX

A COMPILATION OF USEFUL DATA FOR UV, IR, NMR AND MASS SPECTROMETRY

Appendix 1.1 Some spectroscopic terms

Avogadro's number $N = 6.023 \times 10^{23}$ mole^{-1}
Velocity of light $c = 2.998 \times 10^{10}$ cm sec^{-1}
Planck's constant $h = 6.625 \times 10^{-27}$ erg-sec
$= 6.625 \times 10^{-34}$ joule-sec
Energy: 1 electron volt (eV) $= 1.602 \times 10^{-12}$ ergs
$= 1.602 \times 10^{-19}$ joules
1 joule $= 10^7$ ergs $= 2.39 \times 10^{-1}$ calorie
1 calorie $= 4.184$ joules $= 4.184 \times 10^7$ ergs
1 angstrom (Å) $= 10^{-1}$ millimicron $= 10^{-1}$ nanometer
$= 10^{-4}$ micron $= 10^{-8}$ cm
1 micron (μ) $= 1000$ mμ $= 10{,}000$ Å $= 10^{-4}$ cm
1 cm $= 10^4$ μ $= 10^7$ mμ $= 10^7$ nm $= 10^8$ Å

Appendix Table 2.1 Approximate lower cutoff wavelengths for commonly used solvents in UV–visible spectroscopy of organic substances

Solvent	Cutoff wavelength, nm	Solvent	Cutoff wavelength, nm
	200–250		250–300
Acetonitrile	210	Benzene	280
n-Butanol	210	Carbon tetrachloride	265
Chloroform	245	N,N-Dimethylformamide	270
Cyclohexane	210	Methyl formate	260
Decahydronapthalene	200		
1,1-Dichloroethane	235	Tetrachloroethylene	290
Dichloromethane	235	Xylene	295
Dioxane	225		
			300–350
Dodecane	200		
Ethanol	210	Acetone	330
Ethyl ether	210	Benzonitrile	300
Heptane	210	Bromoform	335
		Pyridine	305
Hexane	210		
Methanol	215		
Methylcyclohexane	210		350–400
*iso*octane	210		
*iso*propanol	215	Nitromethane	380
Water	210		

Appendix Table 2.2 Woodward's and Fieser's rules for calculating ultraviolet absorption maxima of substituted dienes (ethanol solution)

i) Basic λ_{max} for an unsubstituted, conjugated, *acyclic* or *heteroannular* diene	214 nm
ii) Basic λ_{max} for an unsubstituted, conjugated *homoannular* diene	253 nm
iii) Extra double bonds in conjugation (for each C=C)	add 30 nm
iv) Exocyclic double bond (effect is two-fold if bond is exocyclic to two rings)	add 5 nm
v) Substituents on vinyl carbons (for each group given below)	
a) O-acyl (—O—CO—R or —O—CO—Ar)	0 nm
b) Simple alkyl (—R)	add 5 nm
c) Halogen (—Cl, —Br)	add 5 nm
d) O-alkyl (—OR)	add 6 nm
e) S-alkyl (—SR)	add 30 nm
f) N-alkyl$_2$ (—NRR′)	add 60 nm

Appendix Table 2.3* Woodward's rules modified to facilitate calculation of ultraviolet absorption maxima of enone derivatives (ethanol solution)

Parent enone (acyclic or rings larger than 5 members)		215 nm
Five-membered cyclic enone		subtract 10
Aldehydes		subtract 5
Extended conjugation (for each double bond)		add 30
Homoannular component		add 39
Exocyclic double bond		add 5
Substituents		
Alkyl	α	add 10
	β	add 12
	γ and higher	add 18
Hydroxyl	α	add 35
	β	add 30
	γ	add 50
Alkoxyl	α	add 35
	β	add 30
	γ	add 17
	δ	add 31
Acetoxyl	α, β, or δ	add 6
Dialkylamino	β	add 95
Chlorine	α	add 15
	β	add 12
Thioalkyl	β	add 85
Bromine	α	add 25
	β	add 30

* From D. J. Pasto and C. R. Johnson, *Organic Structure Determination,* Prentice-Hall, Englewood Cliffs, N.J., 1969.

Appendix Table 2.4* Solvent correction factor for calculating ultraviolet absorption maxima of enones

For solvents other than ethanol, apply the following correction factors to the value calculated on the basis of Table 2.3.

Water	subtract	8 nm
Methanol		0
Chloroform	add	1
Dioxane	add	5
Ether	add	7
Hexane	add	11

* From D. J. Pasto and C. R. Johnson, *Organic Structure Determination,* Prentice-Hall, Englewood Cliffs, N.J., 1969.

Appendix Table 2.5 Near-ultraviolet absorptions of some monosubstituted benzenes

Substituent	202-nm or primary band	ϵ_{max}	255-nm or fine-structure band	ϵ_{max}	Solvent
$-NH_3^+$	203	7500	254	169	2% MeOH in H_2O
—H	203.5	7400	254	204	,,
$-CH_3$	206.5	7000	261	225	,,
—I	207	7000	257	700	,,
—Cl	209.5	7400	263.5	190	,,
—Br	210	7900	261	192	,,
—OH	210.5	6200	270	1450	,,
$-OCH_3$	217	6400	269	1480	,,
$-SO_2NH_2$	217.5	9700	264.5	740	,,
—CN	224	13,000	271	1000	,,
$-COO^-$	224	8700	268	560	,,
—COOH	230	11,600	273	970	,,
$-NH_2$	230	8600	280	1430	,,
$-O^-$	235	9400	287	2600	,,
—C≡CH	236	15,500	278	650	Heptane
$-NHCOCH_3$	238	10,500	—	—	Water
$-CH{=}CH_2$	244	12,000	282	450	EtOH
$-COCH_3$	240	13,000	278	1100	EtOH
—Ph	246	20,000	—	—	Heptane
—CHO	244	15,000	280	1500	EtOH
$-NO_2$	252	10,000	280	1000	Hexane
N=N—Ph (*trans*)	319	19,500	—	—	Chloroform

Appendix Table 2.6 Shifts in the primary bands (203.5 nm bands) of disubstituted ($R_1-C_6H_4-R_2$) benzenes (2% MeOH solution)

R_1	Shift ($\Delta\lambda_1$) for R_1 nm	R_2	Shift ($\Delta\lambda_2$) for R_2 nm	Calculated $\Delta\lambda_1 + \Delta\lambda_2$	Obs. $\Delta\lambda$ for *ortho* compound	Obs. $\Delta\lambda$ for *meta* compound	Obs. $\Delta\lambda$ for *para* compound
CH_3	3.0	CN	20.5	23.5	25.0	26.0	33.5
CH_3	3.0	NO_2	65.0	68.0	62.5	69.5	81.5
Cl	6.0	COOH	26.5	32.5	25.5	28.0	37.5
Cl	6.0	NO_2	65.0	71.0	56.5	60.5	76.5
Br	6.5	COOH	26.5	33.0	—	—	42.0
Br	6.5	$COCH_3$	42.0	48.5	—	—	55.0
OH	7.0	COOH	26.5	33.5	33.5	34.0	51.5
OH	7.0	$COCH_3$	42.0	49.0	49.0	47.0	71.5
OH	7.0	CHO	46.0	53.0	53.0	51.0	80.0
OH	7.0	NO_2	65.0	72.0	75.0	70.0	114.0
OCH_3	13.5	$COCH_3$	42.0	55.5	—	—	73.0
NH_2	26.5	CN	20.5	47.0	—	33.0	66.5
NH_2	26.5	COOH	26.5	53.0	44.5	46.5	80.5
NH_2	26.5	$COCH_3$	42.0	68.5	—	—	108.0
O^-	31.5	$COCH_3$	42.0	73.5	—	—	121.0
O^-	31.5	CHO	46.0	77.5	—	—	127.0
O^-	31.5	NO_2	65.0	96.5	—	—	199.0
$NHCOCH_3$	38.5*	COOH	24.5*	55.0	—	—	84.5*
$NHCOCH_3$	38.5*	NO_2	56.5*	95.0	—	—	112.5*
COOH	26.5	NO_2	65.0	†	—	—	61.0
NO_2	65.0	NO_2	65.0	†	—	38.0	62.5

* In ethanol.
† Not to be added.

Appendix Table 3.1a Characteristic infrared absorption frequencies of common classes of organic compounds

Type	Vibration mode	Frequency, cm^{-1}	Wavelength	Relative intensity*
1. *Alkanes*				
$-CH_3$	C—H str., asym.	2975–2950	3.36–3.39	m.
	C—H str., sym.	2880–2860	3.47–3.50	m.
	C—H def., asym.	1470–1435	6.80–6.97	s.
	C—H def., sym.	1385–1370	7.22–7.30	m., sh.
$-C(CH_3)_2$	C—H def.	1385–1365	7.22–7.33	s., d.
	Skeletal	1175–1140	8.51–8.77	s., br.
		840–790	11.90–12.66	m.
$-C(CH_3)_3$	C—H def.	1395–1365	7.17–7.33	s., d. ratio 1:2
	Skeletal	1255–1200	7.97–8.33	s., br., d.
		750–720	13.33–13.89	s., br.
$-CH_2-$	C—H str., asym.	2940–2915	3.40–3.45	m. } strong if
	C—H str., sym.	2870–2845	3.49–3.52	m. } several
	C—H, scissoring	1480–1440	6.76–6.94	m. } $-CH_2-$
	C—H, twisting and wagging	*ca.* 1250	*ca.* 8.00	m. } present
	Skeletal if $-(CH_2)_4$ or more	750–720	13.33–13.89	s.
CH_2- in cyclopropane	C—H str.	3080–3040	3.25–3.29	v.
	Skeletal	1020–1000	9.80–10.00	m.
$-\overset{\vert}{\underset{\vert}{C}}-H$	C—H str.	2900–2880	3.45–3.47	w.
	C—H def.	*ca.* 1340	*ca.* 7.45	w.
2. *Alkenes*				
See Appendix Table 3.2				
3. *Alkynes and allenes*				
Terminal (RC≡CH)	C—H str.	3310–3300	3.02–3.03	m.
	C≡C str.	2140–2100	4.67–4.76	w.
Nonterminal ($R_1C{\equiv}CR_2$)	C≡C str.	2360–2190	4.43–4.57	v.
Allenes (C=C=C)	C≡C type str.	1970–1950	5.08–5.13	m.
4. *Aromatic hydrocarbons*				
See Appendix Table 3.3				
5. *Alcohols and phenols*				
Free —OH	O—H str.	3670–3580	2.73–2.79	v., sh.
Dimeric	O—H str.	3550–3450	2.82–2.90	v., sh.
Polymeric	O—H str.	3400–3230	2.94–3.10	s., br.
Intramolecular H bonded	O—H str.	3590–3420	2.79–2.92	v., sh.
Chelated	O—H str.	3200–1700	3.13–5.88	w., br.
Deuterated	O—D str.	2780–2400	3.60–4.17	v.
Primary C—O—H	O—H def.	1350–1260	7.40–7.94	s.
	C—O str.	1075–1000	9.30–10.00	s.
Secondary C—O—H	O—H def.	1350–1260	7.44–7.94	s.
	C—O str.	1120–1030	8.93–9.71	s.
Tertiary C—O—H	O—H def.	1410–1310	7.09–7.63	s.
	C—O str.	1170–1100	8.55–9.09	s.
Phenols	O—H def.	1410–1310	7.09–7.63	s.
	C—O str.	1230–1140	8.13–8.77	s.

* str. = stretching; sym. = symmetric; asym. = asymmetric
s. = strong absorption; v. = variable intensity; d. = doublet
m. = medium absorption; sh. = sharp peak; mlt. = multiplet
w. = weak absorption; br. = broad absorption; def. = deformation

Appendix Table 3.1a (*continued*)

Type	Vibration mode	Frequency, cm^{-1}	Wavelength	Relative intensity*
6. *Ethers and epoxides*				
Acyclic—CH_3—O—CH_3	R—O—R str., asym.	1150–1070	8.70–9.35	s.
	C—H str.	2830–2815	3.53–3.55	m.
Aryl and aralkyl	Ar—O—R str., Ar—O—Ar asym.	1260–1200	7.98–8.33	v.–s.
	=C—H str.	3150–3050	3.18–3.28	w.
Conjugated	C—O—C str.	1275–1200	7.84–8.33	v.
t-butyl	$C-C(C)(C)-O$ str.	920–800	10.87–12.50	s.
Cyclic	C—O str.	1140–1070	8.77–9.35	s.
—O—CH_2—O	C—O str.	*ca.* 940	*ca.* 10.65	s.
	C—H str.	*ca.* 2780	*ca.* 3.65	v.
Epoxides	C—O str.	1260–1240	7.94–8.07	s.
trans	C—O str.	950–860	10.53–11.63	v.
cis	C—O str.	865–785	11.56–12.74	m.

7. *Carbonyl compounds*

For C=O stretching vibrations of acid chlorides, acid anhydrides, carboxylic acids, esters, aldehydes, ketones, amides and salts, see Appendix Tables 3.4 and 3.5.

Vibrations other than C=O stretching vibrations

Type	Vibration mode	Frequency, cm^{-1}	Wavelength	Relative intensity*
a) *Anhydrides*				
Cyclic	C—O str.	1310–1210	7.63–8.26	s.
Acyclic	C—O str.	1175–1045	8.51–9.57	s.
b) *Carboxylic acids*				
Free OH	O—H str.	3550–3500	2.82–2.86	m.
Bonded OH	O—H str.	3300–2500	3.00–4.00	w., br.
All OH	O—H def.	955–890	10.47–11.24	v.
Solid fatty acids	CH_2 vib.	1350–1180	7.40–8.48	w. characteristic pattern
—COOH	C—O str. plus O—H def.	1440–1395	6.94–7.17	w.
		1320–1210	7.58–8.26	s.
Carboxylate ion	O=C—O str., asym.	1610–1550	6.21–6.45	s.
	O=C—O str., sym.	1420–1300	7.04–7.69	m.
c) *Esters*				
Formates	C—O str.	1200–1180	8.33–8.48	s.
Acetates	C—O str.	1250–1230	8.00–8.13	s.
Propionates and higher esters	C—O str.	1200–1170	8.33–8.55	s.
Esters of aromatic acids	C—O str.	1300–1250	7.69–8.00	s.
Vinylic and phenolic acetates	C—O str.	1220–1200	8.20–8.33	s.
Esters of α, β unsat. aliphatic acids	C—O str.	1310–1250	7.63–8.00	s.
d) *Aldehydes*				
	C—H str., asym.	2880–2650	3.47–3.77	w.–m., d.
	C—H def.	975–780	10.26–12.82	w.
e) *Ketones*				
CH_3—CO	CH_3 def.	1360–1355	7.35–7.38	s.
—CH_2—CO—	—CH_2 def.	1435–1405	6.97–7.12	s.
	C=O str., overtone	3550–3200	2.82–3.13	w.

Appendix Table 3.1a *(continued)*

Type	Vibration mode	Frequency, cm^{-1}	Wavelength	Relative intensity*
f) *Amides*				
Primary, free NH	N—H str.	3540–3480	2.83–2.88	s.
		3420–3380	2.92–2.96	s.
bonded NH	N—H str.	3360–3320	2.97–3.01	m.
		3220–3180	3.11–3.15	m.
Free *or* bonded NH	N—H def. plus C—N str.	1650–1620 (amide II band)	6.06–6.17	s.
		1620–1590 (amide II band)	6.17–6.31	s.
Secondary, free NH	N—H, *cis* str.	3440–3420	2.91–2.93	s.
	N—H, *trans* str.	3460–3430	2.89–2.91	s.
bonded NH	N—H, *cis* str.	3180–3140	3.15–3.19	m.
	N—H, *trans* str.	3330–3270	3.00–3.06	m.
	N—H, *cis* and *trans* str.	3100–3070	3.23–3.26	w.
	N—H def.	*ca.* 700 (amide V band)	*ca.* 14.30	Conc. dependent
	N—H def. plus C—N str.	1305–1200 (amide III band)	7.67–8.33	m.
Acyclic compound	N—H def. plus C—N str.	1570–1515 (amide II band)	6.37–6.60	s.
		1550–1510 (amide II band)	6.45–6.62	s.
		770–620 (amide IV band)	13.00–16.13	m.
		630–530 (amide VI band)	15.87–18.87	s.

8. ***Amines, amino acids and their salts***

See Appendix Table 3.6.

9. ***Unsaturated nitrogen compounds***

Type	Vibration mode	Frequency, cm^{-1}	Wavelength	Relative intensity*
a) *Nitriles*				
Saturated alkyl	C≡N str.	2260–2240	4.42–4.46	m.
α, β unsaturated alkyl	C≡N str.	2235–2215	4.47–4.51	s.
Iso-	C≡N str.	2185–2120	4.52–4.72	s.
Aryl	C≡N str.	2240–2220	4.46–4.50	m.
		ca. 2145	*ca.* 4.66	s.
b) *Oximes, pyridines, quinolines, purines, pyrimidines, etc.*				
Acyclic saturated	C=N str.	1690–1640	5.92–6.10	v.
Acyclic α, β sat. compound	C=N str.	1665–1630	6.01–6.14	v.
Cyclic α, β unsat. compound	C=N str.	1660–1480	6.02–6.67	v.
Pyridines	C=N str.	1580–1550	6.33–6.45	w.
	=C—H str.	3070–3020	3.26–3.31	s.
	Ring C—H def.	*ca.* 1200	*ca.* 8.33	s.
		1100–1000	9.09–10.00	s.
	C=C str.	1650–1580	6.06–6.35	m.
Pyrimidines and purines	=C—H str.	3060–3010	3.27–3.32	s.
	Ring C—H def.	1000–960	10.00–10.42	m.
		825–775	12.12–12.90	m.
	C=N str.	1580–1520	6.33–6.58	m.
Pyrroles	N—H str.	3440–3400	2.91–2.94	m.
	C=C str.	1565–1500	6.39–6.67	v., d.
Oximes	C=N str.	1690–1620	5.92–6.17	v.
	O—H str.	3650–3500	2.74–2.86	v.

Appendix Table 3.1a *(continued)*

Type	Vibration mode	Frequency, cm^{-1}	Wavelength	Relative intensity*
Azo compounds	—N=N str.	1630–1575	6.14–6.35	v.
Carbodiimides	—N=C=N str.	2155–2130	4.64–4.70	s.
Isocyanates	—N=C=O str.	2275–2240	4.40–4.46	s.,–v.
Azides	—N=N=N str.	2170–2080	4.61–4.81	s.
10. ***Compounds containing nitrogen-oxygen bond***				
Nitrates, nitramines, nitro compounds	NO_2 str., asym.	1590–1500	6.29–6.67	s.
	NO_2 str., sym.	1390–1250	7.20–8.00	w.
	C—N vib.	920–830	10.88–12.05	m.,–s.
Nitroso compounds (R—C—N=O)				
alkyl, aromatic	N=O str.	1550–1500	6.45–6.67	s.
α-halogeno, aliphatic	N=O str.	1620–1560	6.17–6.47	s.
Nitrites (R—O—N=O)				
trans form	N=O str.	1680–1650	5.95–6.05	v.,–s.
	N—O str.	815–750	12.27–13.33	s.
	O—N=O def.	625–565	16.00–17.70	s.
cis form	N=O str.	1625–1610	6.16–6.21	v.,s.
	N—O str.	850–810	11.76–12.35	s.
	O—N=O def.	690–615	14.49–16.26	s.
	Overtone	3360–3220	2.98–3.11	m.
Nitrosamines (R—N—N=O)	N=O str.	1500–1480	6.67–6.76	s. vapor phase
		1460–1440	6.85–6.94	s. solution phase
	N—N str.	*ca.* 1050	*ca.* 9.52	s.
	N—N=O def.	*ca.* 660	*ca.* 15.15	s.
Azoxy comp. (R—N—N—O)	N—O str.	1310–1250	7.63–8.00	m.,–s.
11. ***Sulfur compounds***				
	S—H str.	2600–2550	3.85–3.92	w.
	C—S str.	700–570	14.18–17.54	w.
	C=S str.	1675–1130	5.97–8.85	s.
Covalent sulfates $(RO)_2SO_2$	S=O str., sym.	1440–1350	6.94–7.41	s.
	S=O str., asym.	1230–1150	8.13–8.70	s.
Cov. sulfonates	S=O str., sym.	1420–1330	7.04–7.52	s.
$(R_1—O—SO_2—R_2)$	S=O str., asym.	1200–1145	8.33–8.73	s.
Sulfonyl chlorides	S=O str., sym.	1375–1340	7.27–7.46	s.
$(R—SO_2Cl)$	S=O str., asym.	1190–1160	8.40–8.62	s.
Sulfonamides	S=O str., sym.	1370–1300	7.30–7.69	s.
$(R—SO_2—N)$	S=O str., asym.	1180–1140	8.48–8.77	s.
Sulfones	S=O str., sym.	1350–1300	7.41–7.69	v.,–s.
(R_2SO_2)	S=O str., asym.	1160–1120	8.62–8.93	v.,–s.
Sulfonic acids	S=O str., sym.	1260–1150	7.94–8.70	s.
(RSO_3H)	S=O str., asym.	1080–1010	9.26–9.90	s.
Sulfites $(RO)_2SO$	S=O str.	1220–1170	8.20–8.55	s.
Sulfoxides (R—SO—R)	S=O str.	1070–1030	9.35–9.71	s.
12. ***Phosphorus compounds***				
P—OH	O—H str.	2700–2560	3.70–3.90	w.
	P—H str.	2440–2350	4.10–4.26	m.
	P—C def.	1450–1280	6.90–7.81	m.–s.
	P=O str.	1350–1150	7.41–8.70	v.–s.
	P—O str.	1240–900	8.07–10.10	v.

Appendix Table 3.1a (*continued*)

Type	Vibration mode	Frequency, cm^{-1}	Wavelength	Relative intensity*
13. *Halogen compounds*				
Polyfluorinated	C—F str.	1400–1100	7.14–9.10	v.–s., mlt.
Difluorinated	C—F str.	1250–1050	8.00–9.50	v.–s., d.
Monofluorinated	C—F str.	1110–1000	9.01–10.00	s.
Polychlorinated	C—Cl str.	800–700	12.50–14.30	v.–s.
C—Cl equatorial	C—Cl str.	780–750	12.80–13.33	s.
C—Cl axial, other monochlorinated	C—Cl str.	*ca.* 650	*ca.* 15.40	s.
C—Br equatorial	C—Br str.	750–700	13.33–14.29	s.
C—Br axial	C—Br str.	690–550	14.50–18.20	s.
Other monobromides	C—Br str.	650–560	15.40–17.85	s.
Iodides	C—I str.	*ca.* 500	*ca.* 20.00	s.
14. *Silicon and boron compounds*				
	B—H str.	2220–1600	4.51–6.25	v., mlt.
	B—C def.	1460–1280	6.85–7.81	v.
	Si—H str.	2280–2080	4.39–4.81	v.–s.
	Si—C def.	*ca.* 1260	*ca.* 7.94	v.–s.
	Si—C str.	840–755	11.90–13.25	v.–s.

* str. = stretching; sym. = symmetric; asym. = asymmetric
s. = strong absorption; v. = variable intensity; d. = doublet
m. = medium absorption; sh. = sharp peak; mlt. = multiplet
w. = weak absorption; br. = broad absorption; def. = deformation

Appendix Table 3.1b Correlations of infrared absorption and structure of organic compounds

Frequency range	Wavelength range	Intensity	Type and group	Bond
4000–3001				
3676–3584	2.72–2.79	v., sh.	Alcohols, phenols, free OH	O—H str.
3650–3496	2.74–2.86	v.	Oximes (R—C=NOH)	O—H str.
3595–3425	2.78–2.92	v., sh.	R—OH, Ar—OH, intramolecular hydrogen bonded	O—H str.
3550–3500	2.82–2.86	m.	Carboxylic acids, free OH	O—H str.
3550–3450	2.82–2.90	v., sh.	R—OH, Ar—OH, dimeric	O—H str.
3550–3205	2.82–3.12	w.	Ketones, C=O overtone	C=O str.
3540–3380	2.83–2.96	s., d.	Primary amides, free NH	N—H str.
3500–3300	2.86–3.03	v., d.	Primary amines, free NH secondary amines	N—H str.
3460–3435	2.89–2.91	s.	Sec. amides, free NH (*trans*)	N—H str.
3360–3220	2.98–3.11	m.	Nitrites (R—O—N=O) overtones	N=O str.
3440–3420	2.91–2.93	s.	Sec. amides, free NH (*cis*)	N—H str.
3440–3400	2.91–2.94	m.	Pyrroles	N—H str.
3400–3300	2.94–3.03	v.	Imines	N—H str.
3400–3230	2.94–3.10	s., br.	R—OH, Ar—OH, polymeric	O—H str.
3400–3200	2.94–3.13	m., d.	Amino acid salts	NH_2 str.
3400–3095	2.94–3.23	m.	Amines, imines, associated	N—H str.
3390–3255	2.95–3.07	m.	Amido acids	$N\overset{+}{—}H$ str.
3380–3150	2.96–3.18	m., mlt.	Charged amine derivatives	NH_3^+ str.
3360–3180	2.97–3.15	m., d.	Amides, bonded NH (primary)	N—H str.
3330–3270	3.00–3.06	m.	Sec. amides, bonded (*trans*)	N—H str.

s. = strong absorption; v. = variable intensity; vib. = vibrating; bon. = bonded
m. = medium absorption; sh. = sharp peak; d. = doublet; sat. = saturated
w. = weak absorption; br. = broad absorption; mlt. = multiplet

Appendix Table 3.1b (*continued*)

Frequency range	Wavelength range	Intensity	Type and groups	Bond
3310–3300	3.02–3.03	m.	Alkynes (RC≡CH)	C—H str.
3300–2500	3.00–4.00	w., br.	R—COOH, bonded OH	O—H str.
3200–1700	3.13–5.88	w., br.	R—OH, Ar—OH, chelate	O—H str.
3175–3135	3.15–3.19	m.	Sec. amides, bonded NH (*cis*)	N—H str.
3150–3050	3.18–3.28	w.	Ethers —CH=C—O— and C=CH—O—	C—H str.
3130–3030	3.20–3.30	m.	Amino acids, hydrochlorides	NH_3^+ str.
ca. 3100	*ca.* 3.23	v.	Tropolones	O—H str.
3100–3070	3.23–3.26	w.	Amides, bon. NH (*cis*)	N—H str.
3095–3075	3.23–3.25	m.	Alkenes ($CHR=CH_2$)	RC—H str.
3085–3040	3.24–3.29	v.	Alkanes ($-CH_2-$, cyclopropane)	C—H str.
3085–3030	3.24–3.30	w.–m., mlt.	Aromatic homocyclic (=C—H)	C—H str.
3075–3020	3.25–3.31	s.	Pyridines, quinolines (=C—H)	C—H str.
3060–3010	3.27–3.32	s.	Pyrimidines, purines (=C—H)	C—H str.
3050–2995	3.28–3.34	w.	Ethers, epoxides	OC—H str.
3040–3010	3.29–3.32	m.	Alkenes ($CHR=CH_2$, $CHR_1=CHR_2$, *cis* or *trans*, $CR_1R_2=CH_2$)	C—H str.
3030–2500	3.30–4.00	w., mlt.	Amino acid hydrochlorides	
3000–2001				
2975–2950	3.36–3.39	m.	Alkanes ($-CH_3$)	C—H str.
2940–2915	3.40–3.43	m.	Alkanes ($-CH_2-$)	C—H str.
2905–2875	3.44–3.48	w.	Alkanes (—CH—)	C—H str.
2880–2860	3.47–3.50	m.	Alkanes ($-CH_3$)	C—H str.
2880–2650	3.47–3.77	w.–m., d.	Aldehydes (—CHO)	C—H str.
2870–2845	3.49–3.52	m.	Alkanes ($-CH_2-$)	C—H str.
2835–2815	3.53–3.55	m.	Ethers ($-O-CH_3$)	C—H str.
ca. 2825	*ca.* 3.54	m.	Alkyl acetals	C—H str.
2825–2760	3.54–3.62	m.–s.	Amines (N-methyl)	C—H str.
2780–2400	3.60–4.17	v.	Deuterated R—OH, Ar—OH	O—D str.
ca. 2780	*ca.* 3.60	v.	Ethers ($-O-CH_2-O-$)	C—H str.
2760–2530	3.62–3.95	w.	Amino acids	
ca. 2700	*ca.* 3.70	s.	Charged amine derivatives	NH_2^+ rck.
2700–2560	3.70–3.90	w., b.	Organo-phosphorus compounds	O—H str.
2640–2360	3.79–4.24	w.	Amido acids	
2600–2400	3.85–4.15	v.	Deuterated amines, imines	N—D str.
2590–2550	3.86–3.92	w.	Organo-sulfur compounds	S—H str.
2500–2325	4.00–4.30	s.	Charged amines ($C=NH^+$)	NH^+ str.
2280–2260	4.39–4.43	s.	Diazonium salts $(R-C=N=N)^+$	
2280–2080	4.39–4.81	v.	Organo-silicon compounds	Si—H str.
2275–2240	4.40–4.46	v.	Isocyanates	N=C=O str.
2260–2240	4.43–4.46	w.–m.	Saturated nitriles	C≡N str.
2240–2220	4.46–4.51	m.–s.	Aryl nitriles	C≡N str.
2235–2215	4.47–4.52	s.	Acyclic α, β unsat. nitriles	C≡N str.
2220–1600	4.51–6.25	v., mlt.	Boron compounds	B—H str.
2200–1800	4.55–5.56	w.–m.	Charged amine derivatives	NH^+ vib.
2185–2120	4.58–4.72	s.	Isonitriles	C≡N str.
2160–2120	4.63–4.72	s.	Azides (—N=N=N)	—N=N=N str.
2155–2150	4.64–4.69	s.	Carbodiimides	N=C=N str.
2140–2100	4.67–4.76	w.	Alkynes (RC≡CH)	C≡C str.
2140–2080	4.67–4.81	w.	Amino acids	NH_3^+ str.
ca. 2100	*ca.* 4.76	w.	Deuterated alkanes	C—D str.
2000–1501				
ca. 2000	*ca.* 5.00	w.	Amino acid hydrochlorides	
1970–1950	5.08–5.13	m.	Allenes (C=C=C)	C≡C type str.
1945–1835	5.14–5.45	w.	Amido acids	
1870–1830	5.35–5.46	s.	Acid anhydrides, 5 ring	C=O str.

Appendix Table 3.1b *(continued)*

Frequency range	Wavelength range	Intensity	Type and groups	Bond
1850–1810	5.40–5.53	s.	Acid anhydrides, conj. 5 ring	C=O str.
1850–1800	5.40–5.56	m.	Alkenes ($CHR{=}CH_2$) overtone	C—H str.
1840–1800	5.44–5.56	s.	Acid anhydride, acyclic	C=O str.
1820–1810	5.50–5.53	s.	Acyl peroxides	C=O str.
1820–1780	5.50–5.62	s.	Conj. acyclic acid anhydrides	C=O str.
1815–1785	5.51–5.60	s.	Acid halides	C=O str.
1805–1780	5.54–5.62	s.	Aroyl peroxides, esters, lactone	C=O str.
1800–1780	5.56–5.62	s.	$(R{-}CO{-}O{-})_2$	C=O str.
1800–1780	5.56–5.62	m.	Alkenes ($CR_1R_2{=}CH_2$), overtone	C—H str.
1800–1770	5.56–5.65	s.	Conj. acid halides	C=O str.
1800–1770	5.56–5.65	s.	Vinylic, phenolic esters	C=O str.
1800–1760	5.56–5.68	s.	Acid anhydride, 5 ring	C=O str.
1795–1740	5.57–5.75	s.	Acid anhydrides, conj. 5 ring	C=O str.
1790–1720	5.59–5.81	s.	Ureas (—CO—NH—CO—), amide I	C=O str.
1785–1755	5.60–5.70	s.	$(RCO{-}O{-})_2$	C=O str.
1780–1770	5.62–5.65	s.	Fused-ring β lactams, amide I	C=O str.
1780–1760	5.62–5.68	s.	Ketones, 4 ring, sat. γ lactone	C=O str.
1780–1740	5.62–5.75	s.	Acyclic acid anhydrides	C=O str.
1760–1730	5.68–5.78	s.	Simple β lactams	C=O str.
1760–1720	5.68–5.81	s.	Conj. acyclic anhydrides	C=O str.
1755–1740	5.70–5.75	s.	α keto esters, α diesters	C=O str.
1755–1730	5.70–5.78	s.	α-amino acid hydrochloride	C=O str.
1755–1720	5.70–5.81	s.	Dicarboxylic α-amino acids	C=O str.
1750–1740	5.71–5.75	s.	Ketones, 5 ring	C=O str.
1750–1735	5.71–5.76	s.	γ-keto esters, diesters, nonenolic β-keto esters	C=O str.
1750–1735	5.71–5.76	s.	Sat. aliphatic esters, α-lactones	C=O str.
1745–1725	5.73–5.80	s.	$CO{-}O{-}CH_2{-}CO{-}$	C=O str.
1750–1700	5.71–5.88	s.	Fused-ring γ lactams	C=O str.
1740–1720	5.75–5.81	s.	Sat. aliphatic aldehydes	C=O str.
1740–1715	5.75–5.83	s.	α-halogeno carboxylic acids	C=O str.
1735–1700	5.76–5.88	s.	Urethanes	C=O str.
1730–1710	5.78–5.85	s.	—CO—CO—	C=O str.
1730–1700	5.78–5.88	s.	Amino acid hydrochlorides	C=O str.
1730–1715	5.78–5.83	s.	α,β unsat., aryl esters	C=O str.
1730–1700	5.78–5.88	s.	Dicarboxylic amino acids	C=O str.
1725–1705	5.80–5.87	s.	$-CO{-}CH_2{-}CH_2{-}CO{-}$	C=O str.
1725–1700	5.80–5.88	s.	Sat. aliphatic acids, dimer, acyclic, $CH_2{-}CO{-}CH_2{-}$ ketones	C=O str.
1725–1695	5.80–5.90	s.	α-amido acids	C=O str.
1720–1700	5.81–5.88	s.	Ketones, 6-ring	C=O str.
1715–1700	5.83–5.88	s.	Ketones, 7-ring	C=O str.
1715–1695	5.83–5.90	s.	Aryl aldehydes	C=O str.
1715–1680	5.83–5.95	s.	α,β unsat. acids	C=O str.
1710–1690	5.85–5.92	s.	Carbamates, amide I band	C=O str.
1710–1670	5.85–5.99	s.	—CO—NH—CO—, amide I band	C=O str.
1705–1685	5.78–5.93	s.	α,β unsat. aldehydes	C=O str.
ca. 1700	*ca.* 5.88	s.	Simple γ lactams, amide I band	C=O str.
1700–1680	5.88–5.95	s.	Aryl carboxylic acids, dimer, aryl ketones	C=O str.
1700–1665	5.88–6.01	s.	Secondary amides, amide I band	C=O str.
1695–1660	5.90–6.02	s.	α,β unsat. acyclic or 6-ring ketones	C=O str.
ca. 1690	*ca.* 5.92	s.	Primary amides, amide I band	C=O str.
1690–1670	5.92–5.99	w.	$CR_1R_2{=}CR_3R_4$, alkenes	C=C str.
1690–1670	5.92–5.99	s.	*o*-hydroxy (amino) benzoates	C=O str.
1690–1655	5.92–6.04	s.	Quinones, 2 CO's in same ring	C=O str.
1690–1635	5.92–6.01	v.	Oximes, oxazines, oxazolines, oxazolones, azomethines, acyclic C=N	C=N str.
ca. 1680	*ca.* 5.95	s.	Large-ring cyclic lactams	C=O str.

Appendix Table 3.1b (*continued*)

Frequency range	Wavelength range	Intensity	Type and groups	Bond
1680–1660	5.95–6.02	s.	Conj. polyene aldehydes	C=O str.
1680–1650	5.95–6.06	s.	Intramolecular H-bonded carboxylic acids	C=O str.
1680–1650	5.95–6.06	v.	Nitrites (R—O—N=O *trans*)	N=O str.
1680–1630	5.95–6.14	s.	Sec. amides, amide I band	C=O str.
1680–1620	5.95–6.17	v.	Nonconj. alkenes	C=C str.
1675–1665	5.97–6.00	v.	Alkenes, $CR_1R_2{=}CHR_3$	C=C str.
ca. 1675	*ca.* 5.97	s.	Thioesters	C=S str.
1670–1660	5.99–6.02	s.	Cross-conj. dienones	C=O str.
1670–1645	5.99–6.08	s.	Intramolecular H-bonded (—C(OH)=C—CHO type) aldehydes	C=O str.
1670–1630	5.99–6.14	s.	Tertiary amides, amide I band	C=O str.
1665–1635	6.01–6.12	v.	Alkenes ($CHR_1{=}CHR_2$-*cis*)	C=C str.
1665–1630	6.01–6.14	v.	Oximes, oxazines, etc.	C=N str.
ca. 1660	*ca.* 6.02	s.	Ureas (NH—CO—NH), amide I band	C=O str.
1660–1640	6.02–6.10	v.	Alkenes ($CR_1R_2{=}CH_2$)	C=C str.
1660–1610	6.02–6.21	w.	Amino acids containing NH_2 group, amino acid I band	NH_3^+ def.
1660–1580	6.02–6.33	s.	Alkenes conj. with C=O or C=C	C=C str.
1660–1480	6.02–6.76	v.	Thiazoles (cyclic α, β sat. C=N)	C=N str.
1655–1635	6.04–6.12	s.	Enolic β-keto esters, chelated	C=O str.
1655–1635	6.04–6.12	s.	Quinones, 2 CO's in 2 rings	C=O str.
1655–1610	6.04–6.21	s.	*o*-CO—C_6H_4—OH (or NH_2), H bonded	C=O str.
1655–1610	6.04–6.21	s.	Nitrates ($RONO_2$), asym. vibration	NO_2 str.
ca. 1650	*ca.* 6.06	s.	Primary amides, amide I band	C=O str.
1650–1590	6.06–6.31	s.	Prim. amides, amide II band, combination NH def. + CN str.	NH def. + CN str.
1650–1620	6.06–6.17	s.	Amido acids, amide I band	
1650–1580	6.06–6.33	m.	Pyridines, quinolines	C=C + C=N str.
1650–1580	6.06–6.33	m.–s.	Prim. amines	NH def.
1650–1550	6.06–6.45	w.	Sec. amines	NH def.
1645–1640	6.08–6.10	v.	Alkenes ($CHR{=}CH_2$)	C=C str.
1640–1605	6.10–6.23	s.	Alkyl nitroguanidines, asym. NO_2 vibrations	NO_2 str.
1640–1535	6.10–6.52	s., d.	Ketones (—CO—CH_2—CO or —CO—C=C—OH)	C=O str.
1630–1575	6.14–6.35	v.	Azo compounds	N=N str.
ca. 1625	*ca.* 6.16	s.	Alkenes, phenyl conj. C=C	C=C str.
1625–1610	6.16–6.21	v.	Nitrites (RON=O *cis* form)	N=O str.
1625–1575	6.16–6.35	v.	Aromatic homocyclic comp.	C=C i-p vib.
1620–1600	6.17–6.25	s.	Tropolones	C=O str.
1620–1600	6.17–6.25	s.	Amido acids, amide I band	
1620–1590	6.17–6.31	s.	Prim. amides, amide II band, combination NH def. + CN str.	NH def. + CN str.
1620–1560	6.17–6.41	s.	Nitroso compounds, α halogeno	N=N str.
1620–1560	6.17–6.41	m.–s.	Charged amine derivatives	NH_2^+ def.
1610–1590	6.21–6.29	w.	Amino acid hydrochlorides	NH_3^+ def.
1610–1550	6.21–6.45	s.	Carboxylate ion, asym. str.	
ca. 1600	*ca.* 6.25	m.	Charged amine derivatives	NH_3^+ def.
1600–1575	6.25–6.35	s.	α, α-dihalogenonitro compounds	NO_2 str.
1600–1560	6.25–6.41	s.	Amino acid salts, all amino acids with ionized carboxyl	C=O str.
1590–1575	6.29–6.36	v.	Aromatic homocyclic comp. C=C	i–p str.
1590–1575	6.29–6.36	s.	Nitroureas	NO_2 str.
1585–1530	6.31–6.54	s.	Saturated nitramines	NO_2 str.
1580–1570	6.33–6.37	s.	α-halogenonitro compounds	NO_2 str.
1580–1550	6.33–6.45	w.	Nitrogen heterocycles, combination C=C and C=N str.	C=C + C=N str.
1580–1520	6.33–6.58	m.	Pyrimidines and purines, combination C=C and C=N str.	C=C+C=N str.
1570–1515	6.37–6.60	s.	Sec. acyclic amides, amide II band, combination NH def. and CN str.	NH def. + CN str.
1570–1500	6.37–6.67	s.	All amido acids, amido II band	N—H def.
ca. 1565	*ca.* 6.39	v.	Pyrroles	C=C str.
1565–1545	6.39–6.47	s.	Prim. and sec. nitro compounds	NO_2 str.

Appendix Table 3.1b (*continued*)

Frequency range	Wavelength range	Intensity	Type and groups	Bond
ca. 1550	*ca.* 6.45	s.	Tertiary aliphatic nitroso compounds, vapor phase	N=O str.
1550–1510	6.45–6.62	s.	Aromatic nitro compounds	N=O str.
1550–1485	6.45–6.73	v.	Amino acids cont. NH_2 group, amino acid II band	NH_3^+ def.
1550–1510	6.45–6.62	s.	Sec. acyclic amides, amide II band	NH def. + CN str.
1550–1485	6.45–6.73	v.	Amino acid hydrochlorides	NH_3^+ def.
1545–1530	6.47–6.54	s.	Tert. nitro compounds	NO_2 str.
1530–1510	6.54–6.62	s.	α,β-unsat. nitro compounds	NO_2 str.
1525–1475	6.56–6.78	v.	Aromatic homocyclic comp. C=C	i-p vib.
1510–1480	6.62–6.76	m.	Pyridines, quinolines, combination C=C and C=N str.	C=C + C=N str.
ca. 1500	*ca.* 6.67	v.	Pyrroles	C=C str.
ca. 1500	*ca.* 6.67	s.	Aromatic nitroso compounds	N=O str.
ca. 1500	*ca.* 6.67	s.	Amine salts	NH def.
1500–1001				
1500–1440	6.67–6.94	s.	Nitrosamines (RNN=O)	N=O str.
1485–1445	6.74–6.92	m.	Alkanes ($-CH_2-$)	C—H def.
1470–1430	6.80–7.00	m.	Alkanes ($-CH_3$)	C—H def.
ca. 1467	*ca.* 6.81	m.	Alkanes, CH_2 scissor	C—H def.
ca. 1460	*ca.* 6.85	m.	Alkanes, asym. CH_3	C—H def.
1455 (*ca.*)	*ca.* 6.87	m.	Alkanes, alicyclic CH_2 scissor	C—H def.
ca. 1450	*ca.* 6.90	m.	Aromatic multiple bond	C=C str.
1430–1400	7.0–7.14	v.	Ketones, esters, α-methylene C—H scissor	C—H def.
1430–1350	7.00–7.41	s.	Sulfites	S=O str.
1420–1410	7.04–7.09	s.	Alkenes ($RCH=CH_2$ or $R_1R_2C=CH_2$)	C—H def.
1420–1390	7.04–7.20	w.	Alcohols	O—H def.
1410–1310	7.10–7.60	s.	Phenols, tert. alcohols	O—H def.
1400–1300	7.15–7.69	s.	Carboxylic acids, ionic	
1400–1000	7.15–10.0	s.	Halogen comp., fluorides	C—F str.
1395–1385	7.17–7.22	m., d.	Alkane, tert. butyl	C—H def.
1390–1360	7.20–7.35	m., d.	Alkane, geminal dimethyl, isopropyl, tert. butyl, sym. CH_3 bending	CH_3 def.
1380–1370	7.25–7.30	s.	Alkane, $-CH_3$	C—H def.
1380–1370	7.25–7.30	s.	Aliphatic nitro compounds	NO_2 str.
1370–1340	7.30–7.46	s.	Sulfonyl chlorides	S=O str.
1370–1300	7.30–7.70	s.	Aromatic nitro compounds	NO_2 str.
ca. 1365	*ca.* 7.33	s.	Alkane, tert. butyl	C—H def.
1360–1310	7.35–7.64	s.	Aromatic tert. amines	C—N vib.
1350–1300	7.41–7.69	s.	Sulfones, sulfonamides	S=O str.
1350–1280	7.41–7.81	s.	Aromatic sec. amines	C—N vib.
1350–1260	7.41–7.94	s.	Prim. and sec. alcohols	O—H def.
ca. 1340	*ca.* 7.46	w.	Alkanes	C—H def.
1340–1250	7.46–8.00	s.	Aromatic prim. amines	C—N vib.
1340–1180	7.46–8.48	w.	Azides	$-N_3$ str.
1335–1310	7.39–7.64	v.	Sulfur compounds	S=O str.
1310–1295	7.64–7.72	m.	Alkene ($R_1CH=CHR_2$), *trans*	C—H def.
1300–1250	7.70–8.00	s.	Nitrates	$O-NO_2$ vib.
1275–1200	7.84–8.33	v.	Conj. ethers	ROR str.
ca. 1270	*ca.* 7.88	v.	Aromatic esters	$-C(=O)-O-R$ str.
1260–1200	7.94–8.33	v.	Aromatic ethers	R—O—Ar str.
1257–1232	7.95–8.12	v.	Aliphatic esters, CH_3COOR	
1255–1200	7.97–8.33	s., d.	Alkanes, tert. butyl	Skeletal str.
1240–1190	8.06–8.40	v.	Aromatic phosphorus comp.	P—O str.
1230–1150	8.13–8.70	s.	Sulfites	S=O str.
1210–1150	8.27–8.70	s.	Sulfonic acids	S=O str.
1220–1020	8.20–9.80	w.	Aliphatic amines	C—N vib.

Appendix Table 3.1b (*continued*)

Frequency range	Wavelength range	Intensity	Type and groups	Bond
ca. 1200	*ca.* 8.33	s.	Phenols	C—O str.
1200–1050	8.33–9.52	s.	Sulfur compounds	C=S str.
1200–1190	8.33–8.40	v.	Esters (RCOOR)	
1185–1175	8.44–8.51	v.	Esters (H—COOR)	
1185–1165	8.44–8.59	s.	Sulfonyl chlorides	S=O str.
1180–1140	8.48–8.77	s.	Sulfonamides	S=O str.
1175–1140	8.51–8.77	s.	Alkanes, geminal dimethyl	Skeletal vib.
1175–1155	8.51–8.65	v.	Methyl esters ($R—COOCH_3$)	
1175–1125	8.51–8.89	w.	Substituted benzenes, 1,3-disubstituted or tri-substituted benzenes	C—H def.
1160–1140	8.62–8.77	s.	Sulfones	S=O str.
ca. 1150	*ca.* 8.70	s.	Tertiary alcohols	C—OH str.
1150–1070	8.70–9.35	v.	Aliphatic ethers	R—O—R str.
1120–1100	8.93–9.09	s.	Secondary alcohols	C—OH str.
1070–1030	9.35–9.71	s.	Sulfoxides	S=O str.
ca. 1060	*ca.* 9.40	m.	Allene (C=C=C)	
1060–1030	9.43–9.71	s.	Sulfonic acids	S=O str.
1075–1010	9.30–9.90	v.	Primary alcohols	C—OH str.
1050–990	9.52–10.1	v.	Phosphorus compounds	P—O str.
1000–501				
995–985	10.05–10.15	s.	Monosubstituted alkenes ($RCH=CH_2$)	C—H def.
970–960	10.31–10.42	s.	Disubstituted alkenes ($R_1CH=CHR_2$) *trans*	C—H def.
915–905	10.93–11.05	s.	Monosubstituted alkenes ($RCH=CH_2$)	C—H def.
900–860	11.11–11.63	m.	Tetra- or penta-substituted benzene containing 1 free H	C—H def.
895–885	11.17–11.30	s.	Geminal disubstituted alkene	C—H def.
885–870	11.30–11.50	m.	1,2,4-trisubstituted benzene, another peak at 852–805	C—H def.
ca. 870	11.50 (*ca.*)	m.	Pentasubstituted benzene	C—H def.
870–800	11.50–12.50	v.	Benzene ring containing two adjacent H atoms	C—H def.
840–790	11.90–12.66	s.	Trisubstituted alkenes	C—H def.
810–750	12.34–13.34	v.	Benzene ring with three adjacent H atoms	C—H def.
800–600	12.50–16.67	s.	Halides	C—H def.
770–735	12.98–13.61	v., s.	Benzene ring with four adjacent free H atoms	C—Cl str.
770–730	12.98–13.70	v., s.	Benzene ring with five adjacent free H atoms, second peak at 710–690	C—H def.
ca. 690	*ca.* 14.50	s.	Disubstituted alkenes ($R_1CH=CHR_2$) *cis*	C—H def.
ca. 650	*ca.* 15.40	s.	Sulfonic acids	S=O str.
ca. 630	*ca.* 15.90	s.	Alkynes	C—H def.
600–500	16.60–20.00	s.	Bromides	C—Br str.
ca. 500	*ca.* 20.00	s.	Iodides	C—I str.

s. = strong absorption; m. = medium absorption; w. = weak absorption; v. = variable intensity; sh. = sharp peak; br. = broad absorption; vib. = vibrating; d. = doublet; mlt. = multiplet; bon. = bonded; sat. = saturated

Appendix Table 3.1c Infrared correlation chart

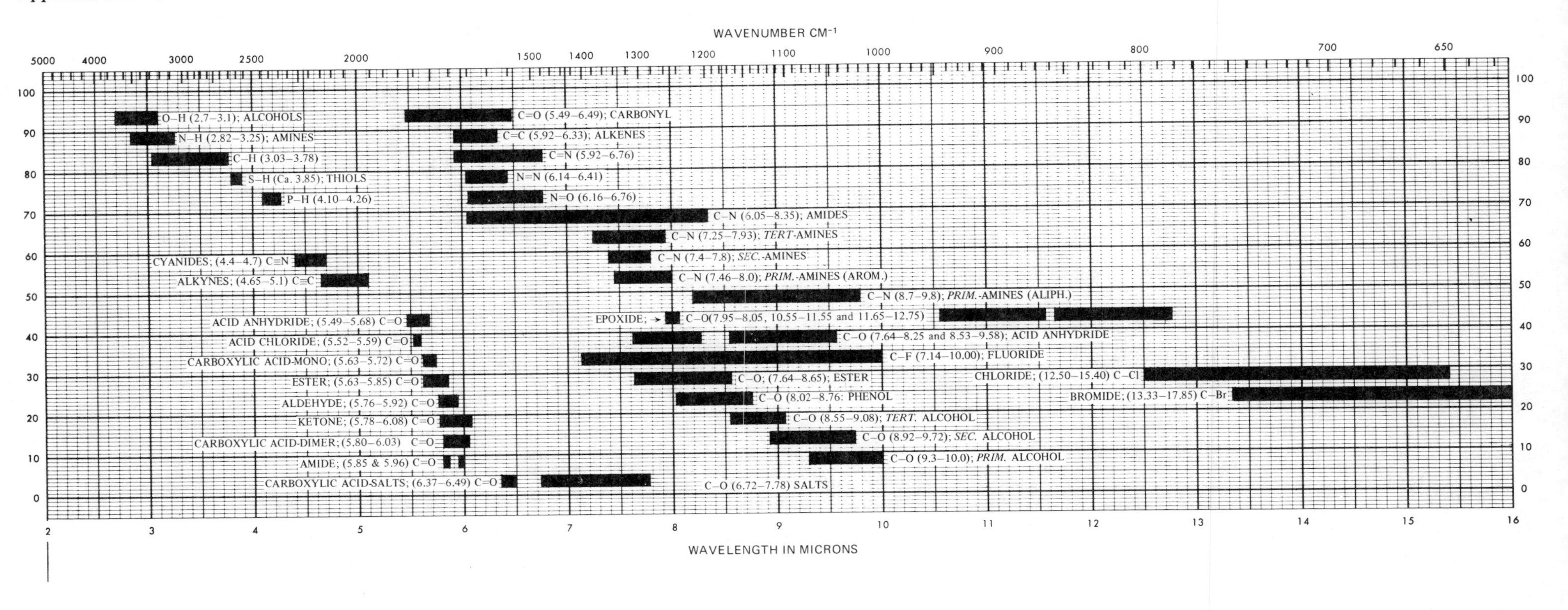

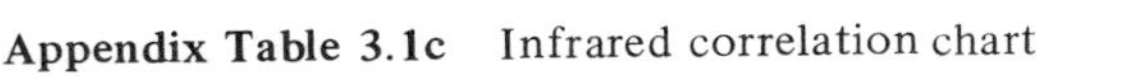

Appendix Table 3.2 Infrared absorption frequencies of alkenes

Alkene type	Vibration mode	Frequency, cm^{-1}	Wavelength, μ	Relative intensity
$RHC{=}CH_2$	C—H str (CH_2)	3095–3075	3.23–3.25	m
	C—H str (CHR)	3040–3010	3.29–3.32	m
	Overtone	1850–1800	5.40–5.56	m
	C=C str	1645–1640	6.08–6.10	v
	CH_2 i-p def	1430–1410	7.00–7.10	w
	CH i-p def	1300–1290	7.69–7.75	v
	CH o-o-p def	995–985	10.05–10.15	m
	CH_2 o-o-p def	915–905	10.93–11.05	s
$R_1HC{=}CHR_2$ (*cis*)	C—H str	3040–3010	3.29–3.32	m
	C=C str	1665–1635	6.01–6.12	v
	C—H i-p def	1430–1400	7.00–7.14	w
	C—H o-o-p def	730–665	13.70–15.04	s
$R_1HC{=}CHR_2$ (*trans*)	C—H str	3040–3010	3.29–3.32	m
	C=C str	1675–1665	5.97–6.01	v
	C—H i-p def	1310–1290	7.63–7.75	w
	C—H o-o-p def	980–960	10.20–10.42	s
$R_1R_2C{=}CH_2$	C—H str	3095–3075	3.23–3.25	m
	Overtone	1800–1780	5.56–5.62	m
	C=C str	1660–1640	6.02–6.10	v
	CH_2 i-p def	1420–1410	7.04–7.09	w
	CH_2 o-o-p def	895–885	11.17–11.30	s
$R_1R_2C{=}CHR_3$	C—H str	3040–3010	3.29–3.32	m
	C=C str	1675–1665	5.97–6.01	v
	C—H o-o-p def	850–790	11.76–12.66	m
$R_1R_2C{=}CR_3R_4$	C=C str	1690–1670	5.92–5.99	w
$Ar{-}HC{=}CH_2$	C=C str	~1625	~6.16	s
C=O or C=C conjugated with C=C	C=C str	1660–1580	6.02–6.33	s

Abbreviations: str = stretching, i-p def = in-plane deformation, o-o-p def = out-of-plane deformation, s = strong, m = medium, w = weak, v = variable.

Appendix Table 3.3 Absorption frequencies typical of aromatic homocyclic compounds

Aromatic compound	=C—H str (multiple bands)		C=C str (ring stretching)		C—H i-p def (w)	C—H o-o-p def
Monosubstituted	3080–3030 cm^{-1}		1604 ± 3 cm^{-1}		1177 ± 6 (2,5 *vs* 3,6)	751 ± 15 s; 5 adj H wag
	3.24–3.33 μ	w–m	6.25–6.22 μ	v	8.45–8.54 μ	13.06–13.58 μ
			1185 ± 3		1156 ± 5 (3,5 *vs* 4)	697 ± 11 s; 5 adj H wag
			6.32–6.30 μ	v	8.61–8.69 μ	14.1–14.58 μ
			1510 → 1480*		1073 ± 4 (2,6 *vs* 3,4,5)	–
			6.62 → 6.75 μ	v	9.28–9.36 μ	
			1452 ± 4		1027 ± 3 (2,3 *vs* 5,6)	–
			6.87–6.91 μ	v	9.71–9.77 μ	
1 : 2-Disubstituted	3080–3030 cm^{-1}		1607 ± 9		1269 ± 17 (all clockwise)	751 ± 7 s; 4 adj H wag
	3.24–3.33 μ	w–m	6.26–6.19 μ	v	7.77–7.99 μ	13.20–13.44 μ
			1577 ± 4		1160 ± 4 (3,5 *vs* 4,6)	–
			6.33–6.36 μ	v	8.59–8.65 μ	
			1510 → 1460		1125 ± 14 (3,6 *vs* 4,5)	–
			6.85 → 6.62 μ	v	8.78–9.0 μ	
			1447 ± 10		1033 ± 11 (3,4 *vs* 5,6)	–
			6.87–6.96 μ	v	9.58–9.78 μ	

pendix **Table 3.3** (*continued*)

omatic compound	=C—H str (multiple bands)		C=C str (ring stretching)		C—H i-p def (w)	C—H o-o-p def
3-Disubstituted	3080–3030 cm^{-1}		1600 → 1620		1278 ± 12 (all clockwise)	900 → 860 m; 1 free H
	3.24–3.33 μ	w–m	6.25 → 6.17 μ	v	7.75–7.90 μ	11.12 → 11.63 μ
			1586 ± 5		1157 ± 5 (2,5 *vs* 4,6)	782 ± 10 s; 3 adj H wag
			6.29–6.32	v	8.60–8.68 μ	12.62–12.96 μ
			1495 → 1470		1096 ± 7 (4 *vs* 6)	725 → 680 m; 3 adj H wag
			6.69 → 6.80 μ	v	9.07–9.18 μ	13.79 → 14.71 μ
			1465 → 1430		1076 ± 7 (2 *vs* 5)	—
			6.83 → 6.99 μ	v	9.23–9.36 μ	
4-Disubstituted	3080–3030 cm^{-1}		1606 ± 6		1258 ± 11 (all clockwise)	817 ± 15 s; 2 adj H wag
	3.24–3.33 μ	w–m	6.20–6.25 μ	v	7.88–8.02 μ	12.02–12.47 μ
			1579 ± 6		1175 ± 6 (2,5 *vs* 3,6)	—
			6.31–6.36 μ	v	8.47–8.56 μ	
			1520 → 1480		1117 ± 7 (2,6 *vs* 3,5)	—
			6.58 → 6.76 μ	v	8.90–9.01 μ	
			1409 ± 8		1013 ± 5 (2,3 *vs* 5,6)	—
			7.05–7.15 μ	v	9.82–9.92 μ	
2 : 3-Trisubstituted	3080–3030 cm^{-1}				1175 → 1125	800 → 770 s; 3 adj H wag
	3.24–3.33 μ	w–m			8.51 → 8.90 μ	12.50 → 12.99 μ
					1110 → 1070	720 → 685 m; 3 adj H wag
					9.01 → 9.35 μ	13.90 → 14.60 μ
					1070 → 1000	—
					9.35 → 10.00 μ	
					1000 → 960	—
					10.00 → 10.42 μ	
: 2 : 4-Trisubstituted	3080–3030 cm^{-1}		1616 ± 8		1225 → 1175	—
	3.24–3.33 μ	w–m	6.15–6.22 μ	v	8.16 → 8.51 μ	
			1577 ± 8		1175 → 1125	900 → 860 m; 1 adj H wag
			6.31–6.37 μ	v	8.51 → 8.89 μ	11.11 → 11.63 μ
			1510 ± 8		1125 → 1090	860 → 800 s; 2 adj H wag
			6.60–6.66 μ	v	8.89 → 9.18 μ	11.63 → 12.50 μ
			1456 ± 1		1070 → 1000	—
			6.86–6.88 μ	v	9.35 → 10.00 μ	
: 3 : 5-Trisubstituted	3080–3030 cm^{-1}				1175 → 1125	900 → 860 m; 1 adj H wag
	3.24–3.33 μ				8.51 → 8.89 μ	11.11 → 11.63 μ
					1070 → 1000	865 → 810 s; 1 adj H wag
					9.35 → 10.00 μ	11.56 → 12.30 μ
						730 → 675 s; 1 adj H wag
						13.70 → 14.82 μ
: 2 : 3 : 4-Tetrasubstituted	3080–3030 cm^{-1}					860 → 800 s; 1 adj H wag
	3.24–3.33 μ					11.63 → 12.50 μ
: 2 : 3 : 5, 1 : 2 : 4 : 5, and	3080–3030 cm^{-1}					900 → 860 m
: 2 : 3 : 4 : 5-Substituted	3.24–3.33 μ					11.11 → 11.63 μ

bbreviations: str = stretching, i-p def = in-plane deformation, o-o-p def = out-of-plane deformation, w = weak, m = medium, s = strong, = variable, adj = adjacent, wag = wagging.

The symbol 1510 → 1480 indicates the range, in cm^{-1}, of absorption of radiation. Usually an *electron-donor* substituent causes bsorption near 1510 cm^{-1}, while an *electron-acceptor* substituent causes absorption near 1480 cm^{-1}.

Appendix Table 3.4 Decreasing order of values of C=O stretching frequency

Type of carbonyl compound	Typical examples	Phase	C=O stretching frequency, cm^{-1}	Value of absorption wavelength, μ
Acid anhydrides (RCOOCOR)			1820 and 1760 (two bands)	5.49 and 5.68
	Acetic anhydride	CCl_4	1825 and 1754	5.48 and 5.70
	Succinic anhydride	$CHCl_3$	1820 and 1776	5.49 and 5.63
	Benzoic anhydride	$CHCl_3$	1818 and 1740	5.50 and 5.75
Acid chlorides (RCOCl)			1812 to 1790	5.52 to 5.59
	Acetyl chloride	CCl_4	1812	5.52
	Isovaleryl chloride	Film	1792	5.58
	Benzoyl chloride	CCl_4	1739	5.75
Carboxylic acids (RCOOH) monomers			1775 to 1750	5.63 to 5.72
	Butyric acid	CCl_4	1775	5.63
	Acetic acid	CCl_4	1760	5.68
Esters (RCOOR)			1780 to 1710	5.62 to 5.85
	Phenyl acetate	Film	1765	5.68
	Vinyl acetate	CCl_4	1765	5.68
	Methyl acetate	CCl_4	1750	5.71
	Methyl propionate	CCl_4	1748	5.72
	Ethyl propionate	CCl_4	1736	5.76
	Propyl formate	CCl_4	1733	5.77
	Ethyl benzoate	CCl_4	1724	5.80
	Benzyl benzoate	Film	1720	5.81
	Ethyl cinnamate	CCl_4	1710	5.85
Aldehydes (RCHO)			1740 to 1690	5.76 to 5.92
	n-butanal	CCl_4	1736	5.76
	Acetaldehyde	CCl_4	1730	5.78
	Valeraldehyde	CCl_4	1730	5.78
	Isovaleraldehyde	Film	1715	5.83
	o-chlorobenzaldehyde	Film	1695	5.90
	p-anisaldehyde	CCl_4	1690	5.92
Ketones (RCOR)			1730 to 1645	5.78 to 6.08
	Norcamphor	Film	1730	5.78
	Butanone	CCl_4	1724	5.80
	Acetone	CCl_4	1720	5.81
	2-Pentanone	CCl_4	1712	5.84
	Methyl isopropyl ketone	Film	1709	5.85
	Methyl phenyl ketone	CCl_4	1695	5.90
	Methyl vinyl ketone	CCl_4	1686	5.93
	Propiophenone	CCl_4	1686	5.93
	Methyl-*p*-tolyl ketone	Film	1675	5.95
	Benzophenone	$CHCl_3$	1669	5.99
	p-Benzoquinone	$CHCl_3$	1645	6.08
Carboxylic acids $(RCOOH)_2$ dimers			1724 to 1665	5.80 to 6.01
	α-chloropropionic acid	Film	1724	5.80
	Butyric acid	Film	1721	5.81
	n-Hexanoic acid	Film	1698	5.89
	Benzoic acid	KBr	1678	5.96
	Salicylic acid	KBr	1665	6.01

Appendix Table 3.4 (*continued*)

Type of carbonyl compound	Typical examples	Phase	C=O stretching frequency, cm^{-1}	Value of absorption wavelength, μ
Amides			1678 and shoulder at 1710	5.96 and 5.85
	Acetanilide	$CHCl_3$	1678 and shoulder at 1701	5.96 and 5.88 (sh.)
($RCONH_2$)	N-methylacetamide	CCl_4	1675 and shoulder at 1705	5.97 and 5.86 (sh.)
(RCONHR)	N,N-dimethylacetamide	CCl_4	1660 and shoulder at 1710	6.02 and 5.85 (sh.)
($RCONR_2$)	Acetamide	$CHCl_3$	1670 and shoulder at 1716	5.97 and 5.83 (sh.)
	Benzamide	$CHCl_3$	1667 and shoulder at 1709	6.00 and 5.85 (sh.)
	Caprolactam	CCl_4	1667 and shoulder at 1701	6.00 and 5.88 (sh.)

Primary and secondary amides display another band in the carbonyl region between 1650 and 1515 cm^{-1} (6.06 and 6.60 μ). This band is called the amide II band.

Type of carbonyl compound	Typical examples	Phase	C=O stretching frequency, cm^{-1}	Value of absorption wavelength, μ
Ions and salts of acids ($R-COO^{\ominus}$)			1570 to 1540	6.37 to 6.49
	Potassium acetate	KBr	1567	6.38
	Potassium stearate	KBr	1543	6.48

Appendix Table 3.5 Variation in frequency of vibrations of carbonyl stretching bands due to inter- and intramolecular factors*

Shift to higher frequency	Add cm^{-1}	Shift to lower frequency	Subtract cm^{-1}
Basic value in CCl_4	1720	Neat—solid or liquid state	10
Solvent		*Solvent*	
Hydrocarbon solvents	7	$CHCl_3$, $CHBr_3$, CH_3CN (partially polar)	15
Ring strain		*Ring strain*	
Angle decreases 6 → 5 ring	35	Angle increases 6 → 7 to 10 ring	10
Bridged systems	15		
Substitution on α-carbon		*Substitution on α-carbon*	
(Field and inductive effects)		Each alkyl group	5
Substituent *cis* oriented and coplanar		*Alkyl groups substituted by amine*	
—Cl, —Br, —OR	20	$-NH_2$ (amide)	5
—OH, and —OAc		—NHMe (monosubstituted amide)	30
Substituent *trans* and		$-NMe_2$ (disubstituted amide)	55
nonplanar	Nil	*Intramolecular hydrogen bonding*	
Alkyl group substituted by		Weak: α or β-OH ketone	10
electronegative atoms or groups		Medium: *o*-OH arylketone	40
—H (aldehyde)	10	Strong: β-diketone	100
—OR (ester)	25	*Intermolecular hydrogen bonding*	
—OH (monomeric acid)	40	Weak: ROH···O=C<	15
—O—C=C (vinyl ester)	50	Strong: RCOOH dimer	45
—Cl (acid chloride)	90	*Conjugation (depends on stereochemistry)*	
—OCOR (anhydrides)	100	First C=C	30
		Second C=C	15
		Third C=C	Nil
		Benzene ring	20
		Vinylogous, —CO—C=C—X (X=H or O)	40

*Adapted by permission of the publishers from J. C. D. Brand and G. Eglington, *Applications of Spectroscopy*, Oldbourne Press, London, 1965.

Appendix Table 3.6 Vibration modes of amines and amino acids (frequencies in cm^{-1})

Compound	N—H stretching		N—H deformation	C—N stretching	C=O stretching	C—O stretching	Other vibrations and remarks
	asym	sym					
Primary amines	3550–3350 (2.82–2.99 μ)	3450–3250 (2.90–3.08 μ)	1650–1580 (6.06–6.33 μ)	1220–1020 ali (8.2–9.8 μ) 1340–1250 aro (7.46–8.0 μ)			Shoulder at 3200 (3.12 μ), overtone of 1610 (6.21 μ) band. N—H wagging strong band at 850–750 (11.76–13.34 μ)
Secondary amines	3550–3350 (2.82–2.99 μ)		1650–1550 (6.06–6.45 μ)	1350–1280 aro (7.41–7.81 μ)			N—H wagging strong band at 750–700 (13.34–14.29 μ). N—H def. band often masked by aromatic band in aromatic compounds.
Tertiary amines				1380–1260 aro (7.24–7.94 μ)			N—Me band in secondary and tertiary amines, observed at 2820–2760 (3.54–3.62 μ)
Amine hydrochloride	About 3380; (2.96 μ) NH_3^+ str		About 1600 (6.25 μ) asym About 1300 (7.7 μ) sym About 800 (12.5 μ) NH_3^+ rock				
Charged amine derivatives	About 3280; (3.05 μ) NH_3^+ str 3350–3150; NH_3^+ str (2.99–3.17 μ)						
Primary amino acids	Solid phase 3125–3030 (3.2–3.3 μ)		1660–1610 band I (6.02–6.21 μ) 1550–1485 band II (6.45–6.74 μ)		1600–1560 ionized (6.25–6.41 μ) carboxyl str.		Weak band at 2760–2530 (3.62–3.96 μ) Medium band at 1300 (7.69 μ)
Dicarboxylic amino acids					1755–1720 (5.7–5.81 μ) unionized carboxyl str		
Amino acid hydrochlorochlorides	3130–3030 (3.2–3.3 μ)		1610–1590 band I (6.21–6.29 μ) 1550–1480 band II (6.45–6.75 μ)		1755–1730 (5.7–5.78 μ) unionized carboxyl	1230–1215 (8.13–8.23 μ)	Series of bands between 3030 and 2500 (3.3 and 4.0 μ)
Amino acid sodium salts	3400–3200 (2.94–3.13 μ) Two bands				1600–1560 (6.25–6.41 μ) ionized carboxyl		

Appendix Table 3.7 Wavelength-wavenumber conversion table

	Wavenumber, cm^{-1}									
λ	0	1	2	3	4	5	6	7	8	9
2.0	**5000**	**4975**	**4950**	**4926**	**4902**	**4878**	**4854**	**4831**	**4808**	**4785**
2.1	4762	4739	4717	4695	4673	4651	4630	4608	4587	4566
2.2	4545	4525	4505	4484	4464	4444	4425	4405	4386	4367
2.3	4348	4329	4310	4292	4274	4255	4237	4219	4202	4184
2.4	4167	4149	4132	4115	4098	4082	4065	4049	4032	4016
2.5	4000	3984	3968	3953	3937	3922	3906	3891	3876	3861
2.6	3846	3831	3817	3802	3788	3774	3759	3745	3731	3717
2.7	3704	3690	3676	3663	3650	3636	3623	3610	3597	3584
2.8	3571	3559	3546	3534	3521	3509	3497	3484	3472	3460
2.9	3448	3436	3425	3413	3401	3390	3378	3367	3356	3344
3.0	**3333**	**3322**	**3311**	**3300**	**3289**	**3279**	**3268**	**3257**	**3247**	**3236**
3.1	3226	3215	3205	3195	3185	3175	3165	3155	3145	3135
3.2	3125	3115	3106	3096	3086	3077	3067	3058	3049	3040
3.3	3030	3021	3012	3003	2994	2985	2976	2967	2959	2950
3.4	2941	2933	2924	2915	2907	2899	2890	2882	2874	2865
3.5	2857	2849	2841	2833	2825	2817	2809	2801	2793	2786
3.6	2778	2770	2762	2755	2747	2740	2732	2725	2717	2710
3.7	2703	2695	2688	2681	2674	2667	2660	2653	2646	2639
3.8	2632	2625	2618	2611	2604	2597	2591	2584	2577	2571
3.9	2564	2558	2551	2545	2538	2532	2525	2519	2513	2506
4.0	**2500**	**2494**	**2488**	**2481**	**2475**	**2469**	**2463**	**2457**	**2451**	**2445**
4.1	2439	2433	2427	2421	2415	2410	2404	2398	2392	2387
4.2	2381	2375	2370	2364	2358	2353	2347	2342	2336	2331
4.3	2326	2320	2315	2309	2304	2299	2294	2288	2283	2278
4.4	2273	2268	2262	2257	2252	2247	2242	2237	2232	2227
4.5	2222	2217	2212	2208	2203	2198	2193	2188	2183	2179
4.6	2174	2169	2165	2160	2155	2151	2146	2141	2137	2132
4.7	2128	2123	2119	2114	2110	2105	2101	2096	2092	2088
4.8	2083	2079	2075	2070	2066	2062	2058	2053	2049	2045
4.9	2041	2037	2033	2028	2024	2020	2016	2012	2008	2004
5.0	**2000**	**1996**	**1992**	**1988**	**1984**	**1980**	**1976**	**1972**	**1969**	**1965**
5.1	1961	1957	1953	1949	1946	1942	1938	1934	1931	1927
5.2	1923	1919	1916	1912	1908	1905	1901	1898	1894	1890
5.3	1887	1883	1880	1876	1873	1869	1866	1862	1859	1855
5.4	1852	1848	1845	1842	1838	1835	1832	1828	1825	1821
5.5	1818	1815	1812	1808	1805	1802	1799	1795	1792	1788
5.6	1786	1783	1779	1776	1773	1770	1767	1764	1761	1757
5.7	1754	1751	1748	1745	1742	1739	1736	1733	1730	1727
5.8	1724	1721	1718	1715	1712	1709	1706	1704	1701	1698
5.9	1695	1692	1689	1686	1684	1681	1678	1675	1672	1669
6.0	**1667**	**1664**	**1661**	**1658**	**1656**	**1653**	**1650**	**1647**	**1645**	**1642**
6.1	1639	1637	1634	1631	1629	1626	1623	1621	1618	1616
6.2	1613	1610	1608	1605	1603	1600	1597	1595	1592	1590
6.3	1587	1585	1582	1580	1577	1575	1572	1570	1567	1565
6.4	1563	1560	1558	1555	1553	1550	1548	1546	1543	1541
6.5	1538	1536	1534	1531	1529	1527	1524	1522	1520	1517
6.6	1515	1513	1511	1508	1506	1504	1502	1499	1497	1495
6.7	1493	1490	1488	1486	1484	1481	1479	1477	1475	1473
6.8	1471	1468	1466	1464	1462	1460	1458	1456	1453	1451
6.9	1449	1447	1445	1443	1441	1439	1437	1435	1433	1431
7.0	**1429**	**1427**	**1425**	**1422**	**1420**	**1418**	**1416**	**1414**	**1412**	**1410**
7.1	1408	1406	1404	1403	1401	1399	1397	1395	1393	1391
7.2	1389	1387	1385	1383	1381	1379	1377	1376	1374	1372
7.3	1370	1368	1366	1364	1362	1361	1359	1357	1355	1353
7.4	1351	1350	1348	1346	1344	1342	1340	1339	1337	1335
7.5	1333	1332	1330	1328	1326	1325	1323	1321	1319	1318
7.6	1316	1314	1312	1311	1309	1307	1305	1304	1302	1300
7.7	1299	1297	1295	1294	1292	1290	1289	1287	1285	1284
7.8	1282	1280	1279	1277	1276	1274	1272	1271	1269	1267
7.9	1266	1264	1263	1261	1259	1258	1256	1255	1253	1252
λ	0	1	2	3	4	5	6	7	8	9

Appendix Table 3.7 (*continued*)

λ	Wavenumber, cm^{-1}									
	0	1	2	3	4	5	6	7	8	9
8.0	**1250**	**1248**	**1247**	**1245**	**1244**	**1242**	**1241**	**1239**	**1238**	**1236**
8.1	1235	1233	1232	1230	1229	1227	1225	1224	1222	1221
8.2	1220	1218	1217	1215	1214	1212	1211	1209	1208	1206
8.3	1205	1203	1202	1200	1199	1198	1196	1195	1193	1192
8.4	1190	1189	1188	1186	1185	1183	1182	1181	1179	1178
8.5	1176	1175	1174	1172	1171	1170	1168	1167	1166	1164
8.6	1163	1161	1160	1159	1157	1156	1155	1153	1152	1151
8.7	1149	1148	1147	1145	1144	1143	1142	1140	1139	1138
8.8	1136	1135	1134	1133	1131	1130	1129	1127	1126	1125
8.9	1124	1122	1121	1120	1119	1117	1116	1115	1114	1112
9.0	**1111**	**1110**	**1109**	**1107**	**1106**	**1105**	**1104**	**1103**	**1101**	**1100**
9.1	1099	1098	1096	1095	1094	1093	1092	1091	1089	1088
9.2	1087	1086	1085	1083	1082	1081	1080	1079	1078	1076
9.3	1075	1074	1073	1072	1071	1070	1068	1067	1066	1065
9.4	1064	1063	1062	1060	1059	1058	1057	1056	1055	1054
9.5	1053	1052	1050	1049	1048	1047	1046	1045	1044	1043
9.6	1042	1041	1040	1038	1037	1036	1035	1034	1033	1032
9.7	1031	1030	1029	1028	1027	1026	1025	1024	1022	1021
9.8	1020	1019	1018	1017	1016	1015	1014	1013	1012	1011
9.9	1010	1009	1008	1007	1006	1005	1004	1003	1002	1001
10.0	**1000**	**999**	**998**	**997**	**996**	**995**	**994**	**993**	**992**	**991**
10.1	990	989	988	987	986	985	984	983	982	981
10.2	980	979	978	978	977	976	975	974	973	972
10.3	971	970	969	968	967	966	965	964	963	962
10.4	962	961	960	959	958	957	956	955	954	953
10.5	952	951	951	950	949	948	947	946	945	944
10.6	943	943	942	941	940	939	938	937	936	935
10.7	935	934	933	932	931	930	929	929	928	927
10.8	926	925	924	923	923	922	921	920	919	918
10.9	917	917	916	915	914	913	912	912	911	910
11.0	**909**	**908**	**907**	**907**	**906**	**905**	**904**	**903**	**903**	**902**
11.1	901	900	899	898	898	897	896	895	894	894
11.2	893	892	891	890	890	889	888	887	887	886
11.3	885	884	883	883	882	881	880	880	879	878
11.4	877	876	876	875	874	873	873	872	871	870
11.5	870	869	868	867	867	866	865	864	864	863
11.6	862	861	861	860	859	858	858	857	856	855
11.7	855	854	853	853	852	851	850	850	849	848
11.8	847	847	846	845	845	844	843	842	842	841
11.9	840	840	839	838	838	837	836	835	835	834
12.0	**833**	**833**	**832**	**831**	**831**	**830**	**829**	**829**	**828**	**827**
12.1	826	826	825	824	824	823	822	822	821	820
12.2	820	819	818	818	817	816	816	815	814	814
12.3	813	812	812	811	810	810	809	808	808	807
12.4	806	806	805	805	804	803	803	802	801	801
12.5	800	799	799	798	797	797	796	796	795	794
12.6	794	793	792	792	791	791	790	789	789	788
12.7	787	787	786	786	785	784	784	783	782	782
12.8	781	781	780	779	779	778	778	777	776	776
12.9	775	775	774	773	773	772	772	771	770	770
13.0	**769**	**769**	**768**	**767**	**767**	**766**	**766**	**765**	**765**	**764**
13.1	763	763	762	762	761	760	760	759	759	758
13.2	758	757	756	756	755	755	754	754	753	752
13.3	752	751	751	750	750	749	749	748	747	747
13.4	746	746	745	745	744	743	743	742	742	741
13.5	741	740	740	739	739	738	737	737	736	736
13.6	735	735	734	734	733	733	732	732	731	730
13.7	730	729	729	728	728	727	727	726	726	725
13.8	725	724	724	723	723	722	722	721	720	720
13.9	719	719	718	718	717	717	716	716	715	715
λ	0	1	2	3	4	5	6	7	8	9

Appendix Table 3.7 (*continued*)

	Wavenumber, cm^{-1}									
λ	0	1	2	3	4	5	6	7	8	9
14.0	**714**	**714**	**713**	**713**	**712**	**712**	**711**	**711**	**710**	**710**
14.1	709	709	708	708	707	707	706	706	705	705
14.2	704	704	703	703	702	702	701	701	700	700
14.3	699	699	698	698	697	697	696	696	695	695
14.4	694	694	693	693	693	692	692	691	691	690
14.5	690	689	689	688	688	687	687	686	686	685
14.6	685	684	684	684	683	683	682	682	681	681
14.7	680	680	679	679	678	678	678	677	677	676
14.8	676	675	675	674	674	673	673	672	672	672
14.9	671	671	670	670	669	669	668	668	668	667
λ	0	1	2	3	4	5	6	7	8	9

Appendix Table 4.1 NMR chart showing correlations of chemical shifts of protons

δ = 0.0 to 2.4	
1 TMS, 10.00	1
2 $-CH_2-$, cyclopropane, 9.78	2
3 CH_3-CN, 8.92–9.12	3
4 CH_3-C- (sat.), 9.05–9.15, (8.7–9.3)	4
5 $CH_3-C-CO-R$, 8.88–9.07	5
6 $CH_3-C-N-CO-R$, 8.80	6
7 $-N-C-$ (bridged by CH_2), 8.52	7
8 $-CH_2-$ (sat.), 8.52–8.80	8
9 $-CH_2-C-O-COR$ and $-CH_2-C-O-Ar$, 8.50	9
10 RSH, 8.5–8.9*	10
11 RNH_2 (conc. less than 1 mole in inert solvent), 8.5–8.9*	11
12 $-CH_2-C-C=C-$, 8.40–8.82	12
13 $-CH_2-CN$, 8.38–8.80	13
14 $-C-H$ (sat.), 8.35–8.60	14
15 $-CH_2-C-Ar$, 8.22–8.40	15
16 $-CH_2-C-O-R$, 8.19–8.79	16
17 $CH_3-C=NOH$, 8.19	17
18 $-CH_2-C-I$, 8.14–8.35	18
19 $-CH_2-C-CO-R$, 8.10–8.40	19
20 $CH_3-C=C$, 8.1–8.4	20
τ = 10.00 to 7.6	

Top scale: δ 2.4, 1.6, 0.8, 0.0 δ; τ 8.0, 8.8, 9.6 τ

Bottom scale: τ 7.6, 8.4, 9.2, 10.00 τ; δ 2.0, 1.2, 0.4 δ

Appendix Table 4.1 (*continued*)

δ = 1.6 to 3.6	δ: 3.2, 2.4, 1.6 — τ: 6.4, 7.2, 8.0	
21	CH_3—C(O—CO—R)=C—, 8.09–8.13	21
22	—CH_2—C=C—O—R, 8.07	22
23	—CH_2—C—Cl, 8.04–8.40	23
24	CH_3—C(COOR or CN)=C—, 7.97–8.06	24
25	—CH_2—C—Br, 7.97–8.32	25
26	CH_3—C=C—CO—R, 7.94–8.07	26
27	—CH_2—C—NO_2, 7.93	27
28	—CH_2—C—SO_2—R, 7.84	28
29	—C—O \ / CH, 7.71	29
30	—CH_2—C=C, 7.69–8.17	30
31	CH_3—N—N—, 7.67	31
32	—CH_2—CO—R, 7.61–7.98‡	32
33	CH_3—SO—R, 7.50	33
34	CH_3—Ar, 7.50–7.75 (7.5–7.9)	34
35	—CH_2—S—R, 7.47–7.61	35
36	CH_3—CO—SR, 7.46–7.67	36
37	—CH_2—C≡N, 7.42	37
38	CH_3—C=O, 7.4–7.9 (7.4–8.1)	38
39	CH_3—S—C≡N, 7.37	39
40	CH_3—CO—C=C or CH_3—CO—Ar, 7.32–8.17	40
41	CH_3—CO—Cl or Br, 7.19–7.34	41
τ = 8.4 to 6.4	τ: 6.8, 7.6, 8.4 — δ: 3.6, 2.8, 2.0	

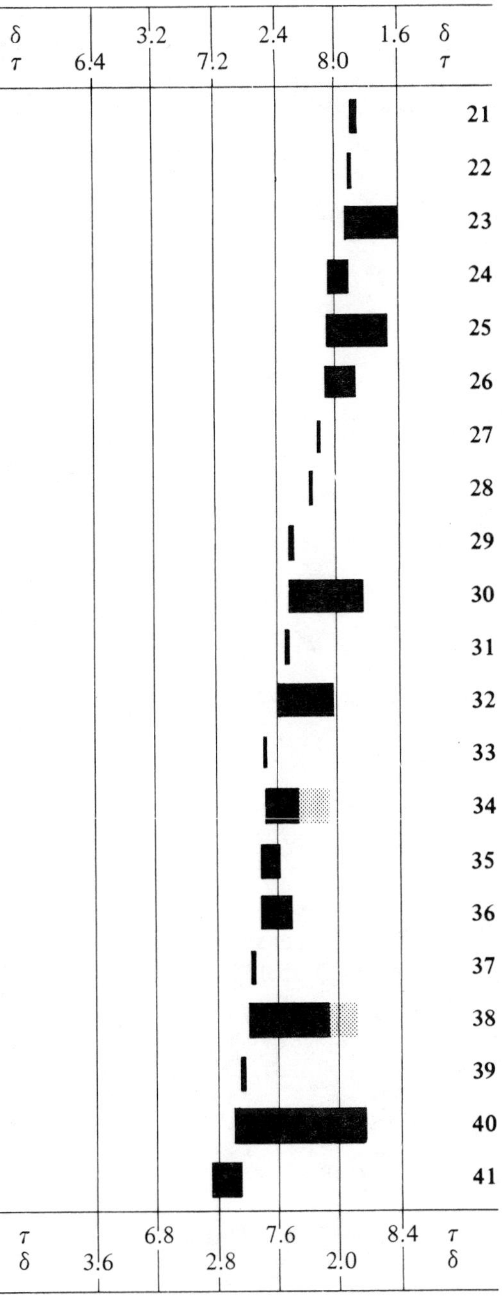

Appendix Table 4.1 (*continued*)

δ = 2.6 to 4.6	
42 $CH_3{-}S{-}$, 7.2–7.9	42
43 $CH_3{-}N$, 7.0–7.9	43
44 $-C{=}C{-}C{\equiv}C{-}H$, 7.13	44
45 $-CH_2{-}SO_2{-}R$, 7.08	45
46 $-C{\equiv}C{-}H$, nonconjugated, 7.35–7.55	46
47 $-C{\equiv}C{-}H$, conjugated, 6.9–7.2	47
48 $Ar{-}C{=}C{-}H$, 6.95	48
49 $-CH_2(C{=}C{-})_2$, 6.95–7.10	49
50 $-CH_2{-}Ar$, 6.94–7.47	50
51 $-CH_2{-}I$, 6.80–6.97	51
52 $-CH_2{-}SO_2F$, 6.72	52
53 $Ar{-}CH_2{-}N$, 6.68	53
54 $-CH_2{-}N{-}Ar$, 6.63–6.72	54
55 $Ar{-}CH_2{-}C{=}C{-}$, 6.62–6.82	55
56 $-CH_2{-}\overset{+}{N}{-}$, 6.60	56
57 $-CH_2{-}Cl$, 6.43–6.65	57
58 $-CH_2{-}O{-}R$, 6.42–7.69	58
59 $CH_3{-}O{-}$, 6.2–6.5 (6.0–6.7)	59
60 $-CH_2{-}Br$, 6.42–6.75	60
61 $CH_3{-}O{-}SO{-}OR$, 6.42	61
62 $-CH_2{-}N{=}C{=}S$, 6.39	62
63 $CH_3{-}SO_2{-}Cl$, 6.36	63
τ = 7.4 to 5.4	

Scale (top): δ 4.6, 3.8, 3.0; τ 5.8, 6.6, 7.4

Scale (bottom): τ 5.4, 6.2, 7.0; δ 4.2, 3.4, 2.6

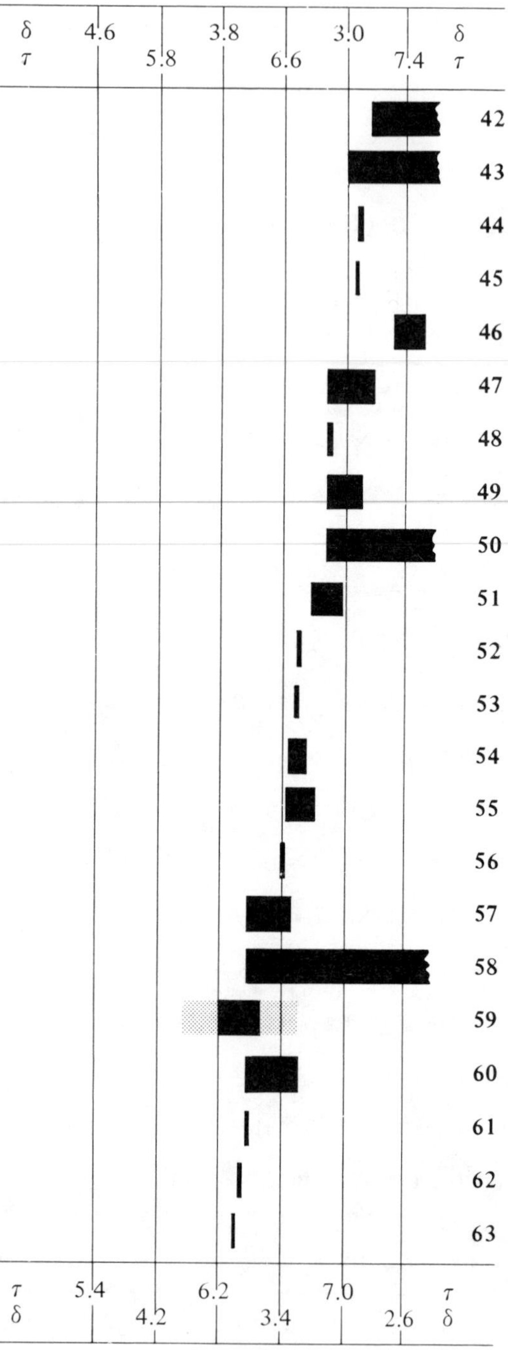

Appendix Table 4.1 (*continued*)

δ = 3.6 to 6.4	
64 Br—CH_2—C≡N, 6.30	64
65 —C≡C—CH_2—Br, 6.18	65
66 Ar—CH_2—Ar, 6.08–6.19	66
67 Ar—NH_2,* Ar—NH—R,* and Ar—NH—Ar,* 6.0–6.6 (5.7–6.7)	67
68 CH_3—O—SO_2—OR, 6.06	68
69 —C=C—CH_2—O—R, 6.03–6.10	69
70 —C=C—CH_2—Cl, 5.96–6.04	70
71 Cl—CH_2—C≡N, 5.93	71
72 H_2—C=C—O—CH_2—C=C, 5.87–6.17	72
73 —C≡C—CH_2—Cl, 5.84–5.91	73
74 —C=C—CH_2—OR, 5.82	74
75 —CH_2—O—CO—R or —CH_2—O—Ar, 5.71–6.02	75
76 —CH_2—NO_2, 5.62	76
77 Ar—CH_2—Br, 5.57–5.59	77
78 Ar—CH_2—OR, 5.51–5.64	78
79 Ar—CH_2—Cl, 5.50	79
80 —C=CH_2, 5.37	80
81 —C=CH—, acyclic, nonconjugated, 4.3–4.9 (4.1–4.9)	81
82 —C=CH—, cyclic, nonconjugated, 4.3–4.8	82
83 —C=CH_2, 4.3–4.7 (3.75–4.8)	83
84 —CH$(OR)_2$, 4.80–5.20	84
85 Ar—CH_2—O—CO—R, 4.74	85
86 R—O—H (conc. less than one mole in inert solvent), 4.8–7.0*	86
τ = 6.4 to 3.6	

Scale: δ 6.4, 5.6, 4.8, 4.0; τ 4.0, 4.8, 5.6, 6.4 (top); τ 4.4, 5.2, 6.0; δ 5.2, 4.4, 3.6 (bottom)

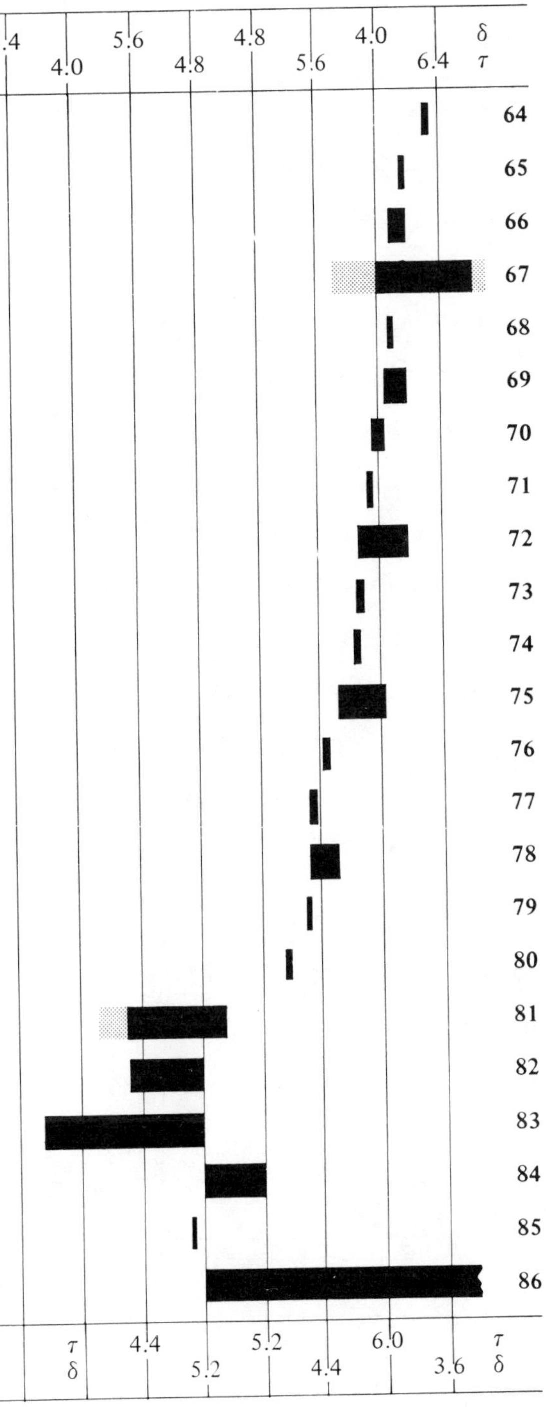

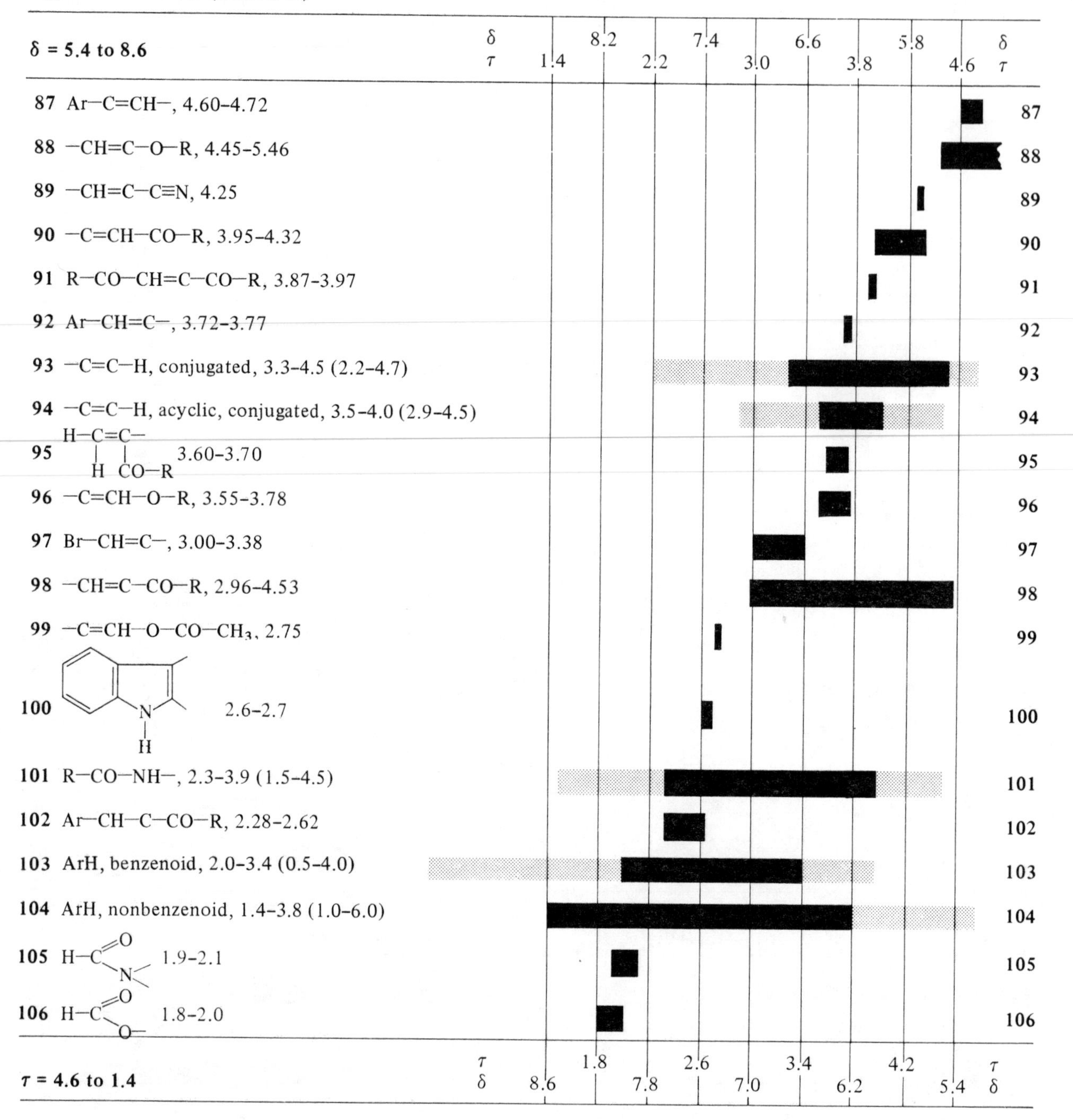

Appendix Table 4.1 (continued)
δ = 5.4 to 8.6
δ τ
8.2 7.4 6.6 5.8
1.4 2.2 3.0 3.8 4.6
87 Ar—C=CH—, 4.60–4.72
88 —CH=C—O—R, 4.45–5.46
89 —CH=C—C≡N, 4.25
90 —C=CH—CO—R, 3.95–4.32
91 R—CO—CH=C—CO—R, 3.87–3.97
92 Ar—CH=C—, 3.72–3.77
93 —C=C—H, conjugated, 3.3–4.5 (2.2–4.7)
94 —C=C—H, acyclic, conjugated, 3.5–4.0 (2.9–4.5)
95 H—C=C— H CO—R 3.60–3.70
96 —C=CH—O—R, 3.55–3.78
97 Br—CH=C—, 3.00–3.38
98 —CH=C—CO—R, 2.96–4.53
99 —C=CH—O—CO—CH3, 2.75
100 N H 2.6–2.7
101 R—CO—NH—, 2.3–3.9 (1.5–4.5)
102 Ar—CH—C—CO—R, 2.28–2.62
103 ArH, benzenoid, 2.0–3.4 (0.5–4.0)
104 ArH, nonbenzenoid, 1.4–3.8 (1.0–6.0)
105 H—C=O N 1.9–2.1
106 H—C=O O— 1.8–2.0
87 88 89 90 91 92 93 94 95 96 97 98 99 100 101 102 103 104 105 106
τ = 4.6 to 1.4
τ δ
1.8 2.6 3.4 4.2
8.6 7.8 7.0 6.2 5.4

Appendix Table 4.1 (*continued*)

δ = 8.0 to 12.4	δ: 12.4, 11.6, 10.8, 10.0, 9.2, 8.4 / τ: −2.0, −1.2, −0.4, 0.4, 1.2, 2.0	
107 —C=C—CHO, aliphatic, α,β unsat. (0.32–0.57)		**107**
108 R—CHO, aliphatic, 0.2–0.3 (0.2–0.5)		**108**
109 Ar—CHO, 0.0–0.3 (−0.1 to 0.5)		**109**
110 R—COOH, −1.52 to −0.97		**110**
111 $-SO_3H$, −2.0 to −1.0		**111**
112 —C=C—COOH, −2.18 to −1.57		**112**
113 R—COOH, dimer, −2.2 to −1.0		**113**
114 Ar—OH, intramolecularly bonded, −2.5 to −0.5 (−5.5 to −0.5)		**114**
115 Ar—OH, polymeric association, +2.3 to 5.5*		**115**
116 Enols, −6.0 to −5.0		**116**

Normally protons resonate within the range shown. In cases in which protons resonate outside the range, the limits are shown by values in parentheses and by shading of the bars.

* The position of these protons depends on concentration, temperature, and the presence of other exchangeable protons. Values for amino protons depend on the basicity of the nitrogen atom.

‡ In these compounds R = H, alkyl, aryl, OH, OR, or NH_2.

Appendix Table 4.2 Proton spin–spin coupling constants J, in Hz

Structure	Coupling constant, Hz or cps	Structure	Coupling constant, Hz or cps
Geminal			
>C(H_A)(H_B)	12–15	>C=C(H_A)(H_B)	0.5–3
H_A(H_B)C=N—OH	7.63–9.95 (solvent dependent)	cyclopropane C(H_A)(H_B)	3.9–8.8
epoxide >C—C(H_A)(H_B) (O)	5.4–6.3	cyclohexane C(H_A)(H_B)	12.6

Appendix Table 4.2 (*continued*)

Structure	Coupling constant, Hz or cps	Structure	Coupling constant, Hz or cps
Vicinal			
CH_3–CH_2–	4.7-9	CH_3CH_2–O–CH_2CH_3	6.97
CH_3–CH_2–X where, if X =	7.1-7.7	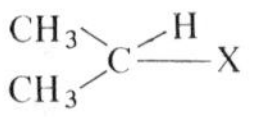 $(CH_3)_2C(H)$–X	6.1-7
–CH_3	7.26	where, if X =	
–Cl	7.23	–OH	6.2
–Br	7.33	–Cl	6.4
–I	7.45	–Br	6.5
–N–Et_2	7.4	–I	6.6
–CN	7.6	–CH_3	6.8
–phenyl	7.62	–phenyl	6.9
		–CHO	7.0
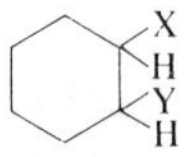 Cyclohexane ring with X, H, Y, H on adjacent carbons	e, e; 2-4 a, e; 2-4 a, a; 5-10	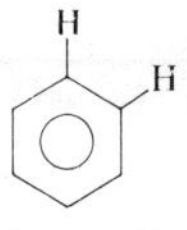 Benzene ring, H and H ortho	6.5-9.4
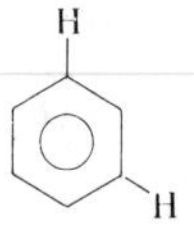 Benzene ring, H and H meta	0.8-3	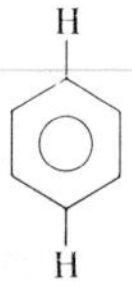 Benzene ring, H and H para	0.4-1
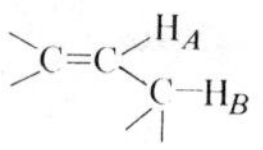 >C=C(H_A)–C–H_B	4-10	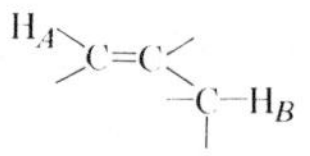 H_A–C=C<–C–H_B	0.5-2.5
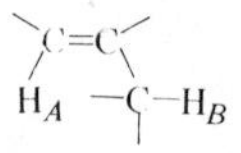 >C=C<, H_A on C=C, –C–H_B	≈ 0	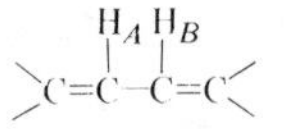 >C=C(H_A)–C(H_B)=C<	9-13
H_AC≡CH_B	9.5-9.8	>CH_A–C≡C–H_B	2-3
H_A–C–C(H_B)=O	1-3	>C=C(H_A)–C(H_B)=O	6-8
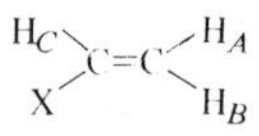 H_C(X)C=CH_AH_B	*cis* H_A–H_C, 4.6-19.3 *trans* H_B–H_C, 12.7-24.0	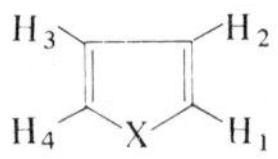 Five-membered ring with X; H_1, H_2, H_3, H_4	1-5.5
where, if X =		where, if X =	
–F	H_A–H_C, 4.6 H_B–H_C, 12.7	O	H_1-H_2, 1-2
–Br	H_A–H_C, 7.1 H_B–H_C, 15.2	N	H_1-H_2, 2-3
–Cl	H_A–H_C, 7.4 H_B–H_C, 14.8	S	H_1-H_2, 5.5
–CH_3	H_A–H_C, 9.6-11.1 H_B–H_C, 16.6-17.4	O, N, S	H_1-H_3, 1-2 H_2-H_3, 3-5
–COOH	H_A–H_C, 10.2 H_B–H_C, 17.2		
–phenyl	H_A–H_C, 10.7 H_B–H_C, 17.5		
–H	H_A–H_C, 11.5 H_B–H_C, 19.0		
–CN	H_A–H_C, 11.7 H_B–H_C, 17.9		

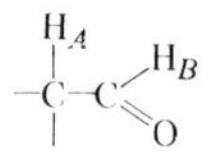

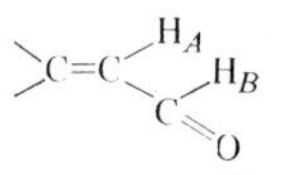

Appendix Table 5.1 Masses and isotopic abundance ratios for various combinations of carbon, hydrogen, nitrogen and oxygen*

	M + 1	M + 2
12		
C	1.08	
13		
CH	1.10	
14		
N	0.38	
CH_2	1.11	
15		
NH	0.40	
CH_3	1.13	
16		
O	0.04	0.20
NH_2	0.41	
CH_4	1.15	
17		
OH	0.06	0.20
NH_3	0.43	
CH_5	1.16	
18		
H_2O	0.07	0.20
NH_4	0.45	
19		
H_3O	0.09	0.20
24		
C_2	2.16	0.01
25		
C_2H	2.18	0.01
26		
CN	1.46	
C_2H_2	2.19	0.01
27		
CHN	1.48	
C_2H_3	2.21	0.01
28		
N_2	0.76	
CO	1.12	0.02
CH_2N	1.49	
C_2H_4	2.23	0.01
29		
N_2H	0.78	
CHO	1.14	0.20
CH_3N	1.51	
C_2H_5	2.24	0.01
30		
NO	0.42	0.20
N_2H_2	0.79	
CH_2O	1.15	0.20
CH_4N	1.53	0.01
C_2H_6	2.26	0.01
31		
NOH	0.44	0.20
N_2H_3	0.81	
CH_3O	1.17	0.20
CH_5N	1.54	
32		
O_2	0.08	0.40
NOH_2	0.45	0.20
N_2H_4	0.83	
CH_4O	1.18	0.20
33		
NOH_3	0.47	0.20
N_2H_5	0.84	
CH_5O	1.12	
34		
N_2H_6	0.86	
36		
C_3	3.24	0.04
37		
C_3H	3.26	0.04
38		
C_2N	2.54	0.02
C_3H_2	3.27	0.04
39		
C_2HN	2.56	0.02
C_3H_3	3.29	0.04
40		
CN_2	1.84	0.01
C_2O	2.20	0.21
C_2H_2N	2.58	0.02
C_3H_4	3.31	0.04
41		
CHN_2	1.86	
C_2HO	2.22	0.21
C_2H_3N	2.59	0.02
C_3H_5	3.32	0.04
42		
CNO	1.50	0.21
CH_2N_2	1.88	0.01
C_2H_2O	2.23	0.21
C_2H_4N	2.61	0.02
C_3H_6	3.34	0.04
43		
CHNO	1.52	0.21
CH_3N_2	1.89	0.01
C_2H_3O	2.25	0.21
C_2H_5N	2.62	0.02
C_3H_7	3.35	0.04
44		
N_2O	0.80	0.20
CO_2	1.16	0.40
CH_2NO	1.53	0.21
CH_4N_2	1.91	0.01
C_2H_4O	2.26	0.21
C_2H_6N	2.64	0.02
C_3H_8	3.37	0.04
45		
HN_2O	0.82	0.20
CHO_2	1.18	0.40
CH_3NO	1.55	0.21
CH_5N_2	1.92	0.01
C_2H_5O	2.28	0.21
C_2H_7N	2.66	0.02
46		
NO_2	0.46	0.40
N_2H_2O	0.83	0.20
CH_2O_2	1.19	0.40
CH_4NO	1.57	0.21
CH_6N_2	1.94	0.01
C_2H_6O	2.30	0.22
C_2H_8N	2.66	0.02
47		
CH_3O_2	1.21	0.40
CH_5NO	1.58	0.21
CH_7N_2	1.96	0.01
C_2H_7O	2.31	0.22
48		
CH_4O_2	1.22	0.40
C_4	4.32	0.07
49		
CH_5O_2	1.24	0.40
C_4H	4.34	0.07
50		
C_4H_2	4.34	0.07
51		
C_4H_3	4.37	0.07
52		
C_2N_2	2.92	0.03
C_3H_2N	3.66	0.05
C_4H_4	4.39	0.07
53		
C_2HN_2	2.94	0.03
C_3HO	3.30	0.24
C_3H_3N	3.67	0.05
C_4H_5	4.40	0.07
54		
C_2NO	2.58	0.22
$C_2H_2N_2$	2.96	0.03
C_3H_2O	3.31	0.24
C_3H_4N	3.69	0.05
C_4H_6	4.42	0.07
55		
C_2HNO	2.60	0.22
$C_2H_3N_2$	2.97	0.03
C_3H_3O	3.33	0.24
C_3H_5N	3.70	0.05
C_4H_7	4.43	0.08

Appendix Table 5.1 (*continued*)

	M + 1	M + 2
	56	
CH_2N_3	2.26	0.02
C_2O_2	2.24	0.41
C_2H_2NO	2.61	0.22
$C_2H_4N_2$	2.99	0.03
C_3H_4O	3.35	0.24
C_3H_6N	3.72	0.05
C_4H_8	4.45	0.08
	57	
CHN_2O	1.90	0.21
CH_3N_3	2.27	0.02
C_2HO_2	2.26	0.41
C_2H_3NO	2.63	0.22
$C_2H_5N_2$	3.00	0.03
C_3H_5O	3.36	0.24
C_3H_7N	3.74	0.05
C_4H_9	4.47	0.08
	58	
CNO_2	1.54	0.41
CH_2N_2O	1.92	0.21
CH_4N_3	2.29	0.02
$C_2H_2O_2$	2.27	0.42
C_2H_4NO	2.65	0.22
$C_2H_6N_2$	3.02	0.03
C_3H_6O	3.38	0.24
C_3H_8N	3.75	0.05
C_4H_{10}	4.48	0.08
	59	
$CHNO_2$	1.56	0.41
CH_3N_2O	1.93	0.21
CH_5N_3	2.31	0.02
$C_2H_3O_2$	2.29	0.42
C_2H_5NO	2.66	0.22
$C_2H_7N_2$	3.04	0.03
C_3H_7O	3.39	0.24
C_3H_9N	3.77	0.05
	60	
CH_2NO_2	1.57	0.41
CH_4N_2O	1.95	0.21
CH_6N_3	2.32	0.02
$C_2H_4O_2$	2.30	0.04
C_2H_6NO	2.68	0.22
$C_2H_8N_2$	3.05	0.03
C_3H_8O	3.41	0.24
	61	
CHO_3	1.21	0.60
CH_3NO_2	1.59	0.41
CH_5N_2O	1.96	0.21
CH_7N_3	2.34	0.02
$C_2H_5O_2$	2.32	0.42
C_2H_7NO	2.69	0.22
C_3H_9O	3.43	0.24
C_5H	5.42	0.12
	62	
CH_2O_3	1.23	0.60
CH_4NO_2	1.60	0.41
CH_6N_2O	1.98	0.21
CH_8N_3	2.35	0.02
$C_2H_6O_2$	2.34	0.42
C_5H_2	5.44	0.12
	63	
CH_3O_3	1.25	0.60
CH_5NO_2	1.62	0.41
C_4HN	4.72	0.09
C_5H_3	5.45	0.12
	64	
CH_4O_3	1.26	0.60
C_4H_2N	4.74	0.09
C_5H_4	5.47	0.12
	65	
C_3HN_2	4.02	0.06
C_4HO	4.38	0.27
C_4H_3N	4.75	0.09
C_5H_5	5.48	0.12
	66	
$C_3H_2N_2$	4.04	0.06
C_4H_2O	4.39	0.27
C_4H_4N	4.77	0.09
C_5H_6	5.50	0.12
	67	
C_2HN_3	3.32	0.04
C_3HNO	3.68	0.25
$C_3H_3N_2$	4.05	0.06
C_4H_3O	4.41	0.27
C_4H_5N	4.78	0.09
C_5H_7	5.52	0.12
	68	
$C_2H_2N_3$	3.34	0.04
C_3O_2	3.32	0.44
C_3H_2NO	3.69	0.25
$C_3H_4N_2$	4.07	0.06
C_4H_4O	4.43	0.28
C_4H_6N	4.80	0.09
C_5H_8	5.53	0.12
	69	
CHN_4	2.62	0.03
C_2HN_2O	2.98	0.23
$C_2H_3N_3$	3.35	0.04
C_3HO_2	3.34	0.44
C_3H_3NO	3.71	0.25
$C_3H_5N_2$	4.09	0.06
C_4H_5O	4.44	0.28
C_4H_7N	4.82	0.09
C_5H_9	5.55	0.12
	70	
CH_2N_4	2.64	0.03
C_2NO_2	2.62	0.42
$C_2H_2N_2O$	3.00	0.23
$C_2H_4N_3$	3.37	0.04
$C_3H_2O_2$	3.35	0.44
C_3H_4NO	3.73	0.25
$C_3H_6N_2$	4.10	0.07
C_4H_6O	4.46	0.28
C_4H_8N	4.83	0.09
C_5H_{10}	5.56	0.13
	71	
CHN_3O	2.28	0.22
CH_3N_4	2.65	0.03
C_2HNO_2	2.64	0.42
$C_2H_3N_2O$	3.01	0.23
$C_3H_5N_3$	3.39	0.04
$C_3H_3O_2$	3.37	0.44
C_3H_5NO	3.74	0.25
$C_3H_7N_2$	4.12	0.07
C_4H_7O	4.47	0.28
C_4H_9N	4.85	0.09
C_5H_{11}	5.58	0.13
	72	
CH_2N_3O	2.30	0.22
CH_4N_4	2.67	0.03
$C_2H_2NO_2$	2.65	0.42
$C_2H_4N_2O$	3.03	0.23
$C_2H_6N_3$	3.40	0.44
$C_3H_4O_2$	3.38	0.44
C_3H_6NO	3.76	0.25
$C_3H_8N_2$	4.13	0.07
C_4H_8O	4.49	0.28
$C_4H_{10}N$	4.86	0.09
C_5H_{12}	5.60	0.13
	73	
CHN_2O_2	1.94	0.41
CH_3N_3O	2.31	0.22
CH_5N_4	2.69	0.03
C_2HO_3	2.30	0.62
$C_2H_3NO_2$	2.67	0.42
$C_2H_5N_2O$	3.04	0.23
$C_2H_7N_3$	3.42	0.04
$C_3H_5O_2$	3.40	0.44
C_3H_7NO	3.77	0.25
$C_3H_9N_2$	4.15	0.07
C_4H_9O	4.51	0.28
$C_4H_{11}N$	4.88	0.10
C_6H	6.50	0.18
	74	
$CH_2N_2O_2$	1.95	0.41
CH_4N_3O	2.33	0.22
CH_6N_4	2.70	0.03
$C_2H_2O_3$	2.31	0.62
$C_2H_4NO_2$	2.69	0.42
$C_2H_6N_2O$	3.06	0.23
$C_2H_8N_3$	3.43	0.05
$C_3H_6O_2$	3.42	0.44
C_3H_8NO	3.79	0.25
$C_3H_{10}N_2$	4.17	0.07
$C_4H_{10}O$	4.52	0.28
C_6H_2	6.52	0.18
	75	
$CHNO_3$	1.60	0.61
$CH_3N_2O_2$	1.97	0.41
CH_5N_3O	2.34	0.22
CH_7N_4	2.72	0.03
$C_2H_3O_3$	2.33	0.62
$C_2H_5NO_2$	2.70	0.43
$C_2H_7N_2O$	3.08	0.23
$C_2H_9N_3$	3.45	0.05
$C_3H_7O_2$	3.43	0.44
C_3H_9NO	3.81	0.25
C_5HN	5.80	0.14
C_6H_3	6.53	0.18
	76	
CH_2NO_3	1.61	0.61
$CH_4N_2O_2$	1.99	0.41
CH_6N_3O	2.36	0.22
CH_8N_4	2.73	0.03
$C_2H_4O_3$	2.34	0.62
$C_2H_6NO_2$	2.72	0.43
$C_2H_8N_2O$	3.09	0.24
$C_3H_8O_2$	3.45	0.44
C_5H_2N	5.82	0.14
C_6H_4	6.55	0.18
	77	
CHO_4	1.25	0.80
CH_3NO_3	1.63	0.61
$CH_5N_2O_2$	2.00	0.41
CH_7N_3O	2.38	0.22
$C_2H_5O_3$	2.39	0.62
$C_2H_7NO_2$	2.73	0.43
C_4HN_2	5.10	0.11
C_5HO	5.45	0.32
C_5H_3N	5.83	0.14
C_6H_5	6.56	0.18
	78	
CH_2O_4	1.27	0.80
CH_4NO_3	1 64	0.61
$CH_6N_2O_2$	2.02	0.41
$C_2H_6O_3$	2.38	0.62
$C_4H_2N_2$	5.12	0.11
C_5H_2O	5.47	0.32
C_5H_4N	5.49	0.14
C_6H_6	6.58	0.18
	79	
CH_3O_4	1.29	0.80
CH_5NO_3	1.66	0.61
C_3HN_3	4.40	0.08
C_4HNO	4.76	0.29
$C_4H_3N_2$	5.13	0.11
C_5H_3O	5.49	0.32
C_5H_5N	5.87	0.14
C_6H_7	6.60	0.18
	80	
CH_4O_4	1.30	0.80
$C_3H_2N_3$	4.42	0.08
C_4H_2NO	4.78	0.29
$C_4H_4N_2$	5.15	0.11
C_5H_4O	5.51	0.32
C_5H_6N	5.88	0.14
C_6H_8	6.61	0 18
	81	
C_2HN_4	3.70	0.05
C_3HN_2O	4.06	0 26

Appendix Table 5.1 (*continued*)

Formula	M + 1	M + 2
$C_3H_3N_3$	4.43	0.08
C_4HO_2	4.42	0.48
C_4H_3NO	4.79	0.29
$C_4H_5N_2$	5.17	0.11
C_5H_5O	5.52	0.32
C_5H_7N	5.90	0.14
C_6H_9	6.63	0.18
82		
$C_2H_2N_4$	3.72	0.05
$C_3H_2N_2O$	4.08	0.36
$C_3H_4N_3$	4.45	0.08
$C_4H_2O_2$	4.43	0.48
C_4H_4NO	4.81	0.29
$C_4H_6N_2$	4.18	0.11
C_5H_6O	5.54	0.32
C_5H_8N	5.91	0.14
C_6H_{10}	6.64	0.19
83		
C_2HN_3O	3.36	0.24
$C_2H_3N_4$	3.74	0.06
C_3HNO_2	3.72	0.45
$C_3H_3N_2O$	4.09	0.27
$C_3H_5N_3$	4.47	0.08
$C_4H_3O_2$	4.45	0.48
C_4H_5NO	4.82	0.29
$C_4H_7N_2$	5.20	0.11
C_5H_7O	5.55	0.33
C_5H_9N	5.93	0.15
C_6H_{11}	6.66	0.19
84		
$C_2H_2N_3O$	3.38	0.24
$C_2H_4N_4$	3.75	0.06
$C_3H_2NO_2$	3.73	0.45
$C_3H_4N_2O$	4.11	0.27
$C_3H_6N_3$	4.48	0.81
$C_4H_4O_2$	4.47	0.48
C_4H_6NO	4.84	0.29
$C_4H_8N_2$	5.21	0.11
C_5H_8O	5.57	0.33
$C_5H_{10}N$	5.95	0.15
C_6H_{12}	6.68	0.19
85		
CHN_4O	2.66	0.23
$C_2HN_2O_2$	3.02	0.43
$C_2H_3N_3O$	3.39	0.24
$C_2H_5N_4$	3.77	0.06
C_3HO_3	3.38	0.64
$C_3H_3NO_2$	3.75	0.45
$C_3H_5N_2O$	4.12	0.27
$C_3H_7N_3$	4.50	0.08
$C_4H_5O_2$	4.48	0.48
C_4H_7NO	4.86	0.29
$C_4H_9N_2$	5.23	0.11
C_5H_9O	5.59	0.33
$C_5H_{11}N$	5.96	0.15
C_6H_{13}	6.69	0.19
C_7H	7.58	0.25
86		
CH_2N_4O	2.68	0.23
$C_2H_2N_2O_2$	3.03	0.43
$C_2H_4N_3O$	3.41	0.24
$C_2H_6N_4$	3.78	0.06
$C_3H_2O_3$	3.39	0.64
$C_3H_4NO_2$	3.77	0.45
$C_3H_6N_2O$	4.14	0.27
$C_3H_8N_3$	4.51	0.08
$C_4H_6O_2$	4.50	0.48
C_4H_8NO	4.87	0.30
$C_4H_{10}N_2$	5.25	0.11
$C_5H_{10}O$	5.60	0.33
$C_5H_{12}N$	5.98	0.15
C_6H_{14}	6.71	0.19
C_7H_2	7.60	0.25
87		
CHN_3O_2	2.32	0.42
CH_3N_4O	2.69	0.23
C_2HNO_3	2.68	0.62
$C_2H_3N_2O_2$	3.05	0.43
$C_2H_5N_3O$	3.43	0.25
$C_2H_7N_4$	3.80	0.06
$C_3H_3O_3$	3.41	0.64
$C_3H_5NO_2$	3.78	0.45
$C_3H_7N_2O$	4.16	0.27
$C_3H_9N_3$	4.53	0.08
$C_4H_7O_2$	4.51	0.48
C_4H_9NO	4.89	0.30
$C_4H_{11}N_2$	5.26	0.11
$C_5H_{11}O$	5.62	0.33
$C_5H_{13}N$	5.99	0.15
C_6HN	6.88	0.20
C_7H_3	7.61	0.25
88		
$CH_2N_3O_2$	2.34	0.42
CH_4N_4O	2.71	0.23
$C_2H_2NO_3$	2.69	0.63
$C_2H_4N_2O_2$	3.07	0.43
$C_2H_6N_3O$	3.44	0.25
$C_2H_8N_4$	3.82	0.06
$C_3H_4O_3$	3.42	0.64
$C_3H_6NO_2$	3.80	0.45
$C_3H_8N_2O$	4.17	0.27
$C_3H_{10}N_3$	4.55	0.08
$C_4H_8O_2$	4.53	0.48
$C_4H_{10}NO$	4.90	0.30
$C_4H_{12}N_2$	5.28	0.11
$C_5H_{12}O$	5.63	0.33
C_6H_2N	6.90	0.20
C_7H_4	7.63	0.25
89		
CHN_2O_3	1.98	0.61
$CH_3N_3O_2$	2.35	0.42
CH_5N_4O	2.73	0.23
C_2HO_4	2.33	0.82
$C_2H_3NO_3$	2.71	0.63
$C_2H_5N_2O_2$	3.08	0.44
$C_2H_7N_3O$	3.46	0.25
$C_2H_9N_4$	3.83	0.06
$C_3H_5O_3$	3.44	0.64
$C_3H_7NO_2$	3.81	0.46
$C_3H_9N_2O$	4.19	0.27
$C_3H_{11}N_3$	4.56	0.84
$C_4H_9O_2$	4.55	0.48
$C_4H_{11}NO$	4.92	0.30
C_5HN_2	6.18	0.16
C_6HO	6.54	0.38
C_6H_3N	6.91	0.20
C_7H_5	7.64	0.25
90		
$CH_2N_2O_3$	1.99	0.61
$CH_4N_3O_2$	2.37	0.42
CH_6N_4O	2.74	0.23
$C_2H_2O_4$	2.35	0.82
$C_2H_4NO_3$	2.72	0.63
$C_2H_6N_2O_2$	3.10	0.44
$C_2H_8N_3O$	3.47	0.25
$C_2H_{10}N_4$	3.85	0.06
$C_3H_6O_3$	3.46	0.64
$C_3H_8NO_2$	3.83	0.46
$C_3H_{10}N_2O$	4.20	0.27
$C_4H_{10}O_2$	4.56	0.48
$C_5H_2N_2$	6.20	0.16
C_6H_2O	6.56	0.38
C_6H_4N	6.93	0.20
C_7H_6	7.66	0.25
91		
$CHNO_4$	1.63	0.81
$CH_3N_2O_3$	2.01	0.61
$CH_5N_3O_2$	2.38	0.42
CH_7N_4O	2.76	0.23
$C_2H_3O_4$	2.37	0.82
$C_2H_5NO_3$	2.74	0.63
$C_2H_7N_2O_2$	3.11	0.44
$C_2H_9N_3O$	3.49	0.25
$C_3H_7O_3$	3.47	0.64
$C_3H_9NO_2$	3.85	0.46
C_4HN_3	5.48	0.12
C_5HNO	5.84	0.34
$C_5H_3N_2$	6.21	0.16
C_6H_3O	6.57	0.38
C_6H_5N	6.95	0.21
C_7H_7	7.68	0.25
92		
CH_2NO_4	1.65	0.81
$CH_4N_2O_3$	2.03	0.61
$CH_6N_3O_2$	2.40	0.42
CH_8N_4O	2.77	0.23
$C_2H_4O_4$	2.38	0.82
$C_2H_6NO_3$	2.76	0.63
$C_2H_8N_2O_2$	3.13	0.44
$C_3H_8O_3$	3.49	0.64
$C_4H_2N_3$	5.50	0.13
C_5H_2NO	5.86	0.34
$C_5H_4N_2$	6.23	0.16
C_6H_4O	6.59	0.38
C_6H_6N	6.96	0.21
C_7H_8	7.69	0.26
N_2O_4	9.19	0.80
93		
CH_3NO_4	1.67	0.81
$CH_5N_2O_3$	2.04	0.61
$CH_7N_3O_2$	2.42	0.42
$C_2H_5O_4$	2.40	0.82
$C_2H_7NO_3$	2.77	0.63
C_3HN_4	4.78	0.09
C_4HN_2O	5.14	0.31
$C_4H_3N_3$	5.52	0.13
C_5HO_2	5.50	0.52
C_5H_3NO	5.87	0.34
$C_5H_5N_2$	6.25	0.16
C_6H_5O	6.60	0.38
C_6H_7N	6.98	0.21
C_7H_9	7.71	0.26
94		
CH_4NO_4	1.68	0.81
$CH_6N_2O_3$	2.06	0.62
$C_2H_6O_4$	2.41	0.82
$C_3H_2N_4$	4.80	0.09
$C_4H_2N_2O$	5.16	0.31
$C_4H_4N_3$	5.53	0.13
$C_5H_2O_2$	5.51	0.52
C_5H_4NO	5.89	0.34
$C_5H_6N_2$	6.26	0.17
C_6H_6O	6.62	0.38
C_6H_8N	6.99	0.21
C_7H_{10}	7.72	0.26
95		
CH_5NO_4	1.70	0.81
C_3HN_3O	4.44	0.28
$C_3H_3N_4$	4.82	0.10
C_4HNO_2	4.80	0.49
$C_4H_3N_2O$	5.17	0.31
$C_4H_5N_3$	5.55	0.13
$C_5H_3O_2$	5.53	0.52
C_5H_5NO	5.90	0.34
$C_5H_7N_2$	6.28	0.17
C_6H_7O	6.64	0.39
C_6H_9N	7.01	0.21
C_7H_{11}	7.74	0.26
96		
$C_3H_2N_3O$	4.46	0.28
$C_3H_4N_4$	4.83	0.10
$C_4H_2NO_2$	4.81	0.49
$C_4H_4N_2O$	5.19	0.31
$C_4H_6N_3$	5.56	0.13
$C_5H_4O_2$	5.55	0.53
C_5H_6NO	5.92	0.35
$C_5H_8N_2$	6.29	0.17
C_6H_8O	6.65	0.39
$C_6H_{10}N$	7.03	0.21
C_7H_{12}	7.76	0.26
97		
C_2HN_4O	3.74	0.26
$C_3HN_2O_2$	4.10	0.47
$C_3H_3N_3O$	4.47	0.28
$C_3H_5N_4$	4.85	0.10
C_4HO_3	4.46	0.68
$C_4H_3NO_2$	4.83	0.49
$C_4H_5N_2O$	5.20	0.31
$C_4H_7N_3$	5.58	0.13
$C_5H_5O_2$	5.56	0.53

Appendix Table 5.1 (*continued*)

	M + 1	M + 2
C_5H_7NO	5.94	0.35
$C_5H_9N_2$	6.31	0.17
C_6H_9O	6.67	0.39
$C_6H_{11}N$	7.04	0.21
C_7H_{13}	7.77	0.26
C_8H	8.66	0.33
98		
$C_2H_2N_4O$	3.76	0.26
$C_3H_2N_2O_2$	4.12	0.47
$C_3H_4N_3O$	4.49	0.28
$C_3H_6N_4$	4.86	0.10
$C_4H_2O_3$	4.47	0.68
$C_4H_4NO_2$	4.85	0.49
$C_4H_6N_2O$	5.22	0.31
$C_4H_8N_3$	5.60	0.13
$C_5H_6O_2$	5.58	0.53
C_5H_8NO	5.95	0.35
$C_5H_{10}N_2$	6.33	0.17
$C_6H_{10}O$	6.68	0.39
$C_6H_{12}N$	7.06	0.21
C_7H_{14}	7.79	0.26
C_8H_2	8.68	0.33
99		
$C_2HN_3O_2$	3.40	0.44
$C_2H_3N_4O$	3.77	0.26
C_3HNO_3	3.76	0.65
$C_3H_3N_2O_2$	4.13	0.47
$C_3H_5N_3O$	4.51	0.28
$C_3H_7N_4$	4.88	0.10
$C_4H_3O_3$	4.49	0.68
$C_4H_5NO_2$	4.86	0.50
$C_4H_7N_2O$	5.24	0.31
$C_4H_9N_3$	5.61	0.13
$C_5H_7O_2$	5.59	0.53
C_5H_9NO	5.97	0.35
$C_5H_{11}N_2$	6.34	0.17
$C_6H_{11}O$	6.70	0.39
$C_6H_{13}N$	7.07	0.21
C_7HN	7.96	0.28
C_7H_{15}	7.80	0.26
C_8H_3	8.69	0.33
100		
$C_2H_2N_3O_2$	3.42	0.45
$C_2H_4N_4O$	3.79	0.26
$C_3H_2NO_3$	3.77	0.65
$C_3H_4N_2O_2$	4.15	0.47
$C_3H_6N_3O$	4.52	0.28
$C_3H_8N_4$	4.90	0.10
$C_4H_4O_3$	4.50	0.68
$C_4H_6NO_2$	4.88	0.50
$C_4H_8N_2O$	5.25	0.31
$C_4H_{10}N_3$	5.63	0.13
$C_5H_8O_2$	5.61	0.53
$C_5H_{10}NO$	5.98	0.35
$C_5H_{12}N_2$	6.36	0.17
$C_6H_{12}O$	6.72	0.39
$C_6H_{14}N$	7.09	0.22
C_7H_2N	7.98	0.28
C_7H_{16}	7.82	0.26
C_8H_4	8.71	0.33
101		
CHN_4O_2	2.70	0.43
$C_2HN_2O_3$	3.06	0.64
$C_2H_3N_3O_2$	3.43	0.45
$C_2H_5N_4O$	3.81	0.26
C_3HO_4	3.41	0.84
$C_3H_3NO_3$	3.79	0.65
$C_3H_5N_2O_2$	4.16	0.47
$C_3H_7N_3O$	4.54	0.28
$C_3H_9N_4$	4.91	0.10
$C_4H_5O_3$	4.52	0.68
$C_4H_7NO_2$	4.89	0.50
$C_4H_9N_2O$	5.27	0.31
$C_4H_{11}N_3$	5.64	0.13
$C_5H_9O_2$	5.63	0.53
$C_5H_{11}NO$	6.00	0.35
$C_5H_{13}N_2$	6.37	0.17
C_6HN_2	7.26	0.23
$C_6H_{13}O$	6.73	0.39
$C_6H_{15}N$	7.11	0.22
C_7HO	7.62	0.45
C_7H_3N	7.99	0.28
C_8H_5	8.73	0.33
102		
$CH_2N_4O_2$	2.72	0.43
$C_2H_2N_2O_3$	3.07	0.64
$C_2H_4N_3O_2$	3.45	0.45
$C_2H_6N_4O$	3.82	0.26
$C_3H_2O_4$	3.43	0.84
$C_3H_4NO_3$	3.81	0.66
$C_3H_6N_2O_2$	4.18	0.47
$C_3H_8N_3O$	4.55	0.28
$C_3H_{10}N_4$	4.93	0.10
$C_4H_6O_3$	4.54	0.68
$C_4H_8NO_2$	4.91	0.50
$C_4H_{10}N_2O$	5.28	0.32
$C_4H_{12}N_3$	5.66	0.13
$C_5H_{10}O_2$	5.64	0.53
$C_5H_{12}NO$	6.02	0.35
$C_5H_{14}N_2$	6.39	0.17
$C_6H_2N_2$	7.28	0.23
$C_6H_{14}O$	6.75	0.39
C_7H_2O	7.64	0.45
C_7H_4N	8.01	0.28
C_8H_6	8.74	0.34
103		
CHN_3O_3	2.36	0.62
$CH_3N_4O_2$	2.73	0.43
C_2HNO_4	2.72	0.83
$C_2H_3N_2O_3$	3.09	0.64
$C_2H_5N_3O_2$	3.46	0.45
$C_2H_7N_4O$	3.84	0.26
$C_3H_3O_4$	3.45	0.84
$C_3H_5NO_3$	3.82	0.66
$C_3H_7N_2O_2$	4.20	0.47
$C_3H_9N_3O$	4.57	0.29
$C_3H_{11}N_4$	4.94	0.10
$C_4H_7O_3$	4.55	0.68
$C_4H_9NO_2$	4.93	0.50
$C_4H_{11}N_2O$	5.30	0.32
$C_4H_{13}N_3$	5.68	0.14
C_5HN_3	6.56	0.18
$C_5H_{11}O_2$	5.66	0.53
$C_5H_{13}NO$	6.03	0.35
C_6HNO	6.92	0.40
$C_6H_3N_2$	7.30	0.23
C_7H_3O	7.65	0.45
C_7H_5N	8.03	0.28
C_8H_7	8.76	0.34
104		
$CH_2N_3O_3$	2.37	0.62
$CH_4N_4O_2$	2.75	0.43
$C_2H_2NO_4$	2.73	0.83
$C_2H_4N_2O_3$	3.11	0.64
$C_2H_6N_3O_2$	3.48	0.45
$C_2H_8N_4O$	3.85	0.26
$C_3H_4O_4$	3.46	0.84
$C_3H_6NO_3$	3.84	0.66
$C_3H_8N_2O_2$	4.21	0.47
$C_3H_{10}N_3O$	4.59	0.29
$C_3H_{12}N_4$	4.96	0.10
$C_4H_8O_3$	4.57	0.68
$C_4H_{10}NO_2$	4.94	0.50
$C_4H_{12}N_2O$	5.32	0.32
$C_5H_2N_3$	6.58	0.19
$C_5H_{12}O_2$	5.67	0.53
C_6H_2NO	6.94	0.41
$C_6H_4N_2$	7.31	0.23
C_7H_4O	7.67	0.45
C_7H_6N	8.04	0.28
C_8H_8	8.77	0.34
105		
CHN_2O_4	2.02	0.81
$CH_3N_3O_3$	2.39	0.62
$CH_5N_4O_2$	2.77	0.43
$C_2H_3NO_4$	2.75	0.83
$C_2H_5N_2O_3$	3.12	0.64
$C_2H_7N_3O_2$	3.50	0.45
$C_2H_9N_4O$	3.87	0.26
$C_3H_5O_4$	3.48	0.84
$C_3H_7NO_3$	3.85	0.66
$C_3H_9N_2O_2$	4.23	0.47
$C_3H_{11}N_3O$	4.60	0.29
C_4HN_4	5.86	0.15
$C_4H_9O_3$	4.58	0.68
$C_4H_{11}NO_2$	4.96	0.50
C_5HN_2O	6.22	0.36
$C_5H_3N_3$	6.60	0.19
C_6HO_2	6.58	0.58
C_6H_3NO	6.95	0.41
$C_6H_5N_2$	7.33	0.23
C_7H_5O	7.68	0.45
C_7H_7N	8.06	0.28
C_8H_9	8.79	0.34
106		
$CH_2N_2O_4$	2.03	0.82
$CH_4N_3O_3$	2.41	0.62
$CH_6N_4O_2$	2.78	0.43
$C_2H_4NO_4$	2.76	0.83
$C_2H_6N_2O_3$	3.14	0.64
$C_2H_8N_3O_2$	3.51	0.45
$C_2H_{10}N_4O$	3.89	0.26
$C_3H_6O_4$	3.49	0.85
$C_3H_8NO_3$	3.87	0.66
$C_3H_{10}N_2O_2$	4.24	0.47
$C_4H_2N_4$	5.88	0.15
$C_4H_{10}O_3$	4.60	0.68
$C_5H_2N_2O$	6.24	0.36
$C_5H_4N_3$	6.61	0.19
$C_6H_2O_2$	6.59	0.58
C_6H_4NO	6.97	0.41
$C_6H_6N_2$	7.34	0.23
C_7H_6O	7.70	0.46
C_7H_8N	8.07	0.28
C_8H_{10}	8.81	0.34
107		
$CH_3N_2O_4$	2.05	0.82
$CH_5N_3O_3$	2.42	0.62
$CH_7N_4O_2$	2.80	0.43
$C_2H_5NO_4$	2.78	0.83
$C_2H_7N_2O_3$	3.15	0.64
$C_2H_9N_3O_2$	3.53	0.45
$C_3H_7O_4$	3.51	0.85
$C_3H_9NO_3$	3.89	0.66
C_4HN_3O	5.52	0.33
$C_4H_3N_4$	5.90	0.15
C_5HNO_2	5.88	0.54
$C_5H_3N_2O$	6.25	0.37
$C_5H_5N_3$	6.63	0.19
$C_6H_3O_2$	6.61	0.58
C_6H_5NO	6.98	0.41
$C_6H_7N_2$	7.36	0.23
C_7H_7O	7.72	0.46
C_7H_9N	8.09	0.29
C_8H_{11}	8.82	0.34
108		
$CH_4N_2O_4$	2.06	0.82
$CH_6N_3O_3$	2.44	0.62
$CH_8N_4O_2$	2.81	0.43
$C_2H_6NO_4$	2.80	0.83
$C_2H_8N_2O_3$	3.17	0.64
$C_3H_8O_4$	3.53	0.85
$C_4H_2N_3O$	5.54	0.33
$C_4H_4N_4$	5.91	0.15
$C_5H_2NO_2$	5.90	0.54
$C_5H_4N_2O$	6.27	0.37
$C_5H_6N_3$	6.64	0.19
$C_6H_4O_2$	6.63	0.59
C_6H_6NO	7.00	0.41
$C_6H_8N_2$	7.38	0.24
C_7H_8O	7.73	0.46
$C_7H_{10}N$	8.11	0.29
C_8H_{12}	8.84	0.34
109		
$CH_5N_2O_4$	2.08	0.82
$CH_7N_3O_3$	2.45	0.62
$C_2H_7NO_4$	2.81	0.83
C_3HN_4O	4.82	0.30
$C_4HN_2O_2$	5.18	0.51
$C_4H_3N_3O$	5.55	0.33
$C_4H_5N_4$	5.93	0.15
C_5HO_3	5.54	0.73

Appendix Table 5.1 (*continued*)

	M + 1	M + 2
$C_5H_3NO_2$	5.91	0.55
$C_5H_5N_2O$	6.29	0.37
$C_5H_7N_3$	6.66	0.19
$C_6H_5O_2$	6.64	0.59
C_6H_7NO	7.02	0.41
$C_6H_9N_2$	7.39	0.24
C_7H_9O	7.75	0.46
$C_7H_{11}N$	8.12	0.29
C_8H_{13}	8.85	0.35
C_9H	9.74	0.42
110		
$CH_6N_2O_4$	2.10	0.82
$C_3H_2N_4O$	4.84	0.30
$C_4H_2N_2O_2$	5.20	0.51
$C_4H_4N_3O$	5.57	0.33
$C_4H_6N_4$	5.94	0.15
$C_5H_2O_3$	5.55	0.73
$C_5H_4NO_2$	5.93	0.55
$C_5H_6N_2O$	6.30	0.37
$C_5H_8N_3$	6.68	0.19
$C_6H_6O_2$	6.66	0.59
C_6H_8NO	7.03	0.41
$C_6H_{10}N_2$	7.41	0.24
$C_7H_{10}O$	7.76	0.46
$C_7H_{12}N$	8.14	0.29
C_8H_{14}	8.87	0.35
C_9H_2	9.76	0.42
111		
$C_3HN_3O_2$	4.48	0.48
$C_3H_3N_4O$	4.86	0.30
C_4HNO_3	4.84	0.69
$C_4H_3N_2O_2$	5.21	0.51
$C_4H_5N_3O$	5.59	0.33
$C_4H_7N_4$	5.96	0.15
$C_5H_3O_3$	5.57	0.73
$C_5H_5NO_2$	5.94	0.55
$C_5H_7N_2O$	6.32	0.37
$C_5H_9N_3$	6.69	0.19
$C_6H_7O_2$	6.67	0.59
C_6H_9NO	7.05	0.41
$C_6H_{11}N_2$	7.42	0.24
$C_7H_{11}O$	7.78	0.46
$C_7H_{13}N$	8.15	0.29
C_8HN	9.04	0.36
C_8H_{15}	8.89	0.35
C_9H_3	9.77	0.43
112		
$C_3H_2N_3O_2$	4.50	0.48
$C_3H_4N_4O$	4.87	0.30
$C_4H_2NO_3$	4.85	0.70
$C_4H_4N_2O_2$	5.23	0.51
$C_4H_6N_3O$	5.60	0.33
$C_4H_8N_4$	5.98	0.15
$C_5H_4O_3$	5.58	0.73
$C_5H_6NO_2$	5.96	0.55
$C_5H_8N_2O$	6.33	0.37
$C_5H_{10}N_3$	6.71	0.19
$C_6H_8O_2$	6.69	0.59
$C_6H_{10}NO$	7.06	0.41
$C_6H_{12}N_2$	7.44	0.24
$C_7H_{12}O$	7.80	0.46

	M + 1	M + 2
$C_7H_{14}N$	8.17	0.29
C_8H_2N	9.06	0.36
C_8H_{16}	8.90	0.35
C_9H_4	9.79	0.43
113		
$C_2HN_4O_2$	3.78	0.46
$C_3HN_2O_3$	4.14	0.67
$C_3H_3N_3O_2$	4.51	0.48
$C_3H_5N_4O$	4.89	0.30
C_4HO_4	4.50	0.88
$C_4H_3NO_3$	4.87	0.70
$C_4H_5N_2O_2$	5.24	0.51
$C_4H_7N_3O$	5.62	0.33
$C_4H_9N_4$	5.99	0.15
$C_5H_5O_3$	5.60	0.73
$C_5H_7NO_2$	5.98	0.55
$C_5H_9N_2O$	6.35	0.37
$C_5H_{11}N_3$	6.72	0.19
$C_6H_9O_2$	6.71	0.59
$C_6H_{11}NO$	7.08	0.42
$C_6H_{13}N_2$	7.46	0.24
C_7HN_2	8.34	0.31
$C_7H_{13}O$	7.81	0.46
$C_7H_{15}N$	8.19	0.29
C_8HO	8.70	0.53
C_8H_3N	9.07	0.36
C_8H_{17}	8.92	0.35
C_9H_5	9.81	0.43
114		
$C_2H_2N_4O_2$	3.80	0.46
$C_3H_2N_2O_3$	4.15	0.67
$C_3H_4N_3O_2$	4.53	0.48
$C_3H_6N_4O$	4.90	0.30
$C_4H_2O_4$	4.51	0.88
$C_4H_4NO_3$	4.89	0.70
$C_4H_6N_2O_2$	5.26	0.51
$C_4H_8N_3O$	5.63	0.33
$C_4H_{10}N_4$	6.01	0.15
$C_5H_6O_3$	5.62	0.73
$C_5H_8NO_2$	5.99	0.55
$C_5H_{10}N_2O$	6.37	0.37
$C_5H_{12}N_3$	6.74	0.20
$C_6H_{10}O_2$	6.72	0.59
$C_6H_{12}NO$	7.10	0.42
$C_6H_{14}N_2$	7.47	0.24
$C_7H_2N_2$	8.36	0.31
$C_7H_{14}O$	7.83	0.47
$C_7H_{16}N$	8.20	0.29
C_8H_2O	8.72	0.53
C_8H_4N	9.09	0.37
C_8H_{18}	8.93	0.35
C_9H_6	9.82	0.43
115		
$C_2HN_3O_3$	3.44	0.65
$C_2H_3N_4O_2$	3.81	0.46
C_3HNO_4	3.80	0.86
$C_3H_3N_2O_3$	4.17	0.67
$C_3H_5N_3O_2$	4.54	0.48
$C_3H_7N_4O$	4.92	0.30
$C_4H_3O_4$	4.53	0.88
$C_4H_5NO_3$	4.90	0.70

	M + 1	M + 2
$C_4H_7N_2O_2$	5.28	0.52
$C_4H_9N_3O$	5.65	0.33
$C_4H_{11}N_4$	6.02	0.16
$C_5H_7O_3$	5.63	0.73
$C_5H_9NO_2$	6.01	0.55
$C_5H_{11}N_2O$	6.38	0.37
$C_5H_{13}N_3$	6.76	0.20
C_6HN_3	7.64	0.25
$C_6H_{11}O_2$	6.74	0.59
$C_6H_{13}NO$	7.11	0.42
$C_6H_{15}N_2$	7.49	0.24
C_7HNO	8.00	0.48
$C_7H_3N_2$	8.38	0.31
$C_7H_{15}O$	7.84	0.47
$C_7H_{17}N$	8.22	0.30
C_8H_3O	8.73	0.54
C_8H_5N	9.11	0.37
C_9H_7	9.84	0.43
116		
$C_2H_2N_3O_3$	3.46	0.65
$C_2H_4N_4O_2$	3.83	0.46
$C_3H_2NO_4$	3.81	0.86
$C_3H_4N_2O_3$	4.19	0.67
$C_3H_6N_3O_2$	4.56	0.49
$C_3H_8N_4O$	4.94	0.30
$C_4H_4O_4$	4.54	0.88
$C_4H_6NO_3$	4.92	0.70
$C_4H_8N_2O_2$	5.29	0.52
$C_4H_{10}N_3O$	5.67	0.34
$C_4H_{12}N_4$	6.04	0.16
$C_5H_8O_3$	5.65	0.73
$C_5H_{10}NO_2$	6.02	0.55
$C_5H_{12}N_2O$	6.40	0.37
$C_5H_{14}N_3$	6.77	0.20
$C_6H_2N_3$	7.66	0.26
$C_6H_{12}O_2$	6.75	0.59
$C_6H_{14}NO$	7.13	0.42
$C_6H_{16}N_2$	7.50	0.24
C_7H_2NO	8.02	0.48
$C_7H_4N_2$	8.39	0.31
$C_7H_{16}O$	7.86	0.47
C_8H_4O	8.75	0.54
C_8H_6N	9.12	0.37
C_9H_8	9.85	0.43
117		
$C_2HN_2O_4$	3.10	0.84
$C_2H_3N_3O_3$	3.47	0.65
$C_2H_5N_4O_2$	3.85	0.46
$C_3H_3NO_4$	3.83	0.86
$C_3H_5N_2O_3$	4.20	0.67
$C_3H_7N_3O_2$	4.58	0.49
$C_3H_9N_4O$	4.95	0.30
$C_4H_5O_4$	4.56	0.88
$C_4H_7NO_3$	4.93	0.70
$C_4H_9N_2O_2$	5.31	0.52
$C_4H_{11}N_3O$	5.68	0.34
$C_4H_{13}N_4$	6.06	0.16
C_5HN_4	6.95	0.21
$C_5H_9O_3$	5.66	0.73
$C_5H_{11}NO_2$	6.04	0.55
$C_5H_{13}N_2O$	6.41	0.38
$C_5H_{15}N_3$	6.79	0.20

	M + 1	M + 2
C_6HN_2O	7.30	0.43
$C_6H_3N_3$	7.68	0.26
$C_6H_{13}O_2$	6.77	0.60
$C_6H_{15}NO$	7.14	0.42
C_7HO_2	7.66	0.65
C_7H_3NO	8.03	0.48
$C_7H_5N_2$	8.41	0.31
C_8H_5O	8.76	0.54
C_8H_7N	9.14	0.37
C_9H_9	9.87	0.43
118		
$C_2H_2N_2O_4$	3.11	0.84
$C_2H_4N_3O_3$	3.49	0.65
$C_2H_6N_4O_2$	3.86	0.46
$C_3H_4NO_4$	3.84	0.86
$C_3H_6N_2O_3$	4.22	0.67
$C_3H_8N_3O_2$	4.59	0.49
$C_3H_{10}N_4O$	4.97	0.30
$C_4H_6O_4$	4.58	0.88
$C_4H_8NO_3$	4.95	0.70
$C_4H_{10}N_2O_2$	5.32	0.52
$C_4H_{12}N_3O$	5.70	0.34
$C_4H_{14}N_4$	6.07	0.16
$C_5H_2N_4$	6.96	0.21
$C_5H_{10}O_3$	5.68	0.73
$C_5H_{12}NO_2$	6.06	0.55
$C_5H_{14}N_2O$	6.43	0.38
$C_6H_2N_2O$	7.32	0.43
$C_6H_4N_3$	7.69	0.26
$C_6H_{14}O_2$	6.79	0.60
$C_7H_2O_2$	7.67	0.65
C_7H_4NO	8.05	0.48
$C_7H_6N_2$	8.42	0.31
C_8H_6O	8.78	0.54
C_8H_8N	9.15	0.37
C_9H_{10}	9.89	0.44
119		
$C_2H_3N_2O_4$	3.13	0.84
$C_2H_5N_3O_3$	3.50	0.65
$C_2H_7N_4O_2$	3.88	0.46
$C_3H_5NO_4$	3.86	0.86
$C_3H_7N_2O_3$	4.23	0.67
$C_3H_9N_3O_2$	4.61	0.49
$C_3H_{11}N_4O$	4.98	0.30
$C_4H_7O_4$	4.59	0.88
$C_4H_9NO_3$	4.97	0.70
$C_4H_{11}N_2O_2$	5.34	0.52
$C_4H_{13}N_3O$	5.71	0.34
C_5HN_3O	6.60	0.39
$C_5H_3N_4$	6.98	0.21
$C_5H_{11}O_3$	5.70	0.73
$C_5H_{13}NO_2$	6.07	0.56
C_6HNO_2	6.96	0.61
$C_6H_3N_2O$	7.33	0.43
$C_6H_5N_3$	7.71	0.26
$C_7H_3O_2$	7.69	0.66
C_7H_5NO	8.07	0.48
$C_7H_7N_2$	8.44	0.31
C_8H_7O	8.80	0.54
C_8H_9N	9.17	0.37
C_9H_{11}	9.90	0.44

Appendix Table 5.1 (*continued*)

	M + 1	M + 2
120		
$C_2H_4N_2O_4$	3.15	0.84
$C_2H_6N_3O_3$	3.52	0.65
$C_2H_8N_4O_2$	3.89	0.46
$C_3H_6NO_4$	3.88	0.86
$C_3H_8N_2O_3$	4.25	0.67
$C_3H_{10}N_3O_2$	4.62	0.49
$C_3H_{12}N_4O$	5.00	0.31
$C_4H_8O_4$	4.61	0.88
$C_4H_{10}NO_3$	4.98	0.70
$C_4H_{12}N_2O_2$	5.36	0.52
$C_5H_2N_3O$	6.62	0.39
$C_5H_4N_4$	6.99	0.21
$C_5H_{12}O_3$	5.71	0.74
$C_6H_2NO_2$	6.98	0.61
$C_6H_4N_2O$	7.35	0.43
$C_6H_6N_3$	7.72	0.26
$C_7H_4O_2$	7.71	0.66
C_7H_6NO	8.08	0.49
$C_7H_8N_2$	8.46	0.32
C_8H_8O	8.81	0.54
$C_8H_{10}N$	9.19	0.37
C_9H_{12}	9.92	0.44
121		
$C_2H_5N_2O_4$	3.16	0.84
$C_2H_7N_3O_3$	3.54	0.65
$C_2H_9N_4O_2$	3.91	0.46
$C_3H_7NO_4$	3.89	0.86
$C_3H_9N_2O_3$	4.27	0.67
$C_3H_{11}N_3O_2$	4.64	0.49
C_4HN_4O	5.90	0.35
$C_4H_9O_4$	4.62	0.89
$C_4H_{11}NO_3$	5.00	0.70
$C_5HN_2O_2$	6.26	0.57
$C_5H_3N_3O$	6.64	0.39
$C_5H_5N_4$	7.01	0.21
C_6HO_3	6.62	0.79
$C_6H_3NO_2$	6.99	0.61
$C_6H_5N_2O$	7.37	0.44
$C_6H_7N_3$	7.74	0.26
$C_7H_5O_2$	7.72	0.66
C_7H_7NO	8.10	0.49
$C_7H_9N_2$	8.47	0.32
C_8H_9O	8.83	0.54
$C_8H_{11}N$	9.20	0.38
C_9H_{13}	9.93	0.44
$C_{10}H$	10.82	0.53
122		
$C_2H_6N_2O_4$	3.18	0.84
$C_2H_8N_3O_3$	3.55	0.65
$C_2H_{10}N_4O_2$	3.93	0.46
$C_3H_8NO_4$	3.91	0.86
$C_3H_{10}N_2O_3$	4.28	0.67
$C_4H_2N_4O$	5.92	0.35
$C_4H_{10}O_4$	4.64	0.89
$C_5H_2N_2O_2$	6.28	0.57
$C_5H_4N_3O$	6.65	0.39
$C_5H_6N_4$	7.03	0.21
$C_6H_2O_3$	6.63	0.79
$C_6H_4NO_2$	7.01	0.61
$C_6H_6N_2O$	7.38	0.44
$C_6H_8N_3$	7.76	0.26
$C_7H_6O_2$	7.74	0.66
C_7H_8NO	8.11	0.49
$C_7H_{10}N_2$	8.49	0.32
$C_8H_{10}O$	8.84	0.54
$C_8H_{12}N$	9.22	0.38
C_9H_{14}	9.95	0.44
$C_{10}H_2$	10.84	0.53
123		
$C_2H_7N_2O_4$	3.19	0.84
$C_2H_9N_3O_3$	3.57	0.65
$C_3H_9NO_4$	3.92	0.86
$C_4HN_3O_2$	5.56	0.53
$C_4H_3N_4O$	5.94	0.35
C_5HNO_3	5.92	0.75
$C_5H_3N_2O_2$	6.29	0.57
$C_5H_5N_3O$	6.67	0.39
$C_5H_7N_4$	7.04	0.22
$C_6H_3O_3$	6.65	0.79
$C_6H_5NO_2$	7.02	0.61
$C_6H_7N_2O$	7.40	0.44
$C_6H_9N_3$	7.77	0.26
$C_7H_7O_2$	7.75	0.66
C_7H_9NO	8.13	0.49
$C_7H_{11}N_2$	8.50	0.32
$C_8H_{11}O$	8.86	0.55
$C_8H_{13}N$	9.23	0.38
C_9HN	10.12	0.46
C_9H_{15}	9.97	0.44
$C_{10}H_3$	10.85	0.53
124		
$C_2H_8N_2O_4$	3.21	0.84
$C_4H_2N_3O_2$	5.58	0.53
$C_4H_4N_4O$	5.95	0.35
$C_5H_2NO_3$	5.93	0.75
$C_5H_4N_2O_2$	6.31	0.57
$C_5H_6N_3O$	6.68	0.39
$C_5H_8N_4$	7.06	0.22
$C_6H_4O_3$	6.67	0.79
$C_6H_6NO_2$	7.04	0.61
$C_6H_8N_2O$	7.41	0.44
$C_6H_{10}N_3$	7.79	0.27
$C_7H_8O_2$	7.77	0.66
$C_7H_{10}NO$	8.15	0.49
$C_7H_{12}N_2$	8.52	0.32
$C_8H_{12}O$	8.88	0.55
$C_8H_{14}N$	9.25	0.38
C_9H_2N	10.14	0.46
C_9H_{16}	9.98	0.45
$C_{10}H_4$	10.87	0.53
125		
$C_3HN_4O_2$	4.86	0.50
$C_4HN_2O_3$	5.22	0.71
$C_4H_3N_3O_2$	5.59	0.53
$C_4H_5N_4O$	5.97	0.35
C_5HO_4	5.58	0.93
$C_5H_3NO_3$	5.95	0.75
$C_5H_5N_2O_2$	6.32	0.57
$C_5H_7N_3O$	6.70	0.39
$C_5H_9N_4$	7.07	0.22
$C_6H_5O_3$	6.68	0.79
$C_6H_7NO_2$	7.06	0.61
$C_6H_9N_2O$	7.43	0.44
$C_6H_{11}N_3$	7.80	0.27
$C_7H_9O_2$	7.79	0.66
$C_7H_{11}NO$	8.16	0.49
$C_7H_{13}N_2$	8.54	0.32
C_8HN_2	9.42	0.40
$C_8H_{13}O$	8.89	0.55
$C_8H_{15}N$	9.27	0.38
C_9HO	9.78	0.63
C_9H_3N	10.16	0.46
C_9H_{17}	10.00	0.45
$C_{10}H_5$	10.89	0.53
126		
$C_3H_2N_4O_2$	4.88	0.50
$C_4H_2N_2O_3$	5.24	0.71
$C_4H_4N_3O_2$	5.61	0.53
$C_4H_6N_4O$	5.98	0.35
$C_5H_2O_4$	5.59	0.93
$C_5H_4NO_3$	5.97	0.75
$C_5H_6N_2O_2$	6.34	0.57
$C_5H_8N_3O$	6.72	0.35
$C_5H_{10}N_4$	7.09	0.22
$C_6H_6O_3$	6.70	0.79
$C_6H_8NO_2$	7.07	0.62
$C_6H_{10}N_2O$	7.45	0.44
$C_6H_{12}N_3$	7.82	0.27
$C_7H_{10}O_2$	7.80	0.66
$C_7H_{12}NO$	8.18	0.49
$C_7H_{14}N_2$	8.55	0.32
$C_8H_2N_2$	9.44	0.40
$C_8H_{14}O$	8.91	0.55
$C_8H_{16}N$	9.28	0.38
C_9H_2O	9.80	0.63
C_9H_4N	10.17	0.46
C_9H_{18}	10.01	0.45
$C_{10}H_6$	10.90	0.54
127		
$C_3HN_3O_3$	4.52	0.68
$C_3H_3N_4O_2$	4.89	0.50
C_4HNO_4	4.88	0.90
$C_4H_3N_2O_3$	5.25	0.71
$C_4H_5N_3O_2$	5.63	0.53
$C_4H_7N_4O$	6.00	0.35
$C_5H_3O_4$	5.61	0.93
$C_5H_5NO_3$	5.98	0.75
$C_5H_7N_2O_2$	6.36	0.57
$C_5H_9N_3O$	6.73	0.40
$C_5H_{11}N_4$	7.11	0.22
$C_6H_7O_3$	6.71	0.79
$C_6H_9NO_2$	7.09	0.62
$C_6H_{11}N_2O$	7.46	0.44
$C_6H_{13}N_3$	7.84	0.27
C_7HN_3	8.73	0.34
$C_7H_{11}O_2$	7.82	0.67
$C_7H_{13}NO$	8.19	0.49
$C_7H_{15}N_2$	8.57	0.32
C_8HNO	9.08	0.57
$C_8H_3N_2$	9.46	0.40
$C_8H_{15}O$	8.92	0.55
$C_8H_{17}N$	9.30	0.38
C_9H_3O	9.81	0.63
C_9H_5N	10.19	0.47
C_9H_{19}	10.03	0.45
$C_{10}H_7$	10.92	0.54
128		
$C_3H_2N_3O_3$	4.54	0.68
$C_3H_4N_4O_2$	4.91	0.50
$C_4H_2NO_4$	4.89	0.90
$C_4H_4N_2O_3$	5.27	0.72
$C_4H_6N_3O_2$	5.64	0.53
$C_4H_8N_4O$	6.02	0.36
$C_5H_4O_4$	5.62	0.93
$C_5H_6NO_3$	6.00	0.75
$C_5H_8N_2O_2$	6.37	0.57
$C_5H_{10}N_3O$	6.75	0.40
$C_5H_{12}N_4$	7.12	0.22
$C_6H_8O_3$	6.73	0.79
$C_6H_{10}NO_2$	7.10	0.62
$C_6H_{12}N_2O$	7.48	0.44
$C_6H_{14}N_3$	7.85	0.27
$C_7H_2N_3$	8.74	0.34
$C_7H_{12}O_2$	7.83	0.67
$C_7H_{14}NO$	8.21	0.50
$C_7H_{16}N_2$	8.58	0.33
C_8H_2NO	9.10	0.57
$C_8H_4N_2$	9.47	0.40
$C_8H_{16}O$	8.94	0.55
$C_8H_{18}N$	9.31	0.39
C_9H_4O	9.83	0.63
C_9H_6N	10.20	0.47
C_9H_{20}	10.05	0.45
$C_{10}H_8$	10.94	0.54
129		
$C_3HN_2O_4$	4.18	0.87
$C_3H_3N_3O_3$	4.55	0.69
$C_3H_5N_4O_2$	4.93	0.50
$C_4H_3NO_4$	4.91	0.90
$C_4H_5N_2O_3$	5.28	0.72
$C_4H_7N_3O_2$	5.66	0.54
$C_4H_9N_4O$	6.03	0.36
$C_5H_5O_4$	5.64	0.93
$C_5H_7NO_3$	6.01	0.75
$C_5H_9N_2O_2$	6.39	0.57
$C_5H_{11}N_3O$	6.76	0.40
$C_5H_{13}N_4$	7.14	0.22
C_6HN_4	8.03	0.28
$C_6H_9O_3$	6.75	0.79
$C_6H_{11}NO_2$	7.12	0.62
$C_6H_{13}N_2O$	7.49	0.44
$C_6H_{15}N_3$	7.87	0.27
C_7HN_2O	8.38	0.51
$C_7H_3N_3$	8.76	0.34
$C_7H_{13}O_2$	7.85	0.67
$C_7H_{15}NO$	8.23	0.50
$C_7H_{17}N_2$	8.60	0.33
C_8HO_2	8.74	0.74
C_8H_3NO	9.11	0.57
$C_8H_5N_2$	9.49	0.40
$C_8H_{17}O$	8.96	0.55
$C_8H_{19}N$	9.33	0.39
C_9H_5O	9.85	0.63
C_9H_7N	10.22	0.47
$C_{10}H_9$	10.95	0.54

Appendix Table 5.1 (*continued*)

	M + 1	M + 2
130		
$C_3H_2N_2O_4$	4.19	0.87
$C_3H_4N_3O_3$	4.57	0.69
$C_3H_6N_4O_2$	4.94	0.50
$C_4H_4NO_4$	4.92	0.90
$C_4H_6N_2O_3$	5.30	0.72
$C_4H_8N_3O_2$	5.67	0.54
$C_4H_{10}N_4O$	6.05	0.36
$C_5H_6O_4$	5.66	0.93
$C_5H_8NO_3$	6.03	0.75
$C_5H_{10}N_2O_2$	6.40	0.58
$C_5H_{12}N_3O$	6.78	0.40
$C_5H_{14}N_4$	7.15	0.22
$C_6H_2N_4$	8.04	0.29
$C_6H_{10}O_3$	6.76	0.79
$C_6H_{12}NO_2$	7.14	0.62
$C_6H_{14}N_2O$	7.51	0.45
$C_6H_{16}N_3$	7.88	0.27
$C_7H_2N_2O$	8.40	0.51
$C_7H_4N_3$	8.77	0.34
$C_7H_{14}O_2$	7.87	0.67
$C_7H_{16}NO$	8.24	0.50
$C_7H_{18}N_2$	8.62	0.33
$C_8H_2O_2$	8.76	0.74
C_8H_4NO	9.13	0.57
$C_8H_6N_2$	9.50	0.40
$C_8H_{18}O$	8.97	0.56
C_9H_6O	9.86	0.63
C_9H_8N	10.24	0.47
$C_{10}H_{10}$	10.97	0.54
131		
$C_3H_3N_2O_4$	4.21	0.87
$C_3H_5N_3O_3$	4.58	0.69
$C_3H_7N_4O_2$	4.96	0.50
$C_4H_5NO_4$	4.94	0.90
$C_4H_7N_2O_3$	5.32	0.72
$C_4H_9N_3O_2$	5.69	0.54
$C_4H_{11}N_4O$	6.06	0.36
$C_5H_7O_4$	5.67	0.93
$C_5H_9NO_3$	6.05	0.75
$C_5H_{11}N_2O_2$	6.42	0.58
$C_5H_{13}N_3O$	6.80	0.40
$C_5H_{15}N_4$	7.17	0.22
C_6HN_3O	7.68	0.46
$C_6H_3N_4$	8.06	0.29
$C_6H_{11}O_3$	6.78	0.80
$C_6H_{13}NO_2$	7.15	0.62
$C_6H_{15}N_2O$	7.53	0.45
$C_6H_{17}N_3$	7.90	0.27
C_7HNO_2	8.04	0.68
$C_7H_3N_2O$	8.41	0.51
$C_7H_5N_3$	8.79	0.34
$C_7H_{15}O_2$	7.88	0.67
$C_7H_{17}NO$	8.26	0.50
$C_8H_3O_2$	8.77	0.74
C_8H_5NO	9.15	0.57
$C_8H_7N_2$	9.52	0.41
C_9H_7O	9.88	0.64
C_9H_9N	10.25	0.47
$C_{10}H_{11}$	10.98	0.54
132		
$C_3H_4N_2O_4$	4.23	0.87
$C_3H_6N_3O_3$	4.60	0.69
$C_3H_8N_4O_2$	4.97	0.50
$C_4H_6NO_4$	4.96	0.90
$C_4H_8N_2O_3$	5.33	0.72
$C_4H_{10}N_3O_2$	5.71	0.54
$C_4H_{12}N_4O$	6.08	0.36
$C_5H_8O_4$	5.69	0.93
$C_5H_{10}NO_3$	6.06	0.76
$C_5H_{12}N_2O_2$	6.44	0.58
$C_5H_{14}N_3O$	6.81	0.40
$C_5H_{16}N_4$	7.19	0.23
$C_6H_2N_3O$	7.70	0.46
$C_6H_4N_4$	8.07	0.29
$C_6H_{12}O_3$	6.97	0.80
$C_6H_{14}NO_2$	7.17	0.62
$C_6H_{16}N_2O$	7.54	0.45
$C_7H_2NO_2$	8.06	0.68
$C_7H_4N_2O$	8.43	0.51
$C_7H_6N_3$	8.81	0.34
$C_7H_{16}O_2$	7.90	0.67
$C_8H_4O_2$	8.79	0.74
C_8H_6NO	9.16	0.57
$C_8H_8N_2$	9.54	0.41
C_9H_8O	9.89	0.64
$C_9H_{10}N$	10.27	0.47
$C_{10}H_{12}$	11.00	0.55
133		
$C_3H_5N_2O_4$	4.24	0.87
$C_3H_7N_3O_3$	4.62	0.69
$C_3H_9N_4O_2$	4.99	0.51
$C_4H_7NO_4$	4.97	0.90
$C_4H_9N_2O_3$	5.35	0.72
$C_4H_{11}N_3O_2$	5.72	0.54
$C_4H_{13}N_4O$	6.10	0.36
C_5HN_4O	6.98	0.41
$C_5H_9O_4$	5.70	0.94
$C_5H_{11}NO_3$	6.08	0.76
$C_5H_{13}N_2O_2$	6.45	0.58
$C_5H_{15}N_3O$	6.83	0.40
$C_6HN_2O_2$	7.34	0.63
$C_6H_3N_3O$	7.72	0.46
$C_6H_5N_4$	8.09	0.29
$C_6H_{13}O_3$	6.81	0.80
$C_6H_{15}NO_2$	7.18	0.62
C_7HO_3	7.70	0.86
$C_7H_3NO_2$	8.07	0.69
$C_7H_5N_2O$	8.45	0.51
$C_7H_7N_3$	8.82	0.35
$C_8H_5O_2$	8.80	0.74
C_8H_7NO	9.18	0.57
$C_8H_9N_2$	9.55	0.41
C_9H_9O	9.91	0.64
$C_9H_{11}N$	10.28	0.48
$C_{10}H_{13}$	11.01	0.55
$C_{11}H$	11.90	0.64
134		
$C_3H_6N_2O_4$	4.26	0.87
$C_3H_8N_3O_3$	4.63	0.69
$C_3H_{10}N_4O_2$	5.01	0.51
$C_4H_8NO_4$	4.99	0.90
$C_4H_{10}N_2O_3$	5.36	0.72
$C_4H_{12}N_3O_2$	5.74	0.54
$C_4H_{14}N_4O$	6.11	0.36
$C_5H_2N_4O$	7.00	0.41
$C_5H_{10}O_4$	5.72	0.94
$C_5H_{12}NO_3$	6.09	0.76
$C_5H_{14}N_2O_2$	6.47	0.58
$C_6H_2N_2O_2$	7.36	0.64
$C_6H_4N_3O$	7.73	0.46
$C_6H_6N_4$	8.11	0.29
$C_6H_{14}O_3$	6.83	0.80
$C_7H_2O_3$	7.71	0.86
$C_7H_4NO_2$	8.09	0.69
$C_7H_6N_2O$	8.46	0.52
$C_7H_8N_3$	8.84	0.35
$C_8H_6O_2$	8.82	0.74
C_8H_8NO	9.19	0.58
$C_8H_{10}N_2$	9.57	0.41
$C_9H_{10}O$	9.93	0.64
$C_9H_{12}N$	10.30	0.48
$C_{10}H_{14}$	11.03	0.55
$C_{11}H_2$	11.92	0.65
135		
$C_3H_7N_2O_4$	4.27	0.87
$C_3H_9N_3O_3$	4.65	0.69
$C_3H_{11}N_4O_2$	5.02	0.51
$C_4H_9NO_4$	5.00	0.90
$C_4H_{11}N_2O_3$	5.38	0.72
$C_4H_{13}N_3O_2$	5.75	0.54
$C_5HN_3O_2$	6.64	0.59
$C_5H_3N_4O$	7.02	0.41
$C_5H_{11}O_4$	5.74	0.94
$C_5H_{13}NO_3$	6.11	0.76
C_6HNO_3	7.00	0.81
$C_6H_3N_2O_2$	7.37	0.64
$C_6H_5N_3O$	7.75	0.46
$C_6H_7N_4$	8.12	0.29
$C_7H_3O_3$	7.73	0.86
$C_7H_5NO_2$	8.10	0.69
$C_7H_7N_2O$	8.48	0.52
$C_7H_9N_3$	8.85	0.35
$C_8H_7O_2$	8.84	0.74
C_8H_9NO	9.21	0.58
$C_8H_{11}N_2$	9.58	0.41
$C_9H_{11}O$	9.94	0.64
$C_9H_{13}N$	10.32	0.48
$C_{10}HN$	11.20	0.57
$C_{10}H_{15}$	11.05	0.55
$C_{11}H_3$	11.94	0.65
136		
$C_3H_8N_2O_4$	4.29	0.87
$C_3H_{10}N_3O_3$	4.66	0.69
$C_3H_{12}N_4O_2$	5.04	0.51
$C_4H_{10}NO_4$	5.02	0.90
$C_4H_{12}N_2O_3$	5.40	0.72
$C_5H_2N_3O_2$	6.66	0.59
$C_5H_4N_4O$	7.03	0.42
$C_5H_{12}O_4$	5.75	0.94
$C_6H_2NO_3$	7.01	0.81
$C_6H_4N_2O_2$	7.39	0.64
$C_6H_6N_3O$	7.76	0.46
$C_6H_8N_4$	8.14	0.29
$C_7H_4O_3$	7.75	0.86
$C_7H_6NO_2$	8.12	0.69
$C_7H_8N_2O$	8.49	0.52
$C_7H_{10}N_3$	8.87	0.35
$C_8H_8O_2$	8.85	0.75
$C_8H_{10}NO$	9.23	0.58
$C_8H_{12}N_2$	9.60	0.41
$C_9H_{12}O$	9.96	0.64
$C_9H_{14}N$	10.33	0.48
$C_{10}H_2N$	11.22	0.57
$C_{10}H_{16}$	11.06	0.55
$C_{11}H_4$	11.95	0.65
137		
$C_3H_9N_2O_4$	4.31	0.88
$C_3H_{11}N_3O_3$	4.68	0.69
$C_4HN_4O_2$	5.94	0.55
$C_4H_{11}NO_4$	5.04	0.90
$C_5HN_2O_3$	6.30	0.77
$C_5H_3N_3O_2$	6.67	0.59
$C_5H_5N_4O$	7.05	0.42
C_6HO_4	6.66	0.99
$C_6H_3NO_3$	7.03	0.81
$C_6H_5N_2O_2$	7.41	0.64
$C_6H_7N_3O$	7.78	0.47
$C_6H_9N_4$	8.15	0.29
$C_7H_5O_3$	7.76	0.86
$C_7H_7NO_2$	8.14	0.69
$C_7H_9N_2O$	8.51	0.52
$C_7H_{11}N_3$	8.89	0.35
$C_8H_9O_2$	8.87	0.75
$C_8H_{11}NO$	9.24	0.58
$C_8H_{13}N_2$	9.62	0.41
C_9HN_2	10.50	0.50
$C_9H_{13}O$	9.97	0.65
$C_9H_{15}N$	10.35	0.48
$C_{10}HO$	10.86	0.73
$C_{10}H_3N$	11.24	0.57
$C_{10}H_{17}$	11.08	0.56
$C_{11}H_5$	11.97	0.65
138		
$C_3H_{10}N_2O_4$	4.32	0.88
$C_4H_2N_4O_2$	5.96	0.55
$C_5H_2N_2O_3$	6.32	0.77
$C_5H_4N_3O_2$	6.69	0.59
$C_5H_6N_4O$	7.06	0.42
$C_6H_2O_4$	6.67	0.99
$C_6H_4NO_3$	7.05	0.81
$C_6H_6N_2O_2$	7.42	0.64
$C_6H_8N_3O$	7.80	0.47
$C_6H_{10}N_4$	8.17	0.30
$C_7H_6O_3$	7.78	0.86
$C_7H_8NO_2$	8.15	0.69
$C_7H_{10}N_2O$	8.53	0.52
$C_7H_{12}N_3$	8.90	0.35
$C_8H_{10}O_2$	8.88	0.75
$C_8H_{12}NO$	9.26	0.58
$C_8H_{14}N_2$	9.63	0.42
$C_9H_2N_2$	10.52	0.50
$C_9H_{14}O$	9.99	0.65
$C_9H_{16}N$	10.36	0.48
$C_{10}H_2O$	10.88	0.73

Appendix Table 5.1 *(continued)*

	M + 1	M + 2
$C_{10}H_4N$	11.25	0.57
$C_{10}H_{18}$	11.09	0.56
$C_{11}H_6$	11.98	0.65
139		
$C_4H_3N_3O_3$	5.60	0.73
$C_4H_3N_4O_2$	5.97	0.55
C_5HNO_4	5.96	0.95
$C_5H_3N_2O_3$	6.33	0.77
$C_5H_5N_3O_2$	6.71	0.59
$C_5H_7N_4O$	7.03	0.42
$C_6H_3O_4$	6.69	0.99
$C_6H_5NO_3$	7.06	0.82
$C_6H_7N_2O_2$	7.44	0.64
$C_6H_9N_3O$	7.81	0.47
$C_6H_{11}N_4$	8.19	0.30
$C_7H_7O_3$	7.79	0.86
$C_7H_9NO_2$	8.17	0.69
$C_7H_{11}N_2O$	8.54	0.52
$C_7H_{13}N_3$	8.92	0.35
C_8HN_3	9.81	0.43
$C_8H_{11}O_2$	8.90	0.75
$C_8H_{13}NO$	9.27	0.58
$C_8H_{15}N_2$	9.65	0.42
C_9HNO	10.16	0.66
$C_9H_3N_2$	10.54	0.50
$C_9H_{15}O$	10.01	0.65
$C_9H_{17}N$	10.38	0.49
$C_{10}H_3O$	10.89	0.74
$C_{10}H_5N$	11.27	0.58
$C_{10}H_{19}$	11.11	0.56
$C_{11}H_7$	12.00	0.66
140		
$C_4H_2N_3O_3$	5.62	0.73
$C_4H_4N_4O_2$	5.99	0.55
$C_5H_2NO_4$	5.97	0.95
$C_5H_4N_2O_3$	6.35	0.77
$C_5H_6N_3O_2$	6.72	0.60
$C_5H_8N_4O$	7.10	0.42
$C_6H_4O_4$	6.70	0.99
$C_6H_6NO_3$	7.08	0.82
$C_6H_8N_2O_2$	7.45	0.64
$C_6H_{10}N_3O$	7.83	0.47
$C_6H_{12}N_4$	8.20	0.30
$C_7H_8O_3$	7.81	0.87
$C_7H_{10}NO_2$	8.18	0.69
$C_7H_{12}N_2O$	8.56	0.52
$C_7H_{14}N_3$	8.93	0.36
$C_8H_2N_3$	9.82	0.43
$C_8H_{12}O_2$	8.92	0.75
$C_8H_{14}NO$	9.29	0.58
$C_8H_{16}N_2$	9.66	0.42
C_9H_2NO	10.18	0.67
$C_9H_4N_2$	10.55	0.50
$C_9H_{16}O$	10.02	0.65
$C_9H_{18}N$	10.40	0.49
$C_{10}H_4O$	10.91	0.74
$C_{10}H_6N$	11.28	0.58
$C_{10}H_{20}$	11.13	0.56
$C_{11}H_8$	12.02	0.66
141		
$C_4HN_2O_4$	5.26	0.92
$C_4H_3N_3O_3$	5.63	0.73
$C_4H_5N_4O_2$	6.01	0.56
$C_5H_3NO_4$	5.99	0.95
$C_5H_5N_2O_3$	6.36	0.77
$C_5H_7N_3O_2$	6.74	0.60
$C_5H_9N_4O$	7.11	0.42
$C_6H_5O_4$	6.72	0.99
$C_6H_7NO_3$	7.09	0.82
$C_6H_9N_2O_2$	7.47	0.64
$C_6H_{11}N_3O$	7.84	0.47
$C_6H_{13}N_4$	8.22	0.30
C_7HN_4	9.11	0.37
$C_7H_9O_3$	7.83	0.87
$C_7H_{11}NO_2$	8.20	0.70
$C_7H_{13}N_2O$	8.57	0.53
$C_7H_{15}N_3$	8.95	0.36
C_8HN_2O	9.46	0.60
$C_8H_3N_3$	9.84	0.44
$C_8H_{13}O_2$	8.93	0.75
$C_8H_{15}NO$	9.31	0.59
$C_8H_{17}N_2$	9.68	0.42
C_9HO_2	9.82	0.83
C_9H_3NO	10.19	0.67
$C_9H_5N_2$	10.57	0.50
$C_9H_{17}O$	10.04	0.65
$C_9H_{19}N$	10.41	0.49
$C_{10}H_5O$	10.93	0.74
$C_{10}H_7N$	11.30	0.58
$C_{10}H_{21}$	11.14	0.56
$C_{11}H_9$	12.03	0.66
142		
$C_4H_2N_2O_4$	5.27	0.92
$C_4H_4N_3O_3$	5.65	0.74
$C_4H_6N_4O_2$	6.02	0.56
$C_5H_4NO_4$	6.01	0.95
$C_5H_6N_2O_3$	6.38	0.77
$C_5H_8N_3O_2$	6.75	0.60
$C_5H_{10}N_4O$	7.13	0.42
$C_6H_6O_4$	6.74	0.99
$C_6H_8NO_3$	7.11	0.82
$C_6H_{10}N_2O_2$	7.49	0.64
$C_6H_{12}N_3O$	7.86	0.47
$C_6H_{14}N_4$	8.23	0.30
$C_7H_2N_4$	9.12	0.37
$C_7H_{10}O_3$	7.84	0.87
$C_7H_{12}NO_2$	8.22	0.70
$C_7H_{14}N_2O$	8.59	0.53
$C_7H_{16}N_3$	8.97	0.36
$C_8H_2N_2O$	9.48	0.60
$C_8H_4N_3$	9.85	0.44
$C_8H_{14}O_2$	8.95	0.75
$C_8H_{16}NO$	9.32	0.59
$C_8H_{18}N_2$	9.70	0.42
$C_9H_2O_2$	9.84	0.83
C_9H_4NO	10.21	0.67
$C_9H_6N_2$	10.58	0.51
$C_9H_{18}O$	10.05	0.65
$C_9H_{20}N$	10.43	0.49
$C_{10}H_6O$	10.94	0.74
$C_{10}H_8N$	11.32	0.58
$C_{10}H_{22}$	11.16	0.56
$C_{11}H_{10}$	12.05	0.66
143		
$C_4H_3N_2O_4$	5.29	0.92
$C_4H_5N_3O_3$	5.66	0.74
$C_4H_7N_4O_2$	6.04	0.56
$C_5H_5NO_4$	6.02	0.95
$C_5H_7N_2O_3$	6.40	0.78
$C_5H_9N_3O_2$	6.77	0.60
$C_5H_{11}N_4O$	7.14	0.42
$C_6H_7O_4$	6.75	0.99
$C_6H_9NO_3$	7.13	0.82
$C_6H_{11}N_2O_2$	7.50	0.65
$C_6H_{13}N_3O$	7.88	0.47
$C_6H_{15}N_4$	8.25	0.30
C_7HN_3O	8.76	0.54
$C_7H_3N_4$	9.14	0.37
$C_7H_{11}O_3$	7.86	0.87
$C_7H_{13}NO_2$	8.23	0.70
$C_7H_{15}N_2O$	8.61	0.53
$C_7H_{17}N_3$	8.98	0.36
C_8HNO_2	9.12	0.77
$C_8H_3N_2O$	9.50	0.60
$C_8H_5N_3$	9.87	0.44
$C_8H_{15}O_2$	8.96	0.76
$C_8H_{17}NO$	9.34	0.59
$C_8H_{19}N_2$	9.71	0.42
$C_9H_3O_2$	9.85	0.83
C_9H_5NO	10.23	0.67
$C_9H_7N_2$	10.60	0.51
$C_9H_{19}O$	10.07	0.65
$C_9H_{21}N$	10.44	0.49
$C_{10}H_7O$	10.96	0.74
$C_{10}H_9N$	11.33	0.58
$C_{11}H_{11}$	12.06	0.66
144		
$C_4H_4N_2O_4$	5.31	0.92
$C_4H_6N_3O_3$	5.68	0.74
$C_4H_8N_4O_2$	6.05	0.56
$C_5H_6NO_4$	6.04	0.95
$C_5H_8N_2O_3$	6.41	0.78
$C_5H_{10}N_3O_2$	6.79	0.60
$C_5H_{12}N_4O$	7.16	0.42
$C_6H_8O_4$	6.77	1.00
$C_6H_{10}NO_3$	7.14	0.82
$C_6H_{12}N_2O_2$	7.52	0.65
$C_6H_{14}N_3O$	7.89	0.47
$C_6H_{16}N_4$	8.27	0.30
$C_7H_2N_3O$	8.78	0.54
$C_7H_4N_4$	9.15	0.38
$C_7H_{12}O_3$	7.87	0.87
$C_7H_{14}NO_2$	8.25	0.70
$C_7H_{16}N_2O$	8.62	0.53
$C_7H_{18}N_3$	9.00	0.36
$C_8H_2NO_2$	9.14	0.77
$C_8H_4N_2O$	9.51	0.60
$C_8H_6N_3$	9.89	0.44
$C_8H_{16}O_2$	8.98	0.76
$C_8H_{18}NO$	9.35	0.59
$C_8H_{20}N_2$	9.73	0.43
$C_9H_4O_2$	9.87	0.84
C_9H_6NO	10.24	0.67
$C_9H_8N_2$	10.62	0.51
$C_9H_{20}O$	10.09	0.66
$C_{10}H_8O$	10.97	0.74
$C_{10}H_{10}N$	11.35	0.58
$C_{11}H_{12}$	12.08	0.67
145		
$C_4H_5N_2O_4$	5.32	0.92
$C_4H_7N_3O_3$	5.70	0.74
$C_4H_9N_4O_2$	6.07	0.56
$C_5H_7NO_4$	6.05	0.96
$C_5H_9N_2O_3$	6.43	0.78
$C_5H_{11}N_3O_2$	6.80	0.60
$C_5H_{13}N_4O$	7.18	0.43
C_6HN_4O	8.07	0.49
$C_6H_9O_4$	6.78	1.00
$C_6H_{11}NO_3$	7.16	0.82
$C_6H_{13}N_2O_2$	7.53	0.65
$C_6H_{15}N_3O$	7.91	0.48
$C_6H_{17}N_4$	8.28	0.31
$C_7HN_2O_2$	8.42	0.71
$C_7H_3N_3O$	8.80	0.54
$C_7H_5N_4$	9.17	0.38
$C_7H_{13}O_3$	7.89	0.87
$C_7H_{15}NO_2$	8.26	0.70
$C_7H_{17}N_2O$	8.64	0.53
$C_7H_{19}N_3$	9.01	0.36
C_8HO_3	8.78	0.94
$C_8H_3NO_2$	9.15	0.77
$C_8H_5N_2O$	9.53	0.61
$C_8H_7N_3$	9.90	0.44
$C_8H_{17}O_2$	9.00	0.76
$C_8H_{19}NO$	9.37	0.59
$C_9H_5O_2$	9.88	0.84
C_9H_7NO	10.26	0.67
$C_9H_9N_2$	10.63	0.51
$C_{10}H_9O$	10.99	0.75
$C_{10}H_{11}N$	11.36	0.59
$C_{11}H_{13}$	12.10	0.67
$C_{12}H$	12.98	0.77
146		
$C_4H_6N_2O_4$	5.34	0.92
$C_4H_8N_3O_3$	5.71	0.74
$C_4H_{10}N_4O_2$	6.09	0.56
$C_5H_8NO_4$	6.07	0.96
$C_5H_{10}N_2O_3$	6.44	0.78
$C_5H_{12}N_3O_2$	6.82	0.60
$C_5H_{14}N_4O$	7.19	0.43
$C_6H_2N_4O$	8.08	0.49
$C_6H_{10}O_4$	6.80	1.00
$C_6H_{12}NO_3$	7.17	0.82
$C_6H_{14}N_2O_2$	7.55	0.65
$C_6H_{16}N_3O$	7.92	0.48
$C_6H_{18}N_4$	8.30	0.31
$C_7H_2N_2O_2$	8.44	0.71
$C_7H_4N_3O$	8.81	0.55
$C_7H_6N_4$	9.19	0.38
$C_7H_{14}O_3$	7.91	0.87
$C_7H_{16}NO_2$	8.28	0.70
$C_7H_{18}N_2O$	8.65	0.53
$C_8H_2O_3$	8.79	0.94
$C_8H_4NO_2$	9.17	0.77
$C_8H_6N_2O$	9.54	0.61
$C_8H_8N_3$	9.92	0.44
$C_8H_{18}O_2$	9.01	0.76
$C_9H_6O_2$	9.90	0.84

Appendix Table 5.1 (*continued*)

Formula	M + 1	M + 2
C_9H_8NO	10.27	0.68
$C_9H_{10}N_2$	10.65	0.51
$C_{10}H_{10}O$	11.01	0.75
$C_{10}H_{12}N$	11.38	0.59
$C_{11}H_{14}$	12.11	0.67
$C_{12}H_2$	13.00	0.77
147		
$C_4H_7N_2O_4$	5.35	0.92
$C_4H_9N_3O_3$	5.73	0.74
$C_4H_{11}N_4O_2$	6.10	0.56
$C_5H_9NO_4$	6.09	0.96
$C_5H_{11}N_2O_3$	6.46	0.78
$C_5H_{13}N_3O_2$	6.83	0.60
$C_5H_{15}N_4O$	7.21	0.43
$C_6HN_3O_2$	7.72	0.66
$C_6H_3N_4O$	8.10	0.49
$C_6H_{11}O_4$	6.82	1.00
$C_6H_{13}NO_3$	7.19	0.82
$C_6H_{15}N_2O_2$	7.57	0.65
$C_6H_{17}N_3O$	7.94	0.48
C_7HNO_3	8.08	0.89
$C_7H_3N_2O_2$	8.45	0.72
$C_7H_5N_3O$	8.83	0.55
$C_7H_7N_4$	9.20	0.38
$C_7H_{15}O_3$	7.92	0.87
$C_7H_{17}NO_2$	8.30	0.70
$C_8H_3O_3$	8.81	0.94
$C_8H_5NO_2$	9.19	0.78
$C_8H_7N_2O$	9.56	0.61
$C_8H_9N_3$	9.93	0.44
$C_9H_7O_2$	9.92	0.84
C_9H_9NO	10.29	0.68
$C_9H_{11}N_2$	10.66	0.51
$C_{10}H_{11}O$	11.02	0.75
$C_{10}H_{13}N$	11.40	0.59
$C_{11}HN$	12.28	0.69
$C_{11}H_{15}$	12.13	0.67
$C_{12}H_3$	13.02	0.78
148		
$C_4H_8N_2O_4$	5.37	0.92
$C_4H_{10}N_3O_3$	5.74	0.74
$C_4H_{12}N_4O_2$	6.12	0.56
$C_5H_{10}NO_4$	6.10	0.96
$C_5H_{12}N_2O_3$	6.48	0.78
$C_5H_{14}N_3O_2$	6.85	0.60
$C_5H_{16}N_4O$	7.22	0.43
$C_6H_2N_3O_2$	7.74	0.66
$C_6H_4N_4O$	8.11	0.49
$C_6H_{12}O_4$	6.83	1.00
$C_6H_{14}NO_3$	7.21	0.83
$C_6H_{16}N_2O_2$	7.58	0.65
$C_7H_2NO_3$	8.10	0.89
$C_7H_4N_2O_2$	8.47	0.72
$C_7H_6N_3O$	8.84	0.55
$C_7H_8N_4$	9.22	0.38
$C_7H_{16}O_3$	7.94	0.88
$C_8H_4O_3$	8.83	0.94
$C_8H_6NO_2$	9.20	0.78
$C_8H_8N_2O$	9.58	0.61
$C_8H_{10}N_3$	9.95	0.45
$C_9H_8O_2$	9.93	0.84
$C_9H_{10}NO$	10.31	0.68

Formula	M + 1	M + 2
$C_9H_{12}N_2$	10.68	0.52
$C_{10}H_{12}O$	11.04	0.75
$C_{10}H_{14}N$	11.41	0.59
$C_{11}H_2N$	12.30	0.69
$C_{11}H_{16}$	12.14	0.67
$C_{12}H_4$	13.03	0.78
149		
$C_4H_9N_2O_4$	5.39	0.92
$C_4H_{11}N_3O_3$	5.76	0.74
$C_4H_{13}N_4O_2$	6.13	0.56
$C_5HN_4O_2$	7.02	0.62
$C_5H_{11}NO_4$	6.12	0.96
$C_5H_{13}N_2O_3$	6.49	0.78
$C_5H_{15}N_3O_2$	6.87	0.61
$C_6HN_2O_3$	7.38	0.84
$C_6H_3N_3O_2$	7.75	0.66
$C_6H_5N_4O$	8.13	0.49
$C_6H_{13}O_4$	6.85	1.00
$C_6H_{15}NO_3$	7.22	0.83
C_7HO_4	7.74	1.06
$C_7H_3NO_3$	8.11	0.89
$C_7H_5N_2O_2$	8.49	0.72
$C_7H_7N_3O$	8.86	0.55
$C_7H_9N_4$	9.23	0.38
$C_8H_5O_3$	8.84	0.95
$C_8H_7NO_2$	9.22	0.78
$C_8H_9N_2O$	9.59	0.61
$C_8H_{11}N_3$	9.97	0.45
$C_9H_9O_2$	9.95	0.84
$C_9H_{11}NO$	10.32	0.68
$C_9H_{13}N_2$	10.70	0.52
$C_{10}HN_2$	11.59	0.61
$C_{10}H_{13}O$	11.05	0.75
$C_{10}H_{15}N$	11.43	0.59
$C_{11}HO$	11.94	0.85
$C_{11}H_3N$	12.32	0.69
$C_{11}H_{17}$	12.16	0.67
$C_{12}H_5$	13.05	0.78
150		
$C_4H_{10}N_2O_4$	5.40	0.92
$C_4H_{12}N_3O_3$	5.78	0.74
$C_4H_{14}N_4O_2$	6.15	0.56
$C_5H_2N_4O_2$	7.04	0.62
$C_5H_{12}NO_4$	6.13	0.96
$C_5H_{14}N_2O_3$	6.51	0.78
$C_6H_2N_2O_3$	7.40	0.84
$C_6H_4N_3O_2$	7.77	0.67
$C_6H_6N_4O$	8.15	0.49
$C_6H_{14}O_4$	6.86	1.00
$C_7H_2O_4$	7.75	1.06
$C_7H_4NO_3$	8.13	0.89
$C_7H_6N_2O_2$	8.50	0.72
$C_7H_8N_3O$	8.88	0.55
$C_7H_{10}N_4$	9.25	0.38
$C_8H_6O_3$	8.86	0.95
$C_8H_8NO_2$	9.23	0.78
$C_8H_{10}N_2O$	9.61	0.61
$C_8H_{12}N_3$	9.98	0.45
$C_9H_{10}O_2$	9.96	0.84
$C_9H_{12}NO$	10.34	0.68
$C_9H_{14}N_2$	10.71	0.52

Formula	M + 1	M + 2
$C_{10}H_2N_2$	11.60	0.61
$C_{10}H_{14}O$	11.07	0.75
$C_{10}H_{16}N$	11.44	0.60
$C_{11}H_2O$	11.96	0.85
$C_{11}H_4N$	12.33	0.70
$C_{11}H_{18}$	12.18	0.68
$C_{12}H_6$	13.06	0.78
151		
$C_4H_{11}N_2O_4$	5.42	0.92
$C_4H_{13}N_3O_3$	5.79	0.74
$C_5HN_3O_3$	6.68	0.79
$C_5H_3N_4O_2$	7.06	0.62
$C_5H_{13}NO_4$	6.15	0.96
C_6HNO_4	7.04	1.01
$C_6H_3N_2O_3$	7.41	0.84
$C_6H_5N_3O_2$	7.79	0.67
$C_6H_7N_4O$	8.16	0.50
$C_7H_3O_4$	7.77	1.06
$C_7H_5NO_3$	8.14	0.89
$C_7H_7N_2O_2$	8.52	0.72
$C_7H_9N_3O$	8.89	0.55
$C_7H_{11}N_4$	9.27	0.39
$C_8H_7O_3$	8.87	0.95
$C_8H_9NO_2$	9.25	0.78
$C_8H_{11}N_2O$	9.62	0.62
$C_8H_{13}N_3$	10.00	0.45
C_9HN_3	10.89	0.54
$C_9H_{11}O_2$	9.98	0.85
$C_9H_{13}NO$	10.36	0.68
$C_9H_{15}N_2$	10.73	0.52
$C_{10}HNO$	11.24	0.77
$C_{10}H_3N_2$	11.62	0.61
$C_{10}H_{15}O$	11.09	0.76
$C_{10}H_{17}N$	11.46	0.60
$C_{11}H_3O$	11.97	0.85
$C_{11}H_5N$	12.35	0.70
$C_{11}H_{19}$	12.19	0.68
$C_{12}H_7$	13.08	0.79
152		
$C_4H_{12}N_2O_4$	5.43	0.92
$C_5H_2N_3O_3$	6.70	0.79
$C_5H_4N_4O_2$	7.07	0.62
$C_6H_2NO_4$	7.05	1.01
$C_6H_4N_2O_3$	7.43	0.84
$C_6H_6N_3O_2$	7.80	0.67
$C_6H_8N_4O$	8.18	0.50
$C_7H_4O_4$	7.79	1.06
$C_7H_6NO_3$	8.16	0.89
$C_7H_8N_2O_2$	8.53	0.72
$C_7H_{10}N_3O$	8.91	0.55
$C_7H_{12}N_4$	9.28	0.39
$C_8H_8O_3$	8.89	0.95
$C_8H_{10}NO_2$	9.27	0.78
$C_8H_{12}N_2O$	9.64	0.62
$C_8H_{14}N_3$	10.01	0.45
$C_9H_2N_3$	10.90	0.54
$C_9H_{12}O_2$	10.00	0.85
$C_9H_{14}NO$	10.37	0.68
$C_9H_{16}N_2$	10.74	0.52
$C_{10}H_2NO$	11.26	0.78
$C_{10}H_4N_2$	11.63	0.62
$C_{10}H_{16}O$	11.10	0.76

Formula	M + 1	M + 2
$C_{10}H_{18}N$	11.48	0.60
$C_{11}H_4O$	11.99	0.86
$C_{11}H_6N$	12.36	0.70
$C_{11}H_{20}$	12.21	0.68
$C_{12}H_8$	13.10	0.79
153		
$C_5HN_2O_4$	6.34	0.97
$C_5H_3N_3O_3$	6.71	0.80
$C_5H_5N_4O_2$	7.09	0.62
$C_6H_3NO_4$	7.07	1.02
$C_6H_5N_2O_3$	7.44	0.84
$C_6H_7N_3O_2$	7.82	0.67
$C_6H_9N_4O$	8.19	0.50
$C_7H_5O_4$	7.80	1.07
$C_7H_7NO_3$	8.18	0.89
$C_7H_9N_2O_2$	8.55	0.72
$C_7H_{11}N_3O$	8.92	0.56
$C_7H_{13}N_4$	9.30	0.39
C_8HN_4	10.19	0.47
$C_8H_9O_3$	8.91	0.95
$C_8H_{11}NO_2$	9.28	0.78
$C_8H_{13}N_2O$	9.66	0.62
$C_8H_{15}N_3$	10.03	0.45
C_9HN_2O	10.54	0.70
$C_9H_3N_3$	10.92	0.54
$C_9H_{13}O_2$	10.01	0.85
$C_9H_{15}NO$	10.39	0.69
$C_9H_{17}N_2$	10.76	0.52
$C_{10}HO_2$	10.90	0.94
$C_{10}H_3NO$	11.28	0.78
$C_{10}H_5N_2$	11.65	0.62
$C_{10}H_{17}O$	11.12	0.76
$C_{10}H_{19}N$	11.49	0.60
$C_{11}H_5O$	12.01	0.86
$C_{11}H_7N$	12.38	0.70
$C_{11}H_{21}$	12.22	0.68
$C_{12}H_9$	13.11	0.79
154		
$C_5H_2N_2O_4$	6.35	0.97
$C_5H_4N_3O_3$	6.73	0.80
$C_5H_6N_4O_2$	7.10	0.62
$C_6H_4NO_4$	7.09	1.02
$C_6H_6N_2O_3$	7.46	0.84
$C_6H_8N_3O_2$	7.83	0.67
$C_6H_{10}N_4O$	8.21	0.50
$C_7H_6O_4$	7.82	1.07
$C_7H_8NO_3$	8.19	0.90
$C_7H_{10}N_2O_2$	8.57	0.73
$C_7H_{12}N_3O$	8.94	0.56
$C_7H_{14}N_4$	9.31	0.39
$C_8H_2N_4$	10.20	0.47
$C_8H_{10}O_3$	8.92	0.95
$C_8H_{12}NO_2$	9.30	0.79
$C_8H_{14}N_2O$	9.67	0.62
$C_8H_{16}N_3$	10.05	0.46
$C_9H_2N_2O$	10.56	0.70
$C_9H_4N_3$	10.93	0.54
$C_9H_{14}O_2$	10.03	0.85
$C_9H_{16}NO$	10.40	0.69
$C_9H_{18}N_2$	10.78	0.53
$C_{10}H_2O_2$	10.92	0.94
$C_{10}H_4NO$	11.29	0.78

Appendix Table 5.1 (*continued*)

	M + 1	M + 2
$C_{10}H_6N_2$	11.67	0.62
$C_{10}H_{18}O$	11.13	0.76
$C_{10}H_{20}N$	11.51	0.60
$C_{11}H_6O$	12.02	0.86
$C_{11}H_8N$	12.40	0.70
$C_{11}H_{22}$	12.24	0.68
$C_{12}H_{10}$	13.13	0.79
	155	
$C_5H_3N_2O_4$	6.37	0.97
$C_5H_5N_3O_3$	6.75	0.80
$C_5H_7N_4O_2$	7.12	0.62
$C_6H_5NO_4$	7.10	1.02
$C_6H_7N_2O_3$	7.48	0.84
$C_6H_9N_3O_2$	7.85	0.67
$C_6H_{11}N_4O$	8.23	0.50
$C_7H_7O_4$	7.83	1.07
$C_7H_9NO_3$	8.21	0.90
$C_7H_{11}N_2O_2$	8.58	0.73
$C_7H_{13}N_3O$	8.96	0.56
$C_7H_{15}N_4$	9.33	0.39
C_8HN_3O	9.84	0.64
$C_8H_3N_4$	10.22	0.47
$C_8H_{11}O_3$	8.94	0.95
$C_8H_{13}NO_2$	9.31	0.79
$C_8H_{15}N_2O$	9.69	0.62
$C_8H_{17}N_3$	10.06	0.46
C_9HNO_2	10.20	0.87
$C_9H_3N_2O$	10.58	0.71
$C_9H_5N_3$	10.95	0.54
$C_9H_{15}O_2$	10.04	0.85
$C_9H_{17}NO$	10.42	0.69
$C_9H_{19}N_2$	10.79	0.53
$C_{10}H_3O_2$	10.93	0.94
$C_{10}H_5NO$	11.31	0.78
$C_{10}H_7N_2$	11.68	0.62
$C_{10}H_{19}O$	11.15	0.76
$C_{10}H_{21}N$	11.52	0.60
$C_{11}H_7O$	12.04	0.86
$C_{11}H_9N$	12.41	0.71
$C_{11}H_{23}$	12.26	0.69
$C_{12}H_{11}$	13.14	0.79
	156	
$C_5H_4N_2O_4$	6.39	0.98
$C_5H_6N_3O_3$	6.76	0.80
$C_5H_8N_4O_2$	7.14	0.62
$C_6H_6NO_4$	7.12	1.02
$C_6H_8N_2O_3$	7.49	0.85
$C_6H_{10}N_3O_2$	7.87	0.67
$C_6H_{12}N_4O$	8.24	0.50
$C_7H_8O_4$	7.85	1.07
$C_7H_{10}NO_3$	8.22	0.90
$C_7H_{12}N_2O_2$	8.60	0.73
$C_7H_{14}N_3O$	8.97	0.56
$C_7H_{16}N_4$	9.35	0.39
$C_8H_2N_3O$	9.86	0.64
$C_8H_4N_4$	10.24	0.47
$C_8H_{12}O_3$	8.95	0.96
$C_8H_{14}NO_2$	9.33	0.79
$C_8H_{16}N_2O$	9.70	0.62
$C_8H_{18}N_3$	10.08	0.46
$C_9H_2NO_2$	10.22	0.87
$C_9H_4N_2O$	10.59	0.71
$C_9H_6N_3$	10.97	0.55
$C_9H_{16}O_2$	10.06	0.85
$C_9H_{18}NO$	10.43	0.69
$C_9H_{20}N_2$	10.81	0.53
$C_{10}H_4O_2$	10.95	0.94
$C_{10}H_6NO$	11.32	0.78
$C_{10}H_8N_2$	11.70	0.62
$C_{10}H_{20}O$	11.17	0.77
$C_{10}H_{22}N$	11.54	0.61
$C_{11}H_8O$	12.05	0.86
$C_{11}H_{10}N$	12.43	0.71
$C_{11}H_{24}$	12.27	0.69
$C_{12}H_{12}$	13.16	0.80
	157	
$C_5H_5N_2O_4$	6.40	0.98
$C_5H_7N_3O_3$	6.78	0.80
$C_5H_9N_4O_2$	7.15	0.62
$C_6H_7NO_4$	7.13	1.02
$C_6H_9N_2O_3$	7.51	0.85
$C_6H_{11}N_3O_2$	7.88	0.67
$C_6H_{13}N_4O$	8.26	0.50
C_7HN_4O	9.15	0.57
$C_7H_9O_4$	7.87	1.07
$C_7H_{11}NO_3$	8.24	0.90
$C_7H_{13}N_2O_2$	8.61	0.73
$C_7H_{15}N_3O$	8.99	0.56
$C_7H_{17}N_4$	9.36	0.39
$C_8HN_2O_2$	9.50	0.80
$C_8H_3N_3O$	9.88	0.64
$C_8H_5N_4$	10.25	0.48
$C_8H_{13}O_3$	8.97	0.96
$C_8H_{15}NO_2$	9.35	0.79
$C_8H_{17}N_2O$	9.72	0.62
$C_8H_{19}N_3$	10.09	0.46
C_9HO_3	9.86	1.03
$C_9H_3NO_2$	10.23	0.87
$C_9H_5N_2O$	10.61	0.71
$C_9H_7N_3$	10.98	0.55
$C_9H_{17}O_2$	10.08	0.86
$C_9H_{19}NO$	10.45	0.69
$C_9H_{21}N_2$	10.82	0.53
$C_{10}H_5O_2$	10.96	0.94
$C_{10}H_7NO$	11.34	0.78
$C_{10}H_9N_2$	11.71	0.63
$C_{10}H_{21}O$	11.18	0.77
$C_{10}H_{23}N$	11.56	0.61
$C_{11}H_9O$	12.07	0.86
$C_{11}H_{11}N$	12.44	0.71
$C_{12}H_{13}$	13.18	0.80
$C_{13}H$	14.06	0.91
	158	
$C_5H_6N_2O_4$	6.42	0.98
$C_5H_8N_3O_3$	6.79	0.80
$C_5H_{10}N_4O_2$	7.17	0.63
$C_6H_8NO_4$	7.15	1.02
$C_6H_{10}N_2O_3$	7.52	0.85
$C_6H_{12}N_3O_2$	7.90	0.68
$C_6H_{14}N_4O$	8.27	0.50
$C_7H_2N_4O$	9.16	0.58
$C_7H_{10}O_4$	7.88	1.07
$C_7H_{12}NO_3$	8.26	0.90
$C_7H_{14}N_2O_2$	8.63	0.73
$C_7H_{16}N_3O$	9.00	0.56
$C_7H_{18}N_4$	9.38	0.40
$C_8H_2N_2O_2$	9.52	0.81
$C_8H_4N_3O$	9.89	0.64
$C_8H_6N_4$	10.27	0.48
$C_8H_{14}O_3$	8.99	0.96
$C_8H_{16}NO_2$	9.36	0.79
$C_8H_{18}N_2O$	9.74	0.63
$C_8H_{20}N_3$	10.11	0.46
$C_9H_2O_3$	9.88	1.04
$C_9H_4NO_2$	10.25	0.87
$C_9H_6N_2O$	10.62	0.71
$C_9H_8N_3$	11.00	0.55
$C_9H_{18}O_2$	10.09	0.86
$C_9H_{20}NO$	10.47	0.69
$C_9H_{22}N_2$	10.84	0.53
$C_{10}H_6O_2$	10.98	0.95
$C_{10}H_8NO$	11.36	0.79
$C_{10}H_{10}N_2$	11.73	0.63
$C_{10}H_{22}O$	11.20	0.77
$C_{11}H_{10}O$	12.09	0.87
$C_{11}H_{12}N$	12.46	0.71
$C_{12}H_{14}$	13.19	0.80
$C_{13}H_2$	14.08	0.92
	159	
$C_5H_7N_2O_4$	6.43	0.98
$C_5H_9N_3O_3$	6.81	0.80
$C_5H_{11}N_4O_2$	7.18	0.63
$C_6H_9NO_4$	7.17	1.02
$C_6H_{11}N_2O_3$	7.54	0.85
$C_6H_{13}N_3O_2$	7.91	0.68
$C_6H_{15}N_4O$	8.29	0.51
$C_7HN_3O_2$	8.80	0.75
$C_7H_3N_4O$	9.18	0.58
$C_7H_{11}O_4$	7.90	1.07
$C_7H_{13}NO_3$	8.27	0.90
$C_7H_{15}N_2O_2$	8.65	0.73
$C_7H_{17}N_3O$	9.02	0.56
$C_7H_{19}N_4$	9.39	0.40
C_8HNO_3	9.16	0.97
$C_8H_3N_2O_2$	9.53	0.81
$C_8H_5N_3O$	9.91	0.64
$C_8H_7N_4$	10.28	0.48
$C_8H_{15}O_3$	9.00	0.96
$C_8H_{17}NO_2$	9.38	0.79
$C_8H_{19}N_2O$	9.75	0.63
$C_8H_{21}N_3$	10.13	0.46
$C_9H_3O_3$	9.89	1.04
$C_9H_5NO_2$	10.27	0.87
$C_9H_7N_2O$	10.64	0.71
$C_9H_9N_3$	11.01	0.55
$C_9H_{19}O_2$	10.11	0.86
$C_9H_{21}NO$	10.48	0.70
$C_{10}H_7O_2$	11.00	0.95
$C_{10}H_9NO$	11.37	0.79
$C_{10}H_{11}N_2$	11.75	0.63
$C_{11}H_{11}O$	12.10	0.87
$C_{11}H_{13}N$	12.48	0.71
$C_{12}HN$	13.37	0.82
$C_{12}H_{15}$	13.21	0.80
$C_{13}H_3$	14.10	0.92
	160	
$C_5H_8N_2O_4$	6.45	0.98
$C_5H_{10}N_3O_3$	6.83	0.80
$C_5H_{12}N_4O_2$	7.20	0.63
$C_6H_{10}NO_4$	7.18	1.02
$C_6H_{12}N_2O_3$	7.56	0.85
$C_6H_{14}N_3O_2$	7.93	0.68
$C_6H_{16}N_4O$	8.31	0.51
$C_7H_2N_3O_2$	8.82	0.75
$C_7H_4N_4O$	9.19	0.58
$C_7H_{12}O_4$	7.91	1.07
$C_7H_{14}NO_3$	8.29	0.90
$C_7H_{16}N_2O_2$	8.66	0.73
$C_7H_{18}N_3O$	9.04	0.57
$C_7H_{20}N_4$	9.41	0.40
$C_8H_2NO_3$	9.18	0.97
$C_8H_4N_2O_2$	9.55	0.81
$C_8H_6N_3O$	9.92	0.64
$C_8H_8N_4$	10.30	0.48
$C_8H_{16}O_3$	9.02	0.96
$C_8H_{18}NO_2$	9.39	0.79
$C_8H_{20}N_2O$	9.77	0.63
$C_9H_4O_3$	9.91	1.04
$C_9H_6NO_2$	10.28	0.88
$C_9H_8N_2O$	10.66	0.71
$C_9H_{10}N_3$	11.03	0.55
$C_9H_{20}O_2$	10.12	0.86
$C_{10}H_8O_2$	11.01	0.95
$C_{10}H_{10}NO$	11.39	0.79
$C_{10}H_{12}N_2$	11.76	0.63
$C_{11}H_{12}O$	12.12	0.87
$C_{11}H_{14}N$	12.49	0.72
$C_{12}H_2N$	13.38	0.82
$C_{12}H_{16}$	13.22	0.80
$C_{13}H_4$	14.11	0.92
	161	
$C_5H_9N_2O_4$	6.47	0.98
$C_5H_{11}N_3O_3$	6.84	0.80
$C_5H_{13}N_4O_2$	7.22	0.63
$C_6HN_4O_2$	8.10	0.69
$C_6H_{11}NO_4$	7.20	1.03
$C_6H_{13}N_2O_3$	7.57	0.85
$C_6H_{15}N_3O_2$	7.95	0.68
$C_6H_{17}N_4O$	8.32	0.51
$C_7HN_2O_3$	8.46	0.92
$C_7H_3N_3O_2$	8.84	0.75
$C_7H_5N_4O$	9.21	0.58
$C_7H_{13}O_4$	7.93	1.08
$C_7H_{15}NO_3$	8.30	0.90
$C_7H_{17}N_2O_2$	8.68	0.74
$C_7H_{19}N_3O$	9.05	0.57
C_8HO_4	8.82	1.14
$C_8H_3NO_3$	9.19	0.98
$C_8H_5N_2O_2$	9.57	0.81
$C_8H_7N_3O$	9.94	0.65
$C_8H_9N_4$	10.32	0.48
$C_8H_{17}O_3$	9.03	0.96
$C_8H_{19}NO_2$	9.41	0.80
$C_9H_5O_3$	9.92	1.04
$C_9H_7NO_2$	10.30	0.88
$C_9H_9N_2O$	10.67	0.72
$C_9H_{11}N_3$	11.05	0.56

Appendix Table 5.1 (*continued*)

	M + 1	M + 2
$C_{10}H_9O_2$	11.03	0.95
$C_{10}H_{11}NO$	11.40	0.79
$C_{10}H_{13}N_2$	11.78	0.63
$C_{11}HN_2$	12.67	0.74
$C_{11}H_{13}O$	12.13	0.87
$C_{11}H_{15}N$	12.51	0.72
$C_{12}HO$	13.02	0.98
$C_{12}H_3N$	13.40	0.83
$C_{12}H_{17}$	13.24	0.81
$C_{13}H_5$	14.13	0.92
162		
$C_5H_{10}N_2O_4$	6.48	0.98
$C_5H_{12}N_3O_3$	6.86	0.81
$C_5H_{14}N_4O_2$	7.23	0.63
$C_6H_2N_4O_2$	8.12	0.69
$C_6H_{12}NO_4$	7.21	1.03
$C_6H_{14}N_2O_3$	7.59	0.85
$C_6H_{16}N_3O_2$	7.96	0.68
$C_6H_{18}N_4O$	8.34	0.51
$C_7H_2N_2O_3$	8.48	0.92
$C_7H_4N_3O_2$	8.85	0.75
$C_7H_6N_4O$	9.23	0.58
$C_7H_{14}O_4$	7.95	1.08
$C_7H_{16}NO_3$	8.32	0.91
$C_7H_{18}N_2O_2$	8.69	0.74
$C_8H_2O_4$	8.83	1.15
$C_8H_4NO_3$	9.21	0.98
$C_8H_6N_2O_2$	9.58	0.81
$C_8H_8N_3O$	9.96	0.65
$C_8H_{10}N_4$	10.33	0.48
$C_8H_{18}O_3$	9.05	0.96
$C_9H_6O_3$	9.94	1.04
$C_9H_8NO_2$	10.31	0.88
$C_9H_{10}N_2O$	10.69	0.72
$C_9H_{12}N_3$	11.06	0.56
$C_{10}H_{10}O_2$	11.04	0.95
$C_{10}H_{12}NO$	11.42	0.79
$C_{10}H_{14}N_2$	11.79	0.64
$C_{11}H_2N_2$	12.68	0.74
$C_{11}H_{14}O$	12.15	0.87
$C_{11}H_{16}N$	12.52	0.72
$C_{12}H_2O$	13.04	0.98
$C_{12}H_4N$	13.41	0.83
$C_{12}H_{18}$	13.26	0.81
$C_{13}H_6$	14.14	0.92
163		
$C_5H_{11}N_2O_4$	6.50	0.98
$C_5H_{13}N_3O_3$	6.87	0.81
$C_5H_{15}N_4O_2$	7.25	0.63
$C_6HN_3O_3$	7.76	0.87
$C_6H_3N_4O_2$	8.14	0.69
$C_6H_{13}NO_4$	7.23	1.03
$C_6H_{15}N_2O_3$	7.60	0.85
$C_6H_{17}N_3O_2$	7.98	0.68
C_7HNO_4	8.12	1.09
$C_7H_3N_2O_3$	8.49	0.92
$C_7H_5N_3O_2$	8.87	0.75
$C_7H_7N_4O$	9.24	0.58
$C_7H_{15}O_4$	7.96	1.08
$C_7H_{17}NO_3$	8.34	0.91
$C_8H_3O_4$	8.85	1.15
$C_8H_5NO_3$	9.22	0.98

	M + 1	M + 2
$C_8H_7N_2O_2$	9.60	0.81
$C_8H_9N_3O$	9.97	0.65
$C_8H_{11}N_4$	10.35	0.49
$C_9H_7O_3$	9.96	1.04
$C_9H_9NO_2$	10.33	0.88
$C_9H_{11}N_2O$	10.70	0.72
$C_9H_{13}N_3$	11.08	0.56
$C_{10}HN_3$	11.97	0.66
$C_{10}H_{11}O_2$	11.06	0.95
$C_{10}H_{13}NO$	11.44	0.80
$C_{10}H_{15}N_2$	11.81	0.64
$C_{11}HNO$	12.32	0.89
$C_{11}H_3N_2$	12.70	0.74
$C_{11}H_{15}O$	12.17	0.88
$C_{11}H_{17}N$	12.54	0.72
$C_{12}H_3O$	13.05	0.98
$C_{12}H_5N$	13.43	0.83
$C_{12}H_{19}$	13.27	0.81
$C_{13}H_7$	14.16	0.93
164		
$C_5H_{12}N_2O_4$	6.51	0.98
$C_5H_{14}N_3O_3$	6.89	0.81
$C_5H_{16}N_4O_2$	7.26	0.63
$C_6H_2N_3O_3$	7.78	0.87
$C_6H_4N_4O_2$	8.15	0.70
$C_6H_{14}NO_4$	7.25	1.03
$C_6H_{16}N_2O_3$	7.62	0.86
$C_7H_2NO_4$	8.13	1.09
$C_7H_4N_2O_3$	8.51	0.92
$C_7H_6N_3O_2$	8.88	0.75
$C_7H_8N_4O$	9.26	0.59
$C_7H_{16}O_4$	7.98	1.08
$C_8H_4O_4$	8.87	1.15
$C_8H_6NO_3$	9.24	0.98
$C_8H_8N_2O_2$	9.61	0.81
$C_8H_{10}N_3O$	9.99	0.65
$C_8H_{12}N_4$	10.36	0.49
$C_9H_8O_3$	9.97	1.05
$C_9H_{10}NO_2$	10.35	0.88
$C_9H_{12}N_2O$	10.72	0.72
$C_9H_{14}N_3$	11.09	0.56
$C_{10}H_2N_3$	11.98	0.66
$C_{10}H_{12}O_2$	11.08	0.96
$C_{10}H_{14}NO$	11.45	0.80
$C_{10}H_{16}N_2$	11.83	0.64
$C_{11}H_2NO$	12.34	0.90
$C_{11}H_4N_2$	12.71	0.74
$C_{11}H_{16}O$	12.18	0.88
$C_{11}H_{18}N$	12.56	0.72
$C_{12}H_4O$	13.07	0.98
$C_{12}H_6N$	13.45	0.83
$C_{12}H_{20}$	13.29	0.81
$C_{13}H_8$	14.18	0.93
165		
$C_5H_{13}N_2O_4$	6.53	0.98
$C_5H_{15}N_3O_3$	6.91	0.81
$C_6HN_2O_4$	7.42	1.04
$C_6H_3N_3O_3$	7.79	0.87
$C_6H_5N_4O_2$	8.17	0.70
$C_6H_{15}NO_4$	7.26	1.03
$C_7H_3NO_4$	8.15	1.09

	M + 1	M + 2
$C_7H_5N_2O_3$	8.52	0.92
$C_7H_7N_3O_2$	8.90	0.75
$C_7H_9N_4O$	9.27	0.59
$C_8H_5O_4$	8.88	1.15
$C_8H_7NO_3$	9.26	0.98
$C_8H_9N_2O_2$	9.63	0.82
$C_8H_{11}N_3O$	10.00	0.65
$C_8H_{13}N_4$	10.38	0.49
C_9HN_4	11.27	0.58
$C_9H_9O_3$	9.99	1.05
$C_9H_{11}NO_2$	10.36	0.88
$C_9H_{13}N_2O$	10.74	0.72
$C_9H_{15}N_3$	11.11	0.56
$C_{10}HN_2O$	11.62	0.82
$C_{10}H_3N_3$	12.00	0.66
$C_{10}H_{13}O_2$	11.09	0.96
$C_{10}H_{15}NO$	11.47	0.80
$C_{10}H_{17}N_2$	11.84	0.64
$C_{11}HO_2$	11.98	1.05
$C_{11}H_3NO$	12.36	0.90
$C_{11}H_5N_2$	12.73	0.74
$C_{11}H_{17}O$	12.20	0.88
$C_{11}H_{19}N$	12.57	0.73
$C_{12}H_5O$	13.09	0.99
$C_{12}H_7N$	13.46	0.84
$C_{12}H_{21}$	13.30	0.81
$C_{13}H_9$	14.19	0.93
166		
$C_5H_{14}N_2O_4$	6.55	0.99
$C_6H_2N_2O_4$	7.44	1.04
$C_6H_4N_3O_3$	7.81	0.87
$C_6H_6N_4O_2$	8.18	0.70
$C_7H_4NO_4$	8.17	1.09
$C_7H_6N_2O_3$	8.54	0.92
$C_7H_8N_3O_2$	8.92	0.76
$C_7H_{10}N_4O$	9.29	0.59
$C_8H_6O_4$	8.90	1.15
$C_8H_8NO_3$	9.27	0.98
$C_8H_{10}N_2O_2$	9.65	0.82
$C_8H_{12}N_3O$	10.02	0.65
$C_8H_{14}N_4$	10.40	0.49
$C_9H_2N_4$	11.28	0.58
$C_9H_{10}O_3$	10.00	1.05
$C_9H_{12}NO_2$	10.38	0.89
$C_9H_{14}N_2O$	10.75	0.72
$C_9H_{16}N_3$	11.13	0.56
$C_{10}H_2N_2O$	11.64	0.82
$C_{10}H_4N_3$	12.01	0.66
$C_{10}H_{14}O_2$	11.11	0.96
$C_{10}H_{16}NO$	11.48	0.80
$C_{10}H_{18}N_2$	11.86	0.64
$C_{11}H_2O_2$	12.00	1.06
$C_{11}H_4NO$	12.37	0.90
$C_{11}H_6N_2$	12.75	0.75
$C_{11}H_{18}O$	12.21	0.88
$C_{11}H_{20}N$	12.59	0.73
$C_{12}H_6O$	13.10	0.99
$C_{12}H_8N$	13.48	0.84
$C_{12}H_{22}$	13.32	0.82
$C_{13}H_{10}$	14.21	0.93
167		
$C_6H_3N_2O_4$	7.45	1.04

	M + 1	M + 2
$C_6H_5N_3O_3$	7.83	0.87
$C_6H_7N_4O_2$	8.20	0.70
$C_7H_5NO_4$	8.18	1.10
$C_7H_7N_2O_3$	8.56	0.93
$C_7H_9N_3O_2$	8.93	0.76
$C_7H_{11}N_4O$	9.31	0.59
$C_8H_7O_4$	8.91	1.15
$C_8H_9NO_3$	9.29	0.99
$C_8H_{11}N_2O_2$	9.66	0.82
$C_8H_{13}N_3O$	10.04	0.66
$C_8H_{15}N_4$	10.41	0.49
C_9HN_3O	10.93	0.74
$C_9H_3N_4$	11.30	0.58
$C_9H_{11}O_3$	10.02	1.05
$C_9H_{13}NO_2$	10.39	0.89
$C_9H_{15}N_2O$	10.77	0.73
$C_9H_{17}N_3$	11.14	0.57
$C_{10}HNO_2$	11.28	0.98
$C_{10}H_3N_2O$	11.66	0.82
$C_{10}H_5N_3$	12.03	0.66
$C_{10}H_{15}O_2$	11.12	0.96
$C_{10}H_{17}NO$	11.50	0.80
$C_{10}H_{19}N_2$	11.87	0.65
$C_{11}H_3O_2$	12.01	1.06
$C_{11}H_5NO$	12.39	0.90
$C_{11}H_7N_2$	12.76	0.75
$C_{11}H_{19}O$	12.23	0.88
$C_{11}H_{21}N$	12.60	0.73
$C_{12}H_7O$	13.12	0.99
$C_{12}H_9N$	13.49	0.84
$C_{12}H_{23}$	13.34	0.82
$C_{13}H_{11}$	14.22	0.94
168		
$C_6H_4N_2O_4$	7.47	1.04
$C_6H_6N_3O_3$	7.84	0.87
$C_6H_8N_4O_2$	8.22	0.70
$C_7H_6NO_4$	8.20	1.10
$C_7H_8N_2O_3$	8.57	0.93
$C_7H_{10}N_3O_2$	8.95	0.76
$C_7H_{12}N_4O$	9.32	0.59
$C_8H_8O_4$	8.93	1.15
$C_8H_{10}NO_3$	9.30	0.99
$C_8H_{12}N_2O_2$	9.68	0.82
$C_8H_{14}N_3O$	10.05	0.66
$C_8H_{16}N_4$	10.43	0.49
$C_9H_2N_3O$	10.94	0.74
$C_9H_4N_4$	11.32	0.58
$C_9H_{12}O_3$	10.04	1.05
$C_9H_{14}NO_2$	10.41	0.89
$C_9H_{16}N_2O$	10.78	0.73
$C_9H_{18}N_3$	11.16	0.57
$C_{10}H_2NO_2$	11.30	0.98
$C_{10}H_4N_2O$	11.67	0.82
$C_{10}H_6N_3$	12.05	0.67
$C_{10}H_{16}O_2$	11.14	0.96
$C_{10}H_{18}NO$	11.52	0.80
$C_{10}H_{20}N_2$	11.89	0.65
$C_{11}H_4O_2$	12.03	1.06
$C_{11}H_6NO$	12.40	0.90
$C_{11}H_8N_2$	12.78	0.75
$C_{11}H_{20}O$	12.25	0.89
$C_{11}H_{22}N$	12.62	0.73

Appendix Table 5.1 (*continued*)

	M + 1	M + 2
$C_{12}H_{8}O$	13.13	0.99
$C_{12}H_{10}N$	13.51	0.84
$C_{12}H_{24}$	13.35	0.82
$C_{13}H_{12}$	14.24	0.94
169		
$C_{6}H_{5}N_{2}O_{4}$	7.48	1.05
$C_{6}H_{7}N_{3}O_{3}$	7.86	0.87
$C_{6}H_{9}N_{4}O_{2}$	8.23	0.70
$C_{7}H_{7}NO_{4}$	8.21	1.10
$C_{7}H_{9}N_{2}O_{3}$	8.59	0.93
$C_{7}H_{11}N_{3}O_{2}$	8.96	0.76
$C_{7}H_{13}N_{4}O$	9.34	0.59
$C_{8}HN_{4}O$	10.23	0.67
$C_{8}H_{9}O_{4}$	8.95	1.16
$C_{8}H_{11}NO_{3}$	9.32	0.99
$C_{8}H_{13}N_{2}O_{2}$	9.69	0.82
$C_{8}H_{15}N_{3}O$	10.07	0.66
$C_{8}H_{17}N_{4}$	10.44	0.50
$C_{9}HN_{2}O_{2}$	10.58	0.91
$C_{9}H_{3}N_{3}O$	10.96	0.75
$C_{9}H_{5}N_{4}$	11.33	0.59
$C_{9}H_{13}O_{3}$	10.05	1.05
$C_{9}H_{15}NO_{2}$	10.43	0.89
$C_{9}H_{17}N_{2}O$	10.80	0.73
$C_{9}H_{19}N_{3}$	11.17	0.57
$C_{10}HO_{3}$	10.94	1.14
$C_{10}H_{3}NO_{2}$	11.31	0.98
$C_{10}H_{5}N_{2}O$	11.69	0.82
$C_{10}H_{7}N_{3}$	12.06	0.67
$C_{10}H_{17}O_{2}$	11.16	0.96
$C_{10}H_{19}NO$	11.53	0.81
$C_{10}H_{21}N_{2}$	11.91	0.65
$C_{11}H_{5}O_{2}$	12.05	1.06
$C_{11}H_{7}NO$	12.42	0.91
$C_{11}H_{9}N_{2}$	12.79	0.75
$C_{11}H_{21}O$	12.26	0.89
$C_{11}H_{23}N$	12.64	0.73
$C_{12}H_{9}O$	13.15	1.00
$C_{12}H_{11}N$	13.53	0.84
$C_{12}H_{25}$	13.37	0.82
$C_{13}H_{13}$	14.26	0.94
$C_{14}H$	15.14	1.07
170		
$C_{6}H_{6}N_{2}O_{4}$	7.50	1.05
$C_{6}H_{8}N_{3}O_{3}$	7.87	0.87
$C_{6}H_{10}N_{4}O_{2}$	8.25	0.70
$C_{7}H_{8}NO_{4}$	8.23	1.10
$C_{7}H_{10}N_{2}O_{3}$	8.60	0.93
$C_{7}H_{12}N_{3}O_{2}$	8.98	0.76
$C_{7}H_{14}N_{4}O$	9.35	0.59
$C_{8}H_{2}N_{4}O$	10.24	0.68
$C_{8}H_{10}O_{4}$	8.96	1.16
$C_{8}H_{12}NO_{3}$	9.34	0.99
$C_{8}H_{14}N_{2}O_{2}$	9.71	0.82
$C_{8}H_{16}N_{3}O$	10.08	0.66
$C_{8}H_{18}N_{4}$	10.46	0.50
$C_{9}H_{2}N_{2}O_{2}$	10.60	0.91
$C_{9}H_{4}N_{3}O$	10.97	0.75
$C_{9}H_{6}N_{4}$	11.35	0.59
$C_{9}H_{14}O_{3}$	10.07	1.06
$C_{9}H_{16}NO_{2}$	10.44	0.89
$C_{9}H_{18}N_{2}O$	10.82	0.73
$C_{9}H_{20}N_{3}$	11.19	0.57
$C_{10}H_{2}O_{3}$	10.96	1.14
$C_{10}H_{4}NO_{2}$	11.33	0.98
$C_{10}H_{6}N_{2}O$	11.70	0.83
$C_{10}H_{8}N_{3}$	12.08	0.67
$C_{10}H_{18}O_{2}$	11.17	0.97
$C_{10}H_{20}NO$	11.55	0.81
$C_{10}H_{22}N_{2}$	11.92	0.65
$C_{11}H_{6}O_{2}$	12.06	1.06
$C_{11}H_{8}NO$	12.44	0.91
$C_{11}H_{10}N_{2}$	12.81	0.75
$C_{11}H_{22}O$	12.28	0.89
$C_{11}H_{24}N$	12.65	0.74
$C_{12}H_{10}O$	13.17	1.00
$C_{12}H_{12}N$	13.54	0.85
$C_{12}H_{26}$	13.38	0.83
$C_{13}H_{14}$	14.27	0.94
$C_{14}H_{2}$	15.16	1.07
171		
$C_{6}H_{7}N_{2}O_{4}$	7.52	1.05
$C_{6}H_{9}N_{3}O_{3}$	7.89	0.88
$C_{6}H_{11}N_{4}O_{2}$	8.26	0.70
$C_{7}H_{9}NO_{4}$	8.25	1.10
$C_{7}H_{11}N_{2}O_{3}$	8.62	0.93
$C_{7}H_{13}N_{3}O_{2}$	9.00	0.76
$C_{7}H_{15}N_{4}O$	9.37	0.60
$C_{8}HN_{3}O_{2}$	9.88	0.84
$C_{8}H_{3}N_{4}O$	10.26	0.68
$C_{8}H_{11}O_{4}$	8.98	1.16
$C_{8}H_{13}NO_{3}$	9.35	0.99
$C_{8}H_{15}N_{2}O_{2}$	9.73	0.83
$C_{8}H_{17}N_{3}O$	10.10	0.66
$C_{8}H_{19}N_{4}$	10.48	0.50
$C_{9}HNO_{3}$	10.24	1.07
$C_{9}H_{3}N_{2}O_{2}$	10.61	0.91
$C_{9}H_{5}N_{3}O$	10.99	0.75
$C_{9}H_{7}N_{4}$	11.36	0.59
$C_{9}H_{15}O_{3}$	10.08	1.06
$C_{9}H_{17}NO_{2}$	10.46	0.89
$C_{9}H_{19}N_{2}O$	10.83	0.73
$C_{9}H_{21}N_{3}$	11.21	0.57
$C_{10}H_{3}O_{3}$	10.97	1.14
$C_{10}H_{5}NO_{2}$	11.35	0.99
$C_{10}H_{7}N_{2}O$	11.72	0.83
$C_{10}H_{9}N_{3}$	12.09	0.67
$C_{10}H_{19}O_{2}$	11.19	0.97
$C_{10}H_{21}NO$	11.56	0.81
$C_{10}H_{23}N_{2}$	11.94	0.65
$C_{11}H_{7}O_{2}$	12.08	1.07
$C_{11}H_{9}NO$	12.45	0.91
$C_{11}H_{11}N_{2}$	12.83	0.76
$C_{11}H_{23}O$	12.29	0.89
$C_{11}H_{25}N$	12.67	0.74
$C_{12}H_{11}O$	13.18	1.00
$C_{12}H_{13}N$	13.56	0.85
$C_{13}HN$	14.45	0.97
$C_{13}H_{15}$	14.29	0.94
$C_{14}H_{3}$	15.18	1.07
172		
$C_{6}H_{8}N_{2}O_{4}$	7.53	1.05
$C_{6}H_{10}N_{3}O_{3}$	7.91	0.88
$C_{6}H_{12}N_{4}O_{2}$	8.28	0.71
$C_{7}H_{10}NO_{4}$	8.26	1.10
$C_{7}H_{12}N_{2}O_{3}$	8.64	0.93
$C_{7}H_{14}N_{3}O_{2}$	9.01	0.76
$C_{7}H_{16}N_{4}O$	9.39	0.60
$C_{8}H_{2}N_{3}O_{2}$	9.90	0.84
$C_{8}H_{4}N_{4}O$	10.27	0.68
$C_{8}H_{12}O_{4}$	8.99	1.16
$C_{8}H_{14}NO_{3}$	9.37	0.99
$C_{8}H_{16}N_{2}O_{2}$	9.74	0.83
$C_{8}H_{18}N_{3}O$	10.12	0.66
$C_{8}H_{20}N_{4}$	10.49	0.50
$C_{9}H_{2}NO_{3}$	10.26	1.07
$C_{9}H_{4}N_{2}O_{2}$	10.63	0.91
$C_{9}H_{6}N_{3}O$	11.01	0.75
$C_{9}H_{8}N_{4}$	11.38	0.59
$C_{9}H_{16}O_{3}$	10.10	1.06
$C_{9}H_{18}NO_{2}$	10.47	0.90
$C_{9}H_{20}N_{2}O$	10.85	0.73
$C_{9}H_{22}N_{3}$	11.22	0.57
$C_{10}H_{4}O_{3}$	10.99	1.15
$C_{10}H_{6}NO_{2}$	11.36	0.99
$C_{10}H_{8}N_{2}O$	11.74	0.83
$C_{10}H_{10}N_{3}$	12.11	0.67
$C_{10}H_{20}O_{2}$	11.20	0.97
$C_{10}H_{22}NO$	11.58	0.81
$C_{10}H_{24}N_{2}$	11.95	0.65
$C_{11}H_{8}O_{2}$	12.09	1.07
$C_{11}H_{10}NO$	12.47	0.91
$C_{11}H_{12}N_{2}$	12.84	0.76
$C_{11}H_{24}O$	12.31	0.89
$C_{12}H_{12}O$	13.20	1.00
$C_{12}H_{14}N$	13.57	0.85
$C_{13}H_{2}N$	14.46	0.97
$C_{13}H_{16}$	14.30	0.95
$C_{14}H_{4}$	15.19	1.07
173		
$C_{6}H_{9}N_{2}O_{4}$	7.55	1.05
$C_{6}H_{11}N_{3}O_{3}$	7.92	0.88
$C_{6}H_{13}N_{4}O_{2}$	8.30	0.71
$C_{7}HN_{4}O_{2}$	9.18	0.78
$C_{7}H_{11}NO_{4}$	8.28	1.10
$C_{7}H_{13}N_{2}O_{3}$	8.65	0.93
$C_{7}H_{15}N_{3}O_{2}$	9.03	0.77
$C_{7}H_{17}N_{4}O$	9.40	0.60
$C_{8}HN_{2}O_{3}$	9.54	1.01
$C_{8}H_{3}N_{3}O_{2}$	9.92	0.84
$C_{8}H_{5}N_{4}O$	10.29	0.68
$C_{8}H_{13}O_{4}$	9.01	1.16
$C_{8}H_{15}NO_{3}$	9.38	0.99
$C_{8}H_{17}N_{2}O_{2}$	9.76	0.83
$C_{8}H_{19}N_{3}O$	10.13	0.66
$C_{8}H_{21}N_{4}$	10.51	0.50
$C_{9}HO_{4}$	9.90	1.24
$C_{9}H_{3}NO_{3}$	10.27	1.08
$C_{9}H_{5}N_{2}O_{2}$	10.65	0.91
$C_{9}H_{7}N_{3}O$	11.02	0.75
$C_{9}H_{9}N_{4}$	11.40	0.59
$C_{9}H_{17}O_{3}$	10.12	1.06
$C_{9}H_{19}NO_{2}$	10.49	0.90
$C_{9}H_{21}N_{2}O$	10.86	0.74
$C_{9}H_{23}N_{3}$	11.24	0.58
$C_{10}H_{5}O_{3}$	11.00	1.15
$C_{10}H_{7}NO_{2}$	11.38	0.99
$C_{10}H_{9}N_{2}O$	11.75	0.83
$C_{10}H_{11}N_{3}$	12.13	0.67
$C_{10}H_{21}O_{2}$	11.22	0.97
$C_{10}H_{23}NO$	11.60	0.81
$C_{11}H_{9}O_{2}$	12.11	1.07
$C_{11}H_{11}NO$	12.48	0.91
$C_{11}H_{13}N_{2}$	12.86	0.76
$C_{12}HN_{2}$	13.75	0.87
$C_{12}H_{13}O$	13.21	1.00
$C_{12}H_{15}N$	13.59	0.85
$C_{13}HO$	14.10	1.12
$C_{13}H_{3}N$	14.48	0.97
$C_{13}H_{17}$	14.32	0.95
$C_{14}H_{5}$	15.21	1.07
174		
$C_{6}H_{10}N_{2}O_{4}$	7.56	1.05
$C_{6}H_{12}N_{3}O_{3}$	7.94	0.88
$C_{6}H_{14}N_{4}O_{2}$	8.31	0.71
$C_{7}H_{2}N_{4}O_{2}$	9.20	0.78
$C_{7}H_{12}NO_{4}$	8.29	1.10
$C_{7}H_{14}N_{2}O_{3}$	8.67	0.93
$C_{7}H_{16}N_{3}O_{2}$	9.04	0.77
$C_{7}H_{18}N_{4}O$	9.42	0.60
$C_{8}H_{2}N_{2}O_{3}$	9.56	1.01
$C_{8}H_{4}N_{3}O_{2}$	9.93	0.85
$C_{8}H_{6}N_{4}O$	10.31	0.68
$C_{8}H_{14}O_{4}$	9.03	1.16
$C_{8}H_{16}NO_{3}$	9.40	1.00
$C_{8}H_{18}N_{2}O_{2}$	9.77	0.83
$C_{8}H_{20}N_{3}O$	10.15	0.67
$C_{8}H_{22}N_{4}$	10.52	0.50
$C_{9}H_{2}O_{4}$	9.91	1.24
$C_{9}H_{4}NO_{3}$	10.29	1.08
$C_{9}H_{6}N_{2}O_{2}$	10.66	0.92
$C_{9}H_{8}N_{3}O$	11.04	0.75
$C_{9}H_{10}N_{4}$	11.41	0.60
$C_{9}H_{18}O_{3}$	10.13	1.06
$C_{9}H_{20}NO_{2}$	10.51	0.90
$C_{9}H_{22}N_{2}O$	10.88	0.74
$C_{10}H_{6}O_{3}$	11.02	1.15
$C_{10}H_{8}NO_{2}$	11.39	0.99
$C_{10}H_{10}N_{2}O$	11.77	0.83
$C_{10}H_{12}N_{3}$	12.14	0.68
$C_{10}H_{22}O_{2}$	11.24	0.97
$C_{11}H_{10}O_{2}$	12.13	1.07
$C_{11}H_{12}NO$	12.50	0.92
$C_{11}H_{14}N_{2}$	12.87	0.76
$C_{12}H_{2}N_{2}$	13.76	0.88
$C_{12}H_{14}O$	13.23	1.01
$C_{12}H_{16}N$	13.61	0.85
$C_{13}H_{2}O$	14.12	1.12
$C_{13}H_{4}N$	14.49	0.97
$C_{13}H_{18}$	14.34	0.95
$C_{14}H_{6}$	15.22	1.08
175		
$C_{6}H_{11}N_{2}O_{4}$	7.58	1.05
$C_{6}H_{13}N_{3}O_{3}$	7.95	0.88
$C_{6}H_{15}N_{4}O_{2}$	8.33	0.71
$C_{7}HN_{3}O_{3}$	8.84	0.95

Appendix Table 5.1 (*continued*)

	M + 1	M + 2
$C_7H_3N_4O_2$	9.22	0.78
$C_7H_{13}NO_4$	8.31	1.11
$C_7H_{15}N_2O_3$	8.68	0.94
$C_7H_{17}N_3O_2$	9.06	0.77
$C_7H_{19}N_4O$	9.43	0.60
C_8HNO_4	9.20	1.18
$C_8H_3N_2O_3$	9.57	1.01
$C_8H_5N_3O_2$	9.95	0.85
$C_8H_7N_4O$	10.32	0.68
$C_8H_{17}NO_3$	9.42	1.00
$C_8H_{19}N_2O_2$	9.79	0.83
$C_8H_{21}N_3O$	10.16	0.67
$C_8H_{15}O_4$	9.04	1.16
$C_9H_3O_4$	9.93	1.24
$C_9H_5NO_3$	10.30	1.08
$C_9H_7N_2O_2$	10.68	0.92
$C_9H_9N_3O$	11.05	0.76
$C_9H_{11}N_4$	11.43	0.60
$C_9H_{19}O_3$	10.15	1.06
$C_9H_{21}NO_2$	10.52	0.90
$C_{10}H_7O_3$	11.04	1.15
$C_{10}H_9NO_2$	11.41	0.99
$C_{10}H_{11}N_2O$	11.78	0.83
$C_{10}H_{13}N_3$	12.16	0.68
$C_{11}HN_3$	13.05	0.78
$C_{11}H_{11}O_2$	12.14	1.07
$C_{11}H_{13}NO$	12.52	0.92
$C_{11}H_{15}N_2$	12.89	0.77
$C_{12}HNO$	13.40	1.03
$C_{12}H_3N_2$	13.78	0.88
$C_{12}H_{15}O$	13.25	1.01
$C_{12}H_{17}N$	13.62	0.86
$C_{13}H_3O$	14.14	1.12
$C_{13}H_5N$	14.51	0.98
$C_{13}H_{19}$	14.35	0.95
$C_{14}H_7$	15.24	1.08
176		
$C_6H_{12}N_2O_4$	7.60	1.05
$C_6H_{14}N_3O_3$	7.97	0.88
$C_6H_{16}N_4O_2$	8.34	0.71
$C_7H_2N_3O_3$	8.86	0.95
$C_7H_4N_4O_2$	9.23	0.78
$C_7H_{14}NO_4$	8.33	1.11
$C_7H_{16}N_2O_3$	8.70	0.94
$C_7H_{18}N_3O_2$	9.08	0.77
$C_7H_{20}N_4O$	9.45	0.60
$C_8H_2NO_4$	9.22	1.18
$C_8H_4N_2O_3$	9.59	1.01
$C_8H_6N_3O_2$	9.96	0.85
$C_8H_8N_4O$	10.34	0.69
$C_8H_{16}O_4$	9.06	1.17
$C_8H_{18}NO_3$	9.43	1.00
$C_8H_{20}N_2O_2$	9.81	0.83
$C_9H_4O_4$	9.95	1.24
$C_9H_6NO_3$	10.32	1.08
$C_9H_8N_2O_2$	10.70	0.92
$C_9H_{10}N_3O$	11.07	0.76
$C_9H_{12}N_4$	11.44	0.60
$C_9H_{20}O_3$	10.16	1.07
$C_{10}H_8O_3$	11.05	1.15
$C_{10}H_{10}NO_2$	11.43	0.99
$C_{10}H_{12}N_2O$	11.80	0.84
$C_{10}H_{14}N_3$	12.17	0.68
$C_{11}H_2N_3$	13.06	0.79
$C_{11}H_{12}O_2$	12.16	1.08
$C_{11}H_{14}NO$	12.53	0.92
$C_{11}H_{16}N_2$	12.91	0.77
$C_{12}H_2NO$	13.42	1.03
$C_{12}H_4N_2$	13.79	0.88
$C_{12}H_{16}O$	13.26	1.01
$C_{12}H_{18}N$	13.64	0.86
$C_{13}H_4O$	14.15	1.13
$C_{13}H_6N$	14.53	0.98
$C_{13}H_{20}$	14.37	0.96
$C_{14}H_8$	15.26	1.08
177		
$C_6H_{13}N_2O_4$	7.61	1.06
$C_6H_{15}N_3O_3$	7.99	0.88
$C_6H_{17}N_4O_2$	8.36	0.71
$C_7HN_2O_4$	8.50	1.12
$C_7H_3N_3O_3$	8.87	0.95
$C_7H_5N_4O_2$	9.25	0.78
$C_7H_{15}NO_4$	8.34	1.11
$C_7H_{17}N_2O_3$	8.72	0.94
$C_7H_{19}N_3O_2$	9.09	0.77
$C_8H_3NO_4$	9.23	1.18
$C_8H_5N_2O_3$	9.61	1.01
$C_8H_7N_3O_2$	9.98	0.85
$C_8H_9N_4O$	10.35	0.69
$C_8H_{17}O_4$	9.07	1.17
$C_8H_{19}NO_3$	9.45	1.00
$C_9H_5O_4$	9.96	1.25
$C_9H_7NO_3$	10.34	1.08
$C_9H_9N_2O_2$	10.71	0.92
$C_9H_{11}N_3O$	11.09	0.76
$C_9H_{13}N_4$	11.46	0.60
$C_{10}HN_4$	12.35	0.70
$C_{10}H_9O_3$	11.07	1.16
$C_{10}H_{11}NO_2$	11.44	1.00
$C_{10}H_{13}N_2O$	11.82	0.84
$C_{10}H_{15}N_3$	12.19	0.68
$C_{11}HN_2O$	12.71	0.94
$C_{11}H_3N_3$	13.08	0.79
$C_{11}H_{13}O_2$	12.17	1.08
$C_{11}H_{15}NO$	12.55	0.92
$C_{11}H_{17}N_2$	12.92	0.77
$C_{12}HO_2$	13.06	1.18
$C_{12}H_3NO$	13.44	1.03
$C_{12}H_5N_2$	13.81	0.88
$C_{12}H_{17}O$	13.28	1.01
$C_{12}H_{19}N$	13.65	0.86
$C_{13}H_5O$	14.17	1.13
$C_{13}H_7N$	14.54	0.98
$C_{13}H_{21}$	14.38	0.96
$C_{14}H_9$	15.27	1.08
178		
$C_6H_{14}N_2O_4$	7.63	1.06
$C_6H_{16}N_3O_3$	8.00	0.88
$C_6H_{18}N_4O_2$	8.38	0.71
$C_7H_2N_2O_4$	8.52	1.12
$C_7H_4N_3O_3$	8.89	0.95
$C_7H_6N_4O_2$	9.26	0.79
$C_7H_{16}NO_4$	8.36	1.11
$C_7H_{18}N_2O_3$	8.73	0.94
$C_8H_4NO_4$	9.25	1.18
$C_8H_6N_2O_3$	9.62	1.02
$C_8H_8N_3O_2$	10.00	0.85
$C_8H_{10}N_4O$	10.37	0.69
$C_8H_{18}O_4$	9.09	1.17
$C_9H_6O_4$	9.98	1.25
$C_9H_8NO_3$	10.35	1.08
$C_9H_{10}N_2O_2$	10.73	0.92
$C_9H_{12}N_3O$	11.10	0.76
$C_9H_{14}N_4$	11.48	0.60
$C_{10}H_2N_4$	12.36	0.70
$C_{10}H_{10}O_3$	11.08	1.16
$C_{10}H_{12}NO_2$	11.46	1.00
$C_{10}H_{14}N_2O$	11.83	0.84
$C_{10}H_{16}N_3$	12.21	0.68
$C_{11}H_2N_2O$	12.72	0.94
$C_{11}H_4N_3$	13.10	0.79
$C_{11}H_{14}O_2$	12.19	1.08
$C_{11}H_{16}NO$	12.56	0.92
$C_{11}H_{18}N_2$	12.94	0.77
$C_{12}H_2O_2$	13.08	1.19
$C_{12}H_4NO$	13.45	1.03
$C_{12}H_6N_2$	13.83	0.88
$C_{12}H_{18}O$	13.29	1.01
$C_{12}H_{20}N$	13.67	0.86
$C_{13}H_6O$	14.18	1.13
$C_{13}H_8N$	14.56	0.98
$C_{13}H_{22}$	14.40	0.96
$C_{14}H_{10}$	15.29	1.09
179		
$C_6H_{15}N_2O_4$	7.64	1.06
$C_6H_{17}N_3O_3$	8.02	0.89
$C_7H_3N_2O_4$	8.53	1.12
$C_7H_5N_3O_3$	8.91	0.95
$C_7H_7N_4O_2$	9.28	0.79
$C_7H_{17}NO_4$	8.37	1.11
$C_8H_5NO_4$	9.26	1.18
$C_8H_7N_2O_3$	9.64	1.02
$C_8H_9N_3O_2$	10.01	0.85
$C_8H_{11}N_4O$	10.39	0.69
$C_9H_7O_4$	9.99	1.25
$C_9H_9NO_3$	10.37	1.09
$C_9H_{11}N_2O_2$	10.74	0.92
$C_9H_{13}N_3O$	11.12	0.76
$C_9H_{15}N_4$	11.49	0.60
$C_{10}HN_3O$	12.01	0.86
$C_{10}H_3N_4$	12.38	0.71
$C_{10}H_{11}O_3$	11.10	1.16
$C_{10}H_{13}NO_2$	11.47	1.00
$C_{10}H_{15}N_2O$	11.85	0.84
$C_{10}H_{17}N_3$	12.22	0.69
$C_{11}HNO_2$	12.36	1.10
$C_{11}H_3N_2O$	12.74	0.95
$C_{11}H_5N_3$	13.11	0.79
$C_{11}H_{15}O_2$	12.21	1.08
$C_{11}H_{17}NO$	12.58	0.93
$C_{11}H_{19}N_2$	12.95	0.77
$C_{12}H_3O_2$	13.09	1.19
$C_{12}H_5NO$	13.47	1.04
$C_{12}H_7N_2$	13.84	0.89
$C_{12}H_{19}O$	13.31	1.02
$C_{12}H_{21}N$	13.69	0.87
$C_{13}H_7O$	14.20	1.13
$C_{13}H_9N$	14.57	0.99
$C_{13}H_{23}$	14.42	0.96
$C_{14}H_{11}$	15.30	1.09
180		
$C_6H_{16}N_2O_4$	7.66	1.06
$C_7H_4N_2O_4$	8.55	1.12
$C_7H_6N_3O_3$	8.92	0.96
$C_7H_8N_4O_2$	9.30	0.79
$C_8H_6NO_4$	9.28	1.18
$C_8H_8N_2O_3$	9.65	1.02
$C_8H_{10}N_3O_2$	10.03	0.85
$C_8H_{12}N_4O$	10.40	0.69
$C_9H_8O_4$	10.01	1.25
$C_9H_{10}NO_3$	10.38	1.09
$C_9H_{12}N_2O_2$	10.76	0.93
$C_9H_{14}N_3O$	11.13	0.77
$C_9H_{16}N_4$	11.51	0.61
$C_{10}H_2N_3O$	12.02	0.86
$C_{10}H_4N_4$	12.40	0.71
$C_{10}H_{12}O_3$	11.12	1.16
$C_{10}H_{14}NO_2$	11.49	1.00
$C_{10}H_{16}N_2O$	11.86	0.84
$C_{10}H_{18}N_3$	12.24	0.69
$C_{11}H_2NO_2$	12.38	1.10
$C_{11}H_4N_2O$	12.75	0.95
$C_{11}H_6N_3$	13.13	0.80
$C_{11}H_{16}O_2$	12.22	1.08
$C_{11}H_{18}NO$	12.60	0.93
$C_{11}H_{20}N_2$	12.97	0.78
$C_{12}H_4O_2$	13.11	1.19
$C_{12}H_6NO$	13.48	1.04
$C_{12}H_8N_2$	13.86	0.89
$C_{12}H_{20}O$	13.33	1.02
$C_{12}H_{22}N$	13.70	0.87
$C_{13}H_8O$	14.22	1.13
$C_{13}H_{10}N$	14.59	0.99
$C_{13}H_{24}$	14.43	0.97
$C_{14}H_{12}$	15.32	1.09
181		
$C_7H_5N_2O_4$	8.56	1.13
$C_7H_7N_3O_3$	8.94	0.96
$C_7H_9N_4O_2$	9.31	0.79
$C_8H_7NO_4$	9.30	1.19
$C_8H_9N_2O_3$	9.67	1.02
$C_8H_{11}N_3O_2$	10.04	0.86
$C_8H_{13}N_4O$	10.42	0.69
C_9HN_4O	11.31	0.78
$C_9H_9O_4$	10.03	1.25
$C_9H_{11}NO_3$	10.40	1.09
$C_9H_{13}N_2O_2$	10.78	0.93
$C_9H_{15}N_3O$	11.15	0.77
$C_9H_{17}N_4$	11.52	0.61
$C_{10}HN_2O_2$	11.66	1.02
$C_{10}H_3N_3O$	12.04	0.86
$C_{10}H_5N_4$	12.41	0.71
$C_{10}H_{13}O_3$	11.13	1.16
$C_{10}H_{15}NO_2$	11.51	1.00
$C_{10}H_{17}N_2O$	11.88	0.85
$C_{10}H_{19}N_3$	12.25	0.69

Appendix Table 5.1 (*continued*)

	M + 1	M + 2
$C_{11}HO_3$	12.02	1.26
$C_{11}H_3NO_2$	12.39	1.10
$C_{11}H_5N_2O$	12.77	0.95
$C_{11}H_7N_3$	13.14	0.80
$C_{11}H_{17}O_2$	12.24	1.09
$C_{11}H_{19}NO$	12.61	0.93
$C_{11}H_{21}N_2$	12.99	0.78
$C_{12}H_5O_2$	13.13	1.19
$C_{12}H_7NO$	13.50	1.04
$C_{12}H_9N_2$	13.87	0.89
$C_{12}H_{21}O$	13.34	1.02
$C_{12}H_{23}N$	13.72	0.87
$C_{13}H_9O$	14.23	1.14
$C_{13}H_{11}N$	14.61	0.99
$C_{13}H_{25}$	14.45	0.97
$C_{14}H_{13}$	15.34	1.09
$C_{15}H$	16.23	1.23
182		
$C_7H_6N_2O_4$	8.58	1.13
$C_7H_8N_3O_3$	8.95	0.96
$C_7H_{10}N_4O_2$	9.33	0.79
$C_8H_8NO_4$	9.31	1.19
$C_8H_{10}N_2O_3$	9.69	1.02
$C_8H_{12}N_3O_2$	10.06	0.86
$C_8H_{14}N_4O$	10.43	0.70
$C_9H_2N_4O$	11.32	0.79
$C_9H_{10}O_4$	10.04	1.25
$C_9H_{12}NO_3$	10.42	1.09
$C_9H_{14}N_2O_2$	10.79	0.93
$C_9H_{16}N_3O$	11.17	0.77
$C_9H_{18}N_4$	11.54	0.61
$C_{10}H_2N_2O_2$	11.68	1.02
$C_{10}H_4N_3O$	12.05	0.87
$C_{10}H_6N_4$	12.43	0.71
$C_{10}H_{14}O_3$	11.15	1.16
$C_{10}H_{16}NO_2$	11.52	1.01
$C_{10}H_{18}N_2O$	11.90	0.85
$C_{10}H_{20}N_3$	12.27	0.69
$C_{11}H_2O_3$	12.04	1.26
$C_{11}H_4NO_2$	12.41	1.11
$C_{11}H_6N_2O$	12.79	0.95
$C_{11}H_8N_3$	13.16	0.80
$C_{11}H_{18}O_2$	12.25	1.09
$C_{11}H_{20}NO$	12.63	0.93
$C_{11}H_{22}N_2$	13.00	0.78
$C_{12}H_6O_2$	13.14	1.19
$C_{12}H_8NO$	13.52	1.04
$C_{12}H_{10}N_2$	13.89	0.89
$C_{12}H_{22}O$	13.36	1.02
$C_{12}H_{24}N$	13.73	0.87
$C_{13}H_{10}O$	14.25	1.14
$C_{13}H_{12}N$	14.62	0.99
$C_{13}H_{26}$	14.46	0.97
$C_{14}H_{14}$	15.35	1.10
$C_{15}H_2$	16.24	1.21
183		
$C_7H_7N_2O_4$	8.60	1.13
$C_7H_9N_3O_3$	8.97	0.96
$C_7H_{11}N_4O_2$	9.34	0.79
$C_8H_9NO_4$	9.33	1.19
$C_8H_{11}N_2O_3$	9.70	1.02
$C_8H_{13}N_3O_2$	10.08	0.86
$C_8H_{15}N_4O$	10.45	0.70
$C_9HN_3O_2$	10.96	0.95
$C_9H_3N_4O$	11.34	0.79
$C_9H_{11}O_4$	10.06	1.26
$C_9H_{13}NO_3$	10.43	1.09
$C_9H_{15}N_2O_2$	10.81	0.93
$C_9H_{17}N_3O$	11.18	0.77
$C_9H_{19}N_4$	11.56	0.61
$C_{10}HNO_3$	11.32	1.18
$C_{10}H_3N_2O_2$	11.70	1.03
$C_{10}H_5N_3O$	12.07	0.87
$C_{10}H_7N_4$	12.44	0.71
$C_{10}H_{15}O_3$	11.16	1.17
$C_{10}H_{17}NO_2$	11.54	1.01
$C_{10}H_{19}N_2O$	11.91	0.85
$C_{10}H_{21}N_3$	12.29	0.69
$C_{11}H_3O_3$	12.05	1.26
$C_{11}H_5NO_2$	12.43	1.11
$C_{11}H_7N_2O$	12.80	0.95
$C_{11}H_9N_3$	13.18	0.80
$C_{11}H_{19}O_2$	12.27	1.09
$C_{11}H_{21}NO$	12.64	0.93
$C_{11}H_{23}N_2$	13.02	0.78
$C_{12}H_7O_2$	13.16	1.20
$C_{12}H_9NO$	13.53	1.05
$C_{12}H_{11}N_2$	13.91	0.90
$C_{12}H_{23}O$	13.37	1.02
$C_{12}H_{25}N$	13.75	0.87
$C_{13}H_{11}O$	14.26	1.14
$C_{13}H_{13}N$	14.64	0.99
$C_{13}H_{27}$	14.48	0.97
$C_{14}HN$	15.53	1.12
$C_{14}H_{15}$	15.37	1.10
$C_{15}H_3$	16.26	1.23
184		
$C_7H_8N_2O_4$	8.61	1.13
$C_7H_{10}N_3O_3$	8.99	0.96
$C_7H_{12}N_4O_2$	9.36	0.80
$C_8H_{10}NO_4$	9.34	1.19
$C_8H_{12}N_2O_3$	9.72	1.03
$C_8H_{14}N_3O_2$	10.09	0.86
$C_8H_{16}N_4O$	10.47	0.70
$C_9H_2N_3O_2$	10.98	0.95
$C_9H_4N_4O$	11.35	0.79
$C_9H_{12}O_4$	10.07	1.26
$C_9H_{14}NO_3$	10.45	1.09
$C_9H_{16}N_2O_2$	10.82	0.93
$C_9H_{18}N_3O$	11.20	0.77
$C_9H_{20}N_4$	11.57	0.61
$C_{10}H_2NO_3$	11.34	1.18
$C_{10}H_4N_2O_2$	11.71	1.03
$C_{10}H_6N_3O$	12.09	0.87
$C_{10}H_8N_4$	12.46	0.71
$C_{10}H_{16}O_3$	11.18	1.17
$C_{10}H_{18}NO_2$	11.55	1.01
$C_{10}H_{20}N_2O$	11.93	0.85
$C_{10}H_{22}N_3$	12.30	0.70
$C_{11}H_4O_3$	12.07	1.27
$C_{11}H_6NO_2$	12.44	1.11
$C_{11}H_8N_2O$	12.82	0.96
$C_{11}H_{10}N_3$	13.19	0.80
$C_{11}H_{20}O_2$	12.29	1.09
$C_{11}H_{22}NO$	12.66	0.94
$C_{11}H_{24}N_2$	13.03	0.78
$C_{12}H_8O_2$	13.17	1.20
$C_{12}H_{10}NO$	13.55	1.05
$C_{12}H_{12}N_2$	13.92	0.90
$C_{12}H_{24}O$	13.39	1.03
$C_{12}H_{26}N$	13.77	0.88
$C_{13}H_{12}O$	14.28	1.14
$C_{13}H_{14}N$	14.65	1.00
$C_{13}H_{28}$	14.50	0.97
$C_{14}H_2N$	15.54	1.13
$C_{14}H_{16}$	15.38	1.10
$C_{15}H_4$	16.27	1.24
185		
$C_7H_9N_2O_4$	8.63	1.13
$C_7H_{11}N_3O_3$	9.00	0.96
$C_7H_{13}N_4O_2$	9.38	0.80
$C_8HN_4O_2$	10.27	0.88
$C_8H_{11}NO_4$	9.36	1.19
$C_8H_{13}N_2O_3$	9.73	1.03
$C_8H_{15}N_3O_2$	10.11	0.86
$C_8H_{17}N_4O$	10.48	0.70
$C_9HN_2O_3$	10.62	1.11
$C_9H_3N_3O_2$	11.00	0.95
$C_9H_5N_4O$	11.37	0.79
$C_9H_{13}O_4$	10.09	1.26
$C_9H_{15}NO_3$	10.46	1.10
$C_9H_{17}N_2O_2$	10.84	0.93
$C_9H_{19}N_3O$	11.21	0.77
$C_9H_{21}N_4$	11.59	0.62
$C_{10}HO_4$	10.98	1.35
$C_{10}H_3NO_3$	11.35	1.19
$C_{10}H_5N_2O_2$	11.73	1.03
$C_{10}H_7N_3O$	12.10	0.87
$C_{10}H_9N_4$	12.48	0.72
$C_{10}H_{17}O_3$	11.20	1.17
$C_{10}H_{19}NO_2$	11.57	1.01
$C_{10}H_{21}N_2O$	11.94	0.85
$C_{10}H_{23}N_3$	12.32	0.70
$C_{11}H_5O_3$	12.08	1.27
$C_{11}H_7NO_2$	12.46	1.11
$C_{11}H_9N_2O$	12.83	0.96
$C_{11}H_{11}N_3$	13.21	0.81
$C_{11}H_{21}O_2$	12.30	1.09
$C_{11}H_{23}NO$	12.68	0.94
$C_{11}H_{25}N_2$	13.05	0.79
$C_{12}H_9O_2$	13.19	1.20
$C_{12}H_{11}NO$	13.56	1.05
$C_{12}H_{13}N_2$	13.94	0.90
$C_{12}H_{25}O$	13.41	1.03
$C_{12}H_{27}N$	13.78	0.88
$C_{13}HN_2$	14.83	1.02
$C_{13}H_{13}O$	14.30	1.15
$C_{13}H_{15}N$	14.67	1.00
$C_{14}HO$	15.18	1.27
$C_{14}H_3N$	15.56	1.13
$C_{14}H_{17}$	15.40	1.10
$C_{15}H_5$	16.29	1.24
186		
$C_7H_{10}N_2O_4$	8.64	1.13
$C_7H_{12}N_3O_3$	9.02	0.97
$C_7H_{14}N_4O_2$	9.39	0.80
$C_8H_2N_4O_2$	10.28	0.88
$C_8H_{12}NO_4$	9.38	1.19
$C_8H_{14}N_2O_3$	9.75	1.03
$C_8H_{16}N_3O_2$	10.12	0.86
$C_8H_{18}N_4O$	10.50	0.70
$C_9H_2N_2O_3$	10.64	1.11
$C_9H_4N_3O_2$	11.01	0.95
$C_9H_6N_4O$	11.39	0.79
$C_9H_{14}O_4$	10.11	1.26
$C_9H_{16}NO_3$	10.48	1.10
$C_9H_{18}N_2O_2$	10.86	0.94
$C_9H_{20}N_3O$	11.23	0.78
$C_9H_{22}N_4$	11.60	0.62
$C_{10}H_2O_4$	10.99	1.35
$C_{10}H_4NO_3$	11.37	1.19
$C_{10}H_6N_2O_2$	11.74	1.03
$C_{10}H_8N_3O$	12.12	0.87
$C_{10}H_{10}N_4$	12.49	0.72
$C_{10}H_{18}O_3$	11.21	1.17
$C_{10}H_{20}NO_2$	11.59	1.01
$C_{10}H_{22}N_2O$	11.96	0.86
$C_{10}H_{24}N_3$	12.33	0.70
$C_{11}H_6O_3$	12.10	1.27
$C_{11}H_8NO_2$	12.47	1.11
$C_{11}H_{10}N_2O$	12.85	0.96
$C_{11}H_{12}N_3$	13.22	0.81
$C_{11}H_{22}O_2$	12.32	1.10
$C_{11}H_{24}NO$	12.69	0.94
$C_{11}H_{26}N_2$	13.07	0.79
$C_{12}H_{10}O_2$	13.21	1.20
$C_{12}H_{12}NO$	13.58	1.05
$C_{12}H_{14}N_2$	13.95	0.90
$C_{12}H_{26}O$	13.42	1.03
$C_{13}H_2N_2$	14.84	1.02
$C_{13}H_{14}O$	14.31	1.15
$C_{13}H_{16}N$	14.69	1.00
$C_{14}H_2O$	15.20	1.27
$C_{14}H_4N$	15.57	1.13
$C_{14}H_{18}$	15.42	1.11
$C_{15}H_6$	16.31	1.24
187		
$C_7H_{11}N_2O_4$	8.66	1.13
$C_7H_{13}N_3O_3$	9.03	0.97
$C_7H_{15}N_4O_2$	9.41	0.80
$C_8HN_3O_3$	9.92	1.04
$C_8H_3N_4O_2$	10.30	0.88
$C_8H_{13}NO_4$	9.39	1.20
$C_8H_{15}N_2O_3$	9.77	1.03
$C_8H_{17}N_3O_2$	10.14	0.87
$C_8H_{19}N_4O$	10.51	0.70
C_9HNO_4	10.28	1.28
$C_9H_3N_2O_3$	10.65	1.11
$C_9H_5N_3O_2$	11.03	0.95
$C_9H_7N_4O$	11.40	0.80
$C_9H_{15}O_4$	10.12	1.26
$C_9H_{17}NO_3$	10.50	1.10
$C_9H_{19}N_2O_2$	10.87	0.94
$C_9H_{21}N_3O$	11.25	0.78
$C_9H_{23}N_4$	11.62	0.62
$C_{10}H_3O_4$	11.01	1.35
$C_{10}H_5NO_3$	11.39	1.19

Appendix Table 5.1 (*continued*)

	M + 1	M + 2
$C_{10}H_7N_2O_2$	11.76	1.03
$C_{10}H_9N_3O$	12.13	0.88
$C_{10}H_{11}N_4$	12.51	0.72
$C_{10}H_{19}O_3$	11.23	1.17
$C_{10}H_{21}NO_2$	11.60	1.01
$C_{10}H_{23}N_2O$	11.98	0.86
$C_{10}H_{25}N_3$	12.35	0.70
$C_{11}H_7O_3$	12.12	1.27
$C_{11}H_9NO_2$	12.49	1.12
$C_{11}H_{11}N_2O$	12.87	0.96
$C_{11}H_{13}N_3$	13.24	0.81
$C_{11}H_{23}O_2$	12.33	1.10
$C_{11}H_{25}NO$	12.71	0.94
$C_{12}HN_3$	14.13	0.93
$C_{12}H_{11}O_2$	13.22	1.20
$C_{12}H_{13}NO$	13.60	1.05
$C_{12}H_{15}N_2$	13.97	0.90
$C_{13}HNO$	14.48	1.17
$C_{13}H_3N_2$	14.86	1.03
$C_{13}H_{15}O$	14.33	1.15
$C_{13}H_{17}N$	14.70	1.00
$C_{14}H_3O$	15.22	1.28
$C_{14}H_5N$	15.59	1.13
$C_{14}H_{19}$	15.43	1.11
$C_{15}H_7$	16.32	1.24
188		
$C_7H_{12}N_2O_4$	8.68	1.14
$C_7H_{14}N_3O_3$	9.05	0.97
$C_7H_{16}N_4O_2$	9.42	0.80
$C_8H_2N_3O_3$	9.94	1.05
$C_8H_4N_4O_2$	10.31	0.88
$C_8H_{14}NO_4$	9.41	1.20
$C_8H_{16}N_2O_3$	9.78	1.03
$C_8H_{18}N_3O_2$	10.16	0.87
$C_8H_{20}N_4O$	10.53	0.71
$C_9H_2NO_4$	10.30	1.28
$C_9H_4N_2O_3$	10.67	1.12
$C_9H_6N_3O_2$	11.04	0.96
$C_9H_8N_4O$	11.42	0.80
$C_9H_{16}O_4$	10.14	1.26
$C_9H_{18}NO_3$	10.51	1.10
$C_9H_{20}N_2O_2$	10.89	0.94
$C_9H_{22}N_3O$	11.26	0.78
$C_9H_{24}N_4$	11.64	0.62
$C_{10}H_4O_4$	11.03	1.35
$C_{10}H_6NO_3$	11.40	1.19
$C_{10}H_8N_2O_2$	11.78	1.03
$C_{10}H_{10}N_3O$	12.15	0.88
$C_{10}H_{12}N_4$	12.52	0.72
$C_{10}H_{20}O_3$	11.24	1.18
$C_{10}H_{22}NO_2$	11.62	1.02
$C_{10}H_{24}N_2O$	11.99	0.86
$C_{11}H_8O_3$	12.13	1.27
$C_{11}H_{10}NO_2$	12.51	1.12
$C_{11}H_{12}N_2O$	12.88	0.96
$C_{11}H_{14}N_3$	13.26	0.81
$C_{11}H_{24}O_2$	12.35	1.10
$C_{12}H_2N_3$	14.14	0.93
$C_{12}H_{12}O_2$	13.24	1.21
$C_{12}H_{14}NO$	13.61	1.06
$C_{12}H_{16}N_2$	13.99	0.91
$C_{13}H_2NO$	14.50	1.18
$C_{13}H_4N_2$	14.88	1.03
$C_{13}H_{16}O$	14.34	1.15
$C_{13}H_{18}N$	14.72	1.01
$C_{14}H_4O$	15.23	1.28
$C_{14}H_6N$	15.61	1.14
$C_{14}H_{20}$	15.45	1.11
$C_{15}H_8$	16.34	1.25
189		
$C_7H_{13}N_2O_4$	8.69	1.14
$C_7H_{15}N_3O_3$	9.07	0.97
$C_7H_{17}N_4O_2$	9.44	0.80
$C_8HN_2O_4$	9.58	1.21
$C_8H_3N_3O_3$	9.95	1.05
$C_8H_5N_4O_2$	10.33	0.88
$C_8H_{15}NO_4$	9.42	1.20
$C_8H_{17}N_2O_3$	9.80	1.03
$C_8H_{19}N_3O_2$	10.17	0.87
$C_8H_{21}N_4O$	10.55	0.71
$C_9H_3NO_4$	10.31	1.28
$C_9H_5N_2O_3$	10.69	1.12
$C_9H_7N_3O_2$	11.06	0.96
$C_9H_9N_4O$	11.43	0.80
$C_9H_{17}O_4$	10.15	1.26
$C_9H_{19}NO_3$	10.53	1.10
$C_9H_{21}N_2O_2$	10.90	0.94
$C_9H_{23}N_3O$	11.28	0.78
$C_{10}H_5O_4$	11.04	1.35
$C_{10}H_7NO_3$	11.42	1.19
$C_{10}H_9N_2O_2$	11.79	1.04
$C_{10}H_{11}N_3O$	12.17	0.88
$C_{10}H_{13}N_4$	12.54	0.72
$C_{10}H_{21}O_3$	11.26	1.18
$C_{10}H_{23}NO_2$	11.63	1.02
$C_{11}HN_4$	13.43	0.83
$C_{11}H_9O_3$	12.15	1.28
$C_{11}H_{11}NO_2$	12.52	1.12
$C_{11}H_{13}N_2O$	12.90	0.97
$C_{11}H_{15}N_3$	13.27	0.81
$C_{12}HN_2O$	13.79	1.08
$C_{12}H_3N_3$	14.16	0.93
$C_{12}H_{13}O_2$	13.25	1.21
$C_{12}H_{15}NO$	13.63	1.06
$C_{12}H_{17}N_2$	14.00	0.91
$C_{13}HO_2$	14.14	1.33
$C_{13}H_3NO$	14.52	1.18
$C_{13}H_5N_2$	14.89	1.03
$C_{13}H_{17}O$	14.36	1.16
$C_{13}H_{19}N$	14.73	1.01
$C_{14}H_5O$	15.25	1.28
$C_{14}H_7N$	15.62	1.14
$C_{14}H_{21}$	15.46	1.11
$C_{15}H_9$	16.35	1.25
190		
$C_7H_{14}N_2O_4$	8.71	1.14
$C_7H_{16}N_3O_3$	9.08	0.97
$C_7H_{18}N_4O_2$	9.46	0.80
$C_8H_2N_2O_4$	9.60	1.21
$C_8H_4N_3O_3$	9.97	1.05
$C_8H_6N_4O_2$	10.35	0.89
$C_8H_{16}NO_4$	9.44	1.20
$C_8H_{18}N_2O_3$	9.81	1.03
$C_8H_{20}N_3O_2$	10.19	0.87
$C_8H_{22}N_4O$	10.56	0.71
$C_9H_4NO_4$	10.33	1.28
$C_9H_6N_2O_3$	10.70	1.12
$C_9H_8N_3O_2$	11.08	0.96
$C_9H_{10}N_4O$	11.45	0.80
$C_9H_{18}O_4$	10.17	1.27
$C_9H_{20}NO_3$	10.54	1.10
$C_9H_{22}N_2O_2$	10.92	0.94
$C_{10}H_6O_4$	11.06	1.35
$C_{10}H_8NO_3$	11.43	1.20
$C_{10}H_{10}N_2O_2$	11.81	1.03
$C_{10}H_{12}N_3O$	12.18	0.88
$C_{10}H_{14}N_4$	12.56	0.73
$C_{10}H_{22}O_3$	11.28	1.18
$C_{11}H_2N_4$	13.44	0.84
$C_{11}H_{10}O_3$	12.16	1.28
$C_{11}H_{12}NO_2$	12.54	1.12
$C_{11}H_{14}N_2O$	12.91	0.97
$C_{11}H_{16}N_3$	13.29	0.82
$C_{12}H_2N_2O$	13.80	1.08
$C_{12}H_4N_3$	14.18	0.93
$C_{12}H_{14}O_2$	13.27	1.21
$C_{12}H_{16}NO$	13.64	1.06
$C_{12}H_{18}N_2$	14.02	0.91
$C_{13}H_2O_2$	14.16	1.33
$C_{13}H_4NO$	14.53	1.18
$C_{13}H_6N_2$	14.91	1.03
$C_{13}H_{18}O$	14.38	1.16
$C_{13}H_{20}N$	14.75	1.01
$C_{14}H_6O$	15.26	1.28
$C_{14}H_8N$	15.64	1.14
$C_{14}H_{22}$	15.48	1.12
$C_{15}H_{10}$	16.37	1.25
191		
$C_7H_{15}N_2O_4$	8.72	1.14
$C_7H_{17}N_3O_3$	9.10	0.97
$C_7H_{19}N_4O_2$	9.47	0.81
$C_8H_3N_2O_4$	9.61	1.22
$C_8H_5N_3O_3$	9.99	1.05
$C_8H_7N_4O_2$	10.36	0.89
$C_8H_{17}NO_4$	9.46	1.20
$C_8H_{19}N_2O_3$	9.83	1.04
$C_8H_{21}N_3O_2$	10.20	0.87
$C_9H_5NO_4$	10.34	1.28
$C_9H_7N_2O_3$	10.72	1.12
$C_9H_9N_3O_2$	11.09	0.96
$C_9H_{11}N_4O$	11.47	0.80
$C_9H_{19}O_4$	10.19	1.27
$C_9H_{21}NO_3$	10.56	1.11
$C_{10}H_7O_4$	11.07	1.36
$C_{10}H_9NO_3$	11.45	1.20
$C_{10}H_{11}N_2O_2$	11.82	1.04
$C_{10}H_{13}N_3O$	12.20	0.88
$C_{10}H_{15}N_4$	12.57	0.73
$C_{11}HN_3O$	13.09	0.99
$C_{11}H_3N_4$	13.46	0.84
$C_{11}H_{11}O_3$	12.18	1.28
$C_{11}H_{13}NO_2$	12.55	1.12
$C_{11}H_{15}N_2O$	12.93	0.97
$C_{11}H_{17}N_3$	13.30	0.82
$C_{12}HNO_2$	13.44	1.23
$C_{12}H_3N_2O$	13.82	1.08
$C_{12}H_5N_3$	14.19	0.93
$C_{12}H_{15}O_2$	13.29	1.21
$C_{12}H_{17}NO$	13.66	1.06
$C_{12}H_{19}N_2$	14.03	0.91
$C_{13}H_3O_2$	14.17	1.33
$C_{13}H_5NO$	14.55	1.18
$C_{13}H_7N_2$	14.92	1.04
$C_{13}H_{19}O$	14.39	1.16
$C_{13}H_{21}N$	14.77	1.01
$C_{14}H_7O$	15.28	1.29
$C_{14}H_9N$	15.65	1.14
$C_{14}H_{23}$	15.50	1.12
$C_{15}H_{11}$	16.39	1.25
192		
$C_7H_{16}N_2O_4$	8.74	1.14
$C_7H_{18}N_3O_3$	9.11	0.97
$C_7H_{20}N_4O_2$	9.49	0.81
$C_8H_4N_2O_4$	9.63	1.22
$C_8H_6N_3O_3$	10.00	1.05
$C_8H_8N_4O_2$	10.38	0.89
$C_8H_{18}NO_4$	9.47	1.20
$C_8H_{20}N_2O_3$	9.85	1.04
$C_9H_6NO_4$	10.36	1.29
$C_9H_8N_2O_3$	10.73	1.12
$C_9H_{10}N_3O_2$	11.11	0.96
$C_9H_{12}N_4O$	11.48	0.80
$C_9H_{20}O_4$	10.20	1.27
$C_{10}H_8O_4$	11.09	1.36
$C_{10}H_{10}NO_3$	11.47	1.20
$C_{10}H_{12}N_2O_2$	11.84	1.04
$C_{10}H_{14}N_3O$	12.21	0.89
$C_{10}H_{16}N_4$	12.59	0.73
$C_{11}H_2N_3O$	13.10	0.99
$C_{11}H_4N_4$	13.48	0.84
$C_{11}H_{12}O_3$	12.20	1.28
$C_{11}H_{14}NO_2$	12.57	1.13
$C_{11}H_{16}N_2O$	12.95	0.97
$C_{11}H_{18}N_3$	13.32	0.82
$C_{12}H_2NO_2$	13.46	1.24
$C_{12}H_4N_2O$	13.83	1.09
$C_{12}H_6N_3$	14.21	0.94
$C_{12}H_{16}O_2$	13.30	1.22
$C_{12}H_{18}NO$	13.68	1.06
$C_{12}H_{20}N_2$	14.05	0.92
$C_{13}H_4O_2$	14.19	1.33
$C_{13}H_6NO$	14.56	1.18
$C_{13}H_8N_2$	14.94	1.04
$C_{13}H_{20}O$	14.41	1.16
$C_{13}H_{22}N$	14.78	1.02
$C_{14}H_8O$	15.30	1.29
$C_{14}H_{10}N$	15.67	1.15
$C_{14}H_{24}$	15.51	1.12
$C_{15}H_{12}$	16.40	1.26
193		
$C_7H_{17}N_2O_4$	8.76	1.14
$C_7H_{19}N_3O_3$	9.13	0.98
$C_8H_5N_2O_4$	9.64	1.22
$C_8H_7N_3O_3$	10.02	1.05
$C_8H_9N_4O_2$	10.39	0.89
$C_8H_{19}NO_4$	9.49	1.20

Appendix Table 5.1 (*continued*)

	M + 1	M + 2
$C_9H_7NO_4$	10.38	1.29
$C_9H_9N_2O_3$	10.75	1.13
$C_9H_{11}N_3O_2$	11.12	0.96
$C_9H_{13}N_4O$	11.50	0.81
$C_{10}HN_4O$	12.39	0.91
$C_{10}H_9O_4$	11.11	1.36
$C_{10}H_{11}NO_3$	11.48	1.20
$C_{10}H_{13}N_2O_2$	11.86	1.04
$C_{10}H_{15}N_3O$	12.23	0.89
$C_{10}H_{17}N_4$	12.60	0.73
$C_{11}HN_2O_2$	12.74	1.15
$C_{11}H_3N_3O$	13.12	0.99
$C_{11}H_5N_4$	13.49	0.84
$C_{11}H_{13}O_3$	12.21	1.28
$C_{11}H_{15}NO_2$	12.59	1.13
$C_{11}H_{17}N_2O$	12.96	0.97
$C_{11}H_{19}N_3$	13.34	0.82
$C_{12}HO_3$	13.10	1.39
$C_{12}H_3NO_2$	13.48	1.24
$C_{12}H_5N_2O$	13.85	1.09
$C_{12}H_7N_3$	14.22	0.94
$C_{12}H_{17}O_2$	13.32	1.22
$C_{12}H_{19}NO$	13.69	1.07
$C_{12}H_{21}N_2$	14.07	0.92
$C_{13}H_5O_2$	14.21	1.33
$C_{13}H_7NO$	14.58	1.19
$C_{13}H_9N_2$	14.96	1.04
$C_{13}H_{21}O$	14.42	1.16
$C_{13}H_{23}N$	14.80	1.02
$C_{14}H_9O$	15.31	1.29
$C_{14}H_{11}N$	15.69	1.15
$C_{14}H_{25}$	15.53	1.12
$C_{15}H_{13}$	16.42	1.26
$C_{16}H$	17.31	1.40
194		
$C_7H_{18}N_2O_4$	8.77	1.14
$C_8H_6N_2O_4$	9.66	1.22
$C_8H_8N_3O_3$	10.03	1.06
$C_8H_{10}N_4O_2$	10.41	0.89
$C_9H_8NO_4$	10.39	1.29
$C_9H_{10}N_2O_3$	10.77	1.13
$C_9H_{12}N_3O_2$	11.14	0.97
$C_9H_{14}N_4O$	11.51	0.81
$C_{10}H_2N_4O$	12.40	0.91
$C_{10}H_{10}O_4$	11.12	1.36
$C_{10}H_{12}NO_3$	11.50	1.20
$C_{10}H_{14}N_2O_2$	11.87	1.05
$C_{10}H_{16}N_3O$	12.25	0.89
$C_{10}H_{18}N_4$	12.62	0.74
$C_{11}H_2N_2O_2$	12.76	1.15
$C_{11}H_4N_3O$	13.13	1.00
$C_{11}H_6N_4$	13.51	0.85
$C_{11}H_{14}O_3$	12.23	1.28
$C_{11}H_{16}NO_2$	12.60	1.13
$C_{11}H_{18}N_2O$	12.98	0.98
$C_{11}H_{20}N_3$	13.35	0.82
$C_{12}H_2O_3$	13.12	1.39
$C_{12}H_4NO_2$	13.49	1.24
$C_{12}H_6N_2O$	13.87	1.09
$C_{12}H_8N_3$	14.24	0.94
$C_{12}H_{18}O_2$	13.33	1.22
$C_{12}H_{20}NO$	13.71	1.07
$C_{12}H_{22}N_2$	14.08	0.92
$C_{13}H_6O_2$	14.22	1.34
$C_{13}H_8NO$	14.60	1.19
$C_{13}H_{10}N_2$	14.97	1.04
$C_{13}H_{22}O$	14.44	1.17
$C_{13}H_{24}N$	14.81	1.02
$C_{14}H_{10}O$	15.33	1.29
$C_{14}H_{12}N$	15.70	1.15
$C_{14}H_{26}$	15.54	1.13
$C_{15}H_{14}$	16.43	1.26
$C_{16}H_2$	17.32	1.41
195		
$C_8H_7N_2O_4$	9.68	1.22
$C_8H_9N_3O_3$	10.05	1.06
$C_8H_{11}N_4O_2$	10.43	0.89
$C_9H_9NO_4$	10.41	1.29
$C_9H_{11}N_2O_3$	10.78	1.13
$C_9H_{13}N_3O_2$	11.16	0.97
$C_9H_{15}N_4O$	11.53	0.81
$C_{10}HN_3O_2$	12.05	1.07
$C_{10}H_3N_4O$	12.42	0.91
$C_{10}H_{11}O_4$	11.14	1.36
$C_{10}H_{13}NO_3$	11.51	1.21
$C_{10}H_{15}N_2O_2$	11.89	1.05
$C_{10}H_{17}N_3O$	12.26	0.89
$C_{10}H_{19}N_4$	12.64	0.74
$C_{11}HNO_3$	12.40	1.31
$C_{11}H_3N_2O_2$	12.78	1.15
$C_{11}H_5N_3O$	13.15	1.00
$C_{11}H_7N_4$	13.52	0.85
$C_{11}H_{15}O_3$	12.24	1.29
$C_{11}H_{17}NO_2$	12.62	1.13
$C_{11}H_{19}N_2O$	12.99	0.98
$C_{11}H_{21}N_3$	13.37	0.83
$C_{12}H_3O_3$	13.13	1.39
$C_{12}H_5NO_2$	13.51	1.24
$C_{12}H_7N_2O$	13.88	1.09
$C_{12}H_9N_3$	14.26	0.94
$C_{12}H_{19}O_2$	13.35	1.22
$C_{12}H_{21}NO$	13.72	1.07
$C_{12}H_{23}N_2$	14.10	0.92
$C_{13}H_7O_2$	14.24	1.34
$C_{13}H_9NO$	14.61	1.19
$C_{13}H_{11}N_2$	14.99	1.05
$C_{13}H_{23}O$	14.46	1.17
$C_{13}H_{25}N$	14.83	1.02
$C_{14}H_{11}O$	15.34	1.30
$C_{14}H_{13}N$	15.72	1.15
$C_{14}H_{27}$	15.56	1.13
$C_{15}HN$	16.61	1.29
$C_{15}H_{15}$	16.45	1.27
$C_{16}H_3$	17.34	1.41
196		
$C_8H_8N_2O_4$	9.69	1.22
$C_8H_{10}N_3O_3$	10.07	1.06
$C_8H_{12}N_4O_2$	10.44	0.90
$C_9H_{10}NO_4$	10.42	1.29
$C_9H_{12}N_2O_3$	10.80	1.13
$C_9H_{14}N_3O_2$	11.17	0.97
$C_9H_{16}N_4O$	11.55	0.81
$C_{10}H_2N_3O_2$	12.06	1.07
$C_{10}H_4N_4O$	12.44	0.91
$C_{10}H_{12}O_4$	11.15	1.37
$C_{10}H_{14}NO_3$	11.53	1.21
$C_{10}H_{16}N_2O_2$	11.90	1.05
$C_{10}H_{18}N_3O$	12.28	0.89
$C_{10}H_{20}N_4$	12.65	0.74
$C_{11}H_2NO_3$	12.42	1.31
$C_{11}H_4N_2O_2$	12.79	1.15
$C_{11}H_6N_3O$	13.17	1.00
$C_{11}H_8N_4$	13.54	0.85
$C_{11}H_{16}O_3$	12.26	1.29
$C_{11}H_{18}NO_2$	12.63	1.13
$C_{11}H_{20}N_2O$	13.01	0.98
$C_{11}H_{22}N_3$	13.38	0.83
$C_{12}H_4O_3$	13.15	1.40
$C_{12}H_6NO_2$	13.52	1.24
$C_{12}H_8N_2O$	13.90	1.09
$C_{12}H_{10}N_3$	14.27	0.95
$C_{12}H_{20}O_2$	13.37	1.22
$C_{12}H_{22}NO$	13.74	1.07
$C_{12}H_{24}N_2$	14.11	0.92
$C_{13}H_8O_2$	14.25	1.34
$C_{13}H_{10}NO$	14.63	1.19
$C_{13}H_{12}N_2$	15.00	1.05
$C_{13}H_{24}O$	14.47	1.17
$C_{13}H_{26}N$	14.85	1.03
$C_{14}H_{12}O$	15.36	1.30
$C_{14}H_{14}N$	15.73	1.16
$C_{14}H_{28}$	15.58	1.13
$C_{15}H_2N$	16.62	1.29
$C_{15}H_{16}$	16.47	1.27
$C_{16}H_4$	17.35	1.41
197		
$C_8H_9N_2O_4$	9.71	1.23
$C_8H_{11}N_3O_3$	10.08	1.06
$C_8H_{13}N_4O_2$	10.46	0.90
$C_9HN_4O_2$	11.35	0.99
$C_9H_{11}NO_4$	10.44	1.29
$C_9H_{13}N_2O_3$	10.81	1.13
$C_9H_{15}N_3O_2$	11.19	0.97
$C_9H_{17}N_4O$	11.56	0.81
$C_{10}HN_2O_3$	11.70	1.23
$C_{10}H_3N_3O_2$	12.08	1.07
$C_{10}H_5N_4O$	12.45	0.91
$C_{10}H_{13}O_4$	11.17	1.37
$C_{10}H_{15}NO_3$	11.55	1.21
$C_{10}H_{17}N_2O_2$	11.92	1.05
$C_{10}H_{19}N_3O$	12.29	0.90
$C_{10}H_{21}N_4$	12.67	0.74
$C_{11}HO_4$	12.06	1.46
$C_{11}H_3NO_3$	12.43	1.31
$C_{11}H_5N_2O_2$	12.81	1.16
$C_{11}H_7N_3O$	13.18	1.00
$C_{11}H_9N_4$	13.56	0.85
$C_{11}H_{17}O_3$	12.28	1.29
$C_{11}H_{19}NO_2$	12.65	1.14
$C_{11}H_{21}N_2O$	13.03	0.98
$C_{11}H_{23}N_3$	13.40	0.83
$C_{12}H_5O_3$	13.16	1.40
$C_{12}H_7NO_2$	13.54	1.25
$C_{12}H_9N_2O$	13.91	1.10
$C_{12}H_{11}N_3$	14.29	0.95
$C_{12}H_{21}O_2$	13.38	1.23
$C_{12}H_{23}NO$	13.76	1.08
$C_{12}H_{25}N_2$	14.13	0.93
$C_{13}H_9O_2$	14.27	1.34
$C_{13}H_{11}NO$	14.64	1.20
$C_{13}H_{13}N_2$	15.02	1.05
$C_{13}H_{25}O$	14.49	1.17
$C_{13}H_{27}N$	14.86	1.03
$C_{14}HN_2$	15.91	1.18
$C_{14}H_{13}O$	15.38	1.30
$C_{14}H_{15}N$	15.75	1.16
$C_{14}H_{29}$	15.59	1.13
$C_{15}HO$	16.26	1.44
$C_{15}H_3N$	16.64	1.30
$C_{15}H_{17}$	16.48	1.27
$C_{16}H_5$	17.37	1.42
198		
$C_8H_{10}N_2O_4$	9.72	1.23
$C_8H_{12}N_3O_3$	10.10	1.06
$C_8H_{14}N_4O$	10.47	0.90
$C_9H_2N_4O_2$	11.36	0.99
$C_9H_{12}NO_4$	10.46	1.30
$C_9H_{14}N_2O_3$	10.83	1.13
$C_9H_{16}N_3O_2$	11.20	0.97
$C_9H_{18}N_4O$	11.58	0.82
$C_{10}H_2N_2O_3$	11.72	1.23
$C_{10}H_4N_3O_2$	12.09	1.07
$C_{10}H_6N_4O$	12.47	0.92
$C_{10}H_{14}O_4$	11.19	1.37
$C_{10}H_{16}NO_3$	11.56	1.21
$C_{10}H_{18}N_2O_2$	11.94	1.05
$C_{10}H_{20}N_3O$	12.31	0.90
$C_{10}H_{22}N_4$	12.68	0.74
$C_{11}H_2O_4$	12.08	1.47
$C_{11}H_4NO_3$	12.45	1.31
$C_{11}H_6N_2O_2$	12.82	1.16
$C_{11}H_8N_3O$	13.20	1.01
$C_{11}H_{10}N_4$	13.57	0.85
$C_{11}H_{18}O_3$	12.29	1.29
$C_{11}H_{20}NO_2$	12.67	1.14
$C_{11}H_{22}N_2O$	13.04	0.99
$C_{11}H_{24}N_3$	13.42	0.83
$C_{12}H_6O_3$	13.18	1.40
$C_{12}H_8NO_2$	13.56	1.25
$C_{12}H_{10}N_2O$	13.93	1.10
$C_{12}H_{12}N_3$	14.30	0.95
$C_{12}H_{22}O_2$	13.40	1.23
$C_{12}H_{24}NO$	13.77	1.08
$C_{12}H_{26}N_2$	14.15	0.93
$C_{13}H_{10}O_2$	14.29	1.35
$C_{13}H_{12}NO$	14.66	1.20
$C_{13}H_{14}N_2$	15.04	1.05
$C_{13}H_{26}O$	14.50	1.18
$C_{13}H_{28}N$	14.88	1.03
$C_{14}H_2N_2$	15.92	1.18
$C_{14}H_{14}O$	15.39	1.30
$C_{14}H_{16}N$	15.77	1.16
$C_{14}H_{30}$	15.61	1.14
$C_{15}H_2O$	16.28	1.44
$C_{15}H_4N$	16.65	1.30
$C_{15}H_{18}$	16.50	1.27
$C_{16}H_6$	17.39	1.42

Appendix Table 5.1 (*continued*)

	M + 1	M + 2
199		
$C_8H_{11}N_2O_4$	9.74	1.23
$C_8H_{13}N_3O_3$	10.11	1.06
$C_8H_{15}N_4O_2$	10.49	0.90
$C_9HN_3O_3$	11.00	1.15
$C_9H_3N_4O_2$	11.38	0.99
$C_9H_{13}NO_4$	10.47	1.30
$C_9H_{15}N_2O_3$	10.85	1.14
$C_9H_{17}N_3O_2$	11.22	0.98
$C_9H_{19}N_4O$	11.59	0.82
$C_{10}HNO_4$	11.36	1.39
$C_{10}H_3N_2O_3$	11.73	1.23
$C_{10}H_5N_3O_2$	12.11	1.07
$C_{10}H_7N_4O$	12.48	0.92
$C_{10}H_{15}O_4$	11.20	1.37
$C_{10}H_{17}NO_3$	11.58	1.21
$C_{10}H_{19}N_2O_2$	11.95	1.01
$C_{10}H_{21}N_3O$	12.33	0.90
$C_{10}H_{23}N_4$	12.70	0.75
$C_{11}H_3O_4$	12.09	1.47
$C_{11}H_5NO_3$	12.47	1.31
$C_{11}H_7N_2O_2$	12.84	1.16
$C_{11}H_9N_3O$	13.21	1.01
$C_{11}H_{11}N_4$	13.59	0.86
$C_{11}H_{19}O_3$	12.31	1.29
$C_{11}H_{21}NO_2$	12.68	1.14
$C_{11}H_{23}N_2O$	13.06	0.99
$C_{11}H_{25}N_3$	13.43	0.84
$C_{12}H_7O_3$	13.20	1.40
$C_{12}H_9NO_2$	13.57	1.25
$C_{12}H_{11}N_2O$	13.95	1.10
$C_{12}H_{13}N_3$	14.32	0.95
$C_{12}H_{23}O_2$	13.41	1.23
$C_{12}H_{25}NO$	13.79	1.08
$C_{12}H_{27}N_2$	14.16	0.93
$C_{13}HN_3$	15.21	1.08
$C_{13}H_{11}O_2$	14.30	1.35
$C_{13}H_{13}NO$	14.68	1.20
$C_{13}H_{15}N_2$	15.05	1.06
$C_{13}H_{27}O$	14.52	1.18
$C_{13}H_{29}N$	14.89	1.03
$C_{14}HNO$	15.57	1.33
$C_{14}H_3N_2$	15.94	1.19
$C_{14}H_{15}O$	15.41	1.31
$C_{14}H_{17}N$	15.78	1.16
$C_{15}H_3O$	16.30	1.44
$C_{15}H_5N$	16.67	1.30
$C_{15}H_{19}$	16.51	1.28
$C_{16}H_7$	17.40	1.42
200		
$C_8H_{12}N_2O_4$	9.76	1.23
$C_8H_{14}N_3O_3$	10.13	1.07
$C_8H_{16}N_4O_2$	10.51	0.90
$C_9H_2N_3O_3$	11.02	1.15
$C_9H_4N_4O_2$	11.39	0.99
$C_9H_{14}NO_4$	10.49	1.30
$C_9H_{16}N_2O_3$	10.86	1.14
$C_9H_{18}N_3O_2$	11.24	0.98
$C_9H_{20}N_4O$	11.61	0.82
$C_{10}H_2NO_4$	11.38	1.39
$C_{10}H_4N_2O_3$	11.75	1.23
$C_{10}H_6N_3O_2$	12.13	1.08

	M + 1	M + 2
$C_{10}H_8N_4O$	12.50	0.92
$C_{10}H_{16}O_4$	11.22	1.37
$C_{10}H_{18}NO_3$	11.59	1.21
$C_{10}H_{20}N_2O_2$	11.97	1.06
$C_{10}H_{22}N_3O$	12.34	0.90
$C_{10}H_{24}N_4$	12.72	0.75
$C_{11}H_4O_4$	12.11	1.47
$C_{11}H_6NO_3$	12.48	1.32
$C_{11}H_8N_2O_2$	12.86	1.16
$C_{11}H_{10}N_3O$	13.23	1.01
$C_{11}H_{12}N_4$	13.60	0.86
$C_{11}H_{20}O_3$	12.32	1.30
$C_{11}H_{22}NO_2$	12.70	1.14
$C_{11}H_{24}N_2O$	13.07	0.99
$C_{11}H_{26}N_3$	13.45	0.84
$C_{12}H_8O_3$	13.21	1.40
$C_{12}H_{10}NO_2$	13.59	1.25
$C_{12}H_{12}N_2O$	13.96	1.10
$C_{12}H_{14}N_3$	13.34	0.96
$C_{12}H_{24}O_2$	13.43	1.23
$C_{12}H_{26}NO$	13.80	1.08
$C_{12}H_{28}N_2$	14.18	0.93
$C_{13}H_2N_3$	15.22	1.08
$C_{13}H_{12}O_2$	14.32	1.35
$C_{13}H_{14}NO$	14.69	1.20
$C_{13}H_{16}N_2$	15.07	1.06
$C_{13}H_{28}O$	14.54	1.18
$C_{14}H_2NO$	15.58	1.33
$C_{14}H_4N_2$	15.96	1.19
$C_{14}H_{16}O$	15.42	1.31
$C_{14}H_{18}N$	15.80	1.17
$C_{15}H_4O$	16.31	1.44
$C_{15}H_6N$	16.69	1.30
$C_{15}H_{20}$	16.53	1.28
$C_{16}H_8$	17.42	1.42
201		
$C_8H_{13}N_2O_4$	9.77	1.23
$C_8H_{15}N_3O_3$	10.15	1.07
$C_8H_{17}N_4O_2$	10.52	0.90
$C_9HN_2O_4$	10.66	1.32
$C_9H_3N_3O_3$	11.04	1.16
$C_9H_5N_4O_2$	11.41	1.00
$C_9H_{15}NO_4$	10.50	1.30
$C_9H_{17}N_2O_3$	10.88	1.14
$C_9H_{19}N_3O_2$	11.25	0.98
$C_9H_{21}N_4O$	11.63	0.82
$C_{10}H_3NO_4$	11.39	1.39
$C_{10}H_5N_2O_3$	11.77	1.23
$C_{10}H_7N_3O_2$	12.14	1.08
$C_{10}H_9N_4O$	12.52	0.92
$C_{10}H_{17}O_4$	11.23	1.37
$C_{10}H_{19}NO_3$	11.61	1.22
$C_{10}H_{21}N_2O_2$	11.98	1.06
$C_{10}H_{23}N_3O$	12.36	0.90
$C_{10}H_{25}N_4$	12.73	0.75
$C_{11}H_5O_4$	12.12	1.47
$C_{11}H_7NO_3$	12.50	1.32
$C_{11}H_9N_2O_2$	12.87	1.16
$C_{11}H_{11}N_3O$	13.25	1.01
$C_{11}H_{13}N_4$	13.62	0.86
$C_{11}H_{21}O_3$	12.34	1.30
$C_{11}H_{23}NO_2$	12.71	1.14

	M + 1	M + 2
$C_{11}H_{25}N_2O$	13.09	0.99
$C_{11}H_{27}N_3$	13.46	0.84
$C_{12}HN_4$	14.51	0.98
$C_{12}H_9O_3$	13.23	1.41
$C_{12}H_{11}NO_2$	13.60	1.26
$C_{12}H_{13}N_2O$	13.98	1.11
$C_{12}H_{15}N_3$	14.35	0.96
$C_{12}H_{25}O_2$	13.45	1.23
$C_{12}H_{27}NO$	13.82	1.08
$C_{13}HN_2O$	14.87	1.23
$C_{13}H_3N_3$	15.24	1.08
$C_{13}H_{13}O_2$	14.33	1.35
$C_{13}H_{15}NO$	14.71	1.21
$C_{13}H_{17}N_2$	15.08	1.06
$C_{14}HO_2$	15.22	1.48
$C_{14}H_3NO$	15.60	1.33
$C_{14}H_5N_2$	15.97	1.19
$C_{14}H_{17}O$	15.44	1.31
$C_{14}H_{19}N$	15.81	1.17
$C_{15}H_5O$	16.33	1.45
$C_{15}H_7N$	16.70	1.31
$C_{15}H_{21}$	16.55	1.28
$C_{16}H_9$	17.43	1.43
202		
$C_8H_{14}N_2O_4$	9.79	1.23
$C_8H_{16}N_3O_3$	10.16	1.07
$C_8H_{18}N_4O_2$	10.54	0.91
$C_9H_2N_2O_4$	10.68	1.32
$C_9H_4N_3O_3$	11.05	1.16
$C_9H_6N_4O_2$	11.43	1.00
$C_9H_{16}NO_4$	10.52	1.30
$C_9H_{18}N_2O_3$	10.89	1.14
$C_9H_{20}N_3O_2$	11.27	0.98
$C_9H_{22}N_4O$	11.64	0.82
$C_{10}H_4NO_4$	11.41	1.39
$C_{10}H_6N_2O_3$	11.78	1.24
$C_{10}H_8N_3O_2$	12.16	1.08
$C_{10}H_{10}N_4O$	12.53	0.92
$C_{10}H_{18}O_4$	11.25	1.38
$C_{10}H_{20}NO_3$	11.63	1.22
$C_{10}H_{22}N_2O_2$	12.00	1.06
$C_{10}H_{24}N_3O$	12.37	0.91
$C_{10}H_{26}N_4$	12.75	0.75
$C_{11}H_6O_4$	12.14	1.47
$C_{11}H_8NO_3$	12.51	1.32
$C_{11}H_{10}N_2O_2$	12.89	1.17
$C_{11}H_{12}N_3O$	13.26	1.01
$C_{11}H_{14}N_4$	13.64	0.86
$C_{11}H_{22}O_3$	12.36	1.30
$C_{11}H_{24}NO_2$	12.73	1.15
$C_{11}H_{26}N_2O$	13.11	0.99
$C_{12}H_2N_4$	14.53	0.98
$C_{12}H_{10}O_3$	13.25	1.41
$C_{12}H_{12}NO_2$	13.62	1.26
$C_{12}H_{14}N_2O$	13.99	1.11
$C_{12}H_{16}N_3$	14.37	0.96
$C_{12}H_{26}O_2$	13.46	1.24
$C_{13}H_2N_2O$	14.88	1.23
$C_{13}H_4N_3$	15.26	1.09
$C_{13}H_{14}O_2$	14.35	1.35
$C_{13}H_{16}NO$	14.72	1.21
$C_{13}H_{18}N_2$	15.10	1.06

	M + 1	M + 2
$C_{14}H_2O_2$	15.24	1.48
$C_{14}H_4NO$	15.61	1.34
$C_{14}H_6N_2$	15.99	1.19
$C_{14}H_{18}O$	15.46	1.31
$C_{14}H_{20}N$	15.83	1.17
$C_{15}H_6O$	16.34	1.45
$C_{15}H_8N$	16.72	1.31
$C_{15}H_{22}$	16.56	1.28
$C_{16}H_{10}$	17.45	1.43
203		
$C_8H_{15}N_2O_4$	9.80	1.23
$C_8H_{17}N_3O_3$	10.18	1.07
$C_8H_{19}N_4O_2$	10.55	0.91
$C_9H_3N_2O_4$	10.69	1.32
$C_9H_5N_3O_3$	11.07	1.16
$C_9H_7N_4O_2$	11.44	1.00
$C_9H_{17}NO_4$	10.54	1.30
$C_9H_{19}N_2O_3$	10.91	1.14
$C_9H_{21}N_3O_2$	11.28	0.98
$C_9H_{23}N_4O$	11.66	0.82
$C_{10}H_5NO_4$	11.42	1.40
$C_{10}H_7N_2O_3$	11.80	1.24
$C_{10}H_9N_3O_2$	12.17	1.08
$C_{10}H_{11}N_4O$	12.55	0.93
$C_{10}H_{19}O_4$	11.27	1.38
$C_{10}H_{21}NO_3$	11.64	1.22
$C_{10}H_{23}N_2O_2$	12.02	1.06
$C_{10}H_{25}N_3O$	12.39	0.91
$C_{11}H_7O_4$	12.16	1.48
$C_{11}H_9NO_3$	12.53	1.32
$C_{11}H_{11}N_2O_2$	12.90	1.17
$C_{11}H_{13}N_3O$	13.28	1.02
$C_{11}H_{15}N_4$	13.65	0.86
$C_{11}H_{23}O_3$	12.37	1.30
$C_{11}H_{25}NO_2$	12.75	1.15
$C_{12}HN_3O$	14.17	1.13
$C_{12}H_3N_4$	14.54	0.98
$C_{12}H_{11}O_3$	13.26	1.41
$C_{12}H_{13}NO_2$	13.64	1.26
$C_{12}H_{15}N_2O$	14.01	1.11
$C_{12}H_{17}N_3$	14.38	0.96
$C_{13}HNO_2$	14.52	1.38
$C_{13}H_3N_2O$	14.90	1.23
$C_{13}H_5N_3$	15.27	1.09
$C_{13}H_{15}O_2$	14.37	1.36
$C_{13}H_{17}NO$	14.74	1.21
$C_{13}H_{19}N_2$	15.12	1.06
$C_{14}H_3O_2$	15.26	1.48
$C_{14}H_5NO$	15.63	1.34
$C_{14}H_7N_2$	16.00	1.20
$C_{14}H_{19}O$	15.47	1.32
$C_{14}H_{21}N$	15.85	1.17
$C_{15}H_7O$	16.36	1.45
$C_{15}H_9N$	16.73	1.31
$C_{15}H_{23}$	16.58	1.29
$C_{16}H_{11}$	17.47	1.43
204		
$C_8H_{16}N_2O_4$	9.82	1.24
$C_8H_{18}N_3O_3$	10.19	1.07
$C_8H_{20}N_4O_2$	10.57	0.91
$C_9H_4N_2O_4$	10.71	1.32

Appendix Table 5.1 (*continued*)

	M + 1	M + 2
$C_9H_6N_3O_3$	11.08	1.16
$C_9H_8N_4O_2$	11.46	1.00
$C_9H_{18}NO_4$	10.55	1.31
$C_9H_{20}N_2O_3$	10.93	1.14
$C_9H_{22}N_3O_2$	11.30	0.98
$C_9H_{24}N_4O$	11.67	0.83
$C_{10}H_6NO_4$	11.44	1.40
$C_{10}H_8N_2O_3$	11.81	1.24
$C_{10}H_{10}N_3O_2$	12.19	1.08
$C_{10}H_{12}N_4O$	12.56	0.93
$C_{10}H_{20}O_4$	11.28	1.38
$C_{10}H_{22}NO_3$	11.66	1.22
$C_{10}H_{24}N_2O_2$	12.03	1.06
$C_{11}H_8O_4$	12.17	1.48
$C_{11}H_{10}NO_3$	12.55	1.32
$C_{11}H_{12}N_2O_2$	12.92	1.17
$C_{11}H_{14}N_3O$	13.29	1.02
$C_{11}H_{16}N_4$	13.67	0.87
$C_{11}H_{24}O_3$	12.39	1.30
$C_{12}H_2N_3O$	14.18	1.13
$C_{12}H_4N_4$	14.56	0.99
$C_{12}H_{12}O_3$	13.28	1.41
$C_{12}H_{14}NO_2$	13.65	1.26
$C_{12}H_{16}N_2O$	14.03	1.11
$C_{12}H_{18}N_3$	14.40	0.96
$C_{13}H_2NO_2$	14.54	1.38
$C_{13}H_4N_2O$	14.91	1.24
$C_{13}H_6N_3$	15.29	1.09
$C_{13}H_{16}O_2$	14.38	1.36
$C_{13}H_{18}NO$	14.76	1.21
$C_{13}H_{20}N_2$	15.13	1.07
$C_{14}H_4O_2$	15.27	1.49
$C_{14}H_6NO$	15.65	1.34
$C_{14}H_8N_2$	16.02	1.20
$C_{14}H_{20}O$	15.49	1.32
$C_{14}H_{22}N$	15.86	1.18
$C_{15}H_8O$	16.38	1.45
$C_{15}H_{10}N$	16.75	1.31
$C_{15}H_{24}$	16.59	1.29
$C_{16}H_{12}$	17.48	1.43
205		
$C_8H_{17}N_2O_4$	9.84	1.24
$C_8H_{19}N_3O_3$	10.21	1.07
$C_8H_{21}N_4O_2$	10.59	0.91
$C_9H_5N_2O_4$	10.73	1.32
$C_9H_7N_3O_3$	11.10	1.16
$C_9H_9N_4O_2$	11.47	1.00
$C_9H_{19}NO_4$	10.57	1.31
$C_9H_{21}N_2O_3$	10.94	1.15
$C_9H_{23}N_3O_2$	11.32	0.99
$C_{10}H_7NO_4$	11.46	1.40
$C_{10}H_9N_2O_3$	11.83	1.24
$C_{10}H_{11}N_3O_2$	12.21	1.09
$C_{10}H_{13}N_4O$	12.58	0.93
$C_{10}H_{21}O_4$	11.30	1.38
$C_{10}H_{23}NO_3$	11.67	1.22
$C_{11}HN_4O$	13.47	1.04
$C_{11}H_9O_4$	12.19	1.48
$C_{11}H_{11}NO_3$	12.56	1.33
$C_{11}H_{13}N_2O_2$	12.94	1.17
$C_{11}H_{15}N_3O$	13.31	1.02
$C_{11}H_{17}N_4$	13.68	0.87
$C_{12}HN_2O_2$	13.82	1.29
$C_{12}H_3N_3O$	14.20	1.14
$C_{12}H_5N_4$	14.57	0.99
$C_{12}H_{13}O_3$	13.29	1.41
$C_{12}H_{15}NO_2$	13.67	1.26
$C_{12}H_{17}N_2O$	14.04	1.11
$C_{12}H_{19}N_3$	14.42	0.97
$C_{13}HO_3$	14.18	1.53
$C_{13}H_3NO_2$	14.56	1.38
$C_{13}H_5N_2O$	14.93	1.24
$C_{13}H_7N_3$	15.30	1.09
$C_{13}H_{17}O_2$	14.40	1.36
$C_{13}H_{19}NO$	14.77	1.21
$C_{13}H_{21}N_2$	15.15	1.07
$C_{14}H_5O_2$	15.29	1.49
$C_{14}H_7NO$	15.66	1.34
$C_{14}H_9N_2$	16.04	1.20
$C_{14}H_{21}O$	15.50	1.32
$C_{14}H_{23}N$	15.88	1.18
$C_{15}H_9O$	16.39	1.46
$C_{15}H_{11}N$	16.77	1.32
$C_{15}H_{25}$	16.61	1.29
$C_{16}H_{13}$	17.50	1.44
$C_{17}H$	18.39	1.59
206		
$C_8H_{18}N_2O_4$	9.85	1.24
$C_8H_{20}N_3O_3$	10.23	1.08
$C_8H_{22}N_4O_2$	10.60	0.91
$C_9H_6N_2O_4$	10.74	1.32
$C_9H_8N_3O_3$	11.12	1.16
$C_9H_{10}N_4O_2$	11.49	1.01
$C_9H_{20}NO_4$	10.58	1.31
$C_9H_{22}N_2O_3$	10.96	1.15
$C_{10}H_8NO_4$	11.47	1.40
$C_{10}H_{10}N_2O_3$	11.85	1.24
$C_{10}H_{12}N_3O_2$	12.22	1.09
$C_{10}H_{14}N_4O$	12.60	0.93
$C_{10}H_{22}O_4$	11.31	1.38
$C_{11}H_2N_4O$	13.48	1.04
$C_{11}H_{10}O_4$	12.20	1.48
$C_{11}H_{12}NO_3$	12.58	1.33
$C_{11}H_{14}N_2O_2$	12.95	1.17
$C_{11}H_{16}N_3O$	13.33	1.02
$C_{11}H_{18}N_4$	13.70	0.87
$C_{12}H_2N_2O_2$	13.84	1.29
$C_{12}H_4N_3O$	14.22	1.14
$C_{12}H_6N_4$	14.59	0.99
$C_{12}H_{14}O_3$	13.31	1.42
$C_{12}H_{16}NO_2$	13.68	1.27
$C_{12}H_{18}N_2O$	14.06	1.12
$C_{12}H_{20}N_3$	14.43	0.97
$C_{13}H_2O_3$	14.20	1.53
$C_{13}H_4NO_2$	14.57	1.39
$C_{13}H_6N_2O$	14.95	1.24
$C_{13}H_8N_3$	15.32	1.10
$C_{13}H_{18}O_2$	14.41	1.36
$C_{13}H_{20}NO$	14.79	1.22
$C_{13}H_{22}N_2$	15.16	1.07
$C_{14}H_6O_2$	15.30	1.49
$C_{14}H_8NO$	15.68	1.35
$C_{14}H_{10}N_2$	16.05	1.21
$C_{14}H_{22}O$	15.52	1.32
$C_{14}H_{24}N$	15.89	1.18
$C_{15}H_{10}O$	16.41	1.46
$C_{15}H_{12}N$	16.78	1.32
$C_{15}H_{26}$	16.63	1.29
$C_{16}H_{14}$	17.51	1.44
$C_{17}H_2$	18.40	1.59
207		
$C_8H_{19}N_2O_4$	9.87	1.24
$C_8H_{21}N_3O_3$	10.24	1.08
$C_9H_7N_2O_4$	10.76	1.33
$C_9H_9N_3O_3$	11.13	1.17
$C_9H_{11}N_4O_2$	11.51	1.01
$C_9H_{21}NO_4$	10.60	1.31
$C_{10}H_9NO_4$	11.49	1.40
$C_{10}H_{11}N_2O_3$	11.86	1.25
$C_{10}H_{13}N_3O_2$	12.24	1.09
$C_{11}H_{15}N_4O$	12.61	0.93
$C_{11}HN_3O_2$	13.13	1.20
$C_{11}H_3N_4O$	13.50	1.04
$C_{11}H_{11}O_4$	12.22	1.48
$C_{11}H_{13}NO_3$	12.59	1.33
$C_{11}H_{15}N_2O_2$	12.97	1.18
$C_{11}H_{17}N_3O$	13.34	1.02
$C_{11}H_{19}N_4$	13.72	0.87
$C_{12}HNO_3$	13.48	1.44
$C_{12}H_3N_2O_2$	13.86	1.29
$C_{12}H_5N_3O$	14.23	1.14
$C_{12}H_7N_4$	14.61	0.99
$C_{12}H_{15}O_3$	13.33	1.42
$C_{12}H_{17}NO_2$	13.70	1.27
$C_{12}H_{19}N_2O$	14.07	1.12
$C_{12}H_{21}N_3$	14.45	0.97
$C_{13}H_3O_3$	14.21	1.54
$C_{13}H_5NO_2$	14.59	1.39
$C_{13}H_7N_2O$	14.96	1.24
$C_{13}H_9N_3$	15.34	1.10
$C_{13}H_{19}O_2$	14.43	1.37
$C_{13}H_{21}NO$	14.80	1.22
$C_{13}H_{23}N_2$	15.18	1.07
$C_{14}H_7O_2$	15.32	1.49
$C_{14}H_9NO$	15.69	1.35
$C_{14}H_{11}N_2$	16.07	1.21
$C_{14}H_{23}O$	15.54	1.33
$C_{14}H_{25}N$	15.91	1.18
$C_{15}H_{11}O$	16.42	1.46
$C_{15}H_{13}N$	16.80	1.32
$C_{15}H_{27}$	16.64	1.30
$C_{16}HN$	17.69	1.47
$C_{16}H_{15}$	17.53	1.44
$C_{17}H_3$	18.42	1.60
208		
$C_8H_{20}N_2O_4$	9.88	1.24
$C_9H_8N_2O_4$	10.77	1.33
$C_9H_{10}N_3O_3$	11.15	1.17
$C_9H_{12}N_4O_2$	11.52	1.01
$C_{10}H_{10}NO_4$	11.50	1.40
$C_{10}H_{12}N_2O_3$	11.88	1.25
$C_{10}H_{14}N_3O_2$	12.25	1.09
$C_{10}H_{16}N_4O$	12.63	0.94
$C_{11}H_2N_3O_2$	13.14	1.20
$C_{11}H_4N_4O$	13.52	1.05
$C_{11}H_{12}O_4$	12.24	1.49
$C_{11}H_{14}NO_3$	12.61	1.33
$C_{11}H_{16}N_2O_2$	12.98	1.18
$C_{11}H_{18}N_3O$	13.36	1.03
$C_{11}H_{20}N_4$	13.73	0.88
$C_{12}H_2NO_3$	13.50	1.44
$C_{12}H_4N_2O_2$	13.87	1.29
$C_{12}H_6N_3O$	14.25	1.14
$C_{12}H_8N_4$	14.62	1.00
$C_{12}H_{16}O_3$	13.34	1.42
$C_{12}H_{18}NO_2$	13.72	1.27
$C_{12}H_{20}N_2O$	14.09	1.12
$C_{12}H_{22}N_3$	14.46	0.97
$C_{13}H_4O_3$	14.23	1.54
$C_{13}H_6NO_2$	14.60	1.39
$C_{13}H_8N_2O$	14.98	1.24
$C_{13}H_{10}N_3$	15.35	1.10
$C_{13}H_{20}O_2$	14.45	1.37
$C_{13}H_{22}NO$	14.82	1.22
$C_{13}H_{24}N_2$	15.20	1.08
$C_{14}H_8O_2$	15.34	1.50
$C_{14}H_{10}NO$	15.71	1.35
$C_{14}H_{12}N_2$	16.08	1.21
$C_{14}H_{24}O$	15.55	1.33
$C_{14}H_{26}N$	15.93	1.19
$C_{15}H_{12}O$	16.44	1.46
$C_{15}H_{14}N$	16.81	1.33
$C_{15}H_{28}$	16.66	1.30
$C_{16}H_2N$	17.70	1.47
$C_{16}H_{16}$	17.55	1.45
$C_{17}H_4$	18.43	1.60
209		
$C_9H_9N_2O_4$	10.79	1.33
$C_9H_{11}N_3O_3$	11.16	1.17
$C_9H_{13}N_4O_2$	11.54	1.01
$C_{10}HN_4O_2$	12.43	1.11
$C_{10}H_{11}NO_4$	11.52	1.41
$C_{10}H_{13}N_2O_3$	11.89	1.25
$C_{10}H_{15}N_3O_2$	12.27	1.09
$C_{10}H_{17}N_4O$	12.64	0.94
$C_{11}HN_2O_3$	12.78	1.35
$C_{11}H_3N_3O_2$	13.16	1.20
$C_{11}H_5N_4O$	13.53	1.05
$C_{11}H_{13}O_4$	12.25	1.49
$C_{11}H_{15}NO_3$	12.63	1.33
$C_{11}H_{17}N_2O_2$	13.00	1.18
$C_{11}H_{19}N_3O$	13.37	1.03
$C_{11}H_{21}N_4$	13.75	0.88
$C_{12}HO_4$	13.14	1.60
$C_{12}H_3NO_3$	13.51	1.44
$C_{12}H_5N_2O_2$	13.89	1.29
$C_{12}H_7N_3O$	14.26	1.15
$C_{12}H_9N_4$	14.64	1.00
$C_{12}H_{17}O_3$	13.36	1.42
$C_{12}H_{19}NO_2$	13.73	1.27
$C_{12}H_{21}N_2O$	14.11	1.12
$C_{12}H_{23}N_3$	14.48	0.98
$C_{13}H_5O_3$	14.25	1.54
$C_{13}H_7NO_2$	14.62	1.39
$C_{13}H_9N_2O$	14.99	1.25
$C_{13}H_{11}N_3$	15.37	1.10

Appendix Table 5.1 (*continued*)

	M + 1	M + 2
$C_{13}H_{21}O_2$	14.46	1.37
$C_{13}H_{23}NO$	14.84	1.22
$C_{13}H_{25}N_2$	15.21	1.08
$C_{14}H_9O_2$	15.35	1.50
$C_{14}H_{11}NO$	15.73	1.35
$C_{14}H_{13}N_2$	16.10	1.21
$C_{14}H_{25}O$	15.57	1.33
$C_{14}H_{27}N$	15.94	1.19
$C_{15}HN_2$	16.99	1.35
$C_{15}H_{13}O$	16.46	1.47
$C_{15}H_{15}N$	16.83	1.33
$C_{15}H_{29}$	16.67	1.30
$C_{16}HO$	17.35	1.61
$C_{16}H_3N$	17.72	1.48
$C_{16}H_{17}$	17.56	1.45
$C_{17}H_5$	18.45	1.60
	210	
$C_9H_{10}N_2O_4$	10.81	1.33
$C_9H_{12}N_3O_3$	11.18	1.17
$C_9H_{14}N_4O_2$	11.55	1.01
$C_{10}H_2N_4O_2$	12.44	1.11
$C_{10}H_{12}NO_4$	11.54	1.41
$C_{10}H_{14}N_2O_3$	11.91	1.25
$C_{10}H_{16}N_3O_2$	12.29	1.09
$C_{10}H_{18}N_4O$	12.66	0.94
$C_{11}H_2N_2O_3$	12.80	1.35
$C_{11}H_4N_3O_2$	13.17	1.20
$C_{11}H_6N_4O$	13.55	1.05
$C_{11}H_{14}O_4$	12.27	1.49
$C_{11}H_{16}NO_3$	12.64	1.34
$C_{11}H_{18}N_2O_2$	13.02	1.18
$C_{11}H_{20}N_3O$	13.39	1.03
$C_{11}H_{22}N_4$	13.76	0.88
$C_{12}H_2O_4$	13.16	1.60
$C_{12}H_4NO_3$	13.53	1.45
$C_{12}H_6N_2O_2$	13.90	1.30
$C_{12}H_8N_3O$	14.28	1.15
$C_{12}H_{10}N_4$	14.65	1.00
$C_{12}H_{18}O_3$	13.37	1.43
$C_{12}H_{20}NO_2$	13.75	1.28
$C_{12}H_{22}N_2O$	14.12	1.13
$C_{12}H_{24}N_3$	14.50	0.98
$C_{13}H_6O_3$	14.26	1.54
$C_{13}H_8NO_2$	14.64	1.40
$C_{13}H_{10}N_2O$	15.01	1.25
$C_{13}H_{12}N_3$	15.38	1.11
$C_{13}H_{22}O_2$	14.48	1.37
$C_{13}H_{24}NO$	14.85	1.23
$C_{13}H_{26}N_2$	15.23	1.08
$C_{14}H_{10}O_2$	15.37	1.50
$C_{14}H_{12}NO$	15.74	1.36
$C_{14}H_{14}N_2$	16.12	1.22
$C_{14}H_{26}O$	15.58	1.33
$C_{14}H_{28}N$	15.96	1.19
$C_{15}H_2N_2$	17.00	1.36
$C_{15}H_{14}O$	16.47	1.47
$C_{15}H_{16}N$	16.85	1.33
$C_{15}H_{30}$	16.69	1.31
$C_{16}H_2O$	17.36	1.61
$C_{16}H_4N$	17.74	1.48
$C_{16}H_{18}$	17.58	1.45
$C_{17}H_6$	18.47	1.61

	M + 1	M + 2
	211	
$C_9H_{11}N_2O_4$	10.82	1.33
$C_9H_{13}N_3O_3$	11.20	1.17
$C_9H_{15}N_4O_2$	11.57	1.01
$C_{10}HN_3O_3$	12.08	1.27
$C_{10}H_3N_4O_2$	12.46	1.12
$C_{10}H_{13}NO_4$	11.55	1.41
$C_{10}H_{15}N_2O_3$	11.93	1.25
$C_{10}H_{17}N_3O_2$	12.30	1.10
$C_{10}H_{19}N_4O$	12.68	0.94
$C_{11}HNO_4$	12.44	1.51
$C_{11}H_3N_2O_3$	12.82	1.36
$C_{11}H_5N_3O_2$	13.19	1.20
$C_{11}H_7N_4O$	13.56	1.05
$C_{11}H_{15}O_4$	12.28	1.49
$C_{11}H_{17}NO_3$	12.66	1.34
$C_{11}H_{19}N_2O_2$	13.03	1.18
$C_{11}H_{21}N_3O$	13.41	1.03
$C_{11}H_{23}N_4$	13.78	0.88
$C_{12}H_3O_4$	13.17	1.60
$C_{12}H_5NO_3$	13.55	1.45
$C_{12}H_7N_2O_2$	13.92	1.30
$C_{12}H_9N_3O$	14.30	1.15
$C_{12}H_{11}N_4$	14.67	1.60
$C_{12}H_{19}O_3$	13.39	1.43
$C_{12}H_{21}NO_2$	13.76	1.28
$C_{12}H_{23}N_2O$	14.14	1.13
$C_{12}H_{25}N_3$	14.51	0.98
$C_{13}H_7O_3$	14.28	1.54
$C_{13}H_9NO_2$	14.65	1.40
$C_{13}H_{11}N_2O$	15.03	1.25
$C_{13}H_{13}N_3$	15.40	1.11
$C_{13}H_{23}O_2$	14.49	1.38
$C_{13}H_{25}NO$	14.87	1.23
$C_{13}H_{27}N_2$	15.24	1.08
$C_{14}HN_3$	16.29	1.24
$C_{14}H_{11}O_2$	15.38	1.50
$C_{14}H_{13}NO$	15.76	1.36
$C_{14}H_{15}N_2$	16.13	1.22
$C_{14}H_{27}O$	15.60	1.34
$C_{14}H_{29}N$	15.97	1.19
$C_{15}HNO$	16.65	1.50
$C_{15}H_3N_2$	17.02	1.36
$C_{15}H_{15}O$	16.49	1.47
$C_{15}H_{17}N$	16.86	1.33
$C_{15}H_{31}$	16.71	1.31
$C_{16}H_3O$	17.38	1.62
$C_{16}H_5N$	17.75	1.48
$C_{16}H_{19}$	17.59	1.45
$C_{17}H_7$	18.48	1.61
	212	
$C_9H_{12}N_2O_4$	10.84	1.34
$C_9H_{14}N_3O_3$	11.21	1.18
$C_9H_{16}N_4O_2$	11.59	1.02
$C_{10}H_2N_3O_3$	12.10	1.27
$C_{10}H_4N_4O_2$	12.47	1.12
$C_{10}H_{14}NO_4$	11.57	1.41
$C_{10}H_{16}N_2O_3$	11.94	1.25
$C_{10}H_{18}N_3O_2$	12.32	1.10
$C_{10}H_{20}N_4O$	12.69	0.94
$C_{11}H_2NO_4$	12.46	1.51
$C_{11}H_4N_2O_3$	12.83	1.36

	M + 1	M + 2
$C_{11}H_6N_3O_2$	13.21	1.21
$C_{11}H_8N_4O$	13.58	1.06
$C_{11}H_{16}O_4$	12.30	1.49
$C_{11}H_{18}NO_3$	12.67	1.34
$C_{11}H_{20}N_2O_2$	13.05	1.19
$C_{11}H_{22}N_3O$	13.42	1.03
$C_{11}H_{24}N_4$	13.80	0.88
$C_{12}H_4O_4$	13.19	1.60
$C_{12}H_6NO_3$	13.56	1.45
$C_{12}H_8N_2O_2$	13.94	1.30
$C_{12}H_{10}N_3O$	14.31	1.15
$C_{12}H_{12}N_4$	14.69	1.01
$C_{12}H_{20}O_3$	13.41	1.43
$C_{12}H_{22}NO_2$	13.78	1.28
$C_{12}H_{24}N_2O$	14.15	1.13
$C_{12}H_{26}N_3$	14.53	0.98
$C_{13}H_8O_3$	14.29	1.55
$C_{13}H_{10}NO_2$	14.67	1.40
$C_{13}H_{12}N_2O$	15.04	1.25
$C_{13}H_{14}N_3$	15.42	1.11
$C_{13}H_{24}O_2$	14.51	1.38
$C_{13}H_{26}NO$	14.88	1.23
$C_{13}H_{28}N_2$	15.26	1.09
$C_{14}H_2N_3$	16.31	1.25
$C_{14}H_{12}O_2$	15.40	1.50
$C_{14}H_{14}NO$	15.77	1.36
$C_{14}H_{16}N_2$	16.15	1.22
$C_{14}H_{28}O$	15.62	1.34
$C_{14}H_{30}N$	15.99	1.20
$C_{15}H_2NO$	16.66	1.50
$C_{15}H_4N_2$	17.04	1.36
$C_{15}H_{16}O$	16.50	1.47
$C_{15}H_{18}N$	16.88	1.34
$C_{15}H_{32}$	16.72	1.31
$C_{16}H_4O$	17.39	1.62
$C_{16}H_6N$	17.77	1.48
$C_{16}H_{20}$	17.61	1.46
$C_{17}H_8$	18.50	1.61
	213	
$C_9H_3N_2O_4$	10.86	1.34
$C_9H_{15}N_3O_3$	11.23	1.18
$C_9H_{17}N_4O_2$	11.60	1.02
$C_{10}HN_2O_4$	11.74	1.43
$C_{10}H_3N_3O_2$	12.12	1.27
$C_{10}H_5N_4O_2$	12.49	1.12
$C_{10}H_{15}NO_4$	11.58	1.41
$C_{10}H_{17}N_2O_3$	11.96	1.26
$C_{10}H_{19}N_3O_2$	12.33	1.10
$C_{10}H_{21}N_4O$	12.71	0.95
$C_{11}H_3NO_4$	12.47	1.51
$C_{11}H_5N_2O_3$	12.85	1.36
$C_{11}H_7N_3O_2$	13.22	1.21
$C_{11}H_9N_4O$	13.60	1.06
$C_{11}H_{17}O_4$	12.32	1.50
$C_{11}H_{19}NO_3$	12.69	1.34
$C_{11}H_{21}N_2O_2$	13.06	1.19
$C_{11}H_{23}N_3O$	13.44	1.04
$C_{11}H_{25}N_4$	13.81	0.89
$C_{12}H_5O_4$	13.20	1.60
$C_{12}H_7NO_3$	13.58	1.45
$C_{12}H_9N_2O_2$	13.95	1.30
$C_{12}H_{11}N_3O$	14.33	1.15

	M + 1	M + 2
$C_{12}H_{13}N_4$	14.70	1.01
$C_{12}H_{21}O_3$	13.42	1.43
$C_{12}H_{23}NO_2$	13.80	1.28
$C_{12}H_{25}N_2O$	14.17	1.13
$C_{12}H_{27}N_3$	14.54	0.99
$C_{13}HN_4$	15.59	1.14
$C_{13}H_9O_3$	14.31	1.55
$C_{13}H_{11}NO_2$	14.68	1.40
$C_{13}H_{13}N_2O$	15.06	1.26
$C_{13}H_{15}N_3$	15.43	1.11
$C_{13}H_{25}O_2$	14.53	1.38
$C_{13}H_{27}NO$	14.90	1.23
$C_{13}H_{29}N_2$	15.28	1.09
$C_{14}HN_2O$	15.95	1.39
$C_{14}H_3N_3$	16.32	1.25
$C_{14}H_{13}O_2$	15.42	1.51
$C_{14}H_{15}NO$	15.79	1.36
$C_{14}H_{17}N_2$	16.16	1.22
$C_{14}H_{29}O$	15.63	1.34
$C_{14}H_{31}N$	16.01	1.20
$C_{15}HO_2$	16.30	1.64
$C_{15}H_3NO$	16.68	1.50
$C_{15}H_5N_2$	17.05	1.36
$C_{15}H_{17}O$	16.52	1.48
$C_{15}H_{19}N$	16.90	1.34
$C_{16}H_5O$	17.41	1.62
$C_{16}H_7N$	17.78	1.49
$C_{16}H_{21}$	17.63	1.46
$C_{17}H_9$	18.51	1.61
	214	
$C_9H_{14}N_2O_4$	10.87	1.34
$C_9H_{16}N_3O_3$	11.24	1.18
$C_9H_{18}N_4O_2$	11.62	1.02
$C_{10}H_2N_2O_4$	11.76	1.43
$C_{10}H_4N_3O_3$	12.13	1.28
$C_{10}H_6N_4O_2$	12.51	1.12
$C_{10}H_{16}NO_4$	11.60	1.42
$C_{10}H_{18}N_2O_3$	11.97	1.26
$C_{10}H_{20}N_3O_2$	12.35	1.10
$C_{10}H_{22}N_4O$	12.72	0.95
$C_{11}H_4NO_4$	12.49	1.52
$C_{11}H_6N_2O_3$	12.86	1.36
$C_{11}H_8N_3O_2$	13.24	1.21
$C_{11}H_{10}N_4O$	13.61	1.06
$C_{11}H_{18}O_4$	12.33	1.50
$C_{11}H_{20}NO_3$	12.71	1.34
$C_{11}H_{22}N_2O_2$	13.08	1.19
$C_{11}H_{24}N_3O$	13.45	1.04
$C_{11}H_{26}N_4$	13.83	0.89
$C_{12}H_6O_4$	13.22	1.61
$C_{12}H_8NO_3$	13.59	1.45
$C_{12}H_{10}N_2O_2$	13.97	1.31
$C_{12}H_{12}N_3O$	14.34	1.16
$C_{12}H_{14}N_4$	14.72	1.01
$C_{12}H_{22}O_3$	13.44	1.43
$C_{12}H_{24}NO_2$	13.81	1.28
$C_{12}H_{26}N_2O$	14.19	1.14
$C_{12}H_{28}N_3$	14.56	0.99
$C_{13}H_2N_4$	15.61	1.14
$C_{13}H_{10}O_3$	14.33	1.55
$C_{13}H_{12}NO_2$	14.70	1.40
$C_{13}H_{14}N_2O$	15.07	1.26

Appendix Table 5.1 (*continued*)

	M + 1	M + 2
$C_{13}H_{16}N_3$	15.45	1.12
$C_{13}H_{26}O_2$	14.54	1.38
$C_{13}H_{28}NO$	14.92	1.24
$C_{13}H_{30}N_2$	15.29	1.09
$C_{14}H_2N_2O$	15.96	1.39
$C_{14}H_4N_3$	16.34	1.25
$C_{14}H_{14}O_2$	15.43	1.51
$C_{14}H_{16}NO$	15.81	1.37
$C_{14}H_{18}N_2$	16.18	1.23
$C_{14}H_{30}O$	15.65	1.34
$C_{15}H_2O_2$	16.32	1.64
$C_{15}H_4NO$	16.69	1.51
$C_{15}H_6N_2$	17.07	1.37
$C_{15}H_{18}O$	16.54	1.48
$C_{15}H_{20}N$	16.91	1.34
$C_{16}H_6O$	17.43	1.63
$C_{16}H_8N$	17.80	1.49
$C_{16}H_{22}$	17.64	1.46
$C_{17}H_{10}$	18.53	1.62
215		
$C_9H_{15}N_2O_4$	10.89	1.34
$C_9H_{17}N_3O_3$	11.26	1.18
$C_9H_{19}N_4O_2$	11.63	1.02
$C_{10}H_3N_2O_4$	11.77	1.44
$C_{10}H_5N_3O_3$	12.15	1.28
$C_{10}H_7N_4O_2$	12.52	1.12
$C_{10}H_{17}NO_4$	11.62	1.42
$C_{10}H_{19}N_2O_3$	11.99	1.26
$C_{10}H_{21}N_3O_2$	12.37	1.10
$C_{10}H_{23}N_4O$	12.74	0.95
$C_{11}H_5NO_4$	12.50	1.52
$C_{11}H_7N_2O_3$	12.88	1.37
$C_{11}H_9N_3O_2$	13.25	1.21
$C_{11}H_{11}N_4O$	13.63	1.06
$C_{11}H_{19}O_4$	12.35	1.50
$C_{11}H_{21}NO_3$	12.72	1.35
$C_{11}H_{23}N_2O_2$	13.10	1.19
$C_{11}H_{25}N_3O$	13.47	1.04
$C_{11}H_{27}N_4$	13.84	0.89
$C_{12}H_7O_4$	13.24	1.61
$C_{12}H_9NO_3$	13.61	1.46
$C_{12}H_{11}N_2O_2$	13.98	1.31
$C_{12}H_{13}N_3O$	14.36	1.16
$C_{12}H_{15}N_4$	14.73	1.01
$C_{12}H_{23}O_3$	13.45	1.44
$C_{12}H_{25}NO_2$	13.83	1.29
$C_{12}H_{27}N_2O$	14.20	1.14
$C_{12}H_{29}N_3$	14.58	0.99
$C_{13}HN_3O$	15.25	1.28
$C_{13}H_3N_4$	15.62	1.14
$C_{13}H_{11}O_3$	14.34	1.55
$C_{13}H_{13}NO_2$	14.72	1.41
$C_{13}H_{15}N_2O$	15.09	1.26
$C_{13}H_{17}N_3$	15.46	1.12
$C_{13}H_{27}O_2$	14.56	1.38
$C_{13}H_{29}NO$	14.93	1.24
$C_{14}HNO_2$	15.60	1.54
$C_{14}H_3N_2O$	15.98	1.39
$C_{14}H_5N_3$	16.35	1.25
$C_{14}H_{15}O_2$	15.45	1.51
$C_{14}H_{17}NO$	15.82	1.37
$C_{14}H_{19}N_2$	16.20	1.23
$C_{15}H_3O_2$	16.34	1.65
$C_{15}H_5NO$	16.71	1.51
$C_{15}H_7N_2$	17.08	1.37
$C_{15}H_{19}O$	16.55	1.48
$C_{15}H_{21}N$	16.93	1.34
$C_{16}H_7O$	17.44	1.63
$C_{16}H_9N$	17.82	1.49
$C_{16}H_{23}$	17.66	1.47
$C_{17}H_{11}$	18.55	1.62
216		
$C_9H_{16}N_2O_4$	10.90	1.34
$C_9H_{18}N_3O_3$	11.28	1.18
$C_9H_{20}N_4O_2$	11.65	1.02
$C_{10}H_4N_2O_4$	11.79	1.44
$C_{10}H_6N_3O_3$	12.16	1.28
$C_{10}H_8N_4O_2$	12.54	1.13
$C_{10}H_{18}NO_4$	11.63	1.42
$C_{10}H_{20}N_2O_3$	12.01	1.26
$C_{10}H_{22}N_3O_2$	12.38	1.11
$C_{10}H_{24}N_4O$	12.76	0.95
$C_{11}H_6NO_4$	12.52	1.52
$C_{11}H_8N_2O_3$	12.90	1.37
$C_{11}H_{10}N_3O_2$	13.27	1.21
$C_{11}H_{12}N_4O$	13.64	1.06
$C_{11}H_{20}O_4$	12.36	1.50
$C_{11}H_{22}NO_3$	12.74	1.35
$C_{11}H_{24}N_2O_2$	13.11	1.19
$C_{11}H_{26}N_3O$	13.49	1.04
$C_{11}H_{28}N_4$	13.86	0.89
$C_{12}H_8O_4$	13.25	1.61
$C_{12}H_{10}NO_3$	13.63	1.46
$C_{12}H_{12}N_2O_2$	14.00	1.31
$C_{12}H_{14}N_3O$	14.38	1.16
$C_{12}H_{16}N_4$	14.75	1.01
$C_{12}H_{24}O_3$	13.47	1.44
$C_{12}H_{26}NO_2$	13.84	1.29
$C_{12}H_{28}N_2O$	14.22	1.14
$C_{13}H_2N_3O$	15.26	1.29
$C_{13}H_4N_4$	15.64	1.14
$C_{13}H_{12}O_3$	14.36	1.56
$C_{13}H_{14}NO_2$	14.73	1.41
$C_{13}H_{16}N_2O$	15.11	1.26
$C_{13}H_{18}N_3$	15.48	1.12
$C_{13}H_{28}O_2$	14.57	1.39
$C_{14}H_2NO_2$	15.62	1.54
$C_{14}H_4N_2O$	15.99	1.40
$C_{14}H_6N_3$	16.37	1.26
$C_{14}H_{16}O_2$	15.46	1.51
$C_{14}H_{18}NO$	15.84	1.37
$C_{14}H_{20}N_2$	16.21	1.23
$C_{15}H_4O_2$	16.35	1.65
$C_{15}H_6NO$	16.73	1.51
$C_{15}H_8N_2$	17.10	1.37
$C_{15}H_{20}O$	16.57	1.49
$C_{15}H_{22}N$	16.94	1.35
$C_{16}H_8O$	17.46	1.63
$C_{16}H_{10}N$	17.83	1.50
$C_{16}H_{24}$	17.67	1.47
$C_{17}H_{12}$	18.56	1.62
217		
$C_9H_{17}N_2O_4$	10.92	1.34
$C_9H_{19}N_3O_3$	11.29	1.18
$C_9H_{21}N_4O_2$	11.67	1.03
$C_{10}H_5N_2O_4$	11.81	1.44
$C_{10}H_7N_3O_3$	12.18	1.28
$C_{10}H_9N_4O_2$	12.55	1.13
$C_{10}H_{19}NO_4$	11.65	1.42
$C_{10}H_{21}N_2O_3$	12.02	1.26
$C_{10}H_{23}N_3O_2$	12.40	1.11
$C_{10}H_{25}N_4O$	12.77	0.95
$C_{11}H_7NO_4$	12.54	1.52
$C_{11}H_9N_2O_3$	12.91	1.37
$C_{11}H_{11}N_3O_2$	13.29	1.22
$C_{11}H_{13}N_4O$	13.66	1.07
$C_{11}H_{21}O_4$	12.38	1.50
$C_{11}H_{23}NO_3$	12.75	1.35
$C_{11}H_{25}N_2O_2$	13.13	1.20
$C_{11}H_{27}N_3O$	13.50	1.05
$C_{12}HN_4O$	14.55	1.19
$C_{12}H_9O_4$	13.27	1.61
$C_{12}H_{11}NO_3$	13.64	1.46
$C_{12}H_{13}N_2O_2$	14.02	1.31
$C_{12}H_{15}N_3O$	14.39	1.16
$C_{12}H_{17}N_4$	14.77	1.02
$C_{12}H_{25}O_3$	13.49	1.44
$C_{12}H_{27}NO_2$	13.86	1.29
$C_{13}HN_2O_2$	14.91	1.43
$C_{13}H_3N_3O$	15.28	1.29
$C_{13}H_5N_4$	15.65	1.15
$C_{13}H_{13}O_3$	14.37	1.56
$C_{13}H_{15}NO_2$	14.75	1.41
$C_{13}H_{17}N_2O$	15.12	1.27
$C_{13}H_{19}N_3$	15.50	1.12
$C_{14}HO_3$	15.26	1.68
$C_{14}H_3NO_2$	15.64	1.54
$C_{14}H_5N_2O$	16.01	1.40
$C_{14}H_7N_3$	16.39	1.26
$C_{14}H_{17}O_2$	15.48	1.52
$C_{14}H_{19}NO$	15.58	1.37
$C_{14}H_{21}N_2$	16.23	1.23
$C_{15}H_5O_2$	16.37	1.65
$C_{15}H_7NO$	16.74	1.51
$C_{15}H_9N_2$	17.12	1.38
$C_{15}H_{21}O$	16.58	1.49
$C_{15}H_{23}N$	16.96	1.35
$C_{16}H_9O$	17.47	1.63
$C_{16}H_{11}N$	17.85	1.50
$C_{16}H_{25}$	17.69	1.47
$C_{17}H_{13}$	18.58	1.63
$C_{18}H$	19.47	1.79
218		
$C_9H_{18}N_2O_4$	10.93	1.35
$C_9H_{20}N_3O_3$	11.31	1.19
$C_9H_{22}N_4O_2$	11.68	1.03
$C_{10}H_6N_2O_4$	11.82	1.44
$C_{10}H_8N_3O_3$	12.20	1.28
$C_{10}H_{10}N_4O_2$	12.57	1.13
$C_{10}H_{20}NO_4$	11.66	1.42
$C_{10}H_{22}N_2O_3$	12.04	1.27
$C_{10}H_{24}N_3O_2$	12.41	1.11
$C_{10}H_{26}N_4O$	12.79	0.96
$C_{11}H_8NO_4$	12.55	1.52
$C_{11}H_{10}N_2O_3$	12.93	1.37
$C_{11}H_{12}N_3O_2$	13.30	1.22
$C_{11}H_{14}N_4O$	13.68	1.07
$C_{11}H_{22}O_4$	12.40	1.51
$C_{11}H_{24}NO_3$	12.77	1.35
$C_{11}H_{26}N_2O_2$	13.14	1.20
$C_{12}H_2N_4O$	14.56	1.19
$C_{12}H_{10}O_4$	13.28	1.61
$C_{12}H_{12}NO_3$	13.66	1.46
$C_{12}H_{14}N_2O_2$	14.03	1.31
$C_{12}H_{16}N_3O$	14.41	1.17
$C_{12}H_{18}N_4$	14.78	1.02
$C_{12}H_{26}O_3$	13.50	1.44
$C_{13}H_2N_2O_2$	14.92	1.44
$C_{13}H_4N_3O$	15.30	1.29
$C_{13}H_6N_4$	15.67	1.15
$C_{13}H_{14}O_3$	14.39	1.56
$C_{13}H_{16}NO_2$	14.76	1.41
$C_{13}H_{18}N_2O$	15.14	1.27
$C_{13}H_{20}N_3$	15.51	1.13
$C_{14}H_2O_3$	15.28	1.69
$C_{14}H_4NO_2$	15.65	1.54
$C_{14}H_6N_2O$	16.03	1.40
$C_{14}H_8N_3$	16.40	1.26
$C_{14}H_{18}O_2$	15.50	1.52
$C_{14}H_{20}NO$	15.87	1.38
$C_{14}H_{22}N_2$	16.24	1.24
$C_{15}H_6O_2$	16.38	1.66
$C_{15}H_8NO$	16.76	1.52
$C_{15}H_{10}N_2$	17.13	1.38
$C_{15}H_{22}O$	16.60	1.49
$C_{15}H_{24}N$	16.98	1.35
$C_{16}H_{10}O$	17.49	1.64
$C_{16}H_{12}N$	17.86	1.50
$C_{16}H_{26}$	17.71	1.47
$C_{17}H_{14}$	18.59	1.63
$C_{18}H_2$	19.48	1.79
219		
$C_9H_{19}N_2O_4$	10.95	1.35
$C_9H_{21}N_3O_3$	11.32	1.19
$C_9H_{23}N_4O_2$	11.70	1.03
$C_{10}H_7N_2O_4$	11.84	1.44
$C_{10}H_9N_3O_3$	12.21	1.29
$C_{10}H_{11}N_4O_2$	12.59	1.13
$C_{10}H_{21}NO_4$	11.68	1.42
$C_{10}H_{23}N_2O_3$	12.05	1.27
$C_{10}H_{25}N_3O_2$	12.43	1.11
$C_{11}H_9NO_4$	12.57	1.53
$C_{11}H_{11}N_2O_3$	12.94	1.37
$C_{11}H_{13}N_3O_2$	13.32	1.22
$C_{11}H_{15}N_4O$	13.69	1.07
$C_{11}H_{23}O_4$	12.41	1.51
$C_{11}H_{25}NO_3$	12.79	1.35
$C_{12}HN_3O_2$	14.21	1.34
$C_{12}H_3N_4O$	14.58	1.19
$C_{12}H_{11}O_4$	13.30	1.62
$C_{12}H_{13}NO_3$	13.67	1.47
$C_{12}H_{15}N_2O_2$	14.05	1.32
$C_{12}H_{17}N_3O$	14.42	1.17
$C_{12}H_{19}N_4$	14.80	1.02
$C_{13}HNO_3$	14.56	1.59
$C_{13}H_3N_2O_2$	14.94	1.44
$C_{13}H_5N_3O$	15.31	1.29

Appendix Table 5.1 (*continued*)

	M + 1	M + 2
$C_{13}H_7N_4$	15.69	1.15
$C_{13}H_{15}O_3$	14.41	1.56
$C_{13}H_{17}NO_2$	14.78	1.42
$C_{13}H_{19}N_2O$	15.15	1.27
$C_{13}H_{21}N_3$	15.53	1.13
$C_{14}H_3O_3$	15.29	1.69
$C_{14}H_5NO_2$	15.67	1.55
$C_{14}H_7N_2O$	16.04	1.40
$C_{14}H_9N_3$	16.42	1.26
$C_{14}H_{19}O_2$	15.51	1.52
$C_{14}H_{21}NO$	15.89	1.38
$C_{14}H_{23}N_2$	16.26	1.24
$C_{15}H_7O_2$	16.40	1.66
$C_{15}H_9NO$	16.77	1.52
$C_{15}H_{11}N_2$	17.15	1.38
$C_{15}H_{23}O$	16.62	1.49
$C_{15}H_{25}N$	16.99	1.36
$C_{16}H_{11}O$	17.51	1.64
$C_{16}H_{13}N$	17.88	1.50
$C_{16}H_{27}$	17.72	1.48
$C_{17}HN$	18.77	1.66
$C_{17}H_{15}$	18.61	1.63
$C_{18}H_3$	19.50	1.80
	220	
$C_9H_{20}N_2O_4$	10.97	1.35
$C_9H_{22}N_3O_3$	11.34	1.19
$C_9H_{24}N_4O_2$	11.71	1.03
$C_{10}H_8N_2O_4$	11.85	1.44
$C_{10}H_{10}N_3O_3$	12.23	1.29
$C_{10}H_{12}N_4O_2$	12.60	1.13
$C_{10}H_{22}NO_4$	11.70	1.43
$C_{10}H_{24}N_2O_3$	12.07	1.27
$C_{11}H_{10}NO_4$	12.58	1.53
$C_{11}H_{12}N_2O_3$	12.96	1.38
$C_{11}H_{14}N_3O_2$	13.33	1.22
$C_{11}H_{16}N_4O$	13.71	1.07
$C_{11}H_{24}O_4$	12.43	1.51
$C_{12}H_2N_3O_2$	14.22	1.34
$C_{12}H_4N_4O$	14.60	1.19
$C_{12}H_{12}O_4$	13.32	1.62
$C_{12}H_{14}NO_3$	13.69	1.47
$C_{12}H_{16}N_2O_2$	14.06	1.32
$C_{12}H_{18}N_3O$	14.44	1.17
$C_{12}H_{20}N_4$	14.81	1.02
$C_{13}H_2NO_3$	14.58	1.59
$C_{13}H_4N_2O_2$	14.95	1.44
$C_{13}H_6N_3O$	15.33	1.30
$C_{13}H_8N_4$	15.70	1.15
$C_{13}H_{16}O_3$	14.42	1.57
$C_{13}H_{18}NO_2$	14.80	1.42
$C_{13}H_{20}N_2O$	15.17	1.27
$C_{13}H_{22}N_3$	15.54	1.13
$C_{14}H_4O_3$	15.31	1.69
$C_{14}H_6NO_2$	15.68	1.55
$C_{14}H_8N_2O$	16.06	1.41
$C_{14}H_{10}N_3$	16.43	1.27
$C_{14}H_{20}O_2$	15.53	1.52
$C_{14}H_{22}NO$	15.90	1.38
$C_{14}H_{24}N_2$	16.28	1.24
$C_{15}H_8O_2$	16.42	1.66
$C_{15}H_{10}NO$	16.79	1.52
$C_{15}H_{12}N_2$	17.16	1.38
$C_{15}H_{24}O$	16.63	1.50
$C_{15}H_{26}N$	17.01	1.36
$C_{16}H_{12}O$	17.52	1.64
$C_{16}H_{14}N$	17.90	1.51
$C_{16}H_{28}$	17.74	1.48
$C_{17}H_2N$	18.78	1.66
$C_{17}H_{16}$	18.63	1.64
$C_{18}H_4$	19.52	1.80
	221	
$C_9H_{21}N_2O_4$	10.98	1.35
$C_9H_{23}N_3O_3$	11.36	1.19
$C_{10}H_9N_2O_4$	11.87	1.45
$C_{10}H_{11}N_3O_3$	12.24	1.29
$C_{10}H_{13}N_4O_2$	12.62	1.14
$C_{10}H_{23}NO_4$	11.71	1.43
$C_{11}HN_4O_2$	13.51	1.25
$C_{11}H_{11}NO_4$	12.60	1.53
$C_{11}H_{13}N_2O_3$	12.98	1.38
$C_{11}H_{15}N_3O_2$	13.35	1.23
$C_{11}H_{17}N_4O$	13.72	1.08
$C_{12}HN_2O_3$	13.86	1.49
$C_{12}H_3N_3O_2$	14.24	1.34
$C_{12}H_5N_4O$	14.61	1.20
$C_{12}H_{13}O_4$	13.33	1.62
$C_{12}H_{15}NO_3$	13.71	1.47
$C_{12}H_{17}N_2O_2$	14.08	1.32
$C_{12}H_{19}N_3O$	14.46	1.17
$C_{12}H_{21}N_4$	14.83	1.03
$C_{13}HO_4$	14.22	1.74
$C_{13}H_3NO_3$	14.60	1.59
$C_{13}H_5N_2O_2$	14.97	1.44
$C_{13}H_7N_3O$	15.34	1.30
$C_{13}H_9N_4$	15.72	1.16
$C_{13}H_{17}O_3$	14.44	1.57
$C_{13}H_{19}NO_2$	14.81	1.42
$C_{13}H_{21}N_2O$	15.19	1.28
$C_{13}H_{23}N_3$	15.56	1.13
$C_{14}H_5O_3$	15.33	1.69
$C_{14}H_7NO_2$	15.70	1.55
$C_{14}H_9N_2O$	16.07	1.41
$C_{14}H_{11}N_3$	16.45	1.27
$C_{14}H_{21}O_2$	15.54	1.53
$C_{14}H_{23}NO$	15.92	1.38
$C_{14}H_{25}N_2$	16.29	1.24
$C_{15}H_9O_2$	16.43	1.66
$C_{15}H_{11}NO$	16.81	1.52
$C_{15}H_{13}N_2$	17.18	1.39
$C_{15}H_{25}O$	16.65	1.50
$C_{15}H_{27}N$	17.02	1.36
$C_{16}HN_2$	18.07	1.54
$C_{16}H_{13}O$	17.54	1.64
$C_{16}H_{15}N$	17.91	1.51
$C_{16}H_{29}$	17.75	1.48
$C_{17}HO$	18.43	1.80
$C_{17}H_3N$	18.80	1.67
$C_{17}H_{17}$	18.64	1.64
$C_{18}H_5$	19.53	1.80
	222	
$C_9H_{22}N_2O_4$	11.00	1.35
$C_{10}H_{10}N_2O_4$	11.89	1.45
$C_{10}H_{12}N_3O_3$	12.26	1.29
$C_{10}H_{14}N_4O_2$	12.63	1.14
$C_{11}H_2N_4O_2$	13.52	1.25
$C_{11}H_{12}NO_4$	12.62	1.53
$C_{11}H_{14}N_2O_3$	12.99	1.38
$C_{11}H_{16}N_3O_2$	13.37	1.23
$C_{11}H_{18}N_4O$	13.74	1.08
$C_{12}H_2N_2O_3$	13.88	1.49
$C_{12}H_4N_3O_2$	14.25	1.34
$C_{12}H_6N_4O$	14.63	1.20
$C_{12}H_{14}O_4$	13.35	1.62
$C_{12}H_{16}NO_3$	13.72	1.47
$C_{12}H_{18}N_2O_2$	14.10	1.32
$C_{12}H_{20}N_3O$	14.47	1.18
$C_{12}H_{22}N_4$	14.85	1.03
$C_{13}H_2O_4$	14.24	1.74
$C_{13}H_4NO_3$	14.61	1.59
$C_{13}H_6N_2O_2$	14.99	1.45
$C_{13}H_8N_3O$	15.36	1.30
$C_{13}H_{10}N_4$	15.73	1.16
$C_{13}H_{18}O_3$	14.45	1.57
$C_{13}H_{20}NO_2$	14.83	1.42
$C_{13}H_{22}N_2O$	15.20	1.28
$C_{13}H_{24}N_3$	15.58	1.14
$C_{14}H_6O_3$	15.34	1.70
$C_{14}H_8NO_2$	15.72	1.55
$C_{14}H_{10}N_2O$	16.09	1.41
$C_{14}H_{12}N_3$	16.47	1.27
$C_{14}H_{22}O_2$	15.56	1.53
$C_{14}H_{24}NO$	15.93	1.39
$C_{14}H_{26}N_2$	16.31	1.25
$C_{15}H_{10}O_2$	16.45	1.67
$C_{15}H_{12}NO$	16.82	1.53
$C_{15}H_{14}N_2$	17.20	1.39
$C_{15}H_{26}O$	16.66	1.50
$C_{15}H_{28}N$	17.04	1.36
$C_{16}H_2N_2$	18.09	1.54
$C_{16}H_{14}O$	17.55	1.65
$C_{16}H_{16}N$	17.93	1.51
$C_{16}H_{30}$	17.77	1.49
$C_{17}H_2O$	18.44	1.80
$C_{17}H_4N$	18.82	1.67
$C_{17}H_{18}$	18.66	1.64
$C_{18}H_6$	19.55	1.81
	223	
$C_{10}H_{11}N_2O_4$	11.90	1.45
$C_{10}H_{13}N_3O_3$	12.28	1.29
$C_{10}H_{15}N_4O_2$	12.65	1.14
$C_{11}HN_3O_3$	13.16	1.40
$C_{11}H_3N_4O_2$	13.54	1.25
$C_{11}H_{13}NO_4$	12.63	1.53
$C_{11}H_{15}N_2O_3$	13.01	1.38
$C_{11}H_{17}N_3O_2$	13.38	1.23
$C_{11}H_{19}N_4O$	13.76	1.08
$C_{12}HNO_4$	13.52	1.65
$C_{12}H_3N_2O_3$	13.90	1.50
$C_{12}H_5N_3O_2$	14.27	1.35
$C_{12}H_7N_4O$	14.64	1.20
$C_{12}H_{15}O_4$	13.36	1.62
$C_{12}H_{17}NO_3$	13.74	1.47
$C_{12}H_{19}N_2O_2$	14.11	1.33
$C_{12}H_{21}N_3O$	14.49	1.18
$C_{12}H_{23}N_4$	14.86	1.03
$C_{13}H_3O_4$	14.25	1.74
$C_{13}H_5NO_3$	14.63	1.59
$C_{13}H_7N_2O_2$	15.00	1.45
$C_{13}H_9N_3O$	15.38	1.30
$C_{13}H_{11}N_4$	15.75	1.16
$C_{13}H_{19}O_3$	14.47	1.57
$C_{13}H_{21}NO_2$	14.84	1.43
$C_{13}H_{23}N_2O$	15.22	1.28
$C_{13}H_{25}N_3$	15.59	1.14
$C_{14}H_7O_3$	15.36	1.70
$C_{14}H_9NO_2$	15.73	1.56
$C_{14}H_{11}N_2O$	16.11	1.41
$C_{14}H_{13}N_3$	16.48	1.27
$C_{14}H_{23}O_2$	15.58	1.53
$C_{14}H_{25}NO$	15.95	1.39
$C_{14}H_{27}N_2$	16.32	1.25
$C_{15}HN_3$	17.37	1.42
$C_{15}H_{11}O_2$	16.46	1.67
$C_{15}H_{13}NO$	16.84	1.53
$C_{15}H_{15}N_2$	17.21	1.39
$C_{15}H_{27}O$	16.68	1.50
$C_{15}H_{29}N$	17.06	1.37
$C_{16}HNO$	17.73	1.68
$C_{16}H_3N_2$	18.10	1.54
$C_{16}H_{15}O$	17.57	1.65
$C_{16}H_{17}N$	17.94	1.52
$C_{16}H_{31}$	17.79	1.49
$C_{17}H_3O$	18.46	1.80
$C_{17}H_5N$	18.83	1.67
$C_{17}H_{19}$	18.67	1.64
$C_{18}H_7$	19.56	1.81
	224	
$C_{10}H_{12}N_2O_4$	11.92	1.45
$C_{10}H_{14}N_3O_3$	12.29	1.30
$C_{10}H_{16}N_4O_2$	12.67	1.14
$C_{11}H_2N_3O_3$	13.18	1.40
$C_{11}H_4N_4O_2$	13.56	1.25
$C_{11}H_{14}NO_4$	12.65	1.54
$C_{11}H_{16}N_2O_3$	13.02	1.38
$C_{11}H_{18}N_3O_2$	13.40	1.23
$C_{11}H_{20}N_4O$	13.77	1.08
$C_{12}H_2NO_4$	13.54	1.65
$C_{12}H_4N_2O_3$	13.91	1.50
$C_{12}H_6N_3O_2$	14.29	1.35
$C_{12}H_8N_4O$	14.66	1.20
$C_{12}H_{16}O_4$	13.38	1.63
$C_{12}H_{18}NO_3$	13.75	1.48
$C_{12}H_{20}N_2O_2$	14.13	1.33
$C_{12}H_{22}N_3O$	14.50	1.18
$C_{12}H_{24}N_4$	14.88	1.03
$C_{13}H_4O_4$	14.27	1.74
$C_{13}H_6NO_3$	14.64	1.60
$C_{13}H_8N_2O_2$	15.02	1.45
$C_{13}H_{10}N_3O$	15.39	1.31
$C_{13}H_{12}N_4$	15.77	1.16
$C_{13}H_{20}O_3$	14.49	1.57
$C_{13}H_{22}NO_2$	14.86	1.43
$C_{13}H_{24}N_2O$	15.23	1.28
$C_{13}H_{26}N_3$	15.61	1.14
$C_{14}H_8O_3$	15.37	1.70
$C_{14}H_{10}NO_2$	15.75	1.56
$C_{14}H_{12}N_2O$	16.12	1.42
$C_{14}H_{14}N_3$	16.50	1.28
$C_{14}H_{24}O_2$	15.59	1.53

Appendix Table 5.1 (*continued*)

	M + 1	M + 2
$C_{14}H_{26}NO$	15.97	1.39
$C_{14}H_{28}N_2$	16.34	1.25
$C_{15}H_2N_3$	17.39	1.42
$C_{15}H_{12}O_2$	16.48	1.67
$C_{15}H_{14}NO$	16.85	1.53
$C_{15}H_{16}N_2$	17.23	1.40
$C_{15}H_{28}O$	16.70	1.51
$C_{15}H_{30}N$	17.07	1.37
$C_{16}H_2NO$	17.74	1.68
$C_{16}H_4N_2$	18.12	1.55
$C_{16}H_{16}O$	17.59	1.65
$C_{16}H_{18}N$	17.96	1.52
$C_{16}H_{32}$	17.80	1.49
$C_{17}H_4O$	18.47	1.81
$C_{17}H_6N$	18.85	1.68
$C_{17}H_{20}$	18.69	1.65
$C_{18}H_8$	19.58	1.81
225		
$C_{10}H_{13}N_2O_4$	11.93	1.45
$C_{10}H_{15}N_3O_3$	12.31	1.30
$C_{10}H_{17}N_4O_2$	12.68	1.14
$C_{11}HN_2O_4$	12.82	1.56
$C_{11}H_3N_3O_3$	13.20	1.41
$C_{11}H_5N_4O_2$	13.57	1.25
$C_{11}H_{15}NO_4$	12.66	1.54
$C_{11}H_{17}N_2O_3$	13.04	1.39
$C_{11}H_{19}N_3O_2$	13.41	1.23
$C_{11}H_{21}N_4O$	13.79	1.08
$C_{12}H_3NO_4$	13.55	1.65
$C_{12}H_5N_2O_3$	13.93	1.50
$C_{12}H_7N_3O_2$	14.30	1.35
$C_{12}H_9N_4O$	14.68	1.20
$C_{12}H_{17}O_4$	13.40	1.63
$C_{12}H_{19}NO_3$	13.77	1.48
$C_{12}H_{21}N_2O_2$	14.14	1.33
$C_{12}H_{23}N_3O$	14.52	1.18
$C_{12}H_{25}N_4$	14.89	1.04
$C_{13}H_5O_4$	14.28	1.75
$C_{13}H_7NO_3$	14.66	1.60
$C_{13}H_9N_2O_2$	15.03	1.45
$C_{13}H_{11}N_3O$	15.41	1.31
$C_{13}H_{13}N_4$	15.78	1.17
$C_{13}H_{21}O_3$	14.50	1.58
$C_{13}H_{23}NO_2$	14.88	1.43
$C_{13}H_{25}N_2O$	15.25	1.29
$C_{13}H_{27}N_3$	15.62	1.14
$C_{14}HN_4$	16.67	1.30
$C_{14}H_9O_3$	15.39	1.70
$C_{14}H_{11}NO_2$	15.76	1.56
$C_{14}H_{13}N_2O$	16.14	1.42
$C_{14}H_{15}N_3$	16.51	1.28
$C_{14}H_{25}O_2$	15.61	1.54
$C_{14}H_{27}NO$	15.98	1.39
$C_{14}H_{29}N_2$	16.36	1.25
$C_{15}HN_2O$	17.03	1.56
$C_{15}H_3N_3$	17.40	1.42
$C_{15}H_{13}O_2$	16.50	1.67
$C_{15}H_{15}NO$	16.87	1.54
$C_{15}H_{17}N_2$	17.24	1.40
$C_{15}H_{29}O$	16.71	1.51
$C_{15}H_{31}N$	17.09	1.37
$C_{16}HO_2$	17.38	1.82

	M + 1	M + 2
$C_{16}H_3NO$	17.76	1.68
$C_{16}H_5N_2$	18.13	1.55
$C_{16}H_{17}O$	17.60	1.66
$C_{16}H_{19}N$	17.98	1.52
$C_{16}H_{33}$	17.82	1.49
$C_{17}H_5O$	18.49	1.81
$C_{17}H_7N$	18.86	1.68
$C_{17}H_{21}$	18.71	1.65
$C_{18}H_9$	19.60	1.81
226		
$C_{10}H_{14}N_2O_4$	11.95	1.46
$C_{10}H_{16}N_3O_3$	12.32	1.30
$C_{10}H_{18}N_4O_2$	12.70	1.15
$C_{11}H_2N_2O_4$	12.84	1.56
$C_{11}H_4N_3O_3$	13.21	1.41
$C_{11}H_6N_4O_2$	13.59	1.26
$C_{11}H_{16}NO_4$	12.68	1.54
$C_{11}H_{18}N_2O_3$	13.06	1.39
$C_{11}H_{20}N_3O_2$	13.43	1.24
$C_{11}H_{22}N_4O$	13.80	1.09
$C_{12}H_4NO_4$	13.57	1.65
$C_{12}H_6N_2O_3$	13.94	1.50
$C_{12}H_8N_3O_2$	14.32	1.35
$C_{12}H_{10}N_4O$	14.69	1.21
$C_{12}H_{18}O_4$	13.41	1.63
$C_{12}H_{20}NO_3$	13.79	1.48
$C_{12}H_{22}N_2O_2$	14.16	1.33
$C_{12}H_{24}N_3O$	14.54	1.18
$C_{12}H_{26}N_4$	14.91	1.04
$C_{13}H_6O_4$	14.30	1.75
$C_{13}H_8NO_3$	14.68	1.60
$C_{13}H_{10}N_2O_2$	15.05	1.46
$C_{13}H_{12}N_3O$	15.42	1.31
$C_{13}H_{14}N_4$	15.80	1.17
$C_{13}H_{22}O_3$	14.52	1.58
$C_{13}H_{24}NO_2$	14.89	1.43
$C_{13}H_{26}N_2O$	15.27	1.29
$C_{13}H_{28}N_3$	15.64	1.15
$C_{14}H_2N_4$	16.69	1.31
$C_{14}H_{10}O_3$	15.41	1.71
$C_{14}H_{12}NO_2$	15.78	1.56
$C_{14}H_{14}N_2O$	16.15	1.42
$C_{14}H_{16}N_3$	16.53	1.28
$C_{14}H_{26}O_2$	15.62	1.54
$C_{14}H_{28}NO$	16.00	1.40
$C_{14}H_{30}N_2$	16.37	1.26
$C_{15}H_2N_2O$	17.04	1.56
$C_{15}H_4N_3$	17.42	1.43
$C_{15}H_{14}O_2$	16.51	1.68
$C_{15}H_{16}NO$	16.89	1.54
$C_{15}H_{18}N_2$	17.26	1.40
$C_{15}H_{30}O$	16.73	1.51
$C_{15}H_{32}N$	17.10	1.37
$C_{16}H_2O_2$	17.40	1.82
$C_{16}H_4NO$	17.77	1.69
$C_{16}H_6N_2$	18.15	1.55
$C_{16}H_{18}O$	17.62	1.66
$C_{16}H_{20}N$	17.99	1.52
$C_{16}H_{34}$	17.83	1.50
$C_{17}H_6O$	18.51	1.81
$C_{17}H_8N$	18.88	1.68
$C_{17}H_{22}$	18.72	1.65

	M + 1	M + 2
$C_{18}H_{10}$	19.61	1.82
227		
$C_{10}H_{15}N_2O_4$	11.97	1.46
$C_{10}H_{17}N_3O_3$	12.34	1.30
$C_{10}H_{19}N_4O_2$	12.71	1.15
$C_{11}H_3N_2O_4$	12.85	1.56
$C_{11}H_5N_3O_3$	13.23	1.41
$C_{11}H_7N_4O_2$	13.60	1.26
$C_{11}H_{17}NO_4$	12.70	1.54
$C_{11}H_{19}N_2O_3$	13.07	1.39
$C_{11}H_{21}N_3O_2$	13.45	1.24
$C_{11}H_{23}N_4O$	13.82	1.09
$C_{12}H_5NO_4$	13.59	1.65
$C_{12}H_7N_2O_3$	13.96	1.50
$C_{12}H_9N_3O_2$	14.33	1.36
$C_{12}H_{11}N_4O$	14.71	1.21
$C_{12}H_{19}O_4$	13.43	1.63
$C_{12}H_{21}NO_3$	13.80	1.48
$C_{12}H_{23}N_2O_2$	14.18	1.33
$C_{12}H_{25}N_3O$	14.55	1.19
$C_{12}H_{27}N_4$	14.93	1.04
$C_{13}H_7O_4$	14.32	1.75
$C_{13}H_9NO_3$	14.69	1.60
$C_{13}H_{11}N_2O_2$	15.07	1.46
$C_{13}H_{13}N_3O$	15.44	1.31
$C_{13}H_{15}N_4$	15.81	1.17
$C_{13}H_{23}O_3$	14.53	1.58
$C_{13}H_{25}NO_2$	14.91	1.44
$C_{13}H_{27}N_2O$	15.28	1.29
$C_{13}H_{29}N_3$	15.66	1.15
$C_{14}HN_3O$	16.33	1.45
$C_{14}H_3N_4$	16.70	1.31
$C_{14}H_{11}O_3$	15.42	1.71
$C_{14}H_{13}NO_2$	15.80	1.57
$C_{14}H_{15}N_2O$	16.17	1.42
$C_{14}H_{17}N_3$	16.55	1.28
$C_{14}H_{27}O_2$	15.64	1.54
$C_{14}H_{29}NO$	16.01	1.40
$C_{14}H_{31}N_2$	16.39	1.26
$C_{15}HNO_2$	16.69	1.70
$C_{15}H_3N_2O$	17.06	1.57
$C_{15}H_5N_3$	17.43	1.43
$C_{15}H_{15}O_2$	16.53	1.68
$C_{15}H_{17}NO$	16.90	1.54
$C_{15}H_{19}N_2$	17.28	1.40
$C_{15}H_{31}O$	16.74	1.51
$C_{15}H_{33}N$	17.12	1.38
$C_{16}H_3O_2$	17.42	1.82
$C_{16}H_5NO$	17.79	1.69
$C_{16}H_7N_2$	18.17	1.55
$C_{16}H_{19}O$	17.63	1.66
$C_{16}H_{21}N$	18.01	1.53
$C_{17}H_7O$	18.52	1.82
$C_{17}H_9N$	18.90	1.69
$C_{17}H_{23}$	18.74	1.66
$C_{18}H_{11}$	19.63	1.82
228		
$C_{10}H_{16}N_2O_4$	11.98	1.46
$C_{10}H_{18}N_3O_3$	12.36	1.30
$C_{10}H_{20}N_4O_2$	12.73	1.15
$C_{11}H_4N_2O_4$	12.87	1.56

	M + 1	M + 2
$C_{11}H_6N_3O_3$	13.24	1.41
$C_{11}H_8N_4O_2$	13.62	1.26
$C_{11}H_{18}NO_4$	12.71	1.55
$C_{11}H_{20}N_2O_3$	13.09	1.39
$C_{11}H_{22}N_3O_2$	13.46	1.24
$C_{11}H_{24}N_4O$	13.84	1.09
$C_{12}H_6NO_4$	13.60	1.66
$C_{12}H_8N_2O_3$	13.98	1.51
$C_{12}H_{10}N_3O_2$	14.35	1.36
$C_{12}H_{12}N_4O$	14.72	1.21
$C_{12}H_{20}O_4$	13.44	1.64
$C_{12}H_{22}NO_3$	13.82	1.49
$C_{12}H_{24}N_2O_2$	14.19	1.34
$C_{12}H_{26}N_3O$	14.57	1.19
$C_{12}H_{28}N_4$	14.94	1.04
$C_{13}H_8O_4$	14.33	1.75
$C_{13}H_{10}NO_3$	14.71	1.61
$C_{13}H_{12}N_2O_2$	15.08	1.46
$C_{13}H_{14}N_3O$	15.46	1.32
$C_{13}H_{16}N_4$	15.83	1.17
$C_{13}H_{24}O_3$	14.55	1.58
$C_{13}H_{26}NO_2$	14.92	1.44
$C_{13}H_{28}N_2O$	15.30	1.29
$C_{13}H_{30}N_3$	15.67	1.15
$C_{14}H_2N_3O$	16.34	1.45
$C_{14}H_4N_4$	16.72	1.31
$C_{14}H_{12}O_3$	15.44	1.71
$C_{14}H_{14}NO_2$	15.81	1.57
$C_{14}H_{16}N_2O$	16.19	1.43
$C_{14}H_{18}N_3$	16.56	1.29
$C_{14}H_{28}O_2$	15.66	1.54
$C_{14}H_{30}NO$	16.03	1.40
$C_{14}H_{32}N_2$	16.40	1.26
$C_{15}H_2NO_2$	16.70	1.71
$C_{15}H_4N_2O$	17.08	1.57
$C_{15}H_6N_3$	17.45	1.43
$C_{15}H_{16}O_2$	16.54	1.68
$C_{15}H_{18}NO$	16.92	1.54
$C_{15}H_{20}N_2$	17.29	1.41
$C_{15}H_{32}O$	16.76	1.52
$C_{16}H_4O_2$	17.43	1.83
$C_{16}H_6NO$	17.81	1.69
$C_{16}H_8N_2$	18.18	1.56
$C_{16}H_{20}O$	17.65	1.66
$C_{16}H_{22}N$	18.02	1.53
$C_{17}H_8O$	18.54	1.82
$C_{17}H_{10}N$	18.91	1.69
$C_{17}H_{24}$	18.75	1.66
$C_{18}H_{12}$	19.64	1.82
229		
$C_{10}H_{17}N_2O_4$	12.00	1.46
$C_{10}H_{19}N_3O_3$	12.37	1.31
$C_{10}H_{21}N_4O_2$	12.75	1.15
$C_{11}H_5N_2O_4$	12.89	1.57
$C_{11}H_7N_3O_3$	13.26	1.41
$C_{11}H_9N_4O_2$	13.64	1.26
$C_{11}H_{19}NO_4$	12.73	1.55
$C_{11}H_{21}N_2O_3$	13.10	1.39
$C_{11}H_{23}N_3O_2$	13.48	1.24
$C_{11}H_{25}N_4O$	13.85	1.09
$C_{12}H_7NO_4$	13.62	1.66
$C_{12}H_9N_2O_3$	13.99	1.51

Appendix Table 5.1 (*continued*)

	M + 1	M + 2
$C_{12}H_{11}N_3O_2$	14.37	1.36
$C_{12}H_{13}N_4O$	14.74	1.21
$C_{12}H_{21}O_4$	13.46	1.64
$C_{12}H_{23}NO_3$	13.83	1.49
$C_{12}H_{25}N_2O_2$	14.21	1.34
$C_{12}H_{27}N_3O$	14.58	1.19
$C_{12}H_{29}N_4$	14.96	1.05
$C_{13}HN_4O$	15.63	1.34
$C_{13}H_9O_4$	14.35	1.76
$C_{13}H_{11}NO_3$	14.72	1.61
$C_{13}H_{13}N_2O_2$	15.10	1.46
$C_{13}H_{15}N_3O$	15.47	1.32
$C_{13}H_{17}N_4$	15.85	1.18
$C_{13}H_{25}O_3$	14.57	1.59
$C_{13}H_{27}NO_2$	14.94	1.44
$C_{13}H_{29}N_2O$	15.31	1.30
$C_{13}H_{31}N_3$	15.69	1.15
$C_{14}HN_2O_2$	15.99	1.60
$C_{14}H_3N_3O$	16.36	1.45
$C_{14}H_5N_4$	16.73	1.32
$C_{14}H_{13}O_3$	15.45	1.71
$C_{14}H_{15}NO_2$	15.83	1.57
$C_{14}H_{17}N_2O$	16.20	1.43
$C_{14}H_{19}N_3$	16.58	1.29
$C_{14}H_{29}O_2$	15.67	1.55
$C_{14}H_{31}NO$	16.05	1.41
$C_{15}HO_3$	16.34	1.85
$C_{15}H_3NO_2$	16.72	1.71
$C_{15}H_5N_2O$	17.09	1.57
$C_{15}H_7N_3$	17.47	1.44
$C_{15}H_{17}O_2$	16.56	1.68
$C_{15}H_{19}NO$	16.93	1.55
$C_{15}H_{21}N_2$	17.31	1.41
$C_{16}H_5O_2$	17.45	1.83
$C_{16}H_7NO$	17.82	1.69
$C_{16}H_9N_2$	18.20	1.56
$C_{16}H_{21}O$	17.67	1.67
$C_{16}H_{23}N$	18.04	1.53
$C_{17}H_9O$	18.55	1.82
$C_{17}H_{11}N$	18.93	1.69
$C_{17}H_{25}$	18.77	1.66
$C_{18}H_{13}$	19.66	1.83
$C_{19}H$	20.55	2.00
	230	
$C_{10}H_{18}N_2O_4$	12.01	1.46
$C_{10}H_{20}N_3O_3$	12.39	1.31
$C_{10}H_{22}N_4O_2$	12.76	1.15
$C_{11}H_6N_2O_4$	12.90	1.57
$C_{11}H_8N_3O_3$	13.28	1.42
$C_{11}H_{10}N_4O_2$	13.65	1.27
$C_{11}H_{20}NO_4$	12.74	1.55
$C_{11}H_{22}N_2O_3$	13.12	1.40
$C_{11}H_{24}N_3O_2$	13.49	1.24
$C_{11}H_{26}N_4O$	13.87	1.09
$C_{12}H_8NO_4$	13.63	1.66
$C_{12}H_{10}N_2O_3$	14.01	1.51
$C_{12}H_{12}N_3O_2$	14.38	1.36
$C_{12}H_{14}N_4O$	14.76	1.22
$C_{12}H_{22}O_4$	13.48	1.64
$C_{12}H_{24}NO_3$	13.85	1.49
$C_{12}H_{26}N_2O_2$	14.22	1.34
$C_{12}H_{28}N_3O$	14.60	1.19
$C_{12}H_{30}N_4$	14.97	1.05
$C_{13}H_2N_4O$	15.65	1.35
$C_{13}H_{10}O_4$	14.36	1.76
$C_{13}H_{12}NO_3$	14.74	1.61
$C_{13}H_{14}N_2O_2$	15.11	1.47
$C_{13}H_{16}N_3O$	15.49	1.32
$C_{13}H_{18}N_4$	15.86	1.18
$C_{13}H_{26}O_3$	14.58	1.59
$C_{13}H_{28}NO_2$	14.96	1.44
$C_{13}H_{30}N_2O$	15.33	1.30
$C_{14}H_2N_2O_2$	16.00	1.60
$C_{14}H_4N_3O$	16.38	1.46
$C_{14}H_6N_4$	16.75	1.32
$C_{14}H_{14}O_3$	15.47	1.72
$C_{14}H_{16}NO_2$	15.84	1.57
$C_{14}H_{18}N_2O$	16.22	1.43
$C_{14}H_{20}N_3$	16.59	1.29
$C_{14}H_{30}O_2$	15.69	1.55
$C_{15}H_2O_3$	16.36	1.85
$C_{15}H_4NO_2$	16.73	1.71
$C_{15}H_6N_2O$	17.11	1.57
$C_{15}H_8N_3$	17.48	1.44
$C_{15}H_{18}O_2$	16.58	1.69
$C_{15}H_{20}NO$	16.95	1.55
$C_{15}H_{22}N_2$	17.32	1.41
$C_{16}H_6O_2$	17.46	1.83
$C_{16}H_8NO$	17.84	1.70
$C_{16}H_{10}N_2$	18.21	1.56
$C_{16}H_{22}O$	17.68	1.67
$C_{16}H_{24}N$	18.06	1.54
$C_{17}H_{10}O$	18.57	1.83
$C_{17}H_{12}N$	18.94	1.69
$C_{17}H_{26}$	18.79	1.67
$C_{18}H_{14}$	19.68	1.83
$C_{\cdot 9}H_2$	20.56	2.00
	231	
$C_{10}H_{19}N_2O_4$	12.03	1.47
$C_{10}H_{21}N_3O_3$	12.40	1.31
$C_{10}H_{23}N_4O_2$	12.78	1.16
$C_{11}H_7N_2O_4$	12.92	1.57
$C_{11}H_9N_3O_3$	13.29	1.42
$C_{11}H_{11}N_4O_2$	13.67	1.27
$C_{11}H_{21}NO_4$	12.76	1.55
$C_{11}H_{23}N_2O_3$	13.14	1.40
$C_{11}H_{25}N_3O_2$	13.51	1.25
$C_{11}H_{27}N_4O$	13.88	1.10
$C_{12}H_9NO_4$	13.65	1.66
$C_{12}H_{11}N_2O_3$	14.02	1.51
$C_{12}H_{13}N_3O_2$	14.40	1.37
$C_{12}H_{15}N_4O$	14.77	1.22
$C_{12}H_{23}O_4$	13.49	1.64
$C_{12}H_{25}NO_3$	13.87	1.49
$C_{12}H_{27}N_2O_2$	14.24	1.34
$C_{12}H_{29}N_3O$	14.62	1.20
$C_{13}HN_3O_2$	15.29	1.49
$C_{13}H_3N_4O$	15.66	1.35
$C_{13}H_{11}O_4$	14.38	1.76
$C_{13}H_{13}NO_3$	14.76	1.61
$C_{13}H_{15}N_2O_2$	15.13	1.47
$C_{13}H_{17}N_3O$	15.50	1.32
$C_{13}H_{19}N_4$	15.88	1.18
$C_{13}H_{27}O_3$	14.60	1.59
$C_{13}H_{29}NO_2$	14.97	1.45
$C_{14}HNO_3$	15.64	1.74
$C_{14}H_3N_2O_2$	16.02	1.60
$C_{14}H_5N_3O$	16.39	1.46
$C_{14}H_7N_4$	16.77	1.32
$C_{14}H_{15}O_3$	15.49	1.72
$C_{14}H_{17}NO_2$	15.86	1.58
$C_{14}H_{19}N_2O$	16.23	1.44
$C_{14}H_{21}N_3$	16.61	1.30
$C_{15}H_3O_3$	16.37	1.85
$C_{15}H_5NO_2$	16.75	1.72
$C_{15}H_7N_2O$	17.12	1.58
$C_{15}H_9N_3$	17.50	1.44
$C_{15}H_{19}O_2$	16.59	1.69
$C_{15}H_{21}NO$	16.97	1.55
$C_{15}H_{23}N_2$	17.34	1.41
$C_{16}H_7O_2$	17.48	1.84
$C_{16}H_9NO$	17.85	1.70
$C_{16}H_{11}N_2$	18.23	1.57
$C_{16}H_{23}O$	17.70	1.67
$C_{16}H_{25}N$	18.07	1.54
$C_{17}H_{11}O$	18.59	1.83
$C_{17}H_{13}N$	18.96	1.70
$C_{17}H_{27}$	18.80	1.67
$C_{18}HN$	19.85	1.86
$C_{18}H_{15}$	19.69	1.83
$C_{19}H_3$	20.58	2.01
	232	
$C_{10}H_{20}N_2O_4$	12.05	1.47
$C_{10}H_{22}N_3O_3$	12.42	1.31
$C_{10}H_{24}N_4O_2$	12.79	1.16
$C_{11}H_8N_2O_4$	12.93	1.57
$C_{11}H_{10}N_3O_3$	13.31	1.42
$C_{11}H_{12}N_4O_2$	13.68	1.27
$C_{11}H_{22}NO_4$	12.78	1.55
$C_{11}H_{24}N_2O_3$	13.15	1.40
$C_{11}H_{26}N_3O_2$	13.53	1.25
$C_{11}H_{28}N_4O$	13.90	1.10
$C_{12}H_{10}NO_4$	13.67	1.66
$C_{12}H_{12}N_2O$	14.04	1.52
$C_{12}H_{14}N_3O_2$	14.41	1.37
$C_{12}H_{16}N_4O$	14.79	1.22
$C_{12}H_{24}O_4$	13.51	1.64
$C_{12}H_{26}NO_3$	13.88	1.49
$C_{12}H_{28}N_2O_2$	14.26	1.35
$C_{13}H_2N_3O_2$	15.30	1.49
$C_{13}H_4N_4O$	15.68	1.35
$C_{13}H_{12}O_4$	14.40	1.76
$C_{13}H_{14}NO_3$	14.77	1.62
$C_{13}H_{16}N_2O_2$	15.15	1.47
$C_{13}H_{18}N_3O$	15.52	1.33
$C_{13}H_{20}N_4$	15.89	1.18
$C_{13}H_{28}O_3$	14.61	1.59
$C_{14}H_2NO_3$	15.66	1.75
$C_{14}H_4N_2O_2$	16.03	1.60
$C_{14}H_6N_3O$	16.41	1.46
$C_{14}H_8N_4$	16.78	1.32
$C_{14}H_{16}O_3$	15.50	1.72
$C_{14}H_{18}NO_2$	15.88	1.58
$C_{14}H_{20}N_2O$	16.25	1.44
$C_{14}H_{22}N_3$	16.63	1.30
$C_{15}H_4O_3$	16.39	1.86
$C_{15}H_6NO_2$	16.77	1.72
$C_{15}H_8N_2O$	17.14	1.58
$C_{15}H_{10}N_3$	17.51	1.44
$C_{15}H_{20}O_2$	16.61	1.69
$C_{15}H_{22}NO$	16.98	1.55
$C_{15}H_{24}N_2$	17.36	1.42
$C_{16}H_8O_2$	17.50	1.84
$C_{16}H_{10}NO$	17.87	1.70
$C_{16}H_{12}N_2$	18.25	1.57
$C_{16}H_{24}O$	17.71	1.68
$C_{16}H_{26}N$	18.09	1.54
$C_{17}H_{12}O$	18.60	1.83
$C_{17}H_{14}N$	18.98	1.70
$C_{17}H_{28}$	18.82	1.67
$C_{18}H_2N$	19.86	1.87
$C_{18}H_{16}$	19.71	1.84
$C_{19}H_4$	20.60	2.01
	233	
$C_{10}H_{21}N_2O_4$	12.06	1.47
$C_{10}H_{23}N_3O_3$	12.44	1.31
$C_{10}H_{25}N_4O_2$	12.81	1.16
$C_{11}H_9N_2O_4$	12.95	1.57
$C_{11}H_{11}N_3O_3$	13.32	1.42
$C_{11}H_{13}N_4O_2$	13.70	1.27
$C_{11}H_{23}NO_4$	12.79	1.56
$C_{11}H_{25}N_2O_3$	13.17	1.40
$C_{11}H_{27}N_3O_2$	13.54	1.25
$C_{12}HN_4O_2$	14.59	1.39
$C_{12}H_{11}NO_4$	13.68	1.67
$C_{12}H_{13}N_2O_3$	14.06	1.52
$C_{12}H_{15}N_3O_2$	14.43	1.37
$C_{12}H_{17}N_4O$	14.80	1.22
$C_{12}H_{25}O_4$	13.52	1.65
$C_{12}H_{27}NO_3$	13.90	1.50
$C_{13}HN_2O_3$	14.94	1.64
$C_{13}H_3N_3O_2$	15.32	1.50
$C_{13}H_5N_4O$	15.69	1.35
$C_{13}H_{13}O_4$	14.41	1.76
$C_{13}H_{15}NO_3$	14.79	1.62
$C_{13}H_{17}N_2O_2$	15.16	1.47
$C_{13}H_{19}N_3O$	15.54	1.33
$C_{13}H_{21}N_4$	15.91	1.19
$C_{14}HO_4$	15.30	1.89
$C_{14}H_3NO_3$	15.68	1.75
$C_{14}H_5N_2O_2$	16.05	1.61
$C_{14}H_7N_3O$	16.42	1.47
$C_{14}H_9N_4$	16.80	1.33
$C_{14}H_{17}O_3$	15.52	1.72
$C_{14}H_{19}NO_2$	15.89	1.58
$C_{14}H_{21}N_2O$	16.27	1.44
$C_{14}H_{23}N_3$	16.64	1.30
$C_{15}H_5O_3$	16.41	1.86
$C_{15}H_7NO_2$	16.78	1.72
$C_{15}H_9N_2O$	17.16	1.58
$C_{15}H_{11}N_3$	17.53	1.45
$C_{15}H_{21}O_2$	16.62	1.70
$C_{15}H_{23}NO$	17.00	1.56
$C_{15}H_{25}N_2$	17.37	1.42
$C_{16}H_9O_2$	17.51	1.84
$C_{16}H_{11}NO$	17.89	1.71
$C_{16}H_{13}N_2$	18.26	1.57
$C_{16}H_{25}O$	17.73	1.68
$C_{16}H_{27}N$	18.10	1.54

Appendix Table 5.1 (*continued*)

	M + 1	M + 2
$C_{17}HN_2$	19.15	1.73
$C_{17}H_{13}O$	18.62	1.83
$C_{17}H_{15}N$	18.99	1.70
$C_{17}H_{29}$	18.83	1.67
$C_{18}HO$	19.51	2.00
$C_{18}H_3N$	19.88	1.87
$C_{18}H_{17}$	19.72	1.84
$C_{19}H_5$	20.61	2.01
	234	
$C_{10}H_{22}N_2O_4$	12.08	1.47
$C_{10}H_{24}N_3O_3$	12.45	1.32
$C_{10}H_{26}N_4O_2$	12.83	1.16
$C_{11}H_{10}N_2O_4$	12.97	1.58
$C_{11}H_{12}N_3O_3$	13.34	1.42
$C_{11}H_{14}N_4O_2$	13.72	1.27
$C_{11}H_{24}NO_4$	12.81	1.56
$C_{11}H_{26}N_2O_3$	13.18	1.40
$C_{12}H_2N_4O_2$	14.60	1.39
$C_{12}H_{12}NO_4$	13.70	1.67
$C_{12}H_{14}N_2O_3$	14.07	1.52
$C_{12}H_{16}N_3O_2$	14.45	1.37
$C_{12}H_{18}N_4O$	14.82	1.23
$C_{12}H_{26}O_4$	13.54	1.65
$C_{13}H_2N_2O_3$	14.96	1.64
$C_{13}H_4N_3O_2$	15.33	1.50
$C_{13}H_6N_4O$	15.71	1.36
$C_{13}H_{14}O_4$	14.43	1.77
$C_{13}H_{16}NO_3$	14.80	1.62
$C_{13}H_{18}N_2O_2$	15.18	1.48
$C_{13}H_{20}N_3O$	15.55	1.33
$C_{13}H_{22}N_4$	15.93	1.19
$C_{14}H_2O_4$	15.32	1.89
$C_{14}H_4NO_3$	15.69	1.75
$C_{14}H_6N_2O_2$	16.07	1.61
$C_{14}H_8N_3O$	16.44	1.47
$C_{14}H_{10}N_4$	16.81	1.33
$C_{14}H_{18}O_3$	15.53	1.73
$C_{14}H_{20}NO_2$	15.91	1.58
$C_{14}H_{22}N_2O$	16.28	1.44
$C_{14}H_{24}N_3$	16.66	1.30
$C_{15}H_6O_3$	16.42	1.86
$C_{15}H_8NO_2$	16.80	1.72
$C_{15}H_{10}N_2O$	17.17	1.59
$C_{15}H_{12}N_3$	17.55	1.45
$C_{15}H_{22}O_2$	16.64	1.70
$C_{15}H_{24}NO$	17.01	1.56
$C_{15}H_{26}N_2$	17.39	1.42
$C_{16}H_{10}O_2$	17.53	1.84
$C_{16}H_{12}NO$	17.90	1.71
$C_{16}H_{14}N_2$	18.28	1.58
$C_{16}H_{26}O$	17.75	1.68
$C_{16}H_{28}N$	18.12	1.55
$C_{17}H_2N_2$	19.17	1.74
$C_{17}H_{14}O$	18.63	1.84
$C_{17}H_{16}N$	19.01	1.71
$C_{17}H_{30}$	18.85	1.68
$C_{18}H_2O$	19.52	2.00
$C_{18}H_4N$	19.90	1.87
$C_{18}H_{18}$	19.74	1.84
$C_{19}H_6$	20.63	2.02
	235	
$C_{10}H_{23}N_2O_4$	12.09	1.47
$C_{10}H_{25}N_3O_3$	12.47	1.32
$C_{11}H_{11}N_2O_4$	12.98	1.58
$C_{11}H_{13}N_3O_3$	13.36	1.43
$C_{11}H_{15}N_4O_2$	13.73	1.28
$C_{11}H_{25}NO_4$	12.82	1.56
$C_{12}HN_3O_3$	14.25	1.54
$C_{12}H_3N_4O_2$	14.62	1.40
$C_{12}H_{13}NO_4$	13.71	1.67
$C_{12}H_{15}N_2O_3$	14.09	1.52
$C_{12}H_{17}N_3O_2$	14.46	1.37
$C_{12}H_{19}N_4O$	14.84	1.23
$C_{13}HNO_4$	14.60	1.79
$C_{13}H_3N_2O_3$	14.98	1.65
$C_{13}H_5N_3O_2$	15.35	1.50
$C_{13}H_7N_4O$	15.73	1.36
$C_{13}H_{15}O_4$	14.44	1.77
$C_{13}H_{17}NO_3$	14.82	1.62
$C_{13}H_{19}N_2O_2$	15.19	1.48
$C_{13}H_{21}N_3O$	15.57	1.33
$C_{13}H_{23}N_4$	15.94	1.19
$C_{14}H_3O_4$	15.33	1.90
$C_{14}H_5NO_3$	15.71	1.75
$C_{14}H_7N_2O_2$	16.08	1.61
$C_{14}H_9N_3O$	16.46	1.47
$C_{14}H_{11}N_4$	16.83	1.33
$C_{14}H_{19}O_3$	15.55	1.73
$C_{14}H_{21}NO_2$	15.92	1.59
$C_{14}H_{23}N_2O$	16.30	1.45
$C_{14}H_{25}N_3$	16.67	1.31
$C_{15}H_7O_3$	16.44	1.86
$C_{15}H_9NO_2$	16.81	1.73
$C_{15}H_{11}N_2O$	17.19	1.59
$C_{15}H_{13}N_3$	17.56	1.44
$C_{15}H_{23}O_2$	16.66	1.70
$C_{15}H_{25}NO$	17.03	1.56
$C_{15}H_{27}N_2$	17.40	1.43
$C_{16}HN_3$	18.45	1.61
$C_{16}H_{11}O_2$	17.54	1.85
$C_{16}H_{13}NO$	17.92	1.71
$C_{16}H_{15}N_2$	18.29	1.58
$C_{16}H_{27}O$	17.76	1.68
$C_{16}H_{29}N$	18.14	1.55
$C_{17}HNO$	18.81	1.87
$C_{17}H_3N_2$	19.18	1.74
$C_{17}H_{15}O$	18.65	1.84
$C_{17}H_{17}N$	19.02	1.71
$C_{17}H_{31}$	18.87	1.68
$C_{18}H_3O$	19.54	2.00
$C_{18}H_3N$	19.91	1.88
$C_{18}H_{19}$	19.76	1.85
$C_{19}H_7$	20.64	2.02
	236	
$C_{10}H_{24}N_2O_4$	12.11	1.48
$C_{11}H_{12}N_2O_4$	13.00	1.58
$C_{11}H_{14}N_3O_3$	13.37	1.43
$C_{11}H_{16}N_4O_2$	13.75	1.28
$C_{12}H_2N_3O_3$	14.26	1.55
$C_{12}H_4N_4O_2$	14.64	1.40
$C_{12}H_{14}NO_4$	13.73	1.67
$C_{12}H_{16}N_2O_3$	14.10	1.52
$C_{12}H_{18}N_3O_2$	14.48	1.38
$C_{12}H_{20}N_4O$	14.85	1.23
$C_{13}H_2NO_4$	14.62	1.79
$C_{13}H_4N_2O_3$	14.99	1.65
$C_{13}H_6N_3O_2$	15.37	1.50
$C_{13}H_8N_4O$	15.74	1.36
$C_{13}H_{16}O_4$	14.46	1.77
$C_{13}H_{18}NO_3$	14.84	1.63
$C_{13}H_{20}N_2O_2$	15.21	1.48
$C_{13}H_{22}N_3O$	15.58	1.34
$C_{13}H_{24}N_4$	15.96	1.19
$C_{14}H_4O_4$	15.35	1.90
$C_{14}H_6NO_3$	15.72	1.76
$C_{14}H_8N_2O_2$	16.10	1.61
$C_{14}H_{10}N_3O$	16.47	1.47
$C_{14}H_{12}N_4$	16.85	1.33
$C_{14}H_{20}O_3$	15.57	1.73
$C_{14}H_{22}NO_2$	15.94	1.59
$C_{14}H_{24}N_2O$	16.31	1.45
$C_{14}H_{26}N_3$	16.69	1.31
$C_{15}H_8O_3$	16.45	1.87
$C_{15}H_{10}NO_2$	16.83	1.73
$C_{15}H_{12}N_2O$	17.20	1.59
$C_{15}H_{14}N_3$	17.58	1.46
$C_{15}H_{24}O_2$	16.67	1.70
$C_{15}H_{26}NO$	17.05	1.56
$C_{15}H_{28}N_2$	17.42	1.43
$C_{16}H_2N_3$	18.47	1.61
$C_{16}H_{12}O_2$	17.56	1.85
$C_{16}H_{14}NO$	17.93	1.71
$C_{16}H_{16}N_2$	18.31	1.58
$C_{16}H_{28}O$	17.78	1.69
$C_{16}H_{30}N$	18.15	1.55
$C_{17}H_2NO$	18.82	1.87
$C_{17}H_4N_2$	19.20	1.74
$C_{17}H_{16}O$	18.67	1.84
$C_{17}H_{18}N$	19.04	1.71
$C_{17}H_{32}$	18.88	1.68
$C_{18}H_4O$	19.55	2.01
$C_{18}H_6N$	19.93	1.88
$C_{18}H_{20}$	19.77	1.85
$C_{19}H_8$	20.66	2.02
	237	
$C_{11}H_{13}N_2O_4$	13.01	1.58
$C_{11}H_{15}N_3O_3$	13.39	1.43
$C_{11}H_{17}N_4O_2$	13.76	1.28
$C_{12}HN_2O_4$	13.90	1.70
$C_{12}H_3N_3O_3$	19.28	1.55
$C_{12}H_5N_4O_2$	14.65	1.40
$C_{12}H_{15}NO_4$	13.75	1.68
$C_{12}H_{17}N_2O_3$	14.12	1.53
$C_{12}H_{19}N_3O_2$	14.49	1.38
$C_{12}H_{21}N_4O$	14.87	1.23
$C_{13}H_3NO_4$	14.63	1.80
$C_{13}H_5N_2O_3$	15.01	1.65
$C_{13}H_7N_3O_2$	15.38	1.51
$C_{13}H_9N_4O$	15.76	1.36
$C_{13}H_{17}O_4$	14.48	1.77
$C_{13}H_{19}NO_3$	14.85	1.63
$C_{13}H_{21}N_2O_2$	15.23	1.48
$C_{13}H_{23}N_3O$	15.60	1.34
$C_{13}H_{25}N_4$	15.97	1.20
$C_{14}H_5O_4$	15.37	1.90
$C_{14}H_7NO_3$	15.74	1.76
$C_{14}H_9N_2O_2$	16.11	1.62
$C_{14}H_{11}N_3O$	16.49	1.48
$C_{14}H_3N_4$	16.86	1.34
$C_{14}H_{21}O_3$	15.58	1.73
$C_{14}H_{23}NO_2$	15.96	1.59
$C_{14}H_{25}N_2O$	16.33	1.45
$C_{14}H_{27}N_3$	16.71	1.31
$C_{15}HN_4$	17.75	1.49
$C_{15}H_9O_3$	16.47	1.87
$C_{15}H_{11}NO_2$	16.85	1.73
$C_{15}H_{13}N_2O$	17.22	1.59
$C_{15}H_{15}N_3$	17.59	1.46
$C_{15}H_{25}O_2$	16.69	1.71
$C_{15}H_{27}NO$	17.06	1.57
$C_{15}H_{29}N_2$	17.44	1.43
$C_{16}HN_2O$	18.11	1.75
$C_{16}H_3N_3$	18.48	1.61
$C_{16}H_{13}O_2$	17.58	1.85
$C_{16}H_{15}NO$	17.95	1.72
$C_{16}H_{17}N_2$	18.33	1.58
$C_{16}H_{29}O$	17.79	1.69
$C_{16}H_{31}N$	18.17	1.56
$C_{17}HO_2$	18.46	2.01
$C_{17}H_3NO$	18.84	1.88
$C_{17}H_5N_2$	19.21	1.75
$C_{17}H_{17}O$	18.68	1.85
$C_{17}H_{19}N$	19.06	1.72
$C_{17}H_{33}$	18.90	1.69
$C_{18}H_5O$	19.57	2.01
$C_{18}H_7N$	19.94	1.88
$C_{18}H_{21}$	19.79	1.85
$C_{19}H_9$	20.68	2.03
	238	
$C_{11}H_{14}N_2O_4$	13.03	1.59
$C_{11}H_{16}N_3O_3$	13.40	1.43
$C_{11}H_{18}N_4O_2$	13.78	1.28
$C_{12}H_2N_2O_4$	13.92	1.70
$C_{12}H_4N_3O_3$	14.29	1.55
$C_{12}H_6N_4O_2$	14.67	1.40
$C_{12}H_{16}NO_4$	13.76	1.68
$C_{12}H_{18}N_2O_3$	14.14	1.53
$C_{12}H_{20}N_3O_2$	14.51	1.38
$C_{12}H_{22}N_4O$	14.88	1.24
$C_{13}H_4NO_4$	14.65	1.80
$C_{13}H_6N_2O_3$	15.02	1.65
$C_{13}H_8N_3O_2$	15.40	1.51
$C_{13}H_{10}N_4O$	15.77	1.37
$C_{13}H_{18}O_4$	14.49	1.78
$C_{13}H_{20}NO_3$	14.87	1.63
$C_{13}H_{22}N_2O_2$	15.24	1.49
$C_{13}H_{24}N_3O$	15.62	1.34
$C_{13}H_{26}N_4$	15.99	1.20
$C_{14}H_6O_4$	15.38	1.90
$C_{14}H_8NO_3$	15.76	1.76
$C_{14}H_{10}N_2O_2$	16.13	1.62
$C_{14}H_{12}N_3O$	16.50	1.48
$C_{14}H_{14}N_4$	16.88	1.34
$C_{14}H_{22}O_3$	15.60	1.74
$C_{14}H_{24}NO_2$	15.97	1.59
$C_{14}H_{26}N_2O$	16.35	1.45
$C_{14}H_{28}N_3$	16.72	1.31
$C_{15}H_2N_4$	17.77	1.49
$C_{15}H_{10}O_3$	16.49	1.87

Appendix Table 5.1 (*continued*)

	M + 1	M + 2
$C_{15}H_{12}NO_2$	16.86	1.73
$C_{15}H_{14}N_2O$	17.24	1.60
$C_{15}H_{16}N_3$	17.61	1.46
$C_{15}H_{26}O_2$	16.70	1.71
$C_{15}H_{28}NO$	17.08	1.57
$C_{15}H_{30}N_2$	17.45	1.43
$C_{16}H_2N_2O$	18.12	1.75
$C_{16}H_4N_3$	18.50	1.62
$C_{16}H_{14}O_2$	17.59	1.85
$C_{16}H_{16}NO$	17.97	1.72
$C_{16}H_{18}N_2$	18.34	1.59
$C_{16}H_{30}O$	17.81	1.69
$C_{16}H_{32}N$	18.18	1.56
$C_{17}H_2O_2$	18.48	2.01
$C_{17}H_4NO$	18.86	1.88
$C_{17}H_6N_2$	19.23	1.75
$C_{17}H_{18}O$	18.70	1.85
$C_{17}H_{20}N$	19.07	1.72
$C_{17}H_{34}$	18.91	1.69
$C_{18}H_6O$	19.59	2.01
$C_{18}H_8N$	19.96	1.89
$C_{18}H_{22}$	19.80	1.86
$C_{19}H_{10}$	20.69	2.03
239		
$C_{11}H_{15}N_2O_4$	13.05	1.59
$C_{11}H_{17}N_3O_3$	13.42	1.44
$C_{11}H_{19}N_4O_2$	13.80	1.29
$C_{12}H_3N_2O_4$	13.93	1.70
$C_{12}H_5N_3O_3$	14.31	1.55
$C_{12}H_7N_4O_2$	14.68	1.41
$C_{12}H_{17}NO_4$	13.78	1.68
$C_{12}H_{19}N_2O_3$	14.15	1.53
$C_{12}H_{21}N_3O_2$	14.53	1.38
$C_{12}H_{23}N_4O$	14.90	1.24
$C_{13}H_5NO_4$	14.67	1.80
$C_{13}H_7N_2O_3$	15.04	1.66
$C_{13}H_9N_3O_2$	15.41	1.51
$C_{13}H_{11}N_4O$	15.79	1.37
$C_{13}H_{19}O_4$	14.51	1.78
$C_{13}H_{21}NO_3$	14.88	1.63
$C_{13}H_{23}N_2O_2$	15.26	1.49
$C_{13}H_{25}N_3O$	15.63	1.34
$C_{13}H_{27}N_4$	16.01	1.20
$C_{14}H_7O_4$	15.40	1.91
$C_{14}H_9NO_3$	15.77	1.76
$C_{14}H_{11}N_2O_2$	16.15	1.62
$C_{14}H_{13}N_3O$	16.52	1.48
$C_{14}H_{15}N_4$	16.89	1.34
$C_{14}H_{23}O_3$	15.61	1.74
$C_{14}H_{25}NO_2$	15.99	1.60
$C_{14}H_{27}N_2O$	16.36	1.46
$C_{14}H_{29}N_3$	16.74	1.32
$C_{15}HN_3O$	17.41	1.63
$C_{15}H_3N_4$	17.78	1.49
$C_{15}H_{11}O_3$	16.50	1.88
$C_{15}H_{13}NO_2$	16.88	1.74
$C_{15}H_{15}N_2O$	17.25	1.60
$C_{15}H_{17}N_3$	17.63	1.46
$C_{15}H_{27}O_2$	16.72	1.71
$C_{15}H_{29}NO$	17.09	1.57
$C_{15}H_{31}N_2$	17.47	1.44
$C_{16}HNO_2$	17.77	1.88
$C_{16}H_3N_2O$	18.14	1.75
$C_{16}H_5N_3$	18.51	1.62
$C_{16}H_{15}O_2$	17.61	1.86
$C_{16}H_{17}NO$	17.98	1.72
$C_{16}H_{19}N_2$	18.36	1.59
$C_{16}H_{31}O$	17.83	1.70
$C_{16}H_{33}N$	18.20	1.56
$C_{17}H_3O_2$	18.50	2.01
$C_{17}H_5NO$	18.87	1.88
$C_{17}H_7N_2$	19.25	1.75
$C_{17}H_{19}O$	18.71	1.85
$C_{17}H_{21}N$	19.09	1.72
$C_{17}H_{35}$	18.93	1.69
$C_{18}H_7O$	19.60	2.02
$C_{18}H_9N$	19.98	1.89
$C_{18}H_{23}$	19.82	1.86
$C_{19}H_{11}$	20.71	2.03
240		
$C_{11}H_{16}N_2O_4$	13.06	1.59
$C_{11}H_{18}N_3O_3$	13.44	1.44
$C_{11}H_{20}N_4O_2$	13.81	1.29
$C_{12}H_4N_2O_4$	13.95	1.70
$C_{12}H_6N_3O_3$	14.33	1.56
$C_{12}H_8N_4O_2$	14.70	1.41
$C_{12}H_{18}NO_4$	13.79	1.68
$C_{12}H_{20}N_2O_3$	14.17	1.53
$C_{12}H_{22}N_3O_2$	14.54	1.39
$C_{12}H_{24}N_4O$	14.92	1.24
$C_{13}H_6NO_4$	14.68	1.80
$C_{13}H_8N_2O_3$	15.06	1.66
$C_{13}H_{10}N_3O_2$	15.43	1.51
$C_{13}H_{12}N_4O$	15.81	1.37
$C_{13}H_{20}O_4$	14.52	1.78
$C_{13}H_{22}NO_3$	14.90	1.63
$C_{13}H_{24}N_2O_2$	15.27	1.49
$C_{13}H_{26}N_3O$	15.65	1.35
$C_{13}H_{28}N_4$	16.02	1.20
$C_{14}H_8O_4$	15.41	1.91
$C_{14}H_{10}NO_3$	15.79	1.77
$C_{14}H_{12}N_2O_2$	16.16	1.62
$C_{14}H_{14}N_3O$	16.54	1.48
$C_{14}H_{16}N_4$	16.91	1.35
$C_{14}H_{24}O_3$	15.63	1.74
$C_{14}H_{26}NO_2$	16.00	1.60
$C_{14}H_{28}N_2O$	16.38	1.46
$C_{14}H_{30}N_3$	16.75	1.32
$C_{15}H_2N_3O$	17.42	1.63
$C_{15}H_4N_4$	17.80	1.49
$C_{15}H_{12}O_3$	16.52	1.88
$C_{15}H_{14}NO_2$	16.89	1.74
$C_{15}H_{16}N_2O$	17.27	1.60
$C_{15}H_{18}N_3$	17.64	1.47
$C_{15}H_{28}O_2$	16.74	1.71
$C_{15}H_{30}NO$	17.11	1.58
$C_{15}H_{32}N_2$	17.48	1.44
$C_{16}H_2NO_2$	17.78	1.89
$C_{16}H_4N_2O$	18.16	1.75
$C_{16}H_6N_3$	18.53	1.62
$C_{16}H_{16}O_2$	17.62	1.86
$C_{16}H_{18}NO$	18.00	1.73
$C_{16}H_{20}N_2$	18.37	1.59
$C_{16}H_{32}O$	17.84	1.70
$C_{16}H_{34}N$	18.22	1.56
$C_{17}H_4O_2$	18.51	2.02
$C_{17}H_6NO$	18.89	1.88
$C_{17}H_8N_2$	19.26	1.75
$C_{17}H_{20}O$	18.73	1.86
$C_{17}H_{22}N$	19.10	1.72
$C_{17}H_{36}$	18.95	1.70
$C_{18}H_8O$	19.62	2.02
$C_{18}H_{10}N$	19.99	1.89
$C_{18}H_{24}$	19.84	1.86
$C_{19}H_{12}$	20.72	2.04
241		
$C_{11}H_{17}N_2O_4$	13.08	1.59
$C_{11}H_{19}N_3O_3$	13.45	1.44
$C_{11}H_{21}N_4O_2$	13.83	1.29
$C_{12}H_5N_2O_4$	13.97	1.71
$C_{12}H_7N_3O_3$	14.34	1.56
$C_{12}H_9N_4O_2$	14.72	1.41
$C_{12}H_{19}NO_4$	13.81	1.68
$C_{12}H_{21}N_2O_3$	14.18	1.54
$C_{12}H_{23}N_3O_2$	14.56	1.39
$C_{12}H_{25}N_4O$	14.93	1.24
$C_{13}H_7NO_4$	14.70	1.81
$C_{13}H_9N_2O_3$	15.07	1.66
$C_{13}H_{11}N_3O_2$	15.45	1.52
$C_{13}H_{13}N_4O$	15.82	1.37
$C_{13}H_{21}O_4$	14.54	1.78
$C_{13}H_{23}NO_3$	14.92	1.64
$C_{13}H_{25}N_2O_2$	15.29	1.49
$C_{13}H_{27}N_3O$	15.66	1.35
$C_{13}H_{29}N_4$	16.04	1.21
$C_{14}HN_4O$	16.71	1.51
$C_{14}H_9O_4$	15.43	1.91
$C_{14}H_{11}NO_3$	15.80	1.77
$C_{14}H_{13}N_2O_2$	16.18	1.63
$C_{14}H_{15}N_3O$	16.55	1.49
$C_{14}H_{17}N_4$	16.93	1.35
$C_{14}H_{25}O_3$	15.65	1.74
$C_{14}H_{27}NO_2$	16.02	1.60
$C_{14}H_{29}N_2O$	16.39	1.46
$C_{14}H_{31}N_3$	16.77	1.32
$C_{15}HN_2O_2$	17.07	1.77
$C_{15}H_3N_3O$	17.44	1.63
$C_{15}H_5N_4$	17.82	1.50
$C_{15}H_{13}O_3$	16.53	1.88
$C_{15}H_{15}NO_2$	16.91	1.74
$C_{15}H_{17}N_2O$	17.28	1.60
$C_{15}H_{19}N_3$	17.66	1.47
$C_{15}H_{29}O_2$	16.75	1.72
$C_{15}H_{31}NO$	17.13	1.58
$C_{15}H_{33}N_2$	17.50	1.44
$C_{16}HO_3$	17.42	2.03
$C_{16}H_3NO_2$	17.80	1.89
$C_{16}H_5N_2O$	18.17	1.76
$C_{16}H_7N_3$	18.55	1.62
$C_{16}H_{17}O_2$	17.64	1.86
$C_{16}H_{19}NO$	18.01	1.73
$C_{16}H_{21}N_2$	18.39	1.60
$C_{16}H_{33}O$	17.86	1.70
$C_{16}H_{35}N$	18.23	1.57
$C_{17}H_5O_2$	18.53	2.02
$C_{17}H_7NO$	18.90	1.89
$C_{17}H_9N_2$	19.28	1.76
$C_{17}H_{21}O$	18.75	1.86
$C_{17}H_{23}N$	19.12	1.73
$C_{18}H_9O$	19.63	2.02
$C_{18}H_{11}N$	20.01	1.90
$C_{18}H_{25}$	19.85	1.87
$C_{19}H_{13}$	20.74	2.04
$C_{20}H$	21.63	2.22
242		
$C_{11}H_{18}N_2O_4$	13.09	1.59
$C_{11}H_{20}N_3O_3$	13.47	1.44
$C_{11}H_{22}N_4O_2$	13.84	1.29
$C_{12}H_6N_2O_4$	13.98	1.71
$C_{12}H_8N_3O_3$	14.36	1.56
$C_{12}H_{10}N_4O_2$	14.73	1.41
$C_{12}H_{20}NO_4$	13.83	1.69
$C_{12}H_{22}N_2O_3$	14.20	1.54
$C_{12}H_{24}N_3O_2$	14.57	1.39
$C_{12}H_{26}N_4O$	14.95	1.24
$C_{13}H_8NO_4$	14.71	1.81
$C_{13}H_{10}N_2O_3$	15.09	1.66
$C_{13}H_{12}N_3O_2$	15.46	1.52
$C_{13}H_{14}N_4O$	15.84	1.38
$C_{13}H_{22}O_4$	14.56	1.79
$C_{13}H_{24}NO_3$	14.93	1.64
$C_{13}H_{26}N_2O_2$	15.31	1.49
$C_{13}H_{28}N_3O$	15.68	1.35
$C_{13}H_{30}N_4$	16.05	1.21
$C_{14}H_2N_4O$	16.73	1.51
$C_{14}H_{10}O_4$	15.45	1.91
$C_{14}H_{12}NO_3$	15.82	1.77
$C_{14}H_{14}N_2O_2$	16.19	1.63
$C_{14}H_{16}N_3O$	16.57	1.49
$C_{14}H_{18}N_4$	16.94	1.35
$C_{14}H_{26}O_3$	15.66	1.75
$C_{14}H_{28}NO_2$	16.04	1.60
$C_{14}H_{30}N_2O$	16.41	1.46
$C_{14}H_{32}N_3$	16.79	1.32
$C_{15}H_2N_2O_2$	17.08	1.77
$C_{15}H_4N_3O$	17.46	1.63
$C_{15}H_6N_4$	17.83	1.50
$C_{15}H_{14}O_3$	16.55	1.88
$C_{15}H_{16}NO_2$	16.93	1.74
$C_{15}H_{18}N_2O$	17.30	1.61
$C_{15}H_{20}N_3$	17.67	1.47
$C_{15}H_{30}O_2$	16.77	1.72
$C_{15}H_{32}NO$	17.14	1.58
$C_{15}H_{34}N_2$	17.52	1.45
$C_{16}H_2O_3$	17.44	2.03
$C_{16}H_4NO_2$	17.81	1.89
$C_{16}H_6N_2O$	18.19	1.76
$C_{16}H_8N_3$	18.56	1.63
$C_{16}H_{18}O_2$	17.66	1.87
$C_{16}H_{20}NO$	18.03	1.73
$C_{16}H_{22}N_2$	18.41	1.60
$C_{16}H_{34}O$	17.87	1.70
$C_{17}H_6O_2$	18.54	2.02
$C_{17}H_8NO$	18.92	1.89
$C_{17}H_{10}N_2$	19.29	1.76
$C_{17}H_{22}O$	18.76	1.86
$C_{17}H_{24}N$	19.14	1.73
$C_{18}H_{10}O$	19.65	2.03
$C_{18}H_{12}N$	20.02	1.90

Appendix Table 5.1 (*continued*)

	M + 1	M + 2
$C_{18}H_{26}$	19.87	1.87
$C_{19}H_{14}$	20.76	2.04
$C_{20}H_{2}$	21.64	2.23
243		
$C_{11}H_{19}N_{2}O_{4}$	13.11	1.60
$C_{11}H_{21}N_{3}O_{3}$	13.48	1.44
$C_{11}H_{23}N_{4}O_{2}$	13.86	1.29
$C_{12}H_{7}N_{2}O_{4}$	14.00	1.71
$C_{12}H_{9}N_{3}O_{3}$	14.37	1.56
$C_{12}H_{11}N_{4}O_{2}$	14.75	1.42
$C_{12}H_{21}NO_{4}$	13.84	1.69
$C_{12}H_{23}N_{2}O_{3}$	14.22	1.54
$C_{12}H_{25}N_{3}O_{2}$	14.59	1.39
$C_{12}H_{27}N_{4}O$	14.96	1.25
$C_{13}H_{9}NO_{4}$	14.73	1.81
$C_{13}H_{11}N_{2}O_{3}$	15.10	1.66
$C_{13}H_{13}N_{3}O_{2}$	15.48	1.52
$C_{13}H_{15}N_{4}O$	15.85	1.38
$C_{13}H_{23}O_{4}$	14.57	1.79
$C_{13}H_{25}NO_{3}$	14.95	1.64
$C_{13}H_{27}N_{2}O_{2}$	15.32	1.50
$C_{13}H_{29}N_{3}O$	15.70	1.35
$C_{13}H_{31}N_{4}$	16.07	1.21
$C_{14}HN_{3}O_{2}$	16.37	1.66
$C_{14}H_{3}N_{4}O$	16.74	1.52
$C_{14}H_{11}O_{4}$	15.46	1.92
$C_{14}H_{13}NO_{3}$	15.84	1.77
$C_{14}H_{15}N_{2}O_{2}$	16.21	1.63
$C_{14}H_{17}N_{3}O$	16.58	1.49
$C_{14}H_{19}N_{4}$	16.96	1.35
$C_{14}H_{27}O_{3}$	15.68	1.75
$C_{14}H_{29}NO_{2}$	16.05	1.61
$C_{14}H_{31}N_{2}O$	16.43	1.47
$C_{14}H_{33}N_{3}$	16.80	1.33
$C_{15}HNO_{3}$	16.72	1.91
$C_{15}H_{3}N_{2}O_{2}$	17.10	1.77
$C_{15}H_{5}N_{3}O$	17.47	1.64
$C_{15}H_{7}N_{4}$	17.85	1.50
$C_{15}H_{15}O_{3}$	16.57	1.89
$C_{15}H_{17}NO_{2}$	16.94	1.75
$C_{15}H_{19}N_{2}O$	17.32	1.61
$C_{15}H_{21}N_{3}$	17.69	1.47
$C_{15}H_{31}O_{2}$	16.78	1.72
$C_{15}H_{33}NO$	17.16	1.58
$C_{16}H_{3}O_{3}$	17.46	2.03
$C_{16}H_{5}NO_{2}$	17.83	1.90
$C_{16}H_{7}N_{2}O$	18.20	1.76
$C_{16}H_{9}N_{3}$	18.58	1.63
$C_{16}H_{19}O_{2}$	17.67	1.87
$C_{16}H_{21}NO$	18.05	1.73
$C_{16}H_{23}N_{2}$	18.42	1.60
$C_{17}H_{7}O_{2}$	18.56	2.02
$C_{17}H_{9}NO$	18.94	1.89
$C_{17}H_{11}N_{2}$	19.31	1.76
$C_{17}H_{23}O$	18.78	1.86
$C_{17}H_{25}N$	19.15	1.73
$C_{18}H_{11}O$	19.67	2.03
$C_{18}H_{13}N$	20.04	1.90
$C_{18}H_{27}$	19.88	1.87
$C_{19}HN$	20.93	2.08
$C_{19}H_{15}$	20.77	2.05
$C_{20}H_{3}$	21.66	2.23
244		
$C_{11}H_{20}N_{2}O_{4}$	13.13	1.60
$C_{11}H_{22}N_{3}O_{3}$	13.50	1.45
$C_{11}H_{24}N_{4}O_{2}$	13.88	1.30
$C_{12}H_{8}N_{2}O_{4}$	14.01	1.71
$C_{12}H_{10}N_{3}O_{3}$	14.39	1.56
$C_{12}H_{12}N_{4}O_{2}$	14.76	1.42
$C_{12}H_{22}NO_{4}$	13.86	1.69
$C_{12}H_{24}N_{2}O_{3}$	14.23	1.54
$C_{12}H_{26}N_{3}O_{2}$	14.61	1.40
$C_{12}H_{28}N_{4}O$	14.98	1.25
$C_{13}H_{10}NO_{4}$	14.75	1.81
$C_{13}H_{12}N_{2}O_{3}$	15.12	1.67
$C_{13}H_{14}N_{3}O_{2}$	15.49	1.52
$C_{13}H_{16}N_{4}O$	15.87	1.38
$C_{13}H_{24}O_{4}$	14.59	1.79
$C_{13}H_{26}NO_{3}$	14.96	1.64
$C_{13}H_{28}N_{2}O_{2}$	15.34	1.50
$C_{13}H_{30}N_{3}O$	15.71	1.36
$C_{13}H_{32}N_{4}$	16.09	1.21
$C_{14}H_{2}N_{3}O_{2}$	16.38	1.66
$C_{14}H_{4}N_{4}O$	16.76	1.52
$C_{14}H_{12}O_{4}$	15.48	1.92
$C_{14}H_{14}NO_{3}$	15.85	1.78
$C_{14}H_{16}N_{2}O_{2}$	16.23	1.63
$C_{14}H_{18}N_{3}O$	16.60	1.49
$C_{14}H_{20}N_{4}$	16.97	1.36
$C_{14}H_{28}O_{3}$	15.69	1.75
$C_{14}H_{30}NO_{2}$	16.07	1.61
$C_{14}H_{32}N_{2}O$	16.44	1.47
$C_{15}H_{2}NO_{3}$	16.74	1.91
$C_{15}H_{4}N_{2}O_{2}$	17.11	1.78
$C_{15}H_{6}N_{3}O$	17.49	1.64
$C_{15}H_{8}N_{4}$	17.86	1.50
$C_{15}H_{16}O_{3}$	16.58	1.89
$C_{15}H_{18}NO_{2}$	16.96	1.75
$C_{15}H_{20}N_{2}O$	17.33	1.61
$C_{15}H_{22}N_{3}$	17.71	1.48
$C_{15}H_{32}O_{2}$	16.80	1.72
$C_{16}H_{4}O_{3}$	17.47	2.03
$C_{16}H_{6}NO_{2}$	17.85	1.90
$C_{16}H_{8}N_{2}O$	18.22	1.77
$C_{16}H_{10}N_{3}$	18.59	1.63
$C_{16}H_{20}O_{2}$	17.69	1.87
$C_{16}H_{22}NO$	18.06	1.74
$C_{16}H_{24}N_{2}$	18.44	1.60
$C_{17}H_{8}O_{2}$	18.58	2.03
$C_{17}H_{10}NO$	18.95	1.90
$C_{17}H_{12}N_{2}$	19.33	1.77
$C_{17}H_{24}O$	18.79	1.87
$C_{17}H_{26}N$	19.17	1.74
$C_{18}H_{12}O$	19.68	2.03
$C_{18}H_{14}N$	20.06	1.91
$C_{18}H_{28}$	19.90	1.87
$C_{19}H_{2}N$	20.95	2.08
$C_{19}H_{16}$	20.79	2.05
$C_{20}H_{4}$	21.68	2.23
245		
$C_{11}H_{21}N_{2}O_{4}$	13.14	1.60
$C_{11}H_{23}N_{3}O_{3}$	13.52	1.45
$C_{11}H_{25}N_{4}O_{2}$	13.89	1.30
$C_{12}H_{9}N_{2}O_{4}$	14.03	1.71
$C_{12}H_{11}N_{3}O_{3}$	14.41	1.57
$C_{12}H_{13}N_{4}O_{2}$	14.78	1.42
$C_{12}H_{23}NO_{4}$	13.87	1.69
$C_{12}H_{25}N_{2}O_{3}$	14.25	1.54
$C_{12}H_{27}N_{3}O_{2}$	14.62	1.40
$C_{12}H_{29}N_{4}O$	15.00	1.25
$C_{13}HN_{4}O_{2}$	15.67	1.55
$C_{13}H_{11}NO_{4}$	14.76	1.81
$C_{13}H_{13}N_{2}O_{3}$	15.14	1.67
$C_{13}H_{15}N_{3}O_{2}$	15.51	1.53
$C_{13}H_{17}N_{4}O$	15.89	1.38
$C_{13}H_{25}O_{4}$	14.60	1.79
$C_{13}H_{27}NO_{3}$	14.98	1.65
$C_{13}H_{29}N_{2}O_{2}$	15.35	1.50
$C_{13}H_{31}N_{3}O$	15.73	1.36
$C_{14}HN_{2}O_{3}$	16.03	1.80
$C_{14}H_{3}N_{3}O_{2}$	16.40	1.66
$C_{14}H_{5}N_{4}O$	16.77	1.52
$C_{14}H_{13}O_{4}$	15.49	1.92
$C_{14}H_{15}NO_{3}$	15.87	1.78
$C_{14}H_{17}N_{2}O_{2}$	16.24	1.64
$C_{14}H_{19}N_{3}O$	16.62	1.50
$C_{14}H_{21}N_{4}$	16.99	1.36
$C_{14}H_{29}O_{3}$	15.71	1.75
$C_{14}H_{31}NO_{2}$	16.08	1.61
$C_{15}HO_{4}$	16.38	2.06
$C_{15}H_{3}NO_{3}$	16.76	1.92
$C_{15}H_{5}N_{2}O_{2}$	17.13	1.78
$C_{15}H_{7}N_{3}O$	17.50	1.64
$C_{15}H_{9}N_{4}$	17.88	1.51
$C_{15}H_{17}O_{3}$	16.60	1.89
$C_{15}H_{19}NO_{2}$	16.97	1.75
$C_{15}H_{21}N_{2}O$	17.35	1.62
$C_{15}H_{23}N_{3}$	17.72	1.48
$C_{16}H_{5}O_{3}$	17.49	2.04
$C_{16}H_{7}NO_{2}$	17.86	1.90
$C_{16}H_{9}N_{2}O$	18.24	1.77
$C_{16}H_{11}N_{3}$	18.61	1.64
$C_{16}H_{21}O_{2}$	17.70	1.87
$C_{16}H_{23}NO$	18.08	1.74
$C_{16}H_{25}N_{2}$	18.45	1.61
$C_{17}H_{9}O_{2}$	18.59	2.03
$C_{17}H_{11}NO$	18.97	1.90
$C_{17}H_{13}N_{2}$	19.34	1.77
$C_{17}H_{25}O$	18.81	1.87
$C_{17}H_{27}N$	19.18	1.74
$C_{18}HN_{2}$	20.23	1.94
$C_{18}H_{13}O$	19.70	2.04
$C_{18}H_{15}N$	20.07	1.91
$C_{18}H_{29}$	19.92	1.88
$C_{19}H_{17}$	20.80	2.05
$C_{19}HO$	20.59	2.21
$C_{19}H_{3}N$	20.96	2.09
$C_{20}H_{5}$	21.69	2.24
246		
$C_{11}H_{22}N_{2}O_{4}$	13.16	1.60
$C_{11}H_{24}N_{3}O_{3}$	13.53	1.45
$C_{11}H_{26}N_{4}O_{2}$	13.91	1.30
$C_{12}H_{10}N_{2}O_{4}$	14.05	1.72
$C_{12}H_{12}N_{3}O_{3}$	14.42	1.57
$C_{12}H_{14}N_{4}O_{2}$	14.80	1.42
$C_{12}H_{24}NO_{4}$	13.89	1.70
$C_{12}H_{26}N_{2}O_{3}$	14.26	1.55
$C_{12}H_{28}N_{3}O_{2}$	14.64	1.40
$C_{12}H_{30}N_{4}O$	15.01	1.25
$C_{13}H_{2}N_{4}O_{2}$	15.68	1.55
$C_{13}H_{12}NO_{4}$	14.78	1.82
$C_{13}H_{14}N_{2}O_{3}$	15.15	1.67
$C_{13}H_{16}N_{3}O_{2}$	15.53	1.53
$C_{13}H_{18}N_{4}O$	15.90	1.39
$C_{13}H_{26}O_{4}$	14.62	1.79
$C_{13}H_{28}NO_{3}$	15.00	1.65
$C_{13}H_{30}N_{2}O_{2}$	15.37	1.50
$C_{14}H_{2}N_{2}O_{3}$	16.04	1.80
$C_{14}H_{4}N_{3}O_{2}$	16.42	1.66
$C_{14}H_{6}N_{4}O$	16.79	1.53
$C_{14}H_{14}O_{4}$	15.51	1.92
$C_{14}H_{16}NO_{3}$	15.88	1.78
$C_{14}H_{18}N_{2}O_{2}$	16.26	1.64
$C_{14}H_{20}N_{3}O$	16.63	1.50
$C_{14}H_{22}N_{4}$	17.01	1.36
$C_{14}H_{30}O_{3}$	15.73	1.76
$C_{15}H_{2}O_{4}$	16.40	2.06
$C_{15}H_{4}NO_{3}$	16.77	1.92
$C_{15}H_{6}N_{2}O_{2}$	17.15	1.78
$C_{15}H_{8}N_{3}O$	17.52	1.65
$C_{15}H_{10}N_{4}$	17.90	1.51
$C_{15}H_{18}O_{3}$	16.61	1.89
$C_{15}H_{20}NO_{2}$	16.99	1.76
$C_{15}H_{22}N_{2}O$	17.36	1.62
$C_{15}H_{24}N_{3}$	17.74	1.48
$C_{16}H_{6}O_{3}$	17.50	2.04
$C_{16}H_{8}NO_{2}$	17.88	1.90
$C_{16}H_{10}N_{2}O$	18.25	1.77
$C_{16}H_{12}N_{3}$	18.63	1.64
$C_{16}H_{22}O_{2}$	17.72	1.88
$C_{16}H_{24}NO$	18.09	1.74
$C_{16}H_{26}N_{2}$	18.47	1.61
$C_{17}H_{10}O_{2}$	18.61	2.03
$C_{17}H_{12}NO$	18.98	1.90
$C_{17}H_{14}N_{2}$	19.36	1.77
$C_{17}H_{26}O$	18.83	1.87
$C_{17}H_{28}N$	19.20	1.74
$C_{18}H_{2}N_{2}$	20.25	1.94
$C_{18}H_{14}O$	19.71	2.04
$C_{18}H_{16}N$	20.09	1.91
$C_{18}H_{30}$	19.93	1.88
$C_{19}H_{2}O$	20.60	2.21
$C_{19}H_{4}N$	20.98	2.09
$C_{19}H_{18}$	20.82	2.06
$C_{20}H_{6}$	21.71	2.24
247		
$C_{11}H_{23}N_{2}O_{4}$	13.17	1.60
$C_{11}H_{25}N_{3}O_{3}$	13.55	1.45
$C_{11}H_{27}N_{4}O_{2}$	13.92	1,30
$C_{12}H_{11}N_{2}O_{4}$	14.06	1.72
$C_{12}H_{13}N_{3}O_{3}$	14.44	1.57
$C_{12}H_{15}N_{4}O_{2}$	14.81	1.42
$C_{12}H_{25}NO_{4}$	13.91	1.70
$C_{12}H_{27}N_{2}O_{3}$	14.28	1.55
$C_{12}H_{29}N_{3}O_{2}$	14.65	1.40
$C_{13}HN_{3}O_{3}$	15.33	1.70
$C_{13}H_{3}N_{4}O_{2}$	15.70	1.55
$C_{13}H_{13}NO_{4}$	14.79	1.82
$C_{13}H_{15}N_{2}O_{3}$	15.17	1.67

Appendix Table 5.1 (*continued*)

	M + 1	M + 2
$C_{13}H_{17}N_3O_2$	15.54	1.53
$C_{13}H_{19}N_4O$	15.92	1.39
$C_{13}H_{27}O_4$	14.64	1.80
$C_{13}H_{29}NO_3$	15.01	1.65
$C_{14}HNO_4$	15.68	1.95
$C_{14}H_3N_2O_3$	16.06	1.81
$C_{14}H_5N_3O_2$	16.43	1.67
$C_{14}H_7N_4O$	16.81	1.53
$C_{14}H_{15}O_4$	15.53	1.93
$C_{14}H_{17}NO_3$	15.90	1.78
$C_{14}H_{19}N_2O_2$	16.27	1.64
$C_{14}H_{21}N_3O$	16.65	1.50
$C_{14}H_{23}N_4$	17.02	1.36
$C_{15}H_3O_4$	16.41	2.06
$C_{15}H_5NO_3$	16.79	1.92
$C_{15}H_7N_2O_2$	17.16	1.78
$C_{15}H_9N_3O$	17.54	1.65
$C_{15}H_{11}N_4$	17.91	1.51
$C_{15}H_{19}O_3$	16.63	1.90
$C_{15}H_{21}NO_2$	17.01	1.76
$C_{15}H_{23}N_2O$	17.38	1.62
$C_{15}H_{25}N_3$	17.75	1.49
$C_{16}H_7O_3$	17.52	2.04
$C_{16}H_9NO_2$	17.89	1.91
$C_{16}H_{11}N_2O$	18.27	1.77
$C_{16}H_{13}N_3$	18.64	1.64
$C_{16}H_{23}O_2$	17.74	1.88
$C_{16}H_{25}NO$	18.11	1.75
$C_{16}H_{27}N_2$	18.49	1.61
$C_{17}HN_3$	19.53	1.81
$C_{17}H_{11}O_2$	18.62	2.04
$C_{17}H_{13}NO$	19.00	1.91
$C_{17}H_{15}N_2$	19.37	1.78
$C_{17}H_{27}O$	18.84	1.88
$C_{17}H_{29}N$	19.22	1.75
$C_{18}HNO$	19.89	2.07
$C_{18}H_3N_2$	20.26	1.95
$C_{18}H_{15}O$	19.73	2.04
$C_{18}H_{17}N$	20.10	1.92
$C_{18}H_{31}$	19.95	1.88
$C_{19}H_3O$	20.62	2.22
$C_{19}H_5N$	20.99	2.09
$C_{19}H_{19}$	20.84	2.06
$C_{20}H_7$	21.72	2.24
248		
$C_{11}H_{24}N_2O_4$	13.19	1.61
$C_{11}H_{26}N_3O_3$	13.56	1.45
$C_{11}H_{28}N_4O_2$	13.94	1.31
$C_{12}H_{12}N_2O_4$	14.08	1.72
$C_{12}H_{14}N_3O_3$	14.45	1.57
$C_{12}H_{16}N_4O_2$	14.83	1.43
$C_{12}H_{26}NO_4$	13.92	1.70
$C_{12}H_{28}N_2O_3$	14.30	1.55

	M + 1	M + 2
$C_{13}H_2N_3O_3$	15.34	1.70
$C_{13}H_4N_4O_2$	15.72	1.56
$C_{13}H_{14}NO_4$	14.81	1.82
$C_{13}H_{16}N_2O_3$	15.18	1.68
$C_{13}H_{18}N_3O_2$	15.56	1.53
$C_{13}H_{20}N_4O$	15.93	1.39
$C_{13}H_{28}O_4$	14.65	1.80
$C_{14}H_2NO_4$	15.70	1.95
$C_{14}H_4N_2O_3$	16.07	1.81
$C_{14}H_6N_3O_2$	16.45	1.67
$C_{14}H_8N_4O$	16.82	1.53
$C_{14}H_{16}O_4$	15.54	1.93
$C_{14}H_{18}NO_3$	15.92	1.79
$C_{14}H_{20}N_2O_2$	16.29	1.64
$C_{14}H_{22}N_3O$	16.66	1.51
$C_{14}H_{24}N_4$	17.04	1.37
$C_{15}H_4O_4$	16.43	2.06
$C_{15}H_6NO_3$	16.80	1.92
$C_{15}H_8N_2O_2$	17.18	1.79
$C_{15}H_{10}N_3O$	17.55	1.65
$C_{15}H_{12}N_4$	17.93	1.52
$C_{15}H_{20}O_3$	16.65	1.90
$C_{15}H_{22}NO_2$	17.02	1.76
$C_{15}H_{24}N_2O$	17.40	1.62
$C_{15}H_{26}N_3$	17.77	1.49
$C_{16}H_8O_3$	17.54	2.05
$C_{16}H_{10}NO_2$	17.91	1.91
$C_{16}H_{12}N_2O$	18.28	1.78
$C_{16}H_{14}N_3$	18.66	1.65
$C_{16}H_{24}O_2$	17.75	1.88
$C_{16}H_{26}NO$	18.13	1.75
$C_{16}H_{28}N_2$	18.50	1.62
$C_{17}H_2N_3$	19.55	1.81
$C_{17}H_{12}O_2$	18.64	2.04
$C_{17}H_{14}NO$	19.02	1.91
$C_{17}H_{16}N_2$	19.39	1.78
$C_{17}H_{28}O$	18.86	1.88
$C_{17}H_{30}N$	19.23	1.75
$C_{18}H_2NO$	19.90	2.08
$C_{18}H_4N_2$	20.28	1.95
$C_{18}H_{16}O$	19.75	2.04
$C_{18}H_{18}N$	20.12	1.92
$C_{18}H_{32}$	19.96	1.89
$C_{19}H_4O$	20.64	2.22
$C_{19}H_6N$	21.01	2.10
$C_{19}H_{20}$	20.85	2.06
$C_{20}H_8$	21.74	2.25
249		
$C_{11}H_{25}N_2O_4$	13.21	1.61
$C_{11}H_{27}N_3O_3$	13.58	1.46
$C_{12}H_{13}N_2O_4$	14.10	1.72
$C_{12}H_{15}N_3O_3$	14.47	1.58
$C_{12}H_{17}N_4O_2$	14.84	1.43

	M + 1	M + 2
$C_{12}H_{27}NO_4$	13.94	1.70
$C_{13}HN_2O_4$	14.98	1.85
$C_{13}H_3N_3O_3$	15.36	1.70
$C_{13}H_5N_4O_2$	15.73	1.56
$C_{13}H_{15}NO_4$	14.83	1.82
$C_{13}H_{17}N_2O_3$	15.20	1.68
$C_{13}H_{19}N_3O_2$	15.57	1.54
$C_{13}H_{21}N_4O$	15.95	1.39
$C_{14}H_3NO_4$	15.71	1.95
$C_{14}H_5N_2O_3$	16.09	1.81
$C_{14}H_7N_3O_2$	16.46	1.67
$C_{14}H_9N_4O$	16.84	1.53
$C_{14}H_{17}O_4$	15.56	1.93
$C_{14}H_{19}NO_3$	15.93	1.79
$C_{14}H_{21}N_2O_2$	16.31	1.65
$C_{14}H_{23}N_3O$	16.68	1.51
$C_{14}H_{25}N_4$	17.05	1.37
$C_{15}H_5O_4$	16.45	2.07
$C_{15}H_7NO_3$	16.82	1.93
$C_{15}H_9N_2O_2$	17.19	1.79
$C_{15}H_{11}N_3O$	17.57	1.65
$C_{15}H_{13}N_4$	17.94	1.52
$C_{15}H_{21}O_3$	16.66	1.90
$C_{15}H_{23}NO_2$	17.04	1.76
$C_{15}H_{25}N_2O$	17.41	1.63
$C_{15}H_{27}N_3$	17.79	1.49
$C_{16}HN_4$	18.83	1.68
$C_{16}H_9O_3$	17.55	2.05
$C_{16}H_{11}NO_2$	17.93	1.91
$C_{16}H_{13}N_2O$	18.30	1.78
$C_{16}H_{15}N_3$	18.67	1.65
$C_{16}H_{25}O_2$	17.77	1.89
$C_{16}H_{27}NO$	18.14	1.75
$C_{16}H_{29}N_2$	18.52	1.62
$C_{17}HN_2O$	19.19	1.94
$C_{17}H_3N_3$	19.56	1.81
$C_{17}H_{13}O_2$	18.66	2.04
$C_{17}H_{15}NO$	19.03	1.91
$C_{17}H_{17}N_2$	19.41	1.78
$C_{17}H_{29}O$	18.87	1.88
$C_{17}H_{31}N$	19.25	1.75
$C_{18}HO_2$	19.55	2.21
$C_{18}H_3NO$	19.92	2.08
$C_{18}H_5N_2$	20.29	1.95
$C_{18}H_{17}O$	19.76	2.05
$C_{18}H_{19}N$	20.14	1.92
$C_{18}H_{33}$	19.98	1.89
$C_{19}H_5O$	20.65	2.22
$C_{19}H_7N$	21.03	2.10
$C_{19}H_{21}$	20.87	2.07
$C_{20}H_9$	21.76	2.25
250		
$C_{11}H_{26}N_2O_4$	13.22	1.61

	M + 1	M + 2
$C_{12}H_{14}N_2O_4$	14.11	1.73
$C_{12}H_{16}N_3O_3$	14.49	1.58
$C_{12}H_{18}N_4O_2$	14.86	1.43
$C_{13}H_2N_2O_4$	15.00	1.85
$C_{13}H_4N_3O_3$	15.37	1.71
$C_{13}H_6N_4O_2$	15.75	1.56
$C_{13}H_{16}NO_4$	14.84	1.83
$C_{13}H_{18}N_2O_3$	15.22	1.68
$C_{13}H_{20}N_3O_2$	15.59	1.54
$C_{13}H_{22}N_4O$	15.97	1.40
$C_{14}H_4NO_4$	15.73	1.96
$C_{14}H_6N_2O_3$	16.11	1.82
$C_{14}H_8N_3O_2$	16.48	1.67
$C_{14}H_{10}N_4O$	16.85	1.54
$C_{14}H_{18}O_4$	15.57	1.93
$C_{14}H_{20}NO_3$	15.95	1.79
$C_{14}H_{22}N_2O_2$	16.32	1.65
$C_{14}H_{24}N_3O$	16.70	1.51
$C_{14}H_{26}N_4$	17.07	1.37
$C_{15}H_6O_4$	16.46	2.07
$C_{15}H_8NO_3$	16.84	1.93
$C_{15}H_{10}N_2O_2$	17.21	1.79
$C_{15}H_{12}N_3O$	17.58	1.66
$C_{15}H_{14}N_4$	17.96	1.52
$C_{15}H_{22}O_3$	16.68	1.90
$C_{15}H_{24}NO_2$	17.05	1.77
$C_{15}H_{26}N_2O$	17.43	1.63
$C_{15}H_{28}N_3$	17.80	1.49
$C_{16}H_2N_4$	18.85	1.68
$C_{16}H_{10}O_3$	17.57	2.05
$C_{16}H_{12}NO_2$	17.94	1.92
$C_{16}H_{14}N_2O$	18.32	1.78
$C_{16}H_{16}N_3$	18.69	1.65
$C_{16}H_{26}O_2$	17.78	1.89
$C_{16}H_{28}NO$	18.16	1.75
$C_{16}H_{30}N_2$	18.53	1.62
$C_{17}H_2N_2O$	19.20	1.94
$C_{17}H_4N_3$	19.58	1.82
$C_{17}H_{14}O_2$	18.67	2.05
$C_{17}H_{16}NO$	19.05	1.91
$C_{17}H_{18}N_2$	19.42	1.79
$C_{17}H_{30}O$	18.89	1.89
$C_{17}H_{32}N$	19.26	1.76
$C_{18}H_2O_2$	19.56	2.21
$C_{18}H_4NO$	19.94	2.08
$C_{18}H_6N_2$	20.31	1.96
$C_{18}H_{18}O$	19.78	2.05
$C_{18}H_{20}N$	20.15	1.92
$C_{18}H_{34}$	20.00	1.89
$C_{19}H_6O$	20.67	2.23
$C_{19}H_8N$	21.04	2.10
$C_{19}H_{22}$	20.88	2.07
$C_{20}H_{10}$	21.77	2.25

* Adapted with permission from:

1. J. H. Beynon, *Mass Spectrometry and Its Applications to Organic Chemistry*, Elsevier Publishing Co., Amsterdam, 1960.
2. R. M. Silverstein and G. C. Basseler, *Spectrometric Identification of Organic Compounds*, Wiley, New York, 1966.

APPENDIX 5.2

RAPID CALCULATION OF MOLECULAR FORMULAS FROM MASS VALUES

The calculation of molecular compositions consistent with a given range of mass values arises particularly in mass spectrometry. Although this can be a trivial exercise on the computer, it has been vexing to do by hand. Published tables, e.g., Beynon and Williams,[1] are bulky, and nevertheless cover a limited range of atom values. The values are also awkward to search, not having been sorted.

The following approach was designed for a desk calculator that ought to be available to any student. As it involves only a few additions and subtractions, it can—*horribilis dictu*—even be done by hand. Furthermore, it lends itself to real time implementation on small computers that lack high precision "divide" instructions in their hardware.

The basis of the calculation is this table (Appendix Table 5.2) which is an ordered list of the mass numbers of the formulas for H from 0 to 10, N from 0 to 5, and O from 0 to 11. *It contains only those compositions whose masses are an integral multiple of 12.* Any number of C's may then be added as required.

The use of the table is best explained by a specific example, say $m = 259.09 \pm 0.001$.

Step 1. Since $259 \equiv 7$ modulo 12, 5 H's (5.03913) will be borrowed to give $m' = m + 5H = 264.129$. This is then divided into $m' = m_i + m_f$; $m_i = 264$ $(= 22 \times 12)$; $m_f = 0.129 \pm 0.001$.

Step 2. The table is searched for entries that correspond to m_f and whose mass does not exceed m_i. (m_i is expressed as $m_i/12$ = C-equivalent.) We find none in this cycle.

Step 3. We therefore remove 12 H's (12.0939) to give $m'' = m' - 12H = 252.035 \pm 0.001$. The table now has entries at 0.034 ($H_8N_4O_8$), 0.035 ($H_{10}NO_9$) and 0.036 ($H_6N_5O_5$). These will be completed in Step 4. 12 H's are again removed until m_f falls below -0.0498,

Appendix Table 5.2. Table of mass fractions for all combinations[a] of H, N, O ⟨H $\leqq$ 0 N $\leqq$ 6 O $\leqq$ 11⟩

Index	$m_f \times 10^6$	H	N	O	=C	Index	$m_f \times 10^6$	H	N	O	=C	Index	$m_f \times 10^6$	H	N	O	=C
−49	−49787	0	2	11	17	0	0	0	0	0	0	31	31537	10	3	11	9
−45	−45765	0	0	9	12	1	510	2	5	6	14	32	32363	4	2	1	14
−38	−38554	0	4	10	18	2	1853	4	2	7	12	34	34216	8	4	8	16
−37	−37211	2	1	11	16	4	4532	2	3	4	9	35	35559	10	1	9	14
−34	−34532	0	2	8	13	5	5875	4	0	5	7	36	36895	6	5	5	13
−30	−30510	0	0	6	8	6	6385	6	5	11	21	38	38238	8	2	6	11
−25	−25978	2	3	10	17	7	7211	0	4	1	6	40	40917	6	3	3	8
−24	−24635	4	0	11	15	8	8554	2	1	2	4	41	42260	8	0	4	6
−23	−23299	0	4	7	14	10	10407	6	3	9	16	42	42770	10	5	10	20
−21	−21956	2	1	8	12	11	11750	8	0	10	14	43	43596	4	4	0	5
−19	−19277	0	2	5	9	13	13086	4	4	6	13	44	44939	6	1	1	3
−15	−15255	0	0	3	4	14	14429	6	1	7	11	46	46792	10	3	8	15
−14	−14745	2	5	9	16	15	15765	2	5	3	10	49	49471	8	4	5	12
−13	−13402	4	2	10	18	17	17108	4	2	4	8	50	50814	10	1	6	10
−10	−10723	2	3	7	13	18	18961	8	4	11	20	52	52150	6	5	2	9
−9	−9380	4	0	8	11	19	19787	2	3	1	5	53	53493	8	2	3	7
−8	−8044	0	4	4	10	20	21130	4	0	2	3	56	56172	6	3	0	4
−6	−6701	2	1	5	8	21	21640	6	5	8	17	57	57515	8	0	1	2
−4	−4022	0	2	2	5	22	22983	8	2	9	15	58	58025	10	5	7	16
−2	−2169	4	4	9	17	25	25662	6	3	6	12	62	62047	10	3	5	11
−1	−826	6	1	10	15	27	27005	8	0	7	10	64	64726	8	4	2	8
						28	28341	4	4	3	9	66	66069	10	1	3	6
						29	29684	6	1	4	7	68	68748	8	2	0	3
						30	31020	2	5	0	6	73	73280	10	5	4	12
												77	77302	10	3	2	7
												81	81324	10	1	0	2
												88	88535	10	5	1	8
(−0.049 to −0.0008)							(0 to 0.03)						(0.03 to 0.088)				

[a] Arranged so that the index for each entry agrees with $1000 \times m_f \pm 1.9$.

Reprinted from: J. Lederberg, *J. Chem. Ed.* **49**, 613 (1972), with permission from the American Chemical Society.

the bottom of the table. In our example, this occurs at the next cycle.

Step 4. The table entries are now completed as follows.

				Add C's to make up m''	Adjust borrowed H's	Check mass (compare 259.0900 ± 0.0010)
34	0.034216	$H_8N_4O_8$	$m_i = C_{16}$	C_5	$C_5H_{15}N_4O_8$	259.089
35	0.035559	$H_{10}NO_9$	$m_i = C_{14}$	C_7	$C_7H_{17}NO_9$	259.090
36	0.036895	$H_6N_5O_5$	$m_i = C_{13}$	C_8	$C_8H_{13}N_5O_5$	259.092

Step 5. Various criteria of chemical plausibility can be used to filter the list. Since the valence rules allow H's to a maximum of 2 + 2C + N, none of these compositions is oversaturated. $C_5H_{15}N_4O_8$ however has an odd number of H's and may therefore represent a free radical.

If wider ranges of hetero atoms are contemplated, adjustments of blocks of 6 N (84.01844) and 12 O (191.9389) can be applied repetitively in a fashion similar to Step 3 so long as the adjusted mass allows.

In fact $m'' = m - 6N - 7H = 168.017 \pm 0.001$ leads to $C_6H_{11}N_8O_4$, $m = 259.090$. Further, $m - 12N - 7H = 83.999 \pm 0.001$. We read this as $m_i = 84$; $m_f = -0.001$ and find two entries in the table: -0.000826 (H_6NO_{10}) and 0.000510 ($H_2N_2O_6$), whose m_i however >84.

The table is arranged so as to illustrate its use in a fast computer program. A linear array with 138 cells, indexed as shown, has entries that never slip more than one position away from the value of the index. The composition values can therefore be accessed by direct lookup, obviating a table search.

This compilation is a greatly shortened form of some tables that were published some time ago.[2]

1 Beynon, J. H., and Williams, A. E., "Mass and Abundance Tables for use in Mass Spectrometry," Elsevier, Amsterdam, 1963.

2 Lederberg, J., "Composition of Molecular Formulas for Mass Spectrometry," Holden-Day, San Francisco, 1964.

Appendix Table 6.1 Four-place logarithms to base 10

N	0	1	2	3	4	5	6	7	8	9	1	2	3	4	5	6	7	8	9
10	0000	0043	0086	0128	0170	0212	0253	0294	0334	0374	4	8	12	17	21	25	29	33	37
11	0414	0453	0492	0531	0569	0607	0645	0682	0719	0755	4	8	11	15	19	23	26	30	34
12	0792	0828	0864	0899	0934	0969	1004	1038	1072	1106	3	7	10	14	17	21	24	28	31
13	1139	1173	1206	1239	1271	1303	1335	1367	1399	1430	3	6	10	13	16	19	23	26	29
14	1461	1492	1523	1553	1584	1614	1644	1673	1703	1732	3	6	9	12	15	18	21	24	27
15	1761	1790	1818	1847	1875	1903	1931	1959	1987	2014	3	6	8	11	14	17	20	22	25
16	2041	2068	2095	2122	2148	2175	2201	2227	2253	2279	3	5	8	11	13	16	18	21	24
17	2304	2330	2355	2380	2405	2430	2455	2480	2504	2529	2	5	7	10	12	15	17	20	22
18	2553	2577	2601	2625	2648	2672	2695	2718	2742	2765	2	5	7	9	12	14	16	19	21
19	2788	2810	2833	2856	2878	2900	2923	2945	2967	2989	2	4	7	9	11	13	16	18	20
20	3010	3032	3054	3075	3096	3118	3139	3160	3181	3201	2	4	6	8	11	13	15	17	19
21	3222	3243	3263	3284	3304	3324	3345	3365	3385	3404	2	4	6	8	10	12	14	16	18
22	3424	3444	3464	3483	3502	3522	3541	3560	3579	3598	2	4	6	8	10	12	14	15	17
23	3617	3636	3655	3674	3692	3711	3729	3747	3766	3784	2	4	6	7	9	11	13	15	17
24	3802	3820	3838	3856	3874	3892	3909	3927	3945	3962	2	4	5	7	9	11	12	14	16
25	3979	3997	4014	4031	4048	4065	4082	4099	4116	4133	2	3	5	7	9	10	12	14	15
26	4150	4166	4183	4200	4216	4232	4249	4265	4281	4298	2	3	5	7	8	10	11	13	15
27	4314	4330	4346	4362	4378	4393	4409	4425	4440	4456	2	3	5	6	8	9	11	13	14
28	4472	4487	4502	4518	4533	4548	4564	4579	4594	4609	2	3	5	6	8	9	11	12	14
29	4624	4639	4654	4669	4683	4698	4713	4728	4742	4757	1	3	4	6	7	9	10	12	13
30	4771	4786	4800	4814	4829	4843	4857	4871	4886	4900	1	3	4	6	7	9	10	11	13
31	4914	4928	4942	4955	4969	4983	4997	5011	5024	5038	1	3	4	6	7	8	10	11	12
32	5051	5065	5079	5092	5105	5119	5132	5145	5159	5172	1	3	4	5	7	8	9	11	12
33	5185	5198	5211	5224	5237	5250	5263	5276	5289	5302	1	3	4	5	6	8	9	10	12
34	5315	5328	5340	5353	5366	5378	5391	5403	5416	5428	1	3	4	5	6	8	9	10	11
35	5441	5453	5465	5478	5490	5502	5514	5527	5539	5551	1	2	4	5	6	7	9	10	11
36	5563	5575	5587	5599	5611	5623	5635	5647	5658	5670	1	2	4	5	6	7	8	10	11
37	5682	5694	5705	5717	5729	5740	5752	5763	5775	5786	1	2	3	5	6	7	8	9	10
38	5798	5809	5821	5832	5843	5855	5866	5877	5888	5899	1	2	3	5	6	7	8	9	10
39	5911	5922	5933	5944	5955	5966	5977	5988	5999	6010	1	2	3	4	5	7	8	9	10
40	6021	6031	6042	6053	6064	6075	6085	6096	6107	6117	1	2	3	4	5	6	8	9	10
41	6128	6138	6149	6160	6170	6180	6191	6201	6212	6222	1	2	3	4	5	6	7	8	9
42	6232	6243	6253	6263	6274	6284	6294	6304	6314	6325	1	2	3	4	5	6	7	8	9
43	6335	6345	6355	6365	6375	6385	6395	6405	6415	6425	1	2	3	4	5	6	7	8	9
44	6435	6444	6454	6464	6474	6484	6493	6503	6513	6522	1	2	3	4	5	6	7	8	9
45	6532	6542	6551	6561	6571	6580	6590	6599	6609	6618	1	2	3	4	5	6	7	8	9
46	6628	6637	6646	6656	6665	6675	6684	6693	6702	6712	1	2	3	4	5	6	7	7	8
47	6721	6730	6739	6749	6758	6767	6776	6785	6794	6803	1	2	3	4	5	5	6	7	8
48	6812	6821	6830	6839	6848	6857	6866	6875	6884	6893	1	2	3	4	4	5	6	7	8
49	6902	6911	6920	6928	6937	6946	6955	6964	6972	6981	1	2	3	4	4	5	6	7	8
50	6990	6998	7007	7016	7024	7033	7042	7050	7059	7067	1	2	3	3	4	5	6	7	8
51	7076	7084	7093	7101	7110	7118	7126	7135	7143	7152	1	2	3	3	4	5	6	7	8
52	7160	7168	7177	7185	7193	7202	7210	7218	7226	7235	1	2	2	3	4	5	6	7	7
53	7243	7251	7259	7267	7275	7284	7292	7300	7308	7316	1	2	2	3	4	5	6	6	7
54	7324	7332	7340	7348	7356	7364	7372	7380	7388	7396	1	2	2	3	4	5	6	6	7

Appendix Table 6.1 (*continued*)

N	0	1	2	3	4	5	6	7	8	9	1	2	3	4	5	6	7	8	9
55	7404	7412	7419	7427	7435	7443	7451	7459	7466	7474	1	2	2	3	4	5	5	6	7
56	7482	7490	7497	7505	7513	7520	7528	7536	7543	7551	1	2	2	3	4	5	5	6	7
57	7559	7566	7574	7582	7589	7597	7604	7612	7619	7627	1	2	2	3	4	5	5	6	7
58	7634	7642	7649	7657	7664	7672	7679	7686	7694	7701	1	1	2	3	4	4	5	6	7
59	7709	7716	7723	7731	7738	7745	7752	7760	7767	7774	1	1	2	3	4	4	5	6	7
60	7782	7789	7796	7803	7810	7818	7825	7832	7839	7846	1	1	2	3	4	4	5	6	6
61	7853	7860	7868	7875	7882	7889	7896	7903	7910	7917	1	1	2	3	4	4	5	6	6
62	7924	7931	7938	7945	7952	7959	7966	7973	7980	7987	1	1	2	3	3	4	5	6	6
63	7993	8000	8007	8014	8021	8028	8035	8041	8048	8055	1	1	2	3	3	4	5	5	6
64	8062	8069	8075	8082	8089	8096	8102	8109	8116	8122	1	1	2	3	3	4	5	5	6
65	8129	8136	8142	8149	8156	8162	8169	8176	8182	8189	1	1	2	3	3	4	5	5	6
66	8195	8202	8209	8215	8222	8228	8235	8241	8248	8254	1	1	2	3	3	4	5	5	6
67	8261	8267	8274	8280	8287	8293	8299	8306	8312	8319	1	1	2	3	3	4	5	5	6
68	8325	8331	8338	8344	8351	8357	8363	8370	8376	8382	1	1	2	3	3	4	4	5	6
69	8388	8395	8401	8407	8414	8420	8426	8432	8439	8445	1	1	2	2	3	4	4	5	6
70	8451	8457	8463	8470	8476	8482	8488	8494	8500	8506	1	1	2	2	3	4	4	5	6
71	8513	8519	8525	8531	8537	8543	8549	8555	8561	8567	1	1	2	2	3	4	4	5	5
72	8573	8579	8585	8591	8597	8603	8609	8615	8621	8627	1	1	2	2	3	4	4	5	5
73	8633	8639	8645	8651	8657	8663	8669	8675	8681	8686	1	1	2	2	3	4	4	5	5
74	8692	8698	8704	8710	8716	8722	8727	8733	8739	8745	1	1	2	2	3	4	4	5	5
75	8751	8756	8762	8768	8774	8779	8785	8791	8797	8802	1	1	2	2	3	3	4	5	5
76	8808	8814	8820	8825	8831	8837	8842	8848	8854	8859	1	1	2	2	3	3	4	5	5
77	8865	8871	8876	8882	8887	8893	8899	8904	8910	8915	1	1	2	2	3	3	4	4	5
78	8921	8927	8932	8938	8943	8949	8954	8960	8965	8971	1	1	2	2	3	3	4	4	5
79	8976	8982	8987	8993	8998	9004	9009	9015	9020	9025	1	1	2	2	3	3	4	4	5
80	9031	9036	9042	9047	9053	9058	9063	9069	9074	9079	1	1	2	2	3	3	4	4	5
81	9085	9090	9096	9101	9106	9112	9117	9122	9128	9133	1	1	2	2	3	3	4	4	5
82	9138	9143	9149	9154	9159	9165	9170	9175	9180	9186	1	1	2	2	3	3	4	4	5
83	9191	9196	9201	9206	9212	9217	9222	9227	9232	9238	1	1	2	2	3	3	4	4	5
84	9243	9248	9253	9258	9263	9269	9274	9279	9284	9289	1	1	2	2	3	3	4	4	5
85	9294	9299	9304	9309	9315	9320	9325	9330	9335	9340	1	1	2	2	3	3	4	4	5
86	9345	9350	9355	9360	9365	9370	9375	9380	9385	9390	1	1	2	2	3	3	4	4	5
87	9395	9400	9405	9410	9415	9420	9425	9430	9435	9440	0	1	1	2	2	3	3	4	4
88	9445	9450	9455	9460	9465	9469	9474	9479	9484	9489	0	1	1	2	2	3	3	4	4
89	9494	9499	9504	9509	9513	9518	9523	9528	9533	9538	0	1	1	2	2	3	3	4	4
90	9542	9547	9552	9557	9652	9566	9571	9576	9581	9586	0	1	1	2	2	3	3	4	4
91	9590	9595	9600	9605	9609	9614	9619	9624	9628	9633	0	1	1	2	2	3	3	4	4
92	9638	9643	9647	9652	9657	9661	9666	9671	9675	9680	0	1	1	2	2	3	3	4	4
93	9685	9689	9694	9699	9703	9708	9713	9717	9722	9727	0	1	1	2	2	3	3	4	4
94	9731	9736	9741	9745	9750	9754	9759	9763	9768	9773	0	1	1	2	2	3	3	4	4
95	9777	9782	9786	9791	9795	9800	9805	9809	9814	9818	0	1	1	2	2	3	3	4	4
96	9823	9827	9832	9836	9841	9845	9850	9854	9859	9863	0	1	1	2	2	3	3	4	4
97	9868	9872	9877	9881	9886	9890	9894	9899	9903	9908	0	1	1	2	2	3	3	4	4
98	9912	9917	9921	9926	9930	9934	9939	9943	9948	9952	0	1	1	2	2	3	3	4	4
99	9956	9961	9965	9969	9974	9978	9983	9987	9991	9996	0	1	1	2	2	3	3	3	4

ANSWERS TO SELECTED PROBLEMS

ANSWERS TO SELECTED PROBLEMS

CHAPTER 1

1.1

Radiation	Wavelength, λ			Wave-number, cm^{-1}	Frequency, ν, Hz
	Å	μ	cm		
Vis., violet	4500	4.5×10^{-1}	4.5×10^{-5}	2.22×10^{4}	6.7×10^{14}
Vis., red	7000	7.0×10^{-1}	7.0×10^{-5}	1.43×10^{4}	4.3×10^{14}
x-rays	10	1.0×10^{-3}	1.0×10^{-7}	1.00×10^{7}	3.0×10^{17}
TV	3.4×10^{10}	3.4×10^{6}	3.4×10^{2}	2.9×10^{-3}	88×10^{6}
UV	3000	3.0×10^{-1}	3.0×10^{-5}	3.33×10^{4}	1.0×10^{15}
Radio	1.88×10^{11}	1.88×10^{7}	1.88×10^{3}	5.34×10^{-4}	1.6×10^{7}

1.2 Vis.–violet 4.44×10^{-12} erg
x-rays 1.99×10^{-9} erg
UV 6.62×10^{-12} erg
Vis.–red 2.85×10^{-12} erg
TV 5.83×10^{-19} erg
Radio 1.05×10^{-19} erg

1.3 a) Radio b) TV c) Radar d) x-rays
e) Visible f) Infrared g) UV h) TV

1.4 a) 2.48×10^{-22} erg b) 1.85×10^{-25} joule
c) 3.97×10^{-17} erg d) 5.96×10^{-16} joule
e) 3.31×10^{-12} erg f) 2.49×10^{-20} joule
g) 9.93×10^{-12} erg h) 1.98×10^{-25} joule

1.5 64.4×10^{10} ergs, 64.4×10^{3} joules, 0.607 eV/photon

1.6 See Appendix 1.1

CHAPTER 2

2.2 Since the concentration is changed from $c_1 = 1$ to $c_2 = 4$, if Beer's law is valid, the ratio between the corresponding absorbances A_1 and A_2 must be

$A_2/A_1 = c_2/c_1 = 4.$

$A_1 = 2 - \log 82 = 0.0862$ | $A_2 = 2 - \log 45.2 = 0.3449$

Therefore $$\frac{A_2}{A_1} = \frac{0.3449}{0.0862} = 4$$

The $\%T$ at concentration $2c = 67.2$.

2.3 Since 2-furoic acid (b) is essentially the *s-cis* isomer of (a), one expects it to display a weaker band at a lower λ_{max} value.

2.4 a) 224 ($\epsilon = 9750$) b) 253 ($\epsilon = 9550$)
c) 235 ($\epsilon = 14{,}000$) d) 248 ($\epsilon = 6890$)

2.5 a) 288 ($\epsilon = 12{,}600$) b) 279 (26,400)
c) 348 ($\epsilon = 11{,}000$)

2.6 *Ortho-para* directing:

$Me < Cl < Br < OH < OMe < NH_2 < O^- < NHCOMe < NMe_2$

Meta directing:

$NH_3^+ < SONH_2 < CN = CO_2^- < COOH < COMe < CHO < NO_2$

NMe_2 and NO_2 *para* to each other produce maximum shift because the inductive effect of Me_2 groups allow *n* electrons of nitrogen to participate readily in the interaction resonance.

2.7 a) 236 (241) b) 295 (283) c) 213 (212)
d) 236.5 (237) e) 236.5 (255) f) 295 (381).
Values in parentheses are experimentally observed values.

2.8 Acetylacetone can undergo keto-enol tautomerism. In the nonpolar solution, the enol form is stabilized by the intramolecular hydrogen bonding, but in the polar solution the keto form is stabilized by hydrogen bonding with the solvent. Therefore a strong band at 269 nm (ϵ = 12,000) corresponding to the enol is observed in cyclohexane, which shifts to 277 nm (ϵ = 1900) in water.

2.9 Resonance interaction in the *para* isomer allows for an intense band at a longer wavelength. On the other hand, appreciable steric interactions in the *ortho* form twist the single bonds away from coplanarity in the conjugated system. This causes considerable decrease in absorption intensity and a small blue shift in λ_{max}.

2.10 The energy per mole of photons of UV radiation in each case can be calculated from the following equation. (See Section 1.11, Problem 1.5.)

$$E = Nh\nu = \frac{Nhc}{\lambda, \text{ in cm}}$$

where E = energy/mole in joule units
N = Avogadro's number = 6.02×10^{23}
h = Planck's constant = 6.62×10^{-34} J-sec
ν = frequency of radiation = c/λ, in cm
1 cal = 4.184 joules

Thus, for cyclohexane solution, λ_{max} = 335 nm or 335×10^{-7} cm.

$$E = \frac{6.02 \times 10^{23} \times 6.62 \times 10^{-34} \times 3 \times 10^{10}}{335 \times 10^{-7}} \text{ joules/mole}$$

$$= \frac{6.02 \times 6.62 \times 3 \times 10^{6}}{4.184 \times 335} \text{ cal/mole}$$

$$= 85.3 \text{ kcal/mole.}$$

Similarly, the energy required for $n \rightarrow \pi^*$ transitions in other solvents is calculated to be:

89.4 kcal/mole for ethanol solution,
91.6 kcal/mole for methanol solution,
95.3 kcal/mole for aqueous solution.

The strengths of hydrogen bonds are therefore:

89.4 – 85.3 = 4.1 kcal/mole in ethanol,
91.6 – 85.3 = 6.3 kcal/mole in methanol,
95.3 – 85.3 = 100 kcal/mole in water.

2.11 a) 218 nm b) 216 nm c) 235 nm
d) 225 nm e) 222 nm f) 231 nm

2.12

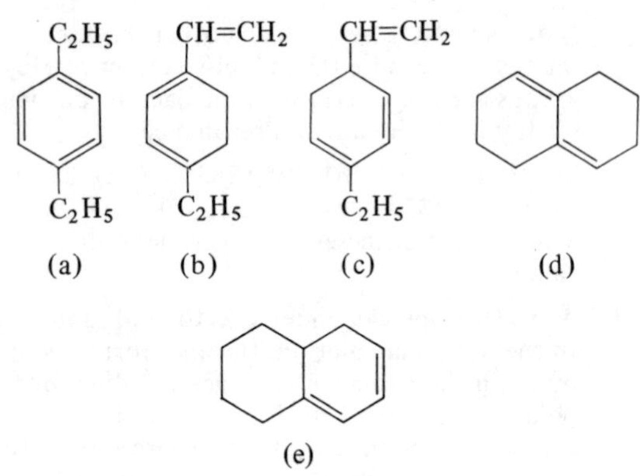

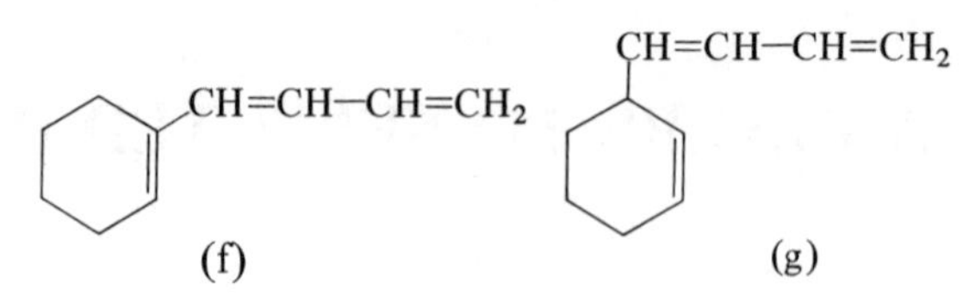

a) An aromatic ring; therefore it has a strong primary band at 207 nm and a weak secondary band at about 260 nm.
b) 253 + (3 x 5) + 30 = 298 nm
c) 253 + (3 x 5) = 268 nm. Third double bond is not conjugated.
d) 214 + (2 x 5) + (4 x 5) = 244 nm. Heteroannular diene with two exocyclic double bonds and four alkyl substituents.
e) 253 + (3 x 5) + 5 = 273 nm. Homoannular diene with three alkyl substituents and one exocyclic bond.
f) 214 + (2 x 5) + 30 = 254 nm
g) 214 + 5 = 219 nm. Third double bond is not conjugated.

2.13 a) 3,3′-dimethylbiphenyl
b) 2,2′-dimethylbiphenyl
c) 4,4′-dimethylbiphenyl

Benzene rings display a primary band at about 203.5 nm and a secondary (fine structure) band at about 255 nm. In biphenyl, these two bands are combined, due to the conjugation between the two rings, to give a broad intense absorption at 245 nm. Loss of this conjugation will again separate the two bands. In 2,2′-dimethylbiphenyl, such a loss of conjugation occurs due to the twist of the central bond caused by the steric interactions between the two methyl groups. Consequently, the allowed primary transition of the benzene rings is observed at 222 nm and the much weaker, forbidden transition at 260 nm. Replacement of *ortho* methyl groups by bulkier *t*-butyl groups would cause a further decrease in intensity and an increase in the blue shift of the primary band. The secondary band, on the other hand, would show increased vibrational resolution.

2.14 a) 256.5 nm (256) b) 227 nm (237)
c) 252.5 nm (258) d) 206.5 nm (207)
e) 217 nm (217) f) 210 nm (207)

CHAPTER 3

3.1 $3.27\mu = 3058 \text{ cm}^{-1}$, $6.90\mu = 1449 \text{ cm}^{-1}$

3.2 HCl: 1.582×10^{-24}, LiH: 1.453×10^{-24}, N_2: 11.63×10^{-24}

3.3 7.163×10^{5} dynes/cm

3.4 $\nu = 5.118 \times 10^{13} \text{ sec}^{-1}$, $\bar{\nu} = 1703 \text{ cm}^{-1}$

3.7 i) O—H str., N—H str.
ii) N—H bending, C≡C str.
iii) C—O str., C—H bending, C—C str., C—N str.
iv) C—H bending

3.8 Identification of the bands:

3030 cm^{-1} C—H str. in terminal $RHC{=}CH_2$ (CHR group)
1765 cm^{-1} C=O str.
1649 cm^{-1} C=C str. (terminal)
1225 cm^{-1} C—O—C str. (asym.)
1140 cm^{-1} C—O—C str. (sym.)

For a compound with the molecular formula $C_4H_6O_2$, only two structures with an -ene and an ester carbonyl group can be written:

a) $CH_2{=}CH{-}C({=}O){-}O{-}CH_3$ methyl acrylate (an α,β-unsaturated ester)

b) $CH_3C({=}O){-}O{-}CH{=}CH_2$ vinyl acetate

The two structures can be distinguished on the basis of their C=O stretching frequencies. We can calculate C=O stretching frequency in each case:

a) Basic value 1720 cm^{-1}
—OR (ester) $+25 \text{ cm}^{-1}$
conjugation (C=C) -30 cm^{-1}
Calculated 1715 cm^{-1}

b) Basic value 1720 cm^{-1}
—O—C=C (vinyl ester) +50
Calculated 1770 cm^{-1}

Since the observed value is 1765 cm^{-1}, the compound is (b).

3.9 Cyclohexanone (O=C in cyclohexane ring)

3.10 Diethylsuccinate $EtO{-}C({=}O){-}CH_2{-}CH_2{-}C({=}O){-}O{-}Et$

3.11 3-methylcyclohexanone (O=C and CH_3 on cyclohexane ring)

3.12 Butyraldehyde $CH_3{-}CH_2{-}CH_2{-}C({=}O){-}H$

3.13 Methyl methacrylate $CH_2{=}C(CH_3){-}C({=}O){-}O{-}CH_3$

3.14 *para*-xylene $CH_3{-}C_6H_4{-}CH_3$

3.15 *para*-anisaldehyde $CH_3O{-}C_6H_4{-}C({=}O){-}H$

3.16 1-nitropropane $CH_3{-}CH_2{-}CH_2{-}NO_2$

3.17 *para*-nitroaniline $NH_2{-}C_6H_4{-}NO_2$

3.18 Ethyl-2-bromobutyrate $CH_3{-}CH_2{-}CHBr{-}C({=}O){-}O{-}Et$

CHAPTER 4

4.1

Radiation	Wavelength	Frequency	Energy	
			eV/photon	cal/mole
UV	200 nm	1.5×10^{15}	6.2	14.29×10^{4}
	800 nm	0.375×10^{15}	1.55	3.57×10^{4}
IR	2000 nm	1.5×10^{14}	6.2×10^{-1}	14.29×10^{3}
	16,000 nm	1.875×10^{13}	7.76×10^{-2}	17.85×10^{2}
Radio	1000 cm	3.0×10^{6}	1.24×10^{-8}	27.58×10^{-5}
	10,000 cm	3.0×10^{5}	1.24×10^{-9}	27.58×10^{-6}

4.2 Half integral spin: ^{23}Na, ^{15}N, ^{17}O, ^{27}Al, ^{35}Cl
Full integral spin: ^{24}Na, ^{10}B
Zero spin: ^{24}Mg

4.3 $^{1}H = 300.4$ MHz, $^{13}C = 75.5$ MHz, $^{19}F = 282.4$ MHz

4.4 $^{1}H = 2.677 \times 10^{4}$ gauss^{-1} sec^{-1}
$^{13}C = 6.726 \times 10^{3}$ gauss^{-1} sec^{-1}
$^{19}F = 2.516 \times 10^{4}$ gauss^{-1} sec^{-1}

4.5 100 MHz

4.6 1.343×10^{-23} erg/gauss

4.8 a) Four
b) Left to right:

Peak no.	Hz	δ	τ
1	484	8.065	1.935
2	425	7.084	2.916
3	140	2.333	7.667
4	0	0	0

c) δ and τ values will remain unchanged, but Hz values for the three peaks from left to right will be 806, 708, 233 and zero, respectively.
d) At 0 ppm
e) 2, 2, 3 (For a detailed discussion of this, see unknown 6.1, Chapter 6.)
f) Doublet, doublet, and singlet. The three-proton singlet must be due to the methyl group. There are no neighboring protons to couple with the methyl protons. The two doublets suggest that the remaining four protons must exist as two nonequivalent groups of two protons each.
g) The subpeaks are not symmetrical around the midpoint. The subpeak on the right-hand side in peak number 1 (1.935τ) is more intense than its counterpart to the left; this indicates that the protons responsible for this peak are coupled with the protons appearing at 2.916τ, and vice versa.
h) 7 Hz; yes.
i) The greater the shielding, the more to the right the protons will appear on the spectrum. Thus the methyl protons giving the signal at 7.667τ are the most shielded. Protons at positions 2 and 6 on the ring are the least shielded, and must give the signal at 1.935τ, whereas the protons at positions 3 and 5 must be responsible for the signal at 2.916τ.

4.9

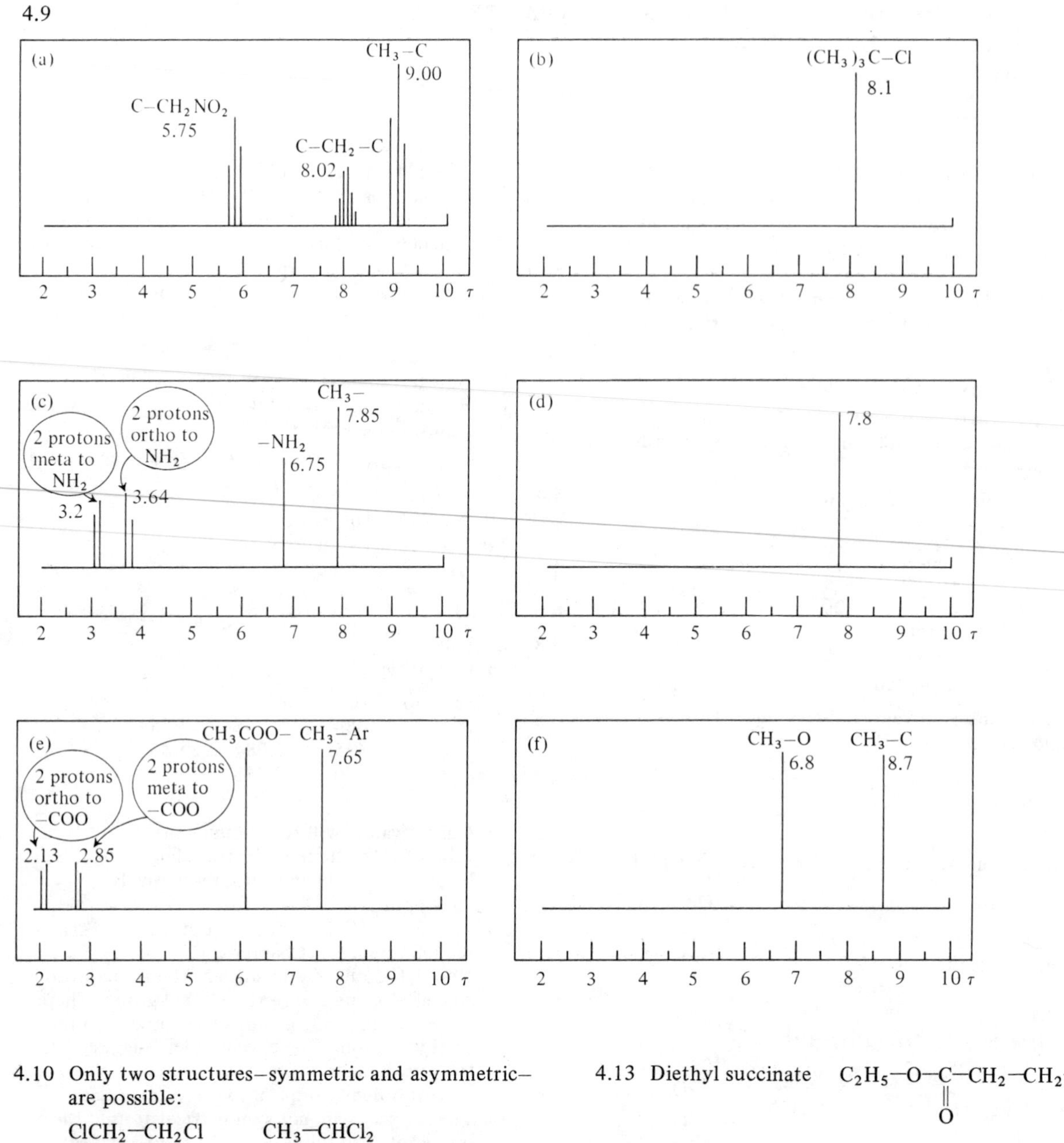

4.10 Only two structures–symmetric and asymmetric–are possible:

$ClCH_2-CH_2Cl$ i) symmetric CH_3-CHCl_2 ii) asymmetric

The symmetric compound will yield only one singlet in the spectrum, perhaps at about 6.0τ. The asymmetric compound will display two peaks: a three-proton doublet at approximately 8.0τ and a one-proton quartet approximately centered at 4.1τ.

4.11 a) AB_3 or AX_3 b) ABX or ABC
c) ABMX or ABCX d) A_3X_2
e) AA′X or A_2X f) AA′BB′ or A_2B_2
g) AMX_6, if the hydroxyl proton is exchanging the system is AX_6
h) A_6B_2

4.12 2,5-hexanedione $CH_3COCH_2CH_2COCH_3$

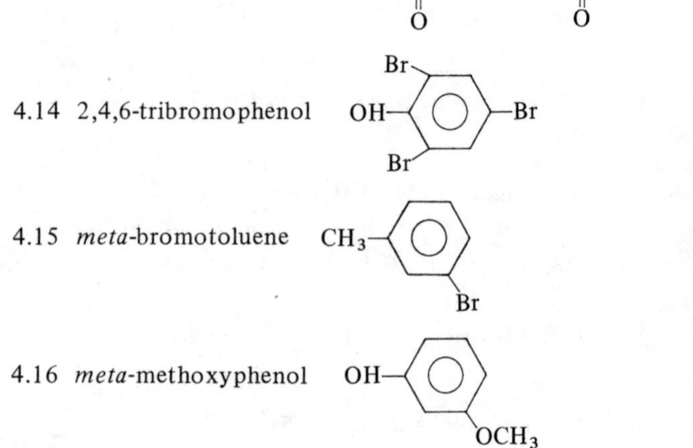

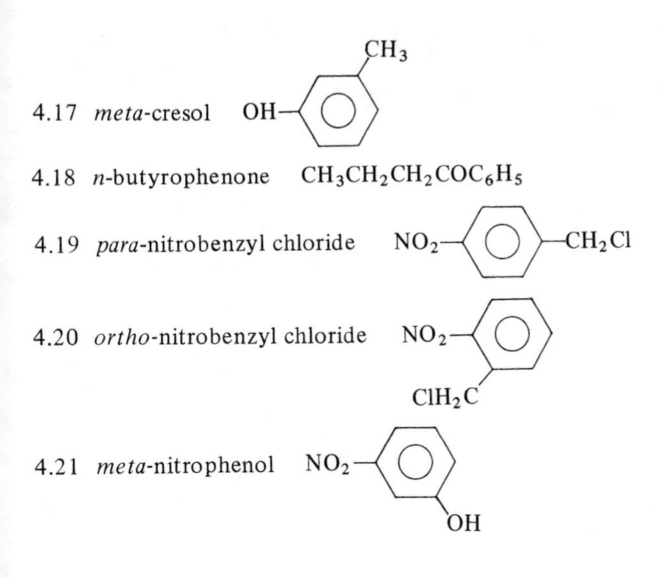

CHAPTER 6

Unknown number	Name and structure
6.1	4-heptanone (di-*n*-propylketone; butyrone) $CH_3-CH_2-CH_2-\overset{O}{\overset{\parallel}{C}}-CH_2-CH_2-CH_3$
6.2	*para*-tolunitrile (4-cyanotoluene) CH_3, CN
6.3	Eugenol (1-allyl-4-hydroxy-3-methoxybenzene) $HO-$, $-CH_2CH{=}CH_2$, OCH_3
6.4	Salicylaldehyde (*ortho*-hydroxybenzaldehyde) O, H, C=O, H
6.5	*n*-hexyl bromide (1-bromohexane) $CH_3-CH_2-CH_2-CH_2-CH_2-CH_2-Br$
6.6	3-pentanone (diethylketone) $CH_3-CH_2-\underset{O}{\underset{\parallel}{C}}-CH_2-CH_3$
6.7	*para*-bromotoluene $Br-$, $-CH_3$
6.8	*para*-tolualdehyde (4-methylbenzaldehyde) O, C, H, $-CH_3$
6.9	*meta*-tolunitrile (*meta*-cyanotoluene) $-CH_3$, CN

Unknown number	Name and structure
6.10	Diethylamine $CH_3{-}CH_2{-}NH{-}CH_2{-}CH_3$
6.11	*para*-bromophenol 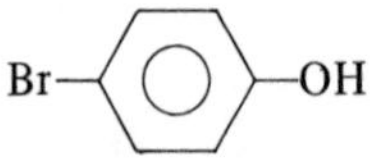 Br–C$_6$H$_4$–OH
6.12	*para*-methoxyphenol (*para*-hydroxyanisol) 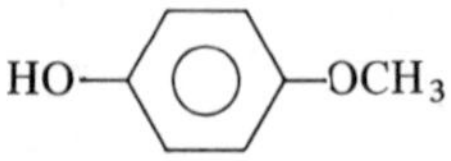 HO–C$_6$H$_4$–OCH_3
6.13	*ortho*-tolunitrile (*ortho*-cyanotoluene) 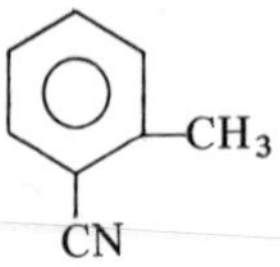 CH_3, CN
6.14	Crotonaldehyde (2-butenal) $CH_3CH{=}CHCHO$
6.15	*ortho*-bromotoluene 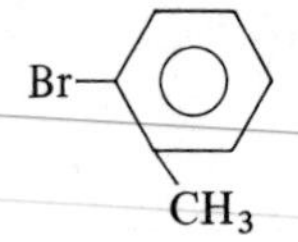 Br, CH_3
6.16	2-heptanone (methyl pentyl ketone) $CH_3{-}CO{-}CH_2{-}CH_2{-}CH_2{-}CH_2{-}CH_3$
6.17	Hydroquinone (quinol; 1,4-dihydroxybenzene)  HO–C$_6$H$_4$–OH
6.18	2-bromopentane (2-amyl bromide) $CH_3{-}CHBr{-}CH_2{-}CH_2{-}CH_3$
6.19	2,6-dimethylpyridine (2,6- or α,α'-lutidine) 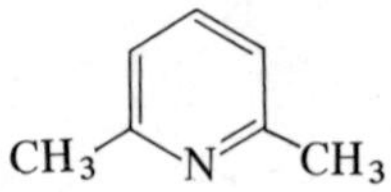 CH_3, N, CH_3
6.20	Iso-safrole (1,2-methylene-dioxy-4-propenyl benzene) 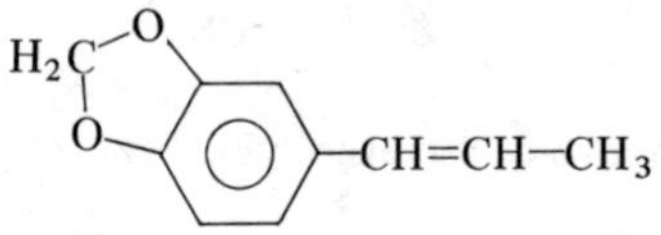 H_2C, O, O, $CH{=}CH{-}CH_3$
6.21	1,3,5-trimethylbenzene (mesitylene) 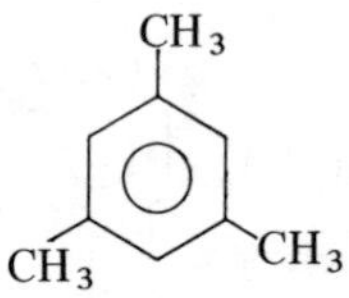 CH_3, CH_3, CH_3
6.22	Cyclohexene 

INDEXES

SUBJECT INDEX

AUTHOR INDEX

*Numbers are listed as follows: boldface-chapter number; italics-page number; lightface-reference number.

INDEX OF COMPOUNDS

Spectra of the following compounds are recorded in the text.